A Brief Table of Laplace Transforms

(See Chapter 7 for more details.)

$f(t)$	$F(s)$	$f(t)$	$F(s)$
1. $h(t)$	$\frac{1}{s}$	**13.** $f(t-\alpha)h(t-\alpha)$	$e^{-\alpha s}F(s)$
2. t^n	$\frac{n!}{s^{n+1}}$	**14.** $h(t-\alpha)$	$\frac{e^{-\alpha s}}{s}$
3. $e^{\alpha t}$	$\frac{1}{s-\alpha}$	**15.** $f'(t)$	$sF(s)-f(0)$
4. $\sin \omega t$	$\frac{\omega}{s^2+\omega^2}$	**16.** $\int_0^t f(u)\,du$	$\frac{F(s)}{s}$
5. $\cos \omega t$	$\frac{s}{s^2+\omega^2}$	**17.** $\sin \omega t - \omega t \cos \omega t$	$\frac{2\omega^3}{(s^2+\omega^2)^2}$
6. $\sinh \alpha t$	$\frac{\alpha}{s^2-\alpha^2}$	**18.** $t \sin \omega t$	$\frac{2\omega s}{(s^2+\omega^2)^2}$
7. $\cosh \alpha t$	$\frac{s}{s^2-\alpha^2}$	**19.** $tf(t)$	$-F'(s)$
8. $e^{\alpha t}f(t)$	$F(s-\alpha)$	**20.** $\frac{1}{t}f(t)$	$\int_s^\infty F(u)\,du$
9. $e^{\alpha t}h(t)$	$\frac{1}{s-\alpha}$	**21.** $f(\alpha t)$	$\frac{1}{\alpha}F\left(\frac{s}{\alpha}\right)$
10. $e^{\alpha t}t^n$	$\frac{n!}{(s-\alpha)^{n+1}}$	**22.** $(f*g)(t)$	$F(s)G(s)$
11. $e^{\alpha t}\sin \omega t$	$\frac{\omega}{(s-\alpha)^2+\omega^2}$	**23.** $f(t+T)=f(t)$	$\frac{\int_0^T e^{-st}f(t)\,dt}{1-e^{-sT}}$
12. $e^{\alpha t}\cos \omega t$	$\frac{(s-\alpha)}{(s-\alpha)^2+\omega^2}$	**24.** $\delta(t-t_0)$	e^{-st_0}

Elementary Differential Equations

Werner Kohler
Virginia Tech

Lee Johnson
Virginia Tech

Boston San Francisco New York
London Toronto Sydney Tokyo Singapore Madrid
Mexico City Munich Paris Cape Town Hong Kong Montreal

Dedication

We especially want to acknowledge our families for making it fun.

Thanks to Abbie, Larry, Tom, Liz, Paul, Maggie, Cathy, Connie, and Luke.
Thanks to Rochelle, Eric, Mark, Ali, Hannah, Quinn, and Casey.

Executive Editor: William Hoffman
Development Editor: David Chelton
Executive Marketing Manager: Yolanda Cossio
Managing Editor: Karen Guardino
Senior Production Supervisor: Peggy McMahon
Production Services: Kathy Diamond
Manufacturing Buyer: Evelyn Beaton
Technical Art Supervisor: Joe Vetere
Composition and Technical Art Rendering: Techsetters, Inc.
Design Supervisor: Barbara T. Atkinson
Interior Design: Andrea Toth
Cover Designer: Barbara T. Atkinson
Cover Photograph: Dominique Serraute/© Getty Images

Library of Congress Cataloging-in-Publication Data

Kohler, W. E. (Werner E.), 1939–
Elementary differential equations / Werner E. Kohler, Lee W. Johnson.
p. cm.
Includes index.
ISBN 0-201-70926-0 (alk. paper)
1. Differential equations. I. Johnson, Lee W. II. Title.

QA371 .K5854 2002
515′.35–dc21

2001056733

1 2 3 4 5 6 7 8 9 10 — CRW — 05040302

Contents

4

Second Order Linear Differential Equations 115

5

Higher Order Linear Differential Equations 221

6

First Order Linear Systems 253

9

Numerical Methods 537

10

Series Solutions of Linear Differential Equations 579

Preface

This book is designed for the sophomore differential equations course taken by students majoring in science and engineering. We assume the reader has had a course in elementary calculus.

We have integrated the underlying theory, the solution procedures, and the numerical and computational aspects of differential equations as seamlessly as possible. For example, when we introduce a new category of problems (such as first order equations, higher order equations, systems of differential equations, etc.), we begin with the basic existence-uniqueness theory since it forms the framework for understanding differential equations and attempting to solve them. However, because we want the text to be easy to read and understand, we discuss the theory as simply as possible and emphasize how to use it.

Linear and nonlinear equations (first order and higher order) are treated in separate chapters. We recognize there is a pedagogical trade off. On the one hand, order is a unifying characteristic of differential equations. On the other hand, linear differential equations are of such importance in terms of applications, theory, and solution techniques that they warrant a strong and separate emphasis. We have opted for the latter approach.

The theory of differential equations has an intrinsic beauty and provides an important tool for understanding the world around us. The interplay of mathematics and science in helping to explain the physical world gives the subject of differential equations much of its vitality. Many readers studying differential equations are preparing for careers in the physical or life sciences. Therefore, when developing models, we try to guide the student carefully through the underlying physical principles leading to the relevant mathematics.

We also emphasize the importance of common sense, intuition, and "back-of-the-envelope" checks. It is important, when solving problems, for students to ask, "Does my answer make sense?" Some of the examples and exercises ask the student to anticipate and interpret the physical content of the solution (for example, "Should we expect an equilibrium solution to exist in this application? If so, why? What should its value be?"). We feel it is particularly important to develop this type of mind-set in the present "computer age," when the temptation is great to accept any computer-generated output as correct.

Some features of this book should be noted:

Readability We have made a determined effort to write a text that is relatively easy for students to read and understand. When we introduce a new topic, such as separable first order equations or the Wronskian, we are careful to give illustrative examples and enough detail so that students can follow the discussion.

A Wide Variety of Exercises The text includes a large collection of exercises ranging from routine drill exercises to interesting applications drawn from a number of different disciplines. It also includes quite a few exercises that ask students to wrestle with the underlying ideas rather than simply turn an algorithmic crank. For example, we may state that the solution of the initial value problem $y'' + ay' + by = 0, y(0) = y_0, y'(0) = y'_0$ is given by $y(t) = e^{-2t}(\cos t - \sin t)$ and then ask the student to determine the constants a, b, y_0, and y'_0.

Answer Key Solutions of the odd-numbered exercises are given at the end of the text.

Computer Awareness Currently, high school courses as well as first-year college courses expose virtually every student to calculators and computers. In their major courses, science and engineering students routinely use computers and computational software. Some of the exercises in this book require some type of electronic computational aid. However, the computational exercises are designed to be generic, not linked to any particular machine or software.

A Multilevel Development of Numerical Methods The basic ideas underlying numerical methods and their use in applications are presented early (in Sections 3.8 and 6.9) in the context of Euler's method. Chapter 9 builds upon this introduction, developing a comprehensive treatment of one-step methods such as Runge-Kutta methods.

Projects There is an Extended Problem at the end of each chapter. Each of these problems can serve as a miniproject. In some cases, the problem brings the concepts introduced in the chapter to bear on a particular application. In other cases, the purpose of the extended problem is to expand the student's horizon, showing how the material in the chapter can be generalized. In certain applications, such as food preservation, the problem exposes students to the mathematical aspects of state-of-the-art work.

An Appendix on Matrix Theory Chapter 6 discusses systems of linear differential equations, $\mathbf{y}' = A(t)\mathbf{y} + \mathbf{g}(t)$. Later, central ideas of systems are revisited in Chapter 7 (Laplace Transforms) and in Chapter 8 (Systems of Nonlinear Equations). However, not all students have the same background in matrix theory. Therefore, we have included an Appendix on Matrix Theory that summarizes the results necessary for understanding systems.

Help for Students The Student Solutions Manual gives detailed solutions of the odd-numbered exercises. The solutions manual also gives examples showing how to use computational software (such as Mathematica, Maple, and MATLAB) as an aid to solving the types of problems we include in the exercises. A web site for the book (http://www.aw.com/kohler) gives examples showing how to use computational software to solve typical differential equations.

Acknowledgments

Many people helped and encouraged us while we wrote this book. Besides the support provided by our families, we are especially thankful for the editorial and developmental assistance of Greg Tobin, Laurie Rosatone, William Hoffman, and Stefanie Borge.

We are very grateful to our reviewers, who made many insightful suggestions that improved the text.

John Baxley, *Wake Forest University*
William Beckner, *University of Texas at Austin*
Jerry L. Bona, *University of Illinois, Chicago*
Thomas Branson, *University of Iowa*
Dennis Brewer, *University of Arkansas*
Peter Brinkmann, *University of Illinois, Urbana-Champagne*
Almut Burchard, *University of Virginia*
Donatella Danielli, *Purdue University*
Charles Friedman, *University of Texas at Austin*
Weimin Han, *University of Iowa*
Harumi Hattori, *West Virginia University*
Yang Kuang, *Arizona State University*
Zhiqin Lu, *University of California at Irvine*
John Neuberger, *Northern Arizona University*
Joan Remski, *University of Michigan at Dearborn*
Sinai Robins, *Temple University*
Robert Snider, *Virginia Polytechnic Institute & State University*
A. Shadi Tahvildar-Zadeh, *Rutgers University*
Jason Whitt, *Rhodes College*
Jennifer Zhao, *University of Michigan at Dearborn*

We also want to express thanks to our developmental editor, David Chelton, who had the task of helping us organize and streamline our material. Our good friend and colleague, George Flick, provided many examples of differential equations applications in the life sciences. We are most appreciative of his encouragement and assistance.

Once our manuscript approached its final form, Addison-Wesley initiated a comprehensive production process to turn the manuscript into a real book. Peggy McMahon was in charge of production, and she made certain we maintained our schedule. In addition, we are grateful to Barbara Atkinson for her design of the text, to Cynthia Benn for her careful copyediting, and to Rena Lam and Mike Lafferty of Techsetters, Inc. who were very thorough and who made helpful suggestions as they typeset the text and produced the art.

To turn a manuscript into a textbook, the manuscript must pass through many stages of copyediting and marking to ensure that it aligns with the design, to make certain the mathematics is displayed properly and consistently, and to check that the figures are rendered correctly. These steps were carried out under the watchful eye of Kathy Diamond, and we are most grateful for her guidance and care.

1

Introduction to Differential Equations

CHAPTER OVERVIEW

Introduction

Scientists and engineers develop mathematical models for physical processes as an aid to understanding and predicting the behavior of the process. In this book we discuss mathematical models that help us understand, among other things, decay of radioactive substances, electrical networks, population dynamics, dispersion of pollutants, and trajectories of moving objects. Modeling a physical process often leads to equations that involve not only the physical quantity of interest but also some of its derivatives. Such equations are referred to as **differential equations**.

In Section 1.1 we give some simple examples that show how mathematical models are derived. We also begin our study of differential equations by introducing the corresponding terminology and by presenting some concrete examples of differential equations. Section 1.2 introduces the idea of a direction field for a differential equation. The concept of direction fields allows us to visualize, in geometric terms, the graphs of solutions of differential equations.

1.1 Examples of Differential Equations

When we apply Newton's second law of motion, $ma = f$, to an object moving in a straight line, we obtain a differential equation of the form

$$my''(t) = f(t, y(t), y'(t)) \tag{1}$$

In equation (1), $y(t)$ represents the position, at time t, of the object. As expressed in equation (1), the product of mass m and acceleration $y''(t)$ is equal to the sum of the applied forces. The applied forces [the right-hand side of equation (1)] often depend on time t, position $y(t)$, and velocity $y'(t)$.

EXAMPLE 1

One of the simplest examples of linear motion is an object falling under the influence of gravity. Let $y(t)$ represent the height of the object above the surface of the earth and let g denote the constant acceleration due to gravity (32 ft/sec^2 or 9.8 m/s^2). See Figure 1.1.

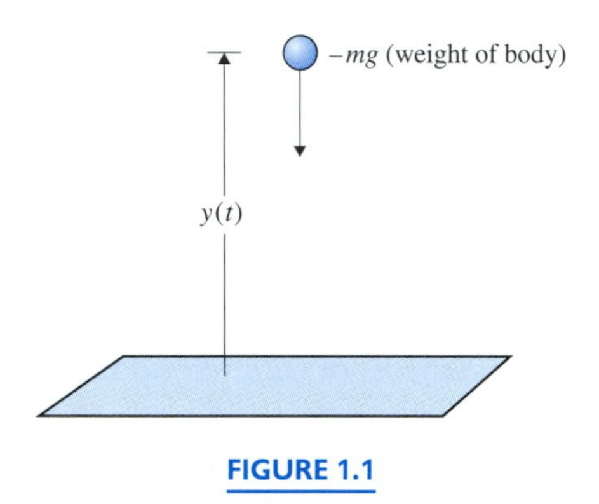

FIGURE 1.1

The only force acting on the falling body is its weight. The body's position, $y(t)$, is governed by the differential equation $y'' = -g$.

Since the only force acting on the body is assumed to be its weight, $W = mg$, equation (1) reduces to $my''(t) = -mg$ or

$$y''(t) = -g. \tag{2}$$

The negative sign appears on the right-hand side of the equation because the acceleration due to gravity is positive downward, while we assumed y to be positive in the upward direction. (Again, see Figure 1.1.)

Equation (2) is solved easily by taking successive antiderivatives. The first antiderivative gives the object's velocity,

$$y'(t) = -gt + C_1.$$

Another antidifferentiation gives the object's position,

$$y(t) = -\frac{1}{2}gt^2 + C_1t + C_2.$$

Here, C_1 and C_2 represent arbitrary constants of integration. ▲

Unless an application suggests otherwise, we normally use t to represent the independent variable and y to represent the dependent variable. Thus, in a typical differential equation, we are searching for a solution $y(t)$.

As is common in a mathematics text, we frequently use a variety of notation to denote derivatives. For instance, we may use d^2y/dt^2 instead of $y''(t)$ or d^4y/dt^4 instead of $y^{(4)}(t)$. In addition, we often suppress the independent variable t and simply write y and y' instead of $y(t)$ and $y'(t)$. An example using this notation is the differential equation

$$y'' + \frac{1}{t}y' + t^3 y = 5.$$

Notice in Example 1 that the solution involves two undetermined constants. This means that, by itself, differential equation (2) does not completely specify the solution $y(t)$. This makes sense physically since, to completely determine the motion, some information about the initial state of the object is also needed. The arbitrary constants of integration that arise are often specified by prescribing velocity and position at some initial time, say $t = 0$. For example, if the object's initial velocity is $y'(0) = v_0$ and its initial position is $y(0) = y_0$, then we obtain a complete description of velocity and position:

$$y'(t) = -gt + v_0, \qquad y(t) = -\frac{1}{2}gt^2 + v_0 t + y_0.$$

EXAMPLE 2

Scientists have observed that radioactive materials have an instantaneous rate of decay (that is, a rate of decrease) that is proportional to the amount of material present. If $Q(t)$ represents the amount of material present at time t, then dQ/dt is proportional to $Q(t)$; that is,

$$\frac{dQ}{dt} = -kQ, \qquad k > 0. \tag{3}$$

The negative sign in equation (3) arises because Q is both positive and decreasing; that is, $Q(t) > 0$ and $Q'(t) < 0$.

Unlike equation (2), differential equation (3) cannot be solved by integrating the right-hand side, $-kQ(t)$, because $Q(t)$ is not known. Instead, equation (3) requires that we somehow find a function $Q(t)$ whose derivative, $Q'(t)$, is a constant multiple of $Q(t)$.

Recall that the exponential function has a derivative that is a constant multiple of itself. For example, if $y = Ce^{-kt}$, then $y' = -kCe^{-kt} = -ky$. Therefore, we see that a solution of equation (3) is

$$Q(t) = Ce^{-kt}, \tag{4}$$

where C can be any constant. ▲

As with Example 1, the differential equation by itself does not completely specify the solution. But, setting $t = 0$ in (4) leads to $Q(0) = C$. Therefore, the quantity $Q(t)$ given in equation (4) is completely determined once the amount of material initially present is specified.

The Form of an *n*th Order Differential Equation

We now state the formal definition of a differential equation and point to some issues that need to be addressed. An equation of the form

$$y^{(n)} = f(t, y, y', \ldots, y^{(n-1)}) \tag{5}$$

is called an ***n*th order ordinary differential equation**.

In equation (5), t is the **independent variable**, while y is the **dependent variable**. A **solution** of the differential equation (5) is any function $y(t)$ that satisfies the equation on our t-interval of interest. For instance, Example 2 showed that $Q(t) = Ce^{-kt}$ is a solution of $Q' = -kQ$ for any value of the constant C; the t-interval of interest for Example 2 is typically the interval $0 \leq t < \infty$.

The **order** of a differential equation is the order of the highest derivative that appears in the equation. For example, $y'' = -g$ is a second order differential equation. Similarly, $Q' = -kQ$ is a first order differential equation.

The form of the nth order ordinary differential equation (5) is not the most general one. In particular, an nth order ordinary differential equation is any equation of the form

$$G(t, y, y', y'', \ldots, y^{(n)}) = 0.$$

For example, the following equation is classified as a second order ordinary differential equation,

$$t^2 \sin y'' + y \ln y'' = 1.$$

Notice that it is not possible to rewrite this equation in the explicit form

$$y'' = f(t, y, y').$$

In our study, however, we usually consider only equations of the form

$$y^{(n)} = f(t, y, y', y'', \ldots, y^{(n-1)}),$$

where the nth derivative is given explicitly in terms of t, y, and lower order derivatives of y.

Differential equation (5) is called **ordinary** because the equation involves just a single independent variable, t. This is in contrast to other equations, called **partial differential equations**, which involve two or more independent variables. An example of a partial differential equation is the one-dimensional wave equation

$$\frac{\partial^2 u(x,t)}{\partial x^2} - \frac{\partial^2 u(x,t)}{\partial t^2} = 0,$$

where the dependent variable u is a function of two independent variables, the spatial coordinate x and time t.

Initial Value Problems

What we have seen about differential equations thus far raises some important questions that we will address throughout this book. One such question is: "What constitutes a properly formulated problem?" Examples 1 and 2 illustrate that auxiliary **initial conditions** are required if the differential equation is to have a single unique solution. The differential equation, together with the proper number of initial conditions, constitutes what is known as an **initial value problem**.

For instance, an initial value problem associated with the falling object in Example 1 consists of the differential equation together with initial conditions specifying the object's initial position and velocity:

$$\frac{d^2y}{dt^2} = -g, \qquad y(0) = y_0, \qquad y'(0) = v_0.$$

Similarly, an initial value problem associated with the radioactive decay process in Example 2 consists of the differential equation together with a specification of the initial amount of the substance:

$$\frac{dQ}{dt} = -kQ, \qquad Q(0) = Q_0.$$

These examples suggest that the number of initial conditions we need to specify must be equal to the order of the differential equation. When we address the question of properly formulating problems, it will be apparent that this is the case. Once we understand how to properly formulate the problem to be solved, the obvious next question is: "How do we go about solving this problem?" Answering the two questions

1. How do we properly formulate the problem?
2. How do we solve the problem?

is central to the study of differential equations.

Solving Initial Value Problems

As Chapters 2, 4, and 5 show, certain special types of differential equations have formulas for the **general solution**. The general solution is an expression containing arbitrary constants (or parameters) that can be adjusted so as to give every solution of the equation. Finding the general solution is often the first step in solving an initial value problem. The next example illustrates this idea.

EXAMPLE 3

Consider the initial value problem

$$y' + 3y = 6t + 5, \qquad y(0) = 3. \tag{6}$$

(a) Show, for any constant C, that

$$y = Ce^{-3t} + 2t + 1 \tag{7}$$

is a solution of the differential equation $y' + 3y = 6t + 5$.

(b) Use expression (7) to solve the initial value problem (6).

Solution:

(a) Inserting expression (7) into the differential equation $y' + 3y = 6t + 5$, we find

$$\begin{aligned} y' + 3y &= (Ce^{-3t} + 2t + 1)' + 3(Ce^{-3t} + 2t + 1) \\ &= (-3Ce^{-3t} + 2) + (3Ce^{-3t} + 6t + 3) \\ &= 6t + 5 \end{aligned}$$

Therefore, for any value C, $y = Ce^{-3t} + 2t + 1$ is a solution of $y' + 3y = 6t + 5$.

(b) Imposing the constraint $y(0) = 3$ on $y(t) = Ce^{-3t} + 2t + 1$ leads to $y(0) = C + 1 = 3$. Therefore, $C = 2$ and a solution of the initial value problem is

$$y = 2e^{-3t} + 2t + 1. \ \blacktriangle$$

We will show later that $y = Ce^{-3t} + 2t + 1$ is the general solution of the differential equation in Example 3. A geometric interpretation is given in Figure 1.2, which shows graphs of the general solution for representative values of C. The solution whose graph passes through the point $(t, y) = (0, 3)$ is the one that solves the initial value problem posed in Example 3.

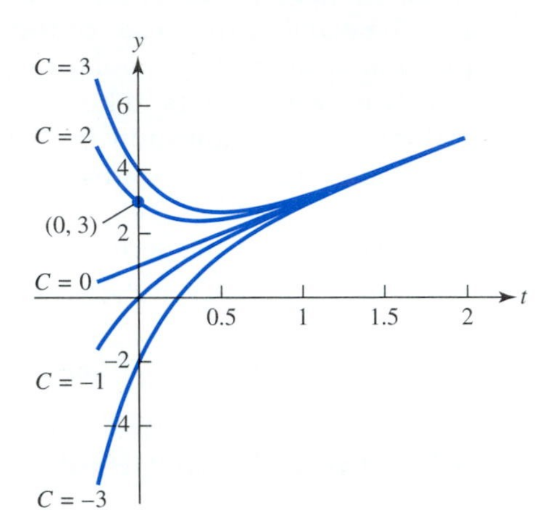

FIGURE 1.2

For any constant C, $y = Ce^{-3t} + 2t + 1$ is a solution of $y' + 3y = 6t + 5$. Solution curves are displayed for several values of C. For $C = 2$, the curve passes through the point $(t, y) = (0, 3)$; this is the solution of the initial value problem posed in Example 3.

EXERCISES

Exercises 1–4:

What is the order of the differential equation?

1. $y'' + 3ty^3 = 1$

2. $t^4y' + y\sin t = 6$

3. $(y')^3 + t^5 \sin y = y^4$

4. $(y''')^4 + \dfrac{t^2}{(y')^4 + 4} = 0$

5. (a) Show that $y(t) = Ce^{t^2}$ is a solution of $y' - 2ty = 0$ for any value of the constant C.

(b) Determine the value of C needed for this solution to satisfy the initial condition $y(1) = 2$.

6. Solve the differential equation $y''' = 2$ by computing successive antiderivatives. What is the order of this differential equation? How many arbitrary constants arise in the antidifferentiation solution process?

7. (a) Show that $y(t) = C_1 \sin 2t + C_2 \cos 2t$ is a solution of the differential equation $y'' + 4y = 0$, where C_1 and C_2 are arbitrary constants.

(b) Find values of the constants C_1 and C_2 so that the solution satisfies the initial conditions $y(\pi/4) = 3, y'(\pi/4) = -2$.

8. Suppose $y(t) = 2e^{-4t}$ is the solution of the initial value problem $y' + ky = 0, y(0) = y_0$. What are the constants k and y_0?

9. Consider $t > 0$. For what value(s) of the constant c, if any, is $y(t) = c/t$ a solution of the differential equation $y' + y^2 = 0$?

10. Let $y(t) = -e^{-t} + \sin t$ be a solution of the initial value problem $y' + y = g(t)$, $y(0) = y_0$. What must the function $g(t)$ and the constant y_0 be?

11. Consider $t > 0$. For what value(s) of the constant r, if any, is $y(t) = t^r$ a solution of the differential equation $t^2y'' - 2ty' + 2y = 0$?

12. Show that $y(t) = C_1e^{2t} + C_2e^{-2t}$ is a solution of the differential equation $y'' - 4y = 0$, where C_1 and C_2 are arbitrary constants.

Exercises 13–14:

Use the result of Exercise 12 to solve the initial value problem.

13. $y'' - 4y = 0, \quad y(0) = 2, \quad y'(0) = 0$

14. $y'' - 4y = 0, \quad y(0) = 1, \quad y'(0) = 2$

Exercises 15–16:

Use the result of Exercise 12 to find a function $y(t)$ that satisfies the given conditions.

15. $y'' - 4y = 0, \quad y(0) = 3, \quad \lim_{t \to \infty} y(t) = 0$

16. $y'' - 4y = 0, \quad y(0) = 10, \quad \lim_{t \to -\infty} y(t) = 0$

Exercises 17–18:

The graph shows the solution to the given initial value problem. In each case, m is an integer. In Exercise 17, determine m, y_0, and $y(t)$. In Exercise 18, determine m, t_0, and $y(t)$.

17. $y'(t) = m + 1, \quad y(1) = y_0$

18. $y'(t) = mt, \quad y(t_0) = -1$

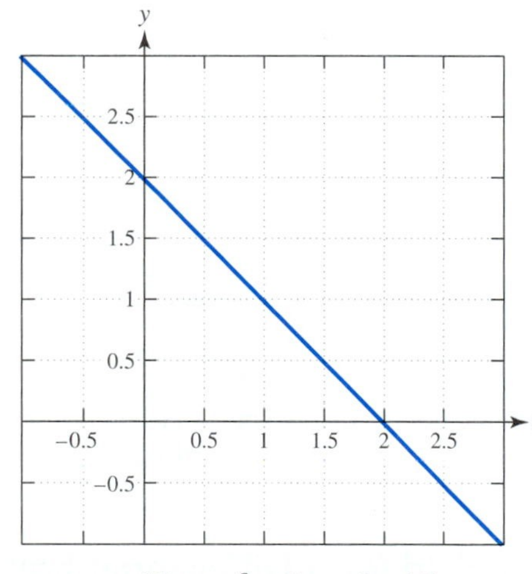

Figure for Exercise 17

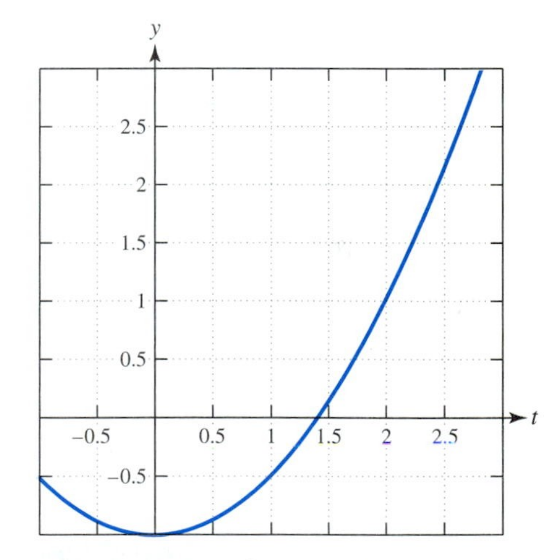

Figure for Exercise 18

19. At time $t = 0$ an object having mass m is released from rest at a height y_0 above the ground. Let g represent the (constant) gravitational acceleration. Derive an expression for the impact time (the time at which the object strikes the ground). What is the velocity with which the object strikes the ground? (Express your answers in terms of the initial height y_0 and the gravitational acceleration g.)

20. A car, initially at rest, begins moving at time $t = 0$ with a constant acceleration down a straight track. If the car achieves a speed of 60 mph (88 ft/sec) at time $t = 8$ sec, what is the car's acceleration? How far down the track will the car have traveled when its speed reaches 60 mph?

21. At time $t = 0$, an object of mass m is released from rest at a height of 252 ft above the floor of an experimental chamber in which gravitational acceleration has been slightly modified. Assume (instead of the usual value of 32 ft/sec^2), that the acceleration has the form $32 - \varepsilon \sin(\pi t/4)$ ft/sec^2, where ε is a constant. In addition, assume that the projectile strikes the ground exactly 4 sec after release. Can this information be used to determine the constant ε? If so, determine ε.

1.2 Direction Fields

Before beginning a systematic study of differential equations, we consider a geometric entity called a direction field, that will assist in understanding the first order differential equation

$$y' = f(t, y).$$

A **direction field** is a way of predicting the *qualitative* behavior of solutions of a differential equation. A good way to understand the idea of a direction field is to recall the "iron filings" experiment that is often done in science classes to illustrate magnetism and lines of magnetic force. In this experiment, iron filings (minute filaments of iron) are sprinkled on a sheet of cardboard, beneath which two magnets of opposite polarity are positioned. When the cardboard sheet is gently tapped, the iron filings align themselves so that their axes are parallel to the magnetic field lines. At a given point on the sheet, the orientation of an iron filing indicates the direction of the magnetic field line. The totality of oriented iron filings gives a good picture of the flow of magnetic field lines connecting the two magnetic poles. Figure 1.3 illustrates this experiment.

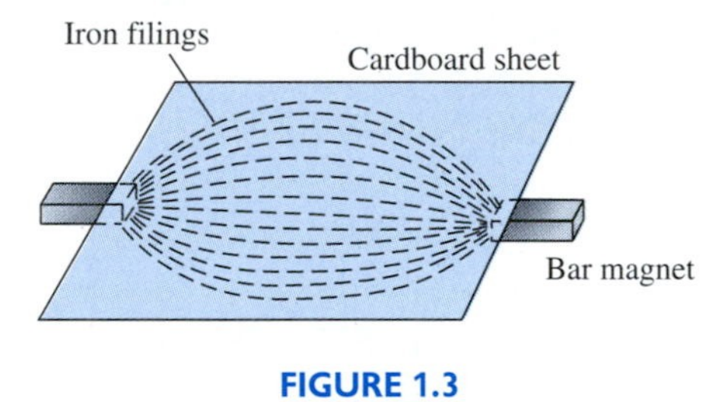

FIGURE 1.3

The orientation of iron filings gives a good picture of the flow of magnetic field lines connecting two poles of a magnet.

The Direction Field for a Differential Equation

What is the connection between the iron filings experiment illustrated in Figure 1.3 and a qualitative understanding of differential equations? From calculus we know that if we graph a differentiable function $y(t)$, the slope of the curve at the point $(t, y(t))$ is $y'(t)$. If $y(t)$ is a solution of a differential equation $y' = f(t, y)$, then we can calculate this slope by simply evaluating the right-hand side $f(t, y)$ at the point $(t, y(t))$.

For example, suppose $y(t)$ is a solution of the equation

$$y' = 1 + 2ty \tag{1}$$

and suppose the graph of $y(t)$ passes through the point $(t, y) = (2, y(2)) = (2, -1)$. For differential equation (1), the right-hand side is $f(t, y) = 1 + 2ty$. Thus we find

$$y'(2) = f(2, y(2)) = f(2, -1) = 1 + 2(2)(-1) = -3.$$

Even though we have not solved $y' = 1 + 2ty$, the preceding calculation has taught us something about the specific solution $y(t)$ passing through $(t, y) = (2, -1)$: it is decreasing (with slope equal to -3) when it passes through the point $(t, y) = (2, -1)$.

To exploit this idea, suppose we systematically evaluate the right-hand side $f(t, y)$ at a large number of points (t, y) throughout a region of interest in the ty-plane. At each point, we evaluate the function $f(t, y)$ to determine the slope of the solution curve passing through that point. We then sketch a tiny line segment at that point, oriented with the given slope $f(t, y)$. The resulting picture, called a direction field, is similar to that illustrated in Figure 1.3. Using such a direction field, we can create a good qualitative picture of the flow of solution curves throughout the region of interest.

EXAMPLE 1

(a) Sketch a direction field for $y' = 1 + 2ty$ in the square $-2 \le t \le 2$, $-2 \le y \le 2$.

(b) Using the direction field, sketch your guess for the solution curve passing through the point $P = (-2, 2)$. Also, using the direction field, sketch your guess for the solution curve passing through the point $Q = (0, -1)$.

Solution: The direction field for $y' = 1 + 2ty$ shown in Figure 1.4(a) was computer generated. There are a number of computer programs available for drawing direction fields, some of which are quite sophisticated. In addition, modern computational software packages have the capability of drawing direction fields. Figure 1.4(b) shows our guesses for the solution of the initial value problems in part (b). ▲

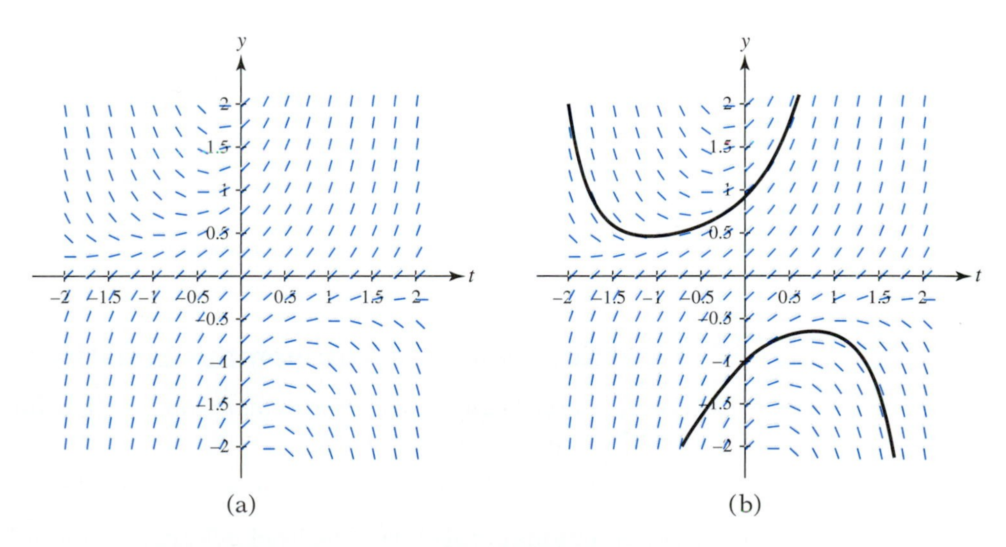

FIGURE 1.4

(a) The direction field for $y' = 1 + 2ty$. (b) Using the direction field, we have drawn our guess for the solution of $y' = 1 + 2ty, y(-2) = 2$ and for the solution of $y' = 1 + 2ty, y(0) = -1$.

Isoclines

The "method of isoclines" is helpful when you need to draw a direction field by hand. An **isocline** of the differential equation $y' = f(t, y)$ is a curve that satisfies the condition

$$f(t, y) = c,$$

where c is a constant. For example, consider the differential equation

$$y' = y - t^2.$$

In this case, curves of the form $y - t^2 = c$ are isoclines of the differential equation. (These curves, $y = t^2 + c$, are parabolas opening upward. Each has its vertex on the y-axis.) Isoclines are useful because, at every point on an isocline, the associated direction field filaments have the same slope, namely $f(t, y) = c$. (In fact, the word "isocline" means "equal inclination" or "equal slope.")

To carry out the method of isoclines we first sketch, for various values of c, the corresponding curves $f(t, y) = c$. Then, at representative points on these curves, we sketch direction field filaments, each having the same slope, $f(t, y) = c$.

EXAMPLE 2

(a) Use the method of isoclines to sketch the direction field for $y' = y - t$. Restrict your direction field to the square defined by $-2 \le t \le 2$, $-2 \le y \le 2$.

(b) Using the direction field, sketch your guess for the solution curve passing through the point $(-1, 1/2)$. Also, using the direction field, sketch your guess for the solution curve passing through the point $(-1, -1/2)$.

Solution:

(a) For the equation $y' = y - t$, lines of the form $y = t + c$ are isoclines. In Figure 1.5(a) we have drawn the isoclines $y = t + 3, y = t + 2, \dots,$ $y = t - 2$. At selected points along an isocline of the form $y = t + c$, we have drawn direction field filaments, each having slope c.

(b) Figure 1.5(b) shows our guesses to the solutions of the initial value problems in part (b) of the example. In addition, note that the line $y = t + 1$ appears to be a solution curve. ▲

Direction Fields for Autonomous Equations

The method of isoclines is particularly well suited for differential equations that have the special form

$$y' = f(y) \tag{2}$$

For equations of this form, the isoclines are *horizontal lines*. That is, if b is any number in the domain of $f(y)$, then the horizontal line $y = b$ is an isocline of the equation $y' = f(y)$. In particular, y' has the same value $f(b)$ all along the horizontal line $y = b$.

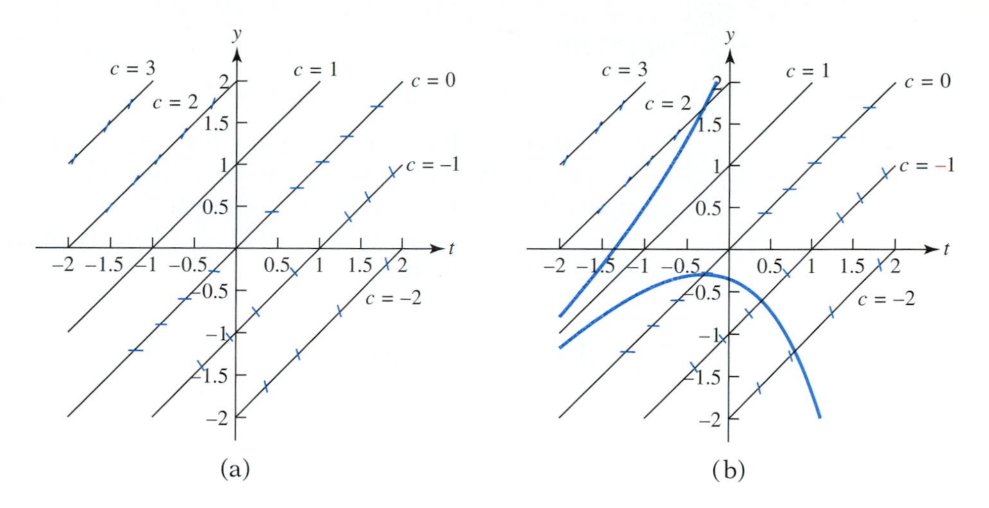

FIGURE 1.5

(a) The method of isoclines was used to sketch the direction field for $y' = y - t$. (b) Using the direction field, we have sketched our guesses to the solutions of the initial value problems in part (b) of Example 2.

Differential equations of the form (2), where the right-hand side does not depend explicitly on t, are called **autonomous differential equations**. An example of an autonomous differential equation is

$$y' = y^2 - 3y.$$

By contrast, the differential equation $y' = y + 2t$ is not autonomous. Autonomous differential equations are quite important in applications, and we study them extensively in Chapter 3.

As noted with respect to the autonomous equation $y' = f(y)$, all the slopes of direction field filaments along the horizontal line $y = b$ are equal. This fact is illustrated in Figure 1.6, which shows the direction field for the differential

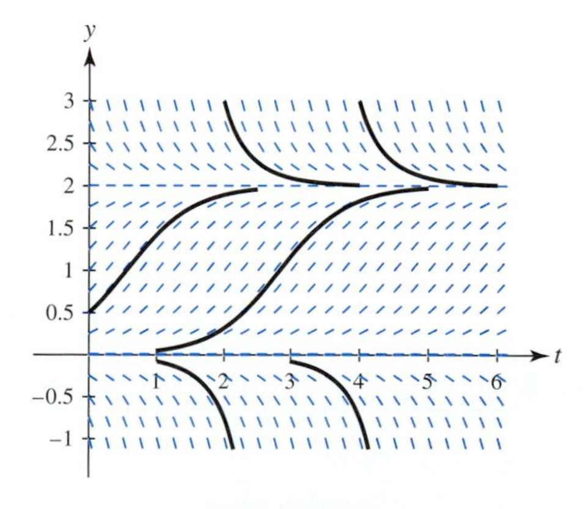

FIGURE 1.6

The direction field for the autonomous equation $y' = y(2 - y)$, together with portions of the graphs of some typical solutions. For an autonomous equation, the slopes are constant along horizontal lines.

equation

$$y' = y(2 - y).$$

For instance, the filaments along the line $y = 1$ all have slope equal to 1. Similarly, the filaments along the line $y = 2$ all have slope equal to 0. In fact, looking at Figure 1.6, the horizontal lines $y = 0$ and $y = 2$ appear to be solution curves for the differential equation $y' = y(2 - y)$. This is indeed the case, as we show in the next subsection.

Equilibrium Solutions

Consider the autonomous differential equation $y' = y(2 - y)$ whose direction field is shown in Figure 1.6. The horizontal lines $y = 0$ and $y = 2$ appear to be solution curves for this differential equation. In fact, by substituting either of the constant functions $y(t) \equiv 0$ or $y(t) \equiv 2$ into the differential equation, we see that it is a solution of $y' = y(2 - y)$.

In general, consider the autonomous differential equation

$$y' = f(y).$$

If the real number β is a root of the equation $f(y) = 0$, then the constant function $y(t) \equiv \beta$ is a solution of $y' = f(y)$. Conversely, if the constant function $y(t) \equiv \beta$ is a solution of $y' = f(y)$, then β must be a root of $f(y) = 0$. Constant solutions of an autonomous differential equation are known as **equilibrium solutions**.

REMARK: It is possible for differential equations that are not autonomous to have constant solutions. For example, $y(t) \equiv 0$ is a solution of $y' = ty + \sin y$ and $y(t) \equiv 1$ is a solution of $y' = (y - 1)t^2$. We will refer to any constant solution of a differential equation (autonomous or not) as an equilibrium solution.

EXAMPLE 3

Find the equilibrium solutions (if any) of

$$y' = y^2 - 4y + 3.$$

Solution: The right-hand side of the differential equation is

$$f(y) = y^2 - 4y + 3 = (y - 1)(y - 3).$$

Therefore, the equilibrium solutions are the constant functions $y(t) \equiv 1$ and $y(t) \equiv 3$. ▲

EXERCISES

Exercises 1–6:

(a) State whether or not the equation is autonomous.

(b) Identify all equilibrium solutions (if any).

(c) Sketch the direction field for the differential equation in the rectangular portion of the ty-plane defined by $-2 \le t \le 2$, $-2 \le y \le 2$.

1. $y' = -y + 1$ **2.** $y' = t - 1$ **3.** $y' = \sin y$

4. $y' = y^2 - y$ **5.** $y' = -1$ **6.** $y' = -ty$

Exercises 7–9:

(a) Determine and sketch the isoclines $f(t, y) = c$ for $c = -1, 0$, and 1.

(b) On each of the isoclines drawn in part (a), add representative direction field filaments.

7. $y' = -y + 1$ **8.** $y' = -y + t$ **9.** $y' = y^2 - t^2$

Exercises 10–13:

Find an autonomous differential equation that possesses the specified properties. [Note: There are many possible solutions for each exercise.]

10. Equilibrium solutions at $y = 0$ and $y = 2$; $y' > 0$ for $0 < y < 2$; $y' < 0$ for $-\infty < y < 0$ and $2 < y < \infty$.

11. An equilibrium solution at $y = 1$; $y' < 0$ for $-\infty < y < 1$ and $1 < y < \infty$.

12. A differential equation with no equilibrium solutions and $y' > 0$ for all y.

13. Equilibrium solutions at $y = n/2, n = 0, \pm 1, \pm 2, \ldots$

Exercises 14–19:

Consider the six direction field plots shown. Associate a direction field with each of the following differential equations.

14. $y' = -y$ **15.** $y' = -t + 1$ **16.** $y' = y^2 - 1$

17. $y' = -\dfrac{1}{2}$ **18.** $y' = y + t$ **19.** $y' = \dfrac{1}{1 + y^2}$

20. For each of the six direction fields shown, assume we are interested in the solution that satisfies the initial condition $y(0) = 0$. Use the graphical information contained in the plots to roughly estimate $y(1)$, the value of the same solution at time $t = 1$.

21. Repeat Exercise 20 with $y(0) = 0$ as before, but this time estimate $y(-1)$.

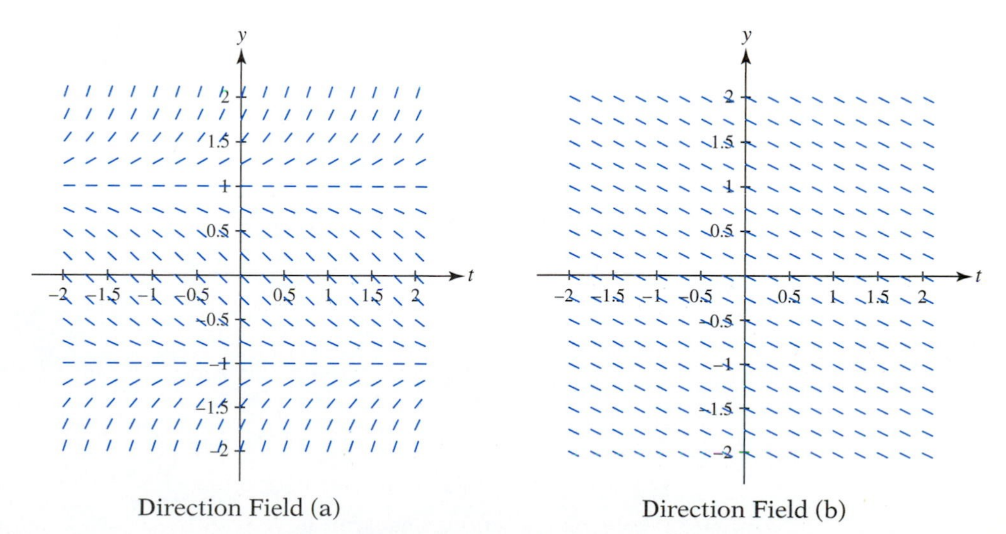

Direction Field (a) Direction Field (b)

Figure for Exercises 14–21

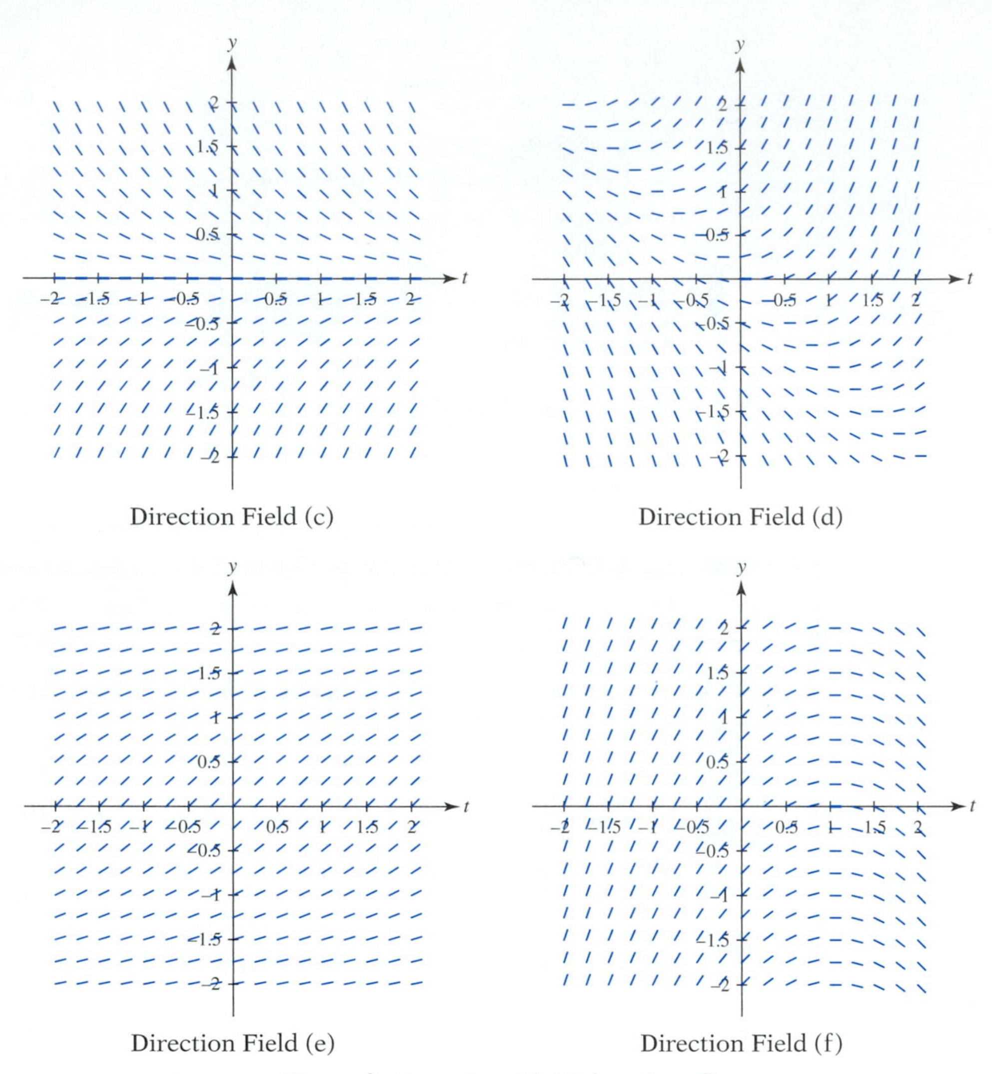

Figure for Exercises 14–21 (continued)

2

First Order Linear Differential Equations

CHAPTER OVERVIEW

Introduction

First order differential equations arise in modeling a wide variety of physical phenomena. In this chapter we develop solution techniques for an important class of first order equations—first order *linear* differential equations. These techniques are then used to study applications such as compound interest, population dynamics, radioactive decay, belt friction, mixing and cooling, and so forth.

The organization of the material in this chapter follows a pattern used in subsequent chapters as well. We begin, in Section 2.1, with a brief statement of the conditions needed to guarantee existence of a unique solution of the problem of interest. This sets the conceptual framework. Once we are assured of the existence of a unique solution, we can go about developing techniques to find the solution.

Solution techniques for first order linear equations are developed in Sections 2.2 and 2.3. In Sections 2.4 and 2.5 we apply these techniques to studying some of the aforementioned applications.

2.1 Existence and Uniqueness

A differential equation of the form

$$y' + p(t)y = g(t) \tag{1}$$

is called a **first order linear differential equation**. In equation (1), $p(t)$ and $g(t)$ are functions defined on some t-interval of interest, $a < t < b$.

If the function $g(t)$ on the right-hand side of (1) is the zero function, then equation (1) is called **homogeneous**. If $g(t)$ is not the zero function, then equation (1) is **nonhomogeneous**.

A first order equation that can be put into the form of equation (1) by algebraic manipulations is also called a first order linear differential equation. For example, the following are first order linear differential equations:

$$\text{(a)}\ \ e^{-t}y' + 3ty = \sin t \qquad \text{(b)}\ \ \frac{1}{y}y' + t^2 = \frac{\ln t}{y}.$$

A first order differential equation that cannot be put into the form of equation (1) is called **nonlinear**. As we will see, it is possible to find an explicit representation for the solution of a first order *linear* equation. By contrast, most first order *nonlinear* equations cannot be solved explicitly. In Chapter 3 we will discuss solution techniques for certain special types of first order nonlinear differential equations.

EXAMPLE 1

Is the differential equation linear or nonlinear? If the equation is linear, decide whether it is homogeneous or nonhomogeneous.

(a) $y' = ty^2$ (b) $y' = t^2y$ (c) $(\cos t)y' + e^ty = \sin t$ (d) $\dfrac{y'}{y} + t^3 = \sin t$

Solution:

(a) This equation is nonlinear because of the presence of the y^2 term.

(b) The equation is linear and homogeneous; it can be put in the form $y' - t^2y = 0$.

(c) This equation can be put into the form of equation (1),

$$y' + \frac{e^t}{\cos t}y = \tan t.$$

Therefore, the equation is linear and nonhomogeneous.

(d) This equation can be put into the form of equation (1),

$$y' + (t^3 - \sin t)\,y = 0.$$

Therefore, the equation is linear and homogeneous. ▲

Existence and Uniqueness for First Order Linear Initial Value Problems

Before looking at how to solve a first order linear equation, we want to address the following question: "What constitutes a properly formulated problem?" This question is answered by the following theorem which we state now and prove later (see Exercises 31–32 in Section 2.3).

Theorem 2.1

Let $p(t)$ and $g(t)$ be continuous functions on the interval (a, b) and let t_0 be in (a, b). Then the initial value problem

$$y' + p(t)y = g(t), \qquad y(t_0) = y_0$$

has a unique solution on the entire interval (a, b).

Notice that the theorem states three conclusions. A solution exists, it is unique, and this unique solution exists on the entire interval (a, b). The interval (a, b) and the initial point (t_0, y_0) referred to in Theorem 2.1 are shown in Figure 2.1. In Chapter 3 we will see that determining intervals of existence is considerably more complicated for nonlinear differential equations.

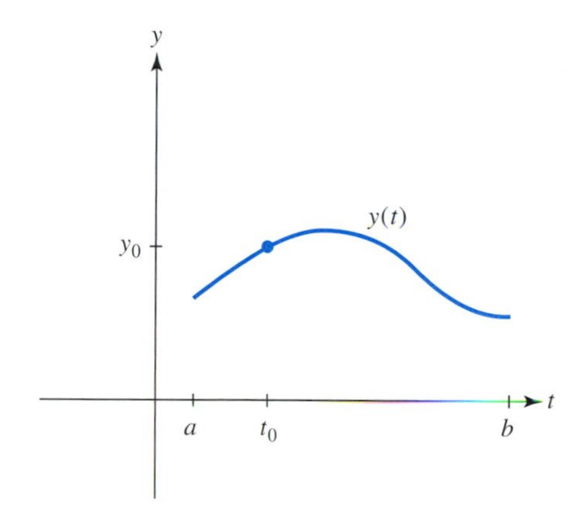

FIGURE 2.1

Let $p(t)$ and $g(t)$ be continuous on (a, b), and let t_0 be in (a, b). Then, by Theorem 2.1, there exists a unique solution, $y(t)$, to the initial value problem $y' + p(t)y = g(t)$, $y(t_0) = y_0$.

The importance of Theorem 2.1 lies in the fact that it defines the framework within which we can construct solutions. In particular, suppose we are given a linear differential equation $y' + p(t)y = g(t)$ with coefficient functions $p(t)$ and $g(t)$ that are continuous on (a, b). If we impose an initial condition of the form $y(t_0) = y_0$, where $a < t_0 < b$, the theorem tells us there is one and only one solution. Therefore, if we are able to construct a solution by using some technique we have discovered, the theorem guarantees that it is the only solution—there is no other solution we might have overlooked, one that is obtainable perhaps by a technique other than the one we are using.

EXAMPLE 2

Consider the initial value problem

$$y' + p(t)y = g(t), \qquad y(3) = 1.$$

Suppose that $p(t)$ and/or $g(t)$ have discontinuities at $t = -2$, $t = 0$, and $t = 5$ but are continuous for all other values of t. What is the largest interval (a, b) on

(continued)

(continued)

which Theorem 2.1 guarantees the existence of a unique solution to the initial value problem?

Solution: Considering only the differential equation, Theorem 2.1 *guarantees* the existence of a unique solution on each of the following t-intervals:

$$(-\infty, -2), \qquad (-2, 0), \qquad (0, 5), \qquad (5, \infty).$$

Since the initial condition for our problem is imposed at $t = 3$, we are guaranteed that a unique solution exists on the interval $0 < t < 5$. (The solution might actually exist over a larger interval, but we cannot be certain without actually solving the initial value problem.) ▲

EXERCISES

Exercises 1–10:

Classify each of the following first order differential equations as linear or nonlinear. If the equation is linear, decide whether it is homogeneous or nonhomogeneous.

1. $y' - \sin t = t^2 y$ **2.** $y' - \sin t = ty^2$ **3.** $\dfrac{y'}{y} - y\cos t = t$

4. $y'\sin y = (t^2+1)y$ **5.** $y'\sin t = \dfrac{t^2+1}{y}$ **6.** $2ty + e^t y' = \dfrac{y}{t^2+4}$

7. $yy' = t^3 + y\sin 3t$ **8.** $2ty + e^y y' = \dfrac{y}{t^2+4}$ **9.** $\dfrac{ty'}{(t^4+2)y} = \cos t + \dfrac{e^{3t}}{y}$

10. $\dfrac{y'}{(t^2+1)y} = \cos t$

Exercises 11–14:

Consider the following first order linear differential equations. For each of the initial conditions, determine the largest interval $a < t < b$ on which Theorem 2.1 guarantees the existence of a unique solution.

11. $y' + \dfrac{t}{t^2+1}y = \sin t;$

(a) $y(-2) = 1,$ (b) $y(0) = \pi,$ (c) $y(\pi) = 0.$

12. $y' + \dfrac{t}{t^2-4}y = 0;$

(a) $y(6) = 2,$ (b) $y(1) = -1,$ (c) $y(0) = 1,$ (d) $y(-6) = 2.$

13. $y' + \dfrac{t}{t^2-4}y = \dfrac{e^t}{t-3};$

(a) $y(5) = 2,$ (b) $y(-3/2) = 1,$ (c) $y(0) = 0,$
(d) $y(-5) = 4,$ (e) $y(3/2) = 3.$

14. $y' + (t-1)y = \dfrac{\ln|t+t^{-1}|}{t-2};$

(a) $y(3) = 0,$ (b) $y(1/2) = -1,$ (c) $y(-1/2) = 1,$ (d) $y(-3) = 2.$

15. If $y(t) = 3e^{t^2}$ is known to be the solution of the initial value problem

$$y' + p(t)y = 0, \qquad y(0) = y_0,$$

what must the function $p(t)$ and the constant y_0 be?

16. (a) For what value of the constant C and exponent r is $y = Ct^r$ the solution of the initial value problem

$$2ty' - 6y = 0, \qquad y(-2) = 8?$$

(b) Determine the largest interval of the form (a, b) on which Theorem 2.1 guarantees the existence of a unique solution.

(c) What is the actual interval of existence for the solution found in part (a)?

17. If $p(t)$ is any function continuous on an interval of the form $a < t < b$ and if t_0 is any point lying within this interval, what is the unique solution of the initial value problem

$$y' + p(t)y = 0, \qquad y(t_0) = 0$$

on this interval? [Hint: If, by inspection, you can identify one solution of the given initial value problem, then Theorem 2.1 tells you that it must be the only solution.]

2.2 First Order Linear Homogeneous Differential Equations

In this section we focus on solving the first order linear homogeneous differential equation

$$y' + p(t)y = 0. \tag{1}$$

In Section 2.3, we build on the results of this section, showing how to solve the nonhomogeneous equation $y' + p(t)y = g(t)$.

Solving the Linear Homogeneous Equation

Consider the homogeneous first order linear equation $y' + p(t)y = 0$, which we rewrite as

$$y' = -p(t)y. \tag{2}$$

In equation (2) we assume that $p(t)$ is continuous on the t-interval of interest.

To solve equation (2), we need to find a function $y(t)$ whose derivative is equal to $-p(t)$ times $y(t)$. Recall from the chain rule of calculus that

$$\frac{d}{dt}e^{-P(t)} = -P'(t)e^{-P(t)}.$$

The function $y = e^{-P(t)}$ has the property that

$$y' = -P'(t)y.$$

Therefore, if we choose a function $P(t)$ such that $P'(t) = p(t)$, then

$$y = e^{-P(t)} \tag{3}$$

satisfies the condition $y' = -p(t)y$.

If $P'(t) = p(t)$, then $P(t)$ is an **antiderivative** of $p(t)$ and is usually denoted by the integral notation, $P(t) = \int p(t)\,dt$. So a solution of $y' = -p(t)y$ can be expressed as

$$y = e^{-\int p(t)\,dt}.$$

Under the hypotheses of Theorem 2.1, $p(t)$ is guaranteed to have antiderivatives since $p(t)$ is continuous on (a, b). In general, let c be any point in (a, b) and define $P(t)$, for each t in (a, b), by the integral

$$P(t) = \int_c^t p(s)\,ds.$$

As defined above, the function $P(t)$ is the particular antiderivative of $p(t)$ that vanishes at the point $t = c$.

Use equation (3) to find a solution of the differential equation

$$y' + 2ty = 0.$$

Solution: For this linear equation, $p(t) = 2t$. For $P(t)$ we can choose any convenient antiderivative of $p(t)$. If we select

$$P(t) = t^2,$$

then, using (3), we obtain the solution

$$y = e^{-t^2}.$$

As a check, let $y = e^{-t^2}$. Then, $y' = -2te^{-t^2} = -2ty$. Thus, we have verified that $y = e^{-t^2}$ is a solution of $y' + 2ty = 0$. Figure 2.2 shows the direction field for this differential equation as well as a graph of the solution. ▲

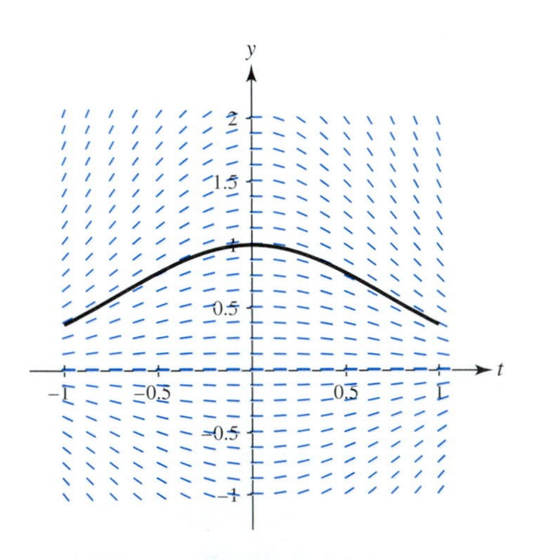

FIGURE 2.2

The direction field for the differential equation in Example 1 and the graph of a solution, $y = e^{-t^2}$.

The General Solution

We can always use equation (3) to find *one* solution of $y' + p(t)y = 0$. But, in order to solve initial value problems, we need to develop a method for finding *all* the solutions of $y' + p(t)y = 0$. To that end, consider the homogeneous equation $y' = -p(t)y$, where $p(t)$ is continuous on (a, b). Let $P(t)$ be some antiderivative for $p(t)$. We know, see equation (3), that one solution of $y' = -p(t)y$ is

$$y = e^{-P(t)}.$$

However, if $y(t)$ solves the homogeneous problem $y' = -p(t)y$, then so does any constant multiple, $u(t) = Cy(t)$. To verify that $u(t)$ is also a solution, observe that

$$u'(t) = \frac{d}{dt}(Cy(t)) = Cy'(t) = -Cp(t)y(t) = -p(t)u(t).$$

In fact (see Section 2.3), Theorem 2.1 can be used to show that *every* solution of $y' = -p(t)y$ can be expressed as

$$y = Ce^{-P(t)} \tag{4}$$

for some constant C. Because every solution of $y' = -p(t)y$ has this form, we call (4) the **general solution**.

EXAMPLE 2

Find the general solution of

$$y' + (\cos t)y = 0.$$

Solution: A convenient antiderivative for $p(t) = \cos t$ is $P(t) = \sin t$. Thus, the general solution is

$$y = Ce^{-\sin t}. \quad \blacktriangle$$

REMARK: Let $P(t)$ be an antiderivative of $p(t)$. From calculus we know that any other antiderivative of $p(t)$ has the form $P(t) + K$, where K is some constant. For instance, in Example 2, we chose $P(t) = \sin t$ as an antiderivative of $p(t) = \cos t$. We could just as well have chosen $P(t) = 2 + \sin t$ as the antiderivative. In that case, the general solution would have had the form

$$y = Ce^{-(2+\sin t)} = Ce^{-2}e^{-\sin t} = C_1 e^{-\sin t}.$$

In this expression, C is an arbitrary constant. Since $C_1 = Ce^{-2}$, we can regard C_1 as an arbitrary constant as well. Thus, no matter which antiderivative we choose, the general solution is still the product of an arbitrary constant and the function $e^{-\sin t}$.

Using the General Solution to Solve Initial Value Problems

An initial value problem for a homogeneous first order linear equation can be solved by first forming the general solution

$$y = Ce^{-P(t)}$$

and then choosing the constant C so as to satisfy the initial condition.

EXAMPLE 3

Solve the initial value problem

$$ty' + 2y = 0, \qquad y(1) = 5.$$

Solution: Notice that the differential equation is not in the standard form for a first order linear equation. In order to use equation (4) to represent the general solution, we need to rewrite the differential equation as

$$y' + \frac{2}{t}y = 0, \qquad y(1) = 5.$$

As rewritten, $p(t) = 2/t$. Therefore, a convenient antiderivative is

$$P(t) = \int \frac{2}{t}\,dt = 2\ln|t| = \ln|t|^2 = \ln t^2.$$

Having an antiderivative $P(t)$, we obtain the general solution,

$$y = Ce^{-P(t)} = Ce^{-\ln t^2} = Ct^{-2}.$$

Using the general solution $y = Ct^{-2}$, we have $y(1) = C$. To satisfy the initial condition, we must choose $C = 5$. Thus, the unique solution of the initial value problem is

$$y = \frac{5}{t^2}. \ \blacktriangle$$

Example 3 illustrates a point about Theorem 2.1. The differential equation has a coefficient function, $p(t) = 2/t$, that is not defined and certainly not continuous at $t = 0$. Therefore, Theorem 2.1 cannot be used to *guarantee* that solutions exist across any interval containing $t = 0$. In fact, for this initial value problem, the solution, $y(t) = 5/t^2$, is not defined at $t = 0$.

However, if we change the initial condition in Example 3 to $y(1) = 0$, we find that the solution is the zero function, $y(t) \equiv 0$ (see Figure 2.3). Thus, even though this particular initial value problem does not satisfy the conditions of

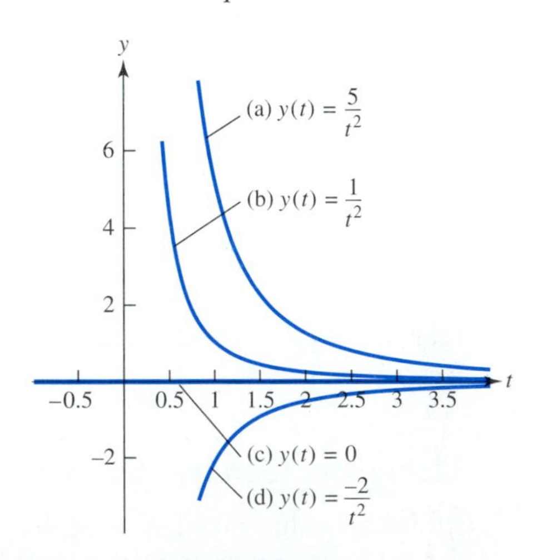

FIGURE 2.3

Some solutions of the problem $ty' + 2y = 0$, $y(1) = y_0$, posed in Example 3. When y_0 is nonzero, the solution is only defined for $t > 0$. But if $y_0 = 0$, the solution is the zero function and is defined for all t.

Theorem 2.1 on $(-\infty, \infty)$, it does in fact have a solution that is defined for all t. It is important to realize that the failure of Theorem 2.1 to apply to an initial value problem does not imply that the solution must necessarily "behave badly." The logical distinction is important. Theorem 2.1 asserts that if the hypotheses are satisfied, "good things will happen." It does *not* assert that when the hypotheses are not satisfied, then "bad things must happen."

EXERCISES

Exercises 1–10:

For each initial value problem

(a) Find the general solution of the differential equation.

(b) Impose the initial condition to obtain the solution of the initial value problem.

1. $y' + 3y = 0, \quad y(0) = -3$

2. $2y' - y = 0, \quad y(-1) = 2$

3. $2ty - y' = 0, \quad y(1) = 3$

4. $ty' - 4y = 0, \quad y(1) = 1$

5. $ty' + 4y = 0, \quad y(1) = 1$

6. $y' + (1 + \sin t)y = 0, \quad y(\pi/2) = 1$

7. $y' - 2(\cos 2t)y = 0, \quad y(\pi) = -2$

8. $(t^2 + 1)y' + 2ty = 0, \quad y(0) = 3$

9. $\dfrac{y'}{(t^2+1)y} = 3, \quad y(1) = 4$

10. $y + e^t y' = 0, \quad y(0) = 2$

11. Consider the three direction fields shown on the next page. Match each of the direction field plots with one of the following differential equations.

(a) $y' + y = 0$ (b) $y' + t^2 y = 0$ (c) $y' - y = 0$

Exercises 12–13:

The graph of the solution of the given initial value problem is known to pass through the (t, y) points listed. Determine the constants α and y_0.

12. $y' + \alpha y = 0, y(0) = y_0$. The graph of the solution passes through the points $(1, 4)$ and $(3, 1)$.

13. $ty' - \alpha y = 0, y(1) = y_0$. The graph of the solution passes through the points $(2, 1)$ and $(4, 4)$.

Exercises 14–16: **A Change of Dependent Variable**

14. Suppose you are given the initial value problem

$$y' = 2y + 4, \qquad y(0) = -1.$$

The differential equation is a first order linear nonhomogeneous equation. We will show how to solve such equations in Section 2.3. In this case, however, since the coefficient functions are constant, we can make a simple change of dependent variable and solve the resulting problem using the ideas of this section.

(a) Note that the differential equation can be rewritten as $y' = 2(y + 2)$. Let $z(t) = y(t) + 2$. Transform the given initial value problem into one for $z(t)$. The new differential equation for $z(t)$ should be a first order linear homogeneous equation. [Don't forget the corresponding initial condition for $z(0)$.]

(b) Solve this new initial value problem for $z(t)$ and use the solution to determine $y(t)$.

15. Apply the ideas of Exercise 14 to solve the initial value problem

$$y' + 2ty = 6t, \qquad y(0) = 4.$$

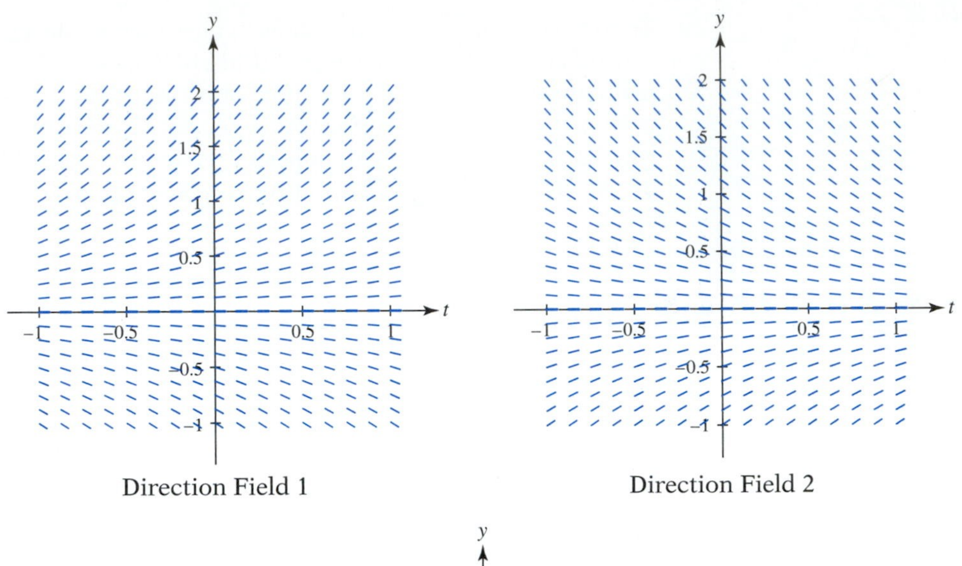

Direction Field 1

Direction Field 2

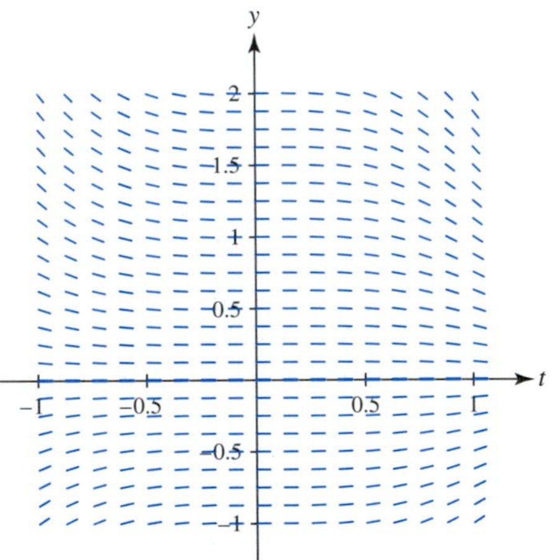

Direction Field 3

Figure for Exercise 11

16. Antioxidants Active oxygen and free radicals are believed to be exacerbating factors in causing cell injury and aging in living tissue.[1] These molecules also accelerate the deterioration of foods. Researchers are therefore interested in understanding the protective role of natural antioxidants. In the study of one such antioxidant (Hsian-tsao leaf gum), the antioxidation activity of the substance has been found to depend upon concentration in the following way:

$$\frac{dA(c)}{dc} = k[A^* - A(c)], \qquad A(0) = 0.$$

In this equation, the dependent variable A is a quantitative measure of antioxidant activity at concentration c. The constant A^* represents a limiting or equilibrium value of this activity and k is a positive rate constant.

(a) Let $B(c) = A(c) - A^*$ and reformulate the given initial value problem in terms of this new dependent variable, B.

(b) Solve the new initial value problem for $B(c)$ and then determine the quantity of interest, $A(c)$. Does the activity $A(c)$ ever exceed the value A^*?

---

[1] Lih-Shiuh Lai, Su-Tze Chou, and Wen-Wan Chao, "Studies on the Antioxidative Activities of Hsian-tsao (*Mesona procumbens Hemsl*) Leaf Gum," *J. Agric. Food Chem.* 49 (2001), pp. 963–968.

(c) Determine the concentration at which 95% of the limiting antioxidation activity is achieved. (Your answer is a function of the rate constant k.)

Exercises 17–18:

For the given initial value problem, a portion of the graph of the solution is shown. Use the information contained in the graph to determine the constants c and y_0. The constants c and y_0 are integers that may be either positive or negative.

17. $y' + cy = 0, \quad y(0) = y_0$

18. $y' + cy = 0, \quad y(1) = y_0$

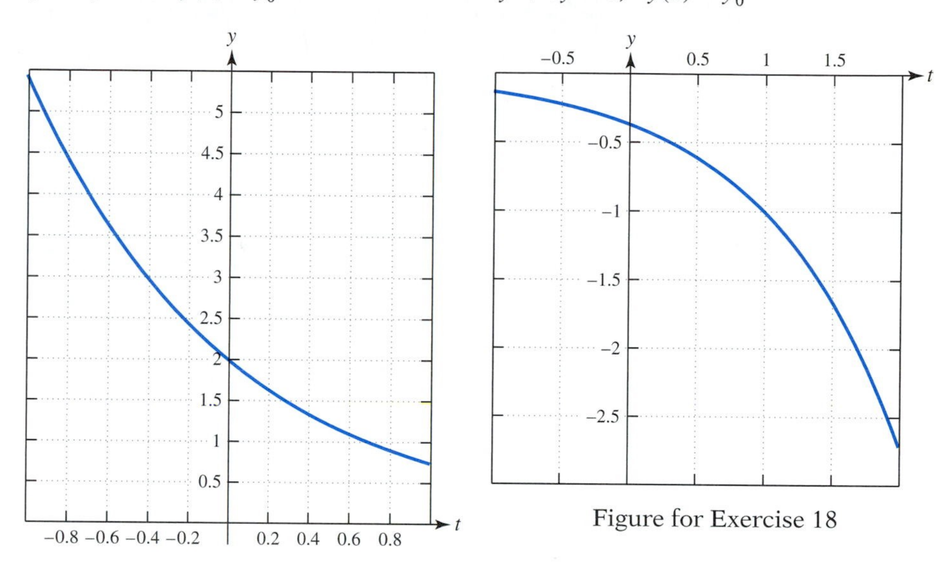

Figure for Exercise 17

Figure for Exercise 18

19. On the next page are four graphs of $\ln[y(t)]$ versus t, $0 \le t \le 4$, corresponding to solutions of the four differential equations, (a)–(d). Match the graphs to the differential equations. For each match, identify the initial condition, $y(0)$. [Caution: Do not forget the graphs are plots of the natural logarithm of the solution versus time.]

(a) $y' + y = 0$ (b) $y' - (\sin 4t + 4t\cos 4t)y = 0$

(c) $y' + ty = 0$ (d) $y' - (1 - 4\cos 4t)y = 0$

20. The figure shown on the next page is the graph of $\ln[y(t)]$ versus t, $0 \le t \le 4$, where $y(t)$ is the solution of the initial value problem

$$y' + p(t)y = 0, \qquad y(0) = y_0.$$

Determine $p(t)$ and y_0.

21. The table below lists values of t and $\ln[y(t)]$ where $y(t)$ is a solution of the initial value problem

$$y' + t^n y = 0, \qquad y(0) = y_0.$$

(a) Determine the nonnegative integer n and the positive constant y_0.

(b) What is $y(-1)$?

t	$\ln[y(t)]$
1	−0.25
2	−4.0
3	−20.25
4	−64.0

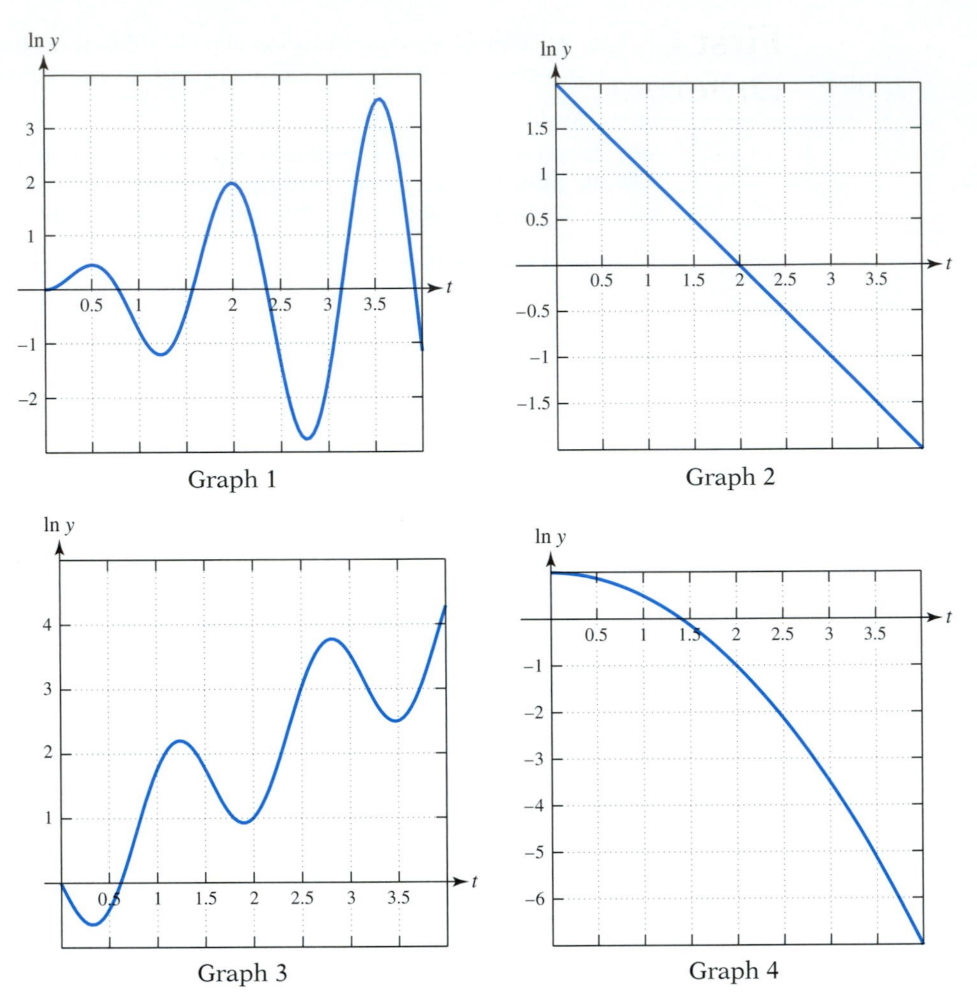

Figure for Exercise 19

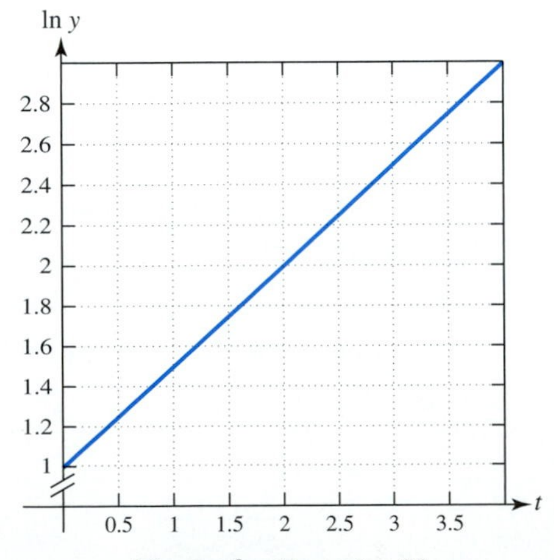

Figure for Exercise 20

2.3 First Order Linear Nonhomogeneous Differential Equations

In the previous section we discussed the homogeneous linear equation

$$y' + p(t)y = 0.$$

In this section we see how to solve the *nonhomogeneous* linear equation

$$y' + p(t)y = g(t).$$

As before, we assume that $p(t)$ and $g(t)$ are continuous on our t-interval of interest.

Integrating Factors

As preparation for solving the nonhomogeneous equation, we consider our previous discussion of the homogeneous equation from a slightly different point of view. In particular, the homogeneous equation has the form

$$y' + p(t)y = 0. \tag{1}$$

Let $P(t)$ be some antiderivative of $p(t)$ and define a new function $\mu(t)$ by

$$\mu(t) = e^{P(t)}. \tag{2}$$

The function $\mu(t) = e^{P(t)}$ is called an **integrating factor**. We will shortly see the reason for this name.

Note from equation (2) that

$$\mu'(t) = P'(t)e^{P(t)} = p(t)\mu(t). \tag{3a}$$

We multiply equation (1) by the integrating factor $\mu(t)$ to obtain a new equation,

$$\mu(t)y' + \mu(t)p(t)y = 0.$$

From (3a), $\mu'(t) = p(t)\mu(t)$, and therefore

$$\mu(t)y' + \mu'(t)y = 0. \tag{3b}$$

The left-hand side of equation (3b) is the derivative of a product and can be rewritten as

$$\frac{d}{dt}(\mu(t)y(t)) = 0. \tag{4}$$

If the derivative of a function is identically zero, then the function must be constant. Therefore, equation (4) implies

$$\mu(t)y(t) = C,$$

where C is a constant. Knowing $\mu(t)y(t) = C$ and knowing that $\mu(t) = e^{P(t)}$ is nonzero, we can solve for $y(t)$,

$$\begin{aligned} y(t) &= \frac{1}{\mu(t)}C \\ &= Ce^{-P(t)}. \end{aligned}$$

The derivation of equation (4) explains why the function $\mu(t) = e^{P(t)}$ is called an "integrating factor." That is, we multiply equation (1) by $\mu(t)$ to obtain the new equation (4), which can be integrated. Also note that this derivation leads to the same general solution for equation (1) that we found in Section 2.2, $y = Ce^{-P(t)}$. The purpose of the derivation is to introduce the concept of an integrating factor.

Using an Integrating Factor to Solve the Nonhomogeneous Equation

Now consider the nonhomogeneous equation

$$y' + p(t)y = g(t). \tag{5}$$

If we multiply equation (5) by the integrating factor $\mu(t) = e^{P(t)}$, we obtain $\mu(t)y' + \mu(t)p(t)y = \mu(t)g(t)$. Since $\mu'(t) = \mu(t)p(t)$, we then obtain

$$\mu(t)y' + \mu'(t)y = \mu(t)g(t)$$

or

$$\frac{d}{dt}(\mu(t)y(t)) = \mu(t)g(t).$$

Integrating both sides,

$$\mu(t)y(t) = \int \mu(t)g(t)\,dt + C,$$

where C is a constant and where $\int \mu(t)g(t)\,dt$ represents some particular antiderivative of $\mu(t)g(t)$. Solving for $y(t)$, we are led to the *general solution* of equation (5):

$$y = e^{-P(t)} \int e^{P(t)}g(t)\,dt + Ce^{-P(t)}. \tag{6}$$

REMARKS:

1. Don't be confused by the notation. In particular, the terms $e^{-P(t)}$ and $e^{P(t)}$ in the general solution do not cancel; $e^{P(t)}$ is part of the function $e^{P(t)}g(t)$ whose antiderivative must be determined. Once this antiderivative has been calculated, it is multiplied by the term $e^{-P(t)}$.
2. Notice that the general solution given by (6) is the sum of two terms, $e^{-P(t)} \int e^{P(t)}g(t)\,dt$ and $Ce^{-P(t)}$. The first term is some *particular* solution of the nonhomogeneous equation, while the second term represents the general solution of the homogeneous equation. We'll see this same solution structure again when we study higher order linear equations and systems of linear equations.
3. Observe that the general solution contains only one arbitrary constant, C. This constant is determined by imposing an initial condition.

4. Although expression (6) is the general solution of the nonhomogeneous equation, you should not try to memorize it. Instead, remember the steps leading to (6). ▮

EXAMPLE 1

Find the general solution and then solve the initial value problem

$$y' + 2ty = 4t, \qquad y(0) = 5.$$

Solution: For this differential equation, $p(t) = 2t$. An antiderivative is $P(t) = t^2$, and so an integrating factor is

$$\mu(t) = e^{t^2}.$$

Multiplying the differential equation by $\mu(t)$, we obtain

$$e^{t^2}y' + 2te^{t^2}y = 4te^{t^2} \quad \text{or} \quad (e^{t^2}y)' = 4te^{t^2}.$$

Integrating both sides with respect to t,

$$e^{t^2}y = 2e^{t^2} + C.$$

Solving for y, we obtain the general solution,

$$y = 2 + Ce^{-t^2}.$$

Imposing the initial condition, we have $y(0) = 5 = 2 + C$ and therefore $C = 3$. The solution of the initial value problem is

$$y = 2 + 3e^{-t^2}.$$

The solution is graphed in Figure 2.4. ▲

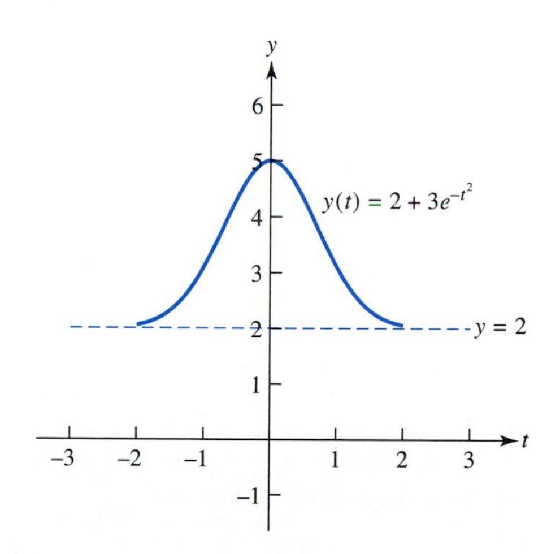

FIGURE 2.4

The solution of the problem posed in Example 1 is $y = 2 + 3e^{-t^2}$.

Example 1 illustrates the second remark following equation (6). The general solution we found (namely $y = 2 + Ce^{-t^2}$) is the sum of some particular solution of the nonhomogeneous equation (namely the constant function $y = 2$) and the general solution of the homogeneous equation (namely $y = Ce^{-t^2}$).

For the differential equation in Example 1, all solutions actually tend to the constant function $y = 2$,

$$\lim_{t\to\infty}(2 + Ce^{-t^2}) = 2.$$

Figure 2.5 shows a portion of the direction field for the equation $y' + 2ty = 4t$ treated in Example 1. As you can see from Figure 2.5, solutions appear to approach the constant solution $y = 2$. Thus, we would have suspected the limit above even if we had not solved the differential equation.

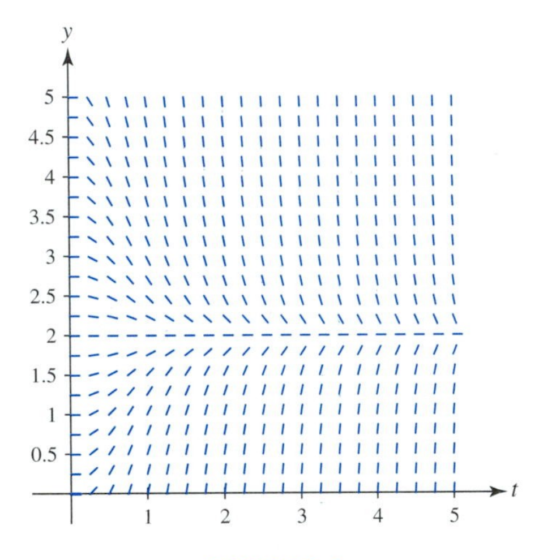

FIGURE 2.5

A portion of the direction field for $y' + 2ty = 4t$. Note that solutions appear to approach the constant solution $y = 2$.

EXAMPLE 2

Solve the initial value problem

$$ty' + 4y = 6t^2, \qquad y(1) = 3, \qquad t > 0.$$

Solution: In order to identify an integrating factor, we first write the differential equation in standard form,

$$y' + \frac{4}{t}y = 6t.$$

Given that $p(t) = 4/t$, an antiderivative is $P(t) = 4\ln|t| = \ln|t|^4 = \ln t^4$. Therefore, an integrating factor is

$$\mu(t) = e^{\ln t^4} = t^4.$$

Multiplying the differential equation by $\mu(t)$, we obtain

$$t^4y' + 4t^3y = 6t^5 \quad \text{or} \quad (t^4y)' = 6t^5.$$

Integrating with respect to t and then solving for y, we find the general solution

$$y = t^2 + \frac{C}{t^4}.$$

Imposing the initial condition, $y(1) = 3 = 1 + C$ and therefore $C = 2$. The solution of the initial value problem is

$$y = t^2 + \frac{2}{t^4}.$$

A portion of the solution is graphed in Figure 2.6. Note that the solution exists on $0 < t < \infty$. [Also note that, without actually solving the differential equation, we could have used Theorem 2.1 to deduce that the initial value problem has a unique solution defined on $(0, \infty)$.] ▲

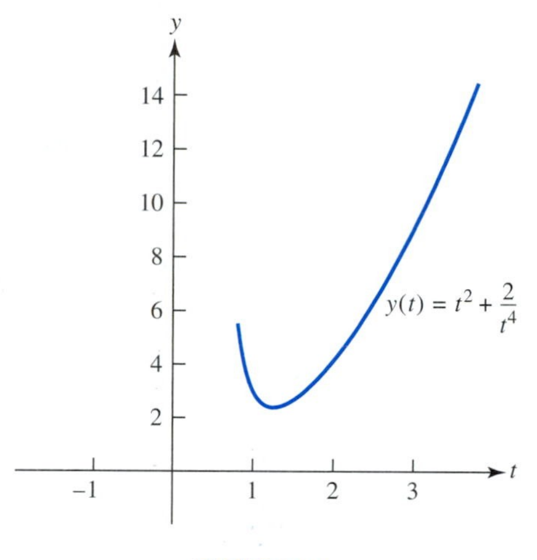

FIGURE 2.6

The solution of the problem posed in Example 2 is $y = t^2 + 2t^{-4}$.

Discontinuous Coefficient Functions

In some applications, physical conditions undergo abrupt changes. For example, a hot metal object might be plunged suddenly into a cooling bath, or we might throw a switch and abruptly change the source voltage in an electrical network.

Such applications frequently involve an initial value problem

$$y' + p(t)y = g(t), \qquad y(a) = y_0, \qquad a \le t_0 \le b$$

where one or both of the functions $p(t)$ and $g(t)$ has a jump discontinuity at some point, say $t = c$, where $a < c < b$. In such cases, even though $y'(t)$ is not continuous at $t = c$, we expect on physical grounds that the solution $y(t)$ is continuous at $t = c$. For these problems we first solve the initial value problem on the interval $a \le t < c$; the solution $y(t)$ will have a one-sided limit,

$$\lim_{t \to c^-} y(t) = y(c^-).$$

To complete the solution, we use the limiting value $y(c^-)$ as the initial condition on the subinterval $[c, b]$ and then solve a second initial value problem on $[c, b]$.[2]

[2]Theorem 2.1 was stated for an open interval of the form (r, s). However, Theorem 2.1 can be extended so that it applies to a closed interval of the form $[r, s]$ or to half-open intervals of the form $[r, s)$ or $(r, s]$.

EXAMPLE 3

Solve the initial value problem on the interval $[0, 2]$,

$$y' - y = g(t), \qquad y(0) = 0, \qquad \text{where } g(t) = \begin{cases} 1, & 0 \le t < 1 \\ -2, & 1 \le t \le 2. \end{cases}$$

Solution: A graph of the function $g(t)$ is shown in Figure 2.7(a); notice the jump discontinuity at $t = 1$. On the interval $[0, 1)$, the differential equation reduces to $y' - y = 1$. An integrating factor is $\mu(t) = e^{-t}$ and the general solution is

$$y(t) = Ce^t - 1.$$

Imposing the initial condition, we find $C = 1$ and hence the solution on $[0, 1)$ is $y(t) = e^t - 1$. As t approaches 1 from the left, $y(t)$ approaches the value $y_1 = e - 1$. To have continuity at $t = 1$, we now solve the initial value problem

$$y' - y = -2, \qquad y(1) = e - 1, \qquad 1 \le t \le 2.$$

The general solution of this initial value problem is

$$y(t) = Ce^t + 2.$$

Imposing the initial condition, $y(1) = e - 1$, we find $C = (1 - 3/e)$.

Having solved these two initial value problems, we obtain the solution $y(t)$ on the entire interval $[0, 2]$. A graph of $y(t)$ is shown in Figure 2.7(b). Note that the solution $y(t)$ is defined in a piecewise fashion:

$$y(t) = \begin{cases} e^t - 1, & 0 \le t < 1 \\ \left(1 - \dfrac{3}{e}\right) e^t + 2, & 1 \le t \le 2. \end{cases}$$

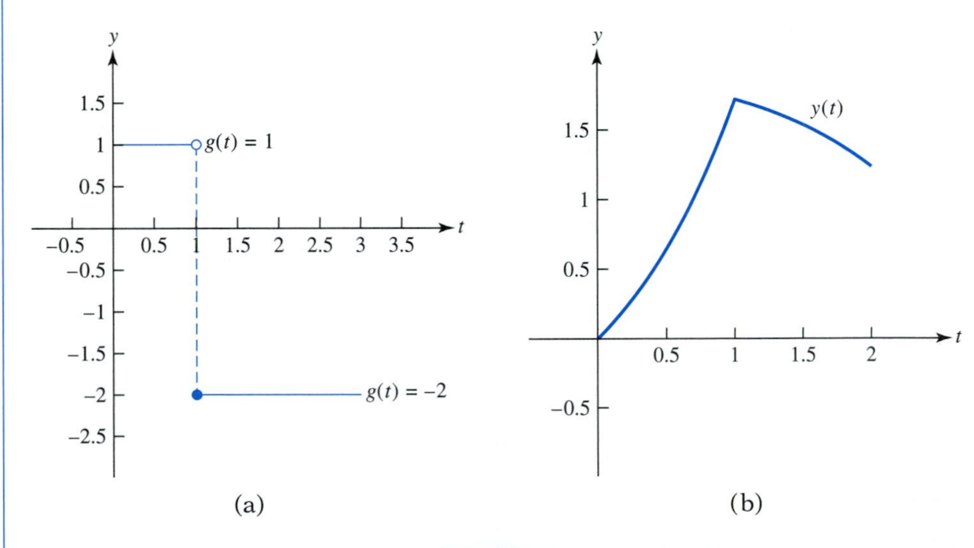

FIGURE 2.7

(a) The coefficient function $g(t)$ in the differential equation $y' - y = g(t)$ of Example 3 has a jump discontinuity at $t = 1$. (b) The solution of the initial value problem $y' - y = g(t)$, $y(0) = 0$ is continuous on $[0, 2]$ but is not differentiable at $t = 1$.

As you can see from this piecewise definition and from the graph, $y(t)$ is continuous on the entire t-interval of interest, $[0, 2]$. However, $y(t)$ is not differentiable at $t = 1$; at $t = 1$, the left-hand derivative of $y(t)$ is e, but the right-hand derivative of $y(t)$ is $e - 3$. Since the one-sided derivatives are not equal at $t = 1$, the solution is not differentiable at $t = 1$. ▲

EXERCISES

Exercises 1–8:

Find the general solution.

1. $y' + 2y = 1$ **2.** $y' + 2y = e^{-t}$ **3.** $y' + 2y = e^{-2t}$

4. $y' + 2ty = t$ **5.** $ty' + 2y = t^2,\ t > 0$ **6.** $(t^2 + 4)y' + 2ty = t^2(t^2 + 4)$

7. $y' + y = t$ **8.** $y' + 2y = \cos 3t$

Exercises 9–15:

Solve the initial value problem.

9. $y' - 3y = 6,\ y(0) = 1$ **10.** $y' - 2y = e^{3t},\ y(0) = 3$

11. $2y' + 3y = e^t,\ y(0) = 0$ **12.** $y' + y = 1 + 2e^{-t}\cos 2t,\ y(\pi/2) = 0$

13. $2y' + (\cos t)y = -3\cos t,\ y(0) = -4$ **14.** $y' + 2y = e^{-t} + t + 1,\ y(-1) = e$

15. $ty' + 3y = t + 1,\ y(-1) = 1/3$ (What is the t-interval on which the solution exists?)

16. The solution of the initial value problem $ty' + 4y = \alpha t^2, y(1) = -1/3$ is known to exist on $-\infty < t < \infty$. What is the constant α?

Exercises 17–19:

In each exercise, the general solution of the differential equation $y' + p(t)y = g(t)$ is given where C is an arbitrary constant. Determine the functions $p(t)$ and $g(t)$.

17. $y(t) = Ce^{-2t} + t + 1$ **18.** $y(t) = Ce^{t^2} + 2$ **19.** $y(t) = Ct^{-1} + 1, \quad t > 0$

Exercises 20–21:

In each exercise, the unique solution of the initial value problem $y' + y = g(t), y(0) = y_0$ is given. Determine the constant y_0 and the function $g(t)$.

20. $y(t) = e^{-t} + t - 1$ **21.** $y(t) = -2e^{-t} + e^t + \sin t$

Exercises 22–23:

In each exercise, discuss the behavior of the solution $y(t)$ as t becomes large. Does $\lim_{t\to\infty} y(t)$ exist? If so, what is the limit?

22. $y' + y + y\cos t = 1 + \cos t,\ y(0) = 3$

23. $\dfrac{y' - e^{-t} + 2}{y} = -2,\ y(0) = -2$

Exercises 24–27:

As in Example 3, find a solution to the initial value problem that is continuous on the given interval $[a, b]$.

24. $y' + \dfrac{1}{t}y = g(t), \quad y(1) = 1; \qquad g(t) = \begin{cases} 3t, & 1 \le t \le 2 \\ 0, & 2 < t \le 3 \end{cases}; \qquad [a, b] = [1, 3]$

25. $y' + (\sin t)y = g(t), \quad y(0) = 3; \quad g(t) = \begin{cases} \sin t, & 0 \le t \le \pi \\ -\sin t, & \pi < t \le 2\pi \end{cases}; \quad [a, b] = [0, 2\pi]$

26. $y' + p(t)y = 2, \quad y(0) = 1; \quad p(t) = \begin{cases} 0, & 0 \le t \le 1 \\ \dfrac{1}{t}, & 1 < t \le 2 \end{cases}; \quad [a, b] = [0, 2]$

27. $y' + p(t)y = 0, \quad y(0) = 3; \quad p(t) = \begin{cases} 2t - 1, & 0 \le t \le 1 \\ 0, & 1 < t \le 3 \\ -\dfrac{1}{t}, & 3 < t \le 4 \end{cases}; \quad [a, b] = [0, 4]$

Exercises 28–29:

In each exercise, we ask you to express the solution in terms of "special functions" [the function $\mathrm{Si}(t)$ in Exercise 28 and $\mathrm{erf}(t)$ in Exercise 29]. Such **special functions** are sufficiently important in applications to warrant giving them names and studying their properties. (A book such as *Handbook of Mathematical Functions* by Abramowitz and Stegun[3] gives the definition for many important special functions, lists their properties, and has tables of their values. Scientific software such as MATLAB, Mathematica, Maple, Derive, and so forth have subroutines for evaluating special functions.)

28. Solve $y' - \dfrac{1}{t}y = \sin t$, $y(1) = 3$. Express your answer in terms of the sine integral, $\mathrm{Si}(t)$, where $\mathrm{Si}(t) = \displaystyle\int_0^t \frac{\sin s}{s}\,ds$. [Note that $\mathrm{Si}(t) = \mathrm{Si}(1) + \displaystyle\int_1^t \frac{\sin s}{s}\,ds$.]

29. Solve $y' - 2ty = 1$, $y(0) = 2$. Express your answer in terms of the error function, $\mathrm{erf}(t)$, where $\mathrm{erf}(t) = \dfrac{2}{\sqrt{\pi}} \displaystyle\int_0^t e^{-s^2}\,ds$.

30. **Superposition** First order linear differential equations possess important superposition properties. Show the following:

(a) If $y_1(t)$ and $y_2(t)$ are any two solutions of the homogeneous equation $y' + p(t)y = 0$ and if c_1 and c_2 are any two constants, then the sum $c_1y_1(t) + c_2y_2(t)$ is also a solution of the homogeneous equation.

(b) If $y_1(t)$ is a solution of the homogeneous equation $y' + p(t)y = 0$ and $y_2(t)$ is a solution of the nonhomogeneous equation $y' + p(t)y = g(t)$ and c is any constant, then the sum $cy_1(t) + y_2(t)$ is also a solution of the nonhomogeneous equation.

(c) If $y_1(t)$ and $y_2(t)$ are any two solutions of the nonhomogeneous equation $y' + p(t)y = g(t)$, then the sum $y_1(t) + y_2(t)$ is *not* a solution of the nonhomogeneous equation.

Exercises 31–32: **Outline of a Proof of Theorem 2.1**

The discussion of integrating factors in this section provides a basis for establishing the existence-uniqueness result stated in Theorem 2.1. In particular, consider the initial value problem $y' + p(t)y = g(t), y(t_0) = y_0$, where $p(t)$ and $g(t)$ are continuous on the interval (a, b) and where t_0 is in the interval (a, b). Let $P(t)$ denote the specific antiderivative of $p(t)$ that vanishes at t_0,

$$P(t) = \int_{t_0}^t p(s)\,ds. \tag{7}$$

Since p is continuous on (a, b), it follows from calculus that $P(t)$ is defined and differ-

[3]Milton Abramowitz and Irene Stegun, *Handbook of Mathematical Functions* (New York: Dover Publications, 1965).

entiable for all t in (a, b). As an instance of equation (6), define $y(t)$ by

$$y(t) = y_0 e^{-P(t)} + e^{-P(t)} \int_{t_0}^{t} e^{P(s)} g(s)\, ds. \tag{8}$$

Since g is continuous on (a, b) and $P(t)$ is differentiable on (a, b), it follows from calculus that $G(t) = \int_{t_0}^{t} e^{P(s)} g(s)\, ds$ is defined and differentiable for all t in (a, b) and that $dG/dt = e^{P(t)} g(t)$.

31. Use the facts above to show that $y(t)$ defined in equation (8) is differentiable on (a, b) and is a solution of the initial value problem $y' + p(t)y = g(t), y(t_0) = y_0$. This explicit construction establishes that at least one solution of the initial value problem exists on the entire interval (a, b).

32. To establish the uniqueness part of Theorem 2.1, assume $y_1(t)$ and $y_2(t)$ are two solutions of the initial value problem $y' + p(t)y = g(t), y(t_0) = y_0$. Define the difference function $w(t) = y_1(t) - y_2(t)$.

(a) Show that $w(t)$ is a solution of the homogeneous linear differential $w' + p(t)w = 0$.

(b) Multiply the differential equation $w' + p(t)w = 0$ by the integrating factor $e^{P(t)}$, where $P(t)$ is defined in equation (7), and deduce that $e^{P(t)} w(t) = C$, where C is a constant.

(c) Evaluate the constant C in part (b) and show that $w(t) \equiv 0$ and (a, b). Therefore, $y_1(t) \equiv y_2(t)$ on (a, b), establishing that the solution of the initial value problem is unique.

2.4 Introduction to Mathematical Models

In Sections 2.4 and 2.5 we begin our study of how differential equations are used in applied science and engineering. Differential equations often serve as mathematical models that can be used to describe and make predictions about physical systems. Sections 2.4 and 2.5 focus on models based on first order linear differential equations. Chapter 3 concerns models involving first order nonlinear differential equations. Later chapters present models based on higher order differential equations and systems of differential equations.

We saw a simple example of a mathematical model in Chapter 1,

$$\frac{d^2y}{dt^2} = -g; \qquad y(0) = y_0, \qquad y'(0) = v_0.$$

This initial value problem is a mathematical model, derived from Newton's[4] second law of motion, for an object falling under the influence of gravity. The starting position of the object is $y(0) = y_0$, and its initial velocity is $y'(0) = v_0$. The solution predicts how the object's position, $y(t)$, and its velocity, $y'(t)$, vary with time as the object falls. In this section we develop models of compound

[4] Sir Isaac Newton (1643–1727) has profoundly influenced the development of mathematics and science. Newton, along with Gottfried Leibniz, is generally credited with laying the foundations of differential and integral calculus. His work, *De Methodis Serierum et Fluxionum,* was completed in 1671 but was not published until 1736. *Optiks,* published in 1704, summarizes Newton's research in the theory of light and color. His greatest work, *Philosophiae naturalis principia mathematica* (or simply *Principia*), was published in 1687. This work summarizes his research in physics and celestial mechanics. It contains his laws of motion and the law of universal gravitation. The *Principia* is arguably the greatest scientific work ever published.

interest, population dynamics, and radioactive decay. The exercises also discuss applications involving flexible belts under conditions of impending slippage. All these models give rise to initial value problems for first order linear differential equations.

Modeling

We can divide the art of mathematical modeling into three phases:

1. ***Formulation*** After observing the physical system, we need to identify the appropriate independent and dependent variables. Then we need to develop a mathematical description of how these variables interact. Often, a differential equation (along with appropriate initial conditions) will serve as a mathematical description of the system.

2. ***Solution*** Once we have formulated the modeling problem, we need to solve it. This involves recognizing the mathematical structure of the problem and bringing the appropriate analytical and/or numerical techniques to bear.

3. ***Validation and Interpretation*** Once we have solved the problem, the solution needs to be examined carefully. Does it make sense? Is it consistent with our physical intuition about what should be expected? The solution needs to be scrutinized for its physical content: What does it say about the physical phenomenon being modeled?

From Discrete to Continuous Models

Many physical systems are inherently discrete in nature. For example, a population is composed of an integer number of individuals. Similarly, the principal in a savings account is some integer number of cents. These quantities, moreover, change in time by undergoing integer jumps in their values. For example, a population changes abruptly whenever an individual is born or dies. As the next two subsections illustrate, however, we can sometimes use differential equations to model discrete systems.

Compound Interest

Suppose an initial amount of money, say A_0 dollars, is deposited into a savings account and is left there over a period of time to accumulate interest and grow. Let r denote the annual interest rate (for example, $r = 0.05$ would correspond to a 5% annual interest rate).

If interest is compounded once per year, the principal $P_1(t)$ in the account after t years is

$$P_1(t) = (1 + r)^t A_0.$$

(The subscript on P denotes the number of compoundings per year.)

If interest is compounded twice per year, the principal $P_2(t)$ increases by a factor of $(1 + (r/2))$ every half year. The principal after t years is

$$P_2(t) = \left(1 + \frac{r}{2}\right)^{2t} A_0.$$

Increasing the compounding rate to monthly (12 times per year) leads, after t years, to

$$P_{12}(t) = \left(1 + \frac{r}{12}\right)^{12t} A_0.$$

In general, if interest is compounded n times per year, the principal after t years amounts to

$$P_n(t) = \left(1 + \frac{r}{n}\right)^{nt} A_0. \tag{1}$$

Does continually increasing the number of times per year that interest is compounded lead to a limiting value for $P_n(t)$? Or, does the principal $P_n(t)$ grow without bound as we compound every hour, then every minute, then every second, and so on? With regard to equation (1), it is logical to ask (for an arbitrary fixed value of t) whether $\lim_{n\to\infty} P_n(t)$ exists. Examining this limit, we find

$$\begin{aligned}\lim_{n\to\infty} P_n(t) &= \lim_{n\to\infty} \left(1 + \frac{r}{n}\right)^{nt} A_0 \\ &= A_0 \left[\lim_{n\to\infty} \left(1 + \frac{r}{n}\right)^{n}\right]^{t} \\ &= A_0 e^{rt}.\end{aligned}$$

The final limit evaluation, leading to $A_0 e^{rt}$, comes from a well-known limit in calculus,

$$\lim_{n\to\infty} \left(1 + \frac{x}{n}\right)^{n} = e^x.$$

In equation (1), we can think of the limit as n tends to infinity as defining **continuous compounding**. Continuous compounding thus leads to the following formula for principal at the end of t years:

$$P(t) = A_0 e^{rt}. \tag{2}$$

Note that $A_0 e^{rt}$ in equation (2) is the solution of the initial value problem

$$P'(t) = rP(t), \qquad P(0) = A_0. \tag{3}$$

Under continuous compounding the instantaneous rate of change of principal is proportional to the principal—the constant of proportionality being the decimal interest rate. Therefore, if compounding frequency is large, we can reasonably approximate growth of principal by continuous compounding, leading to initial value problem (3).

EXAMPLE 1

(a) An investor deposits \$1000 at 6.25%, compounded continuously. How much will the account be worth after five years?

(b) An amount A_0 is deposited at 6.25%, compounded continuously. How long will it take this amount to double?

Solution:

(a) Using equation (2), we find

$$P(5) = e^{(0.0625)(5)}A_0 = e^{0.3125}1000 = \$1366.84.$$

(b) After t years, the initial amount A_0 will have grown to be

$$P(t) = A_0 e^{0.0625t}.$$

We want to find t such that $P(t) = 2A_0$. The equation $P(t) = 2A_0$ reduces to

$$2 = e^{0.0625t}.$$

Solving this equation, we find $t = 11.09$ years or about 11 years and 1 month. ▲

Population Models

Let $P(t)$ represent the population of a species at time t. We assume the population lives in some well-defined environment that we call a *colony*. For this introductory model we will be quite naïve, making no distinction among population members as to age, gender, health, or location within the colony.

Assume that the population can change in time through births, deaths, and by migration in and out of the colony. Also assume the population is sufficiently large to warrant describing its evolution in time by a differential equation based on the following "conservation of population" law,

$$\begin{bmatrix}\text{Rate of change}\\ \text{of population}\end{bmatrix} = \begin{bmatrix}\text{Rate of}\\ \text{population increase}\end{bmatrix} - \begin{bmatrix}\text{Rate of}\\ \text{population decrease}\end{bmatrix}. \tag{4}$$

For our model, population can increase either through births or migration into the colony. Similarly, population decreases either through deaths or migration out of the colony.

To translate the principle in equation (4) into mathematics, we'll need to introduce some notation. Let r_b and r_d be positive constants representing the birth and death rates per unit population. In other words, $r_bP(t)$ represents the *rate* of population increase through births at time t. Similarly, $r_dP(t)$ represents, at time t, the *rate* of population decrease through deaths.

Let $M(t)$ denote the migration rate at time t. Note that $M(t)$ can be positive or negative, depending on whether or not the rate of immigration into the colony exceeds the exodus rate. Combining this notation with the conservation of population principle in equation (4), we obtain the differential equation

$$\frac{dP}{dt} = r_bP - r_dP + M(t)$$

or

$$\frac{dP}{dt} = (r_b - r_d)P + M(t). \tag{5}$$

If there is no migration, then equation (5) takes the form

$$\frac{dP}{dt} = kP. \tag{6}$$

When k is positive in equation (6), we often refer to it as the **growth rate**; if k is negative, we refer to it as the **decay rate**. As the next example illustrates, we can sometimes use data about the population to estimate this rate k.

EXAMPLE 2

For a population of about 100,000 bacteria in a petri dish, we decide to model the growth of this population by the differential equation

$$\frac{dP}{dt} = kP.$$

Suppose, two days later, that the population has grown to about 150,000 bacteria. Find the growth rate k and estimate the bacteria population after seven days.

Solution: The general solution of $P' = kP$ is

$$P(t) = Ce^{kt}$$

where, for this problem, t is measured in days. Imposing the initial condition, $P(0) = 100{,}000$, we obtain $C = 100{,}000$. Thus,

$$P(t) = 100000e^{kt}.$$

Knowing $P(2) = 150{,}000$, we find

$$150000 = 100000e^{2k}.$$

Therefore, $1.5 = e^{2k}$ and hence

$$k = \frac{1}{2}\ln(1.5) \text{ days}^{-1}.$$

Having k, we arrive at an expression for the bacteria population,

$$P(t) = 100000e^{(\ln(1.5)/2)t}.$$

Therefore, at the end of seven days, there will be about 413,000 bacteria. [The formula for $P(t)$ gives $P(7) = 413,351$.] ▲

EXAMPLE 3

An aquaculture firm raises catfish in ponds. At the beginning of the year, the ponds contain approximately 500,000 catfish. The net growth rate coefficient, $r_b - r_d$, is estimated to be about 6.1 per 1000 per week. The firm wants to harvest at a constant rate of R fish per week, but it also wants to increase the population to about 600,000 fish by the end of the year. Find the appropriate harvest rate, R.

Solution: According to the model formulated in equation (5), we have

$$\frac{dP}{dt} = (r_b - r_d)P + M(t).$$

(continued)

(continued)

For our problem, t is measured in weeks. The growth rate coefficient, $r_b - r_d$, is 6.1 per 1000 per week. Therefore,

$$r_b - r_d = \frac{6.1}{1000} = 0.0061 \text{ week}^{-1}.$$

For this problem, the migration rate is the same as the harvest rate. Since the harvest rate is constant, we set $M(t) = -R$, where R is a positive constant. Therefore, we arrive at the following model for catfish population,

$$\frac{dP}{dt} = 0.0061P - R, \qquad P(0) = 500{,}000. \tag{7}$$

Our objective is to choose R so that $P(52) = 600{,}000$. [In the absence of any harvesting, that is, with $R = 0$, the population $P(t)$ would be

$$P(t) = 500000e^{0.0061t}.$$

So, with no harvesting, the population at the end of the year would be about $P(52) = 500000e^{0.3172}$, or about 685,000 fish.]

To determine a reasonable harvest rate R, we first find the general solution for nonhomogeneous differential equation (7) using the techniques of Section 2.3,

$$P(t) = Ce^{0.0061t} + \frac{R}{0.0061}.$$

Imposing the initial condition $P(0) = 500{,}000$, we obtain

$$P(t) = \left(500000 - \frac{R}{0.0061}\right)e^{0.0061t} + \frac{R}{0.0061}$$

or

$$P(t) = \frac{R}{0.0061}\left(1 - e^{0.0061t}\right) + 500000e^{0.0061t}.$$

We want to choose R so that $P(52) = 600{,}000$. Thus, we are led to the requirement

$$600000 = \frac{R}{0.0061}\left(1 - e^{0.3172}\right) + 500000e^{0.3172}.$$

Solving for R, we arrive at a harvest rate of $R = 1416$ per week. Thus, the firm can harvest at a rate of about 1400 fish per week and still see its fish population grow to about 600,000 by year's end. ▲

Radioactive Decay

The process of radioactive decay is, in many respects, like the behavior of a large population in which there are deaths but no births. At the atomic level, individual atoms of a radioactive element spontaneously undergo change, transforming themselves into new material.

At the macroscopic level (the level of continuous modeling), we'll let $Q(t)$ represent the amount of radioactive material present at time t. It has been observed empirically that the rate of decrease of radioactive material is proportional to the amount present. That is, the mathematical model is

$$\frac{dQ}{dt} = -kQ, \qquad k > 0. \tag{8}$$

We can obtain differential equation (8) by invoking the same basic conservation law, equation (4), that was used to derive the population model in equation (5), $P'(t) = (r_b - r_d)P(t) + M(t)$. Here, the birth rate r_b is zero since no radioactive material is being created. The death rate constant r_d has been replaced by k. Likewise, we are tacitly assuming no material is being added or taken away, and therefore the migration rate $M(t)$ is also zero.

EXAMPLE 4

Initially, 50 mg of a radioactive substance is present. Five days later, the quantity has decreased to 43 mg. How much will remain after 30 days?

Solution: The general solution of equation (8) is

$$Q(t) = Ce^{-kt},$$

where t is measured in days. Imposing the initial condition, it follows that $C = 50$ and thus

$$Q(t) = 50e^{-kt}.$$

Next, as in Example 2, we use the fact that $Q(5) = 43$ mg to determine the decay rate k:

$$\begin{aligned} k &= -\frac{1}{5}\ln\frac{43}{50} \\ &= 0.03016\ldots \text{ days}^{-1}. \end{aligned}$$

Thus, at time t, we expect to have about $Q(t)$ mg of the substance where

$$Q(t) = 50e^{-0.0302t}.$$

After 30 days, therefore, we expect to have $Q(30) = 20.21$ mg. ▲

The **half-life** of a radioactive substance is the length of time it takes a given amount of the substance to be reduced to one half of its original amount. Thus, the half-life τ is defined by the equation

$$Q(t + \tau) = \frac{1}{2}Q(t).$$

Since $Q(t) = Ce^{-kt}$, this equation reduces to $e^{-k\tau} = 0.5$ and hence

$$\tau = \frac{\ln 2}{k}.$$

For example, the substance in Example 4 has a half-life of

$$\frac{\ln 2}{0.0302} = 22.95 \text{ days}.$$

If we had 300 mg of the substance at some given time, we would have about 150 mg of the substance 22.95 days later and 75 mg of the substance after 45.9 days.

EXERCISES

1. An investor deposits \$5000 at an annual rate of 5%, compounded continuously. Use equation (2) to decide how much this account is worth after 30 years.

2. An investor deposits \$5000 at an annual rate of 5%, compounded semiannually. How much is this account worth after 30 years? Compare your answer with that of Exercise 1.

3. Using equations (1) and (2), decide how long it will take an investment to double in value if it earns interest at a rate of 6% compounded

   (a) annually (b) semiannually (c) continuously

4. A bond is issued that earns interest at a rate of 5% and is compounded continuously. The bond matures after 10 years. What interest rate is needed if a bond is to reach the same maturity value after only 8 years of continuously compounding interest?

5. An investor is given a choice of two continuous compounding strategies. In Plan A the interest rate is a constant 6%; that is, $r = 0.06$ in equation (3). In Plan B, the interest rate starts at 4% but increases linearly at a rate of 0.4% per year. In other words, $r = r(t) = 0.04 + 0.004t$, where t is measured in years.

   (a) What is the initial value problem modeling how an amount, invested in Plan B, changes with time?

   (b) Solve this initial value problem and determine how an amount, invested in Plan B, increases with time.

   (c) Suppose equal amounts are invested in both plans. How much time must elapse before the principal in Plan B "catches up with" and equals the principal invested in Plan A?

6. A depositor initially places \$1000 in a savings account that earns interest at a rate of 5%, compounded continuously. After 4 years, the depositor withdraws the entire amount and immediately reinvests it in a new savings plan that pays interest at a rate of 7%, compounded continuously. How large is the principal 6 years after this reinvestment?

7. An investor makes an initial deposit of \$1000 into an account that compounds interest continuously. Six years later, she invests another \$1000 into the same account. Twelve years from the date of the original deposit, her total principal has grown to \$4000. What is the constant annual interest rate r of continuous compounding?

Exercises 8–11:

Assume the populations in these exercises evolve according to the differential equation (6).

8. A colony of bacteria initially has 10,000,000 members. After 5 days, the population increases to 11,000,000. Estimate the population after 30 days.

9. How many days will it take the colony in Exercise 8 to double in size?

10. A colony of bacteria is observed to increase in size by 30% over a 2-week period. How long will the colony take to triple its initial size?

11. A colony of bacteria initially has 100,000 members. After 6 days, the population has decreased to 80,000. At that time, 50,000 new organisms are added to replenish its size. How many bacteria will be in the colony after an additional 6 days?

12. The evolution of a population with constant migration rate M is described by the initial value problem

$$\frac{dP}{dt} = kP + M, \qquad P(0) = P_0,$$

where k is net birth rate per unit population (assumed constant).

(a) Solve this initial value problem for $P(t)$.

(b) Examine the solution and determine the relation between the constants k and M that will result in $P(t)$ remaining constant in time and equal to P_0. Explain, on physical grounds, why the two constants k and M must have opposite signs to achieve this constant (equilibrium) solution for $P(t)$.

(c) Suppose we adopt an alternative path of inquiry and ask for what value of the constant A can $P(t) = A$ be a (constant) solution of the differential equation? In other words, what are the equilibrium solutions of this (autonomous) equation? What is the solution of the initial value problem for $P(t)$ if the initial condition $P(0) = P_0$ is set equal to this value A?

13. Assume that the population of fish in an aquaculture farm can be modeled by the differential equation $dP/dt = kP + M(t)$, where k is a positive constant. The manager wants to manage the farm in such a way that the fish population remains constant from year to year. In particular, the following two harvesting strategies are under consideration.

***Strategy* I:** Harvest the fish at a constant and continuous rate so that the population itself remains constant in time. Therefore, $P(t)$ would be a constant and $M(t)$ would be a negative constant; call it $-M$. (Refer to Exercise 12.)

***Strategy* II:** Let the fish population evolve without harvesting throughout the year, and then harvest the excess population at year's end to return the population to its value at the year's beginning.

(a) Determine the number of fish harvested annually for each of the two strategies. Express your answer in terms of the population at year's beginning; call it P_0. (Assume that the units of k are year^{-1}.)

(b) Suppose, as in Example 3, that $P_0 = 500{,}000$ fish and $k = 0.0061 \times 52 = 0.3172$ year^{-1}. Assume further that Strategy I, with its steady harvesting and return, provides the farm with a net profit of \$0.75/fish while Strategy II provides a profit of only \$0.60/fish. Which harvesting strategy will ultimately prove more profitable to the farm?

14. Assume two colonies each have P_0 members at time $t = 0$ and that each evolves with a constant relative birth rate $k = r_b - r_d$. For colony 1, assume that individuals migrate into the colony at a rate of M individuals per unit time. Assume that this immigration occurs for $0 \le t \le 1$ and ceases thereafter. For colony 2, assume a similar migration pattern occurs but is delayed by one unit of time; that is, individuals migrate at a rate of M individuals per unit time, $1 \le t \le 2$. Suppose we are interested in comparing the evolution of these two populations over the time interval $0 \le t \le 2$. The initial value problems governing the two populations are

$$\frac{dP_1}{dt} = kP_1 + M_1(t), \qquad P_1(0) = P_0 \qquad M_1(t) = \begin{cases} 1, & 0 \le t \le 1 \\ 0, & 1 < t \le 2, \end{cases}$$

$$\frac{dP_2}{dt} = kP_2 + M_2(t), \qquad P_2(0) = P_0 \qquad M_2(t) = \begin{cases} 0, & 0 \le t < 1 \\ 1, & 1 \le t \le 2. \end{cases}$$

(a) Solve both problems to determine P_1 and P_2 at time $t = 2$.

(b) Show that $P_1(2) - P_2(2) = (M/k)\left(e^k - 1\right)^2$. If $k > 0$, which population is larger at time $t = 2$? What happens if $k < 0$?

(c) Suppose there is a fixed number of individuals that can be introduced into a population at any time through migration and that the objective is to maximize the population at some fixed future time. Do the calculations performed in this problem suggest a strategy (based upon the relative birth rate) for accomplishing this?

15. After one month of radioactive decay, 100 mg of a radioactive substance was observed to remain. After four months, only 30 mg of this substance was left.

(a) How much of the substance was initially present?

(b) What is the half-life of this radioactive substance?

(c) How long will it take before only 1% of the original amount of this substance remains?

16. **Radiocarbon Dating** Carbon-14 is a radioactive isotope of carbon produced in the upper atmosphere by radiation from the sun. Plants absorb carbon dioxide from the air and living organisms, in turn, eat the plants. The ratio of normal carbon (carbon-12) to carbon-14 in the air and in living things at any given time is nearly constant. When a living creature dies, however, the carbon-14 begins to decrease due to radioactive decay. By comparing the amounts of carbon-14 and carbon-12 present, the amount of carbon-14 that has decayed can therefore be ascertained.

Let $Q(t)$ denote the amount of carbon-14 present at time t after death. If we assume its behavior is modeled by the differential equation $Q'(t) = -kQ(t)$, then $Q(t) = Q(0)e^{-kt}$. Knowing the half-life of carbon-14, we can determine the constant k. Given a specimen to be dated, we can measure its radioactive content and deduce $Q(t)$. Knowing the amount of carbon-12 present enables us to determine $Q(0)$. Therefore, we can use the solution of the differential equation, $Q(t) = Q(0)e^{-kt}$, to deduce the age, t, of the radioactive sample.

(a) The half-life of carbon-14 is nominally 5730 years. Suppose remains have been found in which it is estimated that 30% of the original amount of carbon-14 is present. Estimate the age of the remains.

(b) The half-life of carbon-14 is not known precisely. Let us assume its half-life is 5730 ± 30 years. Determine how this half-life uncertainty affects the age estimate you computed in (a); that is, what is the corresponding uncertainty in the age of the remains?

(c) It is claimed that radiocarbon dating cannot be used to date objects older than about 60,000 years. To appreciate this practical limitation, compute the ratio $Q(60{,}000)/Q(0)$, assuming a half-life of 5730 years.

17. Suppose that 50 mg of a radioactive substance, having a half-life of 3 years, is initially present. More of this material is to be added at a constant rate so that 100 mg of the substance is present at the end of 2 years. At what constant rate must this radioactive material be added?

18. Iodine-131, a fission product created in nuclear reactors and nuclear weapons explosions, has a half-life of 8 days. If 30 micrograms of iodine-131 is detected in a tissue site 3 days after ingestion of the radioactive substance, how much was originally present?

19. U-238, the dominant isotope of natural uranium, has a half-life of roughly 4 billion years. Determine how long it takes for a sample to be reduced in amount by 1% through radioactive decay.

Belt Friction The slippage of flexible belts, cables and ropes over shafts or pulleys of circular cross-section is an important consideration in the design of belt drives. When the frictional contact between the belt and shaft is about to be broken (that is, when slippage is imminent), a belt drive is operating under the most demanding of conditions. The belt tension (force) is not constant along the contact region. Rather it increases along the contact region between belt and shaft in the direction of impending slippage.

Consider the belt drive shown in Figure 2.8. Suppose we ask the following question: "How much greater can we make tension T_2 relative to the opposing tension T_1 before the belt slips over the shaft in the direction of T_2?" The answer obviously depends in part on friction, that is, on the roughness of the belt-shaft contact surface.

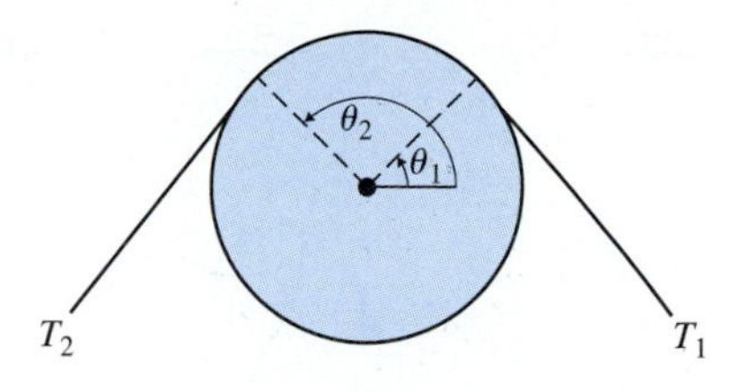

FIGURE 2.8

A belt drive. How much greater can we make tension T_2 relative to the opposing tension T_1 before the belt slips over the shaft in the direction of T_2?

When slippage is imminent, the tension in the belt has been found to satisfy the differential equation[5]

$$\frac{dT(\theta)}{d\theta} = \mu T(\theta).$$

In this equation, the angle θ (in radians) is measured in the direction of the impending slippage over the contact region between belt and shaft. The parameter μ is an empirically determined constant known as the **coefficient of friction**. The larger the value of μ, the rougher the contact surface. In the figure, with slippage impending in the direction of T_2, the value of T_2 is determined relative to T_1 by solving the initial value problem

$$\frac{dT}{d\theta} = \mu T, \qquad T(\theta_1) = T_1, \qquad \theta_1 \le \theta \le \theta_2.$$

The solution gives the critical tension, $T_2 = T(\theta_2) = e^{\mu(\theta_2 - \theta_1)} T_1$. The following exercises ask you to compute the tension under conditions of impending slippage for more complicated belt drive geometries. Often, the task of determining the θ-interval(s) of interest is the most challenging part of the problem.

Exercises 20–23:

Assume that belt slippage is impending in the direction shown by the dashed arrow. Compute the belt tensions at the locations shown for the geometries and coefficients of friction given.

20. **21.**

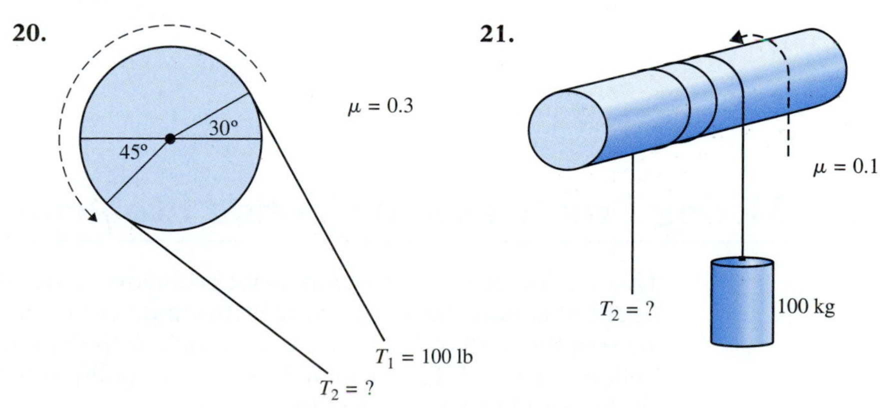

[5]J. L. Meriam and L. G. Kraige, *Engineering Mechanics*, 5th ed., Volume 1, *Statics* (New York: John Wiley and Sons, 2002).

22.

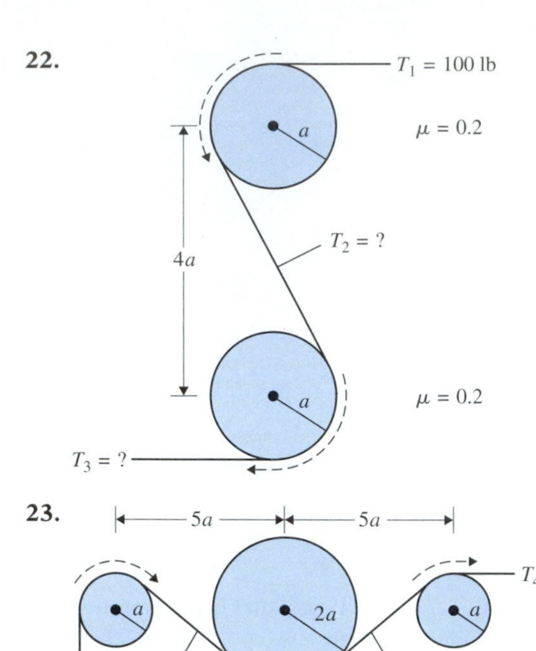

23.

24. In Western movies, a cowboy usually secures his horse by wrapping the reins a few times around the hitching post. Assume the weight of the rein hanging vertically is 1/3 lb and that the coefficient of belt friction is $\mu = 0.4$. If the horse pulls in the direction shown, what force F will cause the rein to slip?

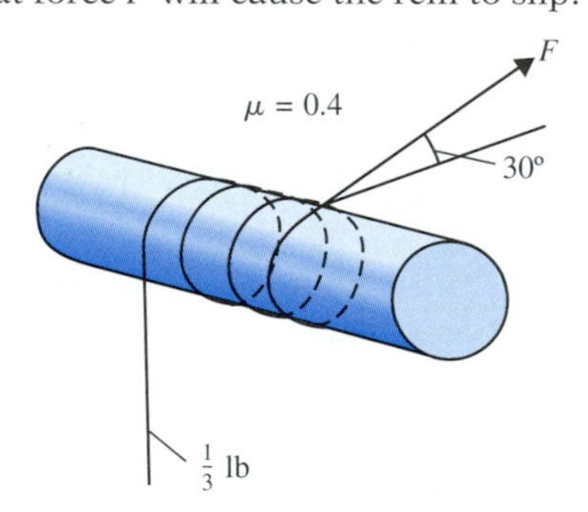

2.5 Mixing Problems and Cooling Problems

This section presents two important problems modeled by first order linear differential equations—mixing problems and cooling problems. These problems arise in the modeling of phenomena such as the mixing of solutes and solvents in flow systems, the spread and removal of pollutants in air and water, and the cooking and sterilization of foods.

Mixing Problems

Consider the fluid mixing problem shown schematically in Figure 2.9. A tank initially contains a volume of fluid, within which is dissolved a certain amount

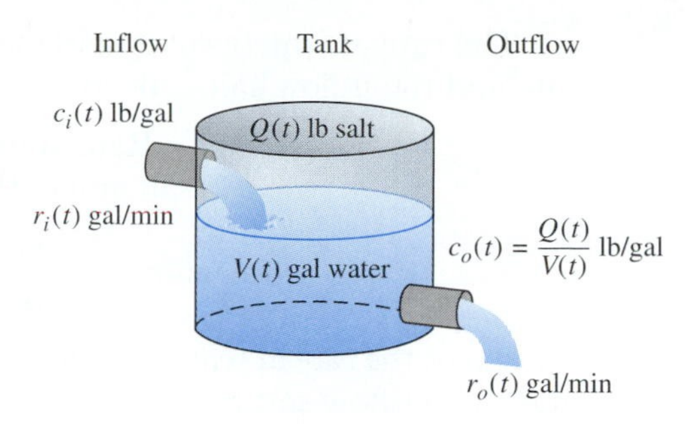

FIGURE 2.9

A salt solution enters the tank at a certain inflow rate and the well-stirred solution leaves the tank at a certain outflow rate. How much salt is in the tank at a given time t?

of solute. For definiteness, consider the liquid to be water, having the units of gallons. Consider the solute to be salt, having the units of pounds. Time is measured in minutes.

At some starting time, say $t = 0$, a salt solution enters the tank at a certain inflow rate and the well-stirred solution flows out of the tank at some outflow rate. The phrase **well stirred** means that the concentration of salt is uniform within the tank; the concentration depends only upon time and not upon spatial location within the tank. In other words, any salt entering the tank is instantaneously dispersed throughout the tank (either through mechanical mixing or diffusion). This is a reasonable approximation if the salt dissolves and disperses into solution very quickly relative to the speed at which the solution enters or leaves the tank.

Our objective in this mixing problem is to determine the amount of salt in the tank, as a function of time.

Modeling the Mixing Problem

To model the dynamics of the mixing process shown in Figure 2.9, we invoke a "conservation of salt" law:

$$\begin{bmatrix}\text{Rate of change of}\\ \text{salt in the tank}\end{bmatrix} = \begin{bmatrix}\text{Rate at which}\\ \text{salt enters the tank}\end{bmatrix} - \begin{bmatrix}\text{Rate at which}\\ \text{salt leaves the tank}\end{bmatrix}. \quad (1)$$

We need to translate the words of equation (1) into mathematics. To that end, let

$$\begin{aligned}
Q(t) &= \text{amount of salt (pounds) in the tank at time } t \text{ (minutes)},\\
V(t) &= \text{volume of water (gallons) in the tank at time } t,\\
c_i(t) &= \text{inflow salt concentration (pounds/gallon) at time } t,\\
c_o(t) &= \text{outflow salt concentration (pounds/gallon) at time } t,\\
r_i(t) &= \text{inflow rate (gallons/minute) at time } t,\\
r_o(t) &= \text{outflow rate (gallons/minute) at time } t.
\end{aligned}$$

Using these definitions, we can convert each term of equation (1) into a mathematical statement and obtain a mathematical model for the mixing process.

The rate at which salt enters the tank is given by the product of the inflow rate and the inflow salt concentration. That is,

$$\begin{bmatrix} \text{Rate at which} \\ \text{salt enters the tank} \end{bmatrix} = r_i(t)c_i(t).$$

$$\left(\text{Dimensionally, } \frac{\text{pounds}}{\text{minute}} = \frac{\text{gallons}}{\text{minute}} \cdot \frac{\text{pounds}}{\text{gallon}}.\right)$$

Similarly, the rate at which salt leaves the tank is the product of the outflow rate and the outflow salt concentration. But, while the inflow salt concentration is a known function, $c_i(t)$, the outflow salt concentration is *not* a known function of t. In particular, outflow salt concentration, $c_o(t)$, is determined by volume $V(t)$ and by how much salt is in the tank at time t,

$$c_o(t) = \frac{Q(t)}{V(t)}.$$

Thus,

$$\begin{bmatrix} \text{Rate at which} \\ \text{salt leaves the tank} \end{bmatrix} = r_o(t)c_o(t) = r_o(t)\frac{Q(t)}{V(t)}.$$

Combining these two calculations, we have a mathematical model for the mixing process,

$$\frac{dQ}{dt} = r_i(t)c_i(t) - r_o(t)\frac{Q}{V(t)}, \qquad Q(0) = Q_0. \tag{2}$$

In equation (2), $Q(0) = Q_0$ gives the amount of salt in the tank at the starting time, $t = 0$.

In equation (2), the volume of water in the tank, $V(t)$, is related to the flow rates by the differential equation

$$\frac{dV}{dt} = r_i(t) - r_o(t).$$

Solving this equation by antidifferentiation, we obtain an expression for $V(t)$,

$$V(t) = V(0) + \int_0^t [r_i(s) - r_o(s)]\, ds.$$

Having $V(t)$, we can solve the first order linear equation (2) for $Q(t)$. [In many cases, the inflow rate and the outflow rate are equal. In such cases, $V(t) = V(0)$ is constant.]

EXAMPLE 1

A tank initially contains 1000 gal of water in which is dissolved 20 lb of salt. A valve is opened and water containing 0.2 lb of salt per gallon flows into the tank at a rate of 5 gal/min. The mixture in the tank is well stirred and drains from the tank at a rate of 5 gal/min.

(a) Find $Q(t)$, the amount of salt in the tank after t minutes.

(b) Find the limiting value: $\lim_{t\to\infty} Q(t)$. Should you expect such a limit to exist?

(c) Let the limit in part (b) be designated as Q_L. How long will it be until $Q(t)$ is within 1% of Q_L?

Solution: Since the inflow rate and the outflow rate are the same, the volume of water in the tank remains constant at 1000 gallons. Equation (2), together with the condition that there was 20 lb of salt in the tank at time $t = 0$, leads to the following initial value problem:

$$\frac{dQ}{dt} = (5)(0.2) - 5\frac{Q}{1000}, \qquad Q(0) = 20.$$

(a) Solving the nonhomogeneous differential equation using the techniques of Section 2.3, we obtain the general solution

$$Q(t) = 200 + Ce^{-t/200}.$$

Imposing the initial condition, we find

$$Q(t) = 200 - 180e^{-t/200} \text{ lb}.$$

(b) From part (a) we see that $Q(t) \to 200$ as $t \to \infty$. The discussion following this example comments on the physical significance of the limit, on why the limit should exist, and why the limit should be $Q_L = 200$ lb.

(c) Given $Q_L = 200$, we need to determine when $Q(t)$ is within 1% of 200. Now, from the solution graphed in Figure 2.10, $Q(t)$ is an increasing function. Thus, $Q(t)$ is within 1% of Q_L when $Q(t)$ has value 198 or more. Solving the equation $Q(t) = 198$, we find

$$-t = 200 \ln\left(\frac{2}{180}\right)$$

or $t = 899.96\ldots$ min. Therefore, after about 900 min (15 hr), there will be at least 198 lb of salt in the tank. ▲

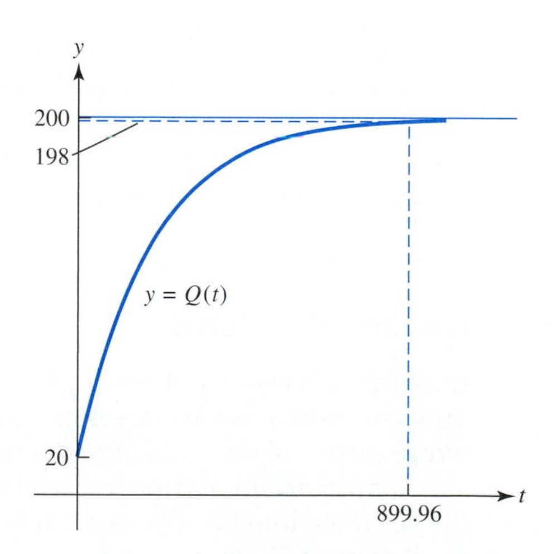

FIGURE 2.10

For the mixing problem in Example 1, there are $Q(t)$ lb of salt in the tank after t min, where $Q(t) = 200 - 180e^{-t/200}$.

The general solution of the differential equation in Example 1 is

$$Q(t) = 200 + Ce^{-t/200}.$$

Thus, $Q(t) \to 200$ as $t \to \infty$, regardless of the value C. Note that the constant function $Q(t) = 200$ is an equilibrium solution of the differential equation. This is the amount of salt needed to make the concentration of salt in the tank equal to the inflow concentration.

This limiting behavior also can be seen in more general situations. For instance, suppose the inflow and outflow rates are constant and equal, say at r gallons/minute. Let the inflow concentration, c_i, be constant as well. Let V_0 denote the volume of water in the tank, and let $Q(0) = Q_0$ denote the amount of salt in the tank at time $t = 0$. Under these conditions, the mixing model becomes

$$\frac{dQ}{dt} = rc_i - r\frac{Q}{V_0}, \qquad Q(0) = Q_0. \tag{3a}$$

The solution of this initial value problem is

$$Q(t) = c_i V_0 + (Q_0 - c_i V_0)e^{-(r/V_0)t}. \tag{3b}$$

Note that the solution $Q(t)$ tends to the same limiting value, $c_i V_0$, regardless of the flow rate r. In fact, the constant function $Q(t) = c_i V_0$ is an equilibrium solution of differential equation (3a). It is the amount of salt needed to make the concentration of the solution in the tank equal to the inflow concentration. Think about flushing out a tank with a salt solution. No matter how much salt is initially in the tank, it is flushed out as time increases and the concentration of salt in the tank approaches the inflow concentration of c_i pounds per gallon. Hence, equation (3b) is consistent with our physical intuition. Increasing or decreasing the flow rate r affects how rapidly the limiting value is approached but does not affect the limiting value itself.

Although we've only talked about tanks and salt, problems of this sort arise in a variety of circumstances, such as in environmental applications where the tank is actually a body of water (such as a lake) and the solute could be some pollutant entering and leaving via connecting streams.

Cooling Problems

Imagine a bowl of hot soup placed on a kitchen table and left there to cool. Suppose you want to develop a mathematical model that predicts how the temperature of the soup changes as time progresses. How would you proceed?

At any instant of time, we would expect the temperature at all points within the soup itself to be approximately the same. (This is the thermal equivalent of "well stirred.") Therefore, we assume that the temperature of the soup is described by a function of time alone. We make the same assumption about the kitchen surroundings, and thus its temperature can be described by a second function of time (quite possibly a constant function). Moreover, the kitchen surroundings are sufficiently large so that kitchen temperature is basically unchanged by introducing the bowl of hot soup.

With this example as a guide, we'll now set up the general framework for **Newton's law of cooling**. Instead of soup and a kitchen, we speak of an **object** and its **surroundings**.

Let $\Theta(t)$ and $S(t)$ denote the temperatures of the object and its surroundings, respectively. The basic assumption underlying Newton's law of cooling is that the rate of change of the object's temperature is proportional to the difference in temperatures between the object and its surroundings. Expressed mathematically, Newton's law of cooling becomes

$$\Theta'(t) = k(S(t) - \Theta(t)). \tag{4}$$

In the model given by equation (4), we assume that the temperature of the surroundings, $S(t)$, is known for all time of interest and is unaffected by the presence of the object.

In equation (4) we also assume the constant of proportionality, k, is positive. Does that make sense to you? Suppose, at some instant in time, that the temperature of the surroundings is less than the temperature of the object. Should the object's temperature be increasing or decreasing at that instant?

EXAMPLE 2

A metal object is heated to 200°C and then placed in a large room to cool. The temperature of the room is held constant at 20°C. After 10 min, the object's temperature is 100°C. How long will it take the object to cool to 25°C?

Solution: Using equation (4), the object's temperature, $\Theta(t)$, is modeled by

$$\Theta'(t) = k(20 - \Theta(t)), \qquad \Theta(0) = 200.$$

For this initial value problem, t is measured in minutes and temperature in degrees Celsius.

The general solution of $\Theta'(t) = k(20 - \Theta(t))$ is

$$\Theta(t) = 20 + Ce^{-kt}.$$

Imposing the initial condition, we get $\Theta(t) = 20 + 180e^{-kt}$. Knowing $\Theta(10) = 100$, we determine the rate constant k, finding $k = 0.08109\ (\text{min})^{-1}$. Thus, $\Theta(t)$ is given (approximately) by

$$\Theta(t) = 20 + 180e^{-0.0811t}.$$

Solving $\Theta(t) = 25$, the metal object cools to 25°C after 44.186... min. ▲

Consider a cooling problem (such as the one in Example 2) where the surrounding temperature $S(t)$ is constant. In particular, let $S(t) = S_0$ for all t of interest. The initial value problem modeling constant-temperature surroundings is

$$\Theta'(t) = k(S_0 - \Theta(t)), \qquad \Theta(0) = \Theta_0. \tag{5}$$

From a mathematical point of view, equations (3a) and (5) are the same. We need only identify k with r/V_0 and S_0 with rc_i. Equation (5) has equilibrium solution $\Theta(t) = S_0$. Therefore, if the object has the same initial temperature as the surroundings (that is, if $\Theta_0 = S_0$), then its temperature will remain constant. If the object's initial temperature is not equal to that of the surroundings, we expect the object's temperature to approach S_0 as $t \to \infty$. These commonsense checks are satisfied by the solution of (5)

$$\Theta(t) = S_0 + (\Theta_0 - S_0)e^{-kt}.$$

You can also gain insight into this behavior by examining the direction field for the differential equation, shown in Figure 2.11. Observe from the direction field that $\Theta(t) = S_0$ is an equilibrium solution. Then, with regard to the derivative $\Theta'(t) = k(S_0 - \Theta(t))$, we see that $\Theta'(t) > 0$ if $\Theta(t) < S_0$ and $\Theta'(t) < 0$ if $\Theta(t) > S_0$. This analysis together with Figure 2.11 clearly suggests that the temperature of the body tends toward the temperature of the surroundings as time evolves.

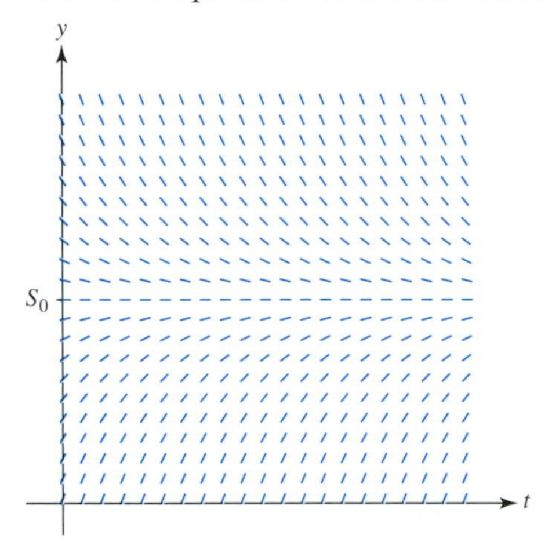

FIGURE 2.11

The direction field for the differential equation $\Theta'(t) = k(S_0 - \Theta(t))$ that models a cooling problem. The constant function $\Theta(t) = S_0$ is an equilibrium solution. The direction field shows that the temperature of the body, $\Theta(t)$, tends toward the temperature of the surroundings, S_0.

EXERCISES

1. A tank originally contains 100 gal of fresh water. At time $t = 0$, a solution containing 0.2 lb of salt per gallon begins to flow into the tank at a rate of 3 gal/min and the well-stirred mixture flows out of the tank at the same rate.

   (a) How much salt is in the tank after 10 min?

   (b) Does the amount of salt approach a limiting value as time increases? If so, what is this limiting value and what is the limiting concentration?

2. An auditorium is 100 m in length, 70 m in width, and 20 m in height. It is ventilated by a system that feeds in fresh air and draws out air at the same rate. Assume that airborne impurities form a well-stirred mixture. The ventilation system is required to reduce air pollutants present at any instant to 1% of their original concentration in 30 min. What inflow (and outflow) rate is required? What fraction of the total auditorium air volume must be vented per minute?

3. A tank originally contains 5 lb of salt dissolved in 200 gal of water. Starting at time $t = 0$, a salt solution containing 0.25 lb of salt per gallon is to be pumped into the tank at a constant rate and the well-stirred mixture is to flow out of the tank at the same rate.

(a) The pumping is to be done so that the tank contains 30 lb of salt after 20 min of pumping. At what rate must the pumping occur in order to achieve this objective?

(b) Suppose the objective was to have 60 lb of salt in the tank after 20 min. Is it possible to achieve this objective? Explain.

4. A 5000-gal aquarium is maintained with a pumping system that circulates 100 gal of water per minute through the tank. To treat a certain fish malady, a soluble antibiotic is introduced into the inflow system. Assume that the inflow concentration of medicine is $10te^{-t/50}$ oz/gal, where t is measured in minutes. The well-stirred mixture flows out of the aquarium at the same rate.

(a) Solve for the amount of medicine in the tank as a function of time.

(b) What is the maximum concentration of medicine achieved by this dosing and when does it occur?

(c) To be effective, the antibiotic concentration must exceed 100 oz/gal for a minimum of 60 min. Was the dosing effective?

5. A tank, having a capacity of 700 gal, initially contains 10 lb of salt dissolved in 100 gal of water. At time $t = 0$, a solution containing 0.5 lb of salt per gallon flows into the tank at a rate of 3 gal/min and the well-stirred mixture flows out of the tank at a rate of 2 gal/min.

(a) How much time will elapse before the tank is filled to capacity?

(b) What is the salt concentration in the tank when it contains 400 gal of solution?

(c) What is the salt concentration at the instant the tank is filled to capacity?

6. A 500-gal aquarium is cleansed by the recirculating filter system schematically shown in the figure. Water, containing impurities is pumped out at a rate of 15 gal/min, filtered and returned to the aquarium at the same rate. Assume that in passing through the filter, the concentration of impurities is reduced by a fractional amount α, as shown in the figure. In other words if the impurity concentration entering the filter is $c(t)$, the exit concentration is $\alpha c(t)$, where $0 < \alpha < 1$.

(a) Apply the basic conservation principle (rate of change = rate in − rate out) to obtain a differential equation for the amount of impurity present in the aquarium at time t. Assume that filtering occurs instantaneously. If the outflow concentration at any time is $c(t)$, assume that the inflow concentration at that same instant is $\alpha c(t)$.

(b) What value of filtering constant α will reduce impurity levels to 1% of their original values in a period of 3 hr?

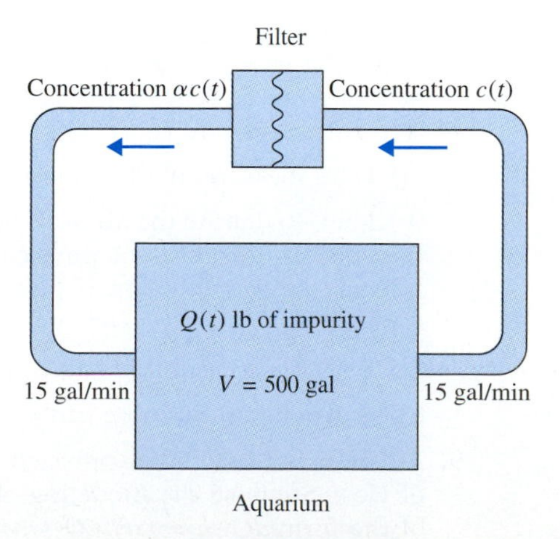

Figure for Exercise 6

7. **Series Connections of Tanks** Consider the sketch shown, where two ponds are connected and fed by a single stream flowing through them. Pond A holds 500,000 gal of water, while Pond B holds 200,000 gal of water. The fresh water stream flows through these ponds at a rate of 1000 gal/hr. Assume at some time, say $t = 0$, that 1,000 lb of a toxin is spilled into Pond A and that it disperses rapidly enough so that a well-stirred assumption is reasonable.

(a) Let $Q_A(t)$ and $Q_B(t)$ denote the amount of toxin in Ponds A and B, respectively, at time t. Apply the "conservation of salt" principle to each pond and formulate initial value problems governing how the amount of toxin in each pond varies with time.

(b) Solve the two initial value problems for $Q_A(t)$ and $Q_B(t)$. (Because of the way the two ponds are connected by the feeder stream, the problem for Pond A can be solved independently of Pond B and the solution, in turn, used to specify the problem for Pond B).

(c) What is the maximum amount of toxin present in Pond B and at what time after the spill is this maximum value reached?

(d) How much time must elapse before the concentration of toxin in both ponds has been reduced to 1 lb per million gallons?

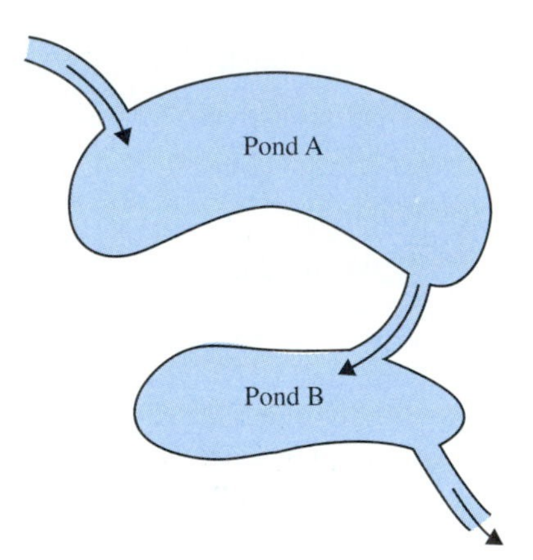

Figure for Exercise 7

8. **Oscillating Flow Rate** A tank initially contains 10 lb of solvent in 200 gal of water. At time $t = 0$ a pulsating or oscillating flow begins. To model this flow, we assume that the input and output flow rates are both equal to $3 + \sin t$ gal/min. Thus, the flow rate oscillates between a maximum of 4 gal/min and a minimum of 2 gal/min; it repeats its pattern every $2\pi \approx 6.28$ min. Assume that the inflow concentration is at a constant rate of 0.5 lb of solvent per gallon.

(a) Does the amount of solution in the tank, V, remain constant or not? Explain.

(b) Let $Q(t)$ denote the amount of solvent (pounds) in the tank at time t (minutes). Explain, on the basis of physical reasoning, whether you expect the amount of solvent in the tank to approach an equilibrium value or not. In other words, do you expect $\lim_{t\to\infty} Q(t)$ to exist and, if so, what is this limit?

(c) Formulate the initial value problem to be solved.

(d) Solve the initial value problem. Determine $\lim_{t\to\infty} Q(t)$ if it exists.

9. **A Change of Clock** We approach the solution of Exercise 8 from a different point of view. Suppose the modeling of a flow problem leads to a differential equation of the form $dQ/dt = f(t)(\alpha Q + \beta)$, where $f(t) > 0$ and α, β are constants. The key

feature that we'll exploit is the fact that the same function $f(t)$ multiplies all terms on the right-hand side. We can change this differential equation to a simpler one by changing the time scale, that is, introducing a new clock. To do this we introduce a new time variable, call it τ, that we relate to time t by the relation $d\tau/dt = f(t)$. Since we want both clocks to start at the same time, set $\tau(0) = 0$. This determines the new time τ as $\tau(t) = \int_0^t f(s)\,ds$. The reason for introducing the new time variable lies in the fact that, using the chain rule, we have

$$\frac{dQ}{dt} = \frac{dQ}{d\tau}\frac{d\tau}{dt} = \frac{dQ}{d\tau}f(t).$$

If we now insert this result into the differential equation and cancel the common positive factor $f(t)$, we obtain $dQ/d\tau = \alpha Q + \beta$. Therefore, the differential equation becomes a constant coefficient equation in the new time variable. Also, since $\tau(0) = 0$, the initial condition at $\tau = 0$ is the same as that at $t = 0$. Use this approach to solve the initial value problem modeling the oscillating flow rate of Exercise 8. What is the function $f(t)$ in this specific problem? Obtain the solution Q first as a function of τ and then substitute $\tau(t) = \int_0^t f(s)\,ds$ to obtain $Q(t)$.

10. Oscillating Inflow Concentration A tank initially contains 10 lb of salt dissolved in 200 gal of water. Assume that a salt solution flows into the tank at a rate of 3 gal/min and the well-stirred mixture flows out at the same rate. Assume that the inflow concentration oscillates in time, however, and is given by $c_i(t) = 0.2(1 + \sin t)$ lb of salt per gallon. Thus, as time evolves, the concentration oscillates back and forth between 0 and 0.4 lb of salt per gallon.

(a) Make a conjecture, on the basis of physical reasoning, as to whether you expect the amount of salt in the tank to reach a constant equilibrium value as time increases. In other words, will $\lim_{t\to\infty} Q(t)$ exist?

(b) Formulate the corresponding initial value problem.

(c) Solve the initial value problem.

(d) Plot $Q(t)$ versus t. How does the amount of salt in the tank vary as time becomes increasingly large? Is this behavior consistent with your intuition?

11. Mixing and Radioactive Decay A lake holding 1,200,000 gal of water is fed by a stream, where 1000 gal/min of water flows into the lake and water flows out at the same rate. At a certain instant, an accidental spill introduces 5 lb of soluble radioactive material, having a half-life of 18 hours, into the lake.

(a) Let $Q(t)$ represent the amount of radioactive material present in the lake at time t, measured from the instant of the spill. Use the basic conservation principle (rate of change = rate in − rate out) to derive a differential equation describing how $Q(t)$ changes with time. Assume the radioactive material, dissolved in the lake water, forms a well-stirred mixture. Add the appropriate initial condition to obtain the initial value problem of interest.

(b) When will the concentration of radioactive material in the lake be reduced to 0.01% of its initial level?

Exercises 12–16:

Assume Newton's law of cooling applies in each of these exercises.

12. A chef removed an apple pie from the oven and allowed it to cool at room temperature (72°F). The pie had a temperature of 350°F when removed from the oven; 10 min later, the pie had cooled to 290°F. How long will it take for the pie to cool to 120°F?

13. Food, initially at a temperature of 40°F was placed in an oven preheated to 350°F. After 10 min in the oven, the food had warmed to 120°F. After 20 min, the food was removed from the oven and allowed to cool at room temperature (72°F). If the ideal serving temperature of the food is 110°F, when should the food be served?

14. **Cooling and Population Dynamics, Sterilization** At a certain instant an uncooked food sample, initially at room temperature (72°F) has its bacterial count measured and is placed in boiling water (212°F). Five minutes later, the temperature of the food sample is found to be 170°F. After 10 min, the sample is removed from the boiling water; at that time the bacterial count has been reduced to 1% of its initial level. A food scientist believes that a suitable model for the bacteria population's change in time is

$$P'(t) = r\left(1 - \frac{\Theta(t)}{140}\right)P(t),$$

where $\Theta(t)$ is the temperature (degrees F) at time t. Assuming the model to be correct, what is the value of the constant r?

15. A student performs the following experiment using two identical cups of water. One cup is removed from a refrigerator at 34°F and allowed to warm in its surroundings to room temperature (72°F). A second cup is simultaneously taken from room temperature surroundings and placed in the refrigerator to cool. The time at which each cup of water reached a temperature of 53°F is recorded. Are the two recorded times the same or not? Explain.

16. At some time, say $t = 0$, a casserole at temperature 40°F is taken from the refrigerator and placed on the table in a kitchen whose room temperature is 72°F. Simultaneously, an oven, set to 300°F, is turned on. Assume that the oven temperature, as it heats up, is described by the equation $S(t) = 72 + 228(1 - e^{-\alpha t})$°F, where t is measured in minutes and α is a positive constant. After 2 min, the casserole, resting on the table, has warmed to 45°F and the oven temperature has risen to 150°F. At that time the casserole is placed in the oven. What is the temperature of the food 8 min after being placed into the oven?

Extended Problem: The Processing of Seafood

Many foods are sterilized by cooking. Sufficient heat kills harmful bacteria and makes the food fit for human consumption. As a specific example, consider the processing of fresh crabmeat. At the time the harvested crabs are taken off the boats, they are heavily laden with bacteria. Therefore, the crabmeat must be taken to a processing facility and steamed so as to reduce the bacteria population to an acceptably low level—the longer the crabmeat is steamed, the lower the final bacteria count. But, steaming forces moisture out of the meat, reducing the amount of crabmeat for sale. Excessive cooking also destroys taste and texture. The processor is therefore faced with a trade-off when choosing an appropriate steaming time.

The basis for a choice of steaming time is the concept of "shelf life." After the steaming treatment is completed, the bacteria population is reduced to a certain low level. The product is then placed in a sterile package and refrigerated. Under refrigeration, the bacterial content in the meat slowly increases. Eventually the bacteria population reaches a size where the crabmeat is no longer suitable for consumption. The time span during which packaged crabmeat is suitable for sale is called the **shelf life** of the product.

We study the following problem:

How long must the crabmeat be steamed to achieve a desired shelf life?

The study involves two of the models we have considered thus far.

Modeling Seafood Shelf Life

The first step in modeling shelf life is to choose a model that describes the population dynamics of the bacteria. For simplicity, use the population growth model introduced in Section 2.4,

$$\frac{dP}{dt} = kP, \tag{1}$$

where $P(t)$ denotes the bacteria population at time t. In equation (1) the parameter k represents the difference between birth and death rates per unit population per unit time. In this application, k is *not* a constant. Rather, it is a function of temperature having the form $k(T)$ where T denotes the temperature of the crabmeat. Thus, equation (1) takes the form

$$\frac{dP}{dt} = k(T)\,P. \tag{2}$$

Note that $k(T)$ is ultimately a function of time since the temperature T of the crabmeat varies with time in the steamer and in the refrigeration case.

Modeling the Bacteria Growth Rate, *k(T)*

We also need to choose a reasonable model for the bacteria growth rate, $k(T)$. We do so by reasoning as follows. At low temperatures (near freezing) the rate of growth of the bacteria population is slow; that's why we refrigerate foods. Mathematically, $k(T)$ is a relatively small positive quantity at these temperatures. As temperature increases, the bacterial growth rate, $k(T)$, first increases, with the most rapid rate of growth occurring near 90°F. Beyond this temperature, the growth rate begins to decrease. Beyond about 145°F, the death rate exceeds the birth rate and the bacteria population begins to decline. Mathematically, $k(T)$ should increase from a small positive value near freezing to a positive maximum near 90°F and then decrease, passing through zero at about 145°F. A simple model for $k(T)$ that captures this qualitative behavior is the quadratic

$$k(T) = k_0 + k_1(T - 34)(140 - T), \tag{3}$$

where k_0 and k_1 are positive constants. (Typically, k_0 and k_1 would be determined by experiment.) In this assumed form for $k(T)$, the value $k(34) = k_0$ represents the relatively slow growth rate at the near-freezing 34°F temperature. Also note that $k(T)$ has a positive maximum at 87°F.

To complete this study of shelf life we need a model that describes the thermal behavior of the crabmeat—how the crabmeat temperature T varies in response to the temperature of the surroundings. Assume that Newton's law of cooling (recall Section 2.5) is a suitable model,

$$\frac{dT}{dt} = \eta(S(t) - T). \tag{4}$$

In equation (4), η is a positive constant and $S(t)$ is the temperature of the surroundings. Note that the surrounding temperature is not constant since the crabmeat is initially in the steamer and then in the refrigeration case.

PROBLEM:

In addition to the modeling assumptions already made, also assume the following:

1. The bacteria are brought into the processing plant at room temperature (75°F), containing (worst case) 10^7 bacteria per cubic centimeter (cm^3).
2. The steam bath is maintained at a constant 250°F temperature.
3. When placed in the steam bath, it has been observed that crabmeat temperature rises from 75°F to 200°F in 5 min.
4. When kept at a constant 34°F temperature, the bacterial count in the crabmeat doubles in 60 hr.
5. Bacterial count in the crabmeat begins to decline once the temperature exceeds 145°F; that is, see equation (3), $k(145) = 0$.
6. A bacterial count of 10^5 bacteria per cubic centimeter governs shelf life. Once this bacterial count is reached, the crabmeat can no longer be offered for sale.

Determine how long the crabmeat must be steamed to achieve a shelf life of 16 days. Assume that the crabmeat goes directly from the 250°F steam bath to the 34°F refrigeration case temperature. Neglect transit time to the store (or assume that it has been factored into the 16-day requirement). Assume that the measurement of shelf life begins the moment the crabmeat is removed from the steam bath.

3

First Order Nonlinear Differential Equations

CHAPTER OVERVIEW

Introduction

In Chapter 2 we studied first order linear differential equations, equations of the form

$$y' + p(t)y = g(t).$$

In this chapter we consider first order *non*linear differential equations. The term "nonlinear differential equation" encompasses all differential equations that are not linear. In particular, a **first order nonlinear differential equation** has the form

$$y' = f(t, y),$$

where $f(t, y) \neq -p(t)y + g(t)$.

The chapter begins with a brief survey of existence and uniqueness results for solutions of nonlinear first order differential equations. This provides us with a perspective of what to expect when solving these equations. In Sections 3.2–3.4, we describe approaches for solving three particular types of nonlinear differential equations—separable equations, exact equations, and Bernoulli equations. Sections 3.5–3.7 apply these solution techniques to problems arising in population dynamics and mechanics.

Sections 3.1–3.3 and the applications of Sections 3.5 and 3.6 form the core of Chapter 3. Section 3.8 introduces Euler's method and uses this simple algorithm to numerically solve some initial value problems that cannot be solved by the analytical techniques described in Sections 3.2–3.4.

As noted above, a nonlinear differential equation has the form $y' = f(t, y)$, where $f(t, y) \neq -p(t)y + g(t)$. Three examples of first order nonlinear differential equations are

$$\text{(a) } y' = t^2 + y^2 \qquad \text{(b) } y' = t + \cos y \qquad \text{(c) } y' = \frac{t}{y}.$$

Nonlinear differential equations arise naturally in many models of physical phenomena, such as population dynamics influenced by environmental constraints or one-dimensional motion in the presence of air resistance. We'll consider such applications later in this chapter.

Because the set of nonlinear differential equations is so diverse, the type of theoretical statements that can be made about the behavior of their solutions is less comprehensive than those made in Chapter 2 for linear equations. In addition, unlike the situation for linear equations, we cannot derive a general solution procedure that applies to the entire class of nonlinear equations. We therefore concentrate on certain subclasses of nonlinear differential equations for which solution procedures do exist. In many cases, however, solving a nonlinear initial value problem requires numerical procedures of the type discussed in Section 3.8 and Chapter 9.

We concentrate on *first order* nonlinear differential equations in this chapter. In Chapter 8, we will discuss higher order nonlinear equations and systems of nonlinear equations. Many of the ideas developed in this chapter have direct generalizations to the more complicated problems of Chapter 8.

3.1 Existence and Uniqueness

We begin our study of nonlinear equations by considering questions of existence and uniqueness for initial value problems. In particular, given the initial value problem

$$y' = f(t, y), \qquad y(t_0) = y_0, \tag{1}$$

what conditions on the function $f(t, y)$ guarantee that problem (1) has a unique solution? On what t-interval does this unique solution exist? The answers to these questions provide a framework within which we can work.

For example, if we use some special technique to find a solution of problem (1), then it is essential to know whether the solution we found is the *only* solution. In fact, if the initial value problem does not have a unique solution, then it probably is not a good mathematical model for the physical phenomenon under consideration.

Existence and uniqueness are also important considerations if we need to use numerical methods to approximate a solution. For example, if a numerical solution "blows up," we want to know whether this behavior arises from inaccuracies in the numerical method or whether it might be inherent in the initial value problem itself.

We now state a theorem that guarantees the existence of a unique solution to an initial value problem. The proof of this theorem is usually studied in a more advanced course in differential equations and we do not give a proof here.

Theorem 3.1

Let R be the open rectangle defined by $a < t < b$, $\alpha < y < \beta$. Let $f(t, y)$ be a function of two variables defined on R, where $f(t, y)$ and the partial derivative $\partial f/\partial y$ are continuous on R. Suppose (t_0, y_0) is a point in R. Then there is an open t-interval (c, d), contained in (a, b) and containing t_0, in which there exists a unique solution of the initial value problem

$$y' = f(t, y), \qquad y(t_0) = y_0.$$

A typical open rectangle R, initial point (t_0, y_0), and interval (c, d) are shown in Figure 3.1. (The rectangle R is called an **open rectangle** because it does not contain the four line segments forming its boundary.)

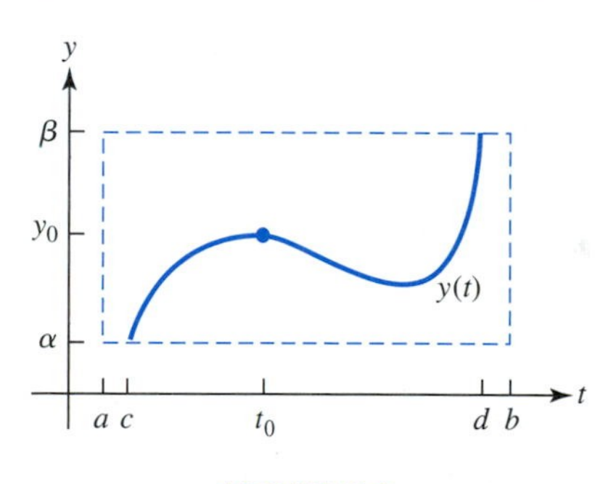

FIGURE 3.1

The open rectangle R, defined by $a < t < b$, $\alpha < y < \beta$, contains the initial point (t_0, y_0). If the hypotheses of Theorem 3.1 hold on R, we are guaranteed a unique solution to the initial value problem on some open interval (c, d).

Although presented in the context of nonlinear differential equations, Theorem 3.1 makes no distinction between linear and nonlinear differential equations. It applies to linear first order equations where

$$f(t, y) = -p(t)y + g(t),$$

as well as to nonlinear first order equations.

Two important observations can be made about Theorem 3.1.

1. The hypotheses of Theorem 3.1 are a natural generalization of those made in Theorem 2.1 for linear differential equations. That is, if $f(t, y) = -p(t)y + g(t)$, then $\partial f/\partial y = -p(t)$. Therefore, requiring $f(t, y)$ and $\partial f/\partial y$ to be continuous on the rectangle R means any linear differential equation satisfying the hypotheses of Theorem 3.1 also satisfies

the hypotheses of Theorem 2.1. Conversely, a linear differential equation satisfying the hypotheses of Theorem 2.1 also satisfies the hypotheses of Theorem 3.1.

2. The conclusions of Theorem 3.1, however, differ substantially from those of Theorem 2.1. Since we have broadened our perspective to encompass nonlinear differential equations, the corresponding conclusions of Theorem 3.1 are therefore weaker than those of Theorem 2.1. Theorem 2.1 guarantees existence and uniqueness on the entire (a, b) interval. Theorem 3.1 guarantees existence and uniqueness on only *some* subinterval (c, d) of (a, b) containing t_0; it does *not* guarantee existence and uniqueness on the entire (a, b) interval. Moreover, Theorem 3.1 gives no insight into how large (c, d) is or how we might go about estimating it.

Although Theorem 3.1 leaves many questions unanswered, it does provide us with the framework we need to study solution techniques for certain classes of nonlinear differential equations; these techniques will be developed in Sections 3.2–3.4. Examining these special cases will give us valuable insight into the behavior of solutions of nonlinear equations.

Autonomous Differential Equations

First order autonomous equations have the form $y' = f(y)$. The right-hand side of the differential equation does not explicitly depend on the independent variable t. Solution curves for an autonomous differential equation have the important geometric property that they can be translated.

As an example, consider the autonomous equation

$$y' = y(2 - y).$$

The direction field for this equation, along with portions of some solution curves, is shown in Figure 3.2. As observed in Section 1.2, the slopes of the

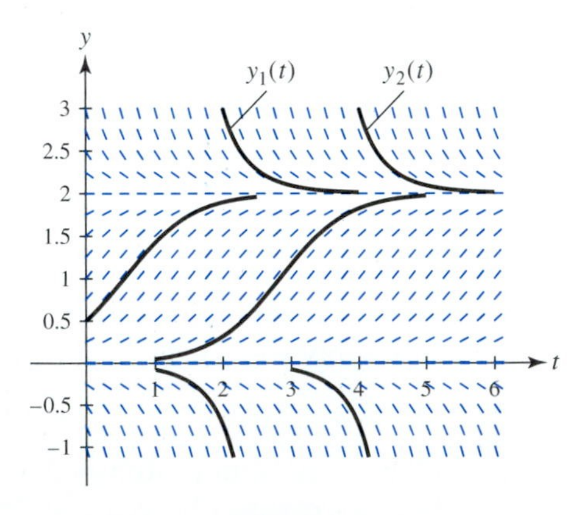

FIGURE 3.2

The direction field for the autonomous equation $y' = y(2 - y)$, together with portions of some typical solutions. Notice that the graph of $y_1(t)$, when translated to the right, looks as though it coincides with the graph of $y_2(t)$.

direction field filaments for an autonomous equation remain constant along horizontal lines. For instance (see Figure 3.2), at every point along the line $y = 1$, the direction field filaments have slope equal to 1. Similarly, at every point along the line $y = 3$, the direction field filaments have slope equal to -3.

Besides illustrating that horizontal lines are isoclines for the autonomous equation, Figure 3.2 also illustrates an important property of solutions to autonomous differential equations. That is, it looks as though the graph of $y_1(t)$, when translated about 2 units to the right, will fall exactly on the graph of $y_2(t)$. This is indeed the case, and we show in Theorem 3.2 that the solution $y_2(t)$ is related to the solution $y_1(t)$ by

$$y_2(t) = y_1(t - c),$$

where c is a constant.

Theorem 3.2

Let the initial value problem

$$y' = f(y), \qquad y(0) = y_0$$

satisfy the conditions of Theorem 3.1 and let $y_1(t)$ be the unique solution, where the interval of existence for $y_1(t)$ is $a < t < b$, with $a < 0 < b$.

Consider the initial value problem

$$y' = f(y), \qquad y(t_0) = y_0. \tag{2}$$

Then the function $y_2(t)$ defined by $y_2(t) = y_1(t - t_0)$ is the unique solution of initial value problem (2), and has an interval of existence

$$t_0 + a < t < t_0 + b.$$

PROOF: Since $y_1(t)$ is defined for $a < t < b$, we know that $y_2(t) = y_1(t - t_0)$ is defined for $a < t - t_0 < b$ and hence for $t_0 + a < t < t_0 + b$. We next observe that $y_2(t)$ satisfies the initial condition of (2), since $y_2(t_0) = y_1(t_0 - t_0) = y_1(0) = y_0$. Therefore, to complete the proof of Theorem 3.2, we need to show that $y_2(t)$ is a solution of the differential equation $y' = f(y)$.

Using the definition of $y_2(t)$, the chain rule, and the fact that $y_1(t)$ solves the differential equation $y' = f(y)$:

$$y_2'(t) = \frac{d}{dt} y_1(t - t_0) = y_1'(t - t_0) \frac{d}{dt}(t - t_0) = y_1'(t - t_0) = f(y_1(t - t_0)) = f(y_2(t)).$$

Therefore, the function $y_2(t) = y_1(t - t_0)$ solves the differential equation $y' = f(y)$ and hence the initial value problem (2). ▮

The important conclusion to be reached from Theorem 3.2 is that the solution of the autonomous initial value problem $y' = f(y), y(t_0) = y_0$ depends on the independent variable t and the initial condition point (t_0, y_0) as a function of the combination $t - t_0$. What matters is time t measured relative to the initial time t_0. As a simple example, recall that the solution of the linear autonomous equation $y' = ky, y(t_0) = y_0$ is

$$y(t) = y_0 e^{k(t - t_0)}.$$

EXERCISES

Exercises 1–8:

For the given initial value problem,

(a) Rewrite the differential equation, if necessary, to obtain the form

$$y' = f(t, y), \qquad y(t_0) = y_0.$$

Identify the function $f(t, y)$.

(b) Compute $\partial f/\partial y$. Determine where in the ty-plane both $f(t, y)$ and $\partial f/\partial y$ are continuous.

(c) Determine the largest open rectangle in the ty-plane that contains the point (t_0, y_0) and in which the hypotheses of Theorem 3.1 are satisfied.

1. $3y' + 2t\cos y = 1, \quad y(\pi/2) = -1$

2. $3ty' + 2\cos y = 1, \quad y(\pi/2) = -1$

3. $2t + (1 + y^2)y' = 0, \quad y(1) = 1$

4. $2t + (1 + y^3)y' = 0, \quad y(1) = 1$

5. $y' + ty^{1/3} = \tan t, \quad y(-1) = 1$

6. $(y^2 - 9)y' + e^{-y} = t^2, \quad y(2) = 2$

7. $(\cos y)y' = 2 + \tan t, \quad y(0) = 0$

8. $(\cos 2t)y' = 2 + \tan y, \quad y(\pi) = 0$

9. Give an example of an initial value problem for which the open rectangle $R = \{(t, y) \mid 0 < t < 4, -1 < y < 2\}$ represents the largest region in the ty-plane where the hypotheses of Theorem 3.1 are satisfied.

10. Consider the initial value problem $t^2y' - y^2 = 0, y(1) = 1$.

(a) Determine the largest open rectangle in the ty-plane, containing the point $(t_0, y_0) = (1, 1)$, in which the hypotheses of Theorem 3.1 are satisfied.

(b) A solution of the initial value problem is $y(t) = t$. This solution exists on $-\infty < t < \infty$. Does this fact contradict Theorem 3.1? Explain your answer.

11. The solution of the initial value problem $y' = f(y), y(0) = 2$ is known to be $y(t) = 2/\sqrt{1-t}$. Let $\overline{y}(t)$ represent the solution of the initial value problem $y' = f(y)$, $y(1) = 2$. What is the value of $\overline{y}(0)$?

12. The solution of the initial value problem $y' = f(y), y(0) = 8$ is known to be $y(t) = (4 + t)^{3/2}$. Let $\overline{y}(t)$ represent the solution of the initial value problem $y' = f(y), y(t_0) = 8$. Suppose we know that $\overline{y}(0) = 1$. What is t_0?

13. The graph shows the solution of $y' = -1/(2y), y(0) = 1$. Use the graph to answer questions (a) and (b).

(a) If $z_1(t)$ is the solution of $z_1' = -1/(2z_1), z_1(-2) = 1$, what is $z_1(-5)$?

(b) If $z_2(t)$ is the solution of $z_2' = -1/(2z_2), z_2(2) = 1$, what is $z_2(3)$?

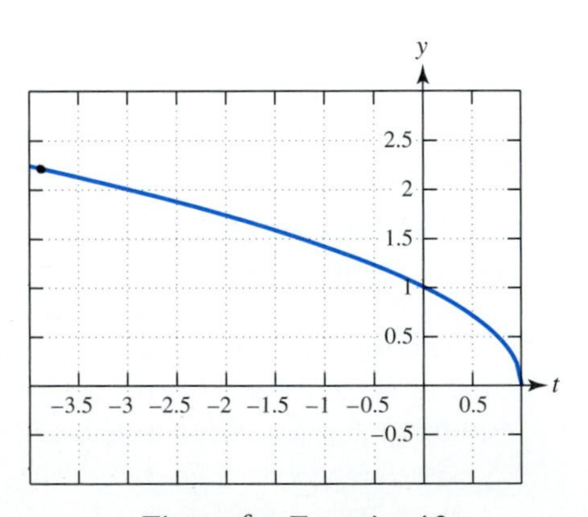

Figure for Exercise 13

3.2 Separable First Order Equations

In Chapter 2 we obtained an explicit representation for the solution of a first order linear differential equation; recall equation (6) in Section 2.3. By contrast, there is no all-encompassing technique that leads to an explicit representation for the solution of a first order nonlinear differential equation.

For certain subclasses of nonlinear equations, however, techniques have been discovered that give us some information about the solution. We study one such subclass in this section—separable differential equations.

Separable Equations

The term **separable differential equation** is used to describe any first order differential equation that can be put into the form

$$n(y)\frac{dy}{dt} + m(t) = 0.$$

For example, the differential equations

$$\text{(a)}\ \frac{dy}{dt} + t^2 \sin y = 0 \quad \text{and} \quad \text{(b)}\ y' + e^{t+y} = e^y \sin t$$

are separable since they can be rewritten (respectively) as

$$\text{(a}'\text{)}\ \csc y\,\frac{dy}{dt} + t^2 = 0 \quad \text{and} \quad \text{(b}'\text{)}\ e^{-y}y' + [e^t - \sin t] = 0.$$

A simple example of a nonseparable differential equation is

$$y' = 2ty^2 + 1.$$

The structure of a separable differential equation,

$$n(y)\frac{dy}{dt} + m(t) = 0,$$

gives the equation its name. The first term is the product of dy/dt and a term $n(y)$ that involves only the dependent variable y. The second term, $m(t)$, involves only the independent variable t. In this sense, the variables "separate."

Solving a Separable Differential Equation

We can get some information about the solution of a separable equation by "reversing the chain rule." We first illustrate this technique in Example 1 and then describe the general procedure.

EXAMPLE 1

Solve the initial value problem

$$y' = 2ty^2, \qquad y(0) = 1.$$

Solution: First, notice that $f(t, y) = 2ty^2$ is continuous on the entire ty-plane, as is the partial derivative $\partial f/\partial y = 4ty$. Therefore, the conditions of Theorem 3.1 are satisfied on the open rectangle R defined by $-\infty < t < \infty$, $-\infty < y < \infty$.

(continued)

(continued)

By Theorem 3.1 we are guaranteed the existence of a unique solution of the initial value problem. But, the theorem provides no insight into the interval of existence of the solution.

The differential equation is separable. It can be rewritten as

$$\frac{y'}{y^2} - 2t = 0.$$

To emphasize the fact that y is the dependent variable, we express the equation as

$$\frac{y'(t)}{y(t)^2} - 2t = 0.$$

Taking antiderivatives, we find

$$\int \frac{y'(t)}{y(t)^2}\,dt - \int 2t\,dt = C.$$

Evaluating the integrals on the left-hand side,

$$\frac{-1}{y(t)} - t^2 = C.$$

Solving for $y(t)$ we obtain a family of solutions

$$y(t) = \frac{-1}{t^2 + C}.$$

Imposing the initial condition, $y(0) = 1$, yields the unique solution of the initial value problem,

$$y(t) = \frac{1}{1 - t^2}.$$

Having determined the solution, we are now able to see that the interval of existence is $-1 < t < 1$. ▲

The process used to solve the initial value problem in Example 1 can be viewed as reversing the chain rule. To explain, we return to the general separable differential equation,

$$n(y)\frac{dy}{dt} + m(t) = 0. \tag{1}$$

Suppose that $N(y)$ is any antiderivative of $n(y)$; that is,

$$\frac{d}{dy}N(y) = n(y).$$

If y is a differentiable function of t, then the chain rule leads to

$$\begin{aligned}\frac{d}{dt}N(y) &= \frac{dN}{dy}\frac{dy}{dt}\\ &= n(y)\frac{dy}{dt}.\end{aligned}$$

This observation shows that the term $n(y)\,dy/dt$ in differential equation (1) can be rewritten as dN/dt.

Similarly, if $M(t)$ represents any antiderivative of $m(t)$, the second term in equation (1) can be written as dM/dt. Combining these two observations, we express the left-hand side of equation (1) as

$$n(y)\frac{dy}{dt}+m(t)=\frac{d}{dt}N(y)+\frac{d}{dt}M(t)=\frac{d}{dt}[N(y)+M(t)].$$

Thus, equation (1) reduces to

$$\frac{d}{dt}[N(y)+M(t)]=0.$$

The term $N(y)+M(t)$ is a function of t [involving the unknown solution $y=y(t)$] whose derivative vanishes identically. Therefore,

$$N(y)+M(t)=C, \tag{2}$$

where C is an arbitrary constant.

This equation provides us with information about the solution $y(t)$. It is not an explicit expression for this solution; it is an equation that the solution must satisfy. An equation in y and t, such as (2), is called an **implicit solution**. Sometimes (as in Example 1 and in Example 2 below) we can "unravel" the implicit solution and solve for $y(t)$ as an explicit function of the independent variable t. In other cases, we must be content with the implicit solution.

Whether or not we can unravel the implicit solution given in equation (2), we can always determine the constant C by imposing the initial condition $y(t_0)=y_0$, finding

$$C=N(y_0)+M(t_0).$$

EXAMPLE 2

Solve the initial value problem

$$\frac{dy}{dt}=-\frac{t}{y},\qquad y(0)=-2.$$

Solution: Separating the variables, we obtain

$$y\frac{dy}{dt}+t=0.$$

Integrating, we find an implicit solution

$$\frac{y^2}{2}+\frac{t^2}{2}=C.$$

Imposing the initial condition, we have $(-2)^2/2+0^2/2=C$, or $C=2$. Thus, an implicit solution of the initial value problem is given by

$$y^2+t^2=4.$$

Suppose we want an explicit solution. Solving the equation above, we find

$$y=\pm\sqrt{4-t^2}.$$

Which root should be taken? To satisfy the initial condition, $y(0)=-2$, we must take the negative root. Thus, the solution is

$$y=-\sqrt{4-t^2}. \tag{3}$$

(continued)

(continued)

This choice of roots is also obvious geometrically, since the graph of $y^2 + t^2 = 4$ is a circle of radius 2 in the ty-plane, as shown in Figure 3.3. The solution of the initial value problem has the lower semicircle as its graph. The function given by equation (3) is defined and continuous on $[-2, 2]$. It is differentiable and satisfies the differential equation on the open interval $(-2, 2)$. ▲

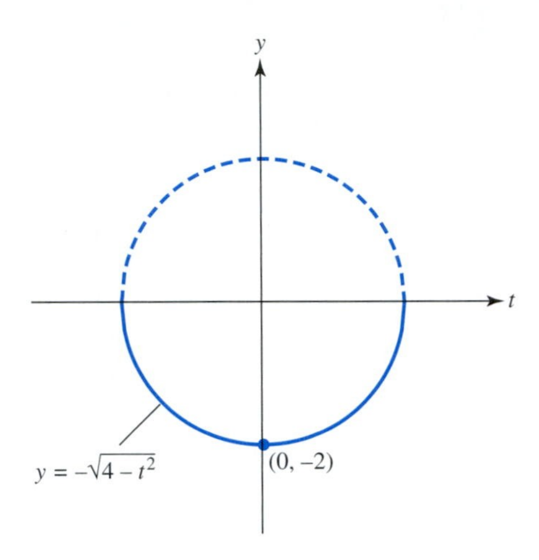

FIGURE 3.3

The implicit solution of the initial value problem in Example 2 is $y^2 + t^2 = 4$; its graph is a circle of radius 2. The explicit solution of the initial value problem is $y = -\sqrt{4 - t^2}$; its graph is the lower semicircle.

EXAMPLE 3

Solve the initial value problem

$$\frac{dy}{dt} = \sqrt{1 - y^2}, \qquad y(0) = 0.$$

Solution: Since the right-hand side of the differential equation has no explicit t-dependence, this differential equation is autonomous. It has two equilibrium solutions, $y(t) = -1$ and $y(t) = 1$.

The differential equation makes sense only in the horizontal strip of the ty-plane defined by $-1 \le y \le 1$, $-\infty < t < \infty$. The hypotheses of Theorem 3.1 are satisfied in the open rectangle R defined by $-1 < y < 1$, $-\infty < t < \infty$. [Note: The partial derivative $\partial f/\partial y$ does not exist along the top and bottom boundaries of R, along the lines $y = -1$ and $y = 1$.]

The differential equation is separable. Separating the variables,

$$\frac{1}{\sqrt{1 - y^2}} \frac{dy}{dt} - 1 = 0,$$

and then integrating, we obtain an implicit solution

$$\sin^{-1} y - t = C.$$

The initial condition $y(0) = 0$ leads to $C = 0$. Therefore, $\sin^{-1} y = t$ and we are tempted to draw the *incorrect* conclusion that $y(t) = \sin t$ is the solution on the *entire* t-axis, $-\infty < t < \infty$.

The situation is, in fact, more complicated. Recall that the range of the arcsine function is $-\pi/2 \le t \le \pi/2$. Therefore, the correct conclusion is

$$\sin^{-1} y = t \quad \text{and hence} \quad y(t) = \sin t, \qquad -\frac{\pi}{2} \le t \le \frac{\pi}{2}.$$

What happens at the endpoints of this t-interval? For definiteness, consider $t = \pi/2$. At this point, $y(\pi/2) = \sin(\pi/2) = 1$ and $y'(\pi/2) = 0$. That is to say, the solution $y(t) = \sin t$ has reached the equilibrium value $y = 1$. Once the solution reaches this value, it stays there. A similar situation holds at $t = -\pi/2$. Therefore, the solution of the initial value problem does indeed exist on $-\infty < t < \infty$. However, its correct description is

$$y(t) = \begin{cases} -1, & -\infty < t < -\pi/2 \\ \sin t, & -\pi/2 \le t \le \pi/2 \\ 1, & \pi/2 < t < \infty \end{cases}.$$

The solution is shown in Figure 3.4.

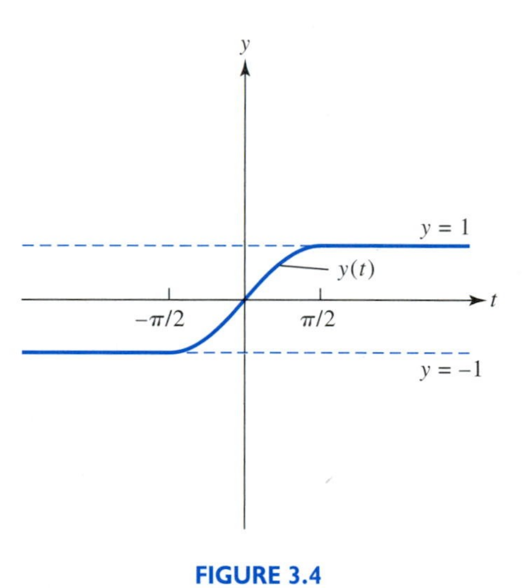

FIGURE 3.4

The solution of the initial value problem posed in Example 3.

To provide another way to understand the issues, we show directly that $y = \sin t$ *cannot* be the solution when, for example, $\pi/2 < t < 3\pi/2$. We do this by inserting $y = \sin t$ into the differential equation and showing that the equation is not satisfied on that interval.

The differential equation is $y' = \sqrt{1 - y^2}$. For $y = \sin t$, we have $y' = \cos t$. However, for $y = \sin t$

$$\sqrt{1 - y^2} = \sqrt{1 - \sin^2 t} = \sqrt{\cos^2 t} = |\cos t| = -\cos t \quad \text{for} \quad \frac{\pi}{2} < t < \frac{3\pi}{2}.$$

Therefore, $y = \sin t$ is not the solution of the initial value problem on this interval. ▲

When we discussed uniqueness in Chapter 2, we noted that one consequence of uniqueness is that solution trajectories do not intersect; they form a streamline flow. In Example 3, we see an example of nonuniqueness. Two distinct solutions, $y(t) = \sin t$ and $y(t) = 1$, intersect at $t = \pi/2$. This, however, does not contradict Theorem 3.1. As we observed earlier, the hypotheses of Theorem 3.1 are not satisfied at the boundaries of the horizontal strip.

Differences between Linear and Nonlinear Differential Equations

We can use Examples 1–3 to make several points about Theorem 3.1 and to illustrate some of the differences between nonlinear and linear differential equations.

1. ***The Interval of Existence May Not Be Obvious.*** If the coefficient functions for a *linear* differential equation are continuous on an interval (a, b) where (a, b) contains the initial point t_0, then a unique solution of the initial value problem $y' + p(t)y = g(t), y(t_0) = y_0$ exists and is defined on all of (a, b).

   By way of contrast, consider the initial value problem in Example 1. The function $f(t, y) = 2ty^2$ is continuous everywhere and is about as nice a nonlinear function as we can expect. However, the solution of the initial value problem has vertical asymptotes at $t = -1$ and at $t = 1$. Note that we cannot predict that the interval of existence is $(-1, 1)$ by simply looking at the equation $y' = 2ty^2$. In fact, if the initial condition is changed from $y(0) = 1$ to $y(0) = -1$, then the interval of existence changes to $(-\infty, \infty)$. (See Exercise 4.)

   Example 2 also provides another illustration. The interval of existence in Example 2 is $-2 < t < 2$. Suppose we leave the initial condition alone and simply change the sign on the right-hand side of the differential equation. This change produces a new initial value problem

$$\frac{dy}{dt} = \frac{t}{y}, \qquad y(0) = -2.$$

   In this case, the solution is defined for all t. (See Exercise 12.)

   In each of these examples, a harmless looking change in the differential equation or in the initial condition leads to a pronounced change in the nature of the solution and in the interval of existence.

2. ***There May Not Be a Single Formula That Gives All Solutions.*** Note that the family of solutions found in Example 1,

$$y(t) = \frac{-1}{t^2 + C}, \tag{4}$$

   does not include the zero function. The zero function is, however, a solution of $y' = 2ty^2$. Thus, the concept of a general solution (which we found so useful for linear equations) is generally not valid for nonlinear equations. In particular, given the initial value problem $y' = 2ty^2, y(0) = 0$, there is no choice for C in equation (4) that yields the unique solution, $y(t) = 0$, that is guaranteed by Theorem 3.1.

3. *We May Have to Be Content with Implicitly Defined Solutions.* In Examples 1 and 2 we were able to find an explicit formula for the solution, $y(t)$, of the initial value problem. But [see Exercises 14 and 15], we may not always be so fortunate and it may not be possible to obtain an explicit formula for the solution. By contrast [see equation (6) in Section 2.3], there is an explicit formula for the solution of any first order *linear* differential equation.

Nonlinear differential equations are often more difficult to solve and analyze than linear differential equations. In later sections, however, we state some useful results that give insight into the behavior of solutions of nonlinear equations.

EXERCISES

Exercises 1–15:

(a) Obtain an implicit solution and, if possible, an explicit solution of the initial value problem.

(b) If an explicit solution of the problem is attainable, determine the t-interval of existence.

1. $y\dfrac{dy}{dt} - \sin t = 0, \quad y(\pi/2) = -2$

2. $\dfrac{dy}{dt} = \dfrac{1}{y^2}, \quad y(1) = 2$

3. $y' + \dfrac{1}{y+1} = 0, \quad y(1) = 0$

4. $y' - 2ty^2 = 0, \quad y(0) = -1$

5. $y' - ty^3 = 0, \quad y(0) = 2$

6. $\dfrac{dy}{dt} + e^y t = e^y \sin t, \quad y(0) = 0$

7. $\dfrac{dy}{dt} = 1 + y^2, \quad y(\pi/4) = -1$

8. $t^2 y' + \sec y = 0, \quad y(-1) = 0$

9. $\dfrac{dy}{dt} = t - ty^2, \quad y(0) = \dfrac{1}{2}$

10. $3y^2\dfrac{dy}{dt} + 2t = 1, \quad y(-1) = -1$

11. $\dfrac{dy}{dt} = e^{t-y}, \quad y(0) = 1$

12. $\dfrac{dy}{dt} = \dfrac{t}{y}, \quad y(0) = -2$

13. $e^t y' + (\cos y)^2 = 0, \quad y(0) = \pi/4$

14. $(2y - \sin y)y' + t = \sin t, \quad y(0) = 0$

15. $e^y y' + \dfrac{t}{y+1} = \dfrac{2}{y+1}, \quad y(1) = 2$

16. For what values of the constants α, y_0, and integer n is the function $y(t) = (4+t)^{-1/2}$ a solution of the initial value problem?

$$y' + \alpha y^n = 0, \qquad y(0) = y_0$$

17. For what values of the constants α, y_0, and integer n is the function $y(t) = 6/(5+t^4)$ a solution of the initial value problem?

$$y' + \alpha t^n y^2 = 0, \qquad y(1) = y_0$$

18. State an initial value problem, with initial condition imposed at $t_0 = 2$, having implicit solution $y^3 + t^2 + \sin y = 4$.

19. State an initial value problem, with initial condition imposed at $t_0 = 0$, having implicit solution $ye^y + t^2 = \sin t$.

20. Consider the initial value problem

$$y' = 2y^2, \qquad y(0) = y_0.$$

For what value(s) y_0 will the solution have a vertical asymptote at $t = 4$, where the t-interval of existence is $-\infty < t < 4$?

21. Let $S(t)$ represent the amount of a chemical reactant present at time $t \geq 0$. Assume that $S(t)$ can be determined by solving the initial value problem

$$S' = -\frac{\alpha S}{K + S}, \qquad S(0) = S_0,$$

where α, K, and S_0 are positive constants. [The differential equation, often referred to as the Michaelis-Menten equation, arises in the study of biochemical reactions. The dependent variable, $S(t)$ represents the amount of chemical substrate present.]

(a) Obtain an implicit solution of the initial value problem.

(b) Suppose the values of time t and corresponding amounts of reactant, $S(t)$, have been measured and tabulated at $t = 0$, $t = 1$, and $t = 6$ (as shown in the table). Determine the constants α, K, and S_0.

t	0	1	6
S(t)	1	3/4	1/8

(c) Determine the time at which $S(t) = 1/50$.

Exercises 22–24:

A differential equation of the form

$$y' = p_1(t) + p_2(t)y + p_3(t)y^2$$

is known as a **Riccati equation**.[1] Equations of this form arise when we model one-dimensional motion with air resistance; see Sections 3.5 and 3.6. In general, this equation is not separable. In certain cases, however, (such as in Exercises 22–24), the equation does assume a separable form.

Solve the given initial value problem and determine the t- interval of existence.

22. $y' = 2 + 2y + y^2, \quad y(0) = 0$

23. $y' = t(5 + 4y + y^2), \quad y(0) = -3$

24. $y' = (y^2 + 2y + 1)\sin t, \quad y(0) = 0$

25. Let $Q(t)$ represent the amount of a certain reactant present at time t. Suppose that the rate of decrease of $Q(t)$ is proportional to $Q^3(t)$. That is, $Q' = -kQ^3$, where k is a positive constant of proportionality. How long will it take for the reactant to be reduced to one half of its original amount? Recall, in problems of radioactive decay where the differential equation has the form $Q' = -kQ$, that the half-life was independent of the amount of material initially present. What happens in this case? Does half-life depend on $Q(0)$, the amount initially present?

26. The rate of decrease of a reactant is proportional to the square of the amount present. During a particular reaction, 40% of the initial amount of this chemical remained after 10 sec. How long will it take before only 25% of the initial amount remains?

[1] Jacopo Riccati (1676–1754) worked on many differential equations, including the one that now bears his name. His work in hydraulics proved useful to his native city of Venice.

27. Consider the differential equation $y' = |y|$.

(a) Is this differential equation linear or nonlinear? Is the differential equation separable?

(b) A student solves the two initial value problems $y' = |y|, y(0) = 1$ and $y' = y$, $y(0) = 1$ and then graphs the two solution curves on the interval $-1 \leq t \leq 1$. Sketch what she observes.

(c) She next solves both problems with initial condition $y(0) = -1$. Sketch what she observes in this case.

28. Consider the following autonomous first order differential equations

$$y' = -y^2, \qquad y' = y^3, \qquad y' = y(4 - y).$$

Match each of these equations with one of the solution curve graphs shown. Note that each solution satisfies the initial condition $y(0) = 1$. Can you match them without solving the differential equations?

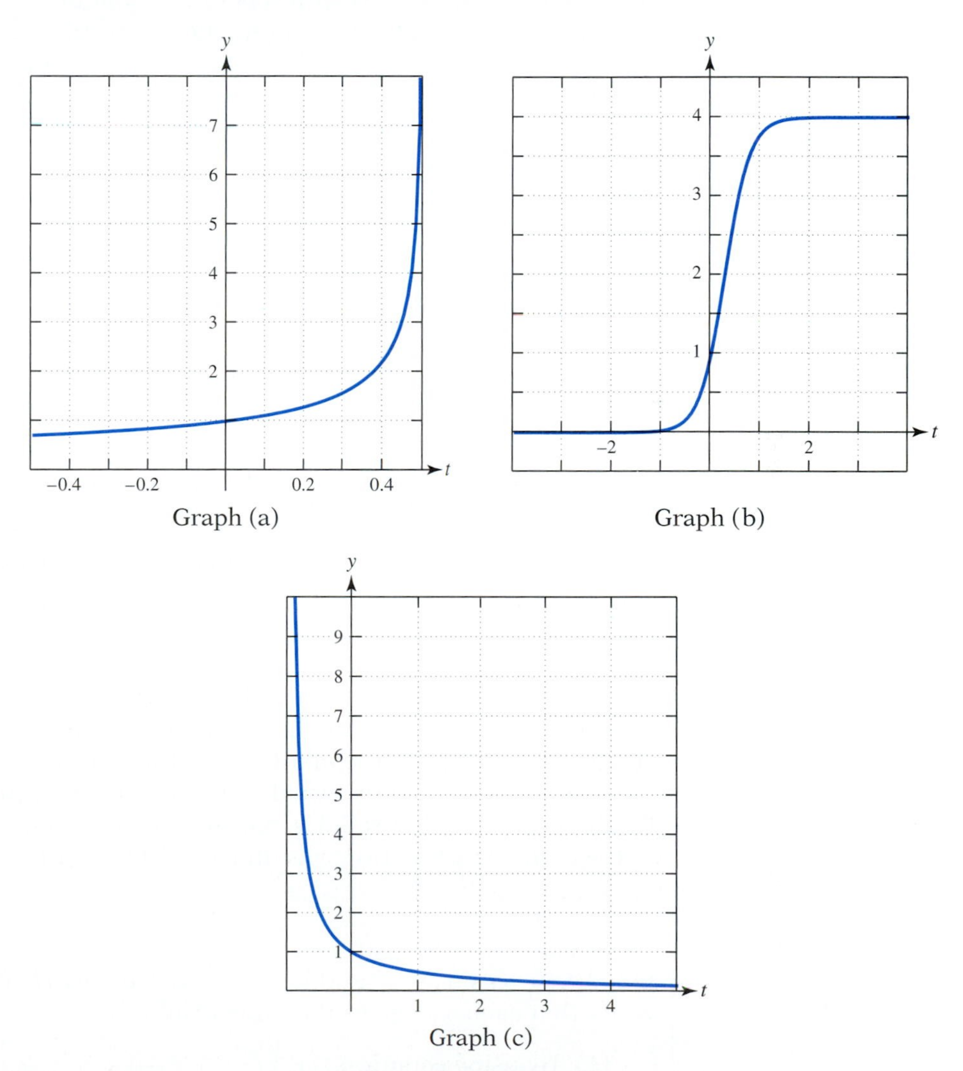

Figure for Exercise 28

29. Exercise 28 shows a graph of the solution of the initial value problem $y' = y(4-y)$, $y(0) = 1$. Use this information to deduce the graph of the solution of the initial value problem $y' = y(4-y), y(3) = 1$.

30. A first order autonomous differential equation has the form $y' = f(y)$. Is such an equation separable? Explain.

31. Assume that $y \sin y - 3t + 3 = 0$ is an implicit solution of the initial value problem $y' = f(y), y(1) = 0$. What is $f(y)$? What is an implicit solution of the initial value problem $y' = t^2 f(y), y(1) = 0$?

3.3 Exact Differential Equations

The class of differential equations referred to as exact includes separable equations as a special case. As with separable equations, the solution procedure for this new class consists of reversing the chain rule. This time, however, we use a chain rule that involves a function of two variables. We begin with a brief review of the chain rule for a function of two variables.

The Extended Chain Rule

Suppose $H(t, y)$ is a function of two independent variables t and y, where $H(t, y)$ has continuous partial derivatives with respect to t and y. If the second independent variable y is replaced with a differentiable function of t, call it $y(t)$, we obtain a composition $H(t, y(t))$ which is now a function of t only. Our hypotheses imply the composite function $H(t, y(t))$ is a differentiable function of t. What is dH/dt?

The appropriate chain rule is

$$\frac{d}{dt} H(t, y(t)) = \frac{\partial H(t, y(t))}{\partial t} + \frac{\partial H(t, y(t))}{\partial y} \frac{dy(t)}{dt}. \tag{1}$$

To understand this equation, note that the partial derivatives on the right-hand side refer to H viewed as a function of the two independent variables t and y. Once these partial derivatives are computed, the variable y is replaced by the function $y(t)$.

Formula (1) is an extension of the chain rule for functions of a single variable. If the function H has the form $H(y)$ so that it is only a function of the single variable y, then the first term on the right-hand side of (1) vanishes and the formula reverts to the usual chain rule for the composite function $H(y(t))$. In the discussion that follows, we will use the notation $H(t, y)$ when referring to the composite function; y is understood to be the function $y(t)$.

EXAMPLE 1

Let $H(t, y) = t \sin y + y^2$ and let $y = t^3$. Calculate dH/dt two different ways and verify that each way gives the same result.

(a) By using equation (1).

(b) By direct differentiation of the composite function $H(t, y) = H(t, t^3)$.

Solution:

(a) Using the chain rule from equation (1), we find

$$\begin{aligned}\frac{d}{dt}H(t,y) &= \frac{\partial H}{\partial t} + \frac{\partial H}{\partial y}\frac{dy}{dt} \\ &= \sin y + (t\cos y + 2y)\frac{dy}{dt} \\ &= \sin t^3 + (t\cos t^3 + 2t^3)(3t^2).\end{aligned}$$

(b) For $y = t^3$, the composite function $H(t, y)$ has the form

$$H(t,y) = t\sin y + y^2 = t\sin t^3 + t^6.$$

Differentiating with respect to t, we find

$$\frac{d}{dt}H(t,y) = \sin t^3 + 3t^3\cos t^3 + 6t^5.$$

The results from parts (a) and (b) agree. ▲

Solving Exact Differential Equations

The basic idea underlying exact differential equations is to reverse the extended chain rule when possible. To that end, let us consider a differential equation of the form

$$M(t,y) + N(t,y)\frac{dy}{dt} = 0. \tag{2}$$

Notice the similarity between the form of differential equation (2) and that of the chain rule (1). To exploit this similarity, suppose there exists some function, call it $H(t, y)$, which satisfies the following two conditions:

$$\frac{\partial H}{\partial t} = M(t,y) \quad \text{and} \quad \frac{\partial H}{\partial y} = N(t,y). \tag{3}$$

Given (3), we can rewrite differential equation (2) as

$$\frac{\partial H}{\partial t} + \frac{\partial H}{\partial y}\frac{dy}{dt} = 0. \tag{4}$$

By the chain rule (1), equation (4) is the same as

$$\frac{d}{dt}H(t,y) = 0.$$

Therefore, we obtain an implicitly defined solution of equation (4),

$$H(t,y) = C. \tag{5}$$

Consider any differential equation of the form (2). If there is a function $H(t, y)$ satisfying the conditions in (3), then the differential equation is called an **exact differential equation**. If we can identify the function $H(t, y)$, then an implicitly defined solution is given by (5).

Recognizing an Exact Differential Equation

As was just shown, once we know a given differential equation is exact and once we can identify a function $H(t, y)$ satisfying the conditions in (3), then we can write down an implicit solution, $H(t, y) = C$, for the differential equation. Two basic questions therefore need to be answered.

1. Given a differential equation of the form $M(t, y) + N(t, y)y' = 0$, how do we know whether or not it is exact? That is, how do we determine if there is a function $H(t, y)$ satisfying the conditions:
$$\frac{\partial H}{\partial t} = M(t, y) \quad \text{and} \quad \frac{\partial H}{\partial y} = N(t, y).$$
2. Suppose we are somehow assured that such a function $H(t, y)$ exists. How do we go about finding $H(t, y)$?

The answer to the first question is given in Theorem 3.3, which is stated without proof. To answer the second question, we will use a process of "anti-partial-differentiation."

Theorem 3.3

Consider the differential equation $M(t, y) + N(t, y)y' = 0$. Let the functions $M, N, \partial M/\partial y$ and $\partial N/\partial t$ be continuous in an open rectangle R of the ty-plane. Then the differential equation is exact in R if and only if
$$\frac{\partial M}{\partial y} = \frac{\partial N}{\partial t} \tag{6}$$
for all points (t, y) in R.

Theorem 3.3 provides an easy test for whether a given differential equation is exact. The theorem does not, however, tell how to construct the implicitly defined solution $H(t, y) = C$. We will discuss finding $H(t, y)$ in the next subsection.

EXAMPLE 2

Which of the following differential equations is (are) exact?

(a) $y + t + ty' = 0$ (b) $y + \sin t + (y \cos t)y' = 0$ (c) $\sin y + (2y + t \cos y)y' = 0$

Solution:

(a) We have in the notation of Theorem 3.3,
$$M(t, y) = y + t \quad \text{and} \quad N(t, y) = t.$$
Calculating the partial derivatives, we find
$$\frac{\partial M}{\partial y} = 1 \quad \text{and} \quad \frac{\partial N}{\partial t} = 1.$$
Therefore, by Theorem 3.3, the differential equation is exact.

(b) For this equation,

$$M(t,y) = y + \sin t \quad \text{and} \quad N(t,y) = y\cos t.$$

Calculating the partial derivatives,

$$\frac{\partial M}{\partial y} = 1 \quad \text{and} \quad \frac{\partial N}{\partial t} = -y\sin t.$$

Since the partial derivatives are not equal, the differential equation is not exact.

(c) Calculating the partial derivatives, we have

$$\frac{\partial M}{\partial y} = \cos y \quad \text{and} \quad \frac{\partial N}{\partial t} = \cos y.$$

Since the partial derivatives are equal, the differential equation is exact. ▲

REMARK: Recall from Section 3.2 that a separable differential equation is one that can be written in the form $n(y)\,dy/dt + m(t) = 0$. Notice that

$$\frac{\partial m}{\partial y} = 0 \quad \text{and} \quad \frac{\partial n}{\partial t} = 0.$$

Thus, by Theorem 3.3, any separable differential equation is also exact.

Anti-Partial-Differentiation

We can use Theorem 3.3 to determine if the differential equation $M(t,y) + N(t,y)y' = 0$ is exact. If it is exact, then we know there must be a function $H(t,y)$ such that

$$\frac{\partial H}{\partial t} = M(t,y) \quad \text{and} \quad \frac{\partial H}{\partial y} = N(t,y).$$

If we can determine $H(t,y)$, then we have an implicitly defined solution,

$$H(t,y) = C.$$

We can use a process of **anti-partial-differentiation** to construct H.

To illustrate anti-partial-differentiation, recall from Example 2 that the following differential equation is exact

$$\sin y + (2y + t\cos y)y' = 0. \tag{7}$$

Since equation (7) is exact, we know there is a function $H(t,y)$ such that

$$\frac{\partial H}{\partial t} = \sin y \quad \text{and} \quad \frac{\partial H}{\partial y} = 2y + t\cos y. \tag{8}$$

Choose one of these equalities, say $\partial H/\partial y = 2y + t\cos y$, and compute an "anti-partial-derivative." Antidifferentiating $2y + t\cos y$ with respect to y, we obtain

$$H(t,y) = y^2 + t\sin y + g(t), \tag{9}$$

where $g(t)$ is an arbitrary function of t. [Note: The "constant of integration" in equation (9) is an arbitrary function of t since t is treated as a constant when the partial derivative with respect to y is computed.] As a check on the form of H in equation (9), the partial derivative of H with respect to y is indeed $N(t, y) = 2y + t\cos y$.

We now determine $g(t)$ so that the representation (9) for H satisfies the first equality in (8). Taking the partial derivative of H with respect to t, we find

$$\begin{aligned}\frac{\partial H}{\partial t} &= \frac{\partial}{\partial t}\left[y^2 + t\sin y + g(t)\right] \\ &= \sin y + \frac{dg}{dt}.\end{aligned}$$

Comparing the preceding result with the first condition of equation (8) it follows that we need

$$\frac{dg}{dt} = 0$$

or $g(t) = C_1$, where C_1 is an arbitrary constant. Thus, from equation (9), we know

$$H(t, y) = y^2 + t\sin y + C_1.$$

We can drop the arbitrary constant C_1 since we will eventually set $H(t, y)$ equal to an arbitrary constant in the implicit solution. Therefore, $H(t, y) = y^2 + t\sin y$.

In the antidifferentiation, we started with the second condition in equation (8). We could just as well have started with the first condition and, had we done so, we would have arrived at the same function H. That is, if we antidifferentiate $\sin y$ with respect to t, we obtain

$$H(t, y) = t\sin y + p(y), \tag{10}$$

where $p(y)$ is an arbitrary function of y. To determine $p(y)$, we turn to the second condition in equation (8). Starting with equation (10), we calculate the partial derivative of $H(t, y)$ with respect to y:

$$\frac{\partial H}{\partial y} = t\cos y + \frac{dp}{dy}.$$

Comparing this partial with the second condition of equation (8), we see that $p(y)$ must be given by $p(y) = y^2 + C_2$. So, from (10), we again find that $H(t, y)$ has the form

$$H(t, y) = y^2 + t\sin y + C_2.$$

We noted earlier that we may as well choose $C_2 = 0$ since the solution of the differential equation is defined implicitly by $H(t, y) = C$.

EXAMPLE 3

Consider the initial value problem

$$1 + y^2 + 2(t + 1)y\frac{dy}{dt} = 0, \qquad y(0) = 1.$$

Verify that the differential equation is exact and solve the initial value problem.

Solution: To verify that the differential equation is exact, we appeal to Theorem 3.3 using $M(t, y) = 1 + y^2$ and $N(t, y) = 2(t + 1)y$. The functions $M, N, \partial M/\partial y,$

and $\partial N/\partial t$ are continuous in the entire ty-plane. Since

$$\frac{\partial M}{\partial y} = 2y \quad \text{and} \quad \frac{\partial N}{\partial t} = 2y,$$

the differential equation is exact.

We now use anti-partial-differentiation to find $H(t, y)$. Since the equation is exact,

$$\frac{\partial H}{\partial t} = 1 + y^2.$$

Antidifferentiating with respect to t,

$$H(t, y) = t(1 + y^2) + g(y) \tag{11}$$

To determine $g(y)$, we differentiate (11) with respect to y, finding

$$\frac{\partial H}{\partial y} = 2ty + \frac{dg}{dy}.$$

Since the differential equation is exact, we know that

$$2ty + \frac{dg}{dy} = 2(t + 1)y.$$

Therefore,

$$g(y) = y^2 + C_1.$$

Without loss of generality, we let $C_1 = 0$ to obtain

$$H(t, y) = t(1 + y^2) + y^2 = (t + 1)y^2 + t.$$

Thus, we have the following implicitly defined solution of the differential equation:

$$(t + 1)y^2 + t = C.$$

Imposing the initial condition, we obtain an implicit solution of the initial value problem

$$(t + 1)y^2 + t = 1.$$

We can solve for y,

$$y^2 = \frac{1 - t}{1 + t} \quad \text{or} \quad y = \pm\sqrt{\frac{1 - t}{1 + t}}.$$

Choosing the positive sign so as to satisfy the initial condition $y(0) = 1$, we arrive at an explicit solution of the initial value problem

$$y = \sqrt{\frac{1 - t}{1 + t}}. \quad \blacktriangle$$

REMARKS:

1. As we observed following Example 2, any separable equation is also exact. Since the differential equation in Example 3 is separable, we can conclude it is exact without appealing to Theorem 3.3. In Exercise 13 you are asked to solve the differential equation by separating the variables.

2. The explicit solution $y = \sqrt{(1-t)/(1+t)}$ has $-1 < t < 1$ as an interval of existence. This should not be surprising in light of existence-uniqueness Theorem 3.1. The differential equation in Example 3 has the form $y' = f(t, y)$, where

$$f(t, y) = -\frac{1+y^2}{2(t+1)y}.$$

Since the initial condition point is $(t_0, y_0) = (0, 1)$, the hypotheses of Theorem 3.1 are satisfied in the infinite rectangle defined by $-1 < t < \infty$, $0 < y < \infty$. Along the lines $t = -1$ and $y = 0$, both f and $\partial f/\partial y$ are undefined. As Figure 3.5 shows, the solution curve has a vertical asymptote at the boundary line $t = -1$ and approaches the boundary line $y = 0$ as $t \to 1$ from the left. ▮

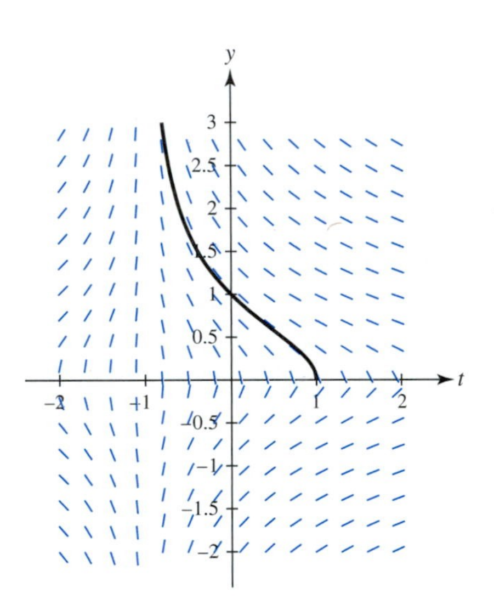

FIGURE 3.5

A portion of the direction field for the differential equation in Example 3. The solution $y = \sqrt{(1-t)/(1+t)}$ is shown as a solid curve.

EXERCISES

Exercises 1–7:

Determine whether the given nonlinear differential equation is exact. (Some algebraic manipulation may be needed.) If the equation is exact, find an implicit solution and (where possible) an explicit solution.

1. $yy' + 3t^2 - 2 = 0, \quad y(-1) = -2$

2. $(t + y^3)y' + y + t^3 = 0, \quad y(0) = -2$

3. $y' = (3t^2 + 1)(y^2 + 1), \quad y(0) = 1$

4. $(6t + y^3)y' + 3t^2 y = 0, \quad y(1) = 2$

5. $(e^{t+y} + 2y)y' + (e^{t+y} + 3t^2) = 0, \quad y(0) = 0$

6. $y' = -\dfrac{y\cos(ty) + 1}{t\cos(ty) + 2ye^{y^2}}, \quad y(\pi) = 0$

7. $(\sin(t+y) + y\cos(t+y) + t + y)y' + (y\cos(t+y) + y + t) = 0, \quad y(1) = -1$

8. For what values of the constants m, n, and α (if any) is the following differential equation exact?

$$t^m y^2 y' + \alpha t^3 y^n = 0$$

9. Assume that $(t^2 + y^2 \sin t)y' + M(t, y) = 0$ is an exact differential equation. Determine the general form for $M(t, y)$.

10. Assume that $N(t, y)y' + t^2 + y^2 \sin t = 0$ is an exact differential equation. Determine the general form for $N(t, y)$.

11. Assume that $t^3 y + e^t + y^2 = 5$ is an implicit solution of the differential equation

$$N(t, y)y' + M(t, y) = 0, \qquad y(0) = y_0.$$

Determine possible functions $M(t, y)$, $N(t, y)$, and the possible value(s) for y_0.

12. Assume that $y = -t - \sqrt{4 - t^2}$ is an explicit solution of the following initial value problem

$$(y + at)y' + (ay + bt) = 0, \qquad y(0) = y_0.$$

Determine values for the constants a, b, and y_0.

13. Verify that the differential equation in Example 3, $1 + y^2 + 2(t+1)yy' = 0$ is separable. Solve the differential equation as a separable equation. Impose the initial condition $y(0) = 1$ and obtain an explicit solution of the initial value problem. For this equation, which approach seems simpler to you?

14. Give an example of a differential equation that is exact but not separable. Solve your equation.

3.4 Bernoulli Equations

Bernoulli differential equations[2] are first order differential equations having the special structure

$$\frac{dy}{dt} + p(t)y = q(t)y^n,$$

where n is an integer. We do not discuss $n = 0$ and $n = 1$ since, in those cases, the Bernoulli equation is a first order linear equation, an equation we examined at length in Chapter 2.

A simple example of a Bernoulli equation is

$$\frac{dy}{dt} + e^{2t}y = y^3 \sin t.$$

Bernoulli equations arise in applications such as population models and models of one-dimensional motion influenced by drag forces. We can use the solution techniques of this section when we discuss these applications in Sections 3.5, 3.6, and the extended problem at the end of this chapter.

[2]Jacob Bernoulli (1654–1705) is one of eight members of the extended Bernoulli family, remembered for their contributions to mathematics and science. While at the University of Basel, Jacob made important contributions to such areas as infinite series, probability theory, geometry, and differential equations. In 1696 he solved the differential equation that now bears his name. Jacob always had a particular fascination for the logarithmic spiral and requested that this curve be carved on his tombstone.

Even though Bernoulli equations are nonlinear (except when $n = 0$ or $n = 1$), there is a procedure for solving them. By making an appropriate change of dependent variable, a Bernoulli equation can be recast as a first order linear differential equation. This first order linear equation can then be solved using the techniques discussed in Chapter 2.

Solving a Bernoulli Equation

Consider a Bernoulli equation

$$\frac{dy}{dt} + p(t)y = q(t)y^n \tag{1}$$

where n is a given integer ($n \neq 0$ and $n \neq 1$). We look for a change of dependent variable of the form $v(t) = y(t)^m$ where m is a constant to be determined. Using the chain rule,

$$\frac{dv}{dt} = my^{m-1}\frac{dy}{dt}$$

and therefore

$$\frac{dy}{dt} = m^{-1}y^{1-m}\frac{dv}{dt} = m^{-1}v^{(1-m)/m}\frac{dv}{dt}.$$

Equation (1) transforms into the following differential equation for $v(t)$:

$$\frac{dv}{dt} + mp(t)v = mq(t)v^{(m+n-1)/m}. \tag{2}$$

At first glance, it may seem that our change of variables has accomplished little. The structure of equation (2) seems similar to what we started with. However, we are free to choose the constant m. In particular, if we select $m = 1 - n$, then equation (2) reduces to the first order linear equation

$$\frac{dv}{dt} + (1-n)p(t)v = (1-n)q(t). \tag{3}$$

We can solve this equation for $v(t)$ and then obtain the desired solution, $y(t) = v(t)^{1/(1-n)}$.

EXAMPLE 1

Solve the initial value problem

$$y' + y = ty^3, \qquad y(0) = 2.$$

Solution: The differential equation is nonlinear, and it is neither separable nor exact. It is a Bernoulli equation, however, with $n = 3$.

We now make the change of dependent variable $v = y^{1-n}$ or, since $n = 3$, $v = y^{-2}$. The initial value problem for $v(t)$ then becomes [recall equation (3)]

$$v' - 2v = -2t, \qquad v(0) = y(0)^{-2} = \frac{1}{4}.$$

The general solution is

$$v = Ce^{2t} + \left(t + \frac{1}{2}\right).$$

Imposing the initial condition, we have

$$v = -\frac{1}{4}e^{2t} + \left(t + \frac{1}{2}\right).$$

Finally, since $y = v^{-1/2}$, we arrive at the desired solution

$$y = \left(-\frac{1}{4}e^{2t} + \left(t + \frac{1}{2}\right)\right)^{-1/2}. \ \blacktriangle$$

EXERCISES

Exercises 1–6:

(a) Determine whether the given Bernoulli equation is also a separable and/or an exact equation.

(b) Solve the initial value problem by

(i) transforming the given Bernoulli differential equation and initial condition into a first order linear differential equation with its corresponding initial condition,

(ii) solving the new initial value problem,

(iii) transforming back to the dependent variable of interest.

(c) Determine the interval of existence.

1. $y' = y(2 - y), \quad y(0) = 1$

2. $y' = 2ty(1 - y), \quad y(0) = -1$

3. $y' = -y + e^t y^2, \quad y(-1) = -1$

4. $y' = y + y^{-1}, \quad y(0) = -1$

5. $ty' + y = t^3 y^{-2}, \quad y(1) = 1$

6. $y' - y = ty^{1/3}, \quad y(0) = -8$

Exercises 7–8:

Solve the given initial value problem by making an appropriate change of dependent variable. Obtain an explicit solution and determine the interval of existence.

7. $y' = 2t^{-1} + e^{-y}, \quad y(1) = 0$

8. $y' = -(y + 1) + t(y + 1)^{-2}, \quad y(0) = 1$

9. The initial value problem $y' + y = q(t)y^2, y(0) = y_0$ is known to have solution

$$y(t) = \frac{3}{(1 - 3t)e^t}$$

on the interval $-\infty < t < 1/3$. Determine the coefficient function $q(t)$ and the initial value y_0.

3.5 The Logistic Population Model

The art of mathematical modeling involves a trade-off between realism and complexity. A model should incorporate enough reality to make it useful as a predictive tool. However, the model must also be tractable mathematically; if it's too complex to analyze and if we cannot deduce any useful information from it, then the model is worthless.

A key assumption of the population model described in Section 2.4, often referred to as the Malthusian[3] model, is that relative birth rate is independent of population. (Recall that the relative birth rate, $r_b - r_d$, is the

[3]Thomas Malthus (1766–1834) was an English political economist whose *Essay on the Principle of Population* had an important influence upon Charles Darwin's thinking. Malthus believed that human population growth, if left unchecked, would inevitably lead to poverty and famine.

difference between birth and death rates per unit population.) The assumption that $r_b - r_d$ is independent of population leads to uninhibited exponential growth or decay of solutions. Such behavior is often a poor approximation of reality.

For a colony possessing limited resources, a more realistic model is one that accounts for the impact of population upon the relative birth rate. When population size is small, resources are relatively plentiful and the population should thrive and grow. When the population becomes larger, however, we expect that resources become scarcer, the population becomes stressed, and the relative birth rate begins to decline (eventually becoming negative). In this section we consider a model that attempts to account for this effect. This model leads to a nonlinear differential equation.

The Verhulst or Logistic Model

The **Verhulst population model**[4] assumes that the population $P(t)$ evolves according to the differential equation

$$\frac{dP}{dt} = r\left(1 - \frac{P}{P_e}\right)P. \tag{1}$$

In equation (1), r and P_e are positive constants. Equation (1) is also known as the **logistic equation**. Comparing equation (1) with the Malthus equation

$$\frac{dP}{dt} = (r_b - r_d)P, \tag{2}$$

we see that the constant relative birth rate $r_b - r_d$ of equation (2) has been replaced by the *population-dependent* relative birth rate,

$$r\left(1 - \frac{P(t)}{P_e}\right),$$

where r is a positive constant. Note that if $P(t) > P_e$, then the relative birth rate is negative, causing the population to decrease. That is, see equation (1), if $P(t) > P_e$, then $dP/dt < 0$. Conversely, if $P(t) < P_e$, then $dP/dt > 0$ and the population is increasing.

The qualitative behavior of solutions of the logistic equation can also be deduced from the direction field. Figure 3.6, for example, shows the direction field for the logistic equation for the parameters $r = 3$ and $P_e = 1$.

The logistic equation has two equilibrium solutions, the trivial solution $P(t) = 0$ and the nontrivial equilibrium solution $P(t) = P_e$. If $P(t_0) = P_e$, then the population remains equal to this value as time evolves. The constant population, P_e, is called an **equilibrium population**. In terms of its structure, the logistic equation is a nonlinear separable differential equation. It is also a Bernoulli equation. We solve it below as a separable equation. Its treatment as a Bernoulli equation is left to the exercises.

[4]Pierre Verhulst (1804–1849) was a Belgian scientist chiefly remembered for his research on population growth. He deduced and studied the nonlinear differential equation named after him.

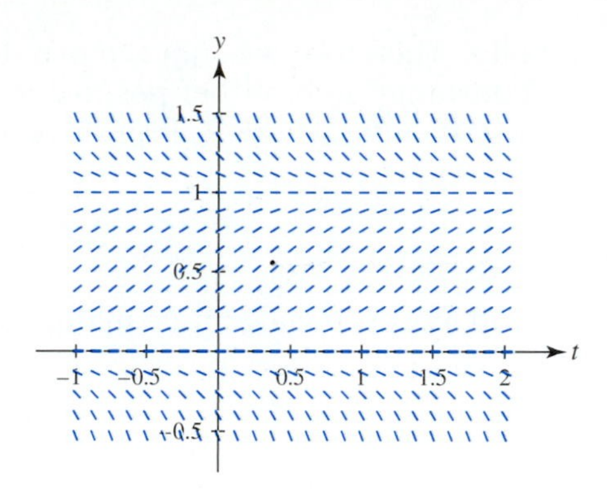

FIGURE 3.6

The direction field for the logistic equation $P' = 3(1 - P)P$. The equilibrium solutions are $P = 0$ and $P = 1$.

Solution of the Logistic Equation

To solve equation (1), we separate variables obtaining

$$\frac{1}{\left(1 - \frac{P}{P_e}\right) P} \frac{dP}{dt} = r.$$

An implicit solution is therefore

$$\int \frac{dP}{\left(1 - \frac{P}{P_e}\right) P} = rt + K, \tag{3}$$

where K is an arbitrary constant. Recalling the method of partial fractions from calculus (also see Section 7.3), we obtain an antiderivative for the integral in (3):

$$\int \frac{dP}{\left(1 - \frac{P}{P_e}\right) P} = \int \left(\frac{1}{P} - \frac{1}{P - P_e}\right) dP = \ln \left| \frac{P}{P - P_e} \right|.$$

Therefore, from equation (3), we find

$$\ln \left| \frac{P}{P - P_e} \right| = rt + K$$

and hence,

$$\left| \frac{P(t)}{P(t) - P_e} \right| = Ce^{rt}, \tag{4}$$

where $C = e^K$ is an arbitrary positive constant. To remove the absolute value signs in equation (4), we argue as follows. The exponential function e^{rt} is never zero. Therefore, the quotient $P(t)/(P(t) - P_e)$ is never zero and, being a continuous function of t, never changes sign; it is either positive or negative for all t. Therefore, we can remove the absolute value signs if we now allow the constant C to be either positive or negative. Assume that the initial population is $P(0) = P_0$, where P_0 is positive and $P_0 \neq P_e$. Then

$$C = \frac{P_0}{P_0 - P_e},$$

and after some algebraic manipulation, we obtain the explicit solution

$$P(t) = \frac{P_0 P_e}{P_0 - (P_0 - P_e)e^{-rt}}. \tag{5}$$

The derivation leading to equation (5) tacitly assumes $P(t)$ never takes on the values 0 or P_e. In particular, the antiderivative for the integral in (3),

$$\ln\left|\frac{P(t)}{P(t) - P_e}\right|,$$

is undefined when $P(t)$ is 0 or P_e. Our final expression (5), however, is valid for any value of P_0 (including the equilibrium values 0 and P_e).

What behavior is predicted by the solution (5)? Since $r > 0$, we see that

$$\lim_{t\to\infty} P(t) = P_e$$

for all positive values of P_0. Therefore, any given nonzero population will tend to the equilibrium population value, P_e, as time increases. Figure 3.6 illustrates this behavior, showing the direction field for the special case $P' = 3(1 - P)P$.

We can ask additional questions. What happens if we allow temporal variations in the relative birth rate by allowing $r = r(t)$ in equation (1)? What happens if migration is introduced into the logistic model? When migration is introduced, equation (1) becomes

$$\frac{dP}{dt} = r\left(1 - \frac{P}{P_e}\right)P + M.$$

This equation has the structure of a Riccati equation (see Exercises 22–24 in Section 3.2). If M is constant, then the question of equilibrium solutions remains relevant. What happens to the equilibria as the constant migration level M is varied? Some of these issues are addressed in the exercises.

Scientists studying the growth of bacteria in food have further refined the logistic model in an attempt to build in more reality. One such model, the Baranyi model, is considered in the extended problem at the end of this chapter.

The Spread of an Infectious Disease

The logistic differential equation also arises in modeling the spread of an infectious disease. Suppose we have a constant population of N individuals and at time t the number of infected members is $P(t)$. The corresponding number of uninfected individuals is then $N - P(t)$. A reasonable assumption is that the rate of spread of the disease at time t is proportional to the product of noninfected and infected individuals. This assumption leads to the differential equation

$$\frac{dP}{dt} = k(N - P)P,$$

where k is the constant of proportionality. If the equation is rewritten as

$$\frac{dP}{dt} = kN\left(1 - \frac{P}{N}\right)P, \tag{6}$$

we can see it is similar to the logistic equation (1). The corresponding initial value problem is completed by specifying the number of infected individuals at some initial time, $P(0) = P_0$.

Note that the differential equation (6) has two equilibrium solutions, $P = 0$ and $P = N$. This certainly makes sense. If no one is infected or if everyone is infected, the disease will not spread. We consider aspects of this infectious disease model in the exercises.

EXERCISES

1. Solve the initial value problem

$$\frac{dP}{dt} = r\left(1 - \frac{P}{P_e}\right)P, \qquad P(0) = P_0$$

by viewing the differential equation as a Bernoulli equation.

2. Let $P(t)$ represent the population of a colony, in millions of individuals. Suppose the colony starts with 100,000 individuals and evolves according to the equation

$$\frac{dP}{dt} = 0.1\left(1 - \frac{P}{3}\right)P$$

with time being measured in years (in particular, $r = 0.1$ year^{-1}). How long will it take the population to reach 90% of its equilibrium value?

3. Consider a population whose dynamics are described by the logistic equation with constant migration,

$$\frac{dP}{dt} = r\left(1 - \frac{P}{P_e}\right)P + M,$$

where r, P_e, and M are constants. Assume that P_e is a fixed positive constant and that we want to understand how the equilibrium solutions of this nonlinear autonomous equation depend upon the parameters r and M. Physically meaningful equilibrium populations are nonnegative constant solutions of this differential equation.

(a) Obtain the roots of the quadratic equation that define the equilibrium population(s) of this differential equation. Note that if $M \neq 0$, the constants 0 and P_e are no longer equilibrium solutions. Does this make sense?

(b) For definiteness, set $P_e = 1$. Plot the equilibrium solutions obtained in (a) as functions of the ratio M/r. How many nonnegative equilibrium populations exist for $M/r > 0$? How many exist for $-1/4 < M/r \leq 0$?

(c) What happens when $M/r = -P_e/4 = -1/4$? What happens when $M/r < -P_e/4 = -1/4$? Are these mathematical results consistent with what one would expect if the migration rate out of the colony were sufficiently large relative to the colony's ability to gain size through reproduction?

4. The direction field shown corresponds to the differential equation

$$\frac{dP}{dt} = r\left(1 - \frac{P}{P_e}\right)P + M,$$

where r, M, and P_e are fixed constants. Following the ideas of Exercise 3, use the figure to determine the constant P_e and the constant ratio M/r.

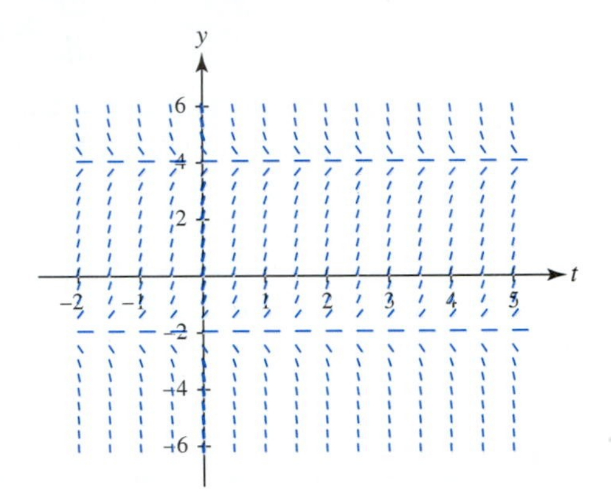

Figure for Exercise 4

Exercises 5–6:

For the given parameters, solve and plot the solution of the initial value problem

$$\frac{dP}{dt} = r\left(1 - \frac{P}{P_e}\right)P + M, \qquad P(0) = P_0$$

5. $r = 1, \quad P_e = 1, \quad M = -\frac{3}{16}, \quad P_0 = \frac{1}{2}$

6. $r = 1, \quad P_e = 1, \quad M = -\frac{1}{4}, \quad P_0 = 1$

Variable Birth Rates In many cases of interest, a population's relative birth rate does not remain constant in time. For example, environmental variations that occur with the change of seasons may play a significant role. Likewise, variations in temperature during storage of food affect the growth of harmful bacteria. Therefore, it is of interest to understand the behavior of the logistic equation when the parameter $r = r(t)$ varies with time.

Exercises 7–9:

Consider the initial value problem

$$\frac{dP}{dt} = r(t)\left(1 - \frac{P}{P_e}\right)P, \qquad P(0) = P_0$$

where $r(t) = r_0[1 + 2\sin 2\pi t]$, with time measured in years. This relative birth rate factor might reflect the impact of seasonal changes on a population's ability to reproduce over the course of a year, achieving a maximum of $3r_0$ when conditions are most favorable and a minimum of $-r_0$ under the least favorable of conditions. Exercises 7–9 examine this problem from three closely-related points of view.

7. Solve the given initial value problem, viewing the differential equation as a separable differential equation. Assume $r_0 = 0.5$ year^{-1} and $P_e = 1$. To understand the behavior of solutions, consider different initial conditions. For example, start with $P(0) = 0.5$; plot the solution $P(t)$ over a 5-year period and then over a 10-year period.

Is the population tending toward an equilibrium value? Do you get similar results if $P(0) = 1.5$?

8. Make the "Change of clock" change of independent variable discussed in Section 2.5, Exercise 9. In other words, define a new independent variable

$$\tau = \int_0^t r(s)\, ds.$$

Transform the initial value problem into a new one with τ as independent variable, use equation (5) to solve this problem in terms of τ and then simply transform the result back into one expressed in terms of independent variable t. Plot the solution $P(t)$ as in Exercise 7.

9. Solve the initial value problem given in Exercise 7, viewing the differential equation as a Bernoulli differential equation. Plot the solution $P(t)$ as in Exercise 7.

10. Let $P(t)$ represent the number of individuals who, at time t, are infected with a certain disease. Let N denote the total number of individuals in the population. Assume that the spread of the disease can be modeled by the initial value problem

$$\frac{dP}{dt} = k(N - P)P, \qquad P(0) = P_0.$$

At time $t = 0$, when 100,000 members of a population of 500,000 are known to be infected, medical authorities intervene with medical treatment. As a consequence of this intervention, the rate factor k is no longer constant but varies with time as $k(t) = 2e^{-t} - 1$, where time t is measured in months and $k(t)$ represents the rate of infection per month per 100,000 individuals.

Initially, as the effects of medical intervention begin to take hold, $k(t)$ remains positive and the disease continues to spread. Eventually, however, the effects of medical treatment cause $k(t)$ to become negative and the number of infected individuals then decreases.

(a) Solve the appropriate initial value problem for the number of infected individuals, $P(t)$, at time t and plot the solution.

(b) From your plot, estimate the maximum number of individuals that are at any time infected with the disease.

(c) How long does it take before the number of infected individuals is reduced to 50,000?

11. Consider a chemical reaction of the form $A + B \to C$, in which the rates of change of the two chemical reactants, A and B, are described by the following two differential equations

$$A' = -kAB, \qquad B' = -kAB,$$

where k is a positive constant. Assume that 5 moles of reactant A and 2 moles of reactant B are present at the beginning of the reaction.

(a) Show that the difference $A(t) - B(t)$ remains constant in time. What is the value of this constant?

(b) Use the observation made in (a) to derive an initial value problem for reactant A.

(c) It was observed, after the reaction had progressed for 1 sec, that 4 moles of reactant A remained. How much of reactants A and B will be left after 4 sec of reaction time?

3.6 One-Dimensional Motion with Air Resistance

In Chapter 1 we considered a falling object acted upon only by gravity. For that case, using $y(t)$ to represent the vertical position of the body at time t and $v(t)$ to represent its velocity, Newton's second law of motion, $ma = F$, reduces to

$$m\frac{dv}{dt} = -mg.$$

Here, m is the mass of the object and g the acceleration due to gravity, nominally 32 ft/sec^2 or 9.8 m/s^2 (seconds is expressed as s in metric notation). The minus sign on the right-hand side is present because we measure $y(t)$ and $v(t)$ as positive upward while the force of gravity acts downward. For the simple model above, we solved for $v(t)$ and $y(t)$ by computing successive antiderivatives.

A more realistic model of one-dimensional motion is one that incorporates the effects of air resistance. As an object moves through air, a retarding aerodynamic force is created by the combination of pressure and frictional forces. The air close to the object exerts a normal pressure force upon it. Likewise, friction creates a tangential force that opposes the motion of air past the object. The combination of these effects creates a drag force that acts to reduce the speed of the moving object.

The drag force depends upon the velocity $v(t)$ of the object and acts upon it in such a way as to reduce its speed, $|v(t)|$. We consider two idealized models of drag force that are consistent with these ideas.

Case 1: Drag Force Proportional to Velocity

Assume that velocity $v(t)$ is positive in the upward direction and that the drag force is proportional to velocity. If k is the positive constant of proportionality, Newton's second law of motion leads to

$$m\frac{dv}{dt} = -mg - kv. \tag{1}$$

Does the model of drag that we have postulated act as we want it to? If the object is moving upward [that is, if $v(t) > 0$] then the drag force $-kv(t)$ is negative and thus acts downward. Conversely, if $v(t) < 0$, then the object is moving downward. In this case, the drag force $-kv(t) = k|v(t)|$ is a positive (upward) force, as it should be. Therefore, whether the object is moving upward or downward, drag acts to slow the object; this drag model is consistent with our ideas of how drag should act. See Figure 3.7.

This drag model leads to equation (1), a first order linear constant coefficient equation; we can solve it using the ideas of Section 2.3. Before we do, however, let's consider the question of equilibrium solutions to see if the model makes sense in that regard. The only constant solution of equation (1) is $v(t) = -mg/k$. (Therefore, the equilibrium condition is one in which the object is falling with speed mg/k.) At the velocity $v(t) = -mg/k$, the drag force and gravitational force acting on the object are equal and opposite. This equilibrium velocity is often referred to as the **terminal velocity** of the object.

The initial value problem corresponding to equation (1) is

$$m\frac{dv}{dt} = -mg - kv, \qquad v(0) = v_0. \tag{2}$$

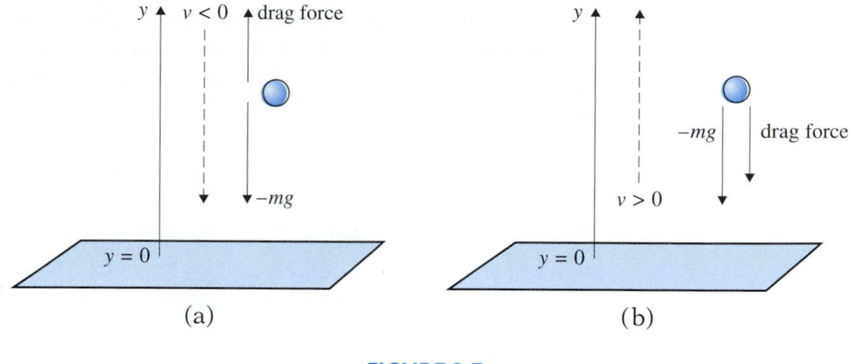

FIGURE 3.7

Assume a drag force of the form $kv(t), k > 0$; see equation (1). (a) When the object is moving downward, drag acts upward and tends to slow the object. (b) When the object is moving upward, drag acts downward and likewise tends to slow the object.

Solving initial value problem (2), we obtain

$$v(t) = -\frac{mg}{k} + \left(v_0 + \frac{mg}{k}\right) e^{-(k/m)t}.$$

From this explicit solution, it is clear that, for any initial velocity v_0, $v(t)$ tends to the terminal velocity

$$\lim_{t\to\infty} v(t) = -\frac{mg}{k}.$$

The concept of terminal velocity is a mathematical abstraction since, in any application, the time interval of interest is finite. However, whether the object is initially moving up or down, it eventually begins to fall and its velocity approaches terminal velocity as the time of falling increases. The velocity that the object actually has the instant before it strikes the ground is called the **impact velocity**. Once impact occurs, the mathematical model (2) is no longer applicable.

Case 2: Drag Force Proportional to the Square of Velocity

A drag force having magnitude $\kappa v^2(t)$ (where the constant of proportionality κ is assumed to be positive) has been found in many cases to be a reasonably good approximation to reality over a range of velocities. However, since it involves an even power of velocity, this model of drag requires more care to incorporate it into the equations of motion.

Drag must act to reduce the speed of the moving object. Therefore, if the object is moving upward, the drag force acts downward and should be $-\kappa v^2(t)$. If the object is moving downward, the drag force acts upward and should be $\kappa v^2(t)$. In other words, when we use this model for drag, Newton's second law leads to

$$m\frac{dv}{dt} = -mg - \kappa v^2, \qquad v(t) \geq 0,$$

$$m\frac{dv}{dt} = -mg + \kappa v^2, \qquad v(t) \leq 0.$$

Here again we can ask about equilibrium solutions and see if the answer makes sense physically. Note that no equilibrium solution exists if $v(t) > 0$ since $-mg - \kappa v^2$ is never zero. When $v(t) < 0$, however, there is an equilibrium solution:

$$v(t) = -\sqrt{\frac{mg}{\kappa}}.$$

This equilibrium solution is again a terminal velocity corresponding to downward motion; at this velocity, drag and gravity exert equal and opposite forces.

Each of the preceding equations is a first order separable equation. If the problem involves a falling object, then velocity is always nonpositive and the second equation is valid for the entire problem (that is, over the entire t-interval of interest). If, however, the problem involves launching a projectile upward, both equations may ultimately be needed. The first equation must be used to model the upward dynamics, the behavior of the projectile from launch until the time t_m it reaches its highest point [that is, when $v(t_m) = 0$]. After time t_m, the projectile begins to fall and the second differential equation [with initial condition $v(t_m) = 0$] is needed to model the descending dynamics.

EXAMPLE 1

Assume a projectile having mass m is launched vertically upward from ground level at time $t = 0$ with initial velocity $v(0) = v_0$. Further assume that the drag force experienced by the projectile is proportional to the square of its velocity, with drag coefficient κ. Determine the maximum height reached by the projectile and the time at which this maximum height is reached.

Solution: The projectile motion being considered involves only upward motion. Therefore,

$$mv' = -mg - \kappa v^2, \qquad v(0) = v_0.$$

Separating variables,

$$\frac{v'}{1 + \dfrac{\kappa}{mg}v^2} = -g.$$

Integrating, we find

$$\int \frac{dv}{1 + \dfrac{\kappa}{mg}v^2} = \sqrt{\frac{mg}{\kappa}} \tan^{-1}\left(\sqrt{\frac{\kappa}{mg}}\, v\right) = -gt + C.$$

Imposing the initial condition,

$$C = \sqrt{\frac{mg}{\kappa}} \tan^{-1}\left(\sqrt{\frac{\kappa}{mg}}\, v_0\right).$$

Finally, after some algebra, we obtain the explicit solution

$$v(t) = \sqrt{\frac{mg}{\kappa}} \tan\left(\tan^{-1}\left(\sqrt{\frac{\kappa}{mg}}\, v_0\right) - \sqrt{\frac{\kappa g}{m}}\, t\right). \tag{3}$$

As a check on (3), note that $v(0)$ does reduce to the given initial velocity v_0.

As a further check, one can show (using L'Hôpital's rule[5]) that for every fixed time t

$$\lim_{\kappa \to 0} v(t) = v_0 - gt,$$

which is the velocity in the absence of drag.

Let t_m denote the time when the maximum height is reached; that is, $v(t_m) = 0$. From equation (3), we see that the maximum height is attained at the first positive value t where the argument of the tangent function is zero. Thus,

$$\tan^{-1}\left(\sqrt{\frac{\kappa}{mg}}\, v_0\right) - \sqrt{\frac{\kappa g}{m}}\, t_m = 0,$$

or,

$$t_m = \sqrt{\frac{m}{\kappa g}} \tan^{-1}\left(\sqrt{\frac{\kappa}{mg}}\, v_0\right). \tag{4}$$

We next want to determine the maximum height, $y(t_m)$, reached by the projectile. To find $y(t_m)$, we need an expression for position, $y(t)$. To determine position, $y(t)$, we integrate velocity:

$$y(t) - y(0) = \int_0^t v(t)\, dt.$$

For our problem, $y(0) = 0$. Therefore,

$$y(t_m) = \int_0^{t_m} v(t)\, dt, \tag{5}$$

where $v(t)$ is given by equation (3). To carry out the integration in equation (5), we use the fact that $\int \tan u\, du = -\ln|\cos u| + C$, finding

$$y(t_m) = \frac{m}{2\kappa} \ln\left(1 + \frac{\kappa v_0^2}{mg}\right).$$

Here again, as a check, one can show using L'Hôpital's rule that

$$\lim_{\kappa \to 0} y(t_m) = \frac{v_0^2}{2g} \tag{6}$$

is the maximum height reached in the absence of drag. Figure 3.8 on the next page shows how the size of the drag constant affects the quantities t_m and $y(t_m)$ for the case of a 2-lb object launched upward from ground level with an initial velocity of 60 mph ($v_0 = 88$ ft/sec). ▲

The computations in Example 1 are relatively complicated. The task of finding maximum projectile height can be simplified by transforming the problem to one in which height rather than time is the independent variable. This transformation is discussed in Section 3.7. In the Exercises of Section 3.7 we ask you to recalculate both the maximum height achieved in Example 1 [see equation (6)] and the impact velocity reached upon descent.

---

[5]Guillaume de L'Hôpital (1661–1704) wrote the first textbook on differential calculus, *Analyse des infiniment petits pour l'intelligence des lignes course*, which appeared in 1692. The book contains the rule that bears his name.

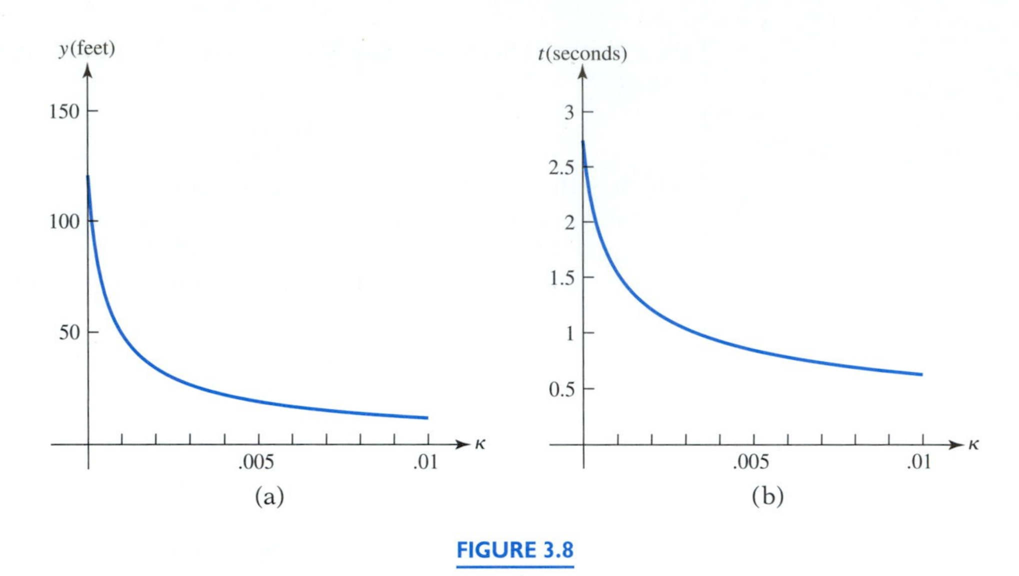

FIGURE 3.8

A 2-lb object is launched upward from ground level with an initial velocity of 88 ft/sec. The graphs show: (a) The maximum altitude as a function of the drag constant κ; see equation (5). (b) The time when the maximum altitude is reached, as a function of the drag constant κ; see equation (4).

EXERCISES

1. An object of mass m is dropped from a high altitude. How long will it take the object to achieve a velocity equal to one-half of its terminal velocity if the drag force is assumed proportional to the velocity?
2. An object of mass m is dropped from a high platform at time $t = 0$. Assume the drag force is proportional to the square of the velocity with drag coefficient κ. As in Example 1, derive an expression for the velocity $v(t)$.
3. Assume that the action of a parachute can be modeled as a drag force proportional to the square of velocity. What drag coefficient κ of the parachute is needed for a 200-lb person to achieve a terminal velocity of 10 mph?
4. A drag chute must be designed to reduce the speed of a 3000-lb dragster from 220 mph to 50 mph in 4 sec. Assume that the drag force is proportional to the velocity.
   (a) What value of the drag coefficient k is needed to accomplish this?
   (b) How far will the dragster travel in the 4-sec interval?
5. Repeat Exercise 4 for the case that the drag force is proportional to the square of the velocity. Determine both the drag coefficient κ and the distance traveled.
6. A projectile of mass m is launched vertically upward from ground level at time $t = 0$ with initial velocity v_0 and is acted upon by gravity and air resistance. Assume the drag force is proportional to velocity, with drag coefficient k. Derive an expression for the time, t_m, when the projectile achieves its maximum height.
7. Derive an expression for the maximum height, $y_m = y(t_m)$, achieved in Exercise 6.
8. A projectile is launched vertically upward from ground level with initial velocity v_0. Neglect air resistance. Show that the time it takes the projectile to reach its maximum height is equal to the time it takes to fall from this maximum height to the ground.

9. On August 24, 1894, Pop Shriver of the Chicago White Stockings caught a baseball dropped (by Clark Griffith) from the top of the Washington Monument. The Washington Monument is 555 ft tall and a baseball weighs 5 1/8 oz.

(a) If we ignore air resistance and assume the baseball was acted upon only by gravity, how fast would the baseball have been traveling when it was 7 ft above the ground?

(b) Suppose we now include air resistance in our model, assuming that the drag force is proportional to velocity with a drag coefficient $k = 0.0018$ lb-sec/ft. How fast is the baseball traveling in this case when it is 7 ft above the ground?

10. A 180-lb skydiver drops from a hot-air balloon. After 10 sec of free fall, a parachute is opened. The parachute immediately introduces a drag force proportional to velocity. After an additional 4 sec, the parachutist reaches the ground. Assume that air resistance is negligible during free fall and that the parachute is designed so that a 200-lb person will reach a terminal velocity of 10 mph.

(a) What is the speed of the skydiver immediately before the parachute is opened?

(b) What is the parachutist's impact velocity?

(c) At what altitude was the parachute opened?

(d) What is the balloon's altitude?

11. When modeling the action of drag chutes and parachutes, we have assumed that the chute opens instantaneously. Real devices take a short amount of time to fully open and deploy.

In this exercise, we try to assess the importance of this distinction. Consider again the assumptions of Exercise 4. A 3000-lb dragster is moving on a straight track at a speed of 220 mph when, at time $t = 0$, the drag chute is opened. If we assume the drag force is proportional to velocity and that the chute opens instantaneously, the differential equation to solve is $mv' = -kv$.

If we assume a short deployment time to open the chute, a reasonable differential equation might be $mv' = -k[\tanh t]v$. Since $\tanh(0) = 0$ and $\tanh(1) \approx 0.76$, it will take about 1 sec for the chute to become 76% deployed in this model.

Assume $k = 25$ lb-sec/ft. Solve the two differential equations and determine in each case how long it takes the vehicle to slow to 50 mph. Which time do you anticipate will be larger? (Explain.) Is the idealization of instantaneous chute deployment realistic?

3.7 One-Dimensional Dynamics with Distance as the Independent Variable

Suppose an object moves along a straight-line path, say along the x-axis. Let $x(t)$ represent the position of the object at time t, measured relative to some origin. Let F denote the sum of the forces acting on the object. Newton's second law of motion, $ma = F$, leads to the differential equation

$$m\frac{dv}{dt} = F, \tag{1}$$

where m is the mass and $v(t) = dx/dt$ is the velocity of the object.

In many applications, velocity is always nonnegative or always nonpositive over the entire time interval of interest. In such cases, position $x(t)$ is a monotonic function of time. If velocity is nonnegative, the position of the object is an increasing function of time; if velocity is nonpositive, then position is a decreasing function of time. As a concrete illustration, suppose velocity is positive but is decreasing from its initial positive value until it reaches zero at some time $t = t_f$. In such a case (which could represent an automobile moving forward but braking to a stop), $x(t)$ increases until the object comes to rest at $t = t_f$.

Changing the Independent Variable from Time to Position

In cases where position $x(t)$ is either always increasing or always decreasing over the time interval of interest, we may be able to simplify the differential equation describing the object's motion if we use position x as the independent variable rather than time t.

Suppose (for definiteness) that position $x(t)$ is an increasing function of time t [a similar analysis is valid if $x(t)$ is decreasing]. Then, an inverse function exists and we can express time as a function of position (whereas we normally view velocity and position as functions of time). Whenever the existence of an inverse function allows us to view time as a function of position, we can ultimately view velocity as a function of position as well. In such cases, when v is a function of x, we can use the chain rule to relate dv/dt to dv/dx,

$$\frac{dv}{dt} = \frac{dv}{dx}\frac{dx}{dt} = v\frac{dv}{dx}. \tag{2}$$

This change of independent variable is useful when the net force acting upon the body is a function of velocity or position or both but does not explicitly depend on time. In such a case, when $F = F(x, v)$, differential equation (1) has the form

$$mv\frac{dv}{dx} = F(x, v). \tag{3}$$

We adopt the customary notation and write $v = v(x)$ when referring to velocity as a function of position and $v = v(t)$ when referring to velocity as a function of time.

Equation (3) is typically supplemented by an initial condition that prescribes velocity at some initial position $v(x_0) = v_0$. Equation (3) and the initial condition define an initial value problem on the underlying x-interval of interest.

An Example That Contrasts the Two Approaches

We now consider a simple example that contrasts the two approaches—using time as the independent variable and using position as the independent variable. The example is deliberately kept simple by assuming the only force acting on the moving object is a drag force proportional to velocity. Since this drag force does not explicitly depend on time, the example leads to a differential equation of the form of (3).

EXAMPLE 1

A boat of mass m is pushed away from a dock in the positive x-direction. The only horizontal force acting on the boat is a drag force that we assume is proportional to the boat's velocity. The velocity at the initial position $x_0 = 0$ is $v_0 > 0$. The initial time is $t_0 = 0$. Find the velocity when the boat has position $x_1 > 0$.

Solution: We solve this problem in two different ways, using time as the independent variable and then using position as the independent variable.

Using Time as the Independent Variable If we solve the problem using time as the independent variable, the steps in the solution are

1. Solve $m\,dv/dt + kv = 0, v(0) = v_0$ for $v(t)$. We obtain
$$v(t) = v_0 e^{-(k/m)t}. \tag{4a}$$
2. Determine the position:
$$x(t) = \int_0^t v(s)\,ds = \int_0^t v_0 e^{-(k/m)s}\,ds = \frac{mv_0}{k}(1 - e^{-(k/m)t}) \tag{4b}$$
Note that position $x(t)$ is an increasing function on $0 \le t < \infty$. Note as well that
$$\lim_{t\to\infty} x(t) = \frac{mv_0}{k}.$$
Thus, if the problem posed is to make sense, $x_1 < mv_0/k$ is required.
3. Use expression (4b) to determine the time, call it t_1, when position is $x = x_1$. Solving the equation $x(t) = x_1$, we find
$$t_1 = -\frac{m}{k}\ln\left[1 - \frac{kx_1}{mv_0}\right].$$
4. Substitute this solution t_1 into the velocity expression (4a) to obtain
$$v(t_1) \equiv v_1 = v_0 - \frac{k}{m}x_1. \tag{5}$$

Using Position as the Independent Variable Transform the problem so as to make position x the independent variable.

1. Solve the transformed problem
$$mv\frac{dv}{dx} + kv = 0, \qquad v(0) = v_0.$$
Here, as noted previously, we now use the symbol $v(x)$ to denote velocity, viewed as a function of position x. The velocity is nonzero over the x-interval of interest so we can cancel velocity in the differential equation to obtain
$$\frac{dv}{dx} = -\frac{k}{m}, \qquad v(0) = v_0.$$
We solve this equation by antidifferentiation, finding
$$v(x) = v_0 - \frac{k}{m}x.$$
2. Evaluate velocity at $x = x_1$ to again obtain
$$v(x_1) = v_0 - \frac{k}{m}x_1.$$
As can be seen, both solutions lead to the same expression for velocity at $x = x_1$. ▲

Impact Velocity

The next example, an object falling through the atmosphere, shows that using position as the independent variable may convert a problem we cannot solve into one that we can solve.

The force of Earth's gravitational attraction diminishes as a body moves higher above the surface. For objects near the surface, the usual assumption of constant gravity is fairly reasonable. However, for a body falling from a great height (such as a satellite reentering the atmosphere) a constant gravity assumption is not accurate and may lead to erroneous predictions of the reentry trajectory.

While most of the realistic gravity models used in aerospace applications are quite complicated, they are all based on Newton's law of gravitation, which states that the force of mutual attraction between two bodies is

$$F = \frac{GmM}{r^2}. \tag{6}$$

In equation (6), $G = 6.673 \times 10^{-11}$ m^3/(kg $\cdot$ s^2) is the universal gravitational constant, m and M are the masses of the two bodies, and r is the distance between the centers of mass of the two bodies. [If the bodies are moving relative to each other, then r varies with time and therefore $r = r(t)$.]

As a first approximation to the force of gravitational attraction for an object falling to the surface of Earth, we assume Earth is a sphere of homogeneous material. Under this assumption, we can use equation (6) to model gravity. The mass of Earth is $M_e = 5.976 \times 10^{24}$ kg and its radius is $R_e = 6371$ km or 6.371×10^6 m.

EXAMPLE 2

Consider an object having mass $m = 100$ kg which is released from rest at an altitude of $h = 200$ km above the surface of Earth. Neglecting drag and considering only the force of gravitational attraction, calculate the impact velocity of the object at Earth's surface.

Solution: In this problem we do not assume that the force of gravitational attraction is constant. Rather, we take into account the variation of this force with the separation distance between the two bodies. See Figure 3.9.

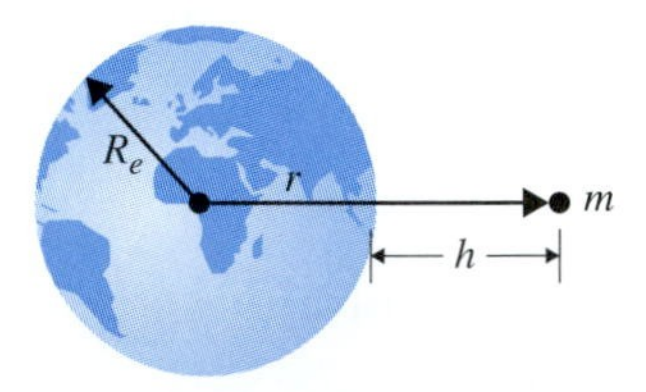

FIGURE 3.9

An object of mass m is released at an altitude of 200 km above the surface of Earth. As the object falls, its distance from the center of Earth is r. Earth has radius R_e and therefore $r = R_e + h$ defines the object's altitude, h, above Earth. The quantity r is positive in the direction of increasing altitude.

As shown in Figure 3.9, separation distance r is measured positive from Earth's center to the 100-kg body. If $v = dr/dt$, then the application of Newton's second law of motion leads to the differential equation

$$m\frac{dv}{dt} = -\frac{GmM_e}{r^2}. \tag{7}$$

If time is retained as the independent variable, we obtain a differential equation for the dependent variable $r(t)$ by using the fact that $v = dr/dt$. The resulting differential equation, however, is second order and nonlinear

$$m\frac{d^2r}{dt^2} = -\frac{GmM_e}{r^2}.$$

Note that the separation distance $r(t)$ is a decreasing function of time. If we transform differential equation (7) into one in which distance r is the independent variable, and use the fact that $v = 0$ when $r = R_e + h$, we obtain the following initial value problem:

$$\begin{aligned} mv\frac{dv}{dr} &= -\frac{GmM_e}{r^2}, \qquad R_e < r < R_e + h \\ v(R_e + h) &= 0. \end{aligned} \tag{8}$$

The differential equation in (8), while nonlinear, is a first order *separable* equation for the quantity of interest, v. Solving, we obtain

$$\frac{v^2}{2} = \frac{GM_e}{r} + C.$$

Imposing the initial condition, we find the implicit solution

$$\frac{v^2}{2} = GM_e\left[\frac{1}{r} - \frac{1}{R_e + h}\right].$$

Since separation distance is decreasing with time, velocity is negative. Therefore, an explicit solution of the problem is

$$v = -\sqrt{2GM_e\left[\frac{1}{r} - \frac{1}{R_e + h}\right]}.$$

The impact velocity is found by evaluating velocity at $r = R_e$:

$$v_{\text{impact}} = -\sqrt{2GM_e\left[\frac{1}{R_e} - \frac{1}{R_e + h}\right]}.$$

Using the values given, we obtain $v_{\text{impact}} = -1952$ m/s (a speed of about 4350 mph). [Note: The impact velocity does not depend on the mass of the object, but only on its distance from Earth when it is released.] ▲

Example 2 is interesting insofar as it illustrates how the transformation to distance as independent variable may be advantageous. The image of a meteor glowing in the night sky tells us, however, that drag due to air resistance is an important effect in high-speed motion through the atmosphere, one whose neglect makes suspect the physical relevance of the answer obtained in Example 2. In the Exercises, we ask you to consider the same problem in the presence of a drag force that is proportional to the square of the velocity. In particular, the governing differential equation becomes

$$m\,\frac{dv}{dt} = -\frac{GmM_e}{r^2} + \kappa v^2. \tag{9}$$

After changing the independent variable from time to distance, the resulting differential equation can be recast as a Bernoulli equation and then solved; see Exercise 10.

EXERCISES

Exercises 1–4:

An object undergoes one-dimensional motion along the x-axis subject to the given decelerating forces. At time $t = 0$, the object's position is $x = 0$ and its velocity is $v = v_0$. In each case, the decelerating force is a function of the object's position $x(t)$ or its velocity $v(t)$ or both. Transform the problem into one having distance x as the independent variable. Determine the position x_f at which the object comes to rest. (If the object does not come to rest, $x_f = \infty$.)

1. $m\,\dfrac{dv}{dt} = -kx^2v$

2. $m\,\dfrac{dv}{dt} = -kxv^2$

3. $m\,\dfrac{dv}{dt} = -ke^{-x}$

4. $m\,\dfrac{dv}{dt} = -\dfrac{kv}{1+x}$

5. A boat having mass m is pushed away from a dock with an initial velocity v_0. Assume the water exerts a drag force on the boat that is proportional to the square of its velocity. Determine the velocity of the boat when it is a distance d from the dock.

6. A projectile having mass m is launched vertically upward from ground level with an initial velocity v_0. Assume that air resistance is proportional to the square of the velocity and that gravitational acceleration g is constant. Determine the maximum height reached by the projectile. [Hint: The equation for dv/dy is a Bernoulli equation.]

7. A block of mass m is pulled over a frictionless (smooth) surface by a cable having a constant tension F; see the figure. The block starts from rest at a horizontal distance D from the base of the pulley. Apply Newton's law of motion ($m\,dv/dt = \sum F$) in the horizontal direction. What is the (horizontal) velocity of the block when $x = D/3$? (Assume the vertical component of the tensile force never exceeds the weight of the block.)

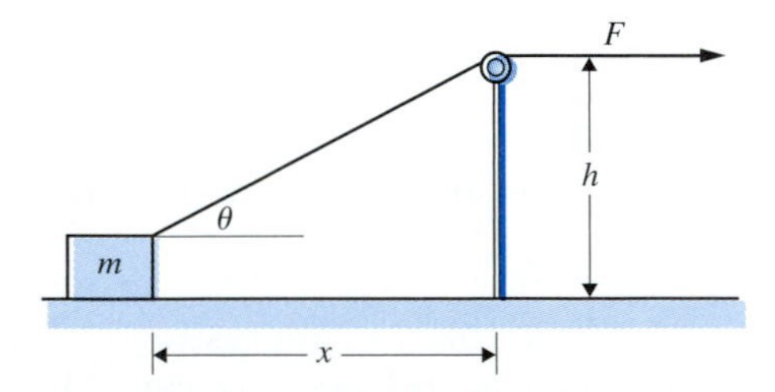

Figure for Exercise 7

8. Power is a measure of the rate of doing work. Work, in turn, is the product of force and distance. Therefore, power can be viewed as the product of force and velocity, $P = Fv$. Assume a car of mass m accelerates on a level road from an initial velocity v_1 to a greater velocity v_2 under conditions where the car's engine delivers a constant power, P. What is the distance traveled by the car during this accelerating motion? Evaluate your answer for the specific case of a 3000-lb car being supplied 200 horsepower while the car is accelerating from 20 mph to 50 mph (1 hp = 550 ft-lb/sec). [Hint: $m\,dv/dt = F = P/v$.]

9. We need to design a ballistics chamber to decelerate test projectiles fired into it. Assume the resistive force encountered by the projectile is proportional to the square of its velocity and neglect gravity. As the figure indicates, the chamber is to be constructed so that the coefficient κ associated with this resistive force is not constant but is, in fact, a linearly increasing function of distance into the chamber. Let $\kappa(x) = \kappa_0 x$, where κ_0 is a constant; the resistive force then has the form $\kappa(x)v^2 = \kappa_0 x v^2$. If we use time t as independent variable, Newton's law of motion leads us to the differential equation

$$m\frac{dv}{dt} + \kappa_0 x v^2 = 0 \quad \text{with} \quad v = \frac{dx}{dt}. \tag{10}$$

(a) Adopt distance x into the chamber as the new independent variable and rewrite differential equation (10) as a first order equation in terms of the new independent variable.

(b) Determine the value κ_0 needed if the chamber is to reduce projectile velocity to 1% of its incoming value within d units of distance.

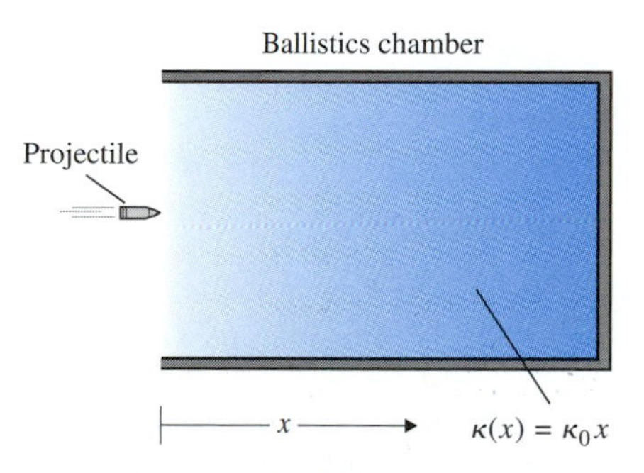

Figure for Exercise 9

10. The motion of a body of mass m, gravitationally attracted to Earth in the presence of a resisting drag force proportional to the square of its velocity, is given by

$$m\frac{dv}{dt} = -\frac{GmM_e}{r^2} + \kappa v^2$$

[recall equation (9)]. In this equation, r is the radial distance of the body from the center of Earth, G is the universal gravitational constant, M_e is the mass of Earth and $v = dr/dt$. Note that the drag force is positive, since it acts in the positive r direction.

(a) Assume that the body is released from rest at an altitude h above the surface of Earth. Recast the differential equation so that distance r is the independent variable. State an appropriate initial condition for the new problem.

(b) Show that the impact velocity can be expressed as

$$v_{\text{impact}} = -\left[2GM_e \int_0^h \frac{e^{-2(\kappa/m)s}}{(R_e + s)^2}\,ds\right]^{1/2}$$

where R_e represents the radius of Earth. (The minus sign reflects the fact that $v = dr/dt < 0$.)

11. A body of mass m is launched upward (that is, radially outward) from the surface of Earth with velocity v_0. Neglect air resistance. Assume that the gravitational attraction between the body and Earth is not constant but rather varies according to Newton's law of universal gravitation, as in equation (7). Assume that the body achieves a maximum height (or radial separation) of 220 km. What was the launch velocity v_0?

Pendulum Motion: Conservation of Energy In Exercises 12 and 13, our goal is to describe the rotational motion of the pendulum shown in Figure 3.10; we neglect the weight of the rod when modeling the motion of this pendulum. The motion about the fixed pivot point O is governed by the equation

$$I\alpha = \sum M.$$

In this equation, I represents the moment of inertia of the body, $\alpha = d^2\theta/dt^2$ is its angular acceleration and $\sum M$ represents the sum of the external moments applied to the body, all referenced to the pivot O. Each moment is the product of an applied force and the perpendicular distance from the pivot to the line of action of the force. The moment is positive if it acts to cause rotation in the positive sense of the angle θ and negative otherwise.

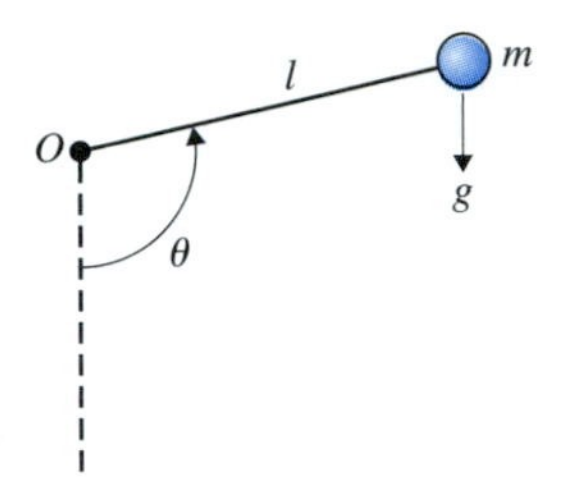

FIGURE 3.10

A pendulum with a bob of mass m at one end of a rod of length l.

Applying the rotational version of Newton's laws to the pendulum and noting that $I = ml^2$, we obtain the second order differential equation

$$ml^2\theta'' = -mgl\sin\theta. \tag{11}$$

In equation (11), the moment caused by the body's weight is negative because it acts to cause clockwise rotation; that is, rotation in the negative θ direction.

12. Suppose at some initial time the pendulum is in the inverted position ($\theta = \pi$) with an angular velocity $d\theta/dt \equiv \omega = -\omega_0$ radians/sec, where ω_0 is a positive constant. The pendulum therefore rotates clockwise.

(a) Equation (11) is a second order differential equation. Rewrite it as a first order separable equation by adopting angle θ as independent variable, using the fact that

$$\theta'' = \frac{d}{dt}\left(\frac{d\theta}{dt}\right) = \frac{d\omega}{dt} = \frac{d\omega}{d\theta}\frac{d\theta}{dt} = \omega\frac{d\omega}{d\theta}.$$

Complete the specification of the initial value problem by specifying an appropriate initial condition.

(b) Obtain the implicit solution

$$ml^2\frac{\omega^2}{2} - mgl\cos\theta = ml^2\frac{\omega_0^2}{2} + mgl. \tag{12}$$

The pendulum is a conservative system; that is, energy is neither created nor destroyed. Equation (12) is a statement of conservation of energy. At a position defined by the angle θ, the quantity $ml^2\omega^2/2$ is the kinetic energy of the pendulum while the term $-mgl\cos\theta$ is the potential energy, referenced to the horizontal position $\theta = \pi/2$. The constant right-hand side is the total energy at the initial position $\theta = \pi$.

(c) Determine the angular velocity at the instant the pendulum reaches the vertically downward position, $\theta = 0$. Express your answer in terms of the constant ω_0.

13. A pendulum, 2 ft in length and initially in the downward position, is launched with an initial angular velocity ω_0. If it achieves a maximum angular displacement of 135 degrees, what is ω_0?

3.8 Euler's Method

Our efforts thus far have focused on the initial value problem

$$y' = f(t, y), \qquad y(t_0) = y_0, \tag{1}$$

on some underlying t-interval of interest, $a < t < b$. In Chapter 2, we discussed the first order linear equation $y' + p(t)y = q(t)$ and developed an explicit representation for the solution of the corresponding initial value problem. This representation, however, was sometimes difficult to evaluate and work with.

In this chapter, we have discussed certain classes of nonlinear differential equations. For these classes, our solution procedures often lead to implicit solutions that cannot be "unraveled" to yield an explicit solution. There are many nonlinear differential equations, moreover, which do not belong to any of the classes considered.

It's clear, therefore, that the analytic methods developed for initial value problems, while important and useful, are not totally adequate. We need a tool that enables us to obtain quantitative information about the solution of (1) in the general case. Numerical methods are such a tool. The application of numerical methods to the solution of problem (1) is a very important topic. At this point we will introduce and discuss the simplest and perhaps most intuitive of numerical methods, Euler's method.[6] This approach will give us some insight into numerical methods. It will also provide us with a tool (albeit a relatively crude one) to analyze initial value problems for which analytical methods are either inadequate or not applicable. Numerical methods are discussed in greater detail in Chapter 9.

---

[6]Leonhard Euler (1707–1783) was one of the most gifted individuals in the history of mathematics and science; his 866 publications (books and articles) make him arguably the most prolific as well. He established the foundations of mathematical analysis, the theory of special functions and analytical mechanics. Complex analysis, number theory, differential equations, differential geometry, lunar theory, elasticity, acoustics, fluid mechanics, and the wave theory of light are some of the other areas to which Euler made important contributions. A backlog of his work continued to be published for nearly 50 years after his death.

Euler's achievements become even more remarkable when one appreciates the circumstances surrounding his work. He was involved in the world about him. In Russia, he worked on state projects involving cartography, science education, magnetism, fire engines, machines, and shipbuilding. Later, in Berlin, he served as an advisor to the government on state lotteries, insurance, annuities, and pensions. He fathered thirteen children, although only five survived infancy. Euler claimed that he made some of his greatest discoveries while holding an infant in his arms with other children playing at his feet. In 1771 Euler became totally blind but he was able to maintain a prodigious output of work until his death.

We begin by asking the most basic question: "What is a numerical solution of initial value problem (1)?" As we will see, a numerical method for (1) typically generates values

$$y_1, y_2, \ldots, y_n$$

that approximate corresponding solution values

$$y(t_1), y(t_2), \ldots, y(t_n),$$

at designated abscissa values $t_0 < t_1 < t_2 \cdots < t_n$. Figure 3.11 illustrates the output of a numerical method, where the solid curve indicates the actual (unknown) solution of initial value problem (1).

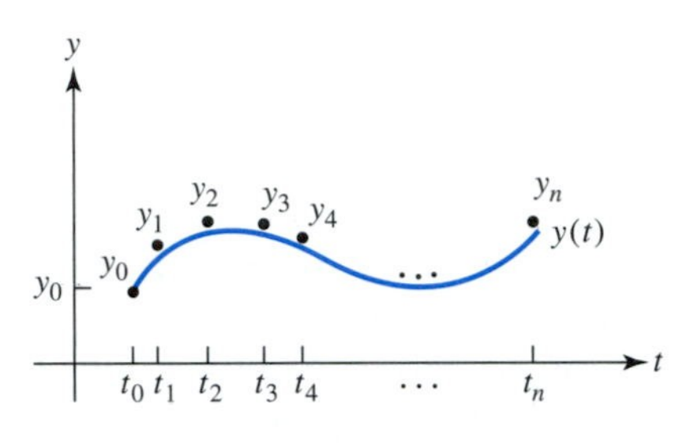

FIGURE 3.11

The solid curve indicates the actual (unknown) solution of initial value problem (1). For $i = 1, 2, \ldots, n$, the points (t_i, y_i) are approximations to corresponding points $(t_i, y(t_i))$ on the actual solution curve. We say that y_i is a numerical approximation to the unknown value $y(t_i)$.

In Figure 3.11, the values $y_1, y_2, y_3, \ldots$ are found sequentially—first y_1, then y_2, and so forth. Note that the starting value $y_0 = y(t_0)$ is known exactly since it is specified in the initial value problem (1). The numerical problem therefore reduces to the question:

> Given (t_0, y_0), how do we determine (t_1, y_1)? And in general, given an approximation (t_k, y_k), how do we determine the next approximation, (t_{k+1}, y_{k+1})?

The simplest answer, the simplest numerical algorithm, is **Euler's method** (also known as the **tangent line method**).

Derivation of Euler's Method

The geometric ideas underlying Euler's method can be understood in terms of direction fields. At the initial condition point (t_0, y_0), the differential equation tells us the slope of the solution curve at (t_0, y_0); it is equal to $f(t_0, y_0)$. Therefore, the line tangent to the solution curve $y(t)$ at the point (t_0, y_0) has equation

$$y = y_0 + f(t_0, y_0)(t - t_0). \tag{2}$$

We follow this tangent line over a short time interval, to $t = t_1$, where $t_0 < t_1$. At $t = t_1$ we reach the point (t_1, y_1) where, by (2),

$$y_1 = y_0 + f(t_0, y_0)(t_1 - t_0).$$

The value found above, y_1, is the Euler's method approximation to the solution value $y(t_1)$. (See Figure 3.12.)

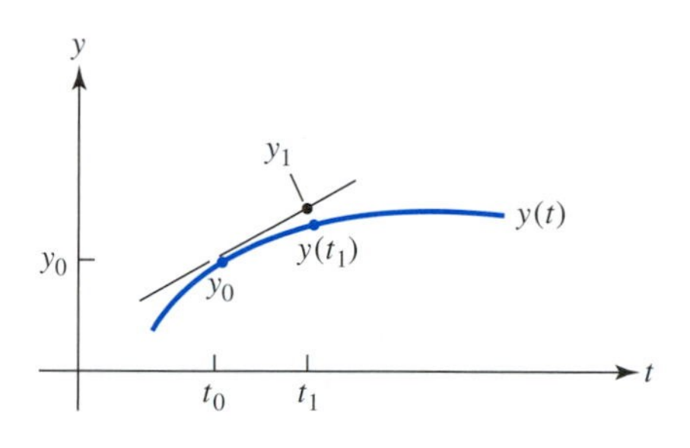

FIGURE 3.12

The line tangent to $y(t)$ at the initial point (t_0, y_0) has slope $f(t_0, y_0)$. Following the tangent line to time t_1, we arrive at the point (t_1, y_1) and have an approximation, y_1, to the solution value, $y(t_1)$.

While the new point (t_1, y_1) does not exactly coincide with the point $(t_1, y(t_1))$, it is generally close to that point (assuming t_1 is sufficiently close to t_0). Moreover, since $f(t, y)$ is continuous, the direction field filament at (t_1, y_1) has nearly the same slope as the filament at $(t_1, y(t_1))$. Hence, although we do not know the exact value of $y(t)$ when $t = t_1$, we are close to it and we do have a good idea of which direction the graph of $y(t)$ is heading. At $t = t_1$, the graph of $y(t)$ has slope $f(t_1, y(t_1))$ which is nearly equal to $f(t_1, y_1)$. So, in an attempt to follow the solution curve $y(t)$, we proceed from (t_1, y_1) along a line having slope $f(t_1, y_1)$:

$$y = y_1 + f(t_1, y_1)(t - t_1).$$

Following this line until $t = t_2$, we reach a point (t_2, y_2) where

$$y_2 = y_1 + f(t_1, y_1)(t_2 - t_1).$$

This process is repeated. In general, starting from a point (t_k, y_k), where y_k is our approximation to $y(t_k)$, we evaluate $f(t_k, y_k)$ to determine the straight line direction in which to proceed. We follow that line,

$$y = y_k + f(t_k, y_k)(t - t_k),$$

until $t = t_{k+1}$, obtaining the next approximation

$$y_{k+1} = y_k + f(t_k, y_k)(t_{k+1} - t_k). \tag{3}$$

Iteration (3) is known as Euler's method and is illustrated in Figure 3.13 on the next page. Euler's method amounts to tracing a polygonal path through the direction field.

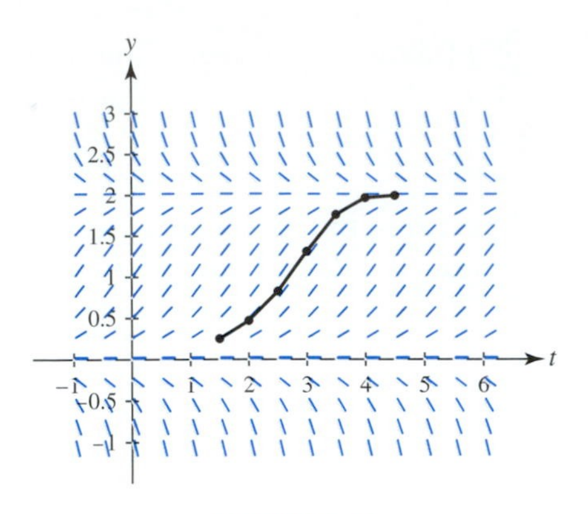

FIGURE 3.13

Starting on the solution curve at (t_0, y_0), Euler's method attempts to track the solution $y(t)$ by following a polygonal path through the direction field. As the path proceeds, its direction is constantly corrected by sampling the direction field, see equation (3).

EXAMPLE 1

Apply Euler's method to the initial value problem

$$y' = t^2 + y, \qquad y(2) = 1.$$

Use $t_1 = 2.1$, $t_2 = 2.2$, and $t_3 = 2.3$. Generate approximations y_1 to $y(2.1)$, y_2 to $y(2.2)$, and y_3 to $y(2.3)$.

Solution: The actual (unknown) solution starts at $(t_0, y_0) = (2, 1)$ and has a starting slope of $f(t_0, y_0) = f(2, 1) = 5$. Following the line of slope 5 passing through $(2, 1)$, we obtain by (3)

$$y_1 = y_0 + f(t_0, y_0)(t_1 - t_0) = 1 + 5(0.1) = 1.5.$$

Having $(t_1, y_1) = (2.1, 1.5)$, we are ready to take the next step to (t_2, y_2). By (3),

$$y_2 = y_1 + f(t_1, y_1)(t_2 - t_1) = 1.5 + 5.91(0.1) = 2.091.$$

Having $(t_2, y_2) = (2.2, 2.091)$, we are ready to take the next step to (t_3, y_3). By (3),

$$y_3 = y_2 + f(t_2, y_2)(t_3 - t_2) = 2.091 + 6.931(0.1) = 2.7841.$$

Note that the differential equation in this example is linear, and hence a formula for the solution can be found. The exact solution is

$$y(t) = 11e^{t-2} - (t^2 + 2t + 2).$$

Figure 3.14 compares the Euler's method approximations with the exact solution at the points $t = 2, 2.1, 2.2$, and 2.3. ▲

Implementing Euler's Method

The simplest way to organize an Euler's method calculation is to choose an appropriate **step size**, h, and then use h to define the equally spaced sample points:

$$t_1 = t_0 + h, \qquad t_2 = t_1 + h,$$

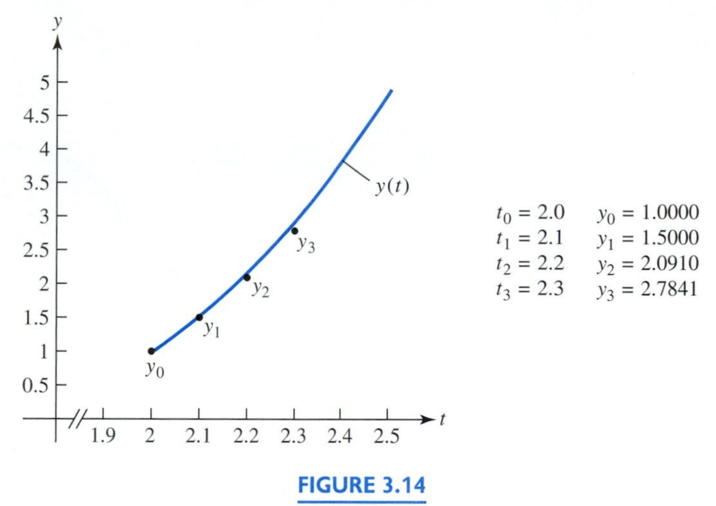

FIGURE 3.14

The values y_1, y_2, and y_3 are the Euler's method approximations found in Example 1.

and, in general,

$$t_{k+1} = t_k + h, \qquad k = 0, 1, \ldots, n-1.$$

For instance, in Example 1 we used a step size of $h = 0.1$ to define sample points $t_0 = 2.0$, $t_1 = 2.1$, $t_2 = 2.2$, and $t_3 = 2.3$.

For a constant step size h, the term $t_{k+1} - t_k$ in (3) is equal to h. Thus, Euler's method takes the form

$$y_{k+1} = y_k + hf(t_k, y_k), \qquad k = 0, 1, \ldots, n-1. \tag{4}$$

We anticipate that Euler's method should become more accurate when we take smaller steps, sampling the direction field more often, and using this information to correct the "Euler path" that is tracking the solution $y(t)$ (see Exercise 12). Using a small step size h, however, may lead to a significant amount of computation. Therefore, numerical methods are usually programmed and run on a computer or programmable calculator.

EXAMPLE 2

Apply Euler's method to the initial value problem

$$y' = y(2-y), \qquad y(0) = 0.1. \tag{5}$$

Use $h = 0.2$ and approximate the solution on the interval $0 \le t \le 4$.

Solution: Using a fixed step size of $h = 0.2$, Euler's method samples the direction field at time values $t_1 = 0.2$, $t_2 = 0.4, \ldots, t_{19} = 3.8$, and $t_{20} = 4.0$. For this test case, initial value problem (5) can be solved since the differential equation is separable. Figure 3.15 on the next page shows a portion of the direction field for $y' = y(2-y)$. The solid curve in Figure 3.15 is the graph of the actual solution of (5), and the circles show the Euler path through the direction field. ▲

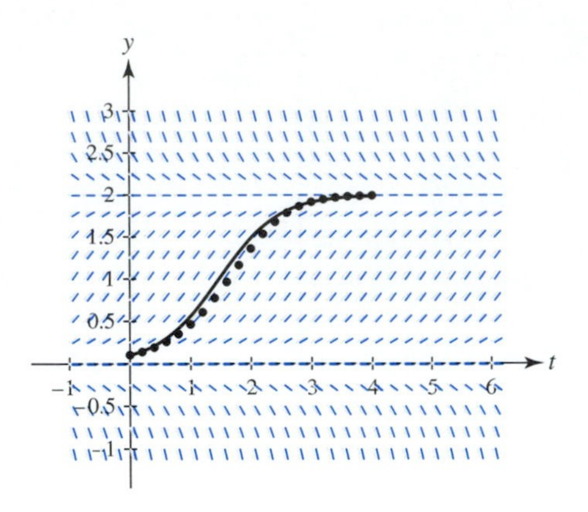

FIGURE 3.15

The direction field for $y' = y(2-y)$. The curve denotes the solution of $y' = y(2-y)$, $y(0) = 0.1$, the initial value problem posed in Example 2. The circles are the points generated by Euler's method, using a step size of $h = 0.2$.

Numerical Issues That Must Be Considered

We have thus far introduced the idea of direction fields and we have used the underlying geometry to motivate a simple numerical method, Euler's method. Several important questions remain to be considered.

1. We expect that the accuracy of Euler's method should improve as the step size is reduced. Although this observation is intuitively plausible, we have not yet presented any mathematical justification.
2. Euler's method seems quite crude—there should be better schemes than tangent line approximations. How do we quantify the notion of "better," and how do we then go about formulating these better schemes? Ultimately, the notion of "better" must combine the characteristics of increased accuracy and ease of implementation.
3. Suppose, for a given problem, that we have determined the t-interval on which the problem is to be solved, chosen the numerical algorithm to be used, and decided on the error that we can tolerate. How do we now select an appropriate step size h?

We will address these important issues in Chapter 9.

EXERCISES

Exercises 1–5:

(a) Solve the problem $y' = f(t, y), y(t_0) = y_0$ analytically, using an appropriate solution technique.

(b) Write the expression $y_{k+1} = y_k + hf(t_k, y_k)$ for the given problem.

(c) Using step size $h = 0.1$, compute the Euler approximations $y_k, k = 1, 2, 3$ at times $t_k = t_0 + kh$.

(d) For $k = 1, 2, 3$, evaluate the exact solution obtained in part (a) at t_k to find $y(t_k)$.

1. $y' = 2t - 1, \quad y(1) = 0$

2. $y' = -y, \quad y(0) = 1$

3. $y' = -ty, \quad y(0) = 1$

4. $y' = -y + t, \quad y(0) = 0$

5. $y' = y^2, \quad y(0) = 1$

Exercises 6–7:

For each exercise the form of an initial value problem is given and a table that corresponds to the Euler's method numerical solution, calculated using a step size of $h = 0.1$. Determine the constants of the initial value problem.

6. Find α, β, t_0, and y_0.
$y' = \alpha t + \beta, \quad y(t_0) = y_0$.

k	t_k	y_k
0	0.0	−1.00
1	0.1	−0.90
2	0.2	−0.81
3	0.3	−0.73

7. Find α, the integer n, t_0, and y_0.
$y' = \alpha + y^n, \quad y(t_0) = y_0$.

k	t_k	y_k
0	1.0	1.00
1	1.1	0.90
2	1.2	0.781
3	1.3	0.6419961

8. Assume we are considering the direction field of an autonomous first order differential equation.

(a) Suppose we can qualitatively establish, by examining this direction field, that all solution curves $y(t)$ in a given region of the ty-plane have one of the following four types of behavior:

(i) increasing, concave up (ii) increasing, concave down
(iii) decreasing, concave up (iv) decreasing, concave down

Suppose we implement an Euler's method approximation to one of the solution curves in the region, using some reasonable step size h. Consider each of the four cases. In each case, how will the Euler approximations y_k compare to the exact solution counterparts, $y(t_k), k = 1, 2, \ldots$? Will the values y_k underestimate the exact values, overestimate the exact values or is it impossible to reach a definite conclusion?

(b) Consider the exact solutions obtained in Exercises 2, 3, and 5. Classify each solution curve as increasing or decreasing, concave up or concave down. Do the computations performed in these exercises support your answers to part (a)?

(c) What do you think will happen if Euler's method is used to approximate an "S-shaped solution curve" similar to the logistic curve shown in Figure 3.15 on page 108. In that case, a solution curve changes from increasing and concave up to increasing and concave down. Are your answers to part (a) consistent with the behavior exhibited by the Euler approximation shown in the figure?

Exercises 9–12:

A programmable calculator or computer is needed for these exercises.

9. Use Euler's method with step size $h = 0.01$ to numerically solve the initial value problem

$$y' - ty = \sin 2\pi t, \qquad y(0) = 1, \qquad 0 \le t \le 1.$$

Graph the numerical solution. [Note: Although the differential equation in this problem is a first order linear equation and we can get an explicit representation for the exact solution, the representation involves antiderivatives that we cannot express in terms of known functions. From a quantitative point of view, the representation itself is of little use.]

10. Assume a tank having a capacity of 200 gal initially contains 90 gal of fresh water. At time $t = 0$, a salt solution begins flowing into the tank at a rate of 6 gal/min and the well-stirred mixture flows out at a rate of 1 gal/min. Assume that the inflow concentration fluctuates in time with the inflow concentration given by $c(t) = 2 - \cos \pi t$ lb/gal, where time t is in minutes. Formulate the appropriate initial value problem for $Q(t)$, the amount of salt (pounds) in the tank at time t. Use Euler's method to approximately determine the amount of salt in the tank when the tank contains 100 gal of liquid. Use a step size of $h = 0.01$.

11. Let $P(t)$ denote the population of a certain colony, measured in millions of members. The colony evolves in time as the solution of the initial value problem,

$$P' = 0.1\left(1 - \frac{P}{3}\right)P + M(t), \qquad P(0) = P_0,$$

where time t is measured in years. Assume that the colony experiences a migration influx that is initially strong but that soon tapers off. Specifically assume that $M(t) = e^{-t}$. Suppose the colony had 500,000 members initially (at time $t = 0$). Use Euler's method to estimate its size after 2 years.

12. In Chapter 9, we will examine how the error in numerical algorithms, such as Euler's method, depends upon step size h. In this exercise, we empirically examine the dependence of errors on step size by studying a particular example,

$$y' = y + 1, \qquad y(0) = 0.$$

(a) Determine the exact solution, $y(t)$.

(b) Use Euler's method to obtain approximate solutions to this initial value problem on the interval $0 \le t \le 1$, using step sizes $h_1 = 0.02$ and $h_2 = 0.01$. You will therefore obtain two sets of points,

$$(t_k^{(1)}, y_k^{(1)}), \qquad k = 0, \ldots, 50$$
$$(t_k^{(2)}, y_k^{(2)}), \qquad k = 0, \ldots, 100$$

where $t_k^{(1)} = 0.02k, k = 0, 1, \ldots, 50$ and $t_k^{(2)} = 0.01k, k = 0, 1, \ldots, 100$.

(c) Print a table of the errors at the common points, $t_k^{(1)}, k = 0, 1, \ldots, 50$,

$$e^{(1)}(t_k^{(1)}) = y(t_k^{(1)}) - y_k^{(1)} \quad \text{and} \quad e^{(2)}(t_k^{(1)}) = y(t_k^{(1)}) - y_{2k}^{(2)}.$$

(d) Note that the approximations $y_k^{(2)}$ were found using a step size equal to one-half of the step size used to obtain the approximations $y_k^{(1)}$; that is, $h_2 = h_1/2$. Compute the corresponding error ratios. In particular, compute

$$\left|\frac{e^{(2)}(t_k^{(1)})}{e^{(1)}(t_k^{(1)})}\right|, \qquad k = 1, \ldots, 50.$$

On the basis of these computations, conjecture how halving the step size affects the error of Euler's method.

13. This exercise treats the simple initial value problem, $y' = \lambda y, y(0) = y_0$, where we can see the behavior of the numerical solution as the step size h approaches zero.

(a) Show that the solution of the initial value problem is $y = e^{\lambda t} y_0$.

(b) Apply Euler's method to the initial value problem, using a step size h. Show that y_n is related to the initial value by the formula $y_n = (1 + h\lambda)^n y_0$.

(c) Consider any fixed time t^*, where $t^* > 0$. For a given value of n, let $h = t^*/n$, so that $t^* = nh$. Show that

$$\lim_{\substack{h \to 0 \\ t^* = nh}} y_n = y(t^*) = e^{\lambda t^*} y_0.$$

14. Consider the initial value problem $y' = y^2, y(0) = 1$.

(a) Obtain an explicit solution of this initial value problem and determine the interval of existence for the solution.

(b) Run Euler's method for this problem on the interval $0 \le t \le 1.2$, using step size $h = 0.1$.

(c) Tabulate the numerical solution and the explicit solution from part (a). Compare the numerical and exact solutions. How accurate is the numerical solution? Is it in any sense misleading? Explain.

EXTENDED PROBLEM: THE BARANYI POPULATION MODEL AND THE STORAGE OF WHOLE MILK

Milk and milk products depend on refrigeration for storage after postpasteurization. Most microorgnisms do not grow at refrigeration temperatures (0°C–7°C). One exception is *Listeria monocytogenes*. This anaerobic pathogen can multiply at refrigeration temperatures and has been responsible for some recent outbreaks of listeriosis, caused by human consumption of contaminated milk products.

Food scientists are interested in developing predictive mathematical models that can accurately model the growth of harmful organisms. These models should be able to relate environmental conditions (such as temperature and pH) to the growth rate of a microbial population. In this regard, the modeler walks a fine line. On the one hand, there is an ongoing need to "build more reality" into the model. On the other hand, the model must be kept simple enough to be mathematically tractable and useful.

In Section 2.4 we introduced the Malthusian population model. This simple model assumed a relative birth rate independent of population and led to predictions of exponential population growth or decay. In Section 3.5 the logistic equation was introduced and viewed as a refinement of the Malthusian model. The logistic model accounted for the influence of population on the relative birth rate; it led, in particular, to predictions of nontrivial equilibrium populations.

A population model currently being studied and used in food science research is the Baranyi population model.[7] The Baranyi model can be viewed as a refinement of the logistic model. It attempts to account for the way certain *critical substances* affect bacterial cell growth. The essence of the model is a pair of initial value problems,

$$\begin{aligned} \frac{dP(t)}{dt} &= \mu \frac{q(t)}{1+q(t)} \left[1 - \left(\frac{P(t)}{P_e} \right)^m \right] P(t), \qquad P(0) = P_0, \\ \frac{dq(t)}{dt} &= \nu q(t), \qquad q(0) = q_0. \end{aligned} \tag{1}$$

In this model, $P(t)$ represents the population of bacteria at time t, while $q(t)$ represents the concentration of critical substance present. The positive constant ν represents the

[7] J. Baranyi, T. A. Roberts, and P. McClure. "A nonautonomous differential equation to model bacterial growth," *Food Microbiology* 10 (1993), pp. 43–59.

growth rate of the critical substance. The parameter μ accounts for the effects of environmental conditions, such as temperature, upon the growth rate of the bacteria. If temperature varies with time, μ is ultimately a prescribed function of time, and this was the case in the work cited. In our initial discussion we shall, for simplicity, assume both μ and ν to be constants. In (1), the initial values P_0 and q_0 represent the bacterial population and the amount of critical substance present at time $t = 0$, respectively. The integer exponent m is introduced into the relative birth rate to allow for greater modeling flexibility. (When $m = 1$, the differential equation reduces to the logistic equation.)

At first glance, the problem posed in (1) may appear to be both complicated and difficult. We can solve it, however, by systematically applying the ideas developed in Chapters 2 and 3.

PROBLEM 1: *Solution of Initial Value Problem (1)*

We first solve the initial value problem for $q(t)$ and then use the solution to fully specify and solve the problem for $P(t)$.

(a) Solve the initial value problem for $q(t)$, obtaining $q(t) = q_0 e^{\nu t}$. For brevity in the initial value problem for $P(t)$, let

$$\alpha(t) = \frac{q(t)}{1 + q(t)} = \frac{q_0 e^{\nu t}}{1 + q_0 e^{\nu t}}.$$

Note that $\alpha(0) = q_0/(1 + q_0) < 1$ and that $\lim_{t\to\infty} \alpha(t) = 1$. The differential equation for $P(t)$ now takes the form

$$\frac{dP(t)}{dt} = \mu\alpha(t)\left[1 - \left(\frac{P(t)}{P_e}\right)^m\right]P(t) \tag{2}$$

As $q(t)$ increases, the relative birth-rate factor $\mu\alpha(t)$ governing the growth of $P(t)$ increases from an initial value of

$$\frac{\mu q_0}{1 + q_0}$$

to a limiting value of μ. In other words, the critical substance can exert only a certain limited effect upon bacterial growth.

(b) Make a "change-of-clock" change of independent variable. Introduce a new independent variable τ by requiring that

$$\frac{d\tau}{dt} = \mu\alpha(t), \qquad \tau(0) = 0.$$

Show that

$$\tau(t) = \mu\int_0^t \alpha(s)\,ds = \mu\nu^{-1}\ln\left[\frac{1 + q_0 e^{\nu t}}{1 + q_0}\right]. \tag{3}$$

Using the chain rule, show that if we introduce the normalized dependent variable $p = P/P_e$ and view p to be a function of τ, then the initial value problem for p becomes

$$\frac{dp(\tau)}{d\tau} = [1 - p^m(\tau)]p(\tau), \qquad p(0) = \frac{P_0}{P_e} \tag{4}$$

(c) Solve initial value problem (4) for $p(\tau)$. Note that this differential equation is both a separable and a Bernoulli equation. For a general integer m, the equation is most easily solved as a Bernoulli equation. In particular, show that

$$p(\tau) = \left[1 + \left(\left[\frac{P_0}{P_e}\right]^{-m} - 1\right)e^{-m\tau}\right]^{-1/m}. \tag{5}$$

(d) Use (3) to express the term $e^{-m\tau}$ in (5) in terms of t, and then use the fact that $P(t) = P_e p(\tau(t))$ to obtain the solution in terms of "actual time" t:

$$P(t) = P_e \left[1 + \left(\left[\frac{P_0}{P_e} \right]^{-m} - 1 \right) \left(\frac{1 + q_0}{1 + q_0 e^{\nu t}} \right)^{(m\mu)/\nu} \right]^{-1/m}. \qquad \textbf{(6)}$$

PROBLEM 2: *Numerical Evaluation of the Results*

Suppose one is interested in applying (6) to study the growth of a pathogen population such as *Listeria monocytogenes* in milk. For simplicity, assume the integer exponent is $m = 1$, so that the first equation in (1) reduces to the logistic model. To test the model predictions against experiments, the food scientist must be able, via independent means, to determine values of μ, ν, and P_e appropriate to the experimental conditions. If the scientist measures P_0 and q_0 at the outset of the experiment, the model is fully specified and the behavior of (6) as a function of time can be compared to the actual growth dynamics of the bacterial culture.

(a) Assume:

$$\mu = \nu = 0.30 \text{ hour}^{-1} \qquad P_e = 5 \times 10^6 \text{ CFU/ml}$$

$$P_0 = 5 \times 10^3 \text{ CFU/ml} \qquad \alpha(0) = \frac{q_0}{1 + q_0} = 0.10$$

and evaluate and plot $\ln[P(t)]$ for $0 \le t \le 10$ hours. (The term CFU stands for "colony forming units." Basically, it refers to the bacterial count of those microbes that are important in the particular study being done.)

(b) Use Euler's method, with $h = 0.02$, to obtain a numerical approximation to the solution of

$$\frac{dP(t)}{dt} = \mu\alpha(t) \left[1 - \frac{P(t)}{P_e} \right] P(t), \qquad P(0) = P_0$$

over the time interval $0 \le t \le 10$ hours. Plot $\ln[P(t)]$ versus t. Does Euler's method provide an adequate approximation to the actual solution over this time span for the step size used? In particular, what is the relative error in $\ln[P(t)]$ at time $t = 10$ hours? What is the corresponding relative error in $P(t)$?

4

Second Order Linear Differential Equations

CHAPTER OVERVIEW

Introduction

In this chapter we discuss initial value problems of the form

$$y'' + p(t)y' + q(t)y = g(t), \qquad y(t_0) = y_0, \qquad y'(t_0) = y'_0, \qquad a < t < b. \tag{1}$$

In equation (1), $p(t)$, $q(t)$, and $g(t)$ are continuous functions on the interval $a < t < b$ and t_0 is some point in this t-interval of interest. Differential equation (1) is called a **second order linear differential equation**.

If $g(t)$ is the zero function, then differential equation (1) is a **homogeneous differential equation**; otherwise the equation is a **nonhomogeneous differential equation**. An initial value problem for a second order equation involves two supplementary or initial conditions, $y(t_0) = y_0$ and $y'(t_0) = y'_0$, where y_0 and y'_0 are given constants.

We begin the study of second order linear differential equations in Section 4.1 with a brief discussion of existence and uniqueness of solutions for initial value problems. The existence and uniqueness theorem, Theorem 4.1, is analogous to Theorem 2.1 for first order linear differential equations.

Sections 4.2 and 4.3 discuss the general solution of the homogeneous equation

$$y'' + p(t)y' + q(t)y = 0, \qquad a < t < b.$$

This discussion of the general solution is based on the concept of a fundamental set of solutions and a determinant known as the Wronskian. In Sections 4.4–4.6 we examine the important special case of homogeneous constant coefficient equations of the form $ay'' + by' + cy = 0$ and develop techniques for finding the general solution of such equations. Section 4.7 discusses applications of the homogeneous equation $ay'' + by' + cy = 0$.

In Section 4.8 we turn our attention to nonhomogeneous equations, describing how the general solution of the nonhomogeneous equation is related to the general solution of the homogeneous equation and some particular solution of the nonhomogeneous equation. Sections 4.9 and 4.10 discuss techniques for finding particular solutions of nonhomogeneous equations. Section 4.11 is devoted to applications and analyzes phenomena such as resonance.

If you wish, you can cover Chapter 5, Higher Order Linear Differential Equations, simultaneously with Chapter 4. The theoretical results and the solution techniques discussed in Chapter 5 are simple generalizations of those for second order linear differential equations and are easily understood as such.

The Bobbing Motion of a Floating Object

Second order differential equations arise frequently when we use Newton's laws of motion to model a physical system. We introduce one such example below and you will see several other applications as we move through this chapter. The applications we consider (such as the bobbing motion of a floating object, the vibrations of a mass-spring system, the operation of a simple centrifuge and the response of *RLC* networks) are all very basic in nature. However, these simple applications embody important physical principles, the same principles that underlie more complicated mechanical and electrical systems. Understanding how to model and analyze the behavior of such simple systems serves as a necessary foundation for learning how to model more complicated systems.

We have all observed a cork, a block of wood, or some other object bobbing up and down in a liquid such as water. How do we mathematically model this bobbing motion?

In its rest or equilibrium state, a floating object is subjected to two equal and opposite forces—the weight of the object is counteracted by an upward buoyant force equal to the weight of the displaced liquid. (This is the law of buoyancy discovered by Archimedes.[1]) If we disturb this equilibrium state by pushing down or pulling up on the object and then releasing it, the object will begin to bob up and down. The physical principle governing the object's movement is Newton's second law of motion, $ma = F$, the product of the mass and acceleration of an object is equal to the sum of the forces acting on it.

As shown in Figure 4.1(a), consider a cylindrical object having uniform mass density ρ, constant cross-sectional area A, and height L. We assume the object is floating in a liquid having density ρ_l; that is, we assume $\rho < \rho_l$. In its rest or equilibrium state, the object sinks into the liquid until the weight of the liquid displaced equals the weight of the object. If, as in Figure 4.1(a), we denote the depth to which the object sinks as Y, then the force-balance equation determining Y is

$$\rho ALg = \rho_l AYg, \quad \text{(2a)}$$

or

$$Y = \frac{\rho}{\rho_l} L. \quad \text{(2b)}$$

In equation (2a), g is the acceleration due to gravity. Notice that equation (2a) simply equates the body's weight, $W = \rho ALg$, to the weight of the displaced liquid, $W_l = \rho_l AYg$. Since the net force acting upon the body is zero, it remains at rest. Also, since $\rho < \rho_l$, we know from equation (2b) that $Y < L$.

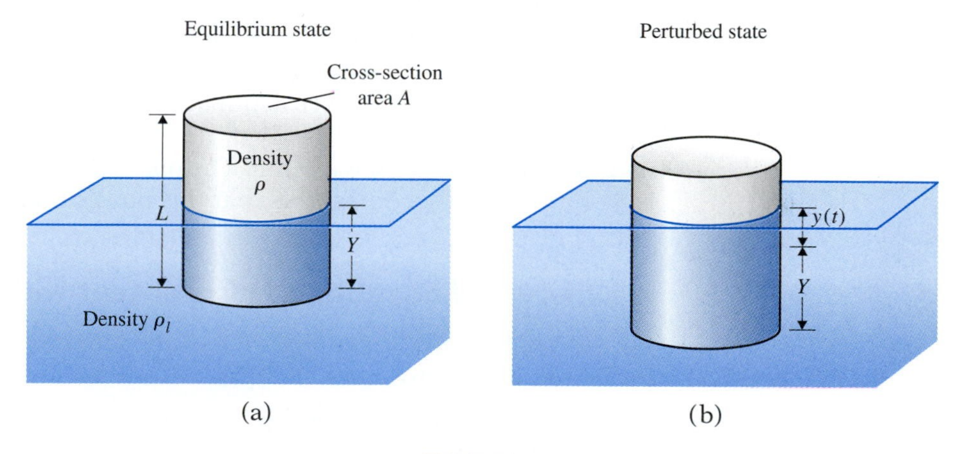

FIGURE 4.1

(a) The floating object is in its equilibrium or rest state when the weight of the displaced liquid is equal to the weight of the object. (b) The object is in a perturbed state when it is displaced from its equilibrium position. At any time t, the quantity $y(t)$ measures how far the object is from its equilibrium position.

---

[1]Archimedes of Syracuse (287–212 B.C.) was a remarkable mathematician and scientist, contributing important results in geometry, mechanics, and hydrostatics. Archimedes developed an early form of integration that empowered his work. He was also an inventor, developing the compound pulley, a pump known as Archimedes' screw, and military machines used to defend his native Syracuse in Sicily from attack by the Romans. Archimedes was killed when Syracuse was captured by the Romans during the Second Punic War.

Suppose now that the object is displaced from its equilibrium state, as illustrated in Figure 4.1(b). Let the depth to which the body is immersed in the liquid at time t be denoted by $Y + y(t)$. Thus, $y(t)$ represents the time-varying perturbation of the object from its equilibrium state. For definiteness, we assume $y(t)$ to be positive in the downward direction. In the perturbed state, the net force acting upon the object is typically nonzero and Newton's second law, $ma = F$, tells us

$$\rho AL \frac{d^2}{dt^2}(Y + y(t)) = \rho ALg - \rho_l A(Y + y(t))g. \tag{3}$$

In equation (3), the right-hand side is the net downward force. In particular, the net downward force is the difference between the weight of the object and the instantaneous upward buoyant force due to the displaced liquid.

Noting that Y is a constant and using equality (2a), equation (3) simplifies to

$$\rho ALy''(t) = -\rho_l Agy(t),$$

or

$$y''(t) + \omega^2 y(t) = 0, \qquad \omega^2 \equiv \frac{\rho_l g}{\rho L}. \tag{4}$$

The perturbation depth, $y(t)$, is thus described by a second order linear homogeneous differential equation.

From a physical point of view, it's clear that we need more than just the differential equation (4) to uniquely characterize the motion of the bobbing object. Specifying the object's depth and velocity at some particular instant of time by specifying $y(t_0) = y_0$ and $y'(t_0) = y'_0$ would seem (on physical grounds) to uniquely characterize the motion. The discussion of existence and uniqueness issues in the next section shows that this physical intuition is, in fact, correct.

Is Differential Equation (4) Consistent with Physical Intuition?

Consider differential equation (4). Assume for the present discussion that the initial value problem

$$y''(t) + \omega^2 y(t) = 0, \qquad y(0) = y_0, \qquad y'(0) = y'_0, \tag{5}$$

has a unique solution on any time interval of interest. For simplicity, we've chosen the initial time to be $t_0 = 0$. Does the solution of initial value problem (5) describe a bobbing or oscillating motion that is consistent with our experience? Answering this question involves many of the basic issues addressed in this chapter.

To provide some insight and a preview of what's to come, note that the functions $y(t) = \sin \omega t$ and $y(t) = \cos \omega t$ are each solutions of differential equation (5). Section 4.6 shows how to obtain these solutions. For now, the assertion can be verified by direct substitution. In fact, for any choice of constants C_1 and C_2, the function

$$y(t) = C_1 \sin \omega t + C_2 \cos \omega t \tag{6}$$

is a solution of $y'' + \omega^2 y = 0$.

You will see later that $y(t) = C_1 \sin \omega t + C_2 \cos \omega t$ is in fact the general solution of the differential equation; that is, any solution of the differential equation can be constructed by making an appropriate choice of the constants C_1 and C_2. For initial value problem (5), imposing the initial conditions upon the general solution (6) leads to the set of equations

$$\begin{aligned} y(0) &= C_1 \sin(0) + C_2 \cos(0) &&= y_0, \\ y'(0) &= C_1 \omega \cos(0) - C_2 \omega \sin(0) &&= y_0'. \end{aligned}$$

Solving this system of equations, we find $C_1 = y_0'/\omega$ and $C_2 = y_0$. The unique solution of initial value problem (5) is therefore

$$y(t) = \frac{y_0'}{\omega} \sin \omega t + y_0 \cos \omega t. \tag{7}$$

If either $y_0 = 0$ or $y_0' = 0$, it's obvious that the solution represents the type of sinusoidal oscillating behavior that is consistent with our experience. In general, as you will see later in this chapter, the solution (7) can always be written as a sinusoid. Figure 4.2, for example, shows the behavior of (7) for the special case $y_0 = y_0' = 1, \omega = 2$. Thus, the mathematical model (5) and its solution (7) do, in fact, predict an oscillatory behavior that is consistent with physical intuition about the motion of a bobbing body.

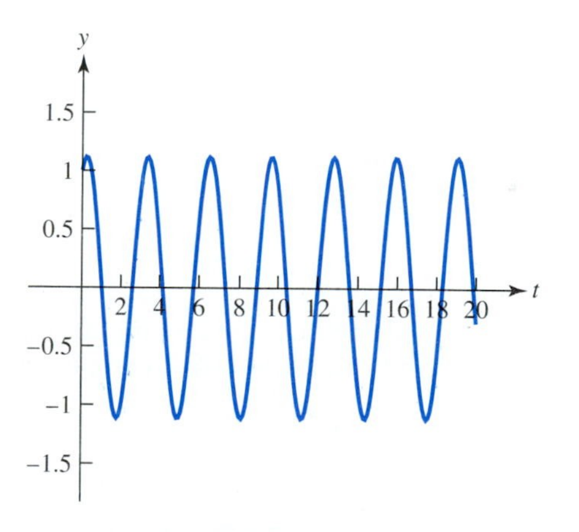

FIGURE 4.2

The graph of the solution of $y'' + 4y = 0, y(0) = 1, y'(0) = 1$. The solution is given by equation (7), using values $\omega = 2, y_0 = 1, y_0' = 1$.

Damping

The solution of the special case of equation (5) graphed in Figure 4.2 is an idealized model of reality. The bobbing motion of real objects does not continue unabated for all time. The presence of drag forces, such as those considered earlier in Section 3.6, causes oscillations to subside as time increases. In later sections of this chapter we will consider some refinements of the model (5), incorporating drag effects as well as applied forces.

EXERCISES

Exercises 1–6:

For these exercises, which are based on the discussion of buoyancy, assume that the mass of water is 1 g/cm^3 and that 1 ft^3 of water weighs 62.4 lb. Assume that $g = 32$ ft/sec^2 or 9.8 m/s^2. Assume that damping is neglected.

1. A cylindrical block of wood has a circular cross-sectional area. The diameter of the base is 1 ft and the height is 2 ft. The wood is hard oak, which weighs 50 lb/ft^3. The block is initially floating at rest in water. At some time, say $t = 0$, it is raised 3 in. and released from rest.

(a) How much of the block's 2-ft height was submerged while the block was initially floating in equilibrium?

(b) Using equation (7), determine $y(t)$, the subsequent displacement from equilibrium (measured positive in the downward direction).

(c) What is the maximum depth to which the block sinks in its subsequent motion? In other words, how much of the block's 2-ft height ultimately gets "wet"?

2. Consider again the cylindrical block of wood described in Exercise 1. Suppose now that the block is perturbed from its rest state by giving it an initial downward velocity, $y'(0) = y'_0$. It is observed that the block, in its subsequent bobbing motion, sinks into the water to the point where it just becomes totally submerged. What was the initial downward velocity y'_0?

3. Since $\sin(\omega t + 2\pi) = \sin \omega t$ and $\cos(\omega t + 2\pi) = \cos \omega t$, the amount of time T it takes a bobbing object to go through one cycle of its motion is determined by the relation $\omega T = 2\pi$ or $T = 2\pi/\omega$. This time T is called the *period* of the motion (see Section 4.7). As the period decreases, the bobbing motion of the floating object becomes more rapid.

(a) Two identically shaped cylindrical drums, made of different material, are floating at rest as shown in Figure (a).

(b) Two cylindrical drums, made of identical material, are floating at rest as shown in Figure (b).

For each case, when the drums are put into motion, is it possible to identify the drum that will bob up and down more rapidly? Explain.

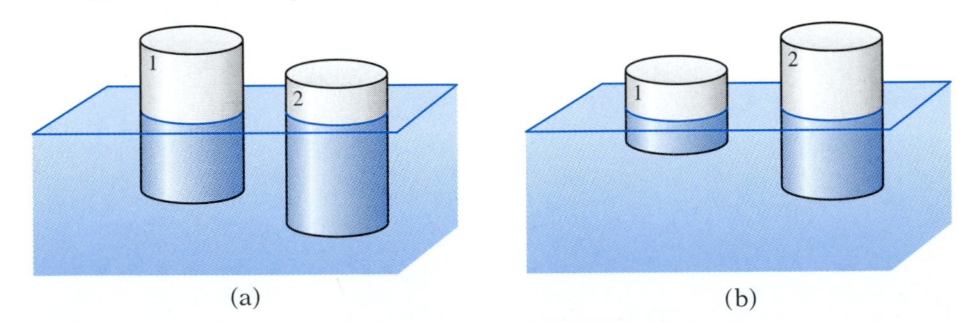

Figure for Exercise 3

4. A buoy having the shape of a right circular cylinder 3 ft in diameter and 5 ft in height is initially floating upright in water. When it was put into motion at time $t = 0$, the following 10-sec record of its displacement from equilibrium, measured in inches positive in the downward direction, was obtained.

(a) Determine the initial displacement y_0 and the period T of the motion (see Exercise 3).

(b) Determine the constant ω and the initial velocity y'_0 of the buoy.

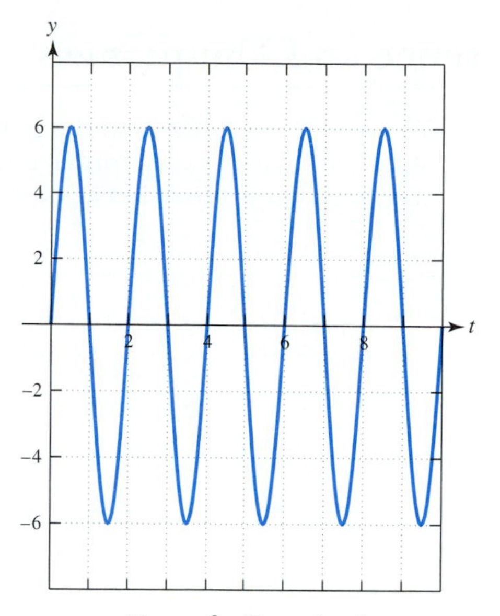

Figure for Exercise 4

(c) The weight of an object having mass density ρ and volume V is $\rho V g$. Use equation (4) and the given information to determine the weight of the buoy.

5. As shown in the figure, an object is fully submerged and $y(t)$ denotes the depth of the object below the water's surface.

(a) Apply Newton's second law of motion to derive a differential equation for the vertical motion of the object beneath the water's surface. Consider only the object's weight and the buoyant force. The resulting differential equation should be a simple one, solvable by antidifferentiation. Is it necessary to restrict your consideration to cylindrical objects?

(b) If the initial depth and velocity of the object are $y(0) = y_0$ and $y'(0) = y'_0$, respectively, obtain the general solution of the initial value problem.

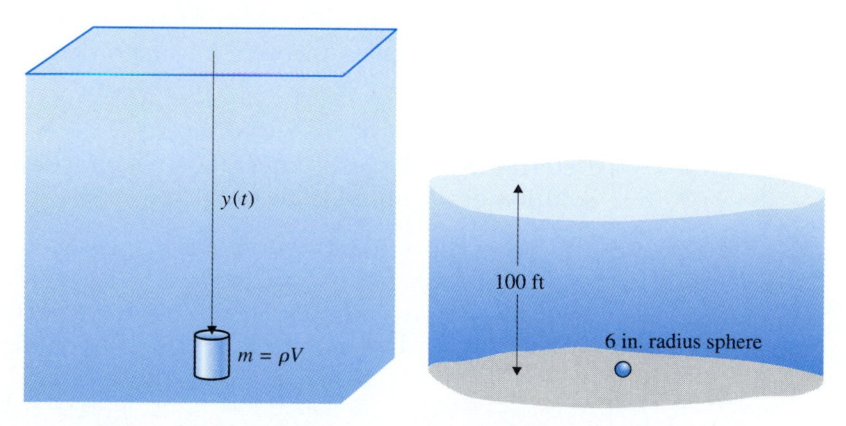

Figure for Exercise 5

Figure for Exercise 6

6. A sphere with 6-in. radius initially rests on a lake bottom 100 ft below the surface of the water as shown in the figure. The sphere is made of soft pine, which weighs 30 lb/ft^3. If the sphere is released from rest, how long will it take before the top of the sphere reaches the surface of the water? Consider only the sphere's weight and the buoyant force.

4.1 Existence and Uniqueness

We now begin to develop the mathematical underpinnings needed to understand and solve second order linear initial value problems. We begin with a theorem that is proved in advanced texts.

Theorem 4.1

Let $p(t)$, $q(t)$, and $g(t)$ be continuous functions on the interval (a, b) and let t_0 be in (a, b). Then the initial value problem

$$y'' + p(t)y' + q(t)y = g(t), \qquad y(t_0) = y_0, \qquad y'(t_0) = y'_0,$$

has a unique solution defined on the entire interval (a, b).

Compare this theorem with Theorem 2.1, which states an analogous existence-uniqueness result for first order linear initial value problems. Both theorems assume that the coefficient functions and the nonhomogeneous term on the right-hand side are continuous on the interval of interest. Both theorems reach the same three conclusions:

1. The solution exists.
2. The solution is unique.
3. The solution exists on the entire interval (a, b).

Theorem 4.1 defines the framework within which we will work. It assures us that, given an initial value problem of the type described, there is one and only one solution. Our job is to find it. The similarity of Theorems 2.1 and 4.1 is no accident. You will see in Chapter 6 that these two theorems, as well as an analogous theorem stated in Chapter 5 for higher order linear initial value problems, can all be viewed as special cases of a single, all-encompassing existence-uniqueness theorem for first order linear systems.

EXAMPLE 1

Determine the largest t-interval on which we can guarantee the existence of a solution of the initial value problem

$$ty'' + (\cos t)y' + t^2y = t, \qquad y(-1) = -1, \qquad y'(-1) = 2.$$

Solution: Before we apply Theorem 4.1, we need to write the differential equation in standard form,

$$y'' + \frac{\cos t}{t}y' + ty = 1.$$

With the equation in this form, we can identify the coefficient functions in the hypotheses of Theorem 4.1. One of the coefficient functions is not continuous at $t = 0$, but there are no other points of discontinuity on the t-axis. Since the initial conditions are posed at the point $t_0 = -1$, it follows from Theorem 4.1 that the given initial value problem is guaranteed to have a unique solution on the interval $-\infty < t < 0$.

Figure 4.3 shows a numerical solution for this initial value problem on the interval $[-1, -0.002]$. As you can see, it appears the solution is not defined at $t = 0$. ▲

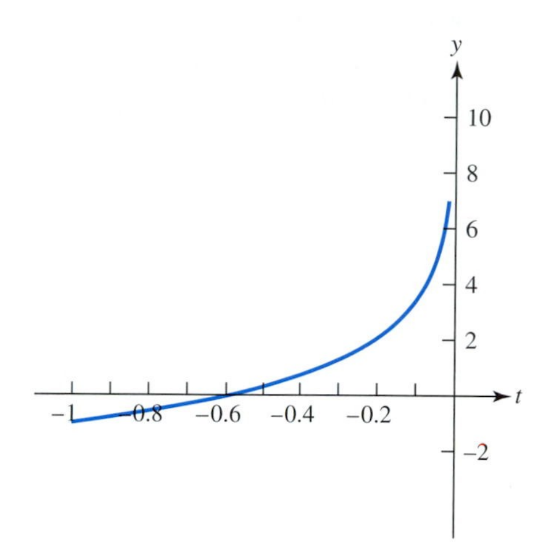

FIGURE 4.3

The graph of a numerical solution of the initial value problem in Example 1. The solution appears to have a vertical asymptote at $t = 0$.

EXERCISES

Exercises 1–4:

For each initial value problem, determine the largest t-interval on which Theorem 4.1 guarantees the existence of a unique solution.

1. $y'' + 3t^2y' + 2y = \sin t, \quad y(1) = 1, \quad y'(1) = -1$

2. $y'' + y' + 3ty = \tan t, \quad y(\pi) = 1, \quad y'(\pi) = -1$

3. $e^t y'' + \dfrac{1}{t^2 - 1} y = \dfrac{4}{t}, \quad y(-2) = 1, \quad y'(-2) = 2$

4. $ty'' + \dfrac{\sin 2t}{t^2 - 9} y' + 2y = 0, \quad y(1) = 0, \quad y'(1) = 1$

Exercises 5–7:

Give an example of an initial value problem of the form

$$y'' + p(t)y' + q(t)y = g(t), \qquad y(t_0) = y_0, \qquad y'(t_0) = y_0'$$

for which the given t-interval is the largest on which Theorem 4.1 guarantees a unique solution.

5. $-\infty < t < \infty$ **6.** $3 < t < \infty$ **7.** $-1 < t < 5$

8. Consider the initial value problem $t^2y'' - ty' + y = 0, y(1) = 1, y'(1) = 1$.

(a) What is the largest interval on which Theorem 4.1 guarantees the existence of a unique solution?

(b) Show by direct substitution that the function $y(t) = t$ is the unique solution of this initial value problem. What is the interval on which this solution actually exists?

(c) Does this example contradict the assertion of Theorem 4.1? Explain.

Exercises 9–10:

Let $y(t)$ denote the solution of the given initial value problem. Is it possible for the corresponding limit to hold? Explain your answer.

9. $y'' + \dfrac{1}{t^2 - 16}y = 0, \quad y(0) = 1, \quad y'(0) = 1 \qquad \lim_{t \to 3^-} y(t) = +\infty$

10. $y'' + 2y' + \dfrac{1}{t - 3}y = 0, \quad y(1) = 1, \quad y'(1) = 2 \qquad \lim_{t \to 0^+} y(t) = +\infty$

Concavity of the Solution Curve In the discussion of direction fields in Section 1.2, you saw how the differential equation indicates the slope of the solution curve at a point in the ty-plane. In particular, given the initial value problem $y' = f(t, y), y(t_0) = y_0$, the slope of the solution curve at initial condition point (t_0, y_0) is $y'(t_0) = f(t_0, y_0)$. In like manner, a second order equation provides direct information about the concavity of the solution curve. Given the initial value problem $y'' = f(t, y, y'), y(t_0) = y_0, y'(t_0) = y'_0$, it follows that the concavity of the solution curve at the initial condition point (t_0, y_0) is $y''(t_0) = f(t_0, y_0, y'_0)$. (What is the slope of the solution curve at that point?)

11. Consider the four graphs shown. Each graph displays a portion of the solution of one of the four initial value problems given. Match each graph with the appropriate initial value problem.

(a) $y'' + y = 2 - \sin t, \quad y(0) = 1, \quad y'(0) = -1$

(b) $y'' + y = -2t, \quad y(0) = 1, \quad y'(0) = -1$

(c) $y'' - y = t^2, \quad y(0) = 1, \quad y'(0) = 1$

(d) $y'' - y = -2\cos t, \quad y(0) = 1, \quad y'(0) = 1$

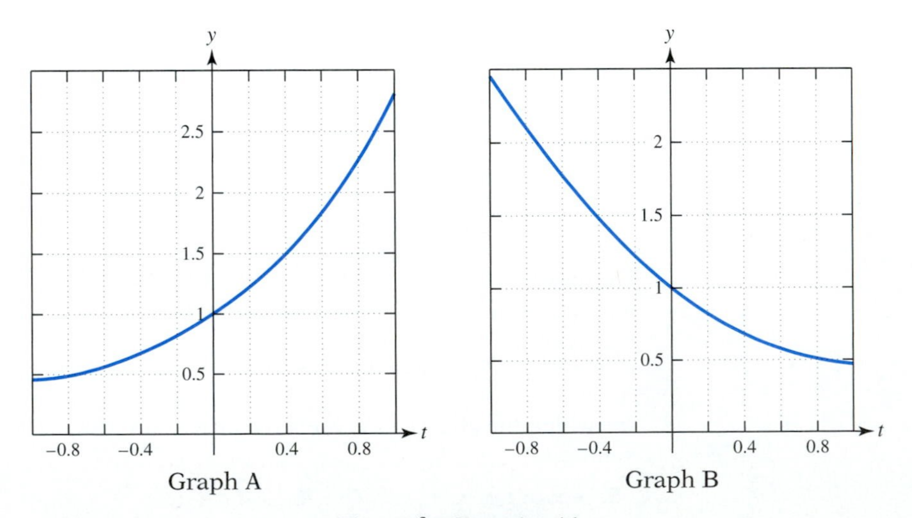

Figure for Exercise 11

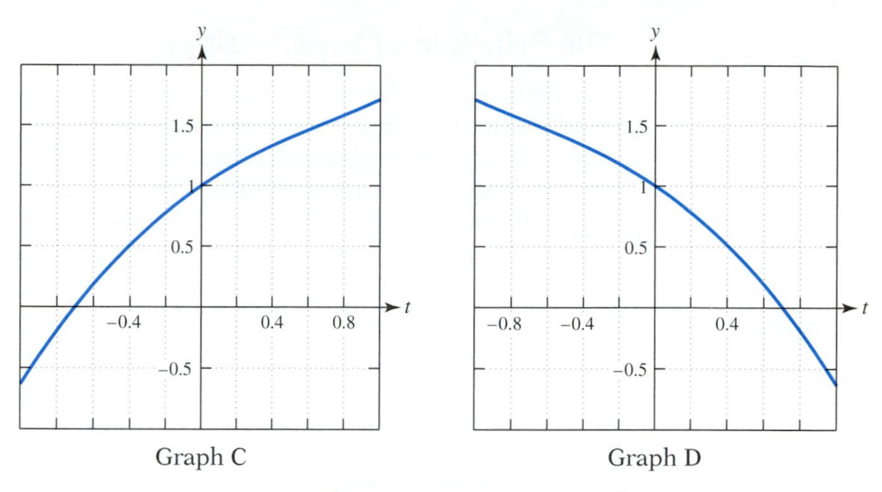

Figure for Exercise 11 (continued)

4.2 The General Solution of Homogeneous Equations

Sections 4.2–4.7 focus on the second order linear homogeneous differential equation

$$y'' + p(t)y' + q(t)y = 0, \qquad a < t < b, \tag{1}$$

where $p(t)$ and $q(t)$ are assumed to be continuous functions defined on (a, b). Sections 4.8–4.11 turn to the general case of the nonhomogeneous equation,

$$y'' + p(t)y' + q(t)y = g(t), \qquad a < t < b.$$

We concentrate first on equation (1) because understanding the solution structure of this equation is the first step in developing methods for solving linear differential equations, whether homogeneous or nonhomogeneous.

Linear Combinations

The general solution of equation (1) can best be expressed as a "linear combination" of functions. In particular, let $f_1(t)$ and $f_2(t)$ be any two functions having a common domain, and let c_1 and c_2 be any two constants. We refer to a function of the form

$$c_1f_1(t) + c_2f_2(t)$$

as a **linear combination of functions** f_1 and f_2. For example, $3\sin t + 8\cos t$ is a linear combination of the functions $\sin t$ and $\cos t$.

In this chapter we are primarily interested in linear combinations of two functions. However, the terminology applies to larger sets of functions as well. If the functions $f_1(t), f_2(t), \ldots, f_n(t)$ have a common domain and if $c_1, c_2, \ldots, c_n$ are constants, we shall refer to the function

$$c_1f_1(t) + c_2f_2(t) + \cdots + c_nf_n(t)$$

as a linear combination of $f_1, f_2, \ldots, f_n$.

The Principle of Superposition

The first result we establish for the homogeneous equation (1) involves linear combinations. This result is frequently referred to as a **superposition principle**. (In engineering and science, the act of forming a linear combination is often called *superposition*.) Theorem 4.2 shows superposing any two solutions of equation (1) (that is, forming a linear combination of the two solutions) results in a function that is also a solution.

Theorem 4.2

Let $y_1(t)$ and $y_2(t)$ be any two solutions of

$$y'' + p(t)y' + q(t)y = 0$$

defined on the interval $a < t < b$, where $p(t)$ and $q(t)$ are continuous on (a, b). Then, for any constants c_1 and c_2, the linear combination

$$y(t) = c_1y_1(t) + c_2y_2(t)$$

is also a solution of this homogeneous differential equation on the interval (a, b).

PROOF: The hypotheses of Theorem 4.2 state that $y_1(t)$ and $y_2(t)$ are each solutions of the homogeneous equation. Therefore, we know

$$y_1'' + p(t)y_1' + q(t)y_1 = 0 \quad \text{and} \quad y_2'' + p(t)y_2' + q(t)y_2 = 0.$$

To prove Theorem 4.2, we need to show that $y'' + p(t)y' + q(t)y = 0$ whenever $y(t)$ is a linear combination of the form $y(t) = c_1y_1(t) + c_2y_2(t)$.

Substituting $y(t) = c_1y_1(t) + c_2y_2(t)$ into the differential equation, we obtain

$$y'' + p(t)y' + q(t)y = (c_1y_1 + c_2y_2)'' + p(t)(c_1y_1 + c_2y_2)' + q(t)(c_1y_1 + c_2y_2). \quad (2)$$

Using basic properties from differential calculus and some algebra, we can rewrite the right-hand side of (2) as

$$c_1[y_1'' + p(t)y_1' + q(t)y_1] + c_2[y_2'' + p(t)y_2' + q(t)y_2] = c_1[0] + c_2[0] = 0.$$

This calculation shows that the linear combination $y(t) = c_1y_1(t) + c_2y_2(t)$ is a solution of the homogeneous linear differential equation on (a, b). ∎

It is important to understand that the superposition principle of Theorem 4.2 is valid for solutions of a *homogeneous* linear equation. In general, a linear combination of solutions of a second order linear *nonhomogeneous* equation is not a solution of the nonhomogeneous equation. Similarly, a linear combination of solutions of a *nonlinear* differential equation is normally not a solution of the nonlinear differential equation.

Fundamental Sets

Theorem 4.2 shows that we can form a linear combination of two solutions of equation (1) and create a new solution. We now turn this idea around and ask: Is it possible to find two solutions, $y_1(t)$ and $y_2(t)$, such that every solution of

$$y'' + p(t)y' + q(t)y = 0, \qquad a < t < b, \quad (3)$$

can be expressed as a linear combination of $y_1(t)$ and $y_2(t)$? In other words, if we are given any solution, $y(t)$, of the homogeneous equation (3), can we determine constants c_1 and c_2 such that

$$y(t) = c_1 y_1(t) + c_2 y_2(t), \qquad a < t < b?$$

If there are two such solutions $y_1(t)$ and $y_2(t)$, we say the set $\{y_1(t), y_2(t)\}$ is a **fundamental set of solutions** for equation (3). The term "fundamental set" is an appropriate one since every solution of equation (3) can be constructed using the basic building blocks $y_1(t)$ and $y_2(t)$.

EXAMPLE 1

Consider the linear homogeneous differential equation

$$y'' + 4y = 0. \tag{4}$$

(a) Show, by direct substitution, that $y_1(t) = \cos 2t$ and $y_2(t) = \sin 2t$ are solutions of differential equation (4).

(b) Show, by direct substitution, that $y(t) = 3\cos(2t + (\pi/4))$ is a solution of (4).

(c) It can be shown (see Example 4) that $y_1(t)$ and $y_2(t)$ form a fundamental set of solutions for equation (4). Find constants c_1 and c_2 such that $y(t) = c_1 \cos 2t + c_2 \sin 2t$ where $y(t)$ is given in (b).

Solution:

(a) Inserting $y_1(t) = \cos 2t$ into equation (4), we obtain

$$y_1'' + 4y_1 = (\cos 2t)'' + 4\cos 2t = -4\cos 2t + 4\cos 2t = 0.$$

This calculation verifies that $y_1(t) = \cos 2t$ is a solution of equation (4). A similar calculation shows that $y_2(t) = \sin 2t$ is also a solution of equation (4).

(b) We leave this part as an exercise.

(c) We want to find constants c_1 and c_2 such that

$$3\cos\left(2t + \frac{\pi}{4}\right) = c_1 \cos 2t + c_2 \sin 2t.$$

Rewriting the left-hand side of this equation using the trigonometric identity

$$\cos(\theta_1 + \theta_2) = \cos\theta_1 \cos\theta_2 - \sin\theta_1 \sin\theta_2,$$

yields

$$\begin{aligned} 3\cos\left(2t + \frac{\pi}{4}\right) &= 3\cos 2t \cos\frac{\pi}{4} - 3\sin 2t \sin\frac{\pi}{4} \\ &= \left[3\cos\frac{\pi}{4}\right]\cos 2t + \left[-3\sin\frac{\pi}{4}\right]\sin 2t. \end{aligned}$$

The constants c_1 and c_2 are therefore

$$c_1 = 3\cos\frac{\pi}{4} = \frac{3\sqrt{2}}{2} \quad \text{and} \quad c_2 = -3\sin\frac{\pi}{4} = -\frac{3\sqrt{2}}{2}. \ \blacktriangle$$

Fundamental Sets and the General Solution

Suppose $y_1(t)$ and $y_2(t)$ form a fundamental set of solutions of the linear homogeneous equation

$$y'' + p(t)y' + q(t)y = 0, \qquad a < t < b, \tag{5}$$

where $p(t)$ and $q(t)$ are continuous on (a, b). If $y(t)$ is any solution of equation (5), then we know we can find constants c_1 and c_2 such that

$$y(t) = c_1y_1(t) + c_2y_2(t), \qquad a < t < b. \tag{6}$$

Expression (6) is called the *general solution* of equation (5).

In Chapter 2 we discussed the general solution for first order linear equations. Similarly, it is important to understand and be able to obtain the general solution for second order equations because we can use it to solve the initial value problem

$$y'' + p(t)y' + q(t)y = 0, \qquad y(t_0) = y_0, \qquad y'(t_0) = y_0'. \tag{7}$$

In particular, if we know that every solution of the differential equation has form (6), then finding the unique solution that satisfies the initial conditions reduces to the problem of choosing constants c_1 and c_2 such that

$$\begin{aligned} y(t_0) &= c_1y_1(t_0) + c_2y_2(t_0) = y_0, \\ y'(t_0) &= c_1y_1'(t_0) + c_2y_2'(t_0) = y_0'. \end{aligned} \tag{8}$$

The next example illustrates how the general solution can be used to solve an initial value problem. You will see shortly that system (8) always has a unique solution whenever $\{y_1, y_2\}$ is a fundamental set.

EXAMPLE 2

Consider the initial value problem

$$y'' - \frac{1}{t}y' - \frac{3}{t^2}y = 0, \qquad y(1) = 4, \qquad y'(1) = 8, \qquad 0 < t < \infty.$$

Since the coefficient functions $p(t) = -t^{-1}$ and $q(t) = -3t^{-2}$ are continuous on $(0, \infty)$, Theorem 4.1 guarantees a unique solution of the initial value problem on this interval. It can be shown, using the result we prove in Theorem 4.3, that $y_1(t) = t^3$ and $y_2(t) = t^{-1}$ form a fundamental set of solutions for this differential equation. Use this fact to solve the initial value problem.

Solution: If $y_1(t)$ and $y_2(t)$ form a fundamental set, then we know the solution $y(t)$ has the form $y(t) = c_1y_1(t) + c_2y_2(t)$ for some constants c_1 and c_2. As in equation (8), we need to determine c_1 and c_2 such that

$$\begin{aligned} y(1) &= c_1y_1(1) + c_2y_2(1) = 4, \\ y'(1) &= c_1y_1'(1) + c_2y_2'(1) = 8. \end{aligned}$$

Since $y_1(t) = t^3$ and $y_2(t) = t^{-1}$, this system reduces to

$$\begin{aligned} c_1 + c_2 &= 4, \\ 3c_1 - c_2 &= 8. \end{aligned}$$

The unique solution of this system is $c_1 = 3$ and $c_2 = 1$. The solution of the initial value problem is $y(t) = 3y_1(t) + y_2(t)$, or

$$y(t) = 3t^3 + \frac{1}{t}, \qquad 0 < t < \infty. \blacktriangle$$

To summarize, we have introduced the related ideas of a fundamental set of solutions and the general solution. These ideas in turn lead to the following questions:

1. How can we identify a fundamental set of solutions? In other words, if we have a particular homogeneous differential equation and two solutions of that differential equation, is there a convenient test that will indicate whether or not these two solutions form a fundamental set of solutions?
2. Is it always possible, in principle, to find a fundamental set of solutions? In other words, does a fundamental set of solutions always exist?
3. For a given homogeneous linear equation, how many fundamental sets of solutions are there? If there is more than just one, how are they related?
4. Assuming a fundamental set of solutions always exists, how do we go about finding it? A significant portion of this chapter is devoted to showing how to find fundamental sets of solutions for the important case of constant coefficient homogeneous equations; that is, equations where $p(t)$ and $q(t)$ in (5) are constants.

We answer question 1 in this section. The other three questions will be answered in Section 4.3.

The Wronskian

A function known as the "Wronskian" is the key to recognizing fundamental sets.[2] Let $f(t)$ and $g(t)$ be functions that are continuous and have a continuous derivative on some interval of interest. The determinant

$$W(t) = \begin{vmatrix} f(t) & g(t) \\ f'(t) & g'(t) \end{vmatrix} = f(t)g'(t) - g(t)f'(t)$$

is called the **Wronskian determinant** or simply the **Wronskian** of f and g. We shall be interested in computing the Wronskian of two solutions of $y'' + p(t)y' + q(t)y = 0$. As Sections 4.2 and 4.3 will show, a nonzero Wronskian implies that these two solutions form a fundamental set of solutions.

EXAMPLE 3

Calculate the Wronskian, $W(t)$, for the solutions in Example 2,

$$y_1(t) = t^3 \quad \text{and} \quad y_2(t) = t^{-1}, \qquad 0 < t < \infty.$$

(continued)

[2]Hoene Wronski (1778–1853) was born Josef Hoene but changed his name just after he married. The determinants we now know as Wronskians were given their name by Muir in 1882.

(continued)

Solution: The Wronskian of y_1 and y_2 is given by

$$W(t) = \begin{vmatrix} t^3 & t^{-1} \\ 3t^2 & -t^{-2} \end{vmatrix} = -t - 3t = -4t.$$

Note that $W(t)$ is nonzero on $(0, \infty)$. ▲

Using the Wronskian to Identify Fundamental Sets

The remainder of this section is devoted to establishing the following simple test for a fundamental set of solutions:

> Suppose $y_1(t)$ and $y_2(t)$ are two solutions of $y'' + p(t)y' + q(t)y = 0$, $a < t < b$. Choose a convenient test point t_0 in (a, b). If the Wronskian of y_1 and y_2 is nonzero at t_0, then $\{y_1, y_2\}$ is a fundamental set of solutions.

This result provides a simple test for determining whether $\{y_1, y_2\}$ is a fundamental set of solutions. We simply calculate the Wronskian, $W(t)$, and evaluate $W(t)$ at some convenient point, $t = t_0$. If $W(t_0)$ is nonzero, then $\{y_1, y_2\}$ is a fundamental set.

In order to establish this test, we begin by proving Theorem 4.3, which says: If we can find *some* point t_0 in (a, b) such that $W(t_0)$ is nonzero, then $\{y_1, y_2\}$ is a fundamental set of solutions for equation (5).

We follow Theorem 4.3 with Theorem 4.4 (Abel's theorem) which states: If $W(t)$ is nonzero at *some* given test point t_0 in (a, b), then $W(t)$ is nonzero at *every* point in (a, b). Therefore, Abel's theorem tells us that we can choose our test point t_0 on the basis of convenience—any test point t_0 will do.

Theorem 4.3

Let $y_1(t)$ and $y_2(t)$ be solutions of the homogeneous linear differential equation

$$y'' + p(t)y' + q(t)y = 0, \qquad a < t < b,$$

where $p(t)$ and $q(t)$ are continuous on (a, b). Let $W(t)$ denote the Wronskian of y_1 and y_2. If there is a point t_0 in (a, b) such that $W(t_0) \neq 0$, then $\{y_1, y_2\}$ is a fundamental set of solutions.

PROOF: Let $\phi(t)$ be any solution of $y'' + p(t)y' + q(t)y = 0$, $a < t < b$. For y_1 and y_2 to form a fundamental set of solutions, we need to show we can find constants c_1 and c_2 such that

$$\phi(t) = c_1 y_1(t) + c_2 y_2(t), \qquad a < t < b. \tag{9a}$$

Because equality (9a) has to hold on the entire interval (a, b) and because all the functions involved are continuously differentiable, satisfying (9a) implies that the following equation, (9b), holds as well,

$$\phi'(t) = c_1 y_1'(t) + c_2 y_2'(t), \qquad a < t < b. \tag{9b}$$

Therefore, if there are constants c_1 and c_2 such that $\phi(t) = c_1y_1(t) + c_2y_2(t)$, $a < t < b$, these constants must satisfy

$$\begin{aligned} c_1y_1(t_0) + c_2y_2(t_0) &= \phi(t_0), \\ c_1y_1'(t_0) + c_2y_2'(t_0) &= \phi'(t_0). \end{aligned} \tag{10a}$$

or in matrix form,

$$\begin{bmatrix} y_1(t_0) & y_2(t_0) \\ y_1'(t_0) & y_2'(t_0) \end{bmatrix} \begin{bmatrix} c_1 \\ c_2 \end{bmatrix} = \begin{bmatrix} \phi(t_0) \\ \phi'(t_0) \end{bmatrix}. \tag{10b}$$

The determinant of the (2×2) coefficient matrix in equation (10b) is equal to $W(t_0)$ and, by hypothesis, $W(t_0) \neq 0$. Therefore, system (10a) has a unique solution for the unknowns c_1 and c_2.

So far, all we have shown is that there are unique constants c_1 and c_2 such that equations (9a) and (9b) hold at the single point t_0. However, we need equation (9a) [and also (9b)] to hold for all t in (a, b). Existence-uniqueness Theorem 4.1 enables us to bridge this gap and arrive at the desired conclusion. Let c_1 and c_2 be the unique solution of equation (10b). Define a function $\overline{y}(t)$ by

$$\overline{y}(t) = c_1y_1(t) + c_2y_2(t), \qquad a < t < b. \tag{11}$$

By Theorem 4.2, $\overline{y}(t)$ is a solution of the differential equation $y'' + p(t)y' + q(t)y = 0$ and, by construction, $\overline{y}(t)$ satisfies the same initial conditions as $\phi(t)$; that is,

$$\overline{y}(t_0) = \phi(t_0) \quad \text{and} \quad \overline{y}'(t_0) = \phi'(t_0).$$

However, Theorem 4.1 asserts that the solution of this initial value problem is unique. Therefore,

$$\phi(t) = \overline{y}(t) = c_1y_1(t) + c_2y_2(t), \qquad a < t < b.$$

Thus, $\{y_1, y_2\}$ is a fundamental set of solutions. ▌

Abel's Theorem

According to Theorem 4.3, we can verify that two solutions of

$$y'' + p(t)y' + q(t)y = 0, \qquad a < t < b,$$

form a fundamental set if we can find at least one point in the interval (a, b) where the Wronskian is nonzero. From this statement alone you might reasonably infer that finding such a point could be a "trial-and-error" procedure. This turns out *not* to be the case. As Theorem 4.4 (Abel's[3] theorem) shows, you can select any convenient point t_0 within (a, b) and evaluate the Wronskian at t_0.

---

[3]Niels Abel (1802–1829) was a Norwegian mathematician whose brief life was marked by poverty and ill health. His mathematical contributions include work on integral equations, elliptic functions, and the quintic equation. One year after his death, Abel, along with Jacobi, was awarded the Grand Prix by the Paris Academy for his outstanding work.

Theorem 4.4

Let $y_1(t)$ and $y_2(t)$ be solutions of the homogeneous linear differential equation

$$y'' + p(t)y' + q(t)y = 0, \qquad a < t < b,$$

where $p(t)$ and $q(t)$ are continuous on (a, b). Let $W(t)$ denote the Wronskian of y_1 and y_2. If t_0 is any point in the interval (a, b), then

$$W(t) = W(t_0)e^{-\int_{t_0}^{t} p(s)\,ds}, \qquad a < t < b.$$

PROOF: The Wronskian of y_1 and y_2 is

$$W(t) = y_1(t)y_2'(t) - y_2(t)y_1'(t).$$

Since the functions $y_1(t)$ and $y_2(t)$ are each twice continuously differentiable on $a < t < b$, we can differentiate $W(t)$ and obtain:

$$\begin{aligned} W'(t) &= [y_1'(t)y_2'(t) + y_1(t)y_2''(t)] - [y_2'(t)y_1'(t) + y_2(t)y_1''(t)] \\ &= y_1(t)y_2''(t) - y_2(t)y_1''(t). \end{aligned}$$

We now use the fact that y_1 and y_2 are both solutions of the differential equation; that is, $y_1'' = -py_1' - qy_1$ and $y_2'' = -py_2' - qy_2$. Making these substitutions in the equation for $W'(t)$ leads to

$$\begin{aligned} W'(t) &= y_1(t)\big(-p(t)y_2'(t) - q(t)y_2(t)\big) - y_2(t)\big(-p(t)y_1'(t) - q(t)y_1(t)\big) \\ &= -p(t)W(t). \end{aligned}$$

Therefore, the Wronskian $W(t)$ is a solution of the initial value problem

$$w' + p(t)w = 0, \qquad w(t_0) = W(t_0). \tag{12}$$

The unique solution of this problem was shown in Section 2.2 to be

$$w(t) = W(t_0)e^{-\int_{t_0}^{t} p(s)\,ds}, \qquad a < t < b. \tag{13}$$

Thus, Theorem 4.4 is proved.

Implications of Abel's Theorem

One consequence of Abel's theorem is that if a Wronskian is formed using two solutions of the homogeneous equation $y'' + p(t)y' + q(t)y = 0, a < t < b$, then this Wronskian will either vanish for all t in the interval (a, b) or it will be nonzero for all t in (a, b).

To understand why this is so, suppose on the one hand that $W(t)$ is zero at some point t_0 in (a, b). Because $W(t_0) = 0$, we see by equation (13) that $W(t)$ is zero for every t in (a, b). On the other hand, suppose $W(t)$ is nonzero at some point t_0 in (a, b). Since $W(t_0) \neq 0$ and since the function $\exp(-\int_{t_0}^{t} p(s)\,ds)$ is positive at every point in (a, b), it follows from equation (13) that $W(t)$ is never zero in (a, b).

It is important to appreciate that Abel's theorem and its consequences apply only to Wronskian determinants formed from solutions of $y'' + p(t)y' + q(t)y = 0$. If one is free to select arbitrary differentiable functions, it is easy to construct Wronskians that vanish at some points and are nonzero at others. For example, if $f(t) = t, g(t) = e^t$, then the Wronskian of f and g is given by

$$W(t) = \begin{vmatrix} t & e^t \\ 1 & e^t \end{vmatrix} = (t-1)e^t. \tag{14}$$

This Wronskian vanishes at $t = 1$ and is nonzero elsewhere.

Since the conclusions of Abel's theorem do not hold for the example in (14), the hypotheses of Theorem 4.4 cannot be true either. We conclude, therefore, that the two functions $f(t) = t$ and $g(t) = e^t$ *cannot* be two solutions of a homogeneous differential equation of the form $y'' + p(t)y' + q(t)y = 0, a < t < b$, where $p(t)$ and $q(t)$ are continuous on an interval (a, b) containing the point $t = 1$. We explore this point further in Exercise 20.

EXAMPLE 4

Example 1 showed that $y_1(t) = \cos 2t$ and $y_2(t) = \sin 2t$ are solutions of the homogeneous linear equation $y'' + 4y = 0, -\infty < t < \infty$. Use Theorem 4.3 to show that $\{y_1, y_2\}$ is a fundamental set of solutions.

Solution: According to the test in Theorem 4.3, if we can find some point where the Wronskian is nonzero, then $\{y_1, y_2\}$ is a fundamental set of solutions. The Wronskian of y_1 and y_2 is

$$W(t) = \begin{vmatrix} \cos 2t & \sin 2t \\ -2\sin 2t & 2\cos 2t \end{vmatrix} = 2\cos^2 2t + 2\sin^2 2t = 2.$$

The Wronskian is nonzero at every point in the interval $(-\infty, \infty)$. Therefore, $\{y_1, y_2\}$ is a fundamental set of solutions and the general solution of $y'' + 4y = 0, -\infty < t < \infty$ is

$$y(t) = c_1 \cos 2t + c_2 \sin 2t. \quad \blacktriangle$$

The fact that the Wronskian W(t) is a constant for the differential equation in Example 4 is consistent with Abel's theorem since $p(t)$ is the zero function.

EXERCISES

Exercises 1–15:

In these exercises, the t-interval of interest is $-\infty < t < \infty$ unless indicated otherwise.

(a) Determine whether the given functions are solutions of the differential equation.

(b) If both functions are solutions, calculate the Wronskian. Does this calculation show that the two functions form a fundamental set of solutions?

(c) If the two functions have been shown in part (b) to form a fundamental set, construct the general solution and determine the unique solution satisfying the given initial conditions.

1. $y'' - 4y = 0;\quad y_1(t) = e^{2t},\ y_2(t) = 2e^{-2t};\quad y(0) = 1,\ y'(0) = -2$

2. $y'' - y = 0;\quad y_1(t) = 2e^t,\ y_2(t) = e^{-t+3};\quad y(-1) = 1,\ y'(-1) = 0$

3. $y'' + y = 0;\quad y_1(t) = \sin t \cos t,\ y_2(t) = \sin t;\quad y(\pi/2) = 1,\ y'(\pi/2) = 1$

4. $y'' + y = 0;\quad y_1(t) = \cos t,\ y_2(t) = \sin t;\quad y(\pi/2) = 1,\ y'(\pi/2) = 1$

5. $y'' - 4y' + 4y = 0;\quad y_1(t) = e^{2t},\ y_2(t) = te^{2t};\quad y(0) = 2,\ y'(0) = 0$

6. $2y'' - y' = 0;\quad y_1(t) = 1,\ y_2(t) = e^{t/2};\quad y(2) = 0,\ y'(2) = 2$

7. $4y'' + y = 0;\quad y_1(t) = \sin((t/2) + (\pi/3)),\ y_2(t) = \sin((t/2) - (\pi/3));$ $y(0) = 0,\ y'(0) = 1$

8. $y'' - 3y' + 2y = 0;\quad y_1(t) = 2e^{t},\ y_2(t) = e^{2t};\quad y(-1) = 1,\ y'(-1) = 0$

9. $ty'' + y' = 0,\ 0 < t < \infty;\quad y_1(t) = \ln t,\ y_2(t) = \ln 3t;\quad y(3) = 0,\ y'(3) = 3$

10. $ty'' + y' = 0,\ 0 < t < \infty;\quad y_1(t) = \ln t,\ y_2(t) = \ln 3;\quad y(1) = 0,\ y'(1) = 3$

11. $t^2y'' - ty' - 3y = 0,\ -\infty < t < 0;\quad y_1(t) = t^3,\ y_2(t) = -t^{-1};\quad y(-1) = 0,$ $y'(-1) = -2$

12. $y'' + 2y' + y = 0;\quad y_1(t) = e^{-t},\ y_2(t) = 2e^{1-t};\quad y(0) = 1,\ y'(0) = 0$

13. $y'' = 0;\quad y_1(t) = t + 1,\ y_2(t) = -t + 2;\quad y(1) = 4,\ y'(1) = -1$

14. $y'' + \pi^2 y = 0;\quad y_1(t) = \sin \pi t + \cos \pi t,\ y_2(t) = \sin \pi t - \cos \pi t;$ $y(1/2) = 1,\ y'(1/2) = 0$

15. $4y'' + 4y' + y = 0;\quad y_1(t) = e^{t/2},\ y_2(t) = te^{t/2};\quad y(1) = 1,\ y'(1) = 0$

Exercises 16–18:

The given pair of functions $\{y_1, y_2\}$ forms a fundamental set of solutions of the differential equation.

(a) Show that the given function $\bar{y}(t)$ is also a solution of the differential equation.

(b) Determine coefficients c_1, c_2 such that $\bar{y}(t) = c_1y_1(t) + c_2y_2(t)$.

16. $y'' + 4y = 0;\quad y_1(t) = 2\cos 2t,\ y_2(t) = \sin 2t;\quad \bar{y}(t) = \sin(2t + (\pi/4))$

17. $t^2y'' - ty' + y = 0,\ 0 < t < \infty;\quad y_1(t) = t,\ y_2(t) = t\ln t;\quad \bar{y}(t) = 2t + t\ln 3t$

18. $4y'' - y = 0;\quad y_1(t) = e^{-t/2},\ y_2(t) = -2e^{t/2};\quad \bar{y}(t) = 2\cosh(t/2)$

19. The functions $y_1(t) = e^{3t}$ and $y_2(t) = e^{-3t}$ are known to be solutions of $y'' + \alpha y' + \beta y = 0$, where α and β are constants. Determine α and β. [Hint: Obtain a system of two equations for the two unknown constants.]

20. The functions $y_1(t) = t$ and $y_2(t) = e^{t}$ are known to be solutions of $y'' + p(t)y' + q(t)y = 0$.

(a) Determine the two functions $p(t)$ and $q(t)$. [Hint: Obtain a system of two equations for these two unknown functions.]

(b) On what t-intervals are the functions $p(t)$ and $q(t)$ continuous?

(c) Compute the Wronskian of these two solutions. On what t-intervals is the Wronskian nonzero?

(d) Are the observations in (b) and (c) consistent with Theorem 4.4 (Abel's theorem)? Is the fact that $y_1(t) = t, y_2(t) = e^{t}$ both exist for $-\infty < t < \infty$ consistent with Theorem 4.1?

21. It is known that two solutions of $y'' + ty' + 2y = 0$ has a Wronskian $W(t)$ that satisfies $W(1) = 4$. What is $W(2)$?

22. The pair of functions $\{y_1, y_2\}$ is known to form a fundamental set of solutions of $y'' + \alpha y' + \beta y = 0$, where α and β are constants. One solution is $y_1(t) = e^{2t}$, and the Wronskian formed by these two solutions is $W(t) = e^{-t}$. Determine the constants α and β.

23. The Wronskian of a pair of solutions of $y'' + t^2y = 0$ satisfies $W(1) = -3$. What is $W(4)$?

24. The Wronskian of a pair of solutions of $y'' + p(t)y' + 3y = 0$ is $W(t) = e^{-t^2}$. What is the coefficient function $p(t)$?

4.3 Fundamental Sets and Linear Independence

In Section 4.2 we introduced the important idea of a fundamental set of solutions for the linear homogeneous differential equation

$$y'' + p(t)y' + q(t)y = 0, \qquad a < t < b. \tag{1}$$

As was shown in Section 4.2, if we can find a fundamental set of solutions $\{y_1, y_2\}$, then we can form the general solution for equation (1),

$$y(t) = c_1 y_1(t) + c_2 y_2(t).$$

An aid to identifying fundamental sets of solutions is the test described in Section 4.2. If y_1 and y_2 are solutions of equation (1) and if y_1 and y_2 have a Wronskian determinant that is nonzero in (a, b), then $\{y_1, y_2\}$ is a fundamental set of solutions.

The discussion in Section 4.2 left unanswered several basic questions about fundamental sets. Do they always exist? How many are there? How are different fundamental sets related? In this section we turn our attention to these questions.

Fundamental Sets of Solutions Always Exist

Up to now, our discussion of fundamental sets of solutions has been predicated on the assumption that we can generate solutions, y_1 and y_2, of differential equation (1) and then test these solutions by forming the Wronskian determinant. You may rightly question whether it is always possible, at least in principle, to obtain a fundamental set of solutions. Theorem 4.5 guarantees that a fundamental set of solutions for equation (1) always exists.

Theorem 4.5

Consider the linear homogeneous differential equation

$$y'' + p(t)y' + q(t)y = 0, \qquad a < t < b,$$

where $p(t)$ and $q(t)$ are continuous on (a, b). A fundamental set of solutions, $\{y_1, y_2\}$, exists for this differential equation.

PROOF: Choose a point t_0 in the interval (a, b). Let $y_1(t)$ and $y_2(t)$ denote the unique solutions of the initial value problems

$$y_1'' + p(t)y_1' + q(t)y_1 = 0, \qquad y_1(t_0) = 1, \qquad y_1'(t_0) = 0,$$
$$y_2'' + p(t)y_2' + q(t)y_2 = 0, \qquad y_2(t_0) = 0, \qquad y_2'(t_0) = 1.$$

Theorem 4.1 guarantees that both these initial value problems have unique solutions on the interval (a, b). To see that $\{y_1, y_2\}$ forms a fundamental set we evaluate the Wronskian at $t = t_0$:

$$W(t_0) = \begin{vmatrix} y_1(t_0) & y_2(t_0) \\ y_1'(t_0) & y_2'(t_0) \end{vmatrix} = \begin{vmatrix} 1 & 0 \\ 0 & 1 \end{vmatrix} = 1.$$

Since the Wronskian is nonzero at a point t_0 in (a, b), we know by Theorem 4.3 that $\{y_1, y_2\}$ is a fundamental set of solutions. ∎

In proving this theorem, the initial conditions were chosen so that evaluating $W(t)$ at $t = t_0$ was particularly simple. However, any choice of initial conditions, say

$$y_1(t_0) = \alpha, \qquad y_1'(t_0) = \beta, \qquad y_2(t_0) = \gamma, \qquad y_2'(t_0) = \delta,$$

also leads to a fundamental set if $\alpha\delta - \beta\gamma \neq 0$.

The Wronskian of a Fundamental Set Is Never Zero

Theorem 4.3 says that if the Wronskian of solutions y_1 and y_2 is nonzero somewhere in the t-interval of interest, then $\{y_1, y_2\}$ is a fundamental set. Abel's theorem, Theorem 4.4, says the Wronskian of solutions is either always nonzero or always zero on this interval. So far, however, we have not logically ruled out the possibility that a fundamental set of solutions might have a Wronskian that is zero everywhere in the interval of interest. Theorem 4.6 rules out this possibility.

Theorem 4.6

Let $\{y_1, y_2\}$ be a fundamental set of solutions for

$$y'' + p(t)y' + q(t)y = 0, \qquad a < t < b,$$

where $p(t)$ and $q(t)$ are continuous on (a, b). Let $W(t)$ denote the Wronskian of y_1 and y_2, and let t_0 be any point in the interval (a, b). Then $W(t_0)$ is nonzero.

PROOF: Our proof of Theorem 4.6 depends on the following fact from matrix theory (see the appendix on matrix theory).

> Consider the equation $A\mathbf{x} = \mathbf{b}$, where A is a (2×2) constant matrix, $\mathbf{b}$ is a (2×1) constant vector, and $\mathbf{x}$ is the (2×1) vector of unknowns. The equation $A\mathbf{x} = \mathbf{b}$ has a unique solution for every right-hand side $\mathbf{b}$ if and only if the determinant of A is nonzero.

Now, let t_0 be any point in (a, b), and let y_0 and y_0' be any two constants. By Theorem 4.1, the initial value problem

$$y'' + p(t)y' + q(t)y = 0, \qquad y(t_0) = y_0, \qquad y'(t_0) = y_0' \tag{2}$$

has a unique solution, $y(t)$. Since $\{y_1, y_2\}$ is a fundamental set of solutions, we know there exist constants c_1 and c_2 such that, for all t in (a, b),

$$\begin{aligned} y(t) &= c_1 y_1(t) + c_2 y_2(t), \\ y'(t) &= c_1 y_1'(t) + c_2 y_2'(t). \end{aligned} \tag{3}$$

Since equation (3) holds at every point in (a, b), it holds at the initial point $t = t_0$ as well. Evaluating (3) at $t = t_0$, we find:

$$\begin{bmatrix} y_1(t_0) & y_2(t_0) \\ y_1'(t_0) & y_2'(t_0) \end{bmatrix} \begin{bmatrix} c_1 \\ c_2 \end{bmatrix} = \begin{bmatrix} y_0 \\ y_0' \end{bmatrix}. \qquad (4)$$

Since equation (4) has a solution and since the right-hand side of (4) was arbitrary, it follows that the coefficient matrix must have a nonzero determinant. The determinant of the coefficient matrix is the Wronskian of y_1 and y_2 evaluated at $t = t_0$; therefore $W(t_0)$ is nonzero. ∎

Combining Theorems 4.3 and 4.6, we obtain the following corollary characterizing a fundamental set of solutions.

Corollary

Let y_1 and y_2 be two solutions of the homogeneous linear differential equation

$$y'' + p(t)y' + q(t)y = 0, \qquad a < t < b,$$

where $p(t)$ and $q(t)$ are continuous on (a, b). Let $W(t)$ denote the Wronskian of y_1 and y_2. Then $\{y_1, y_2\}$ is a fundamental set of solutions if and only if $W(t)$ is never zero in (a, b).

Having developed the related ideas of a fundamental set of solutions and the general solution of the homogeneous linear differential equation, we summarize how these ideas are used to solve the initial value problem

$$y'' + p(t)y' + q(t)y = 0, \qquad a < t < b, \qquad y(t_0) = y_0, \qquad y'(t_0) = y_0'.$$

The solution procedure amounts to the following three steps:

1. Find a fundamental set of solutions, $\{y_1, y_2\}$. The fact that $\{y_1, y_2\}$ forms a fundamental set can be established by computing the Wronskian determinant and showing it is nonzero in (a, b).
2. Form the general solution, $y(t) = c_1y_1(t) + c_2y_2(t)$.
3. Impose the initial conditions and solve the resulting system of equations

$$c_1y_1(t_0) + c_2y_2(t_0) = y_0,$$
$$c_1y_1'(t_0) + c_2y_2'(t_0) = y_0'.$$

We know from step 1 that the determinant of the coefficient matrix of the system in step 3 is nonzero—this determinant is the Wronskian, $W(t_0)$. Therefore, the system of equations in step 3 always has a unique solution for the constants c_1 and c_2.

EXAMPLE 1

It can be shown (see Exercise 5) that $y_1(t) = e^{2t}$ and $y_2(t) = e^{-2t}$ are solutions of the differential equation $y'' - 4y = 0$. Show that $\{y_1, y_2\}$ is a fundamental set of solutions, and use this fact to solve the initial value problem

$$y'' - 4y = 0, \qquad y(0) = 2, \qquad y'(0) = -1.$$

Solution: To verify $\{y_1, y_2\}$ is a fundamental set of solutions, we calculate the Wronskian, finding

$$\begin{vmatrix} y_1(t) & y_2(t) \\ y_1'(t) & y_2'(t) \end{vmatrix} = \begin{vmatrix} e^{2t} & e^{-2t} \\ 2e^{2t} & -2e^{-2t} \end{vmatrix} = -2 - 2 = -4.$$

Since the Wronskian is not zero, we know $\{y_1, y_2\}$ is a fundamental set of solutions and the general solution is

$$y(t) = c_1 e^{2t} + c_2 e^{-2t}.$$

Imposing the initial condition at $t_0 = 0$, we obtain

$$\begin{aligned} c_1 + c_2 &= 2, \\ 2c_1 - 2c_2 &= -1. \end{aligned}$$

Solving this system, we find $c_1 = 3/4$ and $c_2 = 5/4$. The solution of the initial value problem is

$$y(t) = \frac{3}{4}e^{2t} + \frac{5}{4}e^{-2t}. \ ▲$$

Linearly Independent Sets of Functions

We now introduce the concepts of "linearly independent" and "linearly dependent" sets of functions. We then show that a fundamental set of solutions for equation (1) is a linearly independent set of functions.

In this section we restrict our consideration to a pair of functions. However, later in the text, we apply the concepts of linear dependence and independence to sets of more than two functions. We therefore introduce general definitions at this point.

Let $f_1(t), f_2(t), \ldots, f_n(t)$ be n functions defined on a common domain $a < t < b$. The set of functions $\{f_1, f_2, \ldots, f_n\}$ is said to be a **linearly dependent set of functions on (*a*, *b*)** if there exist constants $k_1, k_2, \ldots, k_n$, *not all zero*, such that

$$k_1 f_1(t) + k_2 f_2(t) + \cdots + k_n f_n(t) = 0, \qquad a < t < b. \tag{5}$$

Otherwise, the set of functions $\{f_1, f_2, \ldots, f_n\}$ is said to be a **linearly independent set of functions on (*a*, *b*)**.[4]

Equation (5) holds for any set of functions if we set $k_1 = 0$, $k_2 = 0, \ldots, k_n = 0$. The challenge posed by the definition of linear dependence

[4]We represent the common domain as an open interval, but the definitions also apply to closed and half-open intervals as well as to more complicated domains.

consists of being able to find at least one nonzero coefficient such that the linear combination (5) is nevertheless zero over the entire t-interval. If the only way we can satisfy (5) is by setting $k_1 = 0, k_2 = 0, \ldots, k_n = 0$, then the set of functions $\{f_1, f_2, \ldots, f_n\}$ is linearly independent.

It's not difficult to show that a set of functions $\{f_1, f_2, \ldots, f_n\}$ is linearly dependent on (a, b) if and only if one of the functions can be expressed as a linear combination of the other $n - 1$ functions (see Exercise 23). Therefore $\{f_1, f_2\}$ is linearly dependent if and only if one of the functions is a constant multiple of the other.

Loosely speaking, a linearly dependent set of functions is one in which the constituent functions are not all "basically different" on the common domain of interest. By contrast, the constituent members of a linearly independent set are all "basically different."

Fundamental Sets of Solutions Are Linearly Independent Sets of Functions

We now relate the ideas of linear dependence and independence to fundamental sets of solutions for the second order linear differential equation

$$y'' + p(t)y' + q(t)y = 0, \qquad a < t < b.$$

Theorem 4.7 asserts that a fundamental set of solutions is linearly independent.

Theorem 4.7

(a) Let $\{y_1, y_2\}$ be a fundamental set of solutions for the homogeneous linear differential equation

$$y'' + p(t)y' + q(t)y = 0, \qquad a < t < b,$$

where $p(t)$ and $q(t)$ are continuous on (a, b). Then $\{y_1, y_2\}$ is a linearly independent set of functions on (a, b).

(b) If solutions y_1 and y_2 do not form a fundamental set, then they are a linearly dependent set of functions on (a, b).

PROOF:

(a) Form the linear combination

$$k_1 y_1(t) + k_2 y_2(t) = 0, \qquad a < t < b. \tag{6a}$$

We must show that this equation can be valid only if $k_1 = k_2 = 0$. Since y_1 and y_2 are differentiable on (a, b), it follows from differentiating (6a) that

$$k_1 y_1'(t) + k_2 y_2'(t) = 0, \qquad a < t < b. \tag{6b}$$

Select some point t_0 lying in (a, b). Since equations (6a) and (6b) must be valid for all t, $a < t < b$, they must certainly be valid at the point t_0.

We obtain, therefore,

$$\begin{bmatrix} y_1(t_0) & y_2(t_0) \\ y_1'(t_0) & y_2'(t_0) \end{bmatrix} \begin{bmatrix} k_1 \\ k_2 \end{bmatrix} = \begin{bmatrix} 0 \\ 0 \end{bmatrix}. \tag{7}$$

The determinant of the coefficient matrix in equation (7) is the Wronskian, $W(t)$, evaluated at $t = t_0$. Because $\{y_1, y_2\}$ is a fundamental set, we know from Theorem 4.6 that $W(t_0)$ is nonzero. Since the coefficient matrix in equation (7) is invertible, $k_1 = k_2 = 0$ is the only solution of equation (7) and hence of equation (6a). We conclude that $\{y_1, y_2\}$ is a linearly independent set of functions on (a, b).

(b) Let t_0 be any point in (a, b). Since $\{y_1, y_2\}$ is not a fundamental set of solutions,

$$W(t_0) = \begin{vmatrix} y_1(t_0) & y_2(t_0) \\ y_1'(t_0) & y_2'(t_0) \end{vmatrix} = 0$$

(see the Corollary to Theorem 4.6.). Therefore the (2×2) matrix

$$\begin{bmatrix} y_1(t_0) & y_2(t_0) \\ y_1'(t_0) & y_2'(t_0) \end{bmatrix}$$

is not invertible. Hence, there are constants α and β, *not both zero*, such that

$$\begin{bmatrix} y_1(t_0) & y_2(t_0) \\ y_1'(t_0) & y_2'(t_0) \end{bmatrix} \begin{bmatrix} \alpha \\ \beta \end{bmatrix} = \begin{bmatrix} 0 \\ 0 \end{bmatrix}.$$

It follows that the function $\bar{y}(t) = \alpha y_1(t) + \beta y_2(t)$ is a solution of the given differential equation on (a, b) and satisfies the initial conditions $\bar{y}(t_0) = 0, \bar{y}'(t_0) = 0$. Theorem 4.1, however, asserts that there is only one such solution, namely the function that is identically zero on (a, b). Therefore, $\{y_1, y_2\}$ is linearly dependent. ❚

❚ **REMARK:** It is important to realize that conclusion (b) of Theorem 4.7 follows because the two functions being considered are solutions of an initial value problem that has a unique solution. In particular, Theorem 4.1 plays a crucial role in arriving at conclusion (b). If f and g are two arbitrary differentiable functions, then their Wronskian may be zero at some point t_0 in their common domain and yet the functions f and g may be linearly independent. ❚

How Fundamental Sets of Solutions Are Related

In Section 4.2 we saw that we can construct the general solution of $y'' + p(t)y' + q(t)y = 0$ once we know a fundamental set of solutions. Theorem 4.5 guarantees a fundamental set of solutions always exists. For a given differential equation, however, how many fundamental sets of solutions are there? Is there only one fundamental set or are there, in fact, many? Theorem 4.8 shows that there are many fundamental sets of solutions. The theorem also shows how different fundamental sets of solutions are related to each other.

Theorem 4.8

Let $\{y_1, y_2\}$ be a fundamental set of solutions for the homogeneous differential equation

$$y'' + p(t)y' + q(t)y = 0, \qquad a < t < b, \tag{8}$$

where $p(t)$ and $q(t)$ are continuous on (a, b). Let $\bar{y}_1(t)$ and $\bar{y}_2(t)$ be any solutions of differential equation (8). Then

(a) There is a unique (2×2) constant matrix A,

$$A = \begin{bmatrix} a_{11} & a_{12} \\ a_{21} & a_{22} \end{bmatrix} \tag{9}$$

such that

$$[\bar{y}_1(t), \quad \bar{y}_2(t)] = [y_1(t), \quad y_2(t)]A, \qquad a < t < b. \tag{10}$$

(b) Moreover, $\{\bar{y}_1, \bar{y}_2\}$ forms a fundamental set of solutions for differential equation (8) if and only if the determinant of A is nonzero.

REMARK: Part (b) of this theorem shows that, given one fundamental set of solutions, we can construct others by forming linear combinations of the two solutions [as long as we choose the matrix A in (10) to be invertible]. Parts (a) and (b) together show that every fundamental set of solutions can be obtained in this way; there are no other fundamental sets that we may have overlooked.

PROOF:

(a) Since $\{y_1, y_2\}$ is a fundamental set of solutions and $\bar{y}_1$ and $\bar{y}_2$ are solutions, we can express $\bar{y}_1$ and $\bar{y}_2$ as linear combinations of y_1 and y_2,

$$\begin{aligned} \bar{y}_1(t) &= a_{11}y_1(t) + a_{21}y_2(t) \\ \bar{y}_2(t) &= a_{12}y_1(t) + a_{22}y_2(t) \end{aligned}, \qquad a < t < b. \tag{11}$$

Defining A to be the coefficient matrix of this system, it follows that equation (10) is valid.

Because $\{y_1, y_2\}$ is a fundamental set of solutions for equation (8), we know from Theorem 4.7 that $\{y_1, y_2\}$ is a linearly independent set on (a, b). Linear independence implies (see Exercise 25) that the coefficients a_{11} and a_{21} defining $\bar{y}_1(t)$ in (11) are unique. By the same reasoning, the coefficients a_{12} and a_{22} are also unique. Hence the matrix A in equation (10) is uniquely determined.

(b) In the proof of part (a) we established the validity of equation (10),

$$[\bar{y}_1(t), \quad \bar{y}_2(t)] = [y_1(t), \quad y_2(t)]A, \qquad a < t < b.$$

Differentiating equations (11) and rewriting the result in matrix form, we similarly obtain

$$[\bar{y}_1'(t), \quad \bar{y}_2'(t)] = [y_1'(t), \quad y_2'(t)]A, \qquad a < t < b, \tag{12}$$

Combining (10) and (12), we have

$$\begin{bmatrix} \bar{y}_1(t) & \bar{y}_2(t) \\ \bar{y}_1'(t) & \bar{y}_2'(t) \end{bmatrix} = \begin{bmatrix} y_1(t) & y_2(t) \\ y_1'(t) & y_2'(t) \end{bmatrix} A, \qquad a < t < b. \tag{13}$$

We now make use of the fact from matrix theory that the determinant of a product of square matrices equals the product of the determinants (see the appendix on matrix theory). Let $W(t)$ denote the Wronskian determinant of $y_1(t)$ and $y_2(t)$, and let $\overline{W}(t)$ denote the Wronskian determinant of $\bar{y}_1(t)$ and $\bar{y}_2(t)$. Taking determinants on both sides of equation (13) and using the determinant result, we obtain

$$\overline{W}(t) = W(t)\det(A), \qquad a < t < b. \tag{14}$$

Now, $W(t)$ is nonzero in (a, b) since $\{y_1, y_2\}$ is a fundamental set of solutions for equation (8). Therefore, $\overline{W}(t)$ is nonzero in (a, b) if and only if $\det(A)$ is nonzero. Equivalently, $\{\bar{y}_1, \bar{y}_2\}$ is also a fundamental set of solutions for equation (8) if and only if $\det(A)$ is nonzero. ∎

As shown in Theorem 4.8, fundamental sets of solutions are not unique. If we know one particular fundamental set, however, then every other fundamental set can be obtained by taking appropriate linear combinations of the fundamental set we know. For the differential equation $y'' - 4y = 0$ treated in Example 1, we know one fundamental set,

$$\{y_1, y_2\} = \{e^{2t}, e^{-2t}\}. \tag{15}$$

Another frequently used fundamental set of solutions for $y'' - 4y = 0$ is given in terms of hyperbolic functions:

$$\bar{y}_1(t) = \cosh 2t = \frac{1}{2}e^{2t} + \frac{1}{2}e^{-2t} \tag{16a}$$

and

$$\bar{y}_2(t) = \sinh 2t = \frac{1}{2}e^{2t} - \frac{1}{2}e^{-2t}. \tag{16b}$$

To verify that $\{\bar{y}_1, \bar{y}_2\}$ is also a fundamental set, we can check that the matrix A of linear combination coefficients has a nonzero determinant. From equations (16a) and (16b), the matrix A is given by

$$A = \begin{bmatrix} \frac{1}{2} & \frac{1}{2} \\ \frac{1}{2} & -\frac{1}{2} \end{bmatrix}.$$

Since $\det(A) = -1/4 - 1/4 = -1/2$, part (b) of Theorem 4.8 shows that $\{\bar{y}_1, \bar{y}_2\}$ is also a fundamental set of solutions for $y'' - 4y = 0$.

EXERCISES

Exercises 1–4:

Consider the differential equation $y'' + 2ty' + t^2y = 0$ on the interval $-\infty < t < \infty$. Assume that $y_1(t)$ and $y_2(t)$ are two solutions satisfying the given initial conditions.

(a) Do the solutions form a fundamental set?

(b) Do the two solutions form a linearly independent set of functions on $-\infty < t < \infty$?

1. $y_1(1) = 2,\ y_1'(1) = 2;\quad y_2(1) = -1,\ y_2'(1) = -1$

2. $y_1(-2) = 1,\ y_1'(-2) = 2;\quad y_2(-2) = 0,\ y_2'(-2) = 1$

3. $y_1(0) = 0,\ y_1'(0) = 1;\quad y_2(0) = -1,\ y_2'(0) = 0$

4. $y_1(3) = 0,\ y_1'(3) = 0;\quad y_2(3) = 1,\ y_2'(3) = 2$

Exercises 5–10:

(a) Show that $y_1(t)$ and $y_2(t)$ are solutions of the given differential equation.

(b) Determine the initial conditions satisfied by each function at the specified point t_0.

(c) Determine whether the functions form a fundamental set on $-\infty < t < \infty$.

5. $y'' - 4y = 0;\quad y_1(t) = e^{2t},\ y_2(t) = e^{-2t};\quad t_0 = 1$

6. $4y'' - y = 0;\quad y_1(t) = e^{t/2},\ y_2(t) = -2e^{-t/2};\quad t_0 = -2$

7. $y'' + 9y = 0;\quad y_1(t) = \sin 3(t-1),\ y_2(t) = 2\cos 3(t-1);\quad t_0 = 1$

8. $y'' + 4y' + 5y = 0;\quad y_1(t) = e^{-2t}\cos t,\ y_2(t) = e^{-2t}\sin t;\quad t_0 = 0$

9. $y'' + 2y' - 3y = 0;\quad y_1(t) = e^{-3t},\ y_2(t) = e^{-3(t-2)};\quad t_0 = 2$

10. $y'' - 6y' + 9y = 0;\quad y_1(t) = e^{3(t+2)},\ y_2(t) = te^{3(t+2)};\quad t_0 = -2$

Exercises 11–13:

Assume that $y_1(t)$ and $y_2(t)$ form a fundamental set of solutions of $y'' + p(t)y' + q(t)y = 0$ on the t-interval of interest. Determine whether or not the functions $\bar{y}_1(t)$ and $\bar{y}_2(t)$, formed by the given linear combinations, also form a fundamental set of solutions on the same t-interval.

11. $\bar{y}_1(t) = 2y_1(t) - y_2(t),\ \bar{y}_2(t) = y_1(t) + y_2(t)$

12. $\bar{y}_1(t) = 2y_1(t) - 2y_2(t),\ \bar{y}_2(t) = y_1(t) - y_2(t)$

13. $\bar{y}_1(t) = y_2(t),\ \bar{y}_2(t) = 2y_1(t) - y_2(t)$

Exercises 14–20:

Do the given functions form a linearly independent set on the indicated domain?

14. $f_1(t) = 2,\ f_2(t) = t^2,\quad -\infty < t < \infty$

15. $f_1(t) = \ln t,\ f_2(t) = \ln t^2,\quad 0 < t < \infty$

16. $f_1(t) = e^t,\ f_2(t) = e^{-t},\quad -\infty < t < \infty$

17. $f_1(t) = \sin 2t,\ f_2(t) = -\cos 2t,\quad 0 < t < \pi$

18. $f_1(t) = 2,\ f_2(t) = t,\ f_3(t) = -t^2,\quad -\infty < t < \infty$

19. $f_1(t) = 2,\ f_2(t) = \sin^2 t,\ f_3(t) = 2\cos^2 t,\quad -3 < t < 2$

20. $f_1(t) = e^t,\ f_2(t) = 2e^{-t},\ f_3(t) = \sinh t,\quad -\infty < t < \infty$

21. Consider the graphs of the linear functions shown. In each case, determine if the functions form a linearly independent set of functions on the domain shown.

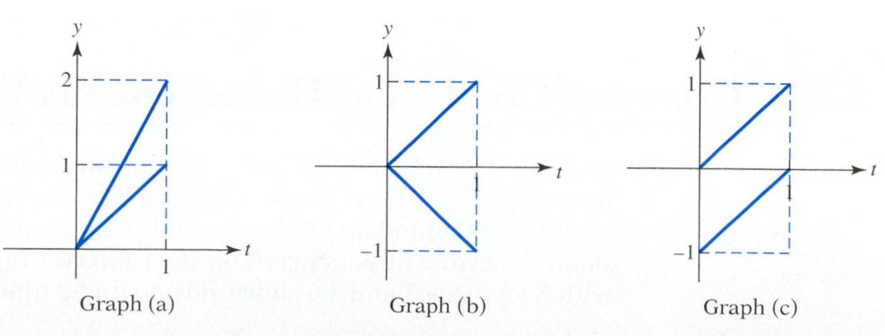

Figure for Exercise 21

Exercises 22–26:

These exercises explore some general properties of linear dependence and independence. Assume that the given functions are defined on a common domain.

22. If a nonzero function, $f_1(t)$, is a constant multiple of another nonzero function, $f_2(t)$, show that the two functions form a linearly dependent set. Conversely, if the two functions form a linearly dependent set, show that one function must be a constant multiple of the other.

23. Consider three functions, $f_1(t), f_2(t)$, and $f_3(t)$. If one of the functions is a linear combination of the other two functions, say $f_2(t) = 3f_1(t) - 2f_3(t)$, show that the set of functions is a linearly dependent set. Conversely, given a set of three linearly dependent functions, show that one of them must be a linear combination of the other two. (This same argument applies to sets of four or more functions.)

24. Consider a set of functions containing the zero function. Can anything be said about whether they form a linearly dependent or linearly independent set? Explain.

25. Suppose, as in the proof of Theorem 4.8, that $\{f_1(t), f_2(t)\}$ is a linearly independent set and that a third function, $f_3(t)$, can be expressed as a linear combination of $f_1(t)$ and $f_2(t)$. That is, $f_3(t) = a_1 f_1(t) + a_2 f_2(t)$. Show that the coefficients, a_1 and a_2, are unique. [Hint: Suppose there are two different representations of $f_3(t)$. Subtract them and deduce a contradiction.]

26. The property of linear dependence or independence depends not only upon the rule defining the functions but also on the domain. To illustrate this fact, show that the pair of functions, $f_1(t) = t, f_2(t) = |t|$, is linearly dependent on the interval $0 < t < \infty$ but is linearly independent on the interval $-\infty < t < \infty$.

Exercises 27–28:

The sets $\{y_1, y_2\}$ and $\{\bar{y}_1, \bar{y}_2\}$ are both fundamental sets of solutions for the given differential equation on the indicated interval. Find a constant (2×2) matrix

$$A = \begin{bmatrix} a_{11} & a_{12} \\ a_{21} & a_{22} \end{bmatrix}$$

such that the matrix equation

$$\begin{bmatrix} \bar{y}_1(t), & \bar{y}_2(t) \end{bmatrix} = \begin{bmatrix} y_1(t), & y_2(t) \end{bmatrix} \begin{bmatrix} a_{11} & a_{12} \\ a_{21} & a_{22} \end{bmatrix}$$

is valid for all t in the given interval.

27. $t^2y'' - 3ty' + 3y = 0,\ 0 < t < \infty;\ \ y_1(t) = t,\ y_2(t) = t^3;\ \ \bar{y}_1(t) = 2t - t^3,\ \bar{y}_2(t) = t^3 + t$

28. $y'' - 4y' + 4y = 0, -\infty < t < \infty;\ \ y_1(t) = e^{2t},\ y_2(t) = te^{2t};\ \ \bar{y}_1(t) = (2t - 1)e^{2t}$, $\bar{y}_2(t) = (t - 3)e^{2t}$

4.4 Constant Coefficient Homogeneous Equations

Sections 4.2–4.3 established the solution structure for second order linear homogeneous equations. As we saw, in order to obtain the general solution, we need to find a fundamental set of solutions—a pair of solutions whose Wronskian determinant is nonzero on the t-interval of interest. This section along with Sections 4.5 and 4.6 shows how to find a fundamental set of solutions for the important case of a constant coefficient equation,

$$ay'' + by' + cy = 0. \tag{1}$$

In equation (1), a, b, and c are constants and we assume that a is nonzero. In later chapters we will discuss methods for variable coefficient equations of the form $y'' + p(t)y' + q(t)y = 0$.

For the discussion of equation (1), we may assume the t-interval of interest is $-\infty < t < \infty$ or any subinterval of $-\infty < t < \infty$ since the coefficient functions $p(t) = b/a$ and $q(t) = c/a$ are constant and hence continuous on every subinterval of $-\infty < t < \infty$.

Finding Solutions of Second Order Constant Coefficient Equations

We look for solutions of the form $y(t) = e^{\lambda t}$, where λ is a constant to be determined. The motivation for assuming this form for a solution comes from observing how the function $e^{\lambda t}$ behaves under repeated differentiation. In particular, we have

$$\frac{d}{dt}e^{\lambda t} = \lambda e^{\lambda t} \quad \text{and} \quad \frac{d^2}{dt^2}e^{\lambda t} = \lambda^2 e^{\lambda t}.$$

Each differentiation of $e^{\lambda t}$ simply multiplies $e^{\lambda t}$ by a power of the constant, λ. Substituting $y(t) = e^{\lambda t}$ into differential equation (1) leads to

$$ay'' + by' + cy = a\lambda^2 e^{\lambda t} + b\lambda e^{\lambda t} + ce^{\lambda t} = e^{\lambda t}(a\lambda^2 + b\lambda + c) = 0. \tag{2}$$

Equation (2) must hold for all t in the interval of interest. Since the factor $e^{\lambda t}$ is never zero for any value of the constant λ, equation (2) is valid only if λ is a root of the polynomial equation

$$a\lambda^2 + b\lambda + c = 0. \tag{3}$$

The quadratic equation (3) is called the **characteristic equation** for $ay'' + by' + cy = 0$, and the polynomial $P(\lambda) = a\lambda^2 + b\lambda + c$ is called the **characteristic polynomial**. The roots of the characteristic equation are exactly those values λ for which $y(t) = e^{\lambda t}$ is a solution of the differential equation $ay'' + by' + cy = 0$.

EXAMPLE 1

Consider the homogeneous linear differential equation

$$y'' + 8y' + 15y = 0.$$

(a) Find all values λ such that $y(t) = e^{\lambda t}$ is a solution of the differential equation.

(b) Do the functions found in part (a) form a fundamental set of solutions for the differential equation? If so, what is the general solution of the differential equation?

(continued)

(continued)

Solution:

(a) The characteristic polynomial for $y'' + 8y' + 15y = 0$ is

$$P(\lambda) = \lambda^2 + 8\lambda + 15 = (\lambda + 5)(\lambda + 3),$$

and so the roots of the characteristic equation are $\lambda_1 = -5$ and $\lambda_2 = -3$. Thus, the trial form $y(t) = e^{\lambda t}$ leads to two solutions of the differential equation:

$$y_1(t) = e^{-5t} \quad \text{and} \quad y_2(t) = e^{-3t}.$$

(b) To decide whether $\{y_1, y_2\}$ is a fundamental set of solutions, we form the Wronskian

$$W(t) = \begin{vmatrix} e^{-5t} & e^{-3t} \\ -5e^{-5t} & -3e^{-3t} \end{vmatrix} = 2e^{-8t}.$$

Since $W(t) = 2e^{-8t}$ is never zero, $\{y_1, y_2\}$ is a fundamental set of solutions for $y'' + 8y' + 15y = 0$ on any t-interval. The general solution of $y'' + 8y' + 15y = 0$ is therefore

$$y(t) = c_1 e^{-5t} + c_2 e^{-3t},$$

where c_1 and c_2 are arbitrary constants. ▲

Roots of the Characteristic Equation

The function $y(t) = e^{\lambda_1 t}$ is a solution of $ay'' + by' + cy = 0$ provided λ_1 is a root of the characteristic equation

$$a\lambda^2 + b\lambda + c = 0. \tag{4}$$

As we know, the quadratic equation $a\lambda^2 + b\lambda + c = 0$ might have two distinct real roots, one real root, or two complex roots. Figure 4.4 shows the graph of $P(\lambda) = a\lambda^2 + b\lambda + c$ versus λ, illustrating each of these three cases.

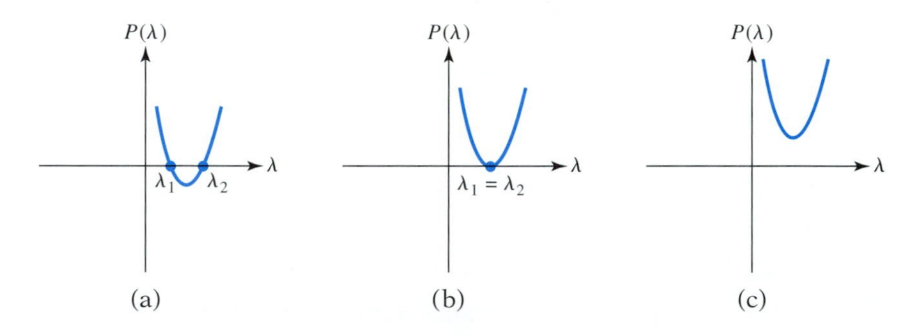

FIGURE 4.4

Three possibilities for the graph of $P(\lambda) = a\lambda^2 + b\lambda + c, a > 0$. (a) The characteristic equation has two real distinct roots, λ_1 and λ_2. (b) The characteristic equation has a single repeated real root, λ_1. (c) The characteristic equation has two complex roots but no real roots.

We can obtain the roots of characteristic equation (4) by using the quadratic formula,

$$\lambda_{1,2} = \frac{-b \pm \sqrt{b^2 - 4ac}}{2a}. \tag{5}$$

As illustrated in Figure 4.4, there are three cases, depending on the value of the discriminant, $b^2 - 4ac$:

(a) $b^2 - 4ac > 0$: In this case, the two roots λ_1 and λ_2 are real and distinct. We obtain two solutions, $y_1(t) = e^{\lambda_1 t}$ and $y_2(t) = e^{\lambda_2 t}$. We show later in this section that these two solutions form a fundamental set of solutions for equation (1).

(b) $b^2 - 4ac = 0$: In this case, the two roots are equal:

$$\lambda_1 = \lambda_2 = \frac{-b}{2a}.$$

Our computation, based on the trial form $y(t) = e^{\lambda t}$, therefore yields only one solution, namely

$$y_1(t) = e^{-(b/2a)t}.$$

Since a fundamental set of solutions consists of two solutions having a nonvanishing Wronskian [and since fundamental sets always exist for equation (1)] there must be another solution having a different functional form. In Section 4.5, we will discuss this real "repeated root" case and show how to obtain the second function needed for a fundamental set.

(c) $b^2 - 4ac < 0$: In this case the roots are complex-valued and we have

$$\lambda_{1,2} = -\frac{b}{2a} \pm i\frac{\sqrt{4ac - b^2}}{2a}.$$

Since a, b, and c are real constants, the roots $\lambda_{1,2}$ form a complex conjugate pair. For brevity, let

$$\alpha = -\frac{b}{2a}, \qquad \beta = \frac{\sqrt{4ac - b^2}}{2a}.$$

Then $\lambda_{1,2} = \alpha \pm i\beta$, where β is nonzero. Several questions arise. What mathematical interpretation do we give to expressions of the form $e^{(\alpha \pm i\beta)t}$? Once we make sense of such expressions mathematically, how do we obtain real-valued, physically meaningful solutions of equation (1)? These issues will be addressed in Section 4.6.

The General Solution When the Characteristic Equation Has Real Distinct Roots

We now take up case (a), where the discriminant $b^2 - 4ac$ is positive. In this case, the two roots λ_1 and λ_2 are real and distinct and $y_1(t) = e^{\lambda_1 t}$ and $y_2(t) = e^{\lambda_2 t}$ are two solutions of $ay'' + by' + cy = 0$. To determine if $\{y_1, y_2\}$ forms a fundamental set of solutions, we calculate the Wronskian,

$$W(t) = \begin{vmatrix} e^{\lambda_1 t} & e^{\lambda_2 t} \\ \lambda_1 e^{\lambda_1 t} & \lambda_2 e^{\lambda_2 t} \end{vmatrix} = (\lambda_2 - \lambda_1)e^{(\lambda_1+\lambda_2)t}.$$

The factor $(\lambda_2 - \lambda_1)$ is nonzero since the roots are distinct. In addition, the exponential function $e^{(\lambda_1+\lambda_2)t}$ is nonzero for all t. Therefore, this calculation establishes, once and for all, that the two exponential solutions obtained in the real, distinct root case form a fundamental set of solutions. There is no need to reestablish this fact for every particular example.

In summary, if the characteristic equation has two real distinct roots, λ_1 and λ_2, then the general solution of $ay'' + by' + cy = 0$ is

$$y(t) = c_1 e^{\lambda_1 t} + c_2 e^{\lambda_2 t}. \tag{6}$$

EXAMPLE 2

Solve the initial value problem

$$y'' + 4y' + 3y = 0, \qquad y(0) = 7, \qquad y'(0) = -17.$$

Solution: The characteristic equation is

$$\lambda^2 + 4\lambda + 3 = 0,$$

or

$$(\lambda + 1)(\lambda + 3) = 0.$$

Therefore, the general solution is

$$y(t) = c_1 e^{-t} + c_2 e^{-3t}.$$

In order to impose the initial conditions, we calculate the derivative

$$y'(t) = -c_1 e^{-t} - 3c_2 e^{-3t}.$$

To satisfy the initial conditions, we need c_1 and c_2 such that

$$\begin{aligned} c_1 + \ c_2 &= \ 7, \\ -c_1 - 3c_2 &= -17. \end{aligned}$$

We find $c_1 = 2$ and $c_2 = 5$. The unique solution of the initial value problem is

$$y(t) = 2e^{-t} + 5e^{-3t}. \ \blacktriangle$$

EXAMPLE 3

Solve the initial value problem

$$y'' + y' - 2y = 0, \qquad y(0) = y_0, \qquad y'(0) = y_0'.$$

For what values of the constants y_0 and y_0' can we guarantee that $\lim_{t\to\infty} y(t) = 0$?

Solution: The characteristic polynomial for $y'' + y' - 2y = 0$ is

$$\lambda^2 + \lambda - 2 = (\lambda + 2)(\lambda - 1).$$

Thus, the general solution is

$$y(t) = c_1 e^{-2t} + c_2 e^{t}.$$

Imposing the initial conditions leads to the system of equations

$$\begin{aligned} y(0) &= \quad c_1 + c_2 = y_0, \\ y'(0) &= -2c_1 + c_2 = y_0'. \end{aligned}$$

The solution of this system of equations is $c_1 = (y_0 - y_0')/3, c_2 = (2y_0 + y_0')/3$. The solution of the initial value problem is therefore

$$y(t) = \left(\frac{y_0 - y_0'}{3}\right) e^{-2t} + \left(\frac{2y_0 + y_0'}{3}\right) e^{t}.$$

Since $\lim_{t\to\infty} e^{-2t} = 0$ and $\lim_{t\to\infty} e^{t} = +\infty$, the solution of the initial value problem will tend to zero as t increases if the coefficient of e^t in the solution is zero. Therefore, $\lim_{t\to\infty} y(t) = 0$ if $y_0' = -2y_0$. ▲

EXERCISES

Exercises 1–15:

(a) Obtain the general solution of the differential equation.

(b) Impose the initial conditions to obtain the unique solution of the initial value problem.

(c) Describe the behavior of the solution $y(t)$ as $t \to -\infty$ and as $t \to \infty$. In each case, does $y(t)$ approach $-\infty$, $+\infty$, or a finite limit?

1. $y'' + y' - 2y = 0, \quad y(0) = 3, \quad y'(0) = -3$

2. $y'' - (1/4)y = 0, \quad y(2) = 1, \quad y'(2) = 0$

3. $y'' - 4y' + 3y = 0, \quad y(0) = -1, \quad y'(0) = 1$

4. $2y'' - 5y' + 2y = 0, \quad y(0) = -1, \quad y'(0) = -5$

5. $y'' - y = 0, \quad y(0) = 1, \quad y'(0) = -1$

6. $y'' + 2y' = 0, \quad y(-1) = 0, \quad y'(-1) = 2$

7. $y'' + 5y' + 6y = 0, \quad y(0) = 1, \quad y'(0) = -1$

8. $y'' - 5y' + 6y = 0, \quad y(0) = 1, \quad y'(0) = -1$

9. $y'' - 4y = 0, \quad y(3) = 0, \quad y'(3) = 0$

10. $8y'' - 6y' + y = 0, \quad y(1) = 4, \quad y'(1) = 3/2$

11. $2y'' - 3y' = 0, \quad y(-2) = 3, \quad y'(-2) = 0$

12. $y'' - 6y' + 8y = 0, \quad y(1) = 2, \quad y'(1) = -8$

13. $y'' + 4y' + 2y = 0, \quad y(0) = 0, \quad y'(0) = 4$

14. $y'' - 4y' - y = 0, \quad y(0) = 1, \quad y'(0) = 2 + \sqrt{5}$

15. $2y'' - y = 0, \quad y(0) = -2, \quad y'(0) = \sqrt{2}$

16. Consider the initial value problem $y'' + \alpha y' + \beta y = 0, y(0) = 1, y'(0) = y_0'$, where α, β, and y_0' are constants. It is known that one solution of the differential equation is $y_1(t) = e^{-3t}$ and that the solution of the initial value problem satisfies $\lim_{t\to\infty} y(t) = 2$. Determine the constants α, β, and y_0'.

17. Consider the initial value problem $y'' + \alpha y' + \beta y = 0, y(0) = 3, y'(0) = 5$. The differential equation has a fundamental set of solutions, $\{y_1(t), y_2(t)\}$. It is known that $y_1(t) = e^{-t}$ and that the Wronskian formed by the two members of the fundamental set is $W(t) = 4e^{2t}$.

(a) Determine the second member of the fundamental set, $y_2(t)$.

(b) Determine the constants α and β.

(c) Solve the initial value problem.

18. The three graphs display solutions of initial value problems on the interval $0 \le t \le 3$. Each solution satisfies the initial conditions $y(0) = 1, y'(0) = -1$. Match the differential equation with the graph of its solution.

(a) $y'' + 2y' = 0$ (b) $6y'' - 5y' + y = 0$ (c) $y'' - y = 0$

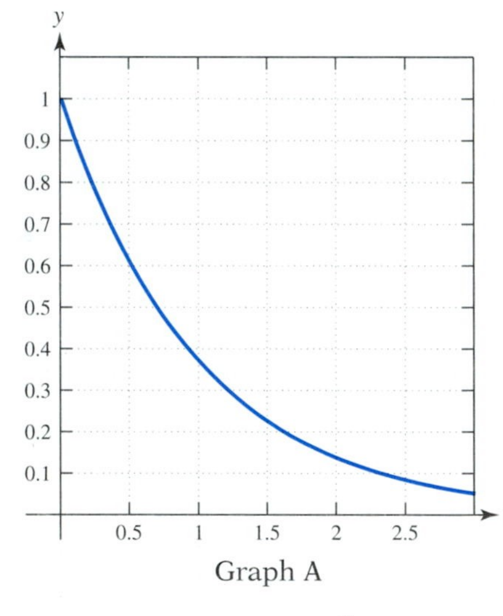

Graph A

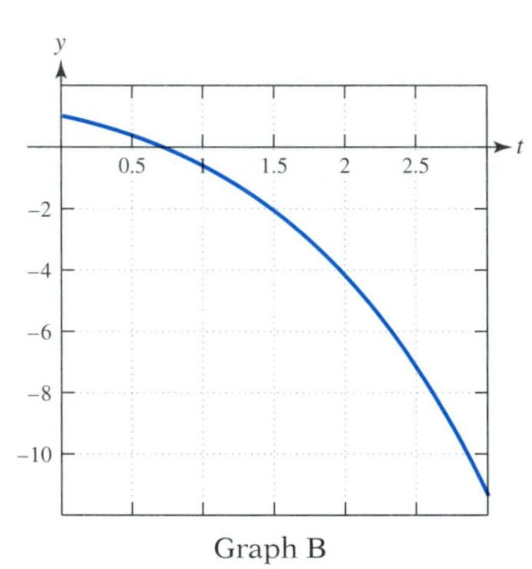

Graph B

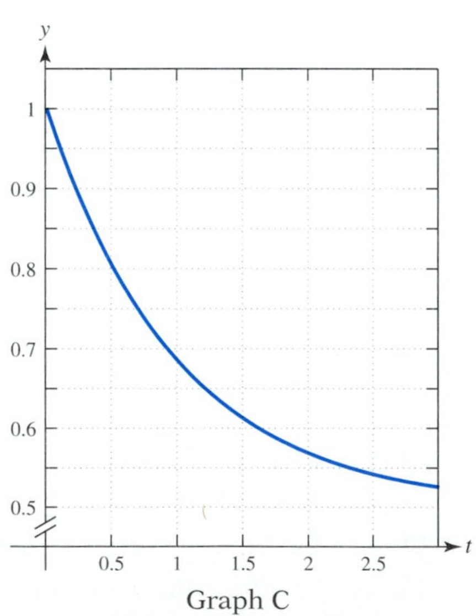

Graph C

Figure for Exercise 18

19. Obtain the general solution of $y''' - 5y'' + 6y' = 0$. [Hint: Make the change of dependent variable $u(t) = y'(t)$, determine $u(t)$, and then antidifferentiate to obtain $y(t)$.]

Rectilinear Motion with a Drag Force In Chapter 3, we considered rectilinear motion in the presence of a drag force proportional to velocity. We solved the first order linear equation for velocity and subsequently antidifferentiated the solution to obtain distance as a function of time. We now consider directly the second order linear differential equation for the distance function.

20. A particle of mass m moves along the x-axis and is acted upon by a drag force proportional to its velocity. The drag constant is denoted by k. If $x(t)$ represents the particle position at time t, Newton's law of motion leads to the differential equation $mx''(t) = -kx'(t)$.

(a) Obtain the general solution of this second order linear differential equation.

(b) Solve the initial value problem if $x(0) = x_0$ and $x'(0) = v_0$.

(c) What is $\lim_{t\to\infty} x(t)$?

A Simple Centrifuge A particle having mass m is initially positioned in a frictionless tube that rotates about a fixed horizontal pivot, as shown in the figure. We anticipate that, as time progresses, the particle's radial distance from the pivot will increase and that the particle will eventually exit the tube. We now analyze this behavior mathematically.

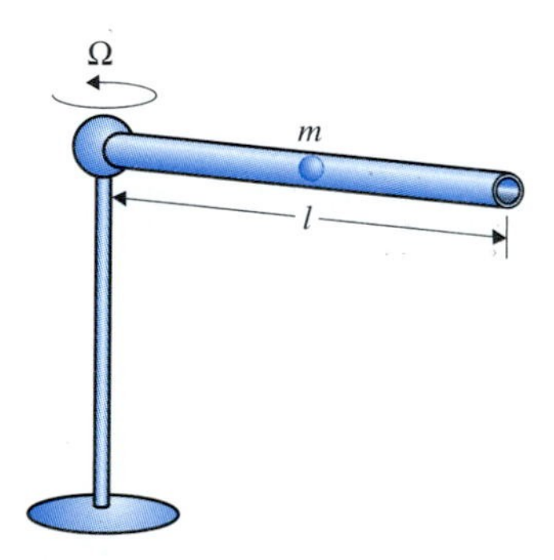

Figure for Exercise 21

The key observation is that at any given instant, the particle experiences no forces in the radial direction. Newton's second law of motion tells us therefore that the component of the acceleration vector in the radial direction must vanish. In an appendix at the end of this section we show that the vanishing of radial acceleration leads to the following differential equation for the radial distance $r(t)$ of the particle from the pivot point,

$$r''(t) - \Omega^2 r(t) = 0,$$

where $\Omega = \theta'(t)$ is the angular velocity of the tube, that is, the rate of change of angle (measured in radians) with respect to time. Specification of the initial radial position and velocity, say $r(0)$ and $r'(0)$, completes the initial value problem. In our present considerations we assume that Ω is constant; in later chapters we consider more complicated scenarios.

21. (a) Solve the initial value problem $r''(t) - \Omega^2 r(t) = 0$, $r(0) = r_0$, $r'(0) = r'_0$, where Ω is a positive constant. As a commonsense check, what is the solution in the case that $r(0) = 0, r'(0) = 0$?

(b) Suppose that at time $t = 0$ a particle is injected into the tube at the pivot point with an initial radial velocity $r'(0) = 1/5$ m/s. Assume that the tube is 3 m long and is rotating at a constant rate of 30 revolutions per minute. How long will it take the particle to exit the tube?

Cloudy Apple Juice An equation similar to the one describing the simple centrifuge in Exercise 21 has been used by food scientists studying cloudy apple juice.[5] Cloudy apple juice consists of minute apple particles suspended in a sweet liquid. As long as the particles remain in suspension, they give apple juice a full-bodied, desirable character. If the particles coagulate and fall out of suspension, however, they form an unattractive sedimentary deposit at the bottom of the container. Since one of the factors influencing coagulation is particle size, scientists study the distribution of particle sizes in the cloudy liquid. One technique used to determine the distribution of particle sizes is to study how the tiny apple particles migrate outward toward the container walls when the cloudy liquid is placed in a centrifuge. We consider a slightly oversimplified version of this problem.

The dynamics governing the radial motion of such particles differs from that considered in Exercise 21 insofar as the particle experiences resistance as it moves radially outward through the liquid. If we model this resistance or damping force as being proportional to radial velocity, then Newton's second law of motion applied in the radial direction yields

$$m[r''(t) - \Omega^2 r(t)] = -kr'(t)$$

where k is a positive constant, m is particle mass, and Ω is the constant angular velocity. The term $-kr'(t)$ is the damping force the particle experiences as it moves radially outward. Assume, at time $t = 0$ a particle is inserted into the rotating system at the pivot location with a positive radial velocity. The initial value problem to be solved is

$$r'' + \frac{k}{m}r' - \Omega^2 r = 0, \qquad r(0) = 0, \qquad r'(0) = r'_0. \tag{7}$$

22. (a) Obtain the solution of initial value problem (7).

(b) Suppose the particles are spheres with mass proportional to volume and drag coefficient proportional to cross-sectional area. Would the coefficient k/m increase or decrease as particle size increased?

(c) Assume the system is rotating at 20 revolutions per minute and that a particle is injected into the system at the pivot location with a radial velocity of 1 cm/s. Let $k/m = 4\ \text{s}^{-1}$. Determine $r(t)$ at $t = 2$ s.

23. At time $t = 0$, a particle is inserted into a rotating tube at its pivot point. The tube is rotating with constant angular velocity Ω and the particle experiences a drag force proportional to its radial velocity. The radial distance of the particle from the pivot, $r(t)$, is therefore the solution of the following initial value problem.

$$r'' + (k/m)r' - \Omega^2 r = 0, \qquad r(0) = 0, \qquad r'(0) = r'_0.$$

Suppose the tube is sufficiently long to infer that

$$\lim_{t\to\infty}\left(\frac{r(t)}{e^{(\Omega/2)t}}\right) = r_\infty,$$

where r_∞ is a positive constant.

(a) Determine the ratio k/m in terms of the angular velocity Ω.

(b) Determine the radial distance $r(t)$ in terms of the constants r'_0 and Ω.

---

[5] D. B. Genovese and J. E. Lozano, "Particle size determination of food suspensions: application to cloudy apple juice," *J. Food Processing Engineering*, 23 (2000), pp. 437–452.

Appendix: Derivation of Radial and Angular Accelerations

The motion of a particle undergoing planar motion can be described in terms of polar coordinates. As shown in the figure below, $x = r\cos\theta$ and $y = r\sin\theta$, where both r and θ vary as functions of time t. Since we are interested in decomposing the acceleration vector into radial and angular components, we introduce orthogonal unit vectors in these directions

$$\mathbf{e}_r = (\cos\theta)\mathbf{i} + (\sin\theta)\mathbf{j}, \quad \mathbf{e}_\theta = (-\sin\theta)\mathbf{i} + (\cos\theta)\mathbf{j}. \tag{8}$$

In contrast to the unit vectors $\mathbf{i}$ and $\mathbf{j}$, along the x and y axes, respectively, the unit vectors $\mathbf{e}_r$ and $\mathbf{e}_\theta$ vary with time. They remain mutually perpendicular and of unit length, but they change direction. In fact, using the chain rule, it follows from (8) that

$$\frac{d}{dt}\mathbf{e}_r = -(\sin\theta)\theta'\mathbf{i} + (\cos\theta)\theta'\mathbf{j} = \theta'\mathbf{e}_\theta \quad \text{and} \quad \frac{d}{dt}\mathbf{e}_\theta = -(\cos\theta)\theta'\mathbf{i} - (\sin\theta)\theta'\mathbf{j} = -\theta'\mathbf{e}_r.$$

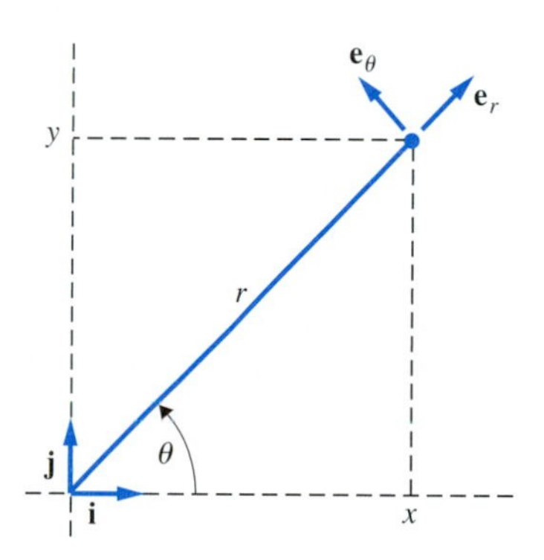

Figure for the Derivation

Since the position vector describing the particle location is $\mathbf{r} = r\mathbf{e}_r$, we need only differentiate this expression twice to obtain the acceleration vector. The velocity vector is given by

$$\mathbf{v} = \frac{d}{dt}\mathbf{r} = \frac{d}{dt}[r\mathbf{e}_r] = r'\mathbf{e}_r + r\frac{d}{dt}\mathbf{e}_r = r'\mathbf{e}_r + r\theta'\mathbf{e}_\theta.$$

The acceleration vector, after using equation (8) and performing some algebra, is found to be

$$\mathbf{a} = \frac{d}{dt}\mathbf{v} = \left(r'' - r\theta'\theta'\right)\mathbf{e}_r + \left(r\theta'' + 2r'\theta'\right)\mathbf{e}_\theta.$$

The radial and angular components of the acceleration vector are therefore

$$a_r = r'' - r\theta'\theta', \qquad a_\theta = r\theta'' + 2r'\theta's.$$

4.5 Real Repeated Roots; Reduction of Order

In Section 4.4 we began discussing how to find the general solution of the constant coefficient differential equation

$$ay'' + by' + cy = 0.$$

When the characteristic equation $a\lambda^2 + b\lambda + c = 0$ has two distinct real roots, λ_1 and λ_2, the general solution is given by

$$y(t) = c_1 e^{\lambda_1 t} + c_2 e^{\lambda_2 t}.$$

In this section, we consider the case where the characteristic equation has a repeated real root (that is, when the discriminant $b^2 - 4ac$ is zero). In this case, looking for solutions of the form $y(t) = e^{\lambda t}$ leads to only one solution since the characteristic equation has only one distinct root. We must somehow find a second solution of $ay'' + by' + cy = 0$ in order to form a fundamental set of solutions.

The Method of Reduction of Order

To obtain a second solution for $ay'' + by' + cy = 0$ in the repeated root case, we use a method called "reduction of order." We apply the method first to the problem at hand,

$$ay'' + by' + cy = 0. \quad (1)$$

Then, at the end of this section, we'll discuss reduction of order as a technique for finding a second solution of the general homogeneous linear equation, $y'' + p(t)y' + q(t)y = 0$, given that we have somehow found one solution, $y_1(t)$, of the equation.

Assume, without loss of generality, that $a = 1$ in equation (1). Then, since $b^2 - 4c = 0$, we know that c is positive. We can represent c as $c = \alpha^2$ and choose $b = -2\alpha$. With these simplifications in notation, differential equation (1) becomes

$$y'' - 2\alpha y' + \alpha^2 y = 0. \quad (2)$$

The characteristic polynomial for equation (2) is

$$\lambda^2 - 2\alpha\lambda + \alpha^2 = (\lambda - \alpha)^2,$$

and therefore one solution is

$$y_1(t) = e^{\alpha t}.$$

To find a second solution, $y_2(t)$, we use the **method of reduction of order**. The basic idea underlying the method is to look for a second solution, $y_2(t)$, of the form

$$y_2(t) = y_1(t)u(t) = e^{\alpha t}u(t). \quad (3)$$

The function $u(t)$ in (3) must be chosen so that $y_2(t)$ is also a solution of equation (2).

At this point there's no obvious reason to believe that this assumed form of the solution provides any simplification. We must substitute (3) into differential equation (2) and see what happens. Substituting, we obtain

$$y_2'' - 2\alpha y_2' + \alpha^2 y_2 = (e^{\alpha t}u)'' - 2\alpha(e^{\alpha t}u)' + \alpha^2(e^{\alpha t}u) \quad (4)$$

which simplifies to

$$y_2'' - 2\alpha y_2' + \alpha^2 y_2 = e^{\alpha t}u''. \quad (5)$$

Since the exponential function is nonzero everywhere, $y_2(t) = e^{\alpha t}u(t)$ is a solution of $y'' - 2\alpha y' + \alpha^2 y = 0$ if and only if $u'' = 0$. The equation $u'' = 0$ can be solved by antidifferentiation to obtain $u(t) = a_1 t + a_2$, where a_1 and a_2 are arbitrary constants.

Thus, the method of reduction of order has led us to a second solution,

$$y_2(t) = e^{\alpha t}(a_1 t + a_2) = a_1 t e^{\alpha t} + a_2 e^{\alpha t}.$$

Notice that the term $a_2 e^{\alpha t}$ is simply a constant multiple of $y_1(t)$. Since the general solution of the differential equation contains $y_1(t)$ multiplied by an arbitrary constant, we lose no generality by setting $a_2 = 0$. We can likewise take $a_1 = 1$ since $y_2(t)$ will also be multiplied by an arbitrary constant in the general solution. With these simplifications, the second solution is

$$y_2(t) = te^{\alpha t}.$$

To verify that $\{y_1, y_2\} = \{e^{\alpha t}, te^{\alpha t}\}$ forms a fundamental set, we compute the Wronskian,

$$W(t) = \begin{vmatrix} e^{\alpha t} & te^{\alpha t} \\ \alpha e^{\alpha t} & (\alpha t + 1)e^{\alpha t} \end{vmatrix} = e^{2\alpha t}.$$

Since the Wronskian is nonzero, we have shown that the general solution is

$$y(t) = c_1 e^{\alpha t} + c_2 t e^{\alpha t}. \tag{6}$$

EXAMPLE 1

Solve the initial value problem

$$4y'' + 4y' + y = 0, \qquad y(2) = 1, \qquad y'(2) = 0.$$

Solution: You can rewrite the differential equation as $y'' + y' + (1/4)y = 0$ if you wish. We'll work with the given form; the results are the same.

Looking for solutions of the form $y(t) = e^{\lambda t}$ leads to the characteristic equation

$$4\lambda^2 + 4\lambda + 1 = (2\lambda + 1)^2 = 0.$$

Therefore, the characteristic equation has real repeated roots $\lambda_1 = \lambda_2 = -1/2$. By (6), the general solution is

$$y(t) = c_1 e^{-t/2} + c_2 t e^{-t/2}.$$

Since

$$y'(t) = -\frac{c_1}{2}e^{-t/2} + c_2\left[1 - \frac{t}{2}\right]e^{-t/2},$$

imposing the initial conditions leads to

$$\begin{aligned} c_1 e^{-1} + c_2 2e^{-1} &= 1, \\ -\frac{c_1}{2}e^{-1} \qquad &= 0. \end{aligned}$$

The solution is $c_1 = 0, c_2 = e/2$. The solution of the initial value problem is

$$y(t) = \frac{e}{2}te^{-t/2} = \frac{t}{2}e^{1-t/2}. \blacktriangle$$

Method of Reduction of Order (General Case)

The method of reduction of order is not restricted to constant coefficient equations. It can be applied to the general second order linear homogeneous equation

$$y'' + p(t)y' + q(t)y = 0, \tag{7}$$

where $p(t)$ and $q(t)$ are continuous functions on the t-interval of interest. Suppose we know one solution of equation (7), call it $y_1(t)$. We again assume that the second unknown member of the fundamental set of solutions, $y_2(t)$, can be expressed as

$$y_2(t) = y_1(t)u(t).$$

We need to determine the unknown function $u(t)$ so that $y_2(t)$ is a solution of equation (7). Substituting the assumed form into (7) leads (after some rearranging of terms) to

$$\begin{aligned}(y_1u)'' + p(t)(y_1u)' + q(t)(y_1u) = y_1u'' &+ [2y_1' + p(t)y_1]u' \\ &+ [y_1'' + p(t)y_1' + q(t)y_1]u = 0.\end{aligned}$$

At first it seems as though this equation offers little improvement. Recall, however, that y_1 is not an arbitrary function; it is a solution of differential equation (7). Therefore, the factor multiplying u in the preceding equation vanishes, and we obtain a considerable simplification:

$$y_1u'' + [2y_1' + p(t)y_1]u' = 0. \tag{8}$$

The structure of equation (8) is what gives the method its name. Although equation (8) has the form of a second order linear differential equation for u, we can define a new dependent variable $v(t) = u'(t)$. Under this change of variables, equation (8) reduces to a first order linear differential equation for v,

$$y_1(t)v' + [2y_1'(t) + p(t)y_1(t)]v = 0. \tag{9}$$

Thus, the task of solving a second order linear differential equation has been replaced by that of solving a first order linear differential equation. Equation (9) can be solved, in principle, using the approach of Section 2.3. Once we have $v = u'$, we can obtain u (and ultimately y_2) by antidifferentiation.

EXAMPLE 2

Observe that $y_1(t) = t$ is a solution of the homogeneous linear differential equation

$$t^2y'' - ty' + y = 0, \qquad 0 < t < \infty. \tag{10}$$

(a) Use reduction of order to obtain a second solution, $y_2(t)$. Does the pair $\{y_1, y_2\}$ form a fundamental set of solutions for this differential equation?

(b) If $\{y_1, y_2\}$ is a fundamental set of solutions, solve the initial value problem

$$t^2y'' - ty' + y = 0, \qquad y(1) = 3, \qquad y'(1) = 8.$$

Solution: Note that the initial value problem, written in standard form, becomes

$$y'' - \frac{1}{t}y' + \frac{1}{t^2}y = 0, \qquad y(1) = 3, \qquad y'(1) = 8. \tag{11}$$

The coefficient functions are not continuous at $t = 0$. Our t-interval of interest, $0 < t < \infty$, is the largest interval containing $t_0 = 1$ on which we are guaranteed the existence of a unique solution of the initial value problem.

(a) Since one solution is known, we apply reduction of order. Assuming $y_2(t) = tu(t)$, we have

$$y_2' = u + tu' \quad \text{and} \quad y_2'' = 2u' + tu''.$$

Substituting these expressions into the differential equation, $t^2y'' - ty' + y = 0$, we find

$$t^2[2u' + tu''] - t[u + tu'] + tu = t^2[tu'' + u'] = 0.$$

Therefore, $tu'' + u' = 0$. Setting $v = u'$, leads to the first order linear equation

$$tv' + v = 0. \tag{12}$$

The general solution of equation (12) is

$$v(t) = \frac{c}{t}.$$

Therefore, since $v(t) = u'(t)$, it follows that $u(t) = c\ln t + d$ and we obtain a second solution

$$y_2(t) = tu(t) = t[c\ln t + d]. \tag{13}$$

[Note that $\ln|t| = \ln t$ since $t > 0$.] Using the same rationale as before, we can take $c = 1, d = 0$, and let

$$y_2(t) = t\ln t.$$

The Wronskian of y_1 and y_2 is $W(t) = t$, which is nonzero on the interval $0 < t < \infty$. Therefore, the general solution is

$$y(t) = c_1y_1(t) + c_2y_2(t) = c_1t + c_2t\ln t, \qquad 0 < t < \infty. \tag{14}$$

(b) For $y(t) = c_1t + c_2t\ln t$, we have $y'(t) = c_1 + c_2[1 + \ln t]$. Imposing the initial conditions $y(1) = 3$ and $y'(1) = 8$, we obtain

$$\begin{aligned} c_1 \qquad &= 3, \\ c_1 + c_2 &= 8. \end{aligned}$$

The solution of the initial value problem is

$$y(t) = 3t + 5t\ln t. \ \blacktriangle$$

EXERCISES

Exercises 1–10:

(a) Obtain the general solution of the differential equation.

(b) Impose the initial conditions to obtain the unique solution of the initial value problem.

(c) Describe the behavior of the solution as $t \to -\infty$ and $t \to \infty$. In each case, does $y(t)$ approach $-\infty$, $+\infty$, or a finite limit?

1. $y'' + 2y' + y = 0, \quad y(1) = 1, \quad y'(1) = 0$

2. $9y'' - 6y' + y = 0, \quad y(3) = -2, \quad y'(3) = -5/3$

3. $y'' + 6y' + 9y = 0, \quad y(0) = 2, \quad y'(0) = -2$

4. $25y'' + 20y' + 4y = 0, \quad y(5) = 4e^{-2}, \quad y'(5) = -(3/5)e^{-2}$
5. $4y'' - 4y' + y = 0, \quad y(1) = -4, \quad y'(1) = 0$
6. $y'' - 4y' + 4y = 0, \quad y(-1) = 2, \quad y'(-1) = 1$
7. $16y'' - 8y' + y = 0, \quad y(0) = -4, \quad y'(0) = 3$
8. $y'' + 2\sqrt{2}y' + 2y = 0, \quad y(0) = 1, \quad y'(0) = 0$
9. $y'' - 5y' + 6.25y = 0, \quad y(-2) = 0, \quad y'(-2) = 1$
10. $3y'' + 2\sqrt{3}y' + y = 0, \quad y(0) = 2\sqrt{3}, \quad y'(0) = 3$

Exercises 11–12:

In each exercise, the graph shown is the solution of $y'' - 2\alpha y' + \alpha^2 y = 0, y(0) = y_0, y'(0) = y_0'$. Determine the constants α, y_0, and y_0' as well as the solution $y(t)$. In Exercise 11, the maximum point shown on the graph has coordinates $(2, 8e^{-1})$.

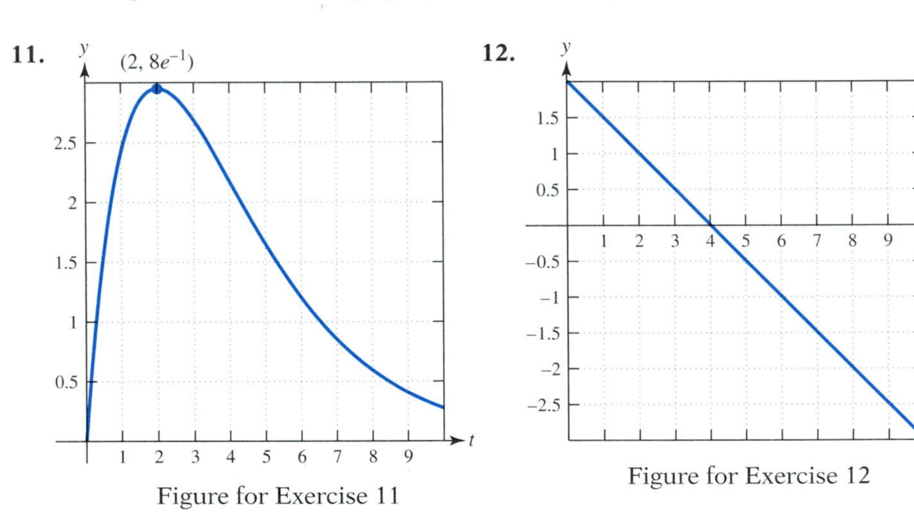

Figure for Exercise 11

Figure for Exercise 12

13. The graph of a solution $y(t)$ of the differential equation $4y'' + 4y' + y = 0$ passes through the points $(1, e^{-1/2})$ and $(2, 0)$. Determine $y(0)$ and $y'(0)$.

Exercises 14–20:

One solution, $y_1(t)$, of the differential equation is given.

(a) Use the method of reduction of order to obtain a second solution, $y_2(t)$.

(b) Compute the Wronskian formed by the solutions $y_1(t)$ and $y_2(t)$. On what interval(s) is the Wronskian continuous and nonzero?

(c) Rewrite the differential equation in the form $y'' + p(t)y' + q(t)y = 0$. On what interval(s) are both $p(t)$ and $q(t)$ continuous? How does this observation compare with the interval(s) determined in part (b)?

14. $ty'' - (2t + 1)y' + (t + 1)y = 0, \quad y_1(t) = e^t$
15. $t^2y'' - ty' + y = 0, \quad y_1(t) = t$
16. $y'' - (2\cot t)y' + (1 + 2\cot^2 t)y = 0, \quad y_1(t) = \sin t$
17. $(t + 1)^2y'' - 4(t + 1)y' + 6y = 0, \quad y_1(t) = (t + 1)^2$
18. $y'' + 4ty' + (2 + 4t^2)y = 0, \quad y_1(t) = e^{-t^2}$
19. $(t - 2)^2y'' + (t - 2)y' - 4y = 0, \quad y_1(t) = (t - 2)^2$
20. $y'' - \left(2 + \dfrac{n-1}{t}\right)y' + \left(1 + \dfrac{n-1}{t}\right)y = 0$, where n is a positive integer, $y_1(t) = e^t$

4.6 Complex Roots

In this section we complete the discussion of the general solution for the differential equation

$$ay'' + by' + cy = 0.$$

In Section 4.4 we found the general solution in the case where the discriminant, $b^2 - 4ac$, is positive. In Section 4.5, we treated the case where the discriminant is zero. We now look for the general solution when the discriminant is negative.

As noted in Section 4.4, looking for solutions of the form $y(t) = e^{\lambda t}$ when the discriminant is negative leads to a characteristic equation $a\lambda^2 + b\lambda + c = 0$ having a complex conjugate pair of roots,

$$\lambda_{1,2} = -\frac{b}{2a} \pm i\frac{\sqrt{4ac - b^2}}{2a}.$$

Using $\alpha = -b/2a$ and $\beta = \sqrt{4ac - b^2}/2a$ for simplicity, the roots are

$$\lambda_{1,2} = \alpha \pm i\beta. \tag{1}$$

The Complex Exponential Function

The approach to solving $ay'' + by' + cy = 0$ has been based on looking for solutions of the form $y(t) = e^{\lambda t}$. When the roots (1) of the characteristic equation are complex, we are led to consider exponential functions with complex arguments:

$$y_1(t) = e^{(\alpha+i\beta)t} \quad \text{and} \quad y_2(t) = e^{(\alpha-i\beta)t}. \tag{2}$$

We need to clarify the mathematical meaning of these two expressions. How is the definition of the exponential function extended or broadened to accommodate complex as well as real arguments? Once such a generalization is understood mathematically, we then need to demonstrate for complex λ that the function $e^{\lambda t}$ is, in fact, a differentiable function of t satisfying the fundamental differentiation formula

$$\frac{d}{dt}e^{\lambda t} = \lambda e^{\lambda t}.$$

This differentiation formula was tacitly assumed in Section 4.4, and it is needed now if we want to show that $y = e^{(\alpha+i\beta)t}$ and $y = e^{(\alpha-i\beta)t}$ are, in fact, solutions of $ay'' + by' + cy = 0$.

A second issue that needs to be addressed is that of physical relevance. We will be able to show that the functions $\{e^{(\alpha+i\beta)t}, e^{(\alpha-i\beta)t}\}$ form a fundamental set of solutions. The physical significance of these solutions is unclear, however. How do we use them to obtain real-valued, physically relevant solutions?

As a case in point, recall the buoyancy example discussed in the Introduction. When we modeled the object's position, $y(t)$, we arrived at a differential equation of the form

$$y'' + \omega^2 y = 0.$$

The characteristic equation for this differential equation is $\lambda^2 + \omega^2 = 0$ and has roots $\lambda_1 = i\omega$ and $\lambda_2 = -i\omega$. How do we relate the two solutions

$$y_1(t) = e^{i\omega t} \quad \text{and} \quad y_2(t) = e^{-i\omega t}$$

to the real-valued and physically relevant general solution

$$y = A \sin \omega t + B \cos \omega t$$

that we used to describe the bobbing motion of the object?

The Definition of the Complex Exponential Function

In calculus, the exponential function $y = e^t$ is encountered in different settings. It is often introduced as the inverse of the natural logarithm function. For our present purposes, however, we want a representation of the exponential function that permits us to generalize from a real argument to a complex argument in a straightforward and natural way. In this regard, the power series representation of the function e^z is very convenient.

From calculus, the Maclaurin[6] series expansion for e^z is the infinite series

$$e^z = 1 + z + \frac{z^2}{2!} + \frac{z^3}{3!} + \cdots = \sum_{n=0}^{\infty} \frac{z^n}{n!}, \tag{3}$$

where z^0 and $0!$ are understood to be equal to 1. For a given value of z, the Maclaurin series (3) is interpreted as the limit of the sequence of partial sums. That is, assuming the limit exists,

$$e^z = \lim_{M\to\infty} \sum_{n=0}^{M} \frac{z^n}{n!}. \tag{4}$$

When limit (4) exists, we say that the series **converges**. It is shown in calculus that the Maclaurin series (3) converges to e^z for every real value of z. Although we do not do so here, it is possible to derive a number of familiar properties of the exponential function from the power series (3). For example, it can be shown that e^z is differentiable and that

$$e^{z_1+z_2} = e^{z_1}e^{z_2}.$$

The importance of the power series representation (3) for e^z is that the representation is not limited to real values of z. It can be shown that power series (3) converges for all complex values of z. In this manner, the function $e^{\lambda t}$ is given meaning even for a complex value of λ. The power series representation can also be used to show that

$$e^{\lambda t} = e^{(\alpha+i\beta)t} = e^{\alpha t+i\beta t} = e^{\alpha t}e^{i\beta t}. \tag{5}$$

---

[6]Colin Maclaurin (1698–1746) was a professor of mathematics at the University of Aberdeen and later at the University of Edinburgh. In a two-volume exposition of Newton's calculus, titled the *Treatise of Fluxions* (published in 1742), Maclaurin used the special form of Taylor series now bearing his name. Maclaurin also is credited with introducing the Integral Test for the convergence of an infinite series.

Euler's Formula

From equation (5),

$$e^{\lambda t} = e^{\alpha t + i\beta t} = e^{\alpha t} e^{i\beta t}.$$

This result simplifies our task of understanding the function $y = e^{\lambda t}$, since we already know the behavior of the factor $e^{\alpha t}$ when α is real. Thus, we focus on the other factor, $e^{i\beta t}$. Using $z = i\beta t$ in power series (3), we obtain

$$\begin{aligned} e^{i\beta t} &= 1 + (i\beta t) + \frac{(i\beta t)^2}{2!} + \frac{(i\beta t)^3}{3!} + \frac{(i\beta t)^4}{4!} + \frac{(i\beta t)^5}{5!} + \cdots \\ &= \left[1 - \frac{(\beta t)^2}{2!} + \frac{(\beta t)^4}{4!} - \cdots\right] + i\left[\beta t - \frac{(\beta t)^3}{3!} + \frac{(\beta t)^5}{5!} - \cdots\right] \\ &= \sum_{n=0}^{\infty} \frac{(-1)^n (\beta t)^{2n}}{(2n)!} + i \sum_{n=0}^{\infty} \frac{(-1)^n (\beta t)^{2n+1}}{(2n+1)!}. \end{aligned} \tag{6}$$

In (6) we used the fact that $i^2 = -1$ and regrouped the terms into real and imaginary parts. The two series on the right-hand side of (6) are Maclaurin series representations of familiar functions:

$$\cos \beta t = 1 - \frac{(\beta t)^2}{2!} + \frac{(\beta t)^4}{4!} - \cdots = \sum_{n=0}^{\infty} \frac{(-1)^n (\beta t)^{2n}}{(2n)!},$$

$$\sin \beta t = \beta t - \frac{(\beta t)^3}{3!} + \frac{(\beta t)^5}{5!} - \cdots = \sum_{n=0}^{\infty} \frac{(-1)^n (\beta t)^{2n+1}}{(2n+1)!}.$$

Using these results in equation (6), we obtain **Euler's formula**,

$$e^{i\beta t} = \cos \beta t + i \sin \beta t. \tag{7}$$

The symmetry properties of the sine and cosine functions ($\cos(-\theta) = \cos\theta$ and $\sin(-\theta) = -\sin\theta$) together with Euler's formula (7), lead to an analogous expression for $e^{-i\beta t}$,

$$e^{-i\beta t} = e^{i(-\beta t)} = \cos(-\beta t) + i \sin(-\beta t) = \cos \beta t - i \sin \beta t. \tag{8}$$

Equations (5), (7), and (8) can be used to express $e^{(\alpha + i\beta)t}$ and $e^{(\alpha - i\beta)t}$ in terms of familiar functions

$$\begin{aligned} e^{(\alpha+i\beta)t} &= e^{\alpha t} e^{i\beta t} &= e^{\alpha t}(\cos \beta t + i \sin \beta t), \\ e^{(\alpha-i\beta)t} &= e^{\alpha t} e^{-i\beta t} &= e^{\alpha t}(\cos \beta t - i \sin \beta t). \end{aligned} \tag{9}$$

We can use equation (9) to show that the Wronskian of $y_1(t) = e^{(\alpha+i\beta)t}$ and $y_2(t) = e^{(\alpha-i\beta)t}$ is $W(t) = -i2\beta e^{2\alpha t}$, which is nonzero for all t since $\beta \neq 0$. The two solutions, y_1 and y_2, therefore form a fundamental set of solutions.

From a mathematical point of view, we are done with the complex roots case since we have found a fundamental set. From a physical point of view, however, if the mathematical problem arises from a (real-valued) physical process and we seek real-valued, physically meaningful answers, the two solutions in (9) are not satisfactory. We want a fundamental set consisting of real-valued solutions.

The Real and Imaginary Parts of a Complex-Valued Solution Are Also Solutions

We now state and prove a theorem that shows how to obtain a fundamental set of real-valued solutions when the characteristic equation has complex roots. Theorem 4.9 states that if we have a complex-valued solution, $y(t) = u(t) + iv(t)$, then the real and imaginary parts, $u(t)$ and $v(t)$, are also solutions.

Theorem 4.9

Let $y(t) = u(t) + iv(t)$ be a solution of the differential equation

$$y'' + p(t)y' + q(t)y = 0,$$

where $p(t)$ and $q(t)$ are real-valued and continuous on $a < t < b$ and where $u(t)$ and $v(t)$ are real-valued functions defined on (a, b). Then $u(t)$ and $v(t)$ are also solutions of the differential equation on this interval.

PROOF: Since $y(t)$ is known to be a solution, we have

$$(u + iv)'' + p(t)(u + iv)' + q(t)(u + iv) = 0.$$

Therefore,

$$u'' + iv'' + p(t)(u' + iv') + q(t)(u + iv) = 0.$$

Collecting real and imaginary parts,

$$[u'' + p(t)u' + q(t)u] + i[v'' + p(t)v' + q(t)v] = 0, \qquad a < t < b.$$

In this identity, $p(t)$, $q(t)$, $u(t)$, and $v(t)$ are real-valued functions defined on the interval (a, b). Since a complex quantity vanishes if and only if its real and imaginary parts both vanish, we know that

$$u'' + p(t)u' + q(t)u = 0, \qquad a < t < b,$$

and

$$v'' + p(t)v' + q(t)v = 0, \qquad a < t < b.$$

Therefore, $u(t)$ and $v(t)$ are both solutions of $y'' + p(t)y' + q(t)y = 0$. ▮

We now apply Theorem 4.9 to the equation $y'' + ay' + by = 0$ in the case where the characteristic equation has complex roots, $\lambda_1 = \alpha + i\beta$ and $\lambda_2 = \alpha - i\beta$. For this case [recall equation (9)], the differential equation has complex-valued solutions

$$y_1(t) = e^{\alpha t}(\cos\beta t + i\sin\beta t) \quad \text{and} \quad y_2(t) = e^{\alpha t}(\cos\beta t - i\sin\beta t).$$

Taking the real and imaginary parts of $y_1(t)$ [or equivalently, of $y_2(t)$] we obtain a pair of real-valued solutions

$$u(t) = e^{\alpha t}\cos\beta t \quad \text{and} \quad v(t) = e^{\alpha t}\sin\beta t.$$

Having established that the two functions $u(t) = e^{\alpha t}\cos\beta t$ and $v(t) = e^{\alpha t}\sin\beta t$ are solutions of $ay'' + by' + cy = 0$, we can calculate the Wronskian and verify that they form a fundamental set of solutions. The Wronskian is

$$W(t) = \begin{vmatrix} e^{\alpha t}\cos\beta t & e^{\alpha t}\sin\beta t \\ e^{\alpha t}[\alpha\cos\beta t - \beta\sin\beta t] & e^{\alpha t}[\alpha\sin\beta t + \beta\cos\beta t] \end{vmatrix} = \beta e^{2\alpha t} \neq 0.$$

Therefore, the general solution of $ay'' + by' + cy = 0$ is

$$y(t) = Ae^{\alpha t}\cos\beta t + Be^{\alpha t}\sin\beta t = e^{\alpha t}[A\cos\beta t + B\sin\beta t], \quad (10)$$

where $\alpha + i\beta$ and $\alpha - i\beta$ are the roots of $a\lambda^2 + b\lambda + c = 0$. In (10) we use A and B to represent the arbitrary constants.

EXAMPLE 1

Find the general solution for the differential equation

$$y'' + 25y = 0.$$

Solution: The characteristic equation is $\lambda^2 + 25 = 0$. The roots are $\lambda_1 = 5i$ and $\lambda_2 = -5i$. Therefore, in equation (10), we have $\alpha = 0$ and $\beta = 5$. The general solution is

$$y(t) = A\cos 5t + B\sin 5t. \quad ▲$$

EXAMPLE 2

Solve the initial value problem

$$y'' + 2y' + 5y = 0, \qquad y(0) = 2, \qquad y'(0) = 2.$$

Solution: The characteristic equation is $\lambda^2 + 2\lambda + 5 = 0$. The roots are

$$\lambda_{1,2} = \frac{-2 \pm \sqrt{4-20}}{2} = -1 \pm 2i.$$

In equation (10), we have $\alpha = -1$ and $\beta = 2$; the general solution of the differential equation is

$$y(t) = e^{-t}(A\cos 2t + B\sin 2t).$$

In order to impose the initial conditions, we differentiate $y(t)$, obtaining

$$y'(t) = -e^{-t}(A\cos 2t + B\sin 2t) + e^{-t}(-2A\sin 2t + 2B\cos 2t).$$

The initial conditions at $t = 0$ lead to the equations

$$\begin{aligned} A \qquad &= 2, \\ -A + 2B &= 2. \end{aligned}$$

Solving this system, we find $A = 2$ and $B = 2$. The unique solution of the initial value problem is

$$y(t) = 2e^{-t}(\cos 2t + \sin 2t). \quad ▲$$

The graph of the solution found in Example 2, $y(t) = 2e^{-t}(\cos 2t + \sin 2t)$, is shown in Figure 4.5 on the next page. Notice that the dashed curves, which are

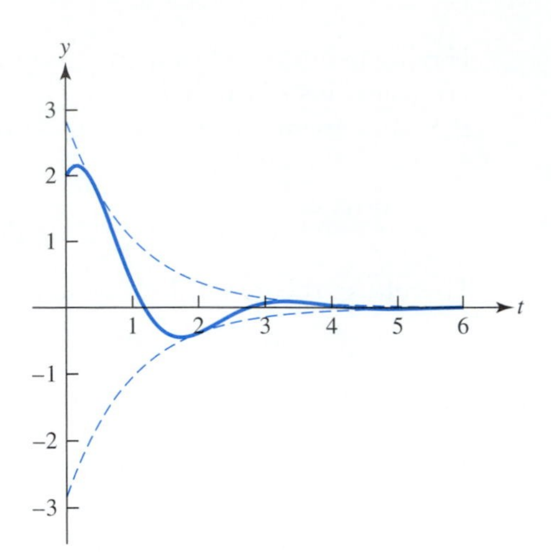

FIGURE 4.5

The graph of the solution, $y(t) = 2e^{-t}(\cos 2t + \sin 2t)$, found in Example 2. The dashed curves, $y = 2\sqrt{2}\,e^{-t}$ and $y = -2\sqrt{2}\,e^{-t}$ constitute an envelope containing the graph of the solution.

actually the graphs of $y = \pm 2\sqrt{2}\,e^{-t}$, represent an envelope that describes the (decreasing) size of the sinusoidal oscillations. We will discuss this envelope in the next subsection.

Amplitude and Phase

Consider an initial value problem whose characteristic equation has complex roots, $\lambda_{1,2} = \alpha \pm i\beta$. As we saw from equation (10), the solution has the form

$$y(t) = e^{\alpha t}[A\cos\beta t + B\sin\beta t], \tag{11}$$

where A and B are constants determined from the given initial conditions. We now show that this solution can also be expressed as

$$y(t) = Re^{\alpha t}\cos(\beta t - \delta), \tag{12}$$

where R and δ are positive constants. In this form, the behavior of the solution as a function of time is more easily understood—it is the product of a sinusoidally oscillating function, $\cos(\beta t - \delta)$, and a term, $Re^{\alpha t}$, that is increasing with time when $\alpha > 0$, constant when $\alpha = 0$, and decreasing with time when $\alpha < 0$.

In (12), the term $Re^{\alpha t}$ is often referred to as the **amplitude** of the oscillations. The constant δ is referred to as the **phase**, the **phase angle**, or the **phase shift**. The term "phase shift" reflects the fact that we obtain the graph of $\cos(\beta t - \delta)$ by shifting the graph of $\cos\beta t$ to the right by an amount $t = \delta/\beta$. To see how equation (11) can be recast as (12), first recall the trigonometric identity

$$\cos(\theta_1 - \theta_2) = \cos\theta_1\cos\theta_2 + \sin\theta_1\sin\theta_2.$$

Using this trigonometric identity on the right-hand side of equation (12), we see that $R\cos(\beta t - \delta) = R\cos\beta t\cos\delta + R\sin\beta t\sin\delta$. Equating expressions (11) and (12), we obtain

$$R[\cos\beta t\cos\delta + \sin\beta t\sin\delta] = A\cos\beta t + B\sin\beta t.$$

Comparing like terms, we see R and δ must be chosen so that

$$R\cos\delta = A \quad \text{and} \quad R\sin\delta = B. \tag{13}$$

From (13) it follows that

$$\begin{aligned} R &= \sqrt{A^2 + B^2}, \\ \tan\delta &= \frac{B}{A}, \qquad A \neq 0. \end{aligned} \tag{14}$$

We need to examine the signs of both $\cos\delta = A/R$ and $\sin\delta = B/R$ in order to determine the quadrant in which the angle δ lies since there are two different choices for an angle δ that satisfies $\tan\delta = B/A$,

$$\delta = \tan^{-1}\left(\frac{B}{A}\right) \quad \text{or} \quad \delta = \tan^{-1}\left(\frac{B}{A}\right) + \pi. \tag{15}$$

EXAMPLE 3

Consider the solution of the initial value problem found in Example 2, $y(t) = 2e^{-t}(\cos 2t + \sin 2t)$. The solution is graphed in Figure 4.5. Rewrite $y(t)$ in the form

$$y(t) = Re^{\alpha t}\cos(\beta t - \delta).$$

Use this form to identify the main features of the graph of $y(t)$.

Solution: Compare the solution, $y(t) = 2e^{-t}(\cos 2t + \sin 2t)$, with expression (11). We see that $A = 2, B = 2, \alpha = -1$, and $\beta = 2$. Therefore,

$$R = 2\sqrt{2} \quad \text{and} \quad \tan\delta = 1.$$

Since $R\cos\delta = 2$ and $R\sin\delta = 2$, it follows that $\cos\delta$ and $\sin\delta$ are both positive. Therefore, δ is in the first quadrant and not in the third quadrant. Having identified the proper quadrant, we obtain $\delta = \tan^{-1}(1) = \pi/4$.

Knowing R, β, and δ, we can rewrite $y(t) = 2e^{-t}(\cos 2t + \sin 2t)$ as

$$y(t) = 2\sqrt{2}e^{-t}\cos\left(2t - \frac{\pi}{4}\right). \tag{16}$$

From expression (16), we can readily deduce the main features of the graph in Figure 4.5. For example, as mentioned earlier, the envelope function $y(t) = 2\sqrt{2}e^{-t}$ governs the amplitude of the oscillations, and this fact is clear from equation (16). The phase angle, $\pi/4$, determines the shift of the cosine function. For instance, in $t > 0, y(t)$ is zero at $t = 3\pi/8, 7\pi/8, 11\pi/8, \ldots$ ▲

EXAMPLE 4

Solve the initial value problem

$$y'' + y = 0, \qquad y(0) = -1, \qquad y'(0) = -\sqrt{3}$$

and put the solution in the form $Re^{\alpha t}\cos(\beta t - \delta)$.

(continued)

(continued)

Solution: The characteristic equation is $\lambda^2 + 1 = 0$ and has roots $\lambda_{1,2} = \pm i$. The general solution is

$$y(t) = A\cos t + B\sin t.$$

Imposing the initial conditions, we obtain the solution

$$y(t) = -\cos t - \sqrt{3}\sin t.$$

Using $R\cos\delta = -1$ and $R\sin\delta = -\sqrt{3}$, we have

$$R = 2 \quad \text{and} \quad \tan\delta = \sqrt{3}.$$

Since $\cos\delta$ and $\sin\delta$ are both negative, the phase angle δ must lie in the third quadrant,

$$\delta = \tan^{-1}\sqrt{3} + \pi = \frac{4\pi}{3}.$$

Hence, see (12),

$$y(t) = 2\cos\left(t - \frac{4\pi}{3}\right).$$

The graph of $y(t)$ is shown in Figure 4.6. It is the graph of $2\cos t$ shifted to the right by $4\pi/3 \approx 4.19$ radians. ▲

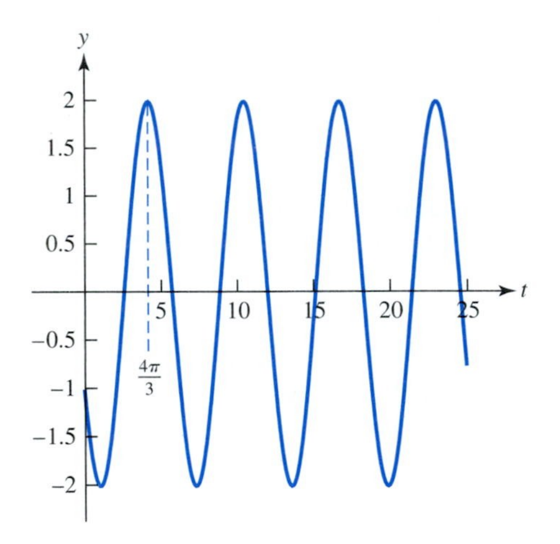

FIGURE 4.6

The graph of the solution found in Example 4, $y(t) = 2\cos(t - (4\pi/3))$.

EXERCISES

The identity $e^{z_1+z_2} = e^{z_1}e^{z_2}$, from which we obtain $(e^z)^n = e^{nz}$, is useful in some of the exercises.

1. Write each of the complex numbers in the form $\alpha + i\beta$, where α and β are real numbers.
   (a) $2e^{i\pi/3}$ (b) $-2\sqrt{2}e^{-i\pi/4}$ (c) $(2-i)e^{i3\pi/2}$
   (d) $\dfrac{1}{2\sqrt{2}}e^{i7\pi/6}$ (e) $\left(\sqrt{2}e^{i\pi/6}\right)^4$

2. Write each of the functions in the form $Ae^{\alpha t}\cos\beta t + iBe^{\alpha t}\sin\beta t$, where α, β, A, and B are real numbers.

(a) $2e^{i\sqrt{2}t}$ (b) $\dfrac{2}{\pi}e^{-(2+3i)t}$ (c) $-\dfrac{1}{2}e^{2t+i(t+\pi)}$

(d) $\left(\sqrt{3}e^{(1+i)t}\right)^3$ (e) $\left(-\dfrac{1}{\sqrt{2}}e^{i\pi t}\right)^3$

Exercises 3–12:

For the given differential equation

(a) Determine the roots of the characteristic equation.

(b) Obtain the general solution as a linear combination of real-valued solutions.

(c) Impose the initial conditions and solve the initial value problem.

3. $y'' + 4y = 0, \quad y(\pi/4) = -2, \ y'(\pi/4) = 1$

4. $y'' + 2y' + 2y = 0, \quad y(0) = 3, \ y'(0) = -1$

5. $9y'' + y = 0, \quad y(\pi/2) = 4, \ y'(\pi/2) = 0$

6. $2y'' - 2y' + y = 0, \quad y(-\pi) = 1, \ y'(-\pi) = -1$

7. $y'' + y' + y = 0, \quad y(0) = -2, \ y'(0) = -2$

8. $y'' + 4y' + 5y = 0, \quad y(\pi/2) = 1/2, \ y'(\pi/2) = -2$

9. $9y'' + 6y' + 2y = 0, \quad y(3\pi) = 0, \ y'(3\pi) = 1/3$

10. $y'' + 4\pi^2 y = 0, \quad y(1) = 2, \ y'(1) = 1$

11. $y'' - 2\sqrt{2}y' + 3y = 0, \quad y(0) = -1/2, \ y'(0) = \sqrt{2}$

12. $9y'' + \pi^2 y = 0, \quad y(3) = 2, \ y'(3) = -\pi$

Exercises 13–17:

The function $y(t)$ is a solution of the initial value problem $y'' + ay' + by = 0$, $y(t_0) = y_0$, $y'(t_0) = y_0'$, where the point t_0 is specified. Determine the constants a, b, y_0, and y_0'.

13. $y(t) = \sin t - \sqrt{2}\cos t, \quad t_0 = \pi/4$

14. $y(t) = 2\sin 2t + \cos 2t, \quad t_0 = \pi/4$

15. $y(t) = e^{-2t}\cos t - e^{-2t}\sin t, \quad t_0 = 0$

16. $y(t) = e^{t-\pi/6}\cos 2t - e^{t-\pi/6}\sin 2t, \quad t_0 = \pi/6$

17. $y(t) = \sqrt{3}\cos \pi t - \sin \pi t, \quad t_0 = 1/2$

Exercises 18–22:

Rewrite the function $y(t)$ in the form $y(t) = Re^{\alpha t}\cos(\beta t - \delta)$, where $0 \le \delta < 2\pi$. Use this representation to sketch a graph of the given function, on a domain sufficiently large to display its main features.

18. $y(t) = \sin t + \cos t$

19. $y(t) = \cos \pi t - \sin \pi t$

20. $y(t) = e^t\cos t + \sqrt{3}e^t \sin t$

21. $y(t) = -e^{-t}\cos t + \sqrt{3}e^{-t}\sin t$

22. $y(t) = e^{-2t}\cos 2t - e^{-2t}\sin 2t$

Exercises 23–25:

In each exercise, the figure shows the graph of the solution of an initial value problem $y'' + ay' + by = 0$, $y(0) = y_0$, $y'(0) = y_0'$. Use the information given to express the solution in the form $y(t) = R\cos(\beta t - \delta)$, where $0 \le \delta < 2\pi$. Determine the constants a, b, y_0, and y_0'.

23.

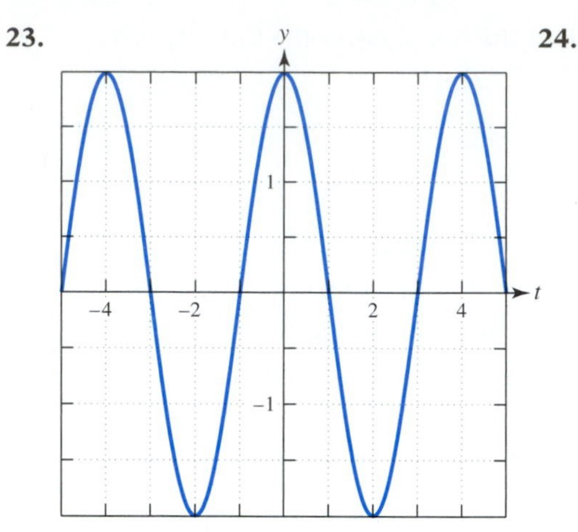

Figure for Exercise 23
The graph has a maximum value at $(0, 2)$ and a t-intercept at $(1, 0)$.

24.

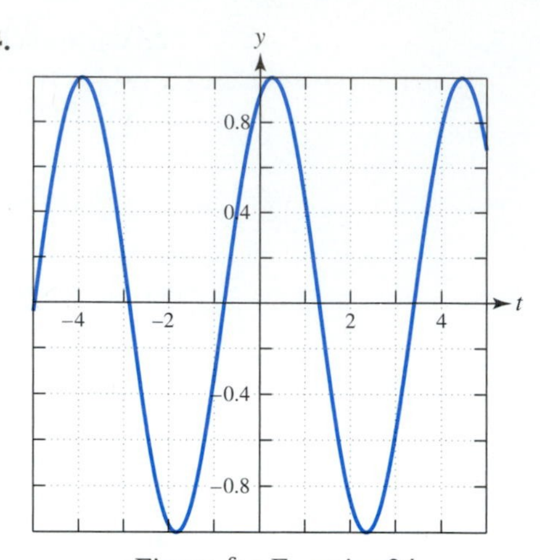

Figure for Exercise 24
The graph has a maximum value at $(\pi/12, 1)$ and a t-intercept at $(5\pi/12, 0)$.

25.

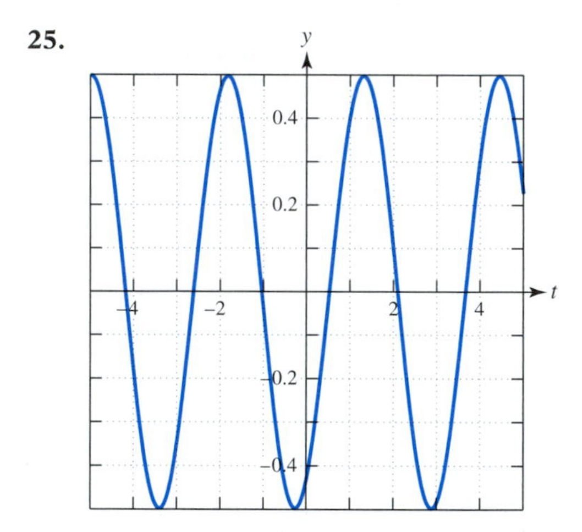

Figure for Exercise 25
The graph has a maximum value at $(5\pi/12, 1/2)$ and a t-intercept at $(\pi/6, 0)$.

26. Buoyancy Problems with Drag Force We discussed modeling the bobbing motion of floating cylindrical objects in the Introduction to Chapter 4. Everyday experience suggests that a bobbing motion will not persist indefinitely; one reason for the reduction or damping of bobbing motion is the drag resistance a floating object experiences as it moves up and down in the liquid. If we assume a drag force proportional to velocity, an application of Newton's second law of motion leads ultimately to the differential equation $y'' + \mu y' + \omega^2 y = 0$, where $y(t)$ is the displacement of the object from its static equilibrium position (measured positive in the downward direction), μ is a positive constant describing the drag force, and ω^2 is a positive constant determined by the mass densities of liquid and object and the vertical extent of the cylindrical object. [See equation (4) in the Introduction.]

(a) Obtain the general solution of this differential equation, assuming that $\mu^2 < 4\omega^2$.

(b) Assume that a cylindrical floating object is initially displaced downward 2 in. and released from rest [so the initial conditions are $y(0) = 2, y'(0) = 0$, with $y(t)$

measured in inches]. Obtain the solution of the initial value problem in terms of μ and ω.

(c) The solution obtained in (b) can be written as $y(t) = Re^{\alpha t}\cos(\beta t - \delta)$. Determine α, β, R, and $\tan\delta$ in terms of μ and ω.

27. For the differential equation $y'' - 2\alpha y' + (\alpha^2 + \beta^2)y = 0$, the complex-valued functions $\{e^{(\alpha+i\beta)t}, e^{(\alpha-i\beta)t}\}$ form a fundamental set of solutions, while $\{e^{\alpha t}\cos\beta t, e^{\alpha t}\sin\beta t\}$ is a fundamental set of real-valued solutions.

(a) Express each of the real-valued solutions, $e^{\alpha t}\cos\beta t$ and $e^{\alpha t}\sin\beta t$, as a linear combination of the complex-valued solutions, $e^{(\alpha+i\beta)t}$ and $e^{(\alpha-i\beta)t}$. [Hint: Use the identities

$$\cos\theta = \frac{e^{i\theta} + e^{-i\theta}}{2} \quad \text{and} \quad \sin\theta = \frac{e^{i\theta} - e^{-i\theta}}{2i}.$$

Some of the constants appearing in the linear combination will be complex numbers.]

(b) Organize the equations obtained in part (a) into a single matrix equation and determine a (2×2) constant matrix

$$A = \begin{bmatrix} a_{11} & a_{12} \\ a_{21} & a_{22} \end{bmatrix}$$

such that

$$[e^{\alpha t}\cos\beta t, e^{\alpha t}\sin\beta t] = [e^{(\alpha+i\beta)t}, e^{(\alpha-i\beta)t}]\begin{bmatrix} a_{11} & a_{12} \\ a_{21} & a_{22} \end{bmatrix}$$

(recall Theorem 4.8 in Section 4.3).

(c) Show that the matrix A obtained in part (b) is invertible. Compute A^{-1} and use the relation

$$[e^{(\alpha+i\beta)t}, e^{(\alpha-i\beta)t}] = [e^{\alpha t}\cos\beta t, e^{\alpha t}\sin\beta t]A^{-1}$$

to obtain equations expressing each of the solutions $e^{(\alpha+i\beta)t}$ and $e^{(\alpha-i\beta)t}$ as linear combinations of the solutions $e^{\alpha t}\cos\beta t$ and $e^{\alpha t}\sin\beta t$.

28. Consider the differential equation $y'' + ay' + 9y = 0$, where a is a real constant. Suppose we know that the Wronskian of a fundamental set of solutions of this differential is constant: $W(t) \equiv 1$, $-\infty < t < \infty$. Find the general solution of this differential equation.

29. This section has focused on the differential equation $ay'' + by' + cy = 0$, where a, b, and c are real constants. The fact that the roots of the characteristic equation occur in conjugate pairs when they are complex is due to the fact that the coefficients a, b, and c are real numbers. To see that this may not be true if the coefficients are allowed to be complex, determine the roots of the characteristic equation for the differential equation

$$y'' + 4iy' + 5y = 0.$$

Find two complex-valued solutions of this equation.

4.7 Unforced Mechanical Vibrations

In this section we model the motion of a simple mechanical system, that of a mass suspended from the end of a hanging spring and subjected to some initial disturbance that causes it to vibrate or oscillate vertically about its rest position. This model possesses features that are similar to the buoyancy model considered in the Introduction.

Hooke's Law

A spring hangs vertically from a ceiling. We assume that the weight of the spring is negligibly small. The natural or unstretched length of the spring is denoted by l, as in Figure 4.7. Suppose we now apply a vertical force to the end of the spring. If the force is directed downward, the spring will stretch. As it stretches, the spring develops an upward restoring force that resists this stretching or elongation. Conversely, if the applied force is directed upward, the spring compresses or shortens in length. In this case, the spring develops a counteracting downward restoring force that tends to resist this compression.

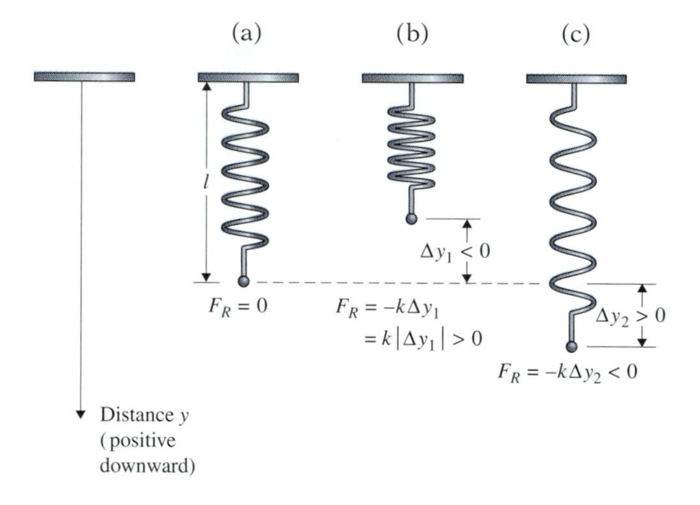

FIGURE 4.7

(a) A spring with natural length l. (b) The restoring force, $F_R = -k\Delta y_1$, is positive (directed downward) when the spring is compressed. (c) The restoring force, $F_R = -k\Delta y_2$, is negative (directed upward) when the spring is stretched.

To describe the restoring force mathematically, we need an equation that relates the restoring force developed by the spring to the amount of elongation or compression that has occurred. The relation we use is **Hooke's law**,[7] which assumes the restoring force developed by the spring is proportional to the amount of stretching or compression that the spring has undergone. Let Δy represent the displacement of the spring end from its undisturbed position. As in Figure 4.7, we assume the downward direction is the positive direction. Therefore, the displacement Δy is positive when the spring is stretched and negative when it is compressed. Whether the spring is stretched or compressed, its length is given by the quantity $l + \Delta y$.

[7]Robert Hooke (1635–1703) served as professor of geometry at Gresham College, London, for 30 years. He worked on problems in elasticity, optics, and simple harmonic motion. Hooke invented the conical pendulum and was the first to build a Gregorian reflecting telescope. He was a competent architect and helped Christopher Wren rebuild London after the Great Fire of 1666.

Hooke's law states that the restoring force is

$$F_R = -k\Delta y, \tag{1}$$

where Δy is the displacement and k is a positive constant of proportionality, called the **spring constant**. The negative sign in equation (1) arises because the restoring force acts to counteract the displacement of the spring end. That is, when the spring is stretched, the restoring force is directed in the upward (negative) direction; when compressed, the spring exerts a downward (positive) restoring force (see Figure 4.7).

The spring constant k in equation (1) has the dimensions of force per unit length. The spring constant represents a measure of spring stiffness. A stiffer spring has a larger value k and the same restoring force arises from a smaller displacement.

Hooke's law is a useful description of reality when the displacement magnitude, $|\Delta y|$, is reasonably small. It cannot remain valid for arbitrarily large $|\Delta y|$; one cannot stretch or compress a spring indefinitely. We will assume in all our modeling and computations that displacement magnitudes are small enough to permit the use of Hooke's law.

A Mathematical Model for the Spring-Mass System

An object, having mass m, is attached to the end of the unstretched spring as in Figure 4.8. The weight of the object is

$$W = mg,$$

where g is the acceleration due to gravity. The spring stretches until it achieves a new rest or equilibrium configuration. Let Y represent the distance the spring stretches to achieve this new equilibrium position. The displacement, Y, is determined by Hooke's law—the spring stretches until the restoring force exactly counteracts the object's weight:

$$W + F_R = mg - kY = 0. \tag{2}$$

It follows from equation (2), that $Y = mg/k$.

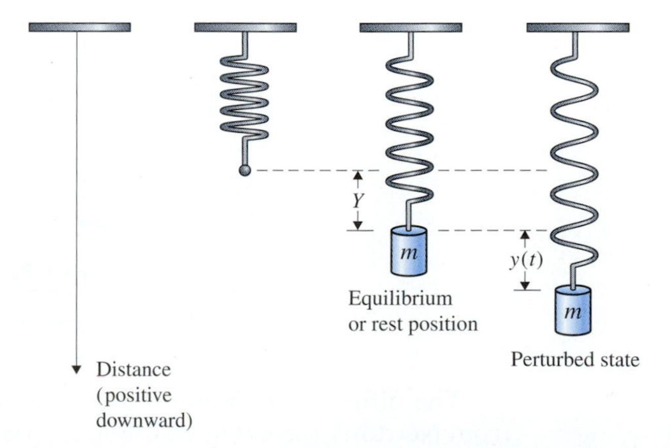

FIGURE 4.8

As the mass moves, the quantity $y(t)$ measures its displacement from the equilibrium position.

What happens when the spring-mass system in Figure 4.8 is perturbed from its equilibrium state? For instance, suppose the mass is pulled down and then released at an initial time $t = 0$. Once the mass is released, our experience suggests that it will begin moving up and down, oscillating back and forth across the equilibrium position. We use Newton's second law of motion to derive a differential equation whose solution describes the position of the mass as a function of time t.

We represent the displacement of the spring end as

$$Y + y(t). \tag{3}$$

With respect to this new dependent variable, y, we see from Figure 4.8 that the restoring force of the spring at time t is given by

$$F_R = -k(Y + y(t)).$$

Finally, we assume a damping mechanism, a dashpot, is attached and suppresses the vibrating motion of the spring-mass system. It is shown schematically in Figure 4.9. The damping force it applies is assumed to be proportional to velocity,

$$F_D = -\gamma \frac{d(Y + y(t))}{dt} = -\gamma \frac{dy(t)}{dt}. \tag{4}$$

In (4), γ is a positive constant of proportionality, referred to as the **damping coefficient**. The negative sign is present because the damping force acts to oppose the motion. For instance, when the mass is moving downward and the velocity of the mass, dy/dt, is positive, the damping force is upward and thus negative. A similar model of velocity damping was assumed in the linear model of projectile motion with air resistance that we discussed in Section 3.5.

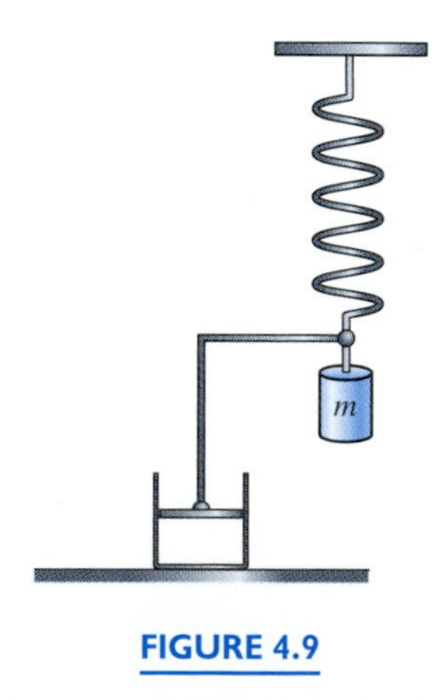

FIGURE 4.9

A spring-mass-dashpot system.

The differential equation describing the motion of the mass is obtained from Newton's second law of motion, $ma = F$:

$$m \frac{d^2y}{dt^2} = W + F_R + F_D = mg - k(Y + y) - \gamma \frac{dy}{dt}. \tag{5}$$

Since $mg - kY = 0$, equation (5) simplifies to

$$m\frac{d^2y}{dt^2} + \gamma\frac{dy}{dt} + ky = 0. \tag{6}$$

Equation (6) is a second order linear constant coefficient differential equation. It is homogeneous because we are considering only the **unforced** or **free vibration** of the system. That is, we are assuming there are no external forces applied to the system.

In Section 4.11, we will consider **forced vibrations**, where equation (6) is modified to include a time-varying applied force, $F(t)$. Inclusion of an applied force leads to a nonhomogeneous equation of the form

$$m\frac{d^2y}{dt^2} + \gamma\frac{dy}{dt} + ky = F(t). \tag{7}$$

Since equation (6) is homogeneous and has constant coefficients, we can solve this equation using the techniques described in Sections 4.4–4.6. In Sections 4.8–4.10, we will derive methods for solving the nonhomogeneous equation (7).

Behavior of the Model

In this subsection we discuss the solutions of equation (6), $my'' + \gamma y' + ky = 0$. These solutions describe how the mass-spring-dashpot system behaves—predicting the position, $y(t)$, and the velocity, $y'(t)$, of the moving mass at any time t. To solve equation (6) we begin with the characteristic equation,

$$m\lambda^2 + \gamma\lambda + k = 0,$$

which has roots

$$\lambda_{1,2} = \frac{-\gamma \pm \sqrt{\gamma^2 - 4mk}}{2m}. \tag{8}$$

The corresponding mass-spring-dashpot system exhibits different behavior, depending on the roots of the characteristic equation. The roots are determined by the relative values of the mass, the spring constant, and the damping coefficient.

Case I: If $\gamma^2 > 4km$ (if damping is relatively strong), the characteristic equation has two negative real roots

$$\lambda_{1,2} = \frac{-\gamma \pm \sqrt{\gamma^2 - 4mk}}{2m}.$$

(Both roots are negative since $\sqrt{\gamma^2 - 4mk}$ is less than γ.) As shown in Section 4.4, the general solution is given by

$$y(t) = c_1e^{\lambda_1 t} + c_2e^{\lambda_2 t}. \tag{9a}$$

Therefore, the strong damping suppresses any vibratory motion of the mass. The general solution is a linear combination of two decreasing exponential functions. This case is referred to as the **overdamped case**.

Case 2: If the three constants are such that $\gamma^2 = 4km$, then the two roots are real and repeated. As we saw in Section 4.5, the general solution is given by

$$y(t) = c_1 e^{\lambda_1 t} + c_2 t e^{\lambda_1 t}, \qquad \textbf{(9b)}$$

where $\lambda_1 = -\gamma/2m$. In this case, known as the **critically damped case**, damping is also sufficiently strong to suppress oscillatory vibrations of the mass.

Case 3: If damping is sufficiently weak, so that $\gamma^2 < 4km$, then the roots are complex conjugates,

$$\lambda_{1,2} = \frac{-\gamma}{2m} \pm i\frac{\sqrt{4mk - \gamma^2}}{2m} = \alpha \pm i\beta.$$

As we saw in Section 4.6, the general solution is given by

$$y(t) = e^{\alpha t}[c_1 \cos \beta t + c_2 \sin \beta t]. \qquad \textbf{(9c)}$$

In this case, known as the **underdamped case**, damping is too weak to totally suppress the vibrations of the mass. Note that the underdamped case also includes the case where there is no damping whatsoever; that is, the case where $\gamma = 0$. Later, we refer to this as the **undamped case**. Finally, recall from Section 4.6 that solution (9c) can be restated in amplitude-phase form as

$$y(t) = Re^{\alpha t} \cos(\beta t - \delta), \qquad \textbf{(10)}$$

where $R = \sqrt{c_1^2 + c_2^2}$, $R \cos \delta = c_1$, and $R \sin \delta = c_2$. If damping is present (that is, if $\gamma > 0$), then $\alpha = -\gamma/2m < 0$ and the motion of the mass described by equation (10) consists of damped vibrations (oscillations that decrease in amplitude as time progresses). If there is no damping, then $\alpha = 0$ since $\gamma = 0$ and the oscillations do not decrease in magnitude.

Representative examples of the motion that occurs in these three cases is shown in Figure 4.10 (a)–(c). When damping is present, it follows from equations (9a)–(9c) that $\lim_{t\to\infty} y(t) = 0$ for any choices of c_1 and c_2. This is to be expected. When there is damping present in the system, energy is dissipated and any initial disturbance will diminish in strength as time increases.

Vibrations and Periodic Functions

As a special case, assume the damping mechanism is absent; that is, $\gamma = 0$. In this case, $\alpha = 0$ and solution (10) reduces to

$$y(t) = R\cos\left(\sqrt{\frac{k}{m}}\,t - \delta\right). \qquad \textbf{(11)}$$

Setting $\omega = \sqrt{k/m}$, equation (11) becomes

$$y(t) = R\cos(\omega t - \delta). \qquad \textbf{(12)}$$

In this case, the amplitude or envelope of the vibrations remains constant and equal to R. The function $y(t)$ in (12) is an example of a "periodic function."

Let $f(t)$ be defined on $-\infty < t < \infty$ or $a \le t < \infty$ for some a. The function $f(t)$ is called a **periodic function** if there exists a positive constant T, called the

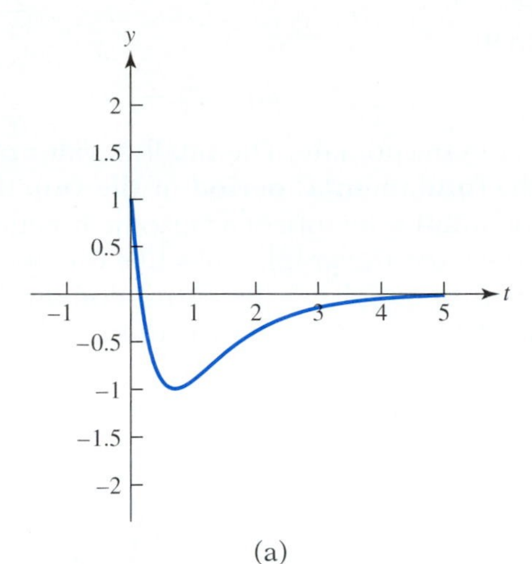

(a)

Equation: $y'' + 4y' + 3y = 0$
$y(0) = 1, \quad y'(0) = -9$
Solution: $y(t) = 4e^{-3t} - 3e^{-t}$

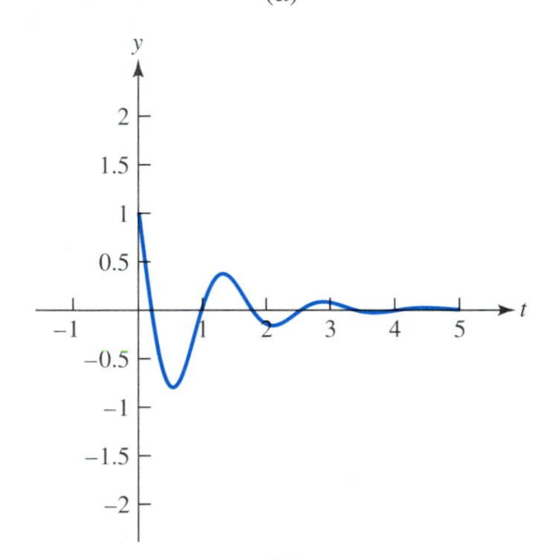

(b)

Equation: $y'' + 2y' + 17y = 0$
$y(0) = 1, \quad y'(0) = -5$
Solution: $y(t) = e^{-t}\,[\cos 4t - \sin 4t]$

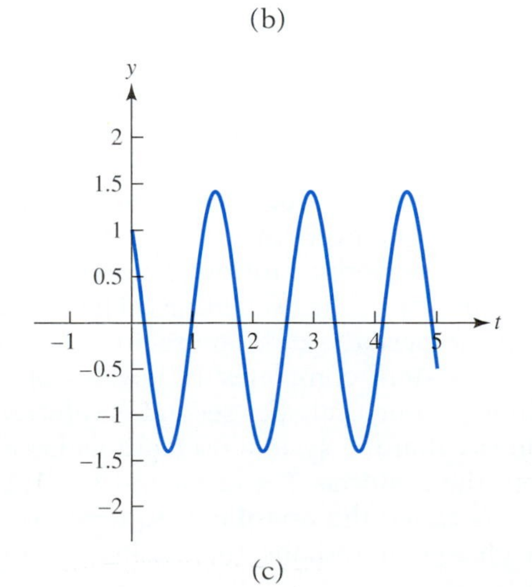

(c)

Equation: $y'' + 16y = 0$
$y(0) = 1, \quad y'(0) = -4$
Solution: $y(t) = \cos 4t - \sin 4t$

FIGURE 4.10

(a) An example of overdamped motion. (b) An example of underdamped motion with nonzero damping. (c) An example of undamped motion.

period, such that

$$f(t+T) = f(t) \tag{13}$$

for all values of t in the domain. The smallest value of the constant T satisfying (13) is called the **fundamental period of the function**.

The basic qualitative feature of a periodic function is that its graph repeats itself. If we know what the graph looks like on any time segment of duration T, we can obtain the graph on the entire domain by simply replicating this segment. Figure 4.11 provides an illustration.

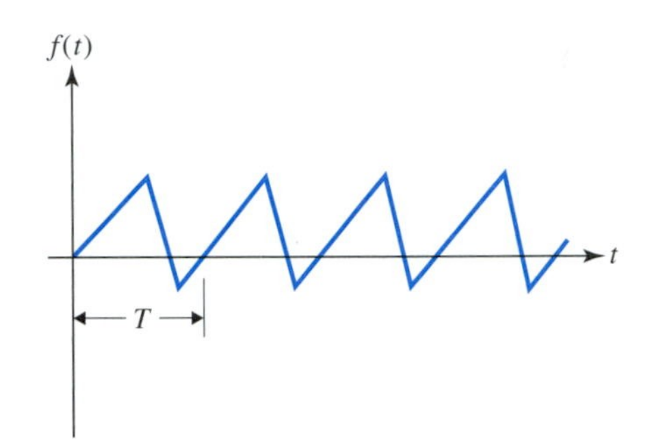

FIGURE 4.11

The graph of a periodic function repeats itself over any time period of duration T, where T is the fundamental period of the function.

Consider again the solution $y(t) = R\cos(\omega t - \delta)$ in equation (12). Since the cosine function repeats itself whenever its argument changes by 2π, $y(t)$ is a periodic function. To find the period for $y(t)$, we set $y(t+T) = y(t)$:

$$\begin{aligned} y(t+T) &= R\cos[\omega(t+T) - \delta] \\ &= R\cos[\omega t + \omega T - \delta] \\ &= y(t), \ \text{if } \omega T = 2n\pi, \qquad n = 1, 2, \ldots . \end{aligned}$$

Therefore, the function $y(t)$ has fundamental period $T = 2\pi/\omega$. In terms of the spring-mass system, $T = 2\pi/\omega$ is referred to as the **fundamental period of the motion** or simply the **period**. The period represents the time required for the mass to execute one cycle of its oscillatory motion. The motion itself is often referred to as **periodic motion**. The reciprocal of the period, $f = 1/T$, is called the **frequency** of the oscillations. The frequency represents the number of cycles of the periodic motion executed per unit time. For example, if $T = 0.01$ sec, the system completes 100 cycles of its motion per second. In current terminology, one cycle per second is referred to as one Hertz.[8] Therefore, we would say that the system oscillations have a frequency of 100 Hertz (100 Hz). From the relations $T = 2\pi/\omega$ and $f = 1/T$, it follows that $\omega = 2\pi f$. The constant ω is called the **angular frequency** or the **radian frequency**. It represents the change, in radians, that $\cos(\omega t - \delta)$ undergoes in one period.

[8] Heinrich Hertz (1857–1894) was a German physicist who confirmed Maxwell's theory of electromagnetism by producing and studying radio waves. He demonstrated that these waves travel at the velocity of light and can be reflected, refracted, and polarized. The unit of frequency was named in his honor.

It's worthwhile to check that the model predictions are consistent with our everyday experience. In the absence of damping, angular frequency is $\omega = \sqrt{k/m}$ and frequency is $f = (1/2\pi)\sqrt{k/m}$. Frequency therefore increases as either k increases or m decreases. Thus, when a given mass is attached in turn to two springs of differing stiffness, the model predicts it will vibrate more rapidly when suspended from the stiffer spring. Likewise, if two bodies of differing mass (and therefore weight) are suspended from the same spring, the smaller mass will vibrate more rapidly than the larger mass.

Now consider what happens when damping is added. The solution $y(t)$ in equation (10) has the form

$$y(t) = Re^{\alpha t}\cos(\beta t - \delta), \tag{14}$$

where

$$\alpha = \frac{-\gamma}{2m} \quad \text{and} \quad \beta = \frac{\sqrt{4km - \gamma^2}}{2m}.$$

Equation (14) predicts that when two different masses are attached to a spring-dashpot system having the same damping coefficient γ and spring constant k, the solution envelope of the larger mass (the heavier body) will decrease more slowly with time because the associated value α is smaller.

Note also that the introduction of damping changes the cosine term in equation (12) from $\cos((\sqrt{k/m})t - \delta)$ when damping is absent, to the term

$$\cos\left(\frac{\sqrt{4km - \gamma^2}}{2m}\,t - \delta\right)$$

in equation (14) when damping is present. Since

$$\sqrt{\frac{k}{m}} > \frac{\sqrt{4km - \gamma^2}}{2m},$$

we see that the introduction of damping causes the vibrations to "slow down" while simultaneously being reduced in amplitude.

Are these model predictions consistent with your everyday experience? What experiments might test these predictions, both qualitatively and quantitatively?

We conclude this section with two examples illustrating the motion of a spring-mass-dashpot system.

EXAMPLE 1

A block weighing 8 lb is attached to the end of a spring, causing the spring to stretch 6 in. beyond its natural length. The block is then pulled down 3 in. and released. Determine the motion of the block, assuming there are no damping forces or external applied forces.

Solution: The motion of the block is governed by equation (6),

$$my'' + \gamma y' + ky = 0,$$

along with the initial conditions of the problem. Taking the gravitational

(continued)

(continued)

constant to be $g = 32$ ft/sec^2 and noting that the weight, W, is given by $W = mg$, the block has mass

$$m = \frac{1}{4}\frac{\text{lb-sec}^2}{\text{ft}}.$$

[Note that 1 lb-sec^2/ft = 1 slug.] By assumption, the damping coefficient γ is zero. The spring constant k can be determined by the fact that an 8-lb force (the weight of the block) causes the spring to stretch 6 in.; by Hooke's law we have 8 lb = (k)(6 in.) or $k = 8/6$ lb/in. = 16 lb/ft. Since 3 in. = 1/4 ft, the initial value problem governing the motion is

$$\frac{1}{4}y'' + 16y = 0, \qquad y(0) = \frac{1}{4}, \qquad y'(0) = 0,$$

or

$$y'' + 64y = 0, \qquad y(0) = \frac{1}{4}, \qquad y'(0) = 0. \tag{15}$$

The units of $y(t)$ are feet and $y'(t)$ are feet per second. The general solution of $y'' + 64y = 0$ is

$$y(t) = c_1 \cos 8t + c_2 \sin 8t.$$

Imposing the initial conditions, we obtain the solution, $y(t) = (1/4)\cos 8t$. A graph of the block's position, $y(t)$, is shown in Figure 4.12(a). ▲

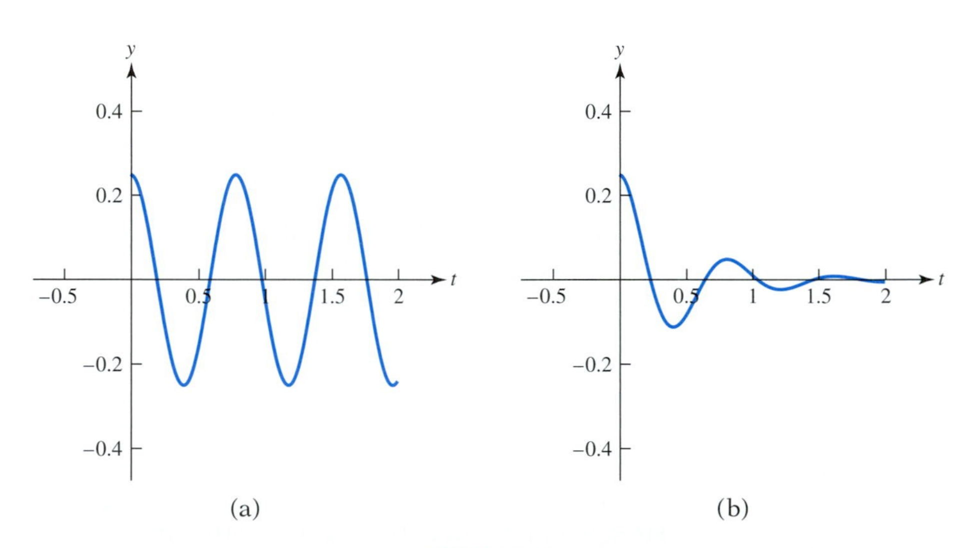

FIGURE 4.12

(a) The position, $y(t)$, of the mass in Example 1 (no damping).
(b) The position, $y(t)$, of the mass in Example 2 (includes damping).

EXAMPLE 2

Consider the spring-mass system in Example 1. Assume there is a damping force associated with the spring and that the damping coefficient is given by $\gamma = 1$ lb-sec/ ft. Determine the motion of the block.

Solution: To account for the assumed damping force, the equation of Example 1 must be modified to include a damping term $\gamma y'$, where $\gamma = 1$. This leads to the equation

$$\frac{1}{4}y'' + y' + 16y = 0.$$

Therefore, the initial value problem governing the motion of the block is

$$y'' + 4y' + 64y = 0, \qquad y(0) = \frac{1}{4}, \qquad y'(0) = 0.$$

The characteristic equation has roots

$$\lambda_{1,2} = \frac{-4 \pm \sqrt{16 - (4)(64)}}{2} = \frac{-4 \pm 4i\sqrt{15}}{2} = -2 \pm 2i\sqrt{15}.$$

The general solution is, therefore,

$$y(t) = e^{-2t}\left[c_1 \cos(2\sqrt{15}\,t) + c_2 \sin(2\sqrt{15}\,t)\right].$$

Imposing the initial conditions $y(0) = 1/4$ and $y'(0) = 0$, we obtain

$$y(t) = \frac{1}{4}e^{-2t}\left[\cos(2\sqrt{15}\,t) + \frac{1}{\sqrt{15}}\sin(2\sqrt{15}\,t)\right].$$

The graph of $y(t)$ is shown in Figure 4.12(b). The block still oscillates about its equilibrium position, but the envelope of the oscillations decreases with time. ▲

EXERCISES

1. The given function $f(t)$ is periodic with fundamental period T; therefore, $f(t+T) = f(t)$. Use the information given to sketch the graph of $f(t)$ over the time interval $0 \le t \le 4T$.

(a) $f(t) = t(2-t), 0 \le t < 2, \quad T = 2$

(b) $f(t) = \begin{cases} 1, & 0 \le t \le \frac{3}{2} \\ 0, & \frac{3}{2} < t < 2 \end{cases}, \quad T = 2$

(c) $f(t) = \begin{cases} t, & 0 \le t \le 2 \\ 4-t, & 2 < t < 4 \end{cases}, \quad T = 4$

(d) $f(t) = -1 + t, 0 \le t < 2, \quad T = 2$

(e) $f(t) = 2e^{-t}, 0 \le t < 1, \quad T = 1$

(f) $f(t) = \sin t, 0 \le t < \pi, \quad T = \pi$

(g) $f(t) = \begin{cases} 2\sin \pi t, & 0 \le t \le 1 \\ 0, & 1 < t < 2 \end{cases}, \quad T = 2$

Exercises 2–10:

Use values 9.8 m/s^2 or 32 ft/sec^2 for the acceleration due to gravity.

2. A 10-kg mass, when attached to the end of a spring hanging vertically, stretches the spring 30 mm. Assume the mass is then pulled down another 70 mm and released (with no initial velocity).

(a) Determine the spring constant k.

(b) State the initial value problem (giving numerical values for all constants) for $y(t)$, where $y(t)$ denotes the displacement (in meters) of the mass from its equilibrium rest position. Assume that y is measured positive in the downward direction.

(c) Solve the initial value problem formulated in part (b).

3. A 20-kg mass was initially at rest, attached to the end of a vertically hanging spring. When given an initial velocity of 2 m/s from its equilibrium rest position, the mass was observed to attain a maximum displacement of 0.2 m from its equilibrium position. What is the value of the spring constant?

4. The graph shows the displacement from equilibrium of a mass-spring system as a function of time after the vertically hanging system was set in motion at $t = 0$. Assume that the units of time and displacement are seconds and centimeters respectively.

(a) What is the period T of the periodic motion?

(b) What is the frequency f (in Hertz)? What is the angular frequency ω (in rad/sec)?

(c) Determine the amplitude R and the phase angle δ (in radians) and express the displacement in the form $y(t) = R\cos(\omega t - \delta)$, with y in meters.

(d) With what initial displacement $y(0)$ and initial velocity $y'(0)$ was the system set into motion?

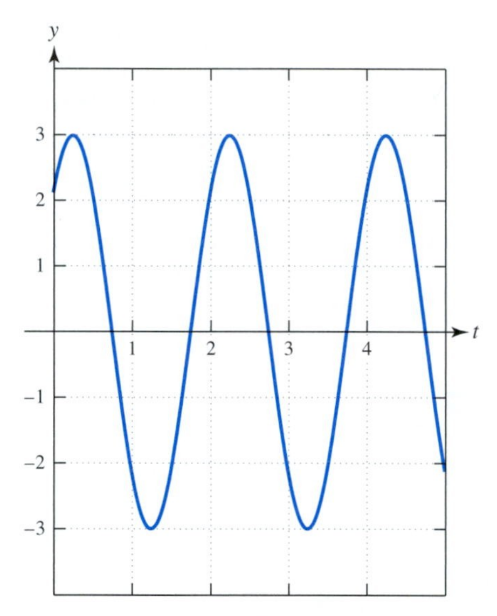

Figure for Exercise 4
The first t-intercept is $(3/4, 0)$, and the first minimum has coordinates $(5/4, -3)$.

5. A spring-mass-dashpot system consists of a 10-kg mass attached to a spring with spring constant $k = 100$ N/m; the dashpot has damping constant 7 kg/s. At time $t = 0$, the system is set into motion by pulling the mass down 0.5 m from its equilibrium rest position while simultaneously giving it an initial downward velocity of 1 m/s.

(a) State the initial value problem to be solved for $y(t)$, the displacement from equilibrium (in meters) measured positive in the downward direction. Give numerical values to all constants involved.

(b) Solve the initial value problem. What is $\lim_{t\to\infty} y(t)$? Explain why your answer for this limit makes sense from a physical perspective.

(c) Plot your solution on a time interval long enough to identify how long it takes for the magnitude of the vibrations to be reduced to 0.1 m. In other words, estimate the smallest time, τ, for which $|y(t)| \le 0.1m$, $\tau \le t < \infty$.

6. A spring and dashpot system is to be designed for a 32-lb weight so that the overall system is critically damped.

(a) How must the damping constant γ and spring constant k be related?

(b) Assume the system is to be designed so that the mass, when given an initial velocity of 4 ft/sec from its rest position, will have a maximum displacement of 6 in. What values of damping constant γ and spring constant k are required?

7. Assume we have a spring-mass-dashpot system that is to be released from rest with an initial displacement of the mass from equilibrium given by $y(0) = y_0$. Consider the following question. "What happens to the displacement $y(t)$ if we keep the values of mass m and spring constant k fixed but increase the damping constant γ?" In particular, select an arbitrary but fixed time $t > 0$, and think of the solution $y(t)$ as being a function of the damping constant γ and then determine $\lim_{\gamma\to\infty} y(t)$. Do you have any physical or intuitive insight as to what the answer should be?

We develop the answer with the following steps.

(a) Show that the roots of the characteristic polynomial are

$$\lambda_1 = -\frac{\gamma}{2m} - \frac{\sqrt{\gamma^2 - 4mk}}{2m}, \qquad \lambda_2 = -\frac{\gamma}{2m} + \frac{\sqrt{\gamma^2 - 4mk}}{2m}.$$

Solve the initial value problem, assuming the system to be overdamped. Express the solution in terms of the two roots λ_1 and λ_2. (Since we are interested in the system's behavior for large values of γ, the overdamped assumption is appropriate.)

(b) Show that $\lim_{\gamma\to\infty} \lambda_1 = -\infty$ and $\lim_{\gamma\to\infty} \lambda_2 = 0$.

(c) Use the results of parts (a) and (b) to determine $\lim_{\gamma\to\infty} y(t)$ (with k, m, and $t > 0$ fixed). What is the physical meaning of your answer? Does it agree with your intuition? Does it make physical sense in retrospect?

8. In this problem we explore computationally the question posed in Exercise 7. Consider the initial value problem

$$y'' + \gamma y' + y = 0, \qquad y(0) = 1, \qquad y'(0) = 0,$$

where, for simplicity, we have given the mass, spring constant, and initial displacement all a numerical value of unity.

(a) Determine γ_{crit}, the damping constant value that makes the given spring-mass-dashpot system critically damped.

(b) Use computational software to plot the solution of the initial value problem for $\gamma = \gamma_{\text{crit}}$, $2\gamma_{\text{crit}}$, and $20\gamma_{\text{crit}}$ over a common time interval sufficiently large to display the main features of each solution. What trend do you observe in the behavior of the solutions as γ increases? Is it consistent with the conclusions reached in Exercise 7?

9. We studied the bobbing motion of a cylindrical object floating in a liquid in the Introduction to Chapter 4. If a drag force is incorporated into the model, the displacement of the object from its equilibrium position (measured positive downward) satisfies the differential equation:

$$y'' + \mu y' + \left(\frac{\rho_l g}{\rho L}\right) y = 0,$$

where μ is a constant of proportionality associated with the drag force, ρ_l and ρ are the mass densities of the liquid and the floating object, respectively ($\rho < \rho_l$), g is acceleration due to gravity, and L is the vertical dimension of the floating object. Structurally, this equation is identical to the differential equation modeling a spring-mass-dashpot system,

$$y'' + \frac{\gamma}{m}y' + \frac{k}{m}y = 0.$$

A scientist is interested in simulating the bobbing motion of a large floating buoy by building a small spring-mass-dashpot system in the laboratory. The buoy is a cylindrical drum that floats and moves in the vertically upright position. The buoy has a diameter of 5 ft and a length of 8 ft; it weighs 6000 lb. Assume that water weighs 62.4 lb per cubic foot, $g = 32$ ft/sec^2, and that $\mu = 0.1$ sec^{-1} has been ascertained. The scientist wants to use a 5-kg mass to simulate the motion. What values of damping constant γ and spring constant k must she use to make the two differential equations identical?

10. Continuous Dependence upon a Parameter Consider a spring-mass-dashpot system that is set into motion from its initial rest state with an initial velocity $y'(0) = y'_0$. The relevant initial value problem is

$$my'' + \gamma y' + ky = 0, \qquad y(0) = 0, \qquad y'(0) = y'_0.$$

Assume that mass m and spring constant k are fixed. We want to study the behavior of the solution as a function of the damping constant γ.

(a) Show that the given initial value problem has the following solutions:

Underdamped case ($\gamma^2 < 4km$): $$y_u(t, \gamma) = y'_0 e^{-(\gamma/2m)t} \frac{\sin\left(\sqrt{\dfrac{k}{m} - \dfrac{\gamma^2}{4m^2}}\, t\right)}{\sqrt{\dfrac{k}{m} - \dfrac{\gamma^2}{4m^2}}}.$$

Critically damped case ($\gamma^2 = 4km$): $$y_c(t, \gamma) = y'_0 t e^{-\sqrt{(k/m)}t}.$$

Overdamped case ($\gamma^2 > 4km$): $$y_o(t, \gamma) = y'_0 e^{-(\gamma/2m)t} \frac{\sinh\left(\sqrt{\dfrac{\gamma^2}{4m^2} - \dfrac{k}{m}}\, t\right)}{\sqrt{\dfrac{\gamma^2}{4m^2} - \dfrac{k}{m}}}.$$

The notation used emphasizes the fact that we are considering the solutions to be functions of the parameter γ.

(b) We expect solutions of the spring-mass-dashpot system to exhibit a continuous dependence upon the parameter γ. In other words, if we change the value of γ continuously, the solution y should not change abruptly. Demonstrate this continuous dependence as follows. Assume that $t > 0$ is arbitrary but fixed. Show that the two limits hold:

$$\lim_{\gamma \to \left(2\sqrt{k/m}\right)^-} y_u(t, \gamma) = y_c(t, \gamma) \quad \text{and} \quad \lim_{\gamma \to \left(2\sqrt{k/m}\right)^+} y_o(t, \gamma) = y_c(t, \gamma).$$

Therefore, if we increase the damping constant up to the critically damped value, the underdamped solution approaches the critically damped solution. If we decrease the damping constant down to the critically damped value, the overdamped solution similarly approaches the critically damped solution.

4.8 The General Solution of the Linear Nonhomogeneous Equation

We consider the linear second order nonhomogeneous differential equation

$$y'' + p(t)y' + q(t)y = g(t), \qquad a < t < b, \tag{1}$$

and ask: "What is the general solution of this equation?"

The General Solution

To understand the structure of the general solution of differential equation (1), we pose a second question: "To what extent can two solutions of equation (1) differ from one another?" Once we can answer this question, we will know how every solution of equation (1) is related to a single particular solution that we may somehow have found.

Assume we have two solutions of nonhomogeneous equation (1); call them $u(t)$ and $v(t)$. Since both are solutions, we know that

$$u'' + p(t)u' + q(t)u = g(t), \quad \text{and} \quad v'' + p(t)v' + q(t)v = g(t), \qquad a < t < b. \tag{2}$$

Subtracting, we obtain

$$[u'' - v''] + p(t)[u' - v'] + q(t)[u - v] = g(t) - g(t) = 0, \qquad a < t < b.$$

Therefore, the difference function, $w(t) = u(t) - v(t)$, is a solution of the associated linear *homogeneous* equation:

$$y'' + p(t)y' + q(t)y = 0. \tag{3}$$

Now, let $y_P(t)$ be a particular solution of equation (1) that we somehow have found. Let $y(t)$ be any solution whatsoever of equation (1). Let $y_1(t)$ and $y_2(t)$ form a fundamental set of solutions for the homogeneous equation (3). As we saw previously, the difference function $y(t) - y_P(t)$ is a solution of equation (3). Since $\{y_1, y_2\}$ is a fundamental set of solutions, there are constants c_1 and c_2 such that

$$y(t) - y_P(t) = c_1 y_1(t) + c_2 y_2(t).$$

Equivalently,

$$y(t) = [c_1 y_1(t) + c_2 y_2(t)] + y_P(t). \tag{4}$$

Since $y(t)$ was any solution whatsoever of equation (1), it follows that the general solution of (1) is given by (4). We can express result (4) in the following schematic form:

The general solution of the nonhomogeneous equation	=	The general solution of the homogeneous equation	+	A particular solution of the nonhomogeneous equation.

Note that the right-hand side of equation (4) contains two arbitrary constants c_1 and c_2 that we can select to satisfy given initial conditions, $y(t_0) = y_0$ and $y'(t_0) = y_0'$; we illustrate this point in Example 1.

We call the general solution of the homogeneous equation the **complementary solution** and denote it y_C. The solution of the nonhomogeneous equation that we somehow have found is called a **particular solution** and is denoted by y_P. If we use y to represent the general solution of nonhomogeneous equation (1), then the preceding schematic statement has the structural form

$$y(t) = y_C(t) + y_P(t). \tag{5}$$

In Sections 4.9 and 4.10 we will discuss methods for finding a particular solution, y_P. For now, we illustrate equation (5) with an example.

EXAMPLE 1

(a) Verify that $y_P(t) = 3t - 4$ is a solution of $y'' - y' - 2y = 5 - 6t$.

(b) Use the result of (a) together with equation (5) to solve the initial value problem

$$y'' - y' - 2y = 5 - 6t, \qquad y(0) = 3, \qquad y'(0) = 11.$$

Solution:

(a) Inserting y_P into the differential equation, we obtain

$$y_P'' - y_P' - 2y_P = (0) - (3) - 2(3t - 4) = 5 - 6t.$$

Therefore, $y_P = 3t - 4$ is a particular solution of the nonhomogeneous differential equation.

(b) The general solution of $y'' - y' - 2y = 0$ is $y_C(t) = c_1e^{-t} + c_2e^{2t}$. Therefore, by equation (5), the general solution of $y'' - y' - 2y = 5 - 6t$ is $y(t) = y_C(t) + y_P(t)$, or

$$y(t) = c_1e^{-t} + c_2e^{2t} + 3t - 4.$$

Imposing the initial conditions, we obtain the system of equations

$$\begin{aligned} c_1 + c_2 - 4 &= 3, \\ -c_1 + 2c_2 + 3 &= 11. \end{aligned}$$

The solution of this system is $c_1 = 2$ and $c_2 = 5$. Thus, the solution of the initial value problem is

$$y(t) = 2e^{-t} + 5e^{2t} + 3t - 4. \ \blacktriangle$$

The Superposition of Particular Solutions

As we noted in Section 4.2, the principle of superposition does not apply to *nonhomogeneous* linear equations. If $u_1(t)$ and $u_2(t)$ are two solutions of the nonhomogeneous equation

$$y'' + p(t)y' + q(t)y = g(t), \tag{6}$$

we do not expect the sum, $w(t) = u_1(t) + u_2(t)$, to be a solution of equation (6).

In fact, as a calculation shows, when $w = u_1 + u_2$ is inserted into the left-hand side of (6) we obtain

$$w'' + p(t)w' + q(t)w = g(t) + g(t) = 2g(t).$$

Therefore, $w = u_1 + u_2$ is not a solution of equation (6) unless g is the zero function [in which case, equation (6) is homogeneous].

There is, however, a different form of superposition that applies to nonhomogeneous equations. We state this result formally as a theorem since we will find it useful in Sections 4.9 and 4.10 which deal with the practical aspects of finding a particular solution, y_P, for equation (6). The proof is left as Exercise 13.

Theorem 4.10

Let $u(t)$ be a solution of $y'' + p(t)y' + q(t)y = g_1(t), a < t < b$. Let $v(t)$ be a solution of $y'' + p(t)y' + q(t)y = g_2(t), a < t < b$. Let a_1 and a_2 be any constants. Then the function $y_P(t) = a_1u(t) + a_2v(t)$ is a particular solution of

$$y'' + p(t)y' + q(t)y = a_1g_1(t) + a_2g_2(t).$$

Theorem 4.10 is a simplifying principle. For example, suppose we need to find a particular solution of

$$y'' + p(t)y' + q(t)y = e^{2t} + \cos t. \tag{7}$$

It often is simpler to separately find a particular solution u_1 that solves

$$y'' + p(t)y' + q(t)y = e^{2t}$$

and then find a particular solution u_2 that solves

$$y'' + p(t)y' + q(t)y = \cos t.$$

The desired particular solution of (7) is $y_P(t) = u_1(t) + u_2(t)$.

EXAMPLE 2

We ask you to show in Exercise 3 that $u(t) = 2e^{4t}$ is a particular solution of $y'' - y' - 2y = 20e^{4t}$. Recall from Example 1 that $v(t) = 3t - 4$ is a particular solution of $y'' - y' - 2y = 5 - 6t$. Find the general solution of

$$y'' - y' - 2y = -5e^{4t} + 20 - 24t.$$

Solution: Applying Theorem 4.10, a particular solution of the equation

$$y'' - y' - 2y = -5e^{4t} + 20 - 24t$$

is

$$y_P(t) = -\frac{1}{4}u(t) + 4v(t) = -\frac{1}{2}e^{4t} + 12t - 16.$$

Therefore, the general solution is

$$y(t) = y_C(t) + y_P(t) = c_1e^{-t} + c_2e^{2t} - \frac{1}{2}e^{4t} + 12t - 16. \blacktriangle$$

EXERCISES

Exercises 1–12:

(a) Verify that the given function, $y_P(t)$, is a particular solution of the differential equation.

(b) Determine the complementary solution, $y_C(t)$.

(c) Form the general solution and impose the initial conditions to obtain the unique solution of the initial value problem.

1. $y'' - 2y' - 3y = -9t - 3, \quad y(0) = 1, y'(0) = 3, \quad y_P(t) = 3t - 1$

2. $y'' - 2y' - 3y = e^{2t}, \quad y(0) = 1, y'(0) = 0, \quad y_P(t) = -e^{2t}/3$

3. $y'' - y' - 2y = 20e^{4t}, \quad y(0) = 0, y'(0) = 1, \quad y_P(t) = 2e^{4t}$

4. $y'' - y' - 2y = 10, \quad y(-1) = 0, y'(-1) = 1, \quad y_P(t) = -5$

5. $y'' + y' = 2t, \quad y(1) = 1, y'(1) = -2, \quad y_P(t) = t^2 - 2t$

6. $y'' + y' = 2e^{-t}, \quad y(0) = 2, y'(0) = 2, \quad y_P(t) = -2te^{-t}$

7. $y'' + y = 2t - 3\cos 2t, \quad y(0) = 0, y'(0) = 0, \quad y_P(t) = 2t + \cos 2t$

8. $y'' + 4y = 10e^{t-\pi}, \quad y(\pi) = 2, y'(\pi) = 0, \quad y_P(t) = 2e^{t-\pi}$

9. $y'' - 2y' + 2y = 10t^2, \quad y(0) = 0, y'(0) = 0, \quad y_P(t) = 5(t+1)^2$

10. $y'' - 2y' + 2y = 5\sin t, \quad y(\pi/2) = 1, y'(\pi/2) = 0, \quad y_P(t) = 2\cos t + \sin t$

11. $y'' - 2y' + y = e^t, \quad y(0) = -2, y'(0) = 2, \quad y_P(t) = (1/2)t^2e^t$

12. $y'' - 2y' + y = t^2 + 4 + 2\sin t, \quad y(0) = 1, y'(0) = 3, \quad y_P(t) = t^2 + 4t + 10 + \cos t$

13. Assume that $u(t)$ and $v(t)$ are (respectively) solutions of the following differential equations,

$$u'' + p(t)u' + q(t)u = g_1(t) \quad \text{and} \quad v'' + p(t)v' + q(t)v = g_2(t),$$

where $p(t), q(t), g_1(t)$, and $g_2(t)$ are continuous on the t-interval of interest. Let a_1 and a_2 be any two constants. Show that the function $y_P(t) = a_1u(t) + a_2v(t)$ is a particular solution of the differential equation

$$y'' + p(t)y' + q(t)y = a_1g_1(t) + a_2g_2(t)$$

on the same t-interval.

Exercises 14–16:

The functions $u_1(t), u_2(t)$, and $u_3(t)$ are solutions of the following differential equations

$$u_1'' + p(t)u_1' + q(t)u_1 = 2e^t + 1, \qquad u_2'' + p(t)u_2' + q(t)u_2 = 2e^{-t} - t - 1,$$
$$u_3'' + p(t)u_3' + q(t)u_3 = 3t.$$

Use the functions of $u_1(t), u_2(t)$, and $u_3(t)$ to construct a particular solution of the given differential equation.

14. $y'' + p(t)y' + q(t)y = e^t + 2t + 1/2$

15. $y'' + p(t)y' + q(t)y = 4e^{-t} - 2$

16. $y'' + p(t)y' + q(t)y = \cosh t$

Exercises 17–21:

The function $y_P(t)$ is a particular solution of the given differential equation. Determine the function $g(t)$.

17. $y'' + y' - y = g(t), \quad y_P(t) = e^{2t} - t^2$ **18.** $y'' - 2y' = g(t), \quad y_P(t) = 3t + \sqrt{t}, t > 0$

19. $ty'' + e^t y' + 2y = g(t), \quad y_P(t) = 3t, t > 0$

20. $y'' + y = g(t), \quad y_P(t) = \ln(1 + t), \quad t > -1$

21. $y'' + (\sin t)y' + 2|t|y = g(t), \quad y_P(t) = \cos t$

Exercises 22–26:

The general solution of the nonhomogeneous differential equation $y'' + \alpha y' + \beta y = g(t)$ is given, where c_1 and c_2 are arbitrary constants. Determine the constants α and β and the function $g(t)$.

22. $y(t) = c_1 e^t + c_2 e^{2t} + 2e^{-2t}$

23. $y(t) = c_1 + c_2 e^{-t} + t^2$

24. $y(t) = c_1 e^t + c_2 t e^t + t^2 e^t$

25. $y(t) = c_1 e^t \cos t + c_2 e^t \sin t + e^t + \sin t$

26. $y(t) = c_1 \sin 2t + c_2 \cos 2t - 1 + \sin t$

4.9 The Method of Undetermined Coefficients

In Section 4.8 we discussed the structure of the general solution for the nonhomogeneous equation

$$y'' + p(t)y' + q(t)y = g(t), \qquad a < t < b.$$

As we saw, the general solution has the form

$$y(t) = y_C(t) + y_P(t), \tag{1}$$

where y_C is the general solution of the homogeneous equation $y'' + p(t)y' + q(t)y = 0$ and y_P is a particular solution of the nonhomogeneous equation $y'' + p(t)y' + q(t)y = g(t)$.

In this section we describe a technique that often can be used to find a particular solution, y_P. The technique is known as the **method of undetermined coefficients**.

We first illustrate the method through a series of examples. Later, we summarize the method in tabular form. In Section 4.10, we discuss a second technique for finding a particular solution, the method of variation of parameters.

Before discussing these two methods for obtaining a particular solution, it's worth stating the procedure that should be followed to obtain the general solution (1):

1. The first step is to find the complementary solution, y_C. As we will see, knowledge of the complementary solution is a prerequisite for using either the method of undetermined coefficients or the method of variation of parameters.
2. Next, use undetermined coefficients or variation of parameters (or anything else that works) to find a particular solution, y_P.
3. Finally, see (1), the general solution is obtained by forming $y_C + y_P$.

If you are solving an initial value problem, the initial conditions are imposed as a last step, step 4.

Introduction to the Method of Undetermined Coefficients

Consider the nonhomogeneous differential equation

$$ay'' + by' + cy = g(t), \tag{2}$$

where a, b, and c are constants. As we will see, we can *guess* the form of a particular solution for certain types of functions $g(t)$. Example 1 introduces some of the main ideas.

EXAMPLE 1

Find the general solution of the nonhomogeneous equation

$$y'' - y' - 2y = 8e^{3t}.$$

Solution: We always begin by finding the complementary solution, y_C. The characteristic polynomial for $y'' - y' - 2y = 0$ is $\lambda^2 - \lambda - 2 = (\lambda + 1)(\lambda - 2)$. Therefore, the complementary solution is

$$y_C(t) = c_1e^{-t} + c_2e^{2t}.$$

We now look for a particular solution, a function $y_P(t)$ such that

$$y_P'' - y_P' - 2y_P = 8e^{3t}. \tag{3}$$

Since all derivatives of $y = e^{3t}$ are again multiples of e^{3t}, it seems reasonable that a particular solution might have the form

$$y_P = Ae^{3t}.$$

Inserting $y_P = Ae^{3t}$ into equation (3), we obtain:

$$9Ae^{3t} - 3Ae^{3t} - 2(Ae^{3t}) = 8e^{3t}.$$

Collecting terms on the left-hand side, $4Ae^{3t} = 8e^{3t}$. Therefore, $A = 2$ and $y_P(t) = 2e^{3t}$ is a particular solution of the nonhomogeneous equation $y'' - y' - 2y = 8e^{3t}$.

Having the complementary solution y_C and a particular solution y_P, we form the general solution of the nonhomogeneous equation,

$$y(t) = y_C(t) + y_P(t) = c_1e^{-t} + c_2e^{2t} + 2e^{3t}. \blacktriangle$$

Example 1 suggests a reasonable approach to finding a particular solution, y_P. If the right-hand side of nonhomogeneous equation (2) is of a certain special type, then it might be possible to guess an appropriate form for y_P. As we shall see, the method of undetermined coefficients amounts to a recipe for choosing the *form* of y_P. This recipe involves unknown (or undetermined) coefficients that must be evaluated by inserting the form, y_P, into the differential equation. (The role that the complementary solution plays in this process will be clarified shortly.)

Trial Forms for the Particular Solution

The method of undetermined coefficients can be applied to differential equations of the form

$$ay'' + by' + cy = g(t),$$

where a, b, and c are constants and where the nonhomogeneous term $g(t)$ is one of several possible types. It's important to understand what types of functions $g(t)$ are suitable and why. To gain insight, we start with some examples.

The first few examples will treat, for various right sides $g(t)$, the differential equation

$$y'' - y' - 2y = g(t). \tag{4}$$

Equation (4) with $g(t) = 8e^{3t}$ was discussed in Example 1; the complementary solution of (4) is

$$y_C(t) = c_1 e^{-t} + c_2 e^{2t}.$$

EXAMPLE 2

Find the general solution of

$$y'' - y' - 2y = 4t^2.$$

Solution: We already know the complementary solution from Example 1. Therefore, we can form the general solution after finding a particular solution.

Following the approach taken in Example 1, we are tempted to look for a solution of the form

$$y_P(t) = At^2,$$

where A is an unknown (undetermined) coefficient. The guess $y_P(t) = At^2$, however, does not work because forming the first and second derivatives of t^2 generates multiples of some new functions, t and 1. In particular, this guess leads to a contradiction when the trial form $y_P(t) = At^2$ is inserted into the differential equation.

Instead, we assume a particular solution of the form

$$y_P(t) = At^2 + Bt + C,$$

where the constants A, B, and C must be chosen so that

$$y_P'' - y_P' - 2y_P = 4t^2.$$

Substituting y_P, we obtain the condition

$$(2A) - (2At + B) - 2(At^2 + Bt + C) = 4t^2$$

or, after collecting terms,

$$-2At^2 - (2A + 2B)t + (2A - B - 2C) = 4t^2. \tag{5}$$

This equality must hold for all t in the interval of interest. Therefore, the coefficients of t^2, t, and 1 on the left-hand side of this equation must equal their counterparts on the right-hand side. We obtain the following three equations for the three unknown coefficients A, B, and C.

$$\begin{aligned} -2A \qquad\qquad &= 4 \\ -2A - 2B \qquad &= 0 \\ 2A - \; B - 2C &= 0. \end{aligned} \tag{6}$$

The solution of this system is $A = -2$, $B = 2$, $C = -3$. Therefore, a particular solution is given by

$$y_P(t) = -2t^2 + 2t - 3.$$

The desired general solution is $y(t) = y_C(t) + y_P(t)$, or

$$y(t) = c_1 e^{-t} + c_2 e^{2t} - 2t^2 + 2t - 3. \quad \blacktriangle$$

REMARKS ABOUT EXAMPLE 2:

1. In retrospect, it is clear why our first guess $y_P(t) = At^2$ failed. Substitution into the nonhomogeneous equation leads [see system (6)] to the contradictory constraints $-2A = 4$, $-2A = 0$, $2A = 0$. The emergence of such contradictions is a clear indicator that the assumed form for the particular solution is incorrect.
2. Note that equation (5) can be rewritten as
$$(-2A - 4)t^2 + (-2A - 2B)t + (2A - B - 2C) = 0, \qquad a < t < b.$$
The assertion that A, B, and C must satisfy equation (6) is equivalent to stating that the set of functions $\{t^2, t, 1\}$ is a linearly independent set on the t-interval of interest, $a < t < b$; recall the discussion of linear independence in Section 4.3.
3. The assumed form of the particular solution, $y_P(t) = At^2 + Bt + C$, was appropriate because of another fact that we did not mention—none of the functions $1, t$, or t^2 is a part of the complementary solution, y_C. If one of these functions *had been* part of the complementary solution, then (as we shall see in Examples 4 and 5) our guess would have failed.

EXAMPLE 3

Find the general solution of
$$y'' - y' - 2y = -20 \sin 2t.$$

Solution: The complementary solution is
$$y_C(t) = c_1 e^{-t} + c_2 e^{2t}.$$
In choosing a guess for the particular solution, we observe that differentiation of the right-hand side,
$$g(t) = -20 \sin 2t,$$
produces a multiple of $\cos 2t$ but that continued differentiation of the set of functions $\{\sin 2t, \cos 2t\}$ simply produces multiples of the functions in the set. In addition, neither of the functions $\sin 2t$ or $\cos 2t$ appears as part of the complementary solution. Therefore, we choose the following trial form for a particular solution:
$$y_P(t) = A \sin 2t + B \cos 2t.$$
Substituting the trial form $y_P(t) = A \sin 2t + B \cos 2t$, we obtain
$$y_P'' - y_P' - 2y_P = -20 \sin 2t,$$
or
$$[-4A \sin 2t - 4B \cos 2t] - [2A \cos 2t - 2B \sin 2t] - 2[A \sin 2t + B \cos 2t] = -20 \sin 2t.$$
Collecting like terms reduces this equation to
$$[-6A + 2B] \sin 2t - [2A + 6B] \cos 2t = -20 \sin 2t.$$
Since this equation must hold for all t in the interval of interest, it follows that
$$\begin{aligned} -6A + 2B &= -20, \\ -2A - 6B &= \quad 0. \end{aligned}$$

The solution of this system is $A = 3$ and $B = -1$, leading to a particular solution

$$y_P(t) = 3\sin 2t - \cos 2t,$$

and the general solution,

$$y(t) = c_1e^{-t} + c_2e^{2t} + 3\sin 2t - \cos 2t. \blacktriangle$$

The Trial Form Cannot Contain Terms Present in the Complementary Solution

Our next two examples illustrate how the trial form for y_P must be modified if portions of $g(t)$ or derivatives of $g(t)$ are present in the complementary solution.

EXAMPLE 4

Find the general solution of

$$y'' - y' - 2y = 4e^{-t}.$$

Solution: In this case, we observe that the function $g(t) = e^{-t}$ is a solution of the homogeneous equation. To illustrate that the trial form $y_P(t) = Ae^{-t}$ is not correct, we substitute it into the differential equation, obtaining

$$Ae^{-t} - [-Ae^{-t}] - 2Ae^{-t} = 4e^{-t}$$

or

$$0Ae^{-t} = 4e^{-t}.$$

Since the condition $0A = 4$ cannot hold for any value A, the assumed form of the trial solution is not correct. This example illustrates why we need first to obtain the complementary solution and then compare our choice for the form of y_P with the complementary solution, y_C.

We obtain the correct form for a particular solution of $y'' - y' - 2y = 4e^{-t}$ if we multiply e^{-t} by t. That is, if we assume a trial solution of the form

$$y_P(t) = Ate^{-t}. \quad (7)$$

At first glance, it may seem surprising that this form is correct. Nevertheless, when we substitute $y_P(t) = Ate^{-t}$, we obtain

$$[Ate^{-t} - 2Ae^{-t}] - [-Ate^{-t} + Ae^{-t}] - 2Ate^{-t} = 4e^{-t} \quad (8)$$

or

$$-3Ae^{-t} = 4e^{-t}.$$

This equation is satisfied by choosing $A = -4/3$, leading to a particular solution

$$y_P(t) = -\frac{4}{3}te^{-t},$$

and the general solution

$$y(t) = y_C(t) + y_P(t) = c_1e^{-t} + c_2e^{2t} - \frac{4}{3}te^{-t}. \blacktriangle$$

REMARK: In retrospect, it should be clear why the te^{-t} terms vanish on the left-hand side of equation (8). After $y_P(t) = Ate^{-t}$ is substituted into the differential equation, the terms that survive as te^{-t} terms are precisely those in which differentiation under the product rule has acted upon the e^{-t} factor and not on the t factor. Such terms ultimately vanish because e^{-t} is a solution of the homogeneous equation.

EXAMPLE 5

Find the general solution of

$$y'' + 2y' + y = 2e^{-t}.$$

Solution: For this problem, the homogeneous equation has characteristic equation $\lambda^2 + 2\lambda + 1 = (\lambda + 1)^2 = 0$. We obtain two real repeated roots, $\lambda_1 = \lambda_2 = -1$, and the complementary solution is

$$y_C(t) = c_1 e^{-t} + c_2 te^{-t}.$$

When looking for the appropriate form for a particular solution, it's clear that a guess

$$y_P(t) = Ae^{-t}$$

will not work because e^{-t} is a solution of the homogeneous equation. Also, trying $y_P(t) = Ate^{-t}$, as in Example 4 will fail for the same reason. It's perhaps not surprising that a guess of the form

$$y_P(t) = At^2e^{-t}$$

does work. Substituting this form of the trial solution leads to

$$[At^2e^{-t} - 4Ate^{-t} + 2Ae^{-t}] + 2[-At^2e^{-t} + 2Ate^{-t}] + At^2e^{-t} = 2e^{-t}.$$

Simplifying, we obtain

$$2Ae^{-t} = 2e^{-t},$$

so that $2A = 2$ and hence $y_P(t) = t^2e^{-t}$. The general solution is

$$y(t) = c_1 e^{-t} + c_2 te^{-t} + t^2 e^{-t}. \quad \blacktriangle$$

A Table Summarizing the Method of Undetermined Coefficients

We summarize the method of undetermined coefficients in Table 4.1. The method applies to the nonhomogeneous linear differential equation

$$ay'' + by' + cy = g(t),$$

where a, b, and c are constants and $g(t)$ has one of the forms listed on the left-hand side of the table. The corresponding form to assume for the particular solution is listed on the right-hand side of the table. The forms listed in Table 4.1 will work; that is, they will always yield a particular solution. In the Exercises, we ask you to solve problems using Table 4.1. The role of the factor t^r in the right-hand column of Table 4.1 is to ensure that no part of the assumed form

TABLE 4.1

The right-hand column gives the proper form to assume for a particular solution of $ay'' + by' + cy = g(t)$. In the right-hand column, choose r to be the smallest nonnegative integer such that no term in the assumed form is a solution of the homogeneous equation $ay'' + by' + cy = 0$. The value of r will be 0, 1, or 2.

Form of $g(t)$	Form to Assume for a Particular Solution $y_P(t)$
$a_n t^n + \cdots + a_1 t + a_0$	$t^r[A_n t^n + \cdots + A_1 t + A_0]$
$[a_n t^n + \cdots + a_1 t + a_0]e^{\alpha t}$	$t^r[A_n t^n + \cdots + A_1 t + A_0]e^{\alpha t}$
$[a_n t^n + \cdots + a_1 t + a_0]\sin\beta t$ or $[a_n t^n + \cdots + a_1 t + a_0]\cos\beta t$	$t^r[(A_n t^n + \cdots + A_1 t + A_0)\sin\beta t + (B_n t^n + \cdots + B_1 t + B_0)\cos\beta t]$
$e^{\alpha t}\sin\beta t$ or $e^{\alpha t}\cos\beta t$	$t^r[Ae^{\alpha t}\sin\beta t + Be^{\alpha t}\cos\beta t]$
$e^{\alpha t}[a_n t^n + \cdots + a_0]\sin\beta t$ or $e^{\alpha t}[a_n t^n + \cdots + a_0]\cos\beta t$	$t^r[(A_n t^n + \cdots + A_0)e^{\alpha t}\sin\beta t + (B_n t^n + \cdots + B_0)e^{\alpha t}\cos\beta t]$

for y_P is present in the complementary solution. You need to choose the proper value for r; the procedure for doing so is described in the table.

EXAMPLE 6

Using Table 4.1, choose an appropriate form for a particular solution of

(a) $y'' + 4y = t^2 e^{3t}$ (b) $y'' + 4y = te^{2t}\cos t$

(c) $y'' + 4y = 2t^2 + 5\sin 2t + e^{3t}$ (d) $y'' + 4y = t^2\cos 2t$

Solution: We note first that the complementary solution for each of parts (a)–(d) is

$$y_C(t) = c_1\sin 2t + c_2\cos 2t.$$

(a) For $g(t) = t^2e^{3t}$, Table 4.1 specifies $y_P(t) = t^r[A_2t^2 + A_1t + A_0]e^{3t}$. If $r = 0$, no term in the assumed form for y_P is present in the complementary solution. So, from Table 4.1, the appropriate form for a trial particular solution is

$$y_P(t) = [A_2t^2 + A_1t + A_0]e^{3t}.$$

(b) For $g(t) = te^{2t}\cos t$, the specified form is

$$y_P(t) = t^r[(A_1t + A_0)e^{2t}\sin t + (B_1t + B_0)e^{2t}\cos t].$$

If $r = 0$, no term in the assumed form for y_P is present in the complementary solution. So, from Table 4.1, the appropriate form for a trial

(continued)

(continued)

particular solution is

$$y_P(t) = (A_1 t + A_0)e^{2t} \sin t + (B_1 t + B_0)e^{2t} \cos t.$$

(c) Note that the nonhomogeneous term, $g(t) = 2t^2 + 5\sin 2t + e^{3t}$ does not match any of the forms listed in Table 4.1. However, we can use the superposition principle described by Theorem 4.10 in Section 4.8. Suppose $u(t)$ is a particular solution of

$$y'' + 4y = 2t^2,$$

$v(t)$ is a particular solution of

$$y'' + 4y = 5\sin 2t,$$

and $w(t)$ is a particular solution of

$$y'' + 4y = e^{3t}.$$

By Theorem 4.9, $y_P(t) = u(t) + v(t) + w(t)$ is a particular solution of

$$y'' + 4y = 2t^2 + 5\sin 2t + e^{3t}. \quad \textbf{(9)}$$

To determine the individual particular solutions $u(t)$, $v(t)$, and $w(t)$, we turn to Table 4.1 to find suitable trial forms. In particular, an appropriate trial form for $y'' + 4y = 2t^2$, is $u(t) = A_2t^2 + A_1t + A_0$. Similarly, from Table 4.1, we find a suitable trial form for $y'' + 4y = 5\sin 2t$ is the function $v(t) = B_0 t\cos 2t + C_0 t\sin 2t$. A suitable trial form for $y'' + 4y = e^{3t}$ is given by $w(t) = D_0 e^{3t}$. (In the first and last cases, $r = 0$. In the second case, $r = 1$.) Therefore,

$$y_P(t) = A_2t^2 + A_1t + A_0 + B_0 t\cos 2t + C_0 t\sin 2t + D_0 e^{3t}.$$

(d) For $g(t) = t^2 \cos 2t$, Table 4.1 prescribes the form

$$y_P(t) = t^r[(A_2t^2 + A_1t + A_0)\sin 2t + (B_2t^2 + B_1t + B_0)\cos 2t].$$

If we set $r = 0$, the assumed form for $y_P(t)$ will contain two terms, $A_0 \sin 2t$ and $B_0 \cos 2t$, that are solutions of the homogeneous equation. Therefore, r cannot be zero. With $r = 1$, we see that no term in the assumed form is a solution of the homogeneous equation. Therefore, the appropriate form is

$$\begin{aligned} y_P(t) &= t[(A_2t^2 + A_1t + A_0)\sin 2t + (B_2t^2 + B_1t + B_0)\cos 2t] \\ &= (A_2t^3 + A_1t^2 + A_0t)\sin 2t + (B_2t^3 + B_1t^2 + B_0t)\cos 2t. \end{aligned}$$ ▲

Although we presented Table 4.1 in the context of discussing nonhomogeneous constant coefficient *second order* linear differential equations, the method of undetermined coefficients and the ideas developed in Table 4.1 are not restricted to second order equations. In Chapter 5, we will see that the method of undetermined coefficients and the ideas of Table 4.1 extend naturally to nonhomogeneous constant coefficient linear equations of order higher than two.

EXERCISES

Exercises 1–15:

For the given differential equation

(a) Determine the complementary solution.

(b) Use the method of undetermined coefficients to find a particular solution.

(c) Form the general solution.

1. $y'' - 4y = 4t^2$

2. $y'' - 4y = \sin 2t$

3. $y'' + y = 8e^t$

4. $y'' + y = e^t \sin t$

5. $y'' - 4y' + 4y = e^{2t}$

6. $y'' - 4y' + 4y = 8 + \sin 2t$

7. $y'' + 2y' + 2y = t^3$

8. $2y'' - 5y' + 2y = te^t$

9. $y'' + 2y' + 2y = \cos t + e^{-t}$

10. $y'' + y' = 6t^2$

11. $2y'' - 5y' + 2y = -6e^{t/2}$

12. $y'' + y' = \cos t$

13. $9y'' - 6y' + y = 9te^{t/3}$

14. $y'' + 4y' + 5y = 5t + e^{-t}$

15. $y'' + 4y' + 5y = 2e^{-2t} + \cos t$

Exercises 16–22:

For the given differential equation,

(a) Determine the complementary solution.

(b) List the form of particular solution prescribed by the method of undetermined coefficients; you need not evaluate the constants in the assumed form. [Hint: In exercises 20 and 22, rewrite the hyperbolic functions in terms of exponential functions. In exercise 21, use trigonometric identities.]

16. $y'' - 2y' - 3y = 2e^{-t}\cos t + t^2 + te^{3t}$

17. $y'' + 9y = t^2 \cos 3t + 4 \sin t$

18. $y'' - y' = t^2(2 + e^t)$

19. $y'' - 2y' + 2y = e^{-t}\sin 2t + 2t + te^{-t}\sin t$

20. $y'' - y = \cosh t + \sinh 2t$

21. $y'' + 4y = \sin t \cos t + \cos^2 2t$

22. $y'' + 4y = 2\sinh t \cosh t + \cosh^2 t$

Exercises 23–27:

Consider the differential equation $y'' + \alpha y' + \beta y = g(t)$. In each exercise, the complementary solution, $y_C(t)$, and nonhomogeneous term, $g(t)$, are given. Determine α and β and then find the general solution of the differential equation.

23. $y_C(t) = c_1e^{-t} + c_2e^{2t}, \quad g(t) = 4t$

24. $y_C(t) = c_1 + c_2e^{-t}, \quad g(t) = t$

25. $y_C(t) = c_1e^{-2t} + c_2te^{-2t}, \quad g(t) = 5\sin t$

26. $y_C(t) = c_1\cos t + c_2\sin t, \; g(t) = t + \sin 2t$

27. $y_C(t) = c_1e^{-t}\cos 2t + c_2e^{-t}\sin 2t, \quad g(t) = 8e^{-t}$

Exercises 28–30:

Consider the differential equation $y'' + \alpha y' + \beta y = g(t)$. In each exercise, the nonhomogeneous term, $g(t)$, and the form of the particular solution prescribed by the method of undetermined coefficients is given. Determine the constants α and β.

28. $g(t) = t + e^{3t}, \quad y_P(t) = A_1t^2 + A_0t + B_0te^{3t}$

29. $g(t) = 3e^{2t} - e^{-2t} + t, \quad y_P(t) = A_0te^{2t} + B_0te^{-2t} + C_1t + C_0$

30. $g(t) = -e^t + \sin 2t + e^t\sin 2t,$

$y_P(t) = A_0e^t + B_0t\cos 2t + C_0t\sin 2t + D_0e^t\cos 2t + E_0e^t\sin 2t$

31. Each of the five graphs is the solution of one of the five differential equations listed. Each solution satisfies the initial conditions $y(0) = 1, y'(0) = 0$. Match each graph with one of the differential equations; justify your answers.

(a) $y'' + 3y' + 2y = \sin t$ (b) $y'' + y = \sin t$ (c) $y'' + y = \sin \dfrac{3t}{2}$

(d) $y'' + 3y' + 2y = e^{t/10}$ (e) $y'' + y' + y = \sin \dfrac{t}{3}$

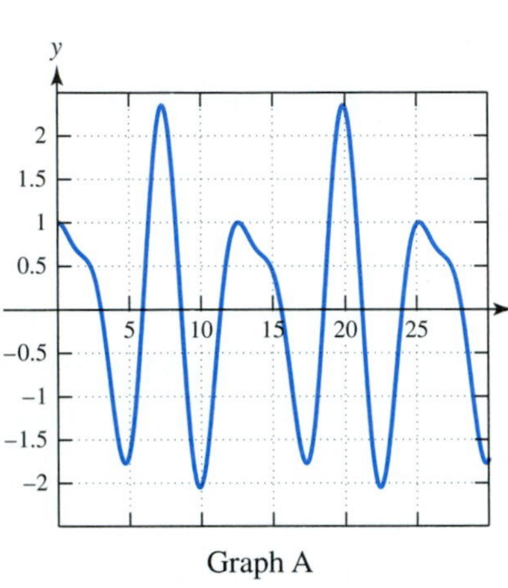

Graph A

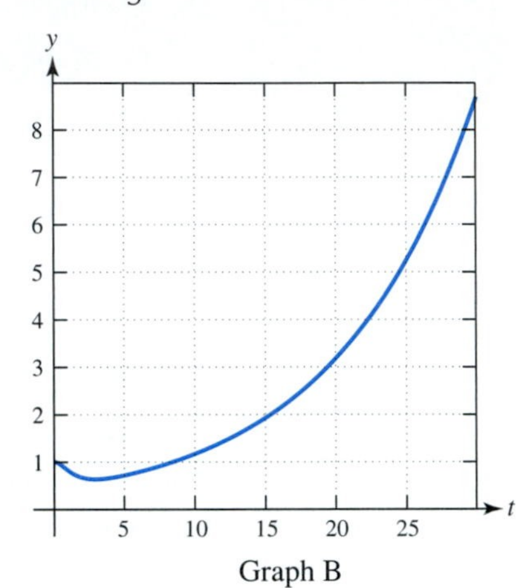

Graph B

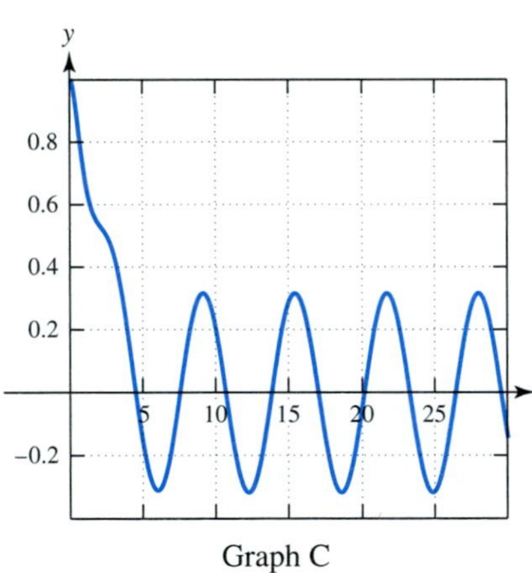

Graph C

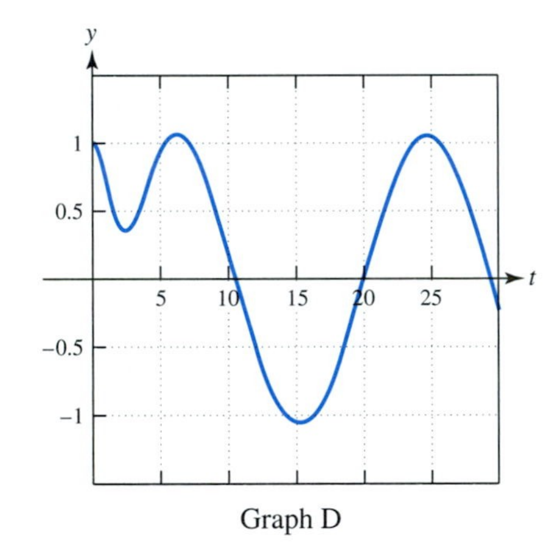

Graph D

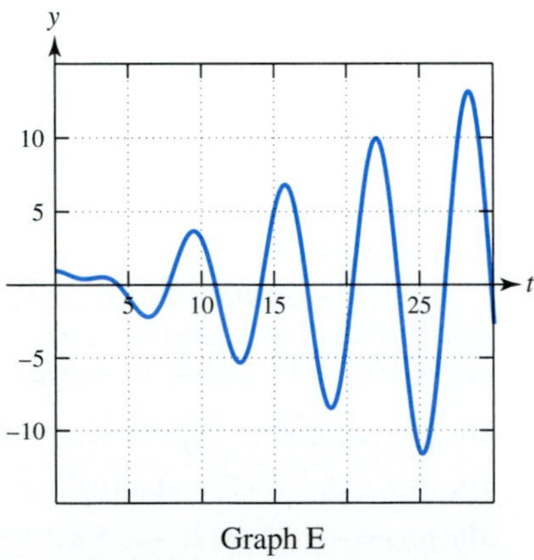

Graph E

Figure for Exercise 31

32. Buoyancy A cube, having a volume of 8 ft^3 and weighing 200 lb, is initially at rest floating in water. At time $t = 0$ a force is applied to the block. Neglecting drag resistance, Newton's second law of motion leads to the following differential equation for $y(t)$,

$$y'' + \omega^2 y = \frac{1}{m_b} F(t).$$

In this equation, $y(t)$ is the vertical displacement of the cube from its equilibrium rest position (measured positive in the downward direction), $\omega^2 = \rho_l g/\rho L$, m_b is the mass of the block, and $F(t)$ is the applied force (measured positive in the downward direction). See the Introduction of Chapter 4 for a discussion of buoyancy problems. Recall that ρ_l and ρ are the mass densities of the liquid and block, respectively, $g = 32$ ft/sec^2 is the acceleration due to gravity, and L is the vertical dimension of the floating object. Let $F(t) = 10 \sin \omega t$ lb. The angular frequency of the applied force is therefore identical to the angular frequency of the solutions forming the complementary solution.

(a) Formulate the initial value problem for $y(t)$.

(b) Solve the initial value problem.

(c) Plot the solution and determine the physically-relevant domain of the problem. Estimate the time at which the displacement $y(t)$ either exceeds the value for which the block is fully submerged or becomes less than that for which the block is completely elevated from the water. Once either event happens, the model ceases to be valid.

Complex-Valued Solutions Although we have emphasized the need to obtain real-valued, physically relevant solutions to problems of interest, it is sometimes computationally convenient to consider differential equations with complex-valued nonhomogeneous terms. The corresponding particular solutions will then likewise be complex-valued functions. Exercises 33 and 34 illustrate some aspects of this type of calculation. Exercises 35–39 provide some additional examples.

33. Consider the differential equation $y'' + p(t)y' + q(t)y = g_1(t) + ig_2(t)$, where $p(t)$, $q(t)$, $g_1(t)$, and $g_2(t)$ are all real-valued functions continuous on some t-interval of interest. Assume that $y_P(t)$ is a particular solution of this equation. Generally, $y_P(t)$ will be a complex-valued function. Let $y_P(t) = u(t) + iv(t)$, where $u(t)$ and $v(t)$ are real-valued functions. Show that

$$u'' + p(t)u' + q(t)u = g_1(t) \quad \text{and} \quad v'' + p(t)v' + q(t)v = g_2(t).$$

That is to say, show that the real and imaginary parts of the complex-valued particular solution, $y_P(t)$, are themselves particular solutions corresponding to the real and imaginary parts of the complex-valued nonhomogeneous term, $g(t)$.

34. Consider the nonhomogeneous differential equation $y'' - y = e^{i2t}$. The complementary solution is $y_C = c_1 e^t + c_2 e^{-t}$. Recall from Euler's formula that $e^{i2t} = \cos 2t + i \sin 2t$. Therefore, the right-hand side is a (complex-valued) linear combination of functions for which the method of undetermined coefficients is applicable.

(a) Assume a particular solution of the form $y_P = Ae^{i2t}$, where A is an undetermined (generally complex) coefficient. Substitute this trial form into the differential equation. Using the fact that $[e^{i2t}]' = i2e^{i2t}$, determine the constant A.

(b) With the constant A as determined in part (a), write $y_P(t) = Ae^{i2t}$ in the form $y_P(t) = u(t) + iv(t)$, where $u(t)$ and $v(t)$ are real-valued functions.

(c) Show that $u(t)$ and $v(t)$ are themselves particular solutions of the following differential equations.

$$u'' - u = \text{Re}[e^{i2t}] = \cos 2t \quad \text{and} \quad v'' - v = \text{Im}[e^{i2t}] = \sin 2t.$$

Therefore, the single computation with the complex-valued nonhomogeneous term yields particular solutions of the differential equation for the two real-valued nonhomogeneous terms forming its real and imaginary parts.

Exercises 35–39:

For each exercise,

(a) Use the indicated trial form for $y_P(t)$ to obtain a (complex-valued) particular solution for the given differential equation with complex-valued nonhomogeneous term, $g(t)$.

(b) Write $y_P(t)$ as $y_P(t) = u(t) + iv(t)$, where $u(t)$ and $v(t)$ are real-valued functions. Show that $u(t)$ and $v(t)$ are particular solutions of the given differential equation with nonhomogeneous terms $\text{Re}[g(t)]$ and $\text{Im}[g(t)]$, respectively.

35. $y'' + 2y' + y = e^{it}, \quad y_P(t) = Ae^{it}$

36. $y'' + 4y = e^{it}, \quad y_P(t) = Ae^{it}$

37. $y'' + 4y = e^{i2t}, \quad y_P(t) = Ate^{i2t}$

38. $y'' + y' = e^{-i2t}, \quad y_P(t) = Ae^{-i2t}$

39. $y'' + y = e^{(1+i)t}, \quad y_P(t) = Ae^{(1+i)t}$

4.10 The Method of Variation of Parameters

Section 4.9 discussed the method of undetermined coefficients as a technique for finding a particular solution of the constant coefficient equation

$$ay'' + by' + cy = g(t). \tag{1}$$

The method of undetermined coefficients can be applied to equation (1) as long as the nonhomogeneous term, $g(t)$, is one of the types listed in Table 4.1 of Section 4.9 or is a linear combination of types listed in the table.

In this section, we consider the general linear second order nonhomogeneous differential equation

$$y'' + p(t)y' + q(t)y = g(t). \tag{2}$$

Unlike the discussion in Section 4.9, we do not insist that this differential equation has constant coefficients or that $g(t)$ belongs to some special class of functions. The only restriction we place on differential equation (2) is that the functions $p(t)$, $q(t)$, and $g(t)$ are continuous on the t-interval of interest.

The technique we discuss, the method of variation of parameters, is one that uses a knowledge of the complementary solution of (2) to construct a corresponding particular solution.

Discussion of the Method

Assume that $\{y_1(t), y_2(t)\}$ is a fundamental set of solutions of the homogeneous equation $y'' + p(t)y' + q(t)y = 0$. In other words, the complementary solution is given by

$$y_C(t) = c_1 y_1(t) + c_2 y_2(t).$$

To obtain a particular solution of (2), we "vary the parameters." That is, we replace the constants c_1 and c_2 by functions $u_1(t)$ and $u_2(t)$ and look for a particular solution of (2) having the form

$$y_P(t) = y_1(t)u_1(t) + y_2(t)u_2(t). \tag{3}$$

In (3) there are two functions, $u_1(t)$ and $u_2(t)$, that we are free to specify. One obvious constraint on $u_1(t)$ and $u_2(t)$ is that they must be chosen so that y_P is, in fact, a solution of nonhomogeneous equation (2). Generally speaking, however, two constraints can be imposed when we want to determine two functions. We will impose a second constraint as we proceed, choosing it so as to simplify the calculations.

Substituting (3) into the left-hand side of equation (2) requires us ultimately to compute the first and second derivatives of (3). Computing the first derivative leads to

$$y_P' = [y_1'u_1 + y_2'u_2] + [y_1u_1' + y_2u_2'].$$

The grouping of terms in this equation is motivated by the fact that we now impose a constraint on the functions $u_1(t)$ and $u_2(t)$. We require

$$y_1u_1' + y_2u_2' = 0. \tag{4}$$

With this constraint, the derivative of y_P becomes

$$y_P' = y_1'u_1 + y_2'u_2, \tag{5}$$

while y_P'' is

$$y_P'' = y_1''u_1 + y_2''u_2 + y_1'u_1' + y_2'u_2'. \tag{6}$$

Notice what requirement (4) has accomplished. The first and second derivatives of y_P have been simplified; in particular, y_P'' does not involve u_1'' or u_2''.

Inserting $y_P = y_1u_1 + y_2u_2$ into the differential equation $y'' + p(t)y' + q(t)y = g(t)$ and using (5) and (6), we find

$$[y_1''u_1 + y_2''u_2 + y_1'u_1' + y_2'u_2'] + p(t)[y_1'u_1 + y_2'u_2] + q(t)[y_1u_1 + y_2u_2] = g(t).$$

Rearranging terms,

$$[y_1'' + p(t)y_1' + q(t)y_1]u_1 + [y_2'' + p(t)y_2' + q(t)y_2]u_2 + [y_1'u_1' + y_2'u_2'] = g(t). \tag{7}$$

Since y_1 and y_2 are solutions of the homogeneous equation $y'' + p(t)y' + q(t)y = 0$, equation (7) reduces to

$$y_1'u_1' + y_2'u_2' = g(t). \tag{8}$$

We therefore obtain two constraints, equations (4) and (8), for the two unknown functions $u_1'(t)$ and $u_2'(t)$. We can combine these two equations into the matrix equation

$$\begin{bmatrix} y_1(t) & y_2(t) \\ y_1'(t) & y_2'(t) \end{bmatrix} \begin{bmatrix} u_1'(t) \\ u_2'(t) \end{bmatrix} = \begin{bmatrix} 0 \\ g(t) \end{bmatrix}. \tag{9}$$

If the (2×2) coefficient matrix has an inverse, then we can solve equation (9) for the unknowns $u_1'(t)$ and $u_2'(t)$. Once they are determined, we can find $u_1(t)$ and $u_2(t)$ by computing antiderivatives.

The coefficient matrix in equation (9) is invertible if and only if its determinant is nonzero. Note, however, that the determinant of this matrix is the Wronskian of the functions y_1 and y_2,

$$W(t) = y_1(t)y_2'(t) - y_1'(t)y_2(t).$$

Since $\{y_1(t), y_2(t)\}$ is a fundamental set of solutions, we are assured that the Wronskian is nonzero for all values of t in our interval of interest. Therefore, solving equation (9) for u_1' and u_2',

$$\begin{bmatrix} u_1'(t) \\ u_2'(t) \end{bmatrix} = \frac{1}{W(t)} \begin{bmatrix} y_2'(t) & -y_2(t) \\ -y_1'(t) & y_1(t) \end{bmatrix} \begin{bmatrix} 0 \\ g(t) \end{bmatrix},$$

or

$$u_1'(t) = \frac{-y_2(t)g(t)}{W(t)} \quad \text{and} \quad u_2'(t) = \frac{y_1(t)g(t)}{W(t)}. \tag{10}$$

Computing antiderivatives to obtain $u_1(t)$ and $u_2(t)$, we form a particular solution,

$$y_P(t) = y_1(t)u_1(t) + y_2(t)u_2(t).$$

EXAMPLE 1

Find the general solution of the differential equation

$$y'' - 2y' + y = e^t \ln t, \qquad t > 0. \tag{11}$$

Solution: Note that the nonhomogeneous term, $g(t) = e^t \ln t$, does not appear in Table 4.1 as a candidate for the method of undetermined coefficients. Since the method of undetermined coefficients is not applicable, we turn to the method of variation of parameters.

For equation (11), the complementary solution is

$$y_C(t) = c_1 y_1(t) + c_2 y_2(t) = c_1 e^t + c_2 t e^t.$$

Using variation of parameters to find a particular solution, we assume

$$y_P(t) = e^t u_1(t) + t e^t u_2(t).$$

Substituting this expression into the nonhomogeneous differential equation (11) and applying the constraint from equation (4), $e^t u_1'(t) + te^t u_2'(t) = 0$, we obtain

$$\begin{bmatrix} e^t & te^t \\ e^t & e^t + te^t \end{bmatrix} \begin{bmatrix} u_1'(t) \\ u_2'(t) \end{bmatrix} = \begin{bmatrix} 0 \\ e^t \ln t \end{bmatrix}.$$

The determinant of the coefficient matrix is $W(t) = e^{2t}$. Solving the matrix equation, as in (10), we find

$$\begin{bmatrix} u_1'(t) \\ u_2'(t) \end{bmatrix} = \begin{bmatrix} -t \ln t \\ \ln t \end{bmatrix}.$$

Computing antiderivatives,

$$u_1(t) = \int u_1'(t)\,dt = -\int t \ln t\,dt = -\frac{t^2}{2} \ln t + \frac{t^2}{4} + K_1$$

and

$$u_2(t) = \int u_2'(t)\,dt = \int \ln t\,dt = t \ln t - t + K_2.$$

We can choose the constants of integration, K_1 and K_2, to suit our convenience because all we need is *some* particular solution, $y_P = y_1u_1 + y_2u_2$. For convenience, we set both K_1 and K_2 equal to zero and obtain the following particular solution:

$$y_P(t) = e^t\left[-\frac{t^2}{2}\ln t + \frac{t^2}{4}\right] + te^t[t\ln t - t]$$
$$= \frac{t^2e^t}{2}\left[\ln t - \frac{3}{2}\right].$$

The general solution of equation (11) is, therefore,

$$y(t) = c_1e^t + c_2te^t + \frac{t^2e^t}{2}\left[\ln t - \frac{3}{2}\right]. \blacktriangle$$

EXAMPLE 2

Observe that $y_1(t) = t$ is a solution of the homogeneous equation

$$t^2y'' - ty' + y = 0, \qquad t > 0.$$

Use this observation to solve the nonhomogeneous initial value problem

$$t^2y'' - ty' + y = t, \qquad y(1) = 1, \qquad y'(1) = 4.$$

Solution: The first step in finding the general solution of the nonhomogeneous equation is determining a fundamental set of solutions $\{y_1, y_2\}$. Thus, we need to find a second solution, $y_2(t)$, to go along with the given solution, $y_1(t) = t$.

The method of reduction of order, described in Section 4.5, can be used to obtain such a second solution, $y_2(t)$. Reduction of order leads to a second solution,

$$y_2(t) = t\ln t.$$

The functions t and $\ln t$ can be shown to form a fundamental set of solutions for the homogeneous equation. It follows that the complementary solution of the nonhomogeneous equation is

$$y_C(t) = c_1t + c_2t\ln t, \qquad t > 0.$$

Since the differential equation has variable coefficients, we cannot use the method of undetermined coefficients to find a particular solution and instead resort to the method of variation of parameters. Assume a particular solution of the form

$$y_P(t) = tu_1(t) + [t\ln t]u_2(t).$$

The simplest approach is to substitute this form into the given nonhomogeneous differential equation to determine one equation for u_1' and u_2', then use the constraint

$$tu_1'(t) + [t\ln t]u_2'(t) = 0$$

to form a second equation for u_1' and u_2'.

As an alternative, we could go directly to equation (9) or equation (10) and use either of these formulas to determine a particular solution. If we proceed in this fashion, however, we have to make certain that the nonhomogeneous

(continued)

(continued)

equation under consideration is in the standard form given by equation (2). Note that since the differential equation in this example does *not* have this form, we need to rewrite it as

$$y'' - \frac{1}{t}y' + \frac{1}{t^2}y = \frac{1}{t}.$$

Doing so, we can identify the term $g(t)$ in equations (9) and (10), $g(t) = 1/t$.

Both approaches lead to the following system of equations for u_1' and u_2':

$$\begin{bmatrix} t & t\ln t \\ 1 & 1+\ln t \end{bmatrix} \begin{bmatrix} u_1' \\ u_2' \end{bmatrix} = \begin{bmatrix} 0 \\ \dfrac{1}{t} \end{bmatrix}.$$

Solving this system, we obtain

$$\begin{bmatrix} u_1' \\ u_2' \end{bmatrix} = \begin{bmatrix} -t^{-1}\ln t \\ t^{-1} \end{bmatrix}.$$

Computing antiderivatives,

$$u_1(t) = -\int \frac{1}{t}\ln t\,dt = -\frac{(\ln t)^2}{2} + K_1,$$

$$u_2(t) = \int \frac{1}{t}\,dt = \ln t + K_2.$$

We can set both of the arbitrary constants equal to zero, obtaining a particular solution

$$y_P(t) = t\left[-\frac{(\ln t)^2}{2}\right] + [t\ln t]\ln t = \frac{t}{2}(\ln t)^2.$$

The general solution of the nonhomogeneous equation is therefore

$$y(t) = c_1 t + c_2 t\ln t + \frac{t}{2}(\ln t)^2, \qquad t > 0.$$

Imposing the initial conditions, we find the solution of the initial value problem is

$$y(t) = t + 3t\ln t + \frac{t}{2}(\ln t)^2, \qquad t > 0. \blacktriangle$$

Figure 4.13 displays the graph of the solution $y(t)$. Note that $\lim_{t\to 0^+} y(t) = 0$. The solution is well behaved near $t = 0$ even though the differential equation has coefficient functions that are not defined at $t = 0$. The differential equation in Example 2 is an example of a nonhomogeneous Euler equation. The solution of Euler equations will be discussed in Chapter 10. Example 2 utilizes two methods we have studied, reduction of order and variation of parameters. Both methods have a common theme—using known information to solve for what is still unknown.

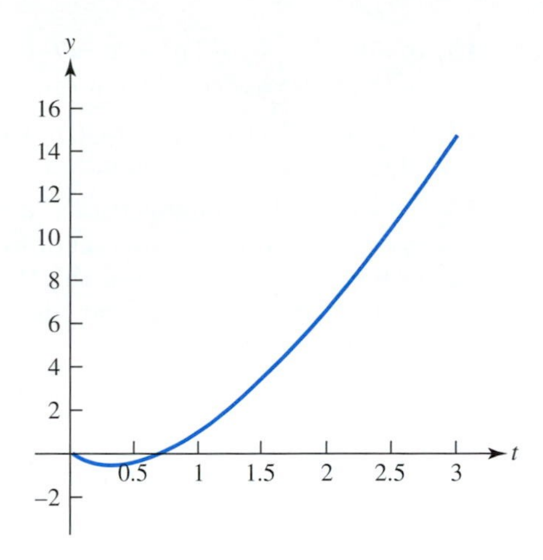

FIGURE 4.13

The solution of the initial value problem in Example 2. Even though some of the coefficient functions are not defined at $t = 0$, the solution $y(t)$ has a limit as t approaches 0 from the right.

EXERCISES

Exercises 1–14:

For the given differential equation,

(a) Determine the complementary solution, $y_C(t) = c_1y_1(t) + c_2y_2(t)$.

(b) Use the method of variation of parameters to construct a particular solution. Then, form the general solution.

(c) If the method of undetermined coefficients is also applicable, use it to find a particular solution. If the general solution formed in this manner is not identical to that found in part (b), reconcile the two answers and show that the general solutions you have found are equivalent.

1. $y'' + 4y = 2\sin 2t$

2. $y'' + y = \sec t, \quad -\pi/2 < t < \pi/2$

3. $y'' - (2+(1/t))y' + (1+(1/t))y = t,\ 0 < t < \infty$. [The functions $y_1(t) = e^t$ and $y_2(t) = t^2e^t$ are solutions of the homogeneous equation.]

4. $y'' - y = \dfrac{1}{1+e^t}$

5. $y'' - y = e^t$

6. $y'' - (2/t)y' + (2/t^2)y = t/(1+t^2),\ 0 < t < \infty$. [The function $y_1(t) = t^2$ is a solution of the homogeneous equation.]

7. $y'' - 2y' + y = e^t$

8. $y'' + 36y = \csc^3(6t)$

9. $y'' - (2\cot t)y' + (2\csc^2 t - 1)y = t\sin t,\ 0 < t < \pi$. [The functions $y_1(t) = \sin t$ and $y_2(t) = t\sin t$ are both solutions of the homogeneous equation.]

10. $t^2y'' - ty' + y = t\ln|t|,\ -\infty < t < \infty$. [The functions $y_1(t) = t$ and $y_2(t) = t\ln|t|$ are both solutions of the homogeneous equation.]

11. $y'' + (t/(1-t))y' - (1/(1-t))y = (t-1)e^t,\ 1 < t < \infty$. [The functions $y_1(t) = t$ and $y_2(t) = e^t$ are both solutions of the homogeneous equation.]

12. $y'' + 4ty' + (2+4t^2)y = t^2e^{-t^2}$. [The functions $y_1(t) = e^{-t^2}$ and $y_2(t) = te^{-t^2}$ are both solutions of the homogeneous equation.]

13. $(t-1)^2y'' - 4(t-1)y' + 6y = t,\ 1 < t < \infty$. [The function $y_1(t) = (t-1)^2$ is a solution of the homogeneous equation.]

14. $y'' - (2 + (2/t))y' + (1 + (2/t))y = e^t,\ 0 < t < \infty$. [The function $y_1(t) = e^t$ is a solution of the homogeneous equation.]

15. Consider the homogeneous differential equation $y'' + p(t)y' + q(t)y = g(t)$. Let $\{y_1, y_2\}$ be a fundamental set of solutions for the corresponding homogeneous equation and let $W(t)$ denote the Wronskian of this fundamental set. Show that the particular solution that vanishes at $t = t_0$ is given by

$$y_P(t) = \int_{t_0}^{t} [y_1(t)y_2(\lambda) - y_2(t)y_1(\lambda)]\frac{g(\lambda)}{W(\lambda)}\,d\lambda.$$

Exercises 16–18:

The given expression is the solution of the initial value problem

$$y'' + \alpha y' + \beta y = g(t), \quad y(0) = y_0, \quad y'(0) = y_0'.$$

Determine the constants α, β, y_0, and y_0'.

16. $y(t) = \dfrac{1}{2}\displaystyle\int_0^t \sin(2(t-\lambda))g(\lambda)\,d\lambda$

17. $y(t) = e^{-t} + \displaystyle\int_0^t \frac{e^{t-\lambda} - e^{-(t-\lambda)}}{2} g(\lambda)\,d\lambda = e^{-t} + \int_0^t \sinh(t-\lambda)g(\lambda)\,d\lambda$

18. $y(t) = t + \displaystyle\int_0^t (t-\lambda)g(\lambda)\,d\lambda$

4.11 Forced Mechanical Vibrations, Electrical Networks, and Resonance

In this section we use what you have learned about solving nonhomogeneous second order linear differential equations to study the behavior of mechanical systems (such as floating objects and spring-mass-dashpot systems) that are subjected to externally applied forces.

We also consider simple electrical networks containing resistors, inductors, and capacitors (called *RLC* networks) that are driven by voltage and current sources. All of these applications ultimately give rise to the same mathematical problem. And, as we shall see, the physical phenomenon of resonance is an important consideration common to all of these applications.

Forced Mechanical Vibrations

(a) ***A Buoyant Body*** Consider again the problem of the buoyant body discussed in the Introduction to this chapter. (See Figure 4.1). A cylindrical block of cross-sectional area A, height L, and mass density ρ is placed in a liquid having mass density ρ_l. Since we assume $\rho < \rho_l$, the block floats in the liquid. In equilibrium, it sinks a depth Y into the liquid; at this depth the weight of the block equals the weight of the liquid displaced. The quantity $y(t)$ represents the instantaneous vertical displacement of the block from its equilibrium position, measured positive downward.

Suppose now that an externally applied vertical force, $F_a(t)$, acts on the buoyant body. Newton's second law of motion ultimately leads to a nonhomogeneous differential equation for the displacement $y(t)$:

$$\frac{d^2y}{dt^2} + \frac{\rho_l g}{\rho L} y = \frac{1}{\rho AL} F_a(t). \tag{1}$$

In equation (1), g denotes gravitational acceleration and the term ρAL is the block's mass. Defining a radian frequency, $\omega_0 = \sqrt{\rho_l g / \rho L}$, and a force per unit mass, $f_a(t) = (1/\rho AL)F_a(t)$, equation (1) can be rewritten as

$$y'' + \omega_0^2 y = f_a(t). \tag{2}$$

Equation (2) is a second order constant coefficient linear nonhomogeneous differential equation. To uniquely prescribe the bobbing motion of the floating body, we would add initial conditions that specify the initial displacement, $y(t_0) = y_0$, and the initial velocity, $y'(t_0) = y_0'$, for the block.

(b) *A Spring-Mass-Dashpot System* In Section 4.7, we used Newton's second law of motion to derive a nonhomogenous differential equation for the displacement of a mass suspended from a spring-dashpot connection. The resulting differential equation is

$$my'' + \gamma y' + ky = F_a(t), \tag{3}$$

where $F_a(t)$ denotes an applied vertical force and the positive constants m, γ, and k represent the mass, damping coefficient, and spring constant of the system, respectively. The dependent variable $y(t)$ measures downward displacement from the equilibrium rest position. (See Figure 4.8.)

If there is no damping (that is, if $\gamma = 0$) and if we define radian frequency $\omega_0 = \sqrt{k/m}$ and force per unit mass $f_a(t) = (1/m)F_a(t)$, equation (3) can be rewritten as

$$y'' + \omega_0^2 y = f_a(t). \tag{4}$$

Note that equation (4), describing a spring-mass system, is identical in structure to equation (2), which describes the bobbing motion of a buoyant body. Both applications lead to the same mathematical problem. As we see shortly, other applications (such as *RLC* networks) also lead to differential equations having exactly the same structure as equation (2). Since these applications all lead to the same mathematical problem, we will discuss equations (2) and (3) in their own right rather than investigate each application separately. The exercises focus on specific applications.

Oscillatory Applied Forces and Resonance

Examples 1 and 2 concern an important special case of equation (2) where the applied force is a sinusoidally varying force,

$$f_a(t) = F \cos \omega_1 t,$$

where F is a constant. For simplicity we assume the system is initially at rest in equilibrium. Thus, for the special case under consideration, the corresponding initial value problem is

$$y'' + \omega_0^2 y = F\cos\omega_1 t, \qquad t > 0,$$
$$y(0) = 0, y'(0) = 0. \tag{5}$$

The complementary solution of (5) is

$$y_C(t) = c_1 \sin\omega_0 t + c_2 \cos\omega_0 t. \tag{6}$$

In order to find a particular solution, we need to consider two separate cases, $\omega_1 \neq \omega_0$ and $\omega_1 = \omega_0$. From a mathematical perspective, the method of undetermined coefficients discussed in Section 4.9 leads us to expect two different types of particular solutions. [If $\omega_1 \neq \omega_0$, the nonhomogeneous term, $g(t) = F\cos\omega_1 t$, is not a solution of the homogeneous equation, whereas $g(t)$ is a solution of the homogeneous equation when $\omega_1 = \omega_0$.]

This mathematical perspective is consistent with the physics of the problem. We should expect different behavior on purely physical grounds. Radian frequency ω_0 (or more properly $f_0 = \omega_0/2\pi$) is called the **natural frequency** of the vibrating system. It represents the frequency at which the system would vibrate if no applied force were present and the system were merely responding to some initial disturbance. The applied force acts on the system with its own applied frequency ω_1. However, in the special case where the natural and applied frequencies are equal, the applied force pushes and pulls on the system with a frequency precisely equal to that at which the system tends naturally to vibrate. This precise reinforcement leads to the phenomenon of **resonance**. For this reason, the natural frequency of the system is also referred to as its **resonant frequency**.

EXAMPLE 1

Assume $\omega_1 \neq \omega_0$. Solve the initial value problem (5).

Solution: Since $\omega_1 \neq \omega_0$, Table 4.1 in Section 4.9 suggests a particular solution of the form

$$y_P(t) = A\cos\omega_1 t + B\sin\omega_1 t.$$

Substituting this form into the nonhomogeneous equation, we obtain [see Exercise 1]

$$y_P(t) = -\frac{F}{\omega_1^2 - \omega_0^2}\cos\omega_1 t.$$

Imposing the initial conditions on the general solution

$$y(t) = c_1 \sin\omega_0 t + c_2\cos\omega_0 t - \frac{F}{\omega_1^2 - \omega_0^2}\cos\omega_1 t,$$

we obtain the solution of initial value problem (5),

$$y(t) = \frac{F}{\omega_1^2 - \omega_0^2}[\cos\omega_0 t - \cos\omega_1 t]. \tag{7}$$

Figure 4.14 shows solution (7) for the special case where $\omega_1 = 12\pi\ \text{s}^{-1}$, $\omega_0 = 10\pi\ \text{s}^{-1}$ and F is chosen so that $F/(\omega_1^2 - \omega_0^2) = 2$ cm. The example there-

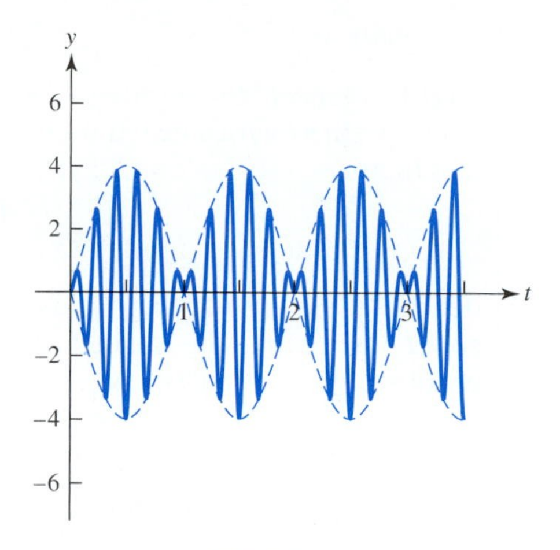

FIGURE 4.14

The solution of initial value problem (5), as given by equation (7). For the case shown, $\omega_1 = 12\pi$, $\omega_0 = 10\pi$, $F = 2(\omega_1^2 - \omega_0^2)$.

fore assumes that the natural frequency of the system is 5 Hz while the applied frequency is 6 Hz; for definiteness, the unit of length is chosen to be the centimeter.

We can better understand Figure 4.14, including the dashed envelope, if we use trigonometric identities to recast equation (7). Suppose we define an average radian frequency, $\overline{\omega}$, and a difference (or **beat**) radian frequency, β, by

$$\overline{\omega} = \frac{\omega_1 + \omega_0}{2}, \qquad \beta = \frac{\omega_1 - \omega_0}{2}.$$

With these definitions, we have $\omega_1 = \overline{\omega} + \beta$ and $\omega_0 = \overline{\omega} - \beta$. Equation (7) becomes

$$y(t) = \frac{F}{4\overline{\omega}\beta}[\cos(\overline{\omega}t - \beta t) - \cos(\overline{\omega}t + \beta t)].$$

Using the trigonometric identities $\cos(\theta_1 \pm \theta_2) = \cos\theta_1 \cos\theta_2 \pm \sin\theta_1 \sin\theta_2$, we can rewrite the second factor as

$$\cos(\overline{\omega}t - \beta t) - \cos(\overline{\omega}t + \beta t) = 2\sin\overline{\omega}t \sin\beta t.$$

Therefore, we obtain an alternative representation for the solution $y(t)$:

$$y(t) = \frac{F\sin\beta t}{2\,\overline{\omega}\beta}\sin\overline{\omega}t = A(t)\sin\overline{\omega}t, \tag{8}$$

where

$$A(t) = \frac{F\sin\beta t}{2\overline{\omega}\beta}.$$

Using equation (8), we can interpret solution (7) as the product of a variable amplitude term, $A(t)$, and a sinusoidal term, $\sin\overline{\omega}t$. In cases where ω_1 and ω_0 are nearly equal, the amplitude term, whose behavior is governed by the factor $\sin\beta t$, is slowly varying relative to the sinusoidal term, $\sin\overline{\omega}t$ [because $|\beta| \ll \overline{\omega}$]. The combination of these disparate rates of variation gives rise to the phenomenon of "beats" seen in Figure 4.14.

(continued)

(continued)

With respect to Figure 4.14, $\overline{\omega} = 11\pi\ \mathrm{s}^{-1}$ and $\beta = \pi\ \mathrm{s}^{-1}$. Therefore, $\sin \beta t = \sin \pi t$ varies much more slowly than $\sin \overline{\omega} t = \sin 11\pi t$. The multiplicative factor

$$A(t) = \frac{F \sin \beta t}{2\overline{\omega}\beta} = 4 \sin \pi t$$

defines a slowly varying sinusoidal envelope for the more rapidly varying $\sin \overline{\omega} t = \sin 11\pi t$. The dashed envelope shown in Figure 4.14 is defined by the graphs of $y = \pm 4 \sin \pi t$. ▲

EXAMPLE 2

Assume $\omega_1 = \omega_0$. Solve initial value problem (5).

Solution: In this case,

$$y'' + \omega_0^2 y = F \cos \omega_0 t.$$

Since the functions $\cos \omega_0 t$ and $\sin \omega_0 t$ are both solutions of the homogeneous equation, Table 4.1 of Section 4.9 prescribes a trial solution of the form

$$y_P(t) = At \cos \omega_0 t + Bt \sin \omega_0 t.$$

Substituting this form leads [see Exercise 1] to the particular solution

$$y_P(t) = \frac{F}{2\omega_0} t \sin \omega_0 t,$$

and the general solution

$$y(t) = c_1 \sin \omega_0 t + c_2 \cos \omega_0 t + \frac{F}{2\omega_0} t \sin \omega_0 t.$$

The initial conditions imply that c_1 and c_2 are zero. The solution of initial value problem (5) is, therefore,

$$y(t) = \frac{F}{2\omega_0} t \sin \omega_0 t. \qquad (9)$$

Figure 4.15 shows solution (9) for the case

$$\omega_0 = 10\pi\ \mathrm{s}^{-1} \quad \text{and} \quad \frac{F}{2\omega_0} = 4\pi\ \mathrm{cm/s}.$$

In this case the applied force reinforces the natural frequency vibrations of the mechanical system and the envelope of the solution (shown by the dashed lines in Figure 4.15) grows linearly with time. This is the phenomenon of resonance.

Does the solution (9) make sense? The vibration amplitude of a real physical system certainly does not continue to grow indefinitely. Therefore, one would expect equation (9) to describe the behavior of a real system for a limited time at best. Once the vibration amplitude becomes sufficiently large, the assumptions made in deriving the mathematical models cease to be valid. For example, equations (1) and (2) for the floating body cease to be valid once the body becomes either totally submerged or completely elevated out of the liquid. Under those conditions, our expression for the buoyant force is no longer correct. Likewise,

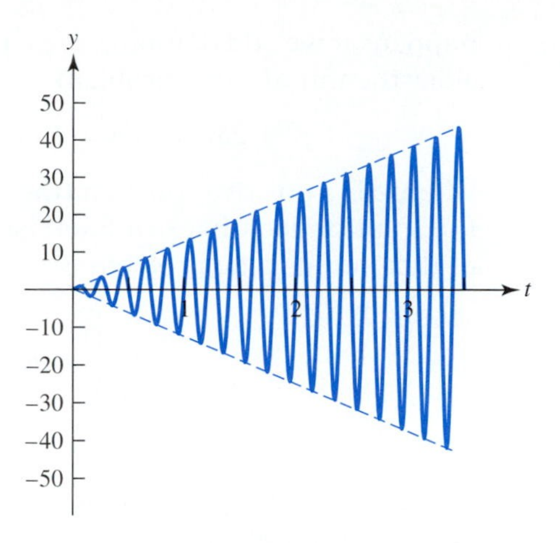

FIGURE 4.15

The solution of initial value problem (5) as given by equation (9). For the case shown, $\omega_0 = 10\pi$ and $F = 8\pi\omega_0$. The solution envelope grows linearly with time, illustrating the phenomenon of resonance.

equation (3) and special case (4) for the spring-mass system assume the validity of Hooke's law. When mass displacement amplitude becomes too large, however, the force-displacement relation becomes more complicated than the simple linear relation embodied in Hooke's law. ▲

REMARK: One property of a well-posed problem is continuous dependence upon the data. Roughly speaking, if an initial value problem is to be a reasonable model of reality, its solution should not change uncontrollably when a parameter (such as a coefficient in the differential equation or an initial condition) is changed slightly. Therefore, we might reasonably ask whether it is possible to see resonant solution (9) emerge from nonresonant solution (7) or (8) in the limit as $\omega_1 \to \omega_0$. Note that (8) can be rewritten as

$$y(t) = \frac{F}{2\overline{\omega}}t\left(\frac{\sin\beta t}{\beta t}\right)\sin\overline{\omega}t. \tag{10}$$

Suppose we fix t and let $\omega_1 \to \omega_0$. Then, from their definitions, $\overline{\omega} \to \omega_0$ and $\beta \to 0$. We know from calculus that

$$\lim_{x\to 0}\frac{\sin x}{x} = 1.$$

Therefore, for any fixed value of t, we do indeed obtain resonant solution (9) from nonresonant expression (8) in the limit as $\omega_1 \to \omega_0$.

The Effect of Damping on Resonance

There are no perpetual motion machines. All physical systems have at least some small loss or damping present. Therefore, it is of interest to see what happens if we add damping to an otherwise resonant system. Suppose we consider the initial value problem

$$y'' + 2\delta y' + \omega_0^2 y = F\cos\omega_0 t, \qquad y(0) = 0, \qquad y'(0) = 0, \tag{11a}$$

where δ is a positive constant (the factor 2 is added for convenience). What does the solution look like? In Exercise 11(a) we ask you to show that the solution of this initial value problem is

$$y(t) = \frac{F}{2\delta}\left[\frac{\sin\omega_0 t}{\omega_0} - \frac{e^{-\delta t}\sin\left(\sqrt{\omega_0^2 - \delta^2}\, t\right)}{\sqrt{\omega_0^2 - \delta^2}}\right]. \tag{11b}$$

[In (11) we tacitly assume that $\omega_0^2 > \delta^2$.] As a check, you can show [see Exercise 12(c)] that for any fixed t and ω_0, expression (11b) reduces to (9) as $\delta \to 0$.

Equation (11b) shows the effect of damping on the otherwise resonant system. If we fix δ at some positive value and let $t \to \infty$, the second term in equation (11b) tends to zero. Therefore, with the addition of damping, displacement does not grow indefinitely as it appears to in Figure 4.15. As time increases, displacement approaches a steady-state behavior given by the first term,

$$\frac{F}{2\delta\omega_0}\sin\omega_0 t.$$

As δ becomes smaller, the amplitude of these steady-state oscillations becomes correspondingly larger. Figure 4.16 shows the variation of displacement for the case

$$\omega_0 = 10\pi \text{ s}^{-1}, \qquad \frac{F}{2\omega_0} = 4\pi \text{ cm/s}, \qquad \delta = 0.5 \text{ s}^{-1}.$$

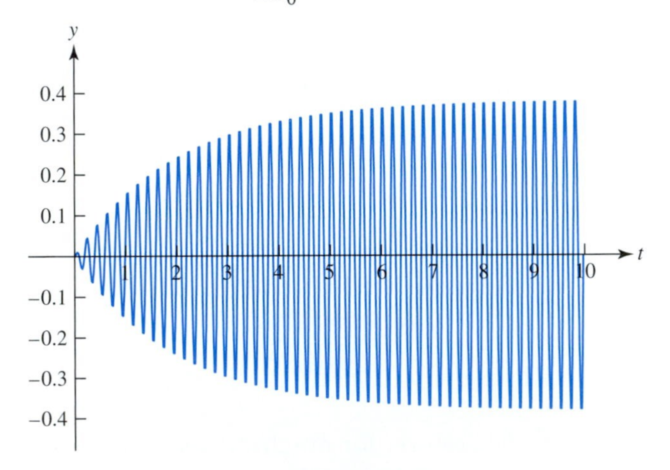

FIGURE 4.16

The solution of $y'' + 2\delta y' + \omega_0^2 y = F\cos\omega_0 t$ for the case $\omega_0 = 10\pi$, $F = 8\pi\omega_0$, and $\delta = 0.5$. As you can see by comparing this graph with the resonant solution graphed in Figure 4.15, the inclusion of damping eliminates the unbounded linear growth in the solution envelope that characterizes the resonant case. As noted, however, the steady-state oscillations, $(F/2\delta\omega_0)\sin\omega_0 t$, have an amplitude proportional to δ^{-1}.

Nonresonant Excitation with Damping Present

Suppose we now change the radian frequency of the applied force in the previous problem to $\omega_1 \neq \omega_0$. In that case the problem becomes

$$y'' + 2\delta y' + \omega_0^2 y = F \cos \omega_1 t, \qquad y(0) = 0, \qquad y'(0) = 0. \tag{12a}$$

This amounts to the addition of a damping force to the problem defined by equation (5). Again assuming that $\omega_0^2 > \delta^2$, the solution of problem (12a) is

$$\begin{aligned} y(t) = {} & \frac{F}{(\omega_0^2 - \omega_1^2)^2 + (2\delta\omega_1)^2}[(\omega_0^2 - \omega_1^2)\cos \omega_1 t + 2\delta\omega_1 \sin \omega_1 t] \\ & - \frac{Fe^{-\delta t}}{(\omega_0^2 - \omega_1^2)^2 + (2\delta\omega_1)^2}\left[(\omega_0^2 - \omega_1^2)\cos\left(\sqrt{\omega_0^2 - \delta^2}\, t\right)\right. \\ & \left. + \frac{\delta(\omega_0^2 + \omega_1^2)}{\sqrt{\omega_0^2 - \delta^2}} \sin\left(\sqrt{\omega_0^2 - \delta^2}\, t\right)\right]. \end{aligned} \tag{12b}$$

This solution seems relatively complicated. However, two checks can be made. At a fixed value of t, the solution should reduce to (11b) in the limit as $\omega_1 \to \omega_0$. Also, the solution should reduce to (7) if we fix ω_1 but let $\delta \to 0$. Exercise 12 asks you not only to derive this solution, but also to make these checks upon its correctness.

The second term in the solution of equation (12a) represents a transient term, one that tends to zero as time increases. The first term is the steady-state portion of the solution. Figure 4.17 shows the behavior of the solution for the case where

$$\omega_0 = 10\pi \text{ s}^{-1}, \qquad \omega_1 = 12\pi \text{ s}^{-1}, \qquad \frac{F}{2\omega_0} = 4\pi \text{ cm/s}, \qquad \delta = 0.5 \text{ s}^{-1}.$$

Compare this behavior with that exhibited in Figures 4.14 and 4.16. When time

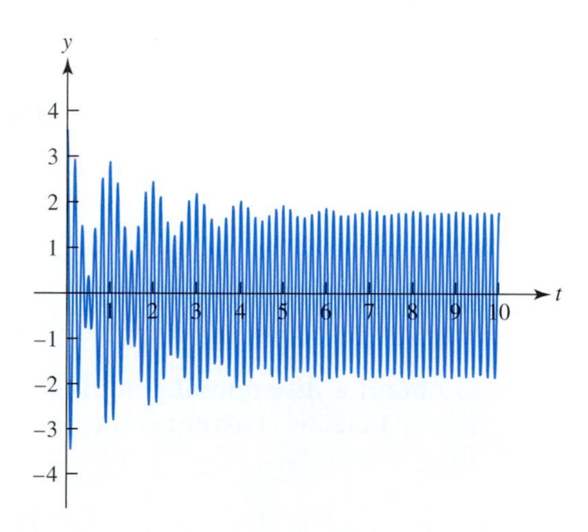

FIGURE 4.17

The solution of equation (12) for representative values of $\omega_0, \omega_1, F, \delta$. Initially, for small values of t, damping is not significant and the motion is similar to that shown in Figure 4.14, exhibiting beats. As t grows, damping diminishes the second term in equation (12) and the motion has a period similar to the applied force.

t is relatively small, before the effects of damping becomes pronounced, the solution exhibits a difference frequency modulation envelope that is qualitatively similar to the behavior shown in Figure 4.14. As time progresses, however, damping eventually diminishes the second term in the solution to a negligibly small contribution and the solution becomes essentially the steady-state portion given by the first term. In this respect, the long-term behavior qualitatively resembles the damping-perturbed resonant case exhibited in Figure 4.16.

RLC Networks

We consider networks containing resistors, inductors, and capacitors. The application of Kirchhoff's[9] voltage law and Kirchhoff's current law to these simple *RLC* networks leads us to second order differential equations.

Consider first the series *RLC* network shown in Figure 4.18. A voltage source $V_S(t)$ having the polarity shown is connected in series with circuit elements having resistance R, inductance L, and capacitance C. A current $I(t)$, assumed positive in the sense shown, flows in the loop. In essence, Kirchhoff's voltage law asserts that the voltage at each point in the network is a well-defined single-valued quantity. Therefore, as we make an excursion around the loop, the sum of the voltage rises must equal the sum of the voltage drops. If we proceed around the loop in Figure 4.18 in a clockwise manner, the voltage rise is the source voltage $V_S(t)$, while the voltage drops are the drops across the three circuit elements. The voltage drop across the resistor is $I(t)R$, the drop across the inductor is $L\,(dI/dt)$ and the drop across the capacitor is $(1/C)Q(t)$, where $Q(t)$ represents the electric charge on the capacitor.

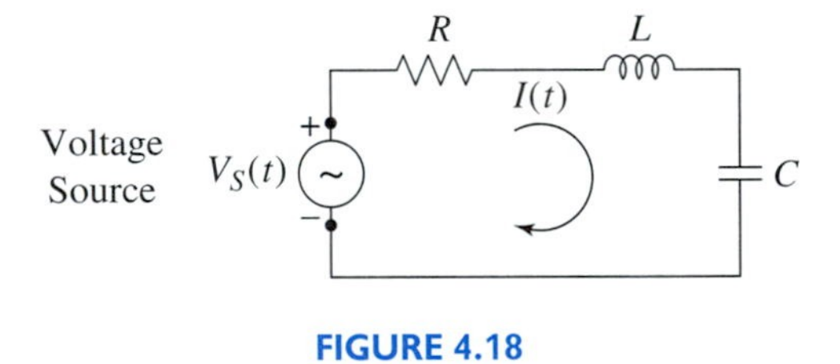

FIGURE 4.18

A series *RLC* network, with voltage source $V_S(t)$ and loop current $I(t)$.

An application of Kirchhoff's voltage law therefore leads to the equation

$$V_S(t) = RI + L\,\frac{dI}{dt} + \frac{1}{C}Q(t). \tag{13a}$$

To obtain a differential equation for a single dependent variable, we use the fact that electric current is the rate of change of electric charge with respect to time,

$$I(t) = \frac{dQ}{dt}.$$

One approach is to rewrite equation (13a) as a second order differential equa-

[9]Gustav Robert Kirchoff (1824–1887) was a German physicist who made important contributions to network theory, elasticity, and our understanding of blackbody radiation. A lunar crater is named in his honor.

tion for the electric charge, obtaining

$$L\frac{d^2Q}{dt^2} + R\frac{dQ}{dt} + \frac{1}{C}Q = V_s(t).$$

This equation, supplemented by initial conditions specifying the charge $Q(t_0)$ and current $Q'(t_0) = I(t_0)$ at some initial time t_0 can be solved for the charge $Q(t)$. Differentiating this solution yields the desired current, $I(t)$. A second approach is simply to differentiate equation (13a), obtaining a second order differential equation for the current $I(t)$. In that case,

$$\frac{d^2I}{dt^2} + \frac{R}{L}\frac{dI}{dt} + \frac{1}{LC}I = \frac{1}{L}\frac{dV_S(t)}{dt}. \tag{13b}$$

Equation (13b) is a nonhomogeneous second order linear differential equation for the unknown loop current $I(t)$. To uniquely prescribe circuit performance, we must add initial conditions $I(t_0) = I_0$ and $I'(t_0) = I'_0$ at some initial time t_0. [Specifying $I'(t_0)$ is tantamount to specifying the voltage drop across the inductor at time t_0.]

As a second example of an *RLC* network, consider the network shown in Figure 4.19. In this case the three circuit elements are connected in parallel with a current source $I_S(t)$, whose current output is assumed to flow in the direction shown. This time the dependent variable of interest is nodal voltage $V(t)$ assumed to have the polarity shown.

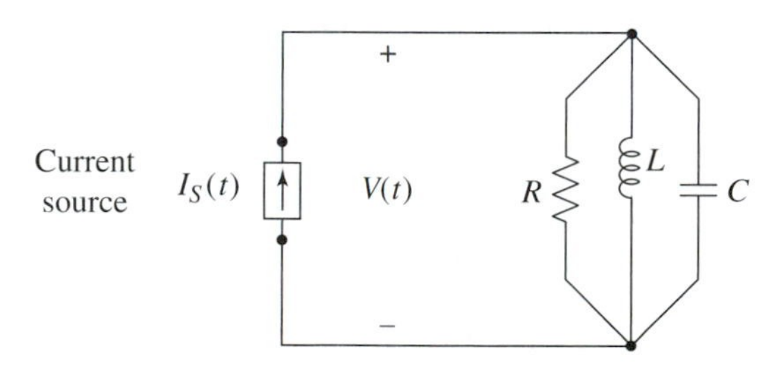

FIGURE 4.19

A parallel *RLC* network, with current source $I_S(t)$ and nodal voltage $V(t)$.

The governing physical principle, Kirchhoff's current law, states (in essence) that electric current does not accumulate at a circuit node. Therefore, the total current flowing into a node must equal the total current flowing out. Consider the upper node. The current flowing in is the source current, while the current flowing out is the current flowing "down" through each of the circuit elements. The current through the resistor is $(1/R)V(t)$, the current through the capacitor is $C(dV/dt)$, and the current through the inductor is $(1/L)\int V(s)\,ds$ (an antiderivative of the nodal voltage). Applying Kirchhoff's current law to the network in Figure 4.19 leads us to the differential equation

$$I_S(t) = \frac{1}{R}V + C\frac{dV}{dt} + \frac{1}{L}\int V(s)\,ds.$$

Upon differentiating and rearranging terms, the equation becomes

$$\frac{d^2V}{dt^2} + \frac{1}{RC}\frac{dV}{dt} + \frac{1}{LC}V = \frac{1}{C}\frac{dI_S(t)}{dt}. \tag{14}$$

Specifying V and V' (that is, the currents through the resistor and capacitor) at some initial time t_0 will uniquely determine circuit performance.

REMARK: If we short circuit the resistor (that is, set $R = 0$) in the series circuit (Figure 4.18) or if we open circuit the resistor (that is, let $R \to \infty$) in the parallel circuit (Figure 4.19), we remove the dissipative (damping) element in each case and obtain a lossless LC circuit. Such circuits can exhibit resonance. Note that equations (13b) and (14) become identical in structure to equations (2) and (4) with a resonant radian frequency defined by

$$\omega_0^2 = \frac{1}{LC}.$$

EXERCISES

1. Consider the differential equation $y'' + \omega_0^2 y = F \cos \omega t$.

(a) Determine the complementary solution of this differential equation.

(b) Use the method of undetermined coefficients to find a particular solution in each of the cases: (i) $\omega = \omega_1 \neq \omega_0$, (ii) $\omega_1 = \omega_0$.

Exercises 2–5:

A 10-kg object suspended from the end of a vertically hanging spring stretches the spring 9.8 cm. At time $t = 0$, the resulting spring-mass system is disturbed from its rest state by an applied force, $F(t)$. The force $F(t)$ is expressed in newtons and is considered positive in the downward direction; time is measured in seconds.

(a) Determine the spring constant, k.

(b) Formulate and solve the initial value problem for $y(t)$, where $y(t)$ is the displacement of the object from its equilibrium rest state, measured positive in the downward direction.

(c) Plot the solution and determine the maximum excursion from equilibrium made by the object on the t-interval $0 \le t < \infty$ or state that there is no such maximum.

2. $F(t) = 20 \cos 10t$

3. $F(t) = 20e^{-t}$

4. $F(t) = 20 \cos 8t$

5. $F(t) = \begin{cases} 20, & 0 \le t \le \pi \\ 0, & \pi < t < \infty \end{cases}$

[Hint: Solve the problem on the t-interval $0 \le t \le \pi$ and then use the fact that position $y(t)$ and velocity $y'(t)$ are both continuous at $t = \pi$ to formulate and solve a second initial value problem on the t-interval $\pi \le t < \infty$.]

6. Consider the initial value problem $my'' + ky = 20 \cos 8\pi t, y(0) = 0, y'(0) = 0$, where $y(t)$ represents the vertical displacement of the mass from equilibrium. This problem models the response of a spring-mass system, initially at rest, to an applied force; assume that the unit of force is the newton. Suppose the motion shown in the figure is recorded and can be described mathematically by the formula $y(t) = 0.1 \sin(\pi t) \sin(7\pi t)$ m. What are the values of mass m and spring constant k for this system? [Hint: Recall the identity $\cos(\alpha \pm \beta) = \cos\alpha \cos\beta \mp \sin\alpha \sin\beta$.]

Exercises 7–10:

Consider the initial value problem

$$my'' + \gamma y' + ky = F(t), \qquad y(0) = 0, \qquad y'(0) = 0,$$

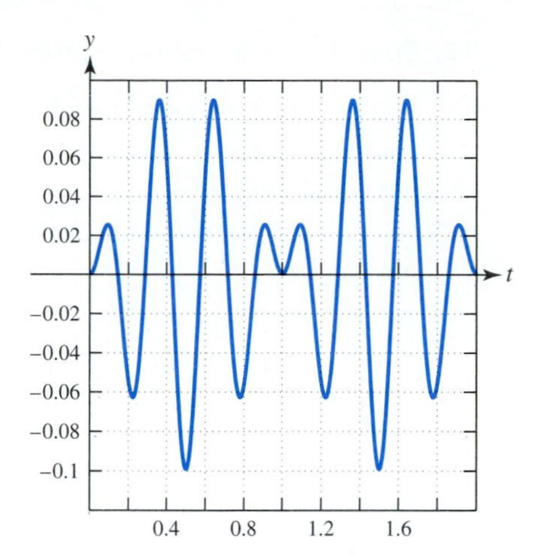

Figure for Exercise 6

which models the motion of a spring-mass-dashpot system initially at rest, and subjected to an applied force $F(t)$, where the unit of force is the newton (N). Assume that $m = 2$ kg, $\gamma = 8\ \mathrm{s}^{-1}$, and $k = 80$ N/m.

(a) Solve the initial value problem for the given applied force. Where appropriate, use the fact that the system displacement $y(t)$ and velocity $y'(t)$ remain continuous at points where the applied force is discontinuous.

(b) Determine the long-time behavior of the system. In particular, is $\lim_{t\to\infty} y(t) = 0$? If not, describe in qualitative terms what the system is doing as $t \to \infty$.

7. $F(t) = 20\cos 8t$

8. $F(t) = 20e^{-t}$

9. $F(t) = 20\sin 6t$

10. $F(t) = \begin{cases} 20, & 0 \le t \le \dfrac{\pi}{2} \\ 0, & \dfrac{\pi}{2} < t < \infty \end{cases}$

11. Consider the initial value problem $my'' + \gamma y' + ky = \overline{F}\cos\sqrt{k/m}\,t, y(0) = 0,$ $y'(0) = 0$. This problem models the response of a spring-mass-dashpot system, initially at rest, to an applied force. If we set $\gamma/m = 2\delta, \omega_0^2 = k/m$, and $\overline{F}/m = F$, we obtain initial value problem (11a). Assume that $\omega_0^2 > 2\delta$. Note that the radian frequency of the applied force is ω_0; this is the resonant radian frequency of the corresponding undamped system.

(a) Derive equation (11b), showing that the solution of this initial value problem is

$$y(t) = \frac{F}{2\delta}\left[\frac{\sin(\omega_0 t)}{\omega_0} - \frac{e^{-\delta t}\sin\left(\sqrt{\omega_0^2 - \delta^2}\,t\right)}{\sqrt{\omega_0^2 - \delta^2}}\right].$$

(b) Show, for any fixed values $t > 0$ and $\omega_0 > 0$, that

$$\lim_{\delta\to 0^+}\left\{\frac{F}{2\delta}\left[\frac{\sin(\omega_0 t)}{\omega_0} - \frac{e^{-\delta t}\sin\left(\sqrt{\omega_0^2 - \delta^2}\,t\right)}{\sqrt{\omega_0^2 - \delta^2}}\right]\right\} = \frac{F}{2\omega_0} t\sin(\omega_0 t).$$

This limit is the response of the undamped spring-mass system to resonant frequency excitation.

(c) Suppose that we know the values of mass m and spring constant k (and $\overline{F}$, the amplitude of the applied force). Explain how we might use our knowledge of the solution in part (a) (observed over a long time interval) to estimate the damping constant δ.

12. Consider the initial value problem given in equation (12a),

$$y'' + 2\delta y' + \omega_0^2 y = F\cos\omega_1 t, \qquad y(0) = 0, \qquad y'(0) = 0,$$

where $\omega_0^2 > \delta$ and $\omega_1 \neq \omega_0$. The radian frequency of the applied force is therefore not equal to ω_0.

(a) Solve the initial value problem for $y(t)$ and verify that equation (12b) represents the solution.

(b) Assume that $t > 0$ and $\delta > 0$ are fixed. Show that

$$\lim_{\omega_1 \to \omega_0} y(t) = \frac{F}{2\delta}\left[\frac{\sin(\omega_0 t)}{\omega_0} - \frac{e^{-\delta t}\sin\left(\sqrt{\omega_0^2 - \delta^2}\, t\right)}{\sqrt{\omega_0^2 - \delta^2}}\right].$$

(c) Assume now that $t > 0$ and ω_1 are fixed. Show that

$$\lim_{\delta \to 0^+} y(t) = \frac{F}{\omega_1^2 - \omega_0^2}\left[\cos(\omega_0 t) - \cos(\omega_1 t)\right].$$

13. The Great Zucchini, daredevil extraordinaire, is a circus performer whose act consists of being "shot from a cannon" to a safety net some distance away. The "cannon" is a frictionless tube containing a large spring, as shown in the figure. The spring constant is $k = 150$ lb/ft and is precompressed 10 ft prior to launching the acrobat. Assume the spring obeys Hooke's law and that Zucchini weighs 150 lb. Neglect the weight of the spring.

(a) Let $x(t)$ represent spring displacement along the tube axis, measured positive in the upward direction. Show that Newton's second law of motion leads to the differential equation $mx'' = -kx - mg\cos(\pi/4), x < 0$, where m is the mass of the daredevil. Specify appropriate initial conditions.

(b) With what speed does he emerge from the tube when the spring is released?

(c) If the safety net is to be placed at the same height as the mouth of the "cannon," how far downrange from the cannon's mouth should the center of the net be placed?

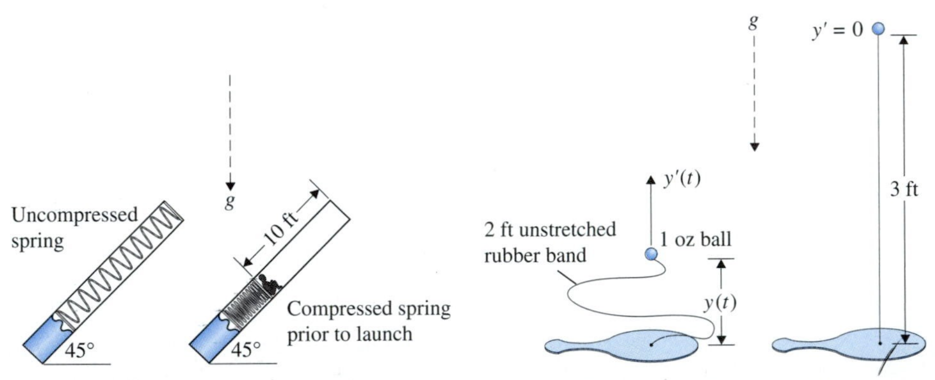

Figure for Exercise 13

Figure for Exercise 14

14. A Change of Independent Variable (see Section 3.7) A popular child's toy consists of a small rubber ball attached to a wooden paddle by a rubber band; see the figure. Assume a 1-oz ball is connected to the paddle by a rubber band having an unstretched length of 2 ft. When the ball is launched vertically upward by the paddle with an initial speed of 30 ft/sec, the rubber band is observed to stretch 1 ft (to a total length of 3 ft) when the ball has reached its highest point. Assume the rubber band behaves like a spring and obeys Hooke's law for this amount of stretching. Our objective is to determine the spring constant k. (Neglect the weight of the rubber band.)

The motion occurs in two phases. Until the ball has risen 2 ft above the paddle, it acts like a projectile influenced only by gravity. Once the height of the ball exceeds 2 ft, it acts like a mass on a spring, acted upon by the elastic restoring force of the rubber band and gravity.

(a) Assume the ball leaves the paddle at time $t = 0$. Let t_2 and t_3 represent the times at which the height of the ball is 2 ft and 3 ft, respectively, and let m denote the mass of the rubber ball. Show that an application of Newton's second law of motion leads to the following two-part description of the problem:

(i) $my'' = -mg, \quad 0 < t < t_2, \;\; y(0) = 0, \;\; y'(0) = 30$

(ii) $my'' = -k(y-2) - mg, \quad t_2 < t < t_3.$

Here, y and y' are assumed to be continuous at $t = t_2$. We also know that $y(t_2) = 2$, $y(t_3) = 3$, and $y'(t_3) = 0$.

If we attempt to solve the problem "directly" with time as independent variable, it is relatively difficult since the times t_2 and t_3 must be determined as part of the problem. Since height $y(t)$ is an increasing function of time t over the interval of interest, however, we can view time t as a function of height, y, and use y as the independent variable.

(b) Let $v = y' = dy/dt$. Adopting y as the independent variable, the acceleration becomes $y'' = dv/dt = (dv/dy)\,(dy/dt) = v(dv/dy)$. Therefore,

(i) $mv\dfrac{dv}{dy} = -mg, \quad 0 < y < 2, \;\; v(0) = 30$

(ii) $mv\dfrac{dv}{dy} = -k(y-2) - mg, \quad 2 < y < 3.$

Here, v is continuous at $y = 2$ and $v|_{y=3} = 0$.

Solve these two separable differential equations, impose the accompanying supplementary conditions, and determine the spring constant k.

Exercises 15–18:

Network Problems Use the following consistent set of scaled units (referred to as the Scaled SI Unit System) in Exercises 15–18.

Quantity	Unit	Symbol
Voltage	volt	V
Current	milliampere (mA)	I
Time	millisecond (ms)	t
Resistance	kilohm kΩ	R
Inductance	Henry (H)	L
Capacitance	microFarad (μF)	C

Exercises 15–16:

Consider the series LC network shown in the figure. Assume that at time $t = 0$, the current and its time rate of change are both zero. For the given source voltage, determine the current, $I(t)$.

Figure for Exercises 15–16

15. $V_S(t) = 5\sin 3t$

16. $V_S(t) = 10te^{-t}$

Exercises 17–18:

Consider the parallel RLC network shown in the figure. Assume that at time $t = 0$, the voltage $V(t)$ and its time rate of change are both zero. For the given source current, determine the voltage $V(t)$.

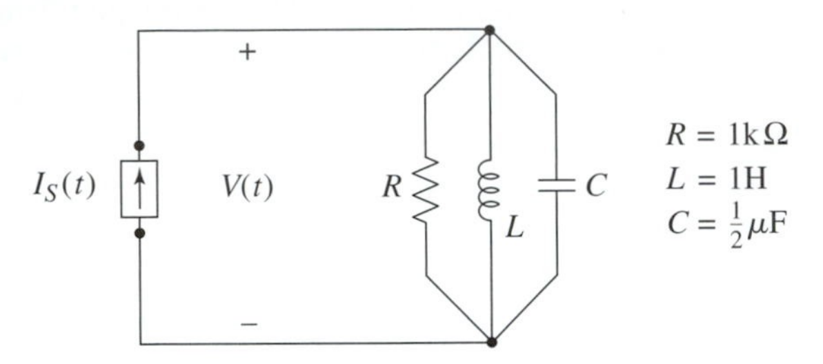

Figure for Exercises 17–18

17. $I_S(t) = 1 - e^{-t}$

18. $I_S(t) = 5\sin t$

Extended Problem: Modeling Buoyant Motion

The primary purpose of this exercise is to derive differential equations rather than solve them. The problems considered involve buoyant motion of bodies having shapes more complicated than those considered earlier in this chapter in Figure 4.1.

In particular, we consider a cone and also a trough having a parabolic cross section. We assume the weights of these bodies are distributed so that the bodies do not tip over in the water. Rather, they bob up and down vertically in a stable fashion. As we shall see, an application of Newton's second law of motion leads to second order nonlinear differential equations that model the bobbing motion. The techniques of Chapter 4 cannot be used to study these equations. Later, in Chapters 8 and 9, we will develop qualitative and numerical approaches that can be used to obtain information about their solutions.

In the last two parts of these problems, however, we consider small disturbances from static equilibrium. When the amplitude of the bobbing motion is small, an approximation can be introduced that reduces the nonlinear differential equations to the type of linear differential equation studied in this chapter.

PROBLEMS 1 and 2: *Cone and Parabolic Trough*

Each problem asks you to perform similar calculations, in one case for a cone-shaped object and in the second case for a parabolic trough. The cone is shown in Figure 4.20 and the parabolic trough (which we might view as a crude model of a boat) is shown in Figure 4.21.

Deriving Equations That Model Bobbing Motion

Let ρ_l denote the mass density of water (liquid) and g the acceleration due to gravity. Then $\rho_l g$ represents the weight of a unit volume of water. Assume that the body weighs one-half the weight of an equal volume of water. Therefore, the object will float when placed in water. Assume that, when placed in water, the body sinks to a level designated as Y_e. At this depth, the body displaces a weight of water equal to its own weight. This is the condition of static equilibrium shown in Figures 4.20(b) and 4.21(b).

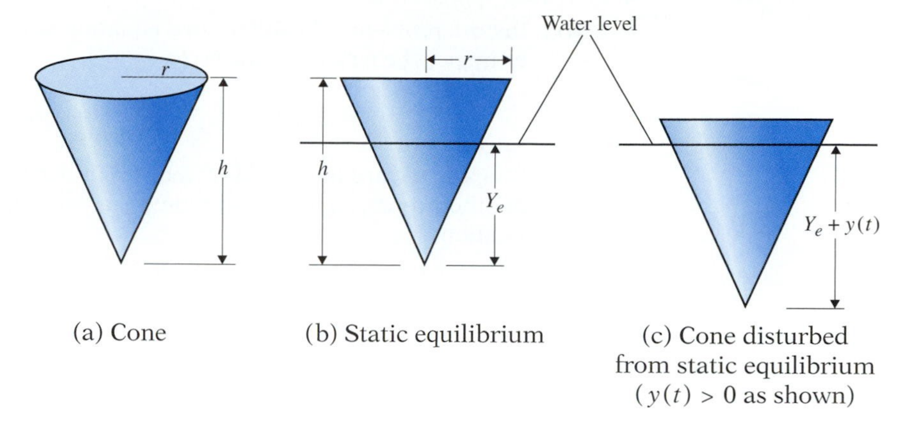

FIGURE 4.20

(a) The dimensions of the cone-shaped object. (b) The cone shown floating in water, in static equilibrium. (c) The cone shown disturbed from its static equilibrium condition.

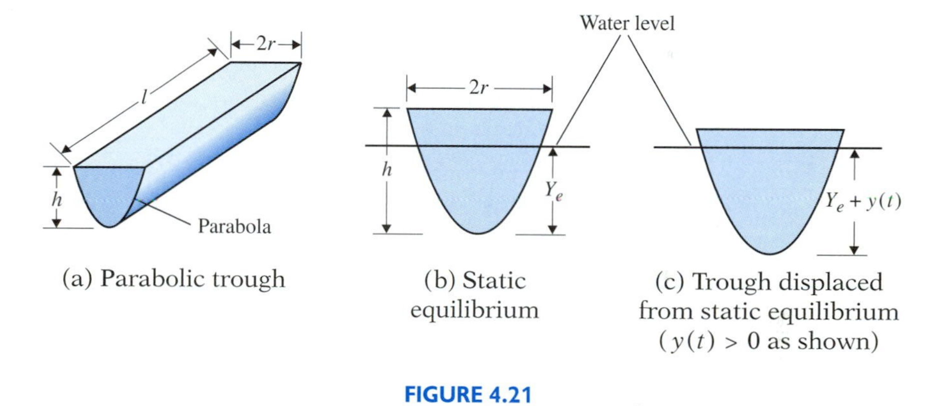

FIGURE 4.21

(a) The dimensions of the parabolic trough. (b) The trough shown floating in water, in static equilibrium. (c) The trough shown disturbed from its static equilibrium condition.

(a) Determine the equilibrium depth, Y_e. (The actual value of Y_e will be different in each of the two problems.)

Assume now that the body is displaced from its static equilibrium position [as in Figures 4.20(c) and 4.21(c)]. The value $y(t)$ represents the additional depth that the body is submerged at time t. We assume $y(t)$ is sufficiently small so that the body is never fully submerged.

(b) Apply Newton's second law of motion in the vertical direction to obtain a differential equation for $y(t)$. The net vertical force, which is positive in the downward direction, is the body weight minus the upward buoyant force.

(c) Is the differential equation that you have derived a nonlinear differential equation as claimed in the introductory comments, or a linear one? Explain your answer. In particular what term(s) in the differential equation make it a nonlinear differential equation?

(d) In both problems, the differential equation governing the buoyant motion can ultimately be written in the form

$$y''(t) = f\left(\frac{y(t)}{h}\right), \quad \text{where} \quad f(0) = 0.$$

Suppose we are interested in very small departures from static equilibrium; in other words, suppose we assume that $|y(t)/h| \ll 1$. In that case, the approximation

$$f\left(\frac{y(t)}{h}\right) \approx f'(0)\frac{y(t)}{h} \tag{1}$$

seems reasonable. [From a mathematical point of view, we are approximating the function $f(y/h)$ by the first nonvanishing term in its Maclaurin series expansion.]

Make the approximation (1) in the differential equation you have obtained. Obtain the general solution of the resulting second order linear differential equation

$$y'' = h^{-1}f'(0)y,$$

using the techniques developed in this chapter.

(e) For both problems, solve the initial value problem

$$y'' = h^{-1}f'(0)y, \qquad y(0) = 1 \text{ in.} \qquad y'(0) = 0,$$

using the values: $h = 3$ ft, $r = 1$ ft, $l = 10$ ft, $g = 32$ ft/sec^2, $\rho_l g = 62.4$ lb/ft^3. Graph your solution.

The approximation discussed in (d) is often referred to as linearization. This approximation will be discussed in more detail in Chapter 8.

5

Higher Order Linear Differential Equations

CHAPTER OVERVIEW

Introduction

In Chapter 4 we discussed second order linear differential equations,

$$y'' + p_1(t)y' + p_0(t)y = g(t).$$

In this chapter we extend the results of Chapter 4 to nth order linear differential equations,

$$y^{(n)} + p_{n-1}(t)y^{(n-1)} + \cdots + p_2(t)y'' + p_1(t)y' + p_0(t)y = g(t).$$

As Sections 5.1–5.3 show, the basic theory for second order linear differential equations carries over with very little change to higher order linear differential equations. Similarly, the techniques described in Chapter 4 for solving constant coefficient second order homogeneous linear equations extend naturally to nth order homogeneous equations (see Section 5.4). Section 5.5 extends the two methods developed in Chapter 4 for finding a particular solution of a nonhomogeneous equation, the method of undetermined coefficients and the method of variation of parameters.

The theory of higher order linear differential equations (and the way in which it generalizes second order linear theory) is aesthetically appealing. In addition, higher order linear equations are important in certain applications. Fourth order linear equations arise, for example, when we model the loading and bending of beams (see the Extended Problem at the end of this chapter).

5.1 Existence and Uniqueness

An ***n*th order linear differential equation** is an equation of the form

$$\frac{d^n y}{dt^n} + p_{n-1}(t)\frac{d^{n-1}y}{dt^{n-1}} + \cdots + p_2(t)\frac{d^2y}{dt^2} + p_1(t)\frac{dy}{dt} + p_0(t)y = g(t),$$

where the functions $p_0(t), p_1(t), \ldots, p_{n-1}(t)$ and $g(t)$ are all defined on some common t-interval, say $a < t < b$. We assume these $n+1$ functions are continuous on the interval (a, b). As before, if the function $g(t)$ is zero throughout the t-interval of interest, the equation is called *homogeneous;* otherwise it is *nonhomogeneous*.

The main purpose of this chapter is to show how the results of Chapter 4 for second order linear differential equations extend to higher order equations. As such, we shall focus much of our attention upon the following initial value problem:

$$\begin{aligned} &y^{(n)} + p_{n-1}(t)y^{(n-1)} + \cdots + p_2(t)y'' + p_1(t)y' + p_0(t)y = g(t), && a < t < b, \\ &y(t_0) = y_0,\ \ y'(t_0) = y_0',\ \ y''(t_0) = y_0'', \ldots,\ \ y^{(n-1)}(t_0) = y_0^{(n-1)}, && a < t_0 < b. \end{aligned} \tag{1}$$

In (1), the values of the solution and its first $n-1$ derivatives are specified at some common point t_0 lying in the t-interval of interest. Note that the number of initial conditions required, n, is again equal to the order of the differential equation.

Existence and Uniqueness for Higher-Order Equations

We now state a theorem that generalizes the existence and uniqueness results given earlier in Theorems 2.1 and 4.1.

Theorem 5.1

Let $p_0(t), p_1(t), \ldots, p_{n-1}(t)$ and $g(t)$ be continuous functions defined on the interval $a < t < b$, and let t_0 be in (a, b). Then the initial value problem

$$\begin{gathered} y^{(n)} + p_{n-1}(t)y^{(n-1)} + \cdots + p_2(t)y'' + p_1(t)y' + p_0(t)y = g(t) \\ y(t_0) = y_0, y'(t_0) = y_0', y''(t_0) = y_0'', \ldots, y^{(n-1)}(t_0) = y_0^{(n-1)} \end{gathered}$$

has a unique solution defined on the entire interval (a, b).

When we compare Theorems 2.1, 4.1, and 5.1, the language and conclusions of the three theorems are virtually identical. Chapter 6 will show that Theorems

2.1, 4.1, and 5.1 can all be viewed as special cases of an overarching existence-uniqueness theorem (Theorem 6.1) for systems of first order linear equations.

EXERCISES

Exercises 1–5:

In each exercise, verify that the given functions satisfy the differential equation.

1. $y''' - y' = 0,\quad y_1(t) = 1,\ y_2(t) = e^t,\ y_3(t) = e^{-t}$

2. $y''' + y'' = 0,\quad y_1(t) = 1,\ y_2(t) = t,\ y_3(t) = e^{-t}$

3. $y^{(4)} + y'' = 0,\quad y_1(t) = 1,\ y_2(t) = t,\ y_3(t) = \cos t,\ y_4(t) = \sin t$

4. $ty''' + 3y'' = 0,\quad t > 0,\quad y_1(t) = 1,\ y_2(t) = t,\ y_3(t) = t^{-1}$

5. $t^2y''' + ty'' - y' = 0,\quad t > 0,\quad y_1(t) = 1,\ y_2(t) = \ln t,\ y_3(t) = t^2$

Exercises 6–9:

For each initial value problem, use Theorem 5.1 to find the largest interval $a < t < b$ in which a unique solution is guaranteed to exist.

6. $y''' - \dfrac{1}{t^2 - 9}y'' + \ln(t+1)y' + (\cos t)y = 0,\quad y(0) = 1,\ y'(0) = 3,\ y''(0) = 0$

7. $y''' + \dfrac{1}{t+1}y' + (\tan t)y = 0,\quad y(0) = 0,\ y'(0) = 1,\ y''(0) = 2$

8. $(t^2 - 16)y^{(4)} + 2y'' + t^2y = \sec t,\quad y(3) = 1,\ y'(3) = 3,\ y''(3) = 0,\ y'''(3) = -1$

9. $y''' - \dfrac{1}{t^2 + 9}y'' + \ln(t^2+1)y' + (\cos t)y = 0,\quad y(0) = 1,\ y'(0) = 3,\ y''(0) = 0$

Exercises 10–13:

Determine all values λ such that $y = e^{\lambda t}$ is a solution of the given differential equation.

10. $y^{(4)} - 5y'' + 4y = 0$

11. $y''' - y' = 0$

12. $y''' - 2y'' - y' + 2y = 0$

13. $y^{(4)} - 2y'' + y = 0$

5.2 The General Solution of nth Order Linear Homogeneous Equations

In this section we consider the nth order linear homogeneous equation

$$y^{(n)} + p_{n-1}(t)y^{(n-1)} + \cdots + p_2(t)y'' + p_1(t)y' + p_0(t)y = 0, \qquad a < t < b, \tag{1}$$

where the coefficient functions $p_0(t), p_1(t), p_2(t), \ldots, p_{n-1}(t)$ are assumed to be continuous on the t-interval (a, b). The theory for nth order linear differential equations parallels the corresponding theory for second order linear differential equations.

The Principle of Superposition

Our first result, Theorem 5.2, is a *superposition principle* for equation (1). It generalizes Theorem 4.2 from Chapter 4.

Theorem 5.2

Let $y_1(t), y_2(t), \ldots, y_r(t)$ be any r solutions of the homogeneous linear equation

$$y^{(n)} + p_{n-1}(t)y^{(n-1)} + \cdots + p_1(t)y' + p_0(t)y = 0, \qquad a < t < b.$$

Then, for any constants $c_1, c_2, \ldots, c_r$, the linear combination

$$y(t) = c_1 y_1(t) + c_2 y_2(t) + \cdots + c_r y_r(t)$$

is also a solution on the t-interval (a, b).

The proof of Theorem 5.2 is a straightforward extension of the proof given for Theorem 4.2 and is left as an exercise. We shall be particularly interested in the case where $r = n$.

EXAMPLE 1

Consider the fourth order linear homogeneous differential equation

$$y^{(4)} - y = 0.$$

You can show, by direct substitution, that each of the four functions $y_1(t) = e^t$, $y_2(t) = e^{-t}$, $y_3(t) = \cos t$, and $y_4(t) = \sin t$ is a solution of this differential equation on the interval $-\infty < t < \infty$. Therefore, Theorem 5.2 asserts that the linear combination

$$y(t) = c_1 e^t + c_2 e^{-t} + c_3 \cos t + c_4 \sin t$$

is also a solution on $-\infty < t < \infty$ for any choice of the constants $c_1, \ldots, c_4$. You can also verify this assertion by direct substitution. ▲

Fundamental Sets and the General Solution

In Chapter 4 we introduced the idea of a fundamental set of solutions for a second order linear homogeneous differential equation. A set of two solutions was called a *fundamental set of solutions* if every solution of the differential equation could be expressed as a linear combination of these two functions. This concept of a fundamental set of solutions extends naturally to nth order linear homogeneous equations.

Let $y_1(t), y_2(t), \ldots, y_n(t)$ denote n solutions of the differential equation

$$y^{(n)} + p_{n-1}(t)y^{(n-1)} + \cdots + p_2(t)y'' + p_1(t)y' + p_0(t)y = 0, \qquad a < t < b.$$

Such a set of n solutions, $\{y_1(t), y_2(t), \ldots, y_n(t)\}$, is called a fundamental set of solutions if any solution of the differential equation can be represented as a linear combination of the form

$$y(t) = c_1 y_1(t) + c_2 y_2(t) + \cdots + c_n y_n(t), \qquad a < t < b,$$

for some choice of constants $c_1, c_2, \ldots, c_n$. Note that the number of solutions comprising a fundamental set is equal to the order of the differential equation.

If $\{y_1(t), y_2(t), \ldots, y_n(t)\}$ is a fundamental set of solutions, the expression

$$y(t) = c_1 y_1(t) + c_2 y_2(t) + \cdots + c_n y_n(t),$$

where $c_1, c_2, \ldots, c_n$ represent arbitrary constants, is called the **general solution of the nth order linear homogeneous equation**. Thus, the members of a fundamental set form basic building blocks in terms of which every solution of the differential equation can be constructed.

EXAMPLE 2

Consider the fourth order differential equation

$$\frac{d^4y}{dt^4} - y = 0, \qquad -\infty < t < \infty.$$

We will later show that the four solutions exhibited in Example 1, $y_1(t) = e^t$, $y_2(t) = e^{-t}$, $y_3(t) = \cos t$, and $y_4(t) = \sin t$ do indeed form a fundamental set of solutions of this equation. Therefore, the general solution is given by

$$y(t) = c_1 e^t + c_2 e^{-t} + c_3 \cos t + c_4 \sin t. \tag{2}$$

(a) Verify, by direct substitution, that the function

$$\bar{y}(t) = \sinh t + \sin\left(t + \frac{\pi}{3}\right)$$

is a solution of the differential equation.

(b) Since (2) is the general solution, there must be constants c_1, c_2, c_3, and c_4 such that

$$\bar{y}(t) = \sinh t + \sin\left(t + \tfrac{\pi}{3}\right) = c_1 e^t + c_2 e^{-t} + c_3 \cos t + c_4 \sin t.$$

What are these constants?

Solution:

(a) Since $\sinh t = (e^t - e^{-t})/2$, we know that $d^4[\sinh t]/dt^4 = \sinh t$. Thus, $u(t) = \sinh t$ is a solution of the differential equation. By a similar calculation, $v(t) = \sin(t + \pi/3)$ is also a solution of the differential equation. By the principle of superposition (Theorem 5.2), it follows that $\bar{y}(t) = \sinh t + \sin(t + \pi/3)$ is a solution of the differential equation.

(b) Using the trigonometric identity $\sin(A + B) = \sin A \cos B + \sin B \cos A$, we obtain

$$\sin\left(t + \frac{\pi}{3}\right) = \sin t \cos\frac{\pi}{3} + \cos t \sin\frac{\pi}{3} = \frac{1}{2}\sin t + \frac{\sqrt{3}}{2}\cos t.$$

Therefore, equation (2) holds for the particular choice of constants

$$c_1 = \frac{1}{2}, \qquad c_2 = -\frac{1}{2}, \qquad c_3 = \frac{\sqrt{3}}{2}, \qquad c_4 = \frac{1}{2}. \ \blacktriangle$$

The Wronskian

In Chapter 4 we defined the Wronskian determinant. In particular, if $y_1(t)$ and $y_2(t)$ are any two differentiable functions, then the Wronskian of y_1 and y_2 is the (2×2) determinant

$$W(t) = \begin{vmatrix} y_1(t) & y_2(t) \\ y_1'(t) & y_2'(t) \end{vmatrix}.$$

The Wronskian proved to be the key tool used in deciding whether two solutions y_1 and y_2 formed a fundamental set of solutions for the second order linear equation

$$y'' + p(t)y' + q(t)y = 0, \qquad a < t < b.$$

We eventually concluded (see the corollary to Theorem 4.6) that $\{y_1, y_2\}$ is a fundamental set of solutions if and only if $W(t)$ is never zero on the t-interval of interest. The definition of the Wronskian and the results leading to this conclusion generalize very naturally from second order to nth order equations.

Let $y_1(t), y_2(t), \ldots, y_n(t)$ be n functions having a common domain $a < t < b$ and having $n-1$ continuous derivatives. The Wronskian of $y_1, y_2, \ldots, y_n$ is the $(n \times n)$ determinant

$$W(t) = \begin{vmatrix} y_1(t) & y_2(t) & \cdots & y_n(t) \\ y_1'(t) & y_2'(t) & & y_n'(t) \\ y_1''(t) & y_2''(t) & & y_n''(t) \\ \vdots & & & \vdots \\ y_1^{(n-1)}(t) & y_2^{(n-1)}(t) & \cdots & y_n^{(n-1)}(t) \end{vmatrix}, \qquad a < t < b.$$

When $n = 2$, this definition reverts to the (2×2) Wronskian defined in Chapter 4.

EXAMPLE 3

Example 1 listed four solutions of the differential equation $y^{(4)} - y = 0$,

$$y_1(t) = e^t, \qquad y_2(t) = e^{-t}, \qquad y_3(t) = \cos t, \qquad y_4(t) = \sin t.$$

Calculate the Wronskian of these four functions.

Solution: The Wronskian is

$$W(t) = \begin{vmatrix} e^t & e^{-t} & \cos t & \sin t \\ e^t & -e^{-t} & -\sin t & \cos t \\ e^t & e^{-t} & -\cos t & -\sin t \\ e^t & -e^{-t} & \sin t & -\cos t \end{vmatrix} = -8[\cos^2 t + \sin^2 t] = -8.$$

(We evaluated this Wronskian using computer software.) ▲

Note, in Example 3, that $W(t)$ is nonvanishing for all t in the interval $-\infty < t < \infty$. This is reminiscent of what we saw in Chapter 4, where the (2×2) Wronskian was either nonvanishing everywhere in the interval of interest, or identically zero on that interval. In this section, we generalize Abel's theorem (see Theorem 5.4). The theorem establishes the vanishing/nonvanishing dichotomy for the higher order case and also shows why we should expect the Wronskian to be constant in Example 3.

A "pencil and paper" evaluation of $(n \times n)$ determinants rapidly becomes tedious as n increases beyond 3. A cofactor expansion of an $(n \times n)$ determinant, for example, ultimately requires us to evaluate $n!/2$ determinants of order 2. Computer software, however, reduces the burden of this task. Moreover, we will see that a symbolic evaluation of the Wronskian is ultimately not required. To test whether a set of solutions is a fundamental set, it is sufficient to evaluate the Wronskian determinant at any convenient point in the t-interval of interest. A judicious choice of an evaluation point may lead to substantial computational simplifications. In any event, all that is required is the evaluation of an $(n \times n)$ determinant of numbers, not the symbolic evaluation of an $(n \times n)$ determinant of functions. We amplify on this point and reconsider the Wronskian calculation in Example 3 after discussing Theorems 5.3 and 5.4 (Abel's theorem).

Using the Wronskian to Identify Fundamental Sets

Theorem 5.3 shows that a set of n solutions of the nth order linear homogeneous equation (1) is a fundamental set of solutions if there exists a point t_0 in (a, b) such that $W(t_0) \neq 0$. The argument establishing this result is a straightforward generalization of the argument used in Theorem 4.3 and the proof is omitted.

Following Theorem 5.3, we state Theorem 5.4, which extends Abel's theorem to the nth order case. Theorem 5.4 shows that if the Wronskian is nonzero at some point t_0 in the interval (a, b), then it is nonzero everywhere in (a, b). Likewise, if the Wronskian vanishes at some point in the interval, then it vanishes everywhere in the interval.

Theorem 5.3

Let $y_1(t), y_2(t), \ldots, y_n(t)$ be solutions of the nth order linear homogeneous differential equation

$$y^{(n)} + p_{n-1}(t)y^{(n-1)} + \cdots + p_1(t)y' + p_0(t)y = 0, \qquad a < t < b,$$

where the coefficient functions $p_0(t)$, $p_1(t), \ldots, p_{n-1}(t)$ are each continuous on the interval (a, b). Let $W(t)$ denote the Wronskian formed by these n solutions. If there is a point t_0 in (a, b) such that $W(t_0) \neq 0$, then $\{y_1, y_2, \ldots, y_n\}$ is a fundamental set of solutions of the differential equation.

Abel's Theorem

We now state, without proof, a generalization of Abel's theorem; see Theorem 4.4 for comparison. This theorem establishes the fact that we can select any point in the t-interval (a, b) as a test point at which to evaluate the Wronskian, $W(t)$.

Theorem 5.4

Let $y_1(t), y_2(t), \ldots, y_n(t)$ denote n solutions of the differential equation

$$y^{(n)} + p_{n-1}(t)y^{(n-1)} + \cdots + p_1(t)y' + p_0(t)y = 0, \qquad a < t < b,$$

where $p_0(t)$, $p_1(t), \ldots, p_{n-1}(t)$ are continuous on (a, b). Let $W(t)$ be the Wronskian of $y_1, y_2, \ldots, y_n$. Then the function $W(t)$ is a solution of the first order linear differential equation

$$W' = -p_{n-1}(t)W.$$

Therefore, if t_0 is any point in the interval (a, b), it follows that

$$W(t) = W(t_0)e^{-\int_{t_0}^{t} p_{n-1}(s)\,ds}, \qquad a < t < b. \tag{3}$$

Note that Theorem 5.4 includes Theorem 4.4 as a special case. The coefficient function $p_{n-1}(t)$ is the coefficient of $y^{(n-1)}$ in the nth order differential equation. When $n = 2$, p_{n-1} becomes the coefficient of y' [in the notation of Theorem 4.4, the coefficient of y' was designated $p(t)$].

The proof of Theorem 5.4 can be found in most advanced texts on differential equations. The basic idea is similar to what we used in proving Theorem 4.4, differentiating the Wronskian. The argument is more complicated because we must now calculate the derivative of an $(n \times n)$ determinant of functions.

The vanishing/nonvanishing dichotomy of the Wronskian is a direct consequence of equation (3). If the Wronskian is nonvanishing at some point in the interval (a, b), we let this point correspond to the point t_0 in equation (3). Since the exponential factor, $\exp(-\int_{t_0}^{t} p_{n-1}(s)\,ds)$, is never zero, it follows from (3) that $W(t)$ is nonvanishing for all t in (a, b). Conversely, if $W(t_0) = 0$, it follows that $W(t) = 0$ for all t in (a, b).

EXAMPLE 4

Let's reconsider the differential equation treated in Examples 1–3,

$$y^{(4)} - y = 0, \qquad -\infty < t < \infty.$$

The functions $y_1(t) = e^t$, $y_2(t) = e^{-t}$, $y_3(t) = \cos t$, $y_4(t) = \sin t$ are four solutions of this equation. Choose a convenient test point t_0, and use Theorem 5.4 to decide if these four functions constitute a fundamental set of solutions.

Solution: According to Theorem 5.3, we can evaluate the Wronskian at any convenient value of t, say $t_0 = 0$, to test whether the four solutions form a fundamental set. The Wronskian is

$$W(t) = \begin{vmatrix} e^t & e^{-t} & \cos t & \sin t \\ e^t & -e^{-t} & -\sin t & \cos t \\ e^t & e^{-t} & -\cos t & -\sin t \\ e^t & -e^{-t} & \sin t & -\cos t \end{vmatrix}.$$

Therefore, at $t = 0$, we find

$$W(0) = \begin{vmatrix} 1 & 1 & 1 & 0 \\ 1 & -1 & 0 & 1 \\ 1 & 1 & -1 & 0 \\ 1 & -1 & 0 & -1 \end{vmatrix}. \tag{4}$$

Such determinants are readily evaluated using a calculator or computer. We can also evaluate it "by hand," using elementary row operations. Using operations of the type where a multiple of one row is added to another row, we can transform the determinant calculation from evaluating (4) to one of evaluating the triangular determinant

$$\begin{vmatrix} 1 & 1 & 1 & 0 \\ 0 & -2 & -1 & 1 \\ 0 & 0 & -2 & 0 \\ 0 & 0 & 0 & -2 \end{vmatrix}. \tag{5}$$

Since determinant (5) is the determinant of a triangular matrix, its value is simply the product of the diagonal elements: $(1)(-2)(-2)(-2) = -8$. Therefore, $W(0) \neq 0$ and we can conclude from Theorem 5.3 that this set of four solutions is a fundamental set. ▲

In Example 3 we found the Wronskian to be constant, $W(t) = -8$. This is to be expected in view of Abel's theorem. In particular, when $n = 4$, the function $p_{n-1}(t) = p_3(t)$ is the function multiplying y''' (see Theorem 5.4). As you can see from the differential equation, $y^{(4)} - y = 0$, the function $p_3(t)$ is zero on the t-interval of interest. Therefore, applying Abel's theorem, we see from equation (3) that $W(t) = W(t_0)$ for all t, $-\infty < t < \infty$.

EXERCISES

1. Consider the differential equation $y''' = 0$.
   (a) Obtain the general solution by taking successive antiderivatives.
   (b) As an alternative, use Theorem 5.3 to obtain the general solution by showing that the functions $1, t, t^2$ form a fundamental set of solutions.

Exercises 2–6:

Exercises 1–5 of Section 5.1 show that the given functions are solutions of the differential equation. Show that they form a fundamental set of solutions for the differential equation by computing the Wronskian determinant.

2. $y''' - y' = 0, \quad y_1(t) = 1, \; y_2(t) = e^t, \; y_3(t) = e^{-t}$
3. $y''' + y'' = 0, \quad y_1(t) = 1, \; y_2(t) = t, \; y_3(t) = e^{-t}$
4. $y^{(4)} + y'' = 0, \quad y_1(t) = 1, \; y_2(t) = t, \; y_3(t) = \cos t, \; y_4(t) = \sin t$
5. $ty''' + 3y'' = 0, t > 0, \quad y_1(t) = 1, \; y_2(t) = t, \; y_3(t) = t^{-1}$
6. $t^2y''' + ty'' - y' = 0, t > 0, \quad y_1(t) = 1, \; y_2(t) = \ln t, \; y_3(t) = t^2$

Exercises 7–11:

Use the fact that the solutions given in Exercises 2–6 form a fundamental set of solutions to solve the following initial value problems.

7. $y''' - y' = 0, \quad y(0) = 3, \; y'(0) = -3, \; y''(0) = 1$
8. $y''' + y'' = 0, \quad y(1) = 4, \; y'(1) = 3, \; y''(1) = 0$
9. $y^{(4)} + y'' = 0, \quad y(\pi/2) = 2 + \pi, \; y'(\pi/2) = 3, \; y''(\pi/2) = -3, \; y'''(\pi/2) = 1$
10. $ty''' + 3y'' = 0, \quad y(2) = -1, \; y'(2) = 3/2, \; y''(2) = -1/2$
11. $t^2y''' + ty'' - y' = 0, \quad y(1) = 1, \; y'(1) = 2, \; y''(1) = -6$

Exercises 12–16:

In each exercise, show that the Wronskian determinant $W(t)$ behaves as predicted by Abel's theorem. That is, for the given value of t_0, show that

$$W(t) = W(t_0)e^{-\int_{t_0}^{t} p_{n-1}(s)\,ds}.$$

12. $W(t)$ found in part (b) of Exercise 2 and $t_0 = -1$.
13. $W(t)$ found in part (b) of Exercise 3 and $t_0 = 0$.
14. $W(t)$ found in part (b) of Exercise 4 and $t_0 = 1$.
15. $W(t)$ found in part (b) of Exercise 5 and $t_0 = 1$.
16. $W(t)$ found in part (b) of Exercise 6 and $t_0 = 2$.

Exercises 17–20:

For each of the following differential equations, assume that a set of solutions exists whose Wronskian determinant, $W(t)$, satisfies the given initial condition. Using the fact (see Theorem 5.4) that $W' = -p_{n-1}(t)W$, determine $W(t)$.

17. $y''' - 3y'' + 3y' - y = 0, \qquad W(1) = 1$

18. $y''' + (\sin t)y'' + (\cos t)y' + 2y = 0, \qquad W(1) = 0$

19. $t^3y''' + t^2y'' - 2y = 0, t > 0, \qquad W(1) = 3$

20. $t^3 y''' - 2y = 0, t > 0, \qquad W(1) = 3$

This chapter will develop techniques for obtaining general solutions of higher order linear constant coefficient equations, such as those in Exercises 2–4. For these particular equations, however, techniques developed in Chapters 2 and 4 can also be adapted. The next three problems explore these ideas.

21. Obtain the general solution of $y''' - y' = 0$, as given in Exercise 2, by making the change of dependent variable, $u(t) = y'(t)$. Determine the general solution of the resulting second order linear constant coefficient equation for $u(t)$ and then obtain $y(t)$ by antidifferentiation.

22. Obtain the general solution of $y''' + y'' = 0$ in two different ways.

(a) By making the change of dependent variable $u(t) = y'(t)$, solving the resulting second-order linear equation for $u(t)$, and obtaining $y(t)$ by antidifferentiation.

(b) By making the change of dependent variable $v(t) = y''(t)$, solving the resulting first-order linear equation for $v(t)$, and obtaining $y(t)$ by two successive antidifferentiations.

23. Obtain the general solution of $y^{(4)} + y'' = 0$ by making the change of dependent variable $v(t) = y''(t)$, solving the resulting second order linear equation for $v(t)$, and then obtaining $y(t)$ by two successive antidifferentiations.

24. Consider the differential equation $y''' + p_2(t)y'' + p_1(t)y' = 0$ on the interval $-1 < t < 1$. Suppose it is known that the coefficient functions $p_2(t)$ and $p_1(t)$ are both continuous functions on $(-1, 1)$. Is it possible that $y(t) = c_1 + c_2t^2 + c_3t^4$ is the general solution for some functions $p_2(t)$ and $p_1(t)$ continuous on $-1 < t < 1$?

(a) Answer this question by considering only the Wronskian of the functions $1, t^2, t^4$ on the given interval.

(b) Explicitly determine functions $p_2(t)$ and $p_1(t)$ such that $y(t) = c_1 + c_2t^2 + c_3t^4$ is the general solution of the differential equation. Use this information, in turn, to provide an alternative answer to the question. [Hint: Substitute the individual solutions into the differential equation and use the resulting information to determine $p_2(t)$ and $p_1(t)$.]

25. Consider the initial value problem

$$y''' - y' = 0, \qquad y(0) = \alpha, \qquad y'(0) = \beta, \qquad y''(0) = 4.$$

The general solution of the differential equation is $y(t) = c_1 + c_2e^t + c_3e^{-t}$.

(a) For what values of α and β will $\lim_{t\to\infty} y(t) = 0$?

(b) For what values of α and β will the solution $y(t)$ be bounded on $0 \le t < \infty$? [Recall that $y(t)$ is bounded on $0 \le t < \infty$ if there is a constant $M > 0$ such that $|y(t)| \le M$ for all t, $0 \le t < \infty$.] Will any values of α and β produce a solution $y(t)$ that is bounded on $-\infty < t < \infty$?

5.3 Fundamental Sets and Linear Independence

In this section, we extend the results of Section 4.3 to nth order linear homogeneous differential equations. Proofs of these generalizations are straightforward extensions of the ideas presented in Section 4.3. We therefore only outline the main ideas.

Fundamental Sets of Solutions Always Exist

We first observe that fundamental sets of solutions always exist.

Theorem 5.5

Consider the linear homogeneous differential equation

$$y^{(n)} + p_{n-1}(t)y^{(n-1)} + \cdots + p_1(t)y' + p_0(t)y = 0, \qquad a < t < b, \tag{1}$$

where the functions $p_0(t), p_1(t), \ldots, p_{n-1}(t)$ are continuous on (a, b). A fundamental set of solutions, $\{y_1, y_2, \ldots, y_n\}$, exists for this differential equation.

OUTLINE OF PROOF: To prove this result, select any point t_0 in (a, b) and let $y_1, y_2, \ldots, y_n$ be n solutions of differential equation (1), satisfying

$$\begin{aligned}
&y_1(t_0) = 1, && y_1'(t_0) = 0, && y_1''(t_0) = 0, \ldots, y_1^{(n-1)}(t_0) = 0,\\
&y_2(t_0) = 0, && y_2'(t_0) = 1, && y_2''(t_0) = 0, \ldots, y_2^{(n-1)}(t_0) = 0,\\
&y_3(t_0) = 0, && y_3'(t_0) = 0, && y_3''(t_0) = 1, \ldots, y_3^{(n-1)}(t_0) = 0,\\
& && && \vdots\\
&y_n(t_0) = 0, && y_n'(t_0) = 0, && y_n''(t_0) = 0, \ldots, y_n^{(n-1)}(t_0) = 1.
\end{aligned}$$

Theorem 5.1 guarantees the existence of a unique solution for each of these n initial value problems. These solutions form a fundamental set because the Wronskian, $W(t)$, formed from these n solutions satisfies the condition $W(t_0) = 1$. ∎

The Wronskian of a Fundamental Set Is Never Zero

We next establish, as in Theorem 4.6, that the Wronskian of a fundamental set of solutions is never zero.

Theorem 5.6

Let $\{y_1, y_2, \ldots, y_n\}$ be a fundamental set of solutions for

$$y^{(n)} + p_{n-1}(t)y^{(n-1)} + \cdots + p_1(t)y' + p_0(t)y = 0, \qquad a < t < b,$$

where the functions $p_{n-1}(t), \ldots, p_1(t), p_0(t)$ are continuous on the interval (a, b). Let $W(t)$ denote the Wronskian formed from these n solutions, and let t_0 be any point in the interval (a, b). Then $W(t_0)$ is nonzero.

OUTLINE OF PROOF: The key result from matrix theory, cited in the proof of Theorem 4.6, holds as well for n linear equations in n unknowns. That is (see the appendix on matrix theory), the matrix equation $A\mathbf{x} = \mathbf{b}$ has a solution for every right-hand side $\mathbf{b}$ if and only if the determinant of A is nonzero. Thus, the proof of Theorem 5.6 parallels exactly the proof of Theorem 4.6. ∎

We can combine Theorems 5.3 and 5.6 to obtain the following corollary.

Corollary

Let $y_1, y_2, \ldots, y_n$ be n solutions of the homogeneous linear differential equation

$$y^{(n)} + p_{n-1}(t)y^{(n-1)} + \cdots + p_1(t)y' + p_0(t)y = 0, \qquad a < t < b,$$

where the coefficient functions $p_0(t), p_1(t), \ldots, p_{n-1}(t)$ are continuous on (a, b). Let $W(t)$ denote the Wronskian formed using these n solutions. Then $\{y_1, y_2, \ldots, y_n\}$ is a fundamental set of solutions if and only if $W(t)$ is never zero on (a, b).

Fundamental Sets of Solutions Are Linearly Independent Sets of Functions

A fundamental set of solutions of a linear homogeneous differential equation is a linearly independent set of functions on the common t-interval of existence, (a, b).

Theorem 5.7

Let $\{y_1, y_2, \ldots, y_n\}$ be a fundamental set of solutions for

$$y^{(n)} + p_{n-1}(t)y^{(n-1)} + \cdots + p_1(t)y' + p_0(t)y = 0, \qquad a < t < b,$$

where the coefficient functions $p_0(t), p_1(t), \ldots, p_{n-1}(t)$ are continuous on (a, b). Then $\{y_1, y_2, \ldots, y_n\}$ is a linearly independent set of functions on (a, b).

Theorem 5.7 can be proved by a straightforward generalization of the ideas of Theorem 4.7. We illustrate the ideas underlying the proof in the following example.

EXAMPLE 1

Consider again the differential equation

$$y^{(4)} - y = 0.$$

We have seen (recall Example 4 in Section 5.2) that the functions $y_1(t) = e^t$, $y_2(t) = e^{-t}$, $y_3(t) = \cos t$, and $y_4(t) = \sin t$ form a fundamental set of solutions on the interval $-\infty < t < \infty$. Show that this set of functions is linearly independent.

Solution: We begin with the equation

$$c_1 e^t + c_2 e^{-t} + c_3 \cos t + c_4 \sin t = 0, \qquad -\infty < t < \infty, \tag{2}$$

and ask whether this equation requires that $c_1 = c_2 = c_3 = c_4 = 0$. If the answer is "Yes," then the set of functions is linearly independent.

Now, if equation (2) holds, then so do the following three equations, obtained by successive differentiation of (2):

$$\begin{aligned} c_1e^t - c_2e^{-t} - c_3\sin t + c_4\cos t &= 0 \\ c_1e^t + c_2e^{-t} - c_3\cos t - c_4\sin t &= 0 \\ c_1e^t - c_2e^{-t} + c_3\sin t - c_4\cos t &= 0, \qquad -\infty < t < \infty. \end{aligned} \tag{3}$$

Equations (2) and (3), taken together, form a homogeneous system of four equations in four unknowns. Written in matrix form, this system is

$$\begin{bmatrix} e^t & e^{-t} & \cos t & \sin t \\ e^t & -e^{-t} & -\sin t & \cos t \\ e^t & e^{-t} & -\cos t & -\sin t \\ e^t & -e^{-t} & \sin t & -\cos t \end{bmatrix} \begin{bmatrix} c_1 \\ c_2 \\ c_3 \\ c_4 \end{bmatrix} = \begin{bmatrix} 0 \\ 0 \\ 0 \\ 0 \end{bmatrix}, \qquad -\infty < t < \infty. \tag{4}$$

We have already observed that the determinant of the (4×4) matrix in (4) is the Wronskian of the solution set. In Example 3 of Section 5.2 this Wronskian was shown to equal -8 for all t. Therefore, for any value of t, homogeneous system (4) has only the trivial solution $c_1 = c_2 = c_3 = c_4 = 0$. It follows that the set of functions is linearly independent. ▲

Fundamental Sets of Solutions Are Not Unique

We showed in Section 4.3 that fundamental sets of solutions of a second order linear homogeneous equation are not unique. In fact, for a given second order equation, there are infinitely many fundamental sets of solutions. Given any one fundamental set, however, we can construct any other set by suitable multiplication by a constant nonsingular matrix.

These results generalize to the nth order linear homogeneous differential equation; they are summarized in the following theorem.

Theorem 5.8

Let $\{y_1, y_2, \ldots, y_n\}$ be a fundamental set of solutions for

$$y^{(n)} + p_{n-1}(t)y^{(n-1)} + \cdots + p_1(t)y' + p_0(t)y = 0, \qquad a < t < b,$$

where the coefficient functions $p_0(t), p_1(t), \ldots, p_{n-1}(t)$ are continuous on (a, b). Let $\bar{y}_1(t), \bar{y}_2(t), \ldots, \bar{y}_n(t)$ be any n solutions of the differential equation.

(a) There is a unique $(n \times n)$ constant matrix A,

$$A = \begin{bmatrix} a_{11} & a_{12} & \cdots & a_{1n} \\ a_{21} & a_{22} & \cdots & a_{2n} \\ \vdots & \vdots & & \vdots \\ a_{n1} & a_{n2} & \cdots & a_{nn} \end{bmatrix},$$

such that

$$[\bar{y}_1(t),\ \bar{y}_2(t),\ \ldots,\ \bar{y}_n(t)] = [y_1(t),\ y_2(t),\ \ldots,\ y_n(t)]A, \quad a < t < b.$$

(b) Moreover, $\{\bar{y}_1, \bar{y}_2, \ldots, \bar{y}_n\}$ is a fundamental set of solutions of the differential equation if and only if the determinant of A is nonzero.

The proof of this theorem is a straightforward generalization of the ideas of Theorem 4.8. The content of the theorem is illustrated in the following example.

EXAMPLE 2

Consider again the differential equation

$$y^{(4)} - y = 0, \qquad -\infty < t < \infty.$$

We know that $y_1(t) = e^t$, $y_2(t) = e^{-t}$, $y_3(t) = \cos t$, $y_4(t) = \sin t$ forms a fundamental set of solutions of this differential equation. We can verify, by direct substitution, that

$$\overline{y}_1(t) = \cosh t, \qquad \overline{y}_2(t) = \sinh t, \qquad \overline{y}_3(t) = \cos(t+\alpha), \qquad \overline{y}_4(t) = \sin(t+\beta)$$

are also solutions of the given differential equation (here, α and β are constant phase angles).

(a) According to Theorem 5.8, there is a unique (4×4) constant matrix A such that

$$[\cosh t, \sinh t, \cos(t+\alpha), \sin(t+\beta)] = [e^t, e^{-t}, \cos t, \sin t]A, \qquad -\infty < t < \infty.$$

Determine the matrix A.

(b) Do the functions $\overline{y}_1(t), \overline{y}_2(t), \overline{y}_3(t), \overline{y}_4(t)$ also form a fundamental set of solutions?

Solution:

(a) To determine the matrix A, we use the fact that

$$\begin{aligned}
\cosh t &= \frac{1}{2}e^t + \frac{1}{2}e^{-t}, \\
\sinh t &= \frac{1}{2}e^t - \frac{1}{2}e^{-t}, \\
\cos(t+\alpha) &= \cos t \cos\alpha - \sin t \sin\alpha, \\
\sin(t+\beta) &= \sin t \cos\beta + \cos t \sin\beta.
\end{aligned}$$

From these equations it follows that the unique matrix A must be

$$A = \begin{bmatrix} \frac{1}{2} & \frac{1}{2} & 0 & 0 \\ \frac{1}{2} & -\frac{1}{2} & 0 & 0 \\ 0 & 0 & \cos\alpha & \sin\beta \\ 0 & 0 & -\sin\alpha & \cos\beta \end{bmatrix}.$$

(b) According to the second part of Theorem 5.8, the functions $\overline{y}_1(t), \overline{y}_2(t), \overline{y}_3(t), \overline{y}_4(t)$ form a fundamental set of solutions on the interval $-\infty < t < \infty$ if and only if the determinant of A is nonzero. Since

$$\det(A) = -\frac{1}{2}[\cos\alpha\cos\beta + \sin\alpha\sin\beta] = -\frac{1}{2}\cos(\alpha - \beta),$$

it follows that $\overline{y}_1(t), \overline{y}_2(t), \overline{y}_3(t), \overline{y}_4(t)$ forms a fundamental set of solutions if $\cos(\alpha - \beta) \neq 0$; that is, if the difference between the two phase angles is not an odd multiple of $\pi/2$. (As an example, consider what happens if $\alpha = \pi/2$, $\beta = 0$. Why would $\overline{y}_1(t), \overline{y}_2(t), \overline{y}_3(t), \overline{y}_4(t)$ not be a fundamental set of solutions in this case?) ▲

EXERCISES

Exercises 1–2:

Find the general solution of $y''' = 0$ by antidifferentiation. Using the general solution and the given point, t_0, construct a fundamental set of solutions $\{y_1(t), y_2(t), y_3(t)\}$ that satisfies the conditions given in the proof outline for Theorem 5.5,

$$\begin{aligned} y_1(t_0) &= 1, & y_1'(t_0) &= 0, & y_1''(t_0) &= 0, \\ y_2(t_0) &= 0, & y_2'(t_0) &= 1, & y_2''(t_0) &= 0, \\ y_3(t_0) &= 0, & y_3'(t_0) &= 0, & y_3''(t_0) &= 1. \end{aligned} \tag{5}$$

1. $t_0 = 0$

2. $t_0 = 1$

Exercises 3–4:

The general solution of $y''' + y'' = 0$ is $c_1 + c_2 t + c_3 e^{-t}$. Proceeding as in Exercises 1–2 for the given point t_0, determine a fundamental set of solutions $\{y_1(t), y_2(t), y_3(t)\}$ that satisfies (5).

3. $t_0 = 0$

4. $t_0 = 1$

Exercises 5–8:

For each differential equation, the corresponding set of functions $\{y_1(t), y_2(t), y_3(t)\}$ forms a fundamental set of solutions.

(a) Determine whether each member of the new set of functions $\{\overline{y}_1(t), \overline{y}_2(t), \overline{y}_3(t)\}$ is likewise a solution. In other words, determine whether $\{\overline{y}_1(t), \overline{y}_2(t), \overline{y}_3(t)\}$ is a *solution set*.

(b) If $\{\overline{y}_1(t), \overline{y}_2(t), \overline{y}_3(t)\}$ is a solution set, find the unique (3×3) matrix $A = (a_{ij})$ such that

$$\left[\overline{y}_1(t), \overline{y}_2(t), \overline{y}_3(t)\right] = \left[y_1(t), y_2(t), y_3(t)\right] \begin{bmatrix} a_{11} & a_{12} & a_{13} \\ a_{21} & a_{22} & a_{23} \\ a_{31} & a_{32} & a_{33} \end{bmatrix}.$$

(c) If $\{\overline{y}_1(t), \overline{y}_2(t), \overline{y}_3(t)\}$ is a solution set, determine whether it is a fundamental set by calculating the determinant of A.

5. $y''' - y' = 0$, $\{y_1(t), y_2(t), y_3(t)\} = \{1, e^t, e^{-t}\}$,
$\{\overline{y}_1(t), \overline{y}_2(t), \overline{y}_3(t)\} = \{\cosh t, 1 - \sinh t, 2 + \sinh t\}$

6. $y''' + y'' = 0$, $\{y_1(t), y_2(t), y_3(t)\} = \{1, t, e^{-t}\}$,
$\{\overline{y}_1(t), \overline{y}_2(t), \overline{y}_3(t)\} = \{1 - 2t, t + 2, e^{-(t+2)}\}$

7. $ty''' + 3y'' = 0, t > 0$, $\{y_1(t), y_2(t), y_3(t)\} = \{1, t, t^{-1}\}$,

$$\{\overline{y}_1(t), \overline{y}_2(t), \overline{y}_3(t)\} = \left\{1 + t, \frac{t+1}{t}, \frac{1}{t+1}\right\}$$

8. $t^2y''' + ty'' - y' = 0, t > 0, \{y_1(t), y_2(t), y_3(t)\} = \{1, \ln t, t^2\},$
$\{\overline{y}_1(t), \overline{y}_2(t), \overline{y}_3(t)\} = \{2t^2 - 1, 3, \ln(t^3)\}$

Exercises 9–11:

Is the given set of functions linearly dependent or linearly independent on the interval $-\infty < t < \infty$?

9. $\{1, t, t^2\}$ **10.** $\{1, 1+t, 1+t+t^2\}$ **11.** $\{\cos^2 t, 2\cos 2t, 2\sin^2 t\}$

12. Consider the set of functions $\{t^2 + 2t, \alpha t + 1, t + \alpha\}$ where α is a constant. For what value(s) α is the given set linearly dependent on the interval $-\infty < t < \infty$?

Exercises 13–15:

Consider the set of functions $\{t|t| + 1, t^2 - 1, t\}$. Determine whether this set of functions is linearly dependent or linearly independent on the given interval.

13. $0 \le t < \infty$ **14.** $-\infty < t \le 0$ **15.** $-\infty < t < \infty$

5.4 Constant Coefficient Homogeneous Equations

In this section we treat nth order linear homogeneous differential equations where all the coefficients are constants,

$$y^{(n)} + a_{n-1}y^{(n-1)} + \cdots + a_1y' + a_0y = 0, \qquad -\infty < t < \infty. \tag{1}$$

The general solution of this differential equation has the form

$$y(t) = c_1y_1(t) + c_2y_2(t) + \cdots + c_ny_n(t),$$

where $c_1, c_2, \ldots, c_n$ are arbitrary constants and $\{y_1(t), y_2(t), \ldots, y_n(t)\}$ is a fundamental set of solutions. We now find these solutions.

The Characteristic Equation

As in Section 4.4, we look for solutions of the form $y(t) = e^{\lambda t}$, where λ is a constant (possibly complex-valued) to be determined. Substituting this form into equation (1) leads to

$$[\lambda^n + a_{n-1}\lambda^{n-1} + \cdots + a_1\lambda + a_0]e^{\lambda t} = 0, \qquad -\infty < t < \infty.$$

The exponential function $e^{\lambda t}$ does not vanish for any value of λ (real or complex). Therefore, if $y(t) = e^{\lambda t}$ is to be a solution, it must be that

$$\lambda^n + a_{n-1}\lambda^{n-1} + \cdots + a_1\lambda + a_0 = 0. \tag{2}$$

The nth degree polynomial in (2) is again called the *characteristic polynomial*, and equation (2) itself is called the *characteristic equation*. The roots of the characteristic equation define solutions of (1) having the form $y(t) = e^{\lambda t}$.

Roots of the Characteristic Equation

In Chapter 4, when considering the case $n = 2$, we were able to classify the three possibilities for the roots of the characteristic polynomial (two distinct real roots, one repeated real root, two distinct complex roots). We cannot make such a short list of possibilities for the general case. We can, however, make some useful observations.

Complex roots occur in complex conjugate pairs. Since $a_0, a_1, \ldots, a_{n-1}$ are real constants, the complex-valued roots of the characteristic equation always occur in complex conjugate pairs. In other words, if $\lambda = \alpha + i\beta$ is a root of (2), then the complex conjugate, $\overline{\lambda} = \alpha - i\beta$, is also a root (here, α and β are real constants). One simple consequence of this observation is that every characteristic polynomial of odd degree has at least one real root.

We noted in Section 4.6 that if $\lambda = \alpha + i\beta$ and $\overline{\lambda} = \alpha - i\beta$ are a complex conjugate pair of roots of the characteristic equation, then a pair of real-valued solutions corresponding to these two roots are the functions:

$$y_1(t) = e^{\alpha t}\cos\beta t \quad \text{and} \quad y_2(t) = e^{\alpha t}\sin\beta t.$$

This result is true for the general nth order equation as well.

EXAMPLE 1

Find the general solution of the third order differential equation

$$y''' + 4y' = 0, \qquad -\infty < t < \infty.$$

Solution: Looking for solutions of the form $e^{\lambda t}$ leads to the characteristic equation $\lambda^3 + 4\lambda = 0$ or $\lambda(\lambda^2 + 4) = \lambda(\lambda + 2i)(\lambda - 2i) = 0$. The three roots are therefore $\lambda_1 = 0, \lambda_2 = 2i$, and $\lambda_3 = \overline{\lambda}_2 = -2i$. The corresponding real-valued solutions are

$$y_1(t) = 1, \qquad y_2(t) = \cos 2t, \qquad y_3(t) = \sin 2t.$$

To show that these three solutions form a fundamental set, we calculate the Wronskian and find

$$W(t) = \begin{vmatrix} 1 & \cos 2t & \sin 2t \\ 0 & -2\sin 2t & 2\cos 2t \\ 0 & -4\cos 2t & -4\sin 2t \end{vmatrix} = 8[\sin^2 2t + \cos^2 2t] = 8.$$

Since the Wronskian is nonzero, these three solutions form a fundamental set of solutions and the general solution is

$$y(t) = c_1 + c_2\cos 2t + c_3\sin 2t, \qquad -\infty < t < \infty. \quad \blacktriangle \tag{3}$$

As an alternative approach in Example 1, you can also obtain (3) by making the change of dependent variable $u(t) = y'(t)$; see Exercise 21, Section 5.2. The resulting differential equation $u'' + 4u = 0$ can be solved using the methods of Section 4.6 and the general solution for $u(t)$ can be antidifferentiated to obtain the general solution for $y(t)$.

If the characteristic polynomial has distinct roots, the corresponding solutions form a fundamental set. In Example 1, the characteristic equation had three distinct roots: the real number $\lambda_1 = 0$ and the complex conjugate pair $\lambda_2 = 2i$ and $\lambda_3 = -2i$. The corresponding set of solutions formed a fundamental set of solutions. In the following theorem, we observe that this situation holds in general; if the characteristic equation has n distinct roots $\lambda_1, \lambda_2, \ldots, \lambda_n$ then the set of solutions $\{e^{\lambda_1 t}, e^{\lambda_2 t}, \ldots, e^{\lambda_n t}\}$ is a fundamental set of solutions.

Theorem 5.9

Assume that the characteristic equation

$$\lambda^n + a_{n-1}\lambda^{n-1} + \cdots + a_1\lambda + a_0 = 0$$

has n distinct roots $\lambda_1, \lambda_2, \ldots, \lambda_n$ (real valued or complex valued). Then the set of solutions $\{e^{\lambda_1 t}, e^{\lambda_2 t}, \ldots, e^{\lambda_n t}\}$ is a fundamental set of solutions for

$$y^{(n)} + a_{n-1}y^{(n-1)} + \cdots + a_1 y' + a_0 y = 0, \qquad -\infty < t < \infty.$$

PROOF: We show that the corresponding Wronskian,

$$W(t) = \begin{vmatrix} e^{\lambda_1 t} & e^{\lambda_2 t} & \cdots & e^{\lambda_n t} \\ \lambda_1 e^{\lambda_1 t} & \lambda_2 e^{\lambda_2 t} & & \lambda_n e^{\lambda_n t} \\ \lambda_1^2 e^{\lambda_1 t} & \lambda_2^2 e^{\lambda_2 t} & & \lambda_n^2 e^{\lambda_n t} \\ \vdots & & & \vdots \\ \lambda_1^{n-1} e^{\lambda_1 t} & \lambda_2^{n-1} e^{\lambda_2 t} & \cdots & \lambda_n^{n-1} e^{\lambda_n t} \end{vmatrix}, \qquad (4)$$

is nonzero.

Recall that if all entries in any row or column of a determinant are multiplied by a common factor, the determinant itself is multiplied by that same factor. Applying this result to each column in (4),

$$W(t) = e^{\lambda_1 t} e^{\lambda_2 t} \cdots e^{\lambda_n t} \begin{vmatrix} 1 & 1 & \cdots & 1 \\ \lambda_1 & \lambda_2 & & \lambda_n \\ \lambda_1^2 & \lambda_2^2 & & \lambda_n^2 \\ \vdots & & & \vdots \\ \lambda_1^{n-1} & \lambda_2^{n-1} & \cdots & \lambda_n^{n-1} \end{vmatrix}. \qquad (5)$$

The product of exponential functions on the right-hand side of (5) is nonzero for all t. The remaining determinant arises in a number of areas of mathematics; it is called the **Vandermonde determinant**. It can be shown that the Vandermonde determinant has value

$$\begin{vmatrix} 1 & 1 & \cdots & 1 \\ \lambda_1 & \lambda_2 & & \lambda_n \\ \lambda_1^2 & \lambda_2^2 & & \lambda_n^2 \\ \vdots & & & \vdots \\ \lambda_1^{n-1} & \lambda_2^{n-1} & \cdots & \lambda_n^{n-1} \end{vmatrix} = \prod_{\substack{i=1 \\ j>i}}^{n-1} (\lambda_j - \lambda_i). \qquad (6)$$

The symbol Π on the right-hand side of (6) is the "product" symbol. The determinant equals the product of all factors of the form $(\lambda_j - \lambda_i)$, where the index i ranges from 1 to $n-1$ and, for each of these values of i, the index j takes on the values $i+1, i+2, \ldots, n$. For instance, if $n = 4$, the Vandermonde determinant equals

$$(\lambda_2 - \lambda_1)(\lambda_3 - \lambda_1)(\lambda_4 - \lambda_1)(\lambda_3 - \lambda_2)(\lambda_4 - \lambda_2)(\lambda_4 - \lambda_3).$$

Returning to equation (6), we note that (since the n roots are distinct) none of the factors in the product on the right-hand side of (6) can be zero. Therefore, the Vandermonde determinant and, in turn, the Wronskian, is nonzero. Thus it follows that the set of functions $\{e^{\lambda_1 t}, e^{\lambda_2 t}, \ldots, e^{\lambda_n t}\}$ is a fundamental set of solutions. ∎

If some of the distinct roots in Theorem 5.8 are complex valued, we can replace the conjugate pair of complex exponentials with the corresponding real-valued pair of solutions and the conclusion remains the same. In Example 1, the complex exponentials, e^{i2t} and e^{-i2t}, were replaced by the pair of real-valued functions, $\cos 2t$ and $\sin 2t$. The next example further illustrates this point.

EXAMPLE 2

Several previous examples involved the differential equation

$$y^{(4)} - y = 0.$$

This differential equation has characteristic polynomial

$$p(\lambda) = \lambda^4 - 1 = (\lambda^2 - 1)(\lambda^2 + 1) = (\lambda - 1)(\lambda + 1)(\lambda - i)(\lambda + i).$$

The real-valued solutions corresponding to the four distinct zeros of $p(\lambda)$ are

$$y_1(t) = e^t, \qquad y_2(t) = e^{-t}, \qquad y_3(t) = \cos t, \qquad y_4(t) = \sin t.$$

Example 4 of Section 5.2 showed that these four solutions do indeed form a fundamental set of solutions. ▲

Roots of the characteristic equation may have multiplicity greater than two and they may be complex. For the second order linear homogeneous equation discussed in Chapter 4, a repeated root of the corresponding quadratic characteristic equation is of necessity real valued. Such a root can be repeated at most once. If $\lambda_1 = \lambda_2 = \alpha$ are repeated roots of the quadratic characteristic equation, then the corresponding solutions

$$y_1(t) = e^{\alpha t} \quad \text{and} \quad y_2(t) = te^{\alpha t}$$

form a fundamental set of solutions.

For higher order differential equations (and their correspondingly higher degree characteristic polynomials) the situation can be more complicated. For example:

(a) ***A root can be repeated more than once.*** In particular, if the characteristic polynomial has the form

$$p(\lambda) = \lambda^n + a_{n-1}\lambda^{n-1} + \cdots + a_1\lambda + a_0 = (\lambda - \lambda_1)^r q(\lambda), \tag{7}$$

where q is a polynomial of degree $n - r$ and $q(\lambda_1) \neq 0$, then we say λ_1 is a **root of multiplicity r**. If λ_1 is a root of multiplicity r, then the functions

$$e^{\lambda_1 t}, \quad te^{\lambda_1 t}, \quad t^2 e^{\lambda_1 t}, \quad \ldots, \quad t^{r-1}e^{\lambda_1 t} \tag{8}$$

form a set of r linearly independent solutions of the differential equations. The remaining $n - r$ solutions needed to complete the fundamental set of solutions are determined by examining the roots of $q(\lambda) = 0$.

(b) ***A repeated root might be complex.*** We recall, however, that complex roots must appear in complex conjugate pairs. Therefore, if $\lambda_1 = \alpha + i\beta$ is a root of multiplicity r, then $\overline{\lambda}_1 = \alpha - i\beta$ is also a root of multiplicity r. In such a case, the characteristic polynomial has the form

$$\lambda^n + a_{n-1}\lambda^{n-1} + \cdots + a_1\lambda + a_0 = (\lambda - \lambda_1)^r(\lambda - \overline{\lambda}_1)^r\hat{q}(\lambda),$$

where $\hat{q}(\lambda)$ is a polynomial of degree $n - 2r$ and where $\hat{q}(\lambda_1) \neq 0$ and $\hat{q}(\overline{\lambda}_1) \neq 0$. In this case, the functions

$$\begin{array}{lllll} e^{\alpha t}\cos\beta t, & te^{\alpha t}\cos\beta t, & t^2e^{\alpha t}\cos\beta t, & \ldots, & t^{r-1}e^{\alpha t}\cos\beta t \\ e^{\alpha t}\sin\beta t, & te^{\alpha t}\sin\beta t, & t^2e^{\alpha t}\sin\beta t, & \ldots, & t^{r-1}e^{\alpha t}\sin\beta t \end{array} \qquad \textbf{(9)}$$

form a set of $2r$ linearly independent, real-valued solutions of differential equation (1).

EXAMPLE 3

(a) Find the general solution of

$$y^{(6)} + 3y^{(5)} + 3y^{(4)} + y''' = 0.$$

(b) Find the general solution of

$$y^{(5)} - y^{(4)} + 2y''' - 2y'' + y' - y = 0.$$

Solution:

(a) The characteristic polynomial is given by

$$p(\lambda) = \lambda^6 + 3\lambda^5 + 3\lambda^4 + \lambda^3 = \lambda^3(\lambda^3 + 3\lambda^2 + 3\lambda + 1) = \lambda^3(\lambda + 1)^3.$$

Therefore, $\lambda = 0$ and $\lambda = -1$ are each roots of multiplicity 3. Noting (8), the set of functions $\{1, t, t^2, e^{-t}, te^{-t}, t^2e^{-t}\}$ forms a fundamental set. The general solution is

$$y(t) = c_1 + c_2t + c_3t^2 + c_4e^{-t} + c_5te^{-t} + c_6t^2e^{-t}.$$

(b) The characteristic polynomial is given by

$$p(\lambda) = \lambda^5 - \lambda^4 + 2\lambda^3 - 2\lambda^2 + \lambda - 1 = (\lambda - 1)(\lambda^2 + 1)^2.$$

This polynomial has one real zero, $\lambda = 1$, and a repeated pair of complex conjugate zeros, $\pm i$. The set of solutions $\{e^t, \cos t, \sin t, t\cos t, t\sin t\}$ forms a fundamental set of solutions. The general solution is

$$y(t) = c_1e^t + c_2\cos t + c_3\sin t + c_4t\cos t + c_5t\sin t. \quad \blacktriangle$$

Finding the Roots of the Characteristic Equation

The differential equations selected for textbook examples and exercises typically have characteristic polynomials that can be factored "by hand." For example, given the characteristic polynomial

$$\lambda^3 + 2\lambda^2 - \lambda - 2,$$

one person might recognize that it can be rewritten and factored as follows:

$$\lambda^2(\lambda + 2) - (\lambda + 2) = (\lambda^2 - 1)(\lambda + 2) = (\lambda - 1)(\lambda + 1)(\lambda + 2).$$

Another person might observe (by trial and error) that substituting $\lambda = 1$ causes the polynomial to vanish. Therefore, we know that $(\lambda - 1)$ is a factor of this polynomial. Using synthetic division we obtain

$$\lambda^3 + 2\lambda^2 - \lambda - 2 = (\lambda - 1)(\lambda^2 + 3\lambda + 2).$$

The quadratic formula now can be used to complete the factorization.

For equations you might encounter in "real life," symbolic and computational computer software is available for solving the characteristic equation. However, such software must be used with care. Two cautionary notes are worth mentioning.

1. Most root-finding algorithms are iterative in nature—they generate estimates that get closer and closer to a root, but only rarely do these estimates reach the root. Moreover, computer arithmetic has finite precision and cannot represent irrational roots such as π or $\sqrt{2}$. Therefore, most numerical estimates generated for roots of the characteristic equation are just that—estimates. However, if the true roots are distinct, then the corresponding "numerical" general solution of the differential equation formed from numerically generated approximate roots will often be close to the true solution, at least on a finite interval of reasonable length.
2. If some of the true roots of the characteristic polynomial are repeated, any small numerical errors made in estimating the roots can be qualitatively misleading and may lead to significant errors even on relatively small t-intervals.

Solving the Differential Equation $y^{(n)} - ay = 0$

We conclude by presenting a procedure for determining the roots of the characteristic equation of

$$y^{(n)} - ay = 0,$$

where a is a nonzero real number. Since the characteristic equation is

$$\lambda^n - a = 0,$$

finding its roots is tantamount to finding the n different nth roots of the real number a.

1. First, we write a in polar form as

   $$a = Re^{i\alpha},$$

   where $R = |a|$ and where $\alpha = 0$ when $a > 0$ and $\alpha = \pi$ when $a < 0$.
2. Recall Euler's formula, $e^{i\theta} = \cos\theta + i\sin\theta$. We observe, for any integer k, that $e^{i2k\pi} = 1$. Therefore, we can write

   $$a = Re^{i(\alpha+2k\pi)}, \qquad k = 0, \pm 1, \pm 2, \ldots$$

   or

   $$a^{1/n} = R^{1/n}e^{i(\alpha+2k\pi)/n}, \qquad k = 0, \pm 1, \pm 2, \ldots. \tag{10}$$

   In equation (10), $R^{1/n}$ is the positive real nth root of $R = |a|$. We generate the n distinct roots of $\lambda^n - a = 0$ by setting $k = 0, 1, \ldots, n-1$ in equation (10). All other values of the integer k simply replicate these n values. Once we determine the n roots, we can again use Euler's formula to rewrite them in the form $x + iy$.

EXAMPLE 4

Find the general solution of

$$y^{(4)} + 16y = 0, \qquad -\infty < t < \infty.$$

Solution: The characteristic equation is

$$\lambda^4 + 16 = 0,$$

so, in this example, $a = -16$ and $n = 4$. Using equation (10) with $R = 16$ and $\alpha = \pi$, the four roots are given by

$$\lambda_k = 2e^{i(\pi+2k\pi)/4}, \qquad k = 0, 1, 2, 3$$

or

$$\begin{aligned}
\lambda_0 &= 2e^{i\pi/4} = \sqrt{2} + i\sqrt{2} \\
\lambda_1 &= 2e^{i3\pi/4} = -\sqrt{2} + i\sqrt{2} \\
\lambda_2 &= 2e^{i5\pi/4} = -\sqrt{2} - i\sqrt{2} \\
\lambda_3 &= 2e^{i7\pi/4} = \sqrt{2} - i\sqrt{2}
\end{aligned}$$

The four roots occur in two sets of complex conjugate pairs. The general solution, expressed in terms of real-valued solutions, is

$$y(t) = e^{\sqrt{2}t}\left[c_1 \cos\sqrt{2}t + c_2 \sin\sqrt{2}t\right] + e^{-\sqrt{2}t}\left[c_3 \cos\sqrt{2}t + c_4 \sin\sqrt{2}t\right]. \ \blacktriangle$$

EXERCISES

Exercises 1–10:

For each of the following problems

(a) Obtain the characteristic polynomial of the given differential equation.

(b) Find the roots of the characteristic equation.

(c) Determine the general solution of the differential equation.

(d) If initial conditions are specified, solve the initial value problem.

1. $y''' - 4y' = 0$

2. $y''' + y'' - y' - y = 0$

3. $y''' + y'' + 4y' + 4y = 0$

4. $16y^{(4)} - 8y'' + y = 0$

5. $16y^{(4)} + 8y'' + y = 0$

6. $y''' - y = 0$

7. $y''' - 2y'' - y' + 2y = 0$

8. $y^{(4)} - 16y = 0$

9. $y''' + 4y' = 0, \quad y(0) = 1, \quad y'(0) = 6, \quad y''(0) = 4$

10. $y''' + 3y'' + 3y' + y = 0, \quad y(0) = 0, \quad y'(0) = 1, \quad y''(0) = 0$

Exercises 11–15:

In each exercise, you are given the general solution of

$$y^{(4)} + a_3y''' + a_2y'' + a_1y' + a_0y = 0,$$

where a_3, a_2, a_1, and a_0 are real constants. Use the general solution to determine the constants a_3, a_2, a_1, and a_0. [Hint: Construct the characteristic equation from the given general solution.]

11. $y(t) = c_1 + c_2t + c_3 \cos 3t + c_4 \sin 3t$

12. $y(t) = c_1 \cos t + c_2 \sin t + c_3 \cos 2t + c_4 \sin 2t$

13. $y(t) = c_1 e^t + c_2 te^t + c_3 e^{-t} + c_4 te^{-t}$

14. $y(t) = c_1 e^{-t} \sin t + c_2 e^{-t} \cos t + c_3 e^t \sin t + c_4 e^t \cos t$

15. $y(t) = c_1 e^t + c_2 te^t + c_3 t^2 e^t + c_4 t^3 e^t$

Exercises 16–20:
Consider the nth order homogeneous linear differential equation

$$y^{(n)} + a_{n-1}y^{(n-1)} \cdots + a_3 y''' + a_2 y'' + a_1 y' + a_0 y = 0,$$

where $a_0, a_1, a_2, \ldots, a_{n-1}$ are real constants. In each exercise, several functions belonging to a fundamental set of solutions for this equation are given.

(a) What is the smallest value n for which the given functions can belong to such a fundamental set?

(b) What is the fundamental set?

16. $y_1(t) = t, \quad y_2(t) = e^t, \quad y_3(t) = \cos t$

17. $y_1(t) = e^t, \quad y_2(t) = e^t \cos 2t, \quad y_3(t) = e^{-t} \cos 2t$

18. $y_1(t) = t^2 \sin t, \quad y_2(t) = e^t \sin t$

19. $y_1(t) = t \sin t, \quad y_2(t) = t^2 e^t$

20. $y_1(t) = t^2, \quad y_2(t) = e^{2t}$

Exercises 21–25:
Consider the nth order differential equation

$$y^{(n)} + ay = 0,$$

where a is a real number. In each exercise, some information is presented about the solutions of this equation. Use the given information to deduce both the order $n (n \geq 1)$ of the differential equation and the value of the constant a. (If more than one answer is possible, determine the smallest possible order n and the corresponding value of a.)

21. $|a| = 1$ and $\lim_{t\to\infty} y(t) = 0$ for all solutions $y(t)$ of the equation.

22. $|a| = 2$ and all nonzero solutions of the differential equation are exponential functions.

23. $y(t) = t^3$ is a solution of the differential equation.

24. $|a| = 4$ and all solutions of the differential equation are bounded functions on the interval $-\infty < t < \infty$.

25. Two solutions are $y_1(t) = e^{3t}$ and $y_2(t) = e^{-(3/2)t} \sin((3\sqrt{3}/2)t)$.

5.5 Nonhomogeneous Linear Equations

We now consider the nth order linear nonhomogeneous differential equation

$$y^{(n)} + p_{n-1}(t)y^{(n-1)} + \cdots + p_1(t)y' + p_0(t)y = g(t), \qquad a < t < b, \tag{1}$$

where the functions $p_0(t), p_1(t), \ldots, p_{n-1}(t)$ and $g(t)$ are assumed to be continuous on the t-interval (a, b). The arguments made in Section 4.8, establishing the solution structure for second order equations, apply as well in the nth order case. The arguments rely upon the fact that the differential equation is linear; they do not depend upon the order of the equation.

The General Solution of the *n*th Order Equation

Let $u(t)$ and $v(t)$ be any two solutions of the nonhomogeneous equation (1). The following calculation shows that the difference function, $u(t) - v(t)$, is a solution of the homogeneous equation:

$$\begin{aligned}
&[u-v]^{(n)} + p_{n-1}(t)[u-v]^{(n-1)} + \cdots + p_1(t)[u-v]' + p_0(t)[u-v] \\
&\quad = [u^{(n)} + p_{n-1}(t)u^{(n-1)} + \cdots + p_0(t)u] - [v^{(n)} + p_{n-1}(t)v^{(n-1)} + \cdots + p_0(t)v] \\
&\quad = g(t) - g(t) = 0, \qquad a < t < b.
\end{aligned}$$

From this observation we conclude, as we did in Section 4.8, that

The general solution of the nonhomogeneous equation	=	The general solution of the homogeneous equation	+	A particular solution of the nonhomogeneous equation.

The general solution of the homogeneous equation is called the *complementary solution* and we designate it as $y_C(t)$. The one solution of the nonhomogeneous equation that we have somehow found is called a *particular solution* and we designate it as $y_P(t)$. If $y(t)$ represents the general solution of nonhomogeneous equation (1), it then follows that

$$y(t) = y_C(t) + y_P(t).$$

Finding a Particular Solution

In Sections 4.9 and 4.10 we discussed two methods, the method of undetermined coefficients and the method of variation of parameters, for constructing particular solutions. Both methods have straightforward generalizations to nth order equations.

Before discussing the methods of undetermined coefficients and variation of parameters, we note that the superposition principle presented in Theorem 4.10 also applies to the nth order equation (1). The generalization is given in Theorem 5.10.

Theorem 5.10

Let $u(t)$ be a solution of

$$y^{(n)} + p_{n-1}(t)y^{(n-1)} + \cdots + p_1(t)y' + p_0(t)y = g_1(t),$$

and let $v(t)$ be a solution of

$$y^{(n)} + p_{n-1}(t)y^{(n-1)} + \cdots + p_1(t)y' + p_0(t)y = g_2(t)$$

on some interval $a < t < b$. Let a_1 and a_2 be constants. Then the function $y_P(t) = a_1u(t) + a_2v(t)$ is a particular solution of

$$y^{(n)} + p_{n-1}(t)y^{(n-1)} + \cdots + p_1(t)y' + p_0(t)y = a_1g_1(t) + a_2g_2(t)$$

on the same interval $a < t < b$.

The Method of Undetermined Coefficients

The method of undetermined coefficients is applicable when the nth order linear differential equation has constant coefficients and the nonhomogeneous term, $g(t)$, belongs to the family of functions we considered in Chapter 4 (see Tables 4.1 and 5.1). In particular, consider the linear differential equation

$$y^{(n)} + a_{n-1}y^{(n-1)} + \cdots + a_1 y' + a_0 y = g(t),$$

where $a_0, a_1, \ldots, a_{n-1}$ are constants. Table 5.1 summarizes the method. Note that this table is basically the same as Table 4.1; the difference lies in the fact that the possible values of the exponent r in the t^r factors now range between the values 0 and n instead of between 0 and 2.

TABLE 5.1

The right-hand column gives the proper form to assume for a particular solution of $y^{(n)} + a_{n-1}y^{(n-1)} + \cdots + a_1 y' + a_0 y = g(t)$. In the right-hand column, choose r to be the smallest nonnegative integer such that no term in the assumed form is a solution of the homogeneous equation $y^{(n)} + a_{n-1}y^{(n-1)} + \cdots + a_1 y' + a_0 y = 0$. The value of r will be an integer such that $0 \le r \le n$.

Form of g(t)	Form to Assume for a Particular Solution $y_P(t)$
$a_k t^k + \cdots + a_1 t + a_0$	$t^r[A_k t^k + \cdots + A_1 t + A_0]$
$[a_k t^k + \cdots + a_1 t + a_0]e^{\alpha t}$	$t^r[A_k t^k + \cdots + A_1 t + A_0]e^{\alpha t}$
$[a_k t^k + \cdots + a_1 t + a_0]\sin\beta t$ or $[a_k t^k + \cdots + a_1 t + a_0]\cos\beta t$	$t^r[(A_k t^k + \cdots + A_1 t + A_0)\sin\beta t + (B_k t^k + \cdots + B_1 t + B_0)\cos\beta t]$
$e^{\alpha t}\sin\beta t$ or $e^{\alpha t}\cos\beta t$	$t^r[Ae^{\alpha t}\sin\beta t + Be^{\alpha t}\cos\beta t]$
$e^{\alpha t}[a_k t^k + \cdots + a_0]\sin\beta t$ or $e^{\alpha t}[a_k t^k + \cdots + a_0]\cos\beta t$	$t^r[(A_k t^k + \cdots + A_0)e^{\alpha t}\sin\beta t + (B_k t^k + \cdots + B_0)e^{\alpha t}\cos\beta t]$

EXAMPLE 1

Choose an appropriate form for a particular solution of

$$y^{(6)} + 3y^{(5)} + 3y^{(4)} + y''' = t + 2te^{-t} + \sin t, \qquad -\infty < t < \infty.$$

Solution: The first step is to find the complementary solution. The characteristic polynomial for the homogeneous equation is

$$\lambda^6 + 3\lambda^5 + 3\lambda^4 + \lambda^3 = \lambda^3[\lambda^3 + 3\lambda^2 + 3\lambda + 1] = \lambda^3[\lambda + 1]^3.$$

(continued)

(continued)

Since $\lambda = 0$ and $\lambda = -1$ are repeated roots, the general solution of the homogeneous equation is

$$y_C(t) = c_1 + c_2 t + c_3 t^2 + c_4 e^{-t} + c_5 t e^{-t} + c_6 t^2 e^{-t}.$$

To find a particular solution, we can use the superposition principle given in Theorem 5.9, where $g_1(t) = t$, $g_2(t) = 2te^{-t}$, and $g_3(t) = \sin t$. Table 5.1 suggests the following particular solutions for $g_1(t)$, $g_2(t)$, and $g_3(t)$:

$$y_{P_1}(t) = t^{r_1}[A_1 t + A_0],$$

$$y_{P_2}(t) = t^{r_2}[B_1 t + B_0]e^{-t},$$

$$y_{P_3}(t) = t^{r_3}[C \sin t + D \cos t].$$

The values of the exponents r_1, r_2, and r_3 are chosen to be the smallest nonnegative integers such that no term in the assumed form for the particular solution is also a solution of the homogeneous equation. So, for this example, $r_1 = 3$, $r_2 = 3$, $r_3 = 0$, and thus the form to assume for the particular solution is

$$y_P(t) = A_1 t^4 + A_0 t^3 + B_1 t^4 e^{-t} + B_0 t^3 e^{-t} + C \cos t + D \sin t. \quad \blacktriangle$$

The Method of Variation of Parameters

The method of variation of parameters, discussed in Section 4.10, can be extended to find a particular solution of a linear nonhomogeneous nth order equation

$$y^{(n)} + p_{n-1}(t)y^{(n-1)} + \cdots + p_1(t)y' + p_0(t)y = g(t), \qquad a < t < b.$$

The basic assumption underlying the method is that we know a fundamental set of solutions of the homogeneous equation. Let $\{y_1(t), y_2(t), \ldots, y_n(t)\}$ denote this set of solutions. The complementary solution (the general solution of the homogeneous equation) is then

$$y_C(t) = c_1 y_1(t) + c_2 y_2(t) + \cdots + c_n y_n(t), \qquad a < t < b.$$

To find a particular solution of the nonhomogeneous equation we "vary the parameters," replacing the constants $c_1, c_2, \ldots, c_n$ with functions $u_1(t)$, $u_2(t), \ldots, u_n(t)$ and assume a particular solution of the form

$$y_P(t) = y_1(t)u_1(t) + y_2(t)u_2(t) + \cdots + y_n(t)u_n(t), \qquad a < t < b. \tag{2}$$

The functions $u_1(t), u_2(t), \ldots, u_n(t)$ must be chosen so that (2) is a solution of nonhomogeneous equation (1). However, since there are n functions in (2), we are free to impose $n - 1$ additional constraints on the n functions. Specifically, we impose the $n - 1$ additional constraints

$$\begin{aligned}
y_1u_1' + y_2u_2' + \cdots + y_nu_n' &= 0 \\
y_1'u_1' + y_2'u_2' + \cdots + y_n'u_n' &= 0 \\
y_1''u_1' + y_2''u_2' + \cdots + y_n''u_n' &= 0 \\
&\vdots \\
y_1^{(n-2)}u_1' + y_2^{(n-2)}u_2' + \cdots + y_n^{(n-2)}u_n' &= 0
\end{aligned} \tag{3}$$

The purpose of (3) is to make successive derivatives of $y_P(t)$ [where $y_P(t)$ is defined by equation (2)] have the following simple forms:

$$\begin{aligned}
y_P' &= y_1'u_1 + y_2'u_2 + \cdots + y_n'u_n \\
y_P'' &= y_1''u_1 + y_2''u_2 + \cdots + y_n''u_n \\
&\vdots \\
y_P^{(n-1)} &= y_1^{(n-1)}u_1 + y_2^{(n-1)}u_2 + \cdots + y_n^{(n-1)}u_n
\end{aligned} \tag{4}$$

We next substitute the representation for y_P given by equation (2) into differential equation (1), use (3) and the fact that each of the functions $y_1, y_2, \ldots, y_n$ is a solution of the homogeneous equation. We obtain

$$y_1^{(n-1)}u_1' + y_2^{(n-1)}u_2' + \cdots + y_n^{(n-1)}u_n' = g. \tag{5}$$

Taken together, equations (3) and (5) form a set of n linear equations for the n unknowns, $u_1', u_2', \ldots, u_n'$. In matrix form this system of equations is

$$\begin{bmatrix} y_1 & y_2 & \cdots & y_n \\ y_1' & y_2' & \cdots & y_n' \\ \vdots & & & \vdots \\ y_1^{(n-1)} & y_2^{(n-1)} & \cdots & y_n^{(n-1)} \end{bmatrix} \begin{bmatrix} u_1' \\ u_2' \\ \vdots \\ u_n' \end{bmatrix} = \begin{bmatrix} 0 \\ 0 \\ \vdots \\ g \end{bmatrix}, \qquad a < t < b. \tag{6}$$

The determinant of the $(n \times n)$ coefficient matrix is the Wronskian of the fundamental set of solutions $\{y_1, y_2, \ldots, y_n\}$. By the corollary to Theorem 5.6, the Wronskian is nonzero for all values t in the interval of interest. Therefore, the system of equations has a unique solution for the unknowns $u_1', u_2', \ldots, u_n'$. Once these n functions are determined, we obtain $u_1, u_2, \ldots, u_n$ by antidifferentiation and then form y_P as prescribed by equation (2).

The method of variation of parameters is, in principle, a very general method. However, the practical limitations noted in Section 4.10 for the second order case also apply to the nth order case. If the differential equation coefficients are not constants, it may be very difficult to determine explicitly a fundamental set of solutions of the homogeneous equation. Even if we know a fundamental set, it may be impossible to express the antiderivatives of $u_1', u_2', \ldots, u_n'$ in terms of known functions. The following example, however, is one in which the entire computational program of variation of parameters can be performed explicitly.

EXAMPLE 2

Consider the nonhomogeneous differential equation

$$t^3y''' - 3t^2y'' + 6ty' - 6y = t, \qquad 0 < t < \infty. \tag{7}$$

(a) Verify that the functions $y_1(t) = t$, $y_2(t) = t^2$, $y_3(t) = t^3$ are solutions of the associated homogeneous equation.

(b) Use variation of parameters to construct a particular solution of the given nonhomogeneous equation.

Solution:

(a) In Section 10.3 we discuss a procedure for solving the homogeneous Euler equation

$$t^3y''' - 3t^2y'' + 6ty' - 6y = 0, \qquad 0 < t < \infty. \tag{8}$$

For now, the fact that these functions are solutions can be verified by direct substitution.

(b) The given set of functions is a fundamental set of solutions of the homogeneous equation since the Wronskian is nonzero on $0 < t < \infty$,

$$W(t) = \begin{vmatrix} t & t^2 & t^3 \\ 1 & 2t & 3t^2 \\ 0 & 2 & 6t \end{vmatrix} = 2t^3.$$

Let us now assume a particular solution of the form

$$y_P = tu_1(t) + t^2u_2(t) + t^3u_3(t).$$

If we want to use equation (6), we must first divide equation (7) by t^3 to put it in standard form. This step is necessary so that we can properly identify the nonhomogeneous term, $g(t)$; for this present example, $g(t) = t^{-2}$. We obtain

$$\begin{bmatrix} t & t^2 & t^3 \\ 1 & 2t & 3t^2 \\ 0 & 2 & 6t \end{bmatrix} \begin{bmatrix} u_1' \\ u_2' \\ u_3' \end{bmatrix} = \begin{bmatrix} 0 \\ 0 \\ t^{-2} \end{bmatrix}. \tag{9}$$

Solving system (9),

$$\begin{bmatrix} u_1' \\ u_2' \\ u_3' \end{bmatrix} = \begin{bmatrix} \dfrac{1}{2t} \\ \dfrac{-1}{t^2} \\ \dfrac{1}{2t^3} \end{bmatrix}.$$

The functions u_1, u_2, u_3 are found by computing convenient antiderivatives: $u_1(t) = (1/2)\ln|t| = (1/2)\ln t$, $u_2(t) = 1/t$, and $u_3(t) = -1/(4t^2)$, where $0 < t < \infty$. Thus, one particular solution is

$$y_P(t) = \frac{t}{2}\ln t + t^2\left[\frac{1}{t}\right] + t^3\left[\frac{-1}{4t^2}\right] = \frac{t}{2}\ln t + \frac{3t}{4}, \qquad 0 < t < \infty.$$

Since the term $3t/4$ is a solution of the homogeneous equation, we can dispense with it and use a simpler particular solution

$$y_P(t) = \frac{t}{2}\ln t \qquad 0 < t < \infty.$$

Adding the complementary solution to the particular solution, we obtain the general solution of equation (7),

$$y(t) = c_1 t + c_2 t^2 + c_3 t^3 + \frac{t}{2}\ln t, \qquad 0 < t < \infty. \ \blacktriangle \qquad (10)$$

EXERCISES

Exercises 1–10:

For each differential equation

(a) Find the complementary solution.

(b) Find a particular solution.

(c) Formulate the general solution.

1. $y''' - y' = e^{2t}$ **2.** $y''' - y' = 4 + 2\cos 2t$ **3.** $y''' - y' = 4t$

4. $y''' - y' = -4e^{t}$ **5.** $y''' + y'' = 6e^{-t}$ **6.** $y''' - y'' = 4e^{-2t}$

7. $y''' - 2y'' + y' = t + 4e^{t}$ **8.** $y''' - 3y'' + 3y' - y = 12e^{t}$

9. $y''' - y = e^{t}$ **10.** $y''' + y = e^{t} + \cos t$

Exercises 11–15:

For each differential equation

(a) Find the complementary solution

(b) Using Theorem 5.9 and Table 5.1, formulate the *appropriate form* for the particular solution suggested by the method of undetermined coefficients. You need not evaluate the undetermined coefficients.

11. $y''' - 4y'' + 4y' = t^3 + 4t^2e^{2t}$ **12.** $y''' - 3y'' + 3y' - y = e^{t} + 4e^{t}\cos 3t + 4$

13. $y^{(4)} - 16y = t\sinh 2t + 4t\cos 2t$ [Hint: Recall that $\sinh 2t = (e^{2t} - e^{-2t})/2$.]

14. $y^{(4)} + 8y'' + 16y = t\cos 2t$ **15.** $y^{(4)} - y = te^{-t} + (3t + 4)\cos t$

Exercises 16–18:

Consider the nonhomogeneous differential equation

$$y''' + ay'' + by' + cy = g(t).$$

In each exercise, the general solution of the differential equation is given, where $c_1, c_2,$ and c_3 represent arbitrary constants. Use this information to determine the constants a, b, c and the function $g(t)$.

16. $y = c_1 + c_2 t + c_3 e^{2t} + 4\sin 2t$ **17.** $y = c_1 \sin 2t + c_2 \cos 2t + c_3 e^{t} + t^2$

18. $y = c_1 + c_2 t + c_3 t^2 - 2t^3$

Exercises 19–20:

Consider the nonhomogeneous differential equation

$$t^3y''' + at^2y'' + bty' + cy = g(t), \qquad t > 0.$$

In each exercise, the general solution of the differential equation is given, where $c_1, c_2,$ and c_3 represent arbitrary constants.. Use this information to determine the constants a, b, c and the function $g(t)$.

19. $y = c_1 + c_2t + c_3t^3 + t^4$ **20.** $y = c_1t + c_2t^2 + c_3t^4 + 2\ln t$

21. (a) Verify that $\{t, t^2, t^4\}$ is a fundamental set of solutions of the differential equation

$$t^3y''' - 4t^2y'' + 8ty' - 8y = 0.$$

(b) Find the general solution of

$$t^3y''' - 4t^2y'' + 8ty' - 8y = 2\sqrt{t}, \qquad t > 0.$$

[Hint: Cramer's rule can be used to solve the system of equations that arises in the method of variation of parameters; see the appendix on matrix theory.]

22. Using the information of Exercise 21(a), find the general solution of

$$t^3y''' - 4t^2y'' + 8ty' - 8y = 2t, \qquad t > 0.$$

23. Using the information of Exercise 21(a), find the general solution of

$$t^3y''' - 4t^2y'' + 8ty' - 8y = t^3 + 2t^2 + 1, \qquad t > 0.$$

Extended Problem: The Loading of a Beam

When a beam is subjected to a load, it bends or deflects. Loads imposed upon beams may be constant in time; such loads are called **static loads**. Loads may also vary in time; these are called **dynamic loads**.

The image of a horizontal plank bending when a heavy weight is placed on it or the image of a diving board flexing as a swimmer begins a dive is familiar to everyone. In this exercise we consider both static and dynamic beam loading. It is important to note that the mathematical problems we ultimately deal with are different from those we have thus far considered; they are two-point boundary value problems rather than initial value problems.

Background

Consider the beam shown in Figure 5.1. The beam is uniform in cross section and composition, has length l, and is clamped at both ends. We neglect the weight of the beam itself.

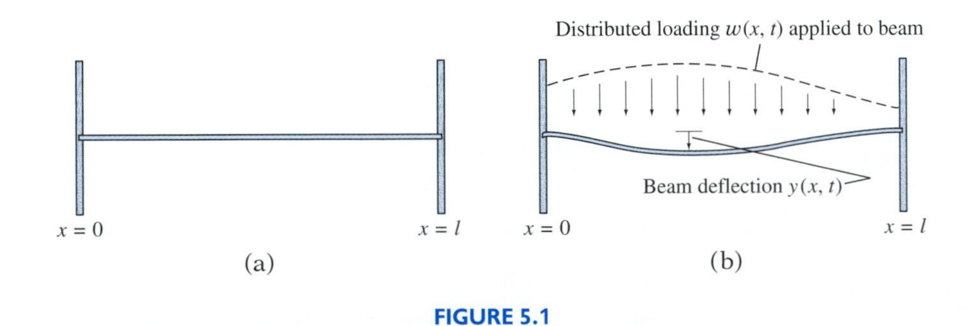

FIGURE 5.1

(a) The beam in the unloaded state. (b) The beam in a loaded state, at time t.

In Figure 5.1(a) no loading is applied—the beam deflection from the horizontal is everywhere zero. Figure 5.1(b) illustrates a distributed dynamic beam loading, $w(x,t)$, applied to the beam. The distributed loading $w(x,t)$ has units of pounds per foot or newtons per meter. As a result of the loading, the beam bends or deflects. As in Figure 5.1(b), we let $y(x,t)$ denote the deflection of the beam at position x along its span and at time t. The loading, $w(x,t)$, and beam deflection, $y(x,t)$, are both assumed to be positive in the downward direction. With a dynamic loading, the load and deflection vary with time and Figure 5.1(b) represents a snapshot taken at time t. We are interested in quantitatively determining how beam deflection varies with the loading.

The Euler-Bernoulli Beam Equation

A mathematical description of how beam position changes under various loadings is needed. A model frequently used to predict beam deflection is the partial differential equation (1), known as the Euler-Bernoulli beam equation

$$\rho\frac{\partial^2 y}{\partial t^2} + EI\frac{\partial^4 y}{\partial x^4} = w(x,t), \qquad 0 < x < l, \qquad 0 < t. \tag{1}$$

As we see, equation (1) involves two independent variables, x and t. Therefore, equation (1) is a partial differential equation. For the two special cases we consider in this problem, equation (1) reduces to a familiar linear ordinary differential equation.

In equation (1), ρ, E, and I are positive constants; ρ is the mass density of the beam per unit length, E is Young's modulus (a constant determined by the beam material), and I is the area moment of inertia of the beam (a constant determined by the beam's cross-sectional geometry).

Because the beam is clamped at its ends, beam displacement and the slope of the beam displacement must both vanish at the ends. Therefore, beam displacement, $y(x,t)$, must satisfy the following four boundary conditions,

$$y(0,t) = 0, \qquad y_x(0,t) = 0, \qquad y(l,t) = 0, \qquad y_x(l,t) = 0, \qquad t \geq 0. \tag{2}$$

To formulate this problem completely, we also need to specify initial conditions at time $t = 0$. These initial conditions describe the beam's initial displacement and its initial velocity at each point along the beam. In other words, the initial conditions specify $y(x,0)$ and $y_t(x,0)$ for $0 \leq x \leq l$.

The derivation of the Euler-Bernoulli equation is beyond the scope of our present discussion. It is important to realize, however, that equation (1) is ultimately a consequence of Newton's law of motion ($ma = F$) applied to a differential element of the beam. We now consider two problems arising as special cases.

PROBLEM 1: *Static Loading*

In this problem, we consider the case where beam loading remains constant in time. In particular, we assume that the loading function, $w(x,t)$ on the right side of equation (1), is given by

$$w(x,t) = w_0 \sin^2\left(\frac{\pi x}{l}\right).$$

In this constant-load case, deflection also remains constant in time and we can therefore look for a time-independent solution of the form $y = y(x)$. Also note, since deflection is independent of time t, the first term on the left-hand side of equation (1), $\rho\partial^2 y/\partial t^2$, is zero.

In a constant-load scenario, therefore, equations (1) and (2) reduce to the following problem, known as a **two-point boundary value problem**:

$$EI\,\frac{d^4y}{dx^4} = w_0 \sin^2\left(\frac{\pi x}{l}\right), \qquad 0 < x < l,$$
$$y(0) = 0, \qquad y'(0) = 0, \qquad y(l) = 0, \qquad y'(l) = 0. \tag{3}$$

Observe how problem (3) differs from the problems we have thus far considered. The four supplementary conditions are not all imposed at the same value of independent variable, x; in fact, two conditions are imposed at $x = 0$ and the other two at $x = l$.

Following steps (a) and (b) below, solve equation (3). For ease of computation, let $E = 1, I = 1, l = 1$, and $w_0 = 2$.

(a) Obtain the general solution of the fourth order differential equation in problem (3) (you can use the characteristic equation techniques of Chapter 5 or you can simply antidifferentiate).

(b) Impose the four boundary conditions, obtaining the solution of this two-point boundary value problem. Note that the four constraints determine the values of the four arbitrary constants appearing in the complementary solution.

(c) Plot the solution and determine the maximum deflection of the beam along its span.

PROBLEM 2: *Dynamic Loading*

Consider now the case where the beam is subjected to a load that varies sinusoidally in time,

$$\begin{aligned} w(x,t) &= w_0 \sin^2\left(\frac{\pi x}{l}\right)\sin(\omega t) \\ &= w_0\left(\frac{1}{2} - \frac{1}{2}\cos\left(\frac{2\pi x}{l}\right)\right)\sin(\omega t). \end{aligned}$$

Here, $\omega = 2\pi f$ is the radian frequency of the periodic-loading time variation. Since $\sin(\omega t)$ varies between ± 1 as time evolves, this type of beam loading takes on both positive and negative values. It corresponds to a loading that sometimes presses down and sometimes pulls up on the beam.

Given appropriate initial conditions, we can find a solution of the form

$$y(x,t) = y(x)\sin(\omega t).$$

The steps are as follows:

(a) Substitute $y(x,t) = y(x)\sin(\omega t)$ into equations (1) and (2). Show that this leads to the following two-point boundary value problem

$$-\rho\omega^2 y(x) + EI\,\frac{d^4y(x)}{dx^4} = w_0\left(\frac{1}{2} - \frac{1}{2}\cos\left(\frac{2\pi x}{l}\right)\right), \qquad 0 < x < l,$$
$$y(0) = 0, \qquad y'(0) = 0, \qquad y(l) = 0, \qquad y'(l) = 0. \tag{4}$$

(b) Obtain the general solution of the fourth order differential equation appearing in (4). Assume that $\rho = \pi^2/4, E = 1, I = 1, l = 1, w_0 = 2$ and $\omega = 2\pi$ (that is, $f = 1$ Hz).

(c) Impose the four boundary conditions and determine the solution of the boundary value problem. (Computer software will be helpful for this computation.)

(d) Plot the beam deflection envelope $y(x)$ and determine the maximum value of this deflection envelope across the span of the beam. How does it compare with the maximum deflection obtained under static loading?

6

First Order Linear Systems

CHAPTER OVERVIEW

Introduction

A linear system of algebraic equations is a familiar concept. For instance, a system of three linear equations in three unknowns has the form

$$\begin{aligned} a_{11}x_1 + a_{12}x_2 + a_{13}x_3 &= b_1, \\ a_{21}x_1 + a_{22}x_2 + a_{23}x_3 &= b_2, \\ a_{31}x_1 + a_{32}x_2 + a_{33}x_3 &= b_3. \end{aligned} \tag{1}$$

In (1), the nine coefficients $a_{11}, a_{12}, \dots, a_{33}$ and the three nonhomogeneous terms b_1, b_2, b_3 are given constants. Solving system (1) entails finding all values x_1, x_2, x_3 that simultaneously satisfy each of the three equations. In most cases these equations cannot be solved "one at a time." Rather, we have to deal with the system as a whole, applying techniques from matrix theory to obtain the solution.

Systems of First Order Linear Differential Equations

In this chapter we consider systems of first order linear differential equations. An example is the system of three linear differential equations:

$$\begin{aligned}\frac{dy_1}{dt} &= p_{11}(t)y_1 + p_{12}(t)y_2 + p_{13}(t)y_3 + g_1(t),\\ \frac{dy_2}{dt} &= p_{21}(t)y_1 + p_{22}(t)y_2 + p_{23}(t)y_3 + g_2(t),\\ \frac{dy_3}{dt} &= p_{31}(t)y_1 + p_{32}(t)y_2 + p_{33}(t)y_3 + g_3(t), \qquad a < t < b.\end{aligned} \tag{2}$$

In this system of equations, the nine coefficient functions $p_{11}(t), p_{12}(t), \ldots, p_{33}(t)$ and the three nonhomogeneous terms $g_1(t)$, $g_2(t)$, $g_3(t)$ are known functions, usually assumed to be continuous on some common t-interval, $a < t < b$.

Solving problem (2) means finding all functions $y_1(t), y_2(t), y_3(t)$ that simultaneously satisfy each of the three differential equations. Here again, as with the algebraic equations in (1), we cannot normally solve the system of differential equations "one at a time." We cannot, for example, use the techniques of Chapter 2 to solve the first equation for $y_1(t)$ because the functions $y_2(t)$ and $y_3(t)$ are not known. Instead, we have to develop techniques to deal with the system of equations as a whole. Not surprisingly, techniques from matrix theory play a central role. (A brief review is given in the appendix on matrix theory.)

Section 6.1 presents a review of matrix functions and the calculus of matrix functions. In Section 6.2 we state the basic existence and uniqueness theorem for systems of differential equations such as (2). We also describe how to rewrite systems of higher order scalar differential equations as a single system of first order differential equations. In Section 6.3 and 6.4 we study the first order homogeneous system $\mathbf{y}' = P(t)\mathbf{y}$. As in Sections 4.3 and 5.3, the discussion centers on the notion of a fundamental set of solutions and the Wronskian.

Section 6.5 parallels Sections 4.4 and 5.4. This section develops methods for solving the constant coefficient problem $\mathbf{y}' = A\mathbf{y}$ using the eigenvalues and eigenvectors of A to define a fundamental set of solutions. Next (Sections 6.6 and 6.7), in parallel with Sections 4.5 and 4.6, we discuss the complications that arise when A has complex eigenvalues or repeated eigenvalues or both. Section 6.8 considers nonhomogeneous linear systems and develops analogs of the method of undetermined coefficients and the method of variation of parameters for linear systems of the form $\mathbf{y}' = P(t)\mathbf{y} + \mathbf{g}(t)$.

Sections 6.9–6.11 are somewhat specialized. Section 6.9 generalizes Euler's method from Section 3.8 so that it applies to systems. Sections 6.10 and 6.11 discuss diagonalization and the exponential matrix, respectively.

An Example

Systems of first order linear differential equations arise frequently in applications. The following example illustrates some of the ideas.

EXAMPLE 1

A Two-Tank Mixing Problem

Consider the two-tank connection shown in Figure 6.1. As in Chapter 2, the solute and solvent are assumed to be salt and water, respectively, and the solutions in both tanks are "well stirred." Assume each tank has a capacity of 500 gal. Initially Tank 1 contains 200 gal of fresh water while Tank 2 has 50 lb of salt dissolved in 300 gal of water. At time $t = 0$, the flow begins at the rates and concentrations shown in Figure 6.1. Let the amounts of salt in the two tanks at time t be denoted by $Q_1(t)$ and $Q_2(t)$, respectively. The problem is to determine $Q_1(t)$ and $Q_2(t)$ on the time interval that is physically relevant (for example, we will stop the flow if one of the tanks becomes completely filled or completely drained). Time t is in minutes.

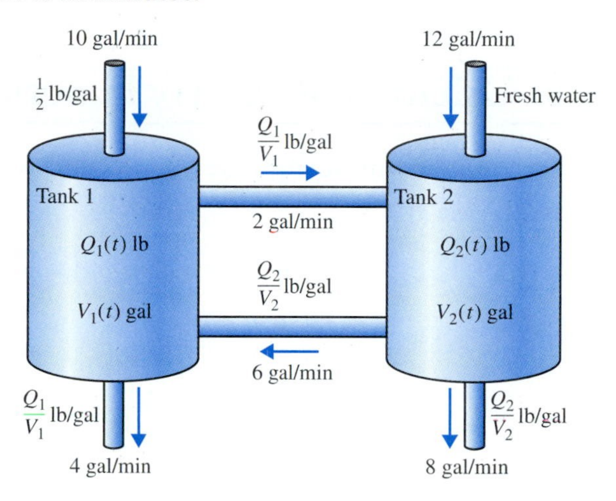

FIGURE 6.1

The two-tank mixing problem described in Example 1.

As a first step, we determine how the volumes of fluid vary in both tanks. Tank 1 has a total of 16 gal of solution entering per minute and 6 gal leaving per minute. Since Tank 1 contains 200 gal at $t = 0$ and since the tank gains 10 gal of fluid per minute, the volume of liquid in Tank 1 is given by $V_1(t) = 200 + 10t$ gal.

Tank 2, on the other hand, gains 14 gal of fluid per minute but also loses 14 gal/min. Therefore, the volume of liquid in Tank 2 remains constant at $V_2(t) = 300$ gal. These considerations of volume also determine the t-interval of interest. Since the capacity of Tank 1 is 500 gal, it will be completely filled at time $t = 30$ min. The interval of interest is therefore $0 \leq t \leq 30$.

We obtain the relevant system of differential equations by applying the principle of "conservation of salt" to each tank. In other words, the rate of change of the amount of salt in a tank is equal to the rate at which salt enters the tank minus the rate at which salt leaves the tank. From Figure 6.1 it follows that:

$$\begin{aligned}\frac{dQ_1}{dt} &= 5 + 6\left[\frac{Q_2}{V_2}\right] - 6\left[\frac{Q_1}{V_1}\right] = -\frac{6}{200+10t}Q_1 + \frac{6}{300}Q_2 + 5,\\ \frac{dQ_2}{dt} &= 2\left[\frac{Q_1}{V_1}\right] - 14\left[\frac{Q_2}{V_2}\right] = \frac{2}{200+10t}Q_1 - \frac{14}{300}Q_2.\end{aligned} \quad (3)$$

The initial value problem, to be solved on the interval $0 \leq t \leq 30$, consists of differential equations (3) together with initial conditions $Q_1(0) = 0, Q_2(0) = 50$. ▲

In Chapter 2 we considered a series connection of two tanks where solution flowed in just one direction between the tanks, from Tank 1 into Tank 2 (see Exercise 7 in Section 2.5). In that case, the techniques of Chapter 2 were used to first determine the amount of salt in Tank 1 as a function of time. We then used this information to solve for the amount of salt in Tank 2. This method *cannot* be used for the flow configuration in Figure 6.1 because solution flows from each tank into the other. In the corresponding mathematical problem, both unknowns, Q_1 and Q_2, appear on the right-hand side of both differential equations. Therefore, we cannot solve the problem by determining the unknowns sequentially; instead, we must address the entire system of differential equations as an entity in itself.

6.1 The Calculus of Matrix Functions

In the Introduction we saw examples of systems of first order linear differential equations, such as the (3×3) system

$$\begin{aligned}\frac{dy_1}{dt} &= p_{11}(t)y_1 + p_{12}(t)y_2 + p_{13}(t)y_3 + g_1(t),\\ \frac{dy_2}{dt} &= p_{21}(t)y_1 + p_{22}(t)y_2 + p_{23}(t)y_3 + g_2(t),\\ \frac{dy_3}{dt} &= p_{31}(t)y_1 + p_{32}(t)y_2 + p_{33}(t)y_3 + g_3(t), \qquad a < t < b.\end{aligned} \tag{1}$$

Techniques from matrix theory play a central role in solving first order systems such as (1). In this section, we present the necessary background in the calculus of matrix functions.

Matrix Functions

Throughout the rest of Chapter 6 we will be concerned with $(m \times n)$ matrices whose entries are real-valued functions of the real variable t. Such functions are called **matrix-valued functions** or simply, **matrix functions**. For example, a (3×2) matrix function has the form

$$A(t) = \begin{bmatrix} a_{11}(t) & a_{12}(t) \\ a_{21}(t) & a_{22}(t) \\ a_{31}(t) & a_{32}(t) \end{bmatrix}. \tag{2}$$

In (2), the entries $a_{ij}(t)$ are real-valued functions defined on a common interval $a < t < b$.

When a matrix function has a single column, such as the (3×1) matrix function

$$\mathbf{y}(t) = \begin{bmatrix} y_1(t) \\ y_2(t) \\ y_3(t) \end{bmatrix},$$

we usually refer to $\mathbf{y}(t)$ as a **vector-valued function** or simply as a **vector function**. (In matrix theory, it is common practice to use boldface type to denote column vectors. Therefore, whenever we refer to a vector function, we will use boldface also.)

The Arithmetic of Matrix Functions

For a fixed value of t, a matrix function is a constant matrix, and thus all the familiar rules of matrix arithmetic hold for matrix functions as well. Rather than stating general formulas for the arithmetic of matrix functions, we illustrate the rules with the following example.

EXAMPLE 1

We consider (2×1) and (2×2) matrix functions for simplicity. All functions are assumed to be defined on some common interval, say $a < t < b$, and all identities listed hold on that interval.

(a) *Equality*

$$\begin{bmatrix} a_1(t) \\ a_2(t) \end{bmatrix} = \begin{bmatrix} b_1(t) \\ b_2(t) \end{bmatrix}$$

if and only if $a_1(t) = b_1(t)$ and $a_2(t) = b_2(t)$

(b) *Addition*

$$\begin{bmatrix} a_1(t) \\ a_2(t) \end{bmatrix} + \begin{bmatrix} b_1(t) \\ b_2(t) \end{bmatrix} = \begin{bmatrix} a_1(t) + b_1(t) \\ a_2(t) + b_2(t) \end{bmatrix}$$

(c) *Scalar Multiplication*

$$f(t) \begin{bmatrix} a_1(t) \\ a_2(t) \end{bmatrix} = \begin{bmatrix} f(t)a_1(t) \\ f(t)a_2(t) \end{bmatrix}$$

(d) *Matrix Multiplication*

$$\begin{bmatrix} a_{11}(t) & a_{12}(t) \\ a_{21}(t) & a_{22}(t) \end{bmatrix} \begin{bmatrix} b_1(t) \\ b_2(t) \end{bmatrix} = \begin{bmatrix} a_{11}(t)b_1(t) + a_{12}(t)b_2(t) \\ a_{21}(t)b_1(t) + a_{22}(t)b_2(t) \end{bmatrix}$$

(e) *Matrix Inversion* Let $A(t)$ be a (2×2) matrix function,

$$A(t) = \begin{bmatrix} a_{11}(t) & a_{12}(t) \\ a_{21}(t) & a_{22}(t) \end{bmatrix}.$$

Then $A^{-1}(t)$ exists for all t such that $\det[A(t)] \neq 0$. Here, $A^{-1}(t)$ is a (2×2) matrix such that $A^{-1}(t)A(t) = I$ where

$$I = \begin{bmatrix} 1 & 0 \\ 0 & 1 \end{bmatrix}.$$

In Example 2, we give a formula for the inverse of a (2×2) matrix. ▲

EXAMPLE 2

Consider the (2×2) matrix function

$$A(t) = \begin{bmatrix} t & 1 \\ 4t & 4t^2 \end{bmatrix}, \qquad -\infty < t < \infty.$$

Determine all values t such that $A(t)$ is invertible and find $A^{-1}(t)$ for those values t.

Solution: The matrix function $A(t)$ is invertible when $\det[A(t)] \neq 0$. Calculating the determinant, we obtain

$$\det[A(t)] = 4t^3 - 4t = 4t(t-1)(t+1).$$

Therefore, $A^{-1}(t)$ exists for all t except $t = -1, 0, 1$. Recall that the inverse of a (2×2) matrix

$$\begin{bmatrix} a & b \\ c & d \end{bmatrix}$$

exists if and only if $ad - bc \neq 0$. Moreover, if $ad - bc \neq 0$, then the inverse is given by

$$\begin{bmatrix} a & b \\ c & d \end{bmatrix}^{-1} = \frac{1}{ad - bc} \begin{bmatrix} d & -b \\ -c & a \end{bmatrix}.$$

Therefore, in our case,

$$A^{-1}(t) = \frac{1}{4t(t^2-1)} \begin{bmatrix} 4t^2 & -1 \\ -4t & t \end{bmatrix} = \begin{bmatrix} \dfrac{t}{t^2-1} & \dfrac{-1}{4t(t^2-1)} \\ \dfrac{-1}{t^2-1} & \dfrac{1}{4(t^2-1)} \end{bmatrix}, \qquad t \neq -1, 0, 1. \; \blacktriangle$$

Limits and Derivatives of Matrix Functions

The concept of the limit of a vector function is familiar from calculus. For example, let $\mathbf{r}(t)$ denote the vector function.

$$\mathbf{r}(t) = f(t)\mathbf{i} + g(t)\mathbf{j} + h(t)\mathbf{k},$$

where the three component functions $f(t)$, $g(t)$, and $h(t)$ are defined in an open interval containing the point $t = a$. The limit of $\mathbf{r}(t)$ as t approaches a is defined to be

$$\lim_{t \to a} \mathbf{r}(t) = \left[\lim_{t \to a} f(t)\right]\mathbf{i} + \left[\lim_{t \to a} g(t)\right]\mathbf{j} + \left[\lim_{t \to a} h(t)\right]\mathbf{k},$$

provided the limits of the three component functions exist.

We take the same approach in defining limits of a matrix function. Let $A(t)$ be an $(m \times n)$ matrix function having component functions $a_{ij}(t)$, all defined in an open interval containing the point $t = a$. To say the limit of $A(t)$ as t approaches a is the $(m \times n)$ matrix L, written

$$\lim_{t \to a} A(t) = L,$$

means that

$$\lim_{t \to a} a_{ij}(t) = l_{ij}, \qquad 1 \le i \le m, \qquad 1 \le j \le n. \tag{3}$$

For instance, if $A(t)$ is a (2×2) matrix function, then the limit

$$\lim_{t \to a} A(t) = L$$

holds if and only if

$$\lim_{t \to a} A(t) = \lim_{t \to a} \begin{bmatrix} a_{11}(t) & a_{12}(t) \\ a_{21}(t) & a_{22}(t) \end{bmatrix} = \begin{bmatrix} \lim_{t \to a} a_{11}(t) & \lim_{t \to a} a_{12}(t) \\ \lim_{t \to a} a_{21}(t) & \lim_{t \to a} a_{22}(t) \end{bmatrix} = \begin{bmatrix} l_{11} & l_{12} \\ l_{21} & l_{22} \end{bmatrix} = L.$$

If one or more of the component function limits does not exist, then the limit of the matrix function does not exist. For example, if

$$B(t) = \begin{bmatrix} t & t^{-1} \\ 0 & e^t \end{bmatrix}, \quad \text{then} \quad \lim_{t \to 2} B(t) \begin{bmatrix} 2 & 1/2 \\ 0 & e^2 \end{bmatrix}.$$

However, $\lim_{t \to 0} B(t)$ does not exist.

As in single-variable calculus, we say the matrix function $A(t)$ is *continuous at* $t = a$ if $A(t)$ is defined in a neighborhood of $t = a$ and if

$$\lim_{t \to a} A(t) = A(a). \tag{4}$$

To define a derivative (an instantaneous rate of change), we are led to a limit of the form

$$A'(t) = \lim_{h \to 0} \frac{1}{h}[A(t+h) - A(t)].$$

For example, let $A(t)$ be a (2×2) matrix function with differentiable component functions. Then

$$\begin{aligned} A'(t) &= \lim_{h \to 0} \frac{1}{h}[A(t+h) - A(t)] \\ &= \begin{bmatrix} \lim_{h \to 0} \dfrac{a_{11}(t+h) - a_{11}(t)}{h} & \lim_{h \to 0} \dfrac{a_{12}(t+h) - a_{12}(t)}{h} \\ \lim_{h \to 0} \dfrac{a_{21}(t+h) - a_{21}(t)}{h} & \lim_{h \to 0} \dfrac{a_{22}(t+h) - a_{22}(t)}{h} \end{bmatrix} \\ &= \begin{bmatrix} a'_{11}(t) & a'_{12}(t) \\ a'_{21}(t) & a'_{22}(t) \end{bmatrix}. \end{aligned}$$

As this special case suggests, the derivative of a matrix function is the matrix of derivatives of its component functions. In general, let $A(t)$ be an $(m \times n)$ matrix function of the form

$$A(t) = \begin{bmatrix} a_{11}(t) & a_{12}(t) & \cdots & a_{1n}(t) \\ a_{21}(t) & a_{22}(t) & & a_{2n}(t) \\ \vdots & & & \vdots \\ a_{m1}(t) & a_{m2}(t) & \cdots & a_{mn}(t) \end{bmatrix},$$

where each of the component functions $a_{11}(t), a_{12}(t), \ldots, a_{mn}(t)$ is differentiable on the interval (a, b). Then the derivative $A'(t)$ exists and is given by

$$A'(t) = \begin{bmatrix} a'_{11}(t) & a'_{12}(t) & \cdots & a'_{1n}(t) \\ a'_{21}(t) & a'_{22}(t) & & a'_{2n}(t) \\ \vdots & & & \vdots \\ a'_{m1}(t) & a'_{m2}(t) & \cdots & a'_{mn}(t) \end{bmatrix}, \qquad a < t < b. \tag{5}$$

We will refer to $A(t)$ as a **differentiable matrix function** or simply a **differentiable matrix**.

Representing Linear Systems in Matrix Terms

Because the derivative of a matrix function has the form (5), we can express systems of linear differential equations in matrix terms. For example, recall the (3×3) system (1)

$$\begin{aligned} \frac{dy_1}{dt} &= p_{11}(t)y_1 + p_{12}(t)y_2 + p_{13}(t)y_3 + g_1(t), \\ \frac{dy_2}{dt} &= p_{21}(t)y_1 + p_{22}(t)y_2 + p_{23}(t)y_3 + g_2(t), \\ \frac{dy_3}{dt} &= p_{31}(t)y_1 + p_{32}(t)y_2 + p_{33}(t)y_3 + g_3(t), \qquad a < t < b. \end{aligned}$$

With regard to this system, let us define $\mathbf{y}(t)$, $P(t)$, and $\mathbf{g}(t)$ as follows:

$$\mathbf{y}(t) = \begin{bmatrix} y_1(t) \\ y_2(t) \\ y_3(t) \end{bmatrix}, \qquad P(t) = \begin{bmatrix} p_{11}(t) & p_{12}(t) & p_{13}(t) \\ p_{21}(t) & p_{22}(t) & p_{23}(t) \\ p_{31}(t) & p_{32}(t) & p_{33}(t) \end{bmatrix}, \qquad \mathbf{g}(t) = \begin{bmatrix} g_1(t) \\ g_2(t) \\ g_3(t) \end{bmatrix}.$$

With these definitions, we can express the (3×3) system in matrix terms as

$$\mathbf{y}'(t) = P(t)\mathbf{y}(t) + \mathbf{g}(t). \tag{6}$$

The simplicity of (6) is more than just shorthand. The notation also helps us focus on the principles of solving systems of linear differential equations because it takes our eyes away from the details of individual equations and allows us to see the system as a single entity. We can view (6) as a single differential equation involving a matrix-valued dependent variable.

Some Useful Formulas

The fact that the derivative of a matrix function is simply the matrix of derivatives can be used to verify the following familiar looking formulas.

Let $A(t)$ and $B(t)$ be two differentiable $(m \times n)$ matrix functions. Then

$$[A(t) + B(t)]' = A'(t) + B'(t). \tag{7}$$

Let $A(t)$ be a differentiable $(m \times n)$ matrix function, and let $f(t)$ be a differentiable scalar function. Then

$$[f(t)A(t)]' = f'(t)A(t) + f(t)A'(t). \tag{8}$$

Let $A(t)$ be a differentiable $(m \times n)$ matrix function and let $B(t)$ be a differentiable $(n \times p)$ matrix function. Then

$$[A(t)B(t)]' = A'(t)B(t) + A(t)B'(t). \tag{9}$$

Formula (9) is the analog of the familiar product formula in calculus. Since the functions involved are matrix functions, however, it is imperative that the order of matrix multiplications be preserved.

To illustrate how identities such as (7)–(9) are established, we verify (9) in the special case where $m = n = 2$ and $p = 1$.

EXAMPLE 3

Verify formula (9) in the case that $A(t)$ is a (2×2) matrix function and $\mathbf{b}(t)$ is a (2×1) vector function.

Solution: Let $A(t)$ and $\mathbf{b}(t)$ be as follows:

$$A(t) = \begin{bmatrix} a_{11}(t) & a_{12}(t) \\ a_{21}(t) & a_{22}(t) \end{bmatrix}, \qquad \mathbf{b}(t) = \begin{bmatrix} b_1(t) \\ b_2(t) \end{bmatrix},$$

where we assume $A(t)$ and $\mathbf{b}(t)$ are matrix functions differentiable on some common interval $a < t < b$.

Computing the matrix product $A(t)\mathbf{b}(t)$ and using the fact that the derivative of this resulting (2×1) matrix function is the matrix of derivatives,

$$[A(t)\mathbf{b}(t)]' = \begin{bmatrix} a_{11}(t)b_1(t) + a_{12}(t)b_2(t) \\ a_{21}(t)b_1(t) + a_{22}(t)b_2(t) \end{bmatrix}' = \begin{bmatrix} [a_{11}(t)b_1(t) + a_{12}(t)b_2(t)]' \\ [a_{21}(t)b_1(t) + a_{22}(t)b_2(t)]' \end{bmatrix}.$$

We now use familiar calculus formulas to compute the derivatives of the two components of the matrix on the right. With some additional rearranging of terms we obtain

$$\begin{bmatrix} [a_{11}(t)b_1(t) + a_{12}(t)b_2(t)]' \\ [a_{21}(t)b_1(t) + a_{22}(t)b_2(t)]' \end{bmatrix}$$

$$= \begin{bmatrix} [a'_{11}(t)b_1(t) + a'_{12}(t)b_2(t)] + [a_{11}(t)b'_1(t) + a_{12}(t)b'_2(t)] \\ [a'_{21}(t)b_1(t) + a'_{22}(t)b_2(t)] + [a_{21}(t)b'_1(t) + a_{22}(t)b'_2(t)] \end{bmatrix}$$

$$= \begin{bmatrix} a'_{11}(t)b_1(t) + a'_{12}(t)b_2(t) \\ a'_{21}(t)b_1(t) + a'_{22}(t)b_2(t) \end{bmatrix} + \begin{bmatrix} a_{11}(t)b'_1(t) + a_{12}(t)b'_2(t) \\ a_{21}(t)b'_1(t) + a_{22}(t)b'_2(t) \end{bmatrix}$$

$$= \begin{bmatrix} a'_{11}(t) & a'_{12}(t) \\ a'_{21}(t) & a'_{22}(t) \end{bmatrix} \begin{bmatrix} b_1(t) \\ b_2(t) \end{bmatrix} + \begin{bmatrix} a_{11}(t) & a_{12}(t) \\ a_{21}(t) & a_{22}(t) \end{bmatrix} \begin{bmatrix} b'_1(t) \\ b'_2(t) \end{bmatrix}.$$

As you can see, the term on the right is equal to $A'(t)\mathbf{b}(t) + A(t)\mathbf{b}'(t)$ and establishes formula (9) for this example. ▲

Antiderivatives of Matrix Functions

Since the derivative of a matrix function reduces to the matrix of derivatives, it should not be surprising that antiderivatives of matrix functions can be evaluated by performing the corresponding antidifferentiation operations upon each component of the matrix function. That is, if $A(t)$ is the $(m \times n)$ matrix function

$$A(t) = \begin{bmatrix} a_{11}(t) & a_{12}(t) & \cdots & a_{1n}(t) \\ a_{21}(t) & a_{22}(t) & & a_{2n}(t) \\ \vdots & & & \vdots \\ a_{m1}(t) & a_{m2}(t) & \cdots & a_{mn}(t) \end{bmatrix},$$

then the antiderivative of $A(t)$ is the $(m \times n)$ matrix function

$$\int A(t)\,dt = \begin{bmatrix} \int a_{11}(t)\,dt & \int a_{12}(t)\,dt & \cdots & \int a_{1n}(t)\,dt \\ \int a_{21}(t)\,dt & \int a_{22}(t)\,dt & & \int a_{2n}(t)\,dt \\ \vdots & & & \vdots \\ \int a_{m1}(t)\,dt & \int a_{m2}(t)\,dt & \cdots & \int a_{mn}(t)\,dt \end{bmatrix}.$$

EXAMPLE 4

Determine the matrix function $A(t)$ if

$$A'(t) = \begin{bmatrix} 2e^{2t} & 2t \\ 0 & -1 \end{bmatrix}.$$

Solution: Since the antiderivative of a matrix function is the matrix of antiderivatives,

$$A(t) = \begin{bmatrix} e^{2t} + c_{11} & t^2 + c_{12} \\ c_{21} & -t + c_{22} \end{bmatrix} = \begin{bmatrix} e^{2t} & t^2 \\ 0 & -t \end{bmatrix} + \begin{bmatrix} c_{11} & c_{12} \\ c_{21} & c_{22} \end{bmatrix} = \begin{bmatrix} e^{2t} & t^2 \\ 0 & -t \end{bmatrix} + C. \blacktriangle$$

As Example 4 illustrates, when we calculate the antiderivative of a matrix function we need to allow for different arbitrary constants in each component function antidifferentiation. Therefore, the general antiderivative of a matrix function is a matrix of convenient antiderivatives *plus* an arbitrary constant matrix of comparable dimensions.

EXERCISES

Exercises 1–5:

For the given matrix functions $A(t)$, $B(t)$ and $\mathbf{c}(t)$, make the indicated calculations

$$A(t) = \begin{bmatrix} t-1 & t^2 \\ 2 & 2t+1 \end{bmatrix}, \qquad B(t) = \begin{bmatrix} t & -1 \\ 0 & t+2 \end{bmatrix}, \qquad \mathbf{c}(t) = \begin{bmatrix} t+1 \\ -1 \end{bmatrix}.$$

1. $2A(t) - 3tB(t)$ **2.** $A(t)B(t) - B(t)A(t)$ **3.** $A(t)\mathbf{c}(t)$

4. $\det[tA(t)]$ **5.** $\det[B(t)A(t)]$

Exercises 6–9:

Determine all values t such that $A(t)$ is invertible and, for those t-values, find $A^{-1}(t)$.

6. $A(t) = \begin{bmatrix} t+1 & t \\ t & t+1 \end{bmatrix}$

7. $A(t) = \begin{bmatrix} t & 2 \\ 2 & t-3 \end{bmatrix}$

8. $A(t) = \begin{bmatrix} \sin t & -\cos t \\ \sin t & \cos t \end{bmatrix}$

9. $A(t) = \begin{bmatrix} e^t & e^{3t} \\ e^{2t} & e^{4t} \end{bmatrix}$

Exercises 10–11:

Find $\lim_{t\to 0} A(t)$ or state that the limit does not exist.

10. $A(t) = \begin{bmatrix} \dfrac{\sin t}{t} & t\cos t & \dfrac{3}{t+1} \\ e^{3t} & \sec t & \dfrac{2t}{t^2-1} \end{bmatrix}$

11. $A(t) = \begin{bmatrix} te^{-t} & \tan t \\ t^2-2 & e^{\sin t} \end{bmatrix}$

Exercises 12–13:

Find $A'(t)$ and $A''(t)$. For what values of t are the matrices $A(t)$, $A'(t)$, and $A''(t)$ defined?

12. $A(t) = \begin{bmatrix} \sin t & 3t \\ t^2+2 & 5 \end{bmatrix}$

13. $A(t) = \begin{bmatrix} 7 & \ln|t| \\ \sqrt{1-t} & e^{3t} \end{bmatrix}$

Exercises 14–15:

Each of the systems of linear differential equations can be expressed in the form $\mathbf{y}' = P(t)\mathbf{y} + \mathbf{g}(t)$. Determine $P(t)$ and $\mathbf{g}(t)$.

14. $y_1' = t^2 y_1 + 3y_2 + \sec t$
$y_2' = (\sin t)y_1 + ty_2 - 5$

15. $y_1' = t^{-1}y_1 + (t^2+1)y_2 + t$
$y_2' = 4y_1 + t^{-1}y_2 + 8t\ln t$

Exercises 16–18:

Determine $A(t)$ where

16. $A'(t) = \begin{bmatrix} 2t & 1 \\ \cos t & 3t^2 \end{bmatrix}$ and $A(0) = \begin{bmatrix} 2 & 5 \\ 1 & -2 \end{bmatrix}$

17. $A'(t) = \begin{bmatrix} t^{-1} & 4t \\ 5 & 3t^2 \end{bmatrix}$ and $A(1) = \begin{bmatrix} 2 & 5 \\ 1 & -2 \end{bmatrix}$

18. $A''(t) = \begin{bmatrix} 1 & t \\ 0 & 0 \end{bmatrix}$, $A(0) = \begin{bmatrix} 1 & 1 \\ -2 & 1 \end{bmatrix}$, and $A(1) = \begin{bmatrix} -1 & 2 \\ -2 & 3 \end{bmatrix}$

Exercises 19–20:

Calculate $A(t) = \int_0^t B(s)\,ds$.

19. $B(s) = \begin{bmatrix} 2s & \cos s & 2 \\ 5 & (s+1)^{-1} & 3s^2 \end{bmatrix}$

20. $B(s) = \begin{bmatrix} e^s & 6s \\ \cos 2\pi s & \sin 2\pi s \end{bmatrix}$

21. Let $A(t)$ be an $(n \times n)$ matrix function. We use the notation $A^2(t)$ to mean the matrix function $A(t)A(t)$.

(a) Construct an explicit (2×2) differentiable matrix function to show that

$$\frac{d}{dt}(A^2(t)) \quad \text{and} \quad 2A(t)\frac{d}{dt}(A(t))$$

are generally not equal.

(b) What is the correct formula relating the derivative of $A^2(t)$ to the matrices $A(t)$ and $A'(t)$?

22. Construct an example of a (2×2) matrix function $A(t)$ such that $A^2(t)$ is a constant matrix but $A(t)$ is not a constant matrix.

23. Let $A(t)$ be an $(n \times n)$ matrix function that is both differentiable and invertible on some t-interval of interest. It can be shown that $A^{-1}(t)$ is likewise differentiable on this interval. Differentiate the matrix identity $A^{-1}(t)A(t) = I$ to obtain the following formula:

$$\frac{d}{dt}(A^{-1}(t)) = -A^{-1}(t)A'(t)A^{-1}(t).$$

[Hint: Recall the product rule, equation (9). Notice that the formula you derive is **not** the same as the power rule of single-variable calculus.]

24. Consider the matrix function

$$A(t) = \begin{bmatrix} t & t^3 \\ 0 & 2t \end{bmatrix}.$$

Explicitly calculate both $(d/dt)(A^{-1}(t))$ and $-A^{-1}(t)A'(t)A^{-1}(t)$ for this special case and illustrate the formula derived in Exercise 23.

6.2 Existence and Uniqueness

In Section 6.1 we introduced matrix and vector functions and saw how to deal with the calculus of such functions. We now return to the main focus of Chapter 6, studying initial value problems involving systems of first order differential equations. Specifically, we consider the initial value problem

$$\begin{aligned}
\frac{dy_1}{dt} &= p_{11}(t)y_1 + p_{12}(t)y_2 + \cdots + p_{1n}(t)y_n + g_1(t),\\
\frac{dy_2}{dt} &= p_{21}(t)y_1 + p_{22}(t)y_2 + \cdots + p_{2n}(t)y_n + g_2(t),\\
&\vdots\\
\frac{dy_n}{dt} &= p_{n1}(t)y_1 + p_{n2}(t)y_2 + \cdots + p_{nn}(t)y_n + g_n(t), \qquad a < t < b,\\
y_1(t_0) &= y_1^0,\ \ y_2(t_0) = y_2^0, \ldots,\ \ y_n(t_0) = y_n^0,
\end{aligned} \tag{1}$$

where $y_1^0, y_2^0, \ldots, y_n^0$ are n constants specified at some point t_0 in the t-interval of interest, (a, b). The n^2 coefficient functions $p_{11}(t), p_{12}(t), \ldots, p_{nn}(t)$ and the n functions $g_1(t), g_2(t), \ldots, g_n(t)$ are given functions, defined on $a < t < b$.

As Section 6.1 showed, we can recast problem (1) in matrix form as

$$\mathbf{y}'(t) = P(t)\mathbf{y}(t) + \mathbf{g}(t), \qquad \mathbf{y}(t_0) = \mathbf{y}_0, \tag{2}$$

where, for $a < t < b$,

$$\mathbf{y}(t) = \begin{bmatrix} y_1(t) \\ y_2(t) \\ \vdots \\ y_n(t) \end{bmatrix}, \qquad P(t) = \begin{bmatrix} p_{11}(t) & p_{12}(t) & \cdots & p_{1n}(t) \\ p_{21}(t) & p_{22}(t) & & p_{2n}(t) \\ \vdots & & & \vdots \\ p_{n1}(t) & p_{n2}(t) & \cdots & p_{nn}(t) \end{bmatrix},$$

$$\mathbf{g}(t) = \begin{bmatrix} g_1(t) \\ g_2(t) \\ \vdots \\ g_n(t) \end{bmatrix}, \qquad \mathbf{y}_0 = \begin{bmatrix} y_1^0 \\ y_2^0 \\ \vdots \\ y_n^0 \end{bmatrix}.$$

The differential equation in (2) is referred to as a **first order system of linear differential equations** or, more simply, a **first order linear system**. If the $(n \times 1)$ vector function $\mathbf{g}(t)$ is the $(n \times 1)$ zero vector, $\mathbf{g}(t) = \mathbf{0}$, $a < t < b$, then the system is called a **first order homogeneous linear system**; if $\mathbf{g}(t)$ is not identically zero on the interval of interest, the system is a **first order nonhomogeneous linear system**.

We sometimes need to distinguish the differential equation in (2), where the dependent variable $\mathbf{y}(t)$ is a vector-valued function, from the differential equations considered in Chapters 1–5, where the dependent variable $y(t)$ is a single real-valued function. We will refer to the differential equations studied in Chapters 1–5 as **scalar differential equations**. In particular, Chapters 2, 4, and 5 dealt (respectively) with first order, second order, and nth order linear scalar differential equations.

An Existence and Uniqueness Theorem

We now discuss the conditions needed to ensure that initial value problem (2) has a unique solution. In Chapters 2, 4, and 5 (respectively) we looked at initial value problems for scalar linear differential equations. A general theme emerged from the existence/uniqueness results given in Theorems 2.1, 4.1, and 5.1, namely:

> If (a, b) denotes the interval of interest, then continuity of the coefficient functions of a linear differential equation, together with continuity of the nonhomogeneous term $g(t)$, is sufficient to guarantee existence of a unique solution of the initial value problem on the entire interval (a, b).

Theorem 6.1 continues this theme. The theory of linear differential equations—whether scalar equations or first order systems—has an underlying conceptual unity; as one might suspect, this underlying unity is no accident. The scalar initial value problems dealt with in Theorems 2.1, 4.1, and 5.1 can, in fact, be recast into the framework of a first order linear system. Theorem 6.1 is therefore an overarching result, including Theorems 2.1, 4.1, and 5.1 as special cases.

Theorem 6.1

Consider the initial value problem

$$\mathbf{y}'(t) = P(t)\mathbf{y}(t) + \mathbf{g}(t), \qquad \mathbf{y}(t_0) = \mathbf{y}_0,$$

where $\mathbf{y}(t)$, $P(t)$, $\mathbf{g}(t)$, and $\mathbf{y}_0$ are defined as in equation (2). Let the n^2 components of $P(t)$ and the n components of $\mathbf{g}(t)$ be continuous on the interval (a, b), and let t_0 be in (a, b). Then the initial value problem has a unique solution that exists on the entire interval $a < t < b$.

EXAMPLE 1

Consider the initial value problem

$$y_1' = (\sin 2t)y_1 + \frac{t}{t^2 - 2t - 8}y_2 + 4, \qquad y_1(1) = 2,$$

$$y_2' = (\ln|t+1|)y_1 + e^{-2t}y_2 + \sec t, \qquad y_2(1) = 0.$$

(continued)

(continued)

Determine the largest t-interval on which Theorem 6.1 guarantees the existence of a unique solution of this problem.

Solution: The given problem can be rewritten as

$$\mathbf{y}'(t) = P(t)\mathbf{y}(t) + \mathbf{g}(t), \qquad a < t < b,$$
$$\mathbf{y}(t_0) = \mathbf{y}_0,$$

where

$$\mathbf{y}(t) = \begin{bmatrix} y_1(t) \\ y_2(t) \end{bmatrix}, \qquad P(t) = \begin{bmatrix} \sin 2t & \dfrac{t}{t^2 - 2t - 8} \\ \ln|t+1| & e^{-2t} \end{bmatrix},$$
$$\mathbf{g}(t) = \begin{bmatrix} 4 \\ \sec t \end{bmatrix}, \qquad t_0 = 1, \qquad \mathbf{y}_0 = \begin{bmatrix} 2 \\ 0 \end{bmatrix}.$$

According to Theorem 6.1, a unique solution is guaranteed to exist on the largest interval (a, b) containing the point $t_0 = 1$, in which the four components of $P(t)$ and the two components of $\mathbf{g}(t)$ are continuous.

The functions $p_{11}(t) = \sin 2t$, $p_{22}(t) = e^{-2t}$ and $g_1(t) = 4$ are continuous for all t, $-\infty < t < \infty$. The function

$$p_{12}(t) = \frac{t}{t^2 - 2t - 8} = \frac{t}{(t-4)(t+2)}$$

has discontinuities at $t = -2$ and $t = 4$. Similarly, $p_{21}(t) = \ln|t+1|$ is discontinuous at $t = -1$ while $g_2(t) = \sec t$ has discontinuities at odd multiples of $\pi/2$. The largest interval containing $t_0 = 1$ on which all six functions are continuous is $-1 < t < \pi/2$. Theorem 6.1 guarantees the existence of a unique solution on the interval $-1 < t < \pi/2$. ▲

Rewriting an *n*th Order Scalar Linear Equation as a First Order Linear System

Consider the initial value problem

$$\begin{aligned} &y^{(n)} + p_{n-1}(t)y^{(n-1)} + \cdots + p_1(t)y' + p_0(t)y = g(t), \qquad a < t < b \\ &y(t_0) = y_0, \quad y'(t_0) = y'_0, \ldots, \quad y^{(n-1)}(t_0) = y_0^{(n-1)}. \end{aligned} \tag{3}$$

We now show that it is possible to rewrite the scalar problem (3) as an initial value problem for a first order linear system. (Following Example 2, we comment on why it is often useful to rewrite a scalar problem as a first order system.) To begin, let $\mathbf{Y}(t)$ be the $(n \times 1)$ vector function

$$\mathbf{Y}(t) = \begin{bmatrix} Y_1(t) \\ Y_2(t) \\ \vdots \\ Y_n(t) \end{bmatrix},$$

where the component functions $Y_i(t)$ are defined by

$$Y_1(t) = y(t), \quad Y_2(t) = y'(t), \quad Y_3(t) = y''(t), \ldots, \quad Y_n(t) = y^{(n-1)}(t). \tag{4}$$

If we calculate $\mathbf{Y}'(t)$, use the relations in (4) and the fact [see equation (3)] that $y^{(n)} = -p_{n-1}(t)y^{(n-1)} - \cdots - p_1(t)y' - p_0(t)y + g(t)$, we obtain

$$\mathbf{Y}'(t) = \begin{bmatrix} Y_1'(t) \\ Y_2'(t) \\ Y_3'(t) \\ \vdots \\ Y_n'(t) \end{bmatrix} = \begin{bmatrix} y'(t) \\ y''(t) \\ y'''(t) \\ \vdots \\ y^{(n)}(t) \end{bmatrix} = \begin{bmatrix} Y_2(t) \\ Y_3(t) \\ Y_4(t) \\ \vdots \\ -p_{n-1}(t)Y_n - \cdots - p_0(t)Y_1 + g(t) \end{bmatrix}. \tag{5}$$

Equation (5) can be written as the matrix equation

$$\begin{bmatrix} Y_1'(t) \\ Y_2'(t) \\ Y_3'(t) \\ \vdots \\ Y_n'(t) \end{bmatrix} = \begin{bmatrix} 0 & 1 & 0 & \cdots & 0 \\ 0 & 0 & 1 & & 0 \\ 0 & 0 & 0 & & 0 \\ \vdots & & & & \vdots \\ -p_0(t) & -p_1(t) & -p_2(t) & \cdots & -p_{n-1}(t) \end{bmatrix} \begin{bmatrix} Y_1(t) \\ Y_2(t) \\ Y_3(t) \\ \vdots \\ Y_n(t) \end{bmatrix} + \begin{bmatrix} 0 \\ 0 \\ 0 \\ \vdots \\ g(t) \end{bmatrix}. \tag{6}$$

Matrix equation (6) has the form

$$\mathbf{Y}' = P(t)\mathbf{Y} + \mathbf{G}(t),$$

where $P(t)$ is an $(n \times n)$ matrix function and $\mathbf{G}(t)$ is an $(n \times 1)$ vector function. The initial conditions in (3) likewise transform into an initial condition for the function $\mathbf{Y}(t)$,

$$\mathbf{Y}(t_0) = \begin{bmatrix} Y_1(t_0) \\ Y_2(t_0) \\ Y_3(t_0) \\ \vdots \\ Y_n(t_0) \end{bmatrix} = \begin{bmatrix} y(t_0) \\ y'(t_0) \\ y''(t_0) \\ \vdots \\ y^{(n-1)}(t_0) \end{bmatrix} = \begin{bmatrix} y_0 \\ y_0' \\ y_0'' \\ \vdots \\ y_0^{(n-1)} \end{bmatrix} \equiv \mathbf{Y}_0.$$

Thus, we have transformed initial value problem (3) into

$$\mathbf{Y}' = P(t)\mathbf{Y} + \mathbf{G}(t), \qquad a < t < b, \qquad \mathbf{Y}(t_0) = \mathbf{Y}_0.$$

Example 2 illustrates this process for a specific scalar problem.

EXAMPLE 2

Rewrite the following scalar initial value problem

$$y''' + t^2y'' + e^ty' + 3y = \sin 2t, \qquad y(0) = 2, \qquad y'(0) = -1, \qquad y''(0) = 0$$

as an initial value problem for a first order system.

(continued)

(continued)

Solution: Define the vector function $\mathbf{Y}(t)$,

$$\mathbf{Y}(t) = \begin{bmatrix} Y_1(t) \\ Y_2(t) \\ Y_3(t) \end{bmatrix} = \begin{bmatrix} y(t) \\ y'(t) \\ y''(t) \end{bmatrix}.$$

Calculating $\mathbf{Y}'(t)$, we find

$$\mathbf{Y}'(t) = \begin{bmatrix} Y_2(t) \\ Y_3(t) \\ -t^2 Y_3(t) - e^t Y_2(t) - 3Y_1(t) + \sin 2t \end{bmatrix} = \begin{bmatrix} 0 & 1 & 0 \\ 0 & 0 & 1 \\ -3 & -e^t & -t^2 \end{bmatrix} \begin{bmatrix} Y_1(t) \\ Y_2(t) \\ Y_3(t) \end{bmatrix} + \begin{bmatrix} 0 \\ 0 \\ \sin 2t \end{bmatrix}. \tag{7}$$

Thus, the third-order scalar equation can be rewritten as the first order system:

$$\mathbf{Y}' = P(t)\mathbf{Y} + \mathbf{G}(t), \qquad -\infty < t < \infty.$$

The initial conditions for the scalar problem transform into a corresponding initial condition for the vector function $\mathbf{Y}(t)$:

$$\mathbf{Y}(0) = \begin{bmatrix} Y_1(0) \\ Y_2(0) \\ Y_3(0) \end{bmatrix} = \begin{bmatrix} y(0) \\ y'(0) \\ y''(0) \end{bmatrix} = \begin{bmatrix} 2 \\ -1 \\ 0 \end{bmatrix}. \tag{8}$$

▲

The ability to rewrite an nth order scalar equation as a first order linear system leads to an important conceptual unity for the theory of differential equations. As we present the theory of first order linear systems in this chapter, we will point out how the results being developed relate to analogous results from Chapters 2, 4, and 5.

What are the practical implications of this relationship? On one hand, the techniques we have seen in the prior chapters for solving nth order scalar problems have not been made obsolete; they remain as relevant as ever. As you will see, if an initial value problem for a scalar differential equation can be solved using the techniques of Chapters 4 or 5, that process is usually easier than rewriting the equation as a first order system and then solving the resulting matrix problem.

On the other hand, the ability to recast higher order scalar problems as first order systems is very useful, for example when applying numerical algorithms to solve these problems. Consider, for instance, the scalar differential equation in Example 2. Although Theorem 5.1 assures us that the problem has a unique solution on $-\infty < t < \infty$, we have no method to explicitly construct the solution; the third order nonhomogeneous differential equation, although linear, has variable coefficients. Therefore, we might solve this problem numerically. As we will see in Section 6.9, Euler's method (developed in Section 3.8 for scalar problems) extends very naturally to initial value problems for first order systems such as equation (7). In Chapter 9 we will develop algorithms more accurate than Euler's method, which are also easily extended to first order systems. In Chapter 8, we discuss similar techniques for rewriting a higher order scalar nonlinear equation as a first order matrix system of nonlinear equations.

EXERCISES

Exercises 1–4:

Find the largest interval $a < t < b$ such that a unique solution of the given initial value problem is guaranteed to exist.

1. $y_1' = t^{-1}y_1 + (\tan t)y_2, \quad y_1(3) = 0$
$y_2' = (\ln|t|)y_1 + e^t y_2, \quad y_2(3) = 1$

2. $y_1' = y_1 + (\tan t)y_2 + (t+1)^{-2} \quad y_1(0) = 0$
$y_2' = (t^2 - 2)y_1 + 4y_2, \quad y_2(0) = 0$

3. $t^2 y_1' = (\cos t)y_1 + y_2 + 1, \quad y_1(1) = 0$
$y_2' = 2y_1 + 4ty_2 + \sec t, \quad y_2(1) = 2$

4. $(t+2)y_1' = 3ty_1 + 5y_2, \quad y_1(1) = 0$
$(t-2)y_2' = 2y_1 + 4ty_2, \quad y_2(1) = 2$

Exercises 5–6:

Verify, for any values c_1 and c_2, that the functions $y_1(t)$ and $y_2(t)$ satisfy the given system of linear differential equations.

5. $y_1' = 4y_1 + y_2, \quad y_1(t) = c_1 e^{5t} + c_2 e^{3t}$
$y_2' = y_1 + 4y_2, \quad y_2(t) = c_1 e^{5t} - c_2 e^{3t}$

6. $y_1' = y_1 + y_2, \quad y_1(t) = c_1 e^t \cos t + c_2 e^t \sin t$
$y_2' = -y_1 + y_2, \quad y_2(t) = -c_1 e^t \sin t + c_2 e^t \cos t$

Exercises 7–8:

For each of the exercises

(a) Rewrite the equations from the given exercise in vector form as $\mathbf{y}'(t) = A\mathbf{y}(t)$, identifying the constant matrix A.

(b) Rewrite the solution of the equations in part (a) in vector form as $\mathbf{y}(t) = c_1\mathbf{y}_1(t) + c_2\mathbf{y}_2(t)$.

7. The equations in Exercise 5.

8. The equations in Exercise 6.

Exercises 9–10:

Each exercise lists a candidate for the solution, $\mathbf{y}(t)$, of the equation $\mathbf{y}' = A\mathbf{y}$, where A is the given constant matrix. Verify that $\mathbf{y}(t)$ is indeed a solution for any choice of the constants c_1 and c_2. Find values of c_1 and c_2 such that $\mathbf{y}(t)$ solves the given initial value problem. [According to Theorem 6.1, you have found the unique solution of $\mathbf{y}' = A\mathbf{y}$, $\mathbf{y}(0) = \mathbf{y}_0$.]

9. $\mathbf{y}' = A\mathbf{y}, \ \mathbf{y}(0) = \begin{bmatrix} 4 \\ -3 \end{bmatrix}$; where $A = \begin{bmatrix} 1 & -2 \\ 1 & 4 \end{bmatrix}$ and $\mathbf{y}(t) = c_1 e^{2t}\begin{bmatrix} 2 \\ -1 \end{bmatrix} + c_2 e^{3t}\begin{bmatrix} 1 \\ -1 \end{bmatrix}$

10. $\mathbf{y}' = A\mathbf{y}, \ \mathbf{y}(0) = \begin{bmatrix} -1 \\ 8 \end{bmatrix}$; where $A = \begin{bmatrix} 3 & 2 \\ 4 & 1 \end{bmatrix}$ and $\mathbf{y}(t) = c_1 e^{5t}\begin{bmatrix} 1 \\ 1 \end{bmatrix} + c_2 e^{-t}\begin{bmatrix} -1 \\ 2 \end{bmatrix}$

Exercises 11–14:

As in Example 2, rewrite the scalar linear differential equation as a system of first order linear differential equations of the form $\mathbf{Y}' = P(t)\mathbf{Y} + \mathbf{G}(t)$. Identify the matrix function $P(t)$ and the vector function $\mathbf{G}(t)$.

11. $y'' + t^2y' + 4y = \sin t$

12. $(\cos t)y'' - 3ty' + \sqrt{t}\,y = t^2 + 1$

13. $e^t y''' + 5y'' + t^{-1}y' + (\tan t)y = 1$

14. $2y'' + ty + e^{3t} = y''' + (\cos t)y'$

Exercises 15–17:

Each initial value problem was obtained from an initial value problem for a higher order scalar differential equation. What is the corresponding scalar initial value problem?

15. $\mathbf{y}' = \begin{bmatrix} 0 & 1 \\ -3 & 2 \end{bmatrix}\mathbf{y} + \begin{bmatrix} 0 \\ 2\cos 2t \end{bmatrix}, \quad \mathbf{y}(-1) = \begin{bmatrix} 1 \\ 4 \end{bmatrix}$

16. $\mathbf{y}' = \begin{bmatrix} y_2 \\ y_3 \\ -2y_1 + 4y_3 + e^{3t} \end{bmatrix}, \quad \mathbf{y}(0) = \begin{bmatrix} 1 \\ -2 \\ 3 \end{bmatrix}$

17. $\mathbf{y}' = \begin{bmatrix} y_2 \\ y_3 \\ y_4 \\ y_2 + y_3\sin(y_1) + y_3^2 \end{bmatrix}, \quad \mathbf{y}(1) = \begin{bmatrix} 0 \\ 0 \\ -1 \\ 2 \end{bmatrix}$

Exercises 18–21:

Exercises 11–17 dealt with rewriting a single scalar equation as a first order system. Frequently, however, we need to convert systems of higher order equations into a single first order system. In each exercise, rewrite the given system of two second order equations as a system of four first order linear equations of the form $\mathbf{Y}' = P(t)\mathbf{Y} + \mathbf{G}(t)$. In each exercise, use the following change of variables and identify $P(t)$ and $\mathbf{G}(t)$,

$$\mathbf{Y}(t) = \begin{bmatrix} Y_1(t) \\ Y_2(t) \\ Y_3(t) \\ Y_4(t) \end{bmatrix} = \begin{bmatrix} y(t) \\ y'(t) \\ z(t) \\ z'(t) \end{bmatrix}.$$

18. $y'' = tz' + y' + z$
$z'' = y' + z' + 2ty$

19. $y'' = t^{-1}y' + 4y - tz + (\sin t)z' + e^{2t}$
$z'' = y - 5z'$

20. $y'' = 7y' + 4y - 8z + 6z' + t^2$
$z'' = 5z' + 2z - 6y' + 3y - \sin t$

21. $15z + 9y' + 3y'' = 12y - 6z' + 3t^2$
$z' + 5y - z'' = 2z - 6y' + t$

6.3 Homogeneous Linear Systems

Our previous discussions of linear differential equations, in Chapters 2, 4, and 5, began with homogeneous equations. We use the same approach here. Consider the system of n homogeneous linear differential equations,

$$\begin{aligned} y_1' &= p_{11}(t)y_1 + p_{12}(t)y_2 + \cdots + p_{1n}(t)y_n \\ y_2' &= p_{21}(t)y_1 + p_{22}(t)y_2 + \cdots + p_{2n}(t)y_n \\ &\ \vdots \\ y_n' &= p_{n1}(t)y_1 + p_{n2}(t)y_2 + \cdots + p_{nn}(t)y_n, \qquad a < t < b. \end{aligned} \tag{1}$$

The **first order homogeneous linear system** (1) can be written in matrix form as

$$\mathbf{y}' = P(t)\mathbf{y}, \quad a < t < b, \tag{2}$$

where

$$\mathbf{y}(t) = \begin{bmatrix} y_1(t) \\ y_2(t) \\ \vdots \\ y_n(t) \end{bmatrix} \quad \text{and} \quad P(t) = \begin{bmatrix} p_{11}(t) & p_{12}(t) & \cdots & p_{1n}(t) \\ p_{21}(t) & p_{22}(t) & & p_{2n}(t) \\ \vdots & & & \vdots \\ p_{n1}(t) & p_{n2}(t) & \cdots & p_{nn}(t) \end{bmatrix}.$$

The Principle of Superposition

Our first goal is to establish a superposition principle for the solutions of equation (2). As a start, we extend the concept of "linear combination" to vector functions. In particular, let $\mathbf{f}_1(t), \mathbf{f}_2(t), \ldots, \mathbf{f}_r(t)$ be a collection of r vector functions of size $(n \times 1)$ defined on some common domain $a < t < b$,

$$\mathbf{f}_1(t) = \begin{bmatrix} f_{1,1}(t) \\ f_{2,1}(t) \\ \vdots \\ f_{n,1}(t) \end{bmatrix}, \quad \mathbf{f}_2(t) = \begin{bmatrix} f_{1,2}(t) \\ f_{2,2}(t) \\ \vdots \\ f_{n,2}(t) \end{bmatrix}, \ldots, \quad \mathbf{f}_r(t) = \begin{bmatrix} f_{1,r}(t) \\ f_{2,r}(t) \\ \vdots \\ f_{n,r}(t) \end{bmatrix}.$$

Let $c_1, c_2, \ldots, c_r$ denote r arbitrary constants. The vector function

$$\mathbf{f}(t) = c_1\mathbf{f}_1(t) + c_2\mathbf{f}_2(t) + \cdots + c_r\mathbf{f}_r(t), \qquad a < t < b$$

is called a **linear combination** of $\mathbf{f}_1(t), \mathbf{f}_2(t), \ldots, \mathbf{f}_r(t)$.

We now state a superposition theorem, Theorem 6.2, showing that a linear combination of solutions of (2) is again a solution of (2); compare Theorem 6.2 with Theorems 4.2 and 5.2.

Theorem 6.2

Let $\mathbf{y}_1(t), \mathbf{y}_2(t), \ldots, \mathbf{y}_r(t)$ be any r solutions of the homogeneous linear equation

$$\mathbf{y}' = P(t)\mathbf{y}, \qquad a < t < b.$$

Then, for any r constants $c_1, c_2, \ldots, c_r$, the linear combination

$$\mathbf{y}(t) = c_1\mathbf{y}_1(t) + c_2\mathbf{y}_2(t) + \cdots + c_r\mathbf{y}_r(t)$$

is also a solution on the t-interval $a < t < b$.

PROOF: The proof of Theorem 6.2 is conceptually the same as the proofs of Theorems 4.2 and 5.2. We simply differentiate $\mathbf{y}(t)$, obtaining

$$\mathbf{y}' = (c_1\mathbf{y}_1 + c_2\mathbf{y}_2 + \cdots c_r\mathbf{y}_r)' = c_1\mathbf{y}_1' + c_2\mathbf{y}_2' + \cdots c_r\mathbf{y}_r'. \tag{3}$$

Since $\mathbf{y}_i' = P(t)\mathbf{y}_i$ for $1 \le i \le r$, we can rewrite equation (3) as

$$\mathbf{y}' = c_1P(t)\mathbf{y}_1 + c_2P(t)\mathbf{y}_2 + \cdots + c_rP(t)\mathbf{y}_r = P(t)(c_1\mathbf{y}_1 + c_2\mathbf{y}_2 + \cdots + c_r\mathbf{y}_r) = P(t)\mathbf{y}.$$

Fundamental Sets and the General Solution

We are mainly concerned with Theorem 6.2 in the case $r = n$; that is, when the number of solutions of (2) is equal to the size of the square matrix $P(t)$. In this context, we again introduce the concept of a fundamental set of solutions.

Let $\{\mathbf{y}_1(t), \mathbf{y}_2(t), \ldots, \mathbf{y}_n(t)\}$ be a set of n solutions of the linear system (2). This set of solutions is called a **fundamental set of solutions** if every solution of (2) can be written as a linear combination of the form $c_1\mathbf{y}_1(t) + c_2\mathbf{y}_2(t) + \cdots + c_n\mathbf{y}_n(t)$ for some choice of constants $c_1, c_2, \ldots, c_n$. If $\{\mathbf{y}_1(t), \mathbf{y}_2(t), \ldots, \mathbf{y}_n(t)\}$ is a fundamental set of solutions of $\mathbf{y}' = P(t)\mathbf{y}$, then the expression

$$\mathbf{y}(t) = c_1\mathbf{y}_1(t) + c_2\mathbf{y}_2(t) + \cdots + c_n\mathbf{y}_n(t)$$

is called the *general solution*.

EXAMPLE 1

Consider the first order linear homogeneous system $\mathbf{y}' = A\mathbf{y}$, $-\infty < t < \infty$, where

$$A = \begin{bmatrix} 0 & 1 & 0 \\ -6 & 5 & 0 \\ 0 & 0 & 1 \end{bmatrix}.$$

We can verify by direct substitution that each of the vector functions

$$\mathbf{y}_1(t) = \begin{bmatrix} e^{2t} \\ 2e^{2t} \\ 0 \end{bmatrix}, \qquad \mathbf{y}_2(t) = \begin{bmatrix} e^{3t} \\ 3e^{3t} \\ 0 \end{bmatrix}, \qquad \mathbf{y}_3(t) = \begin{bmatrix} 0 \\ 0 \\ e^{t} \end{bmatrix},$$

is a solution of $\mathbf{y}' = A\mathbf{y}$. For example, to verify that $\mathbf{y}_1(t)$ is a solution, we calculate

$$\mathbf{y}_1'(t) = \begin{bmatrix} 2e^{2t} \\ 4e^{2t} \\ 0 \end{bmatrix} \quad \text{and} \quad A\mathbf{y}_1(t) = \begin{bmatrix} 0 & 1 & 0 \\ -6 & 5 & 0 \\ 0 & 0 & 1 \end{bmatrix} \begin{bmatrix} e^{2t} \\ 2e^{2t} \\ 0 \end{bmatrix} = \begin{bmatrix} 2e^{2t} \\ 4e^{2t} \\ 0 \end{bmatrix}.$$

From Theorem 6.2, it follows that the linear combination

$$\mathbf{y}(t) = c_1\mathbf{y}_1(t) + c_2\mathbf{y}_2(t) + c_3\mathbf{y}_3(t) \tag{4}$$

is also a solution for any choice of constants c_1, c_2, c_3. Is (4) the general solution of $\mathbf{y}' = A\mathbf{y}$? We will show in Example 4 that the answer is "Yes." Also, in Section 6.5, we show how to obtain these three solutions $\mathbf{y}_1(t)$, $\mathbf{y}_2(t)$, and $\mathbf{y}_3(t)$. ▲

An Important Identity

Before continuing, we take note of a simple but important matrix identity known as the **column form for matrix-vector multiplication**. (See the appendix on matrix theory.) Let A be an $(m \times n)$ matrix and let $\mathbf{x}$ be an $(n \times 1)$ vector. Let us represent the matrix A in **column form** as

$$A = [\mathbf{A}_1, \mathbf{A}_2, \ldots, \mathbf{A}_n].$$

[In this column form representation, the terms $\mathbf{A}_1, \mathbf{A}_2, \ldots, \mathbf{A}_n$ are the columns of A. Note that each $\mathbf{A}_i$ is an $(m \times 1)$ vector.] Finally, let the vector $\mathbf{x}$ be given by

$$\mathbf{x} = \begin{bmatrix} x_1 \\ x_2 \\ \vdots \\ x_n \end{bmatrix}.$$

Then the matrix-vector product $A\mathbf{x}$ is equal to the linear combination

$$A\mathbf{x} = x_1\mathbf{A}_1 + x_2\mathbf{A}_2 + \cdots + x_n\mathbf{A}_n. \tag{5}$$

For instance, consider the linear combination (4) in Example 1, $\mathbf{y}(t) = c_1\mathbf{y}_1(t) + c_2\mathbf{y}_2(t) + c_3\mathbf{y}_3(t)$. Using identity (5), we can rewrite this linear combination in the form $\mathbf{y}(t) = \Psi(t)\mathbf{c}$ where

$$\Psi(t) = [\mathbf{y}_1(t), \mathbf{y}_2(t), \mathbf{y}_3(t)] = \begin{bmatrix} e^{2t} & e^{3t} & 0 \\ 2e^{2t} & 3e^{3t} & 0 \\ 0 & 0 & e^t \end{bmatrix} \quad \text{and} \quad \mathbf{c} = \begin{bmatrix} c_1 \\ c_2 \\ c_3 \end{bmatrix}. \ \blacktriangle$$

The Wronskian

Let $\mathbf{y}_1(t), \mathbf{y}_2(t), \ldots, \mathbf{y}_n(t)$ be a set of n solutions of $\mathbf{y}' = P(t)\mathbf{y}$, where

$$\mathbf{y}_1(t) = \begin{bmatrix} y_{1,1}(t) \\ y_{2,1}(t) \\ \vdots \\ y_{n,1}(t) \end{bmatrix}, \quad \mathbf{y}_2(t) = \begin{bmatrix} y_{1,2}(t) \\ y_{2,2}(t) \\ \vdots \\ y_{n,2}(t) \end{bmatrix}, \quad \ldots, \quad \mathbf{y}_n(t) = \begin{bmatrix} y_{1,n}(t) \\ y_{2,n}(t) \\ \vdots \\ y_{n,n}(t) \end{bmatrix}.$$

We can form an $(n \times n)$ matrix $\Psi(t)$ using these n solutions as columns, obtaining the matrix $\Psi(t) = [\mathbf{y}_1(t), \mathbf{y}_2(t), \ldots, \mathbf{y}_n(t)]$. In terms of its components,

$$\Psi(t) = \begin{bmatrix} y_{1,1}(t) & y_{1,2}(t) & \cdots & y_{1,n}(t) \\ y_{2,1}(t) & y_{2,2}(t) & & y_{2,n}(t) \\ \vdots & & & \vdots \\ y_{n,1}(t) & y_{n,2}(t) & \cdots & y_{n,n}(t) \end{bmatrix}. \tag{6}$$

The determinant of the matrix $\Psi(t)$ is defined to be the Wronskian, $W(t)$, of $\{\mathbf{y}_1(t), \mathbf{y}_2(t), \ldots, \mathbf{y}_n(t)\}$,

$$W(t) = \det[\Psi(t)].$$

EXAMPLE 2

Recall the linear system $\mathbf{y}' = A\mathbf{y}$ treated in Example 1. As given in Example 1, the solutions were $\mathbf{y}_1(t)$, $\mathbf{y}_2(t)$, $\mathbf{y}_3(t)$. Calculate the Wronskian, $W(t)$, for these solutions.

(continued)

(continued)

Solution: The solutions $\mathbf{y}_1(t)$, $\mathbf{y}_2(t)$, $\mathbf{y}_3(t)$ given in Example 1 are

$$\mathbf{y}_1(t) = \begin{bmatrix} e^{2t} \\ 2e^{2t} \\ 0 \end{bmatrix}, \qquad \mathbf{y}_2(t) = \begin{bmatrix} e^{3t} \\ 3e^{3t} \\ 0 \end{bmatrix}, \qquad \mathbf{y}_3(t) = \begin{bmatrix} 0 \\ 0 \\ e^{t} \end{bmatrix}.$$

The corresponding matrix $\Psi(t)$ is

$$\Psi(t) = \begin{bmatrix} e^{2t} & e^{3t} & 0 \\ 2e^{2t} & 3e^{3t} & 0 \\ 0 & 0 & e^{t} \end{bmatrix}.$$

Using a cofactor expansion along the third row (or the third column):

$$W(t) = \det[\Psi(t)] = (3e^{5t} - 2e^{5t})(e^{t}) = e^{6t}. \ ▲$$

Using the Wronskian to Identify Fundamental Sets

Theorem 6.3 shows that a set of n solutions is a fundamental set if there is a point t_0 within the t-interval of interest where the Wronskian is nonzero. Compare Theorem 6.3 with Theorems 4.3 and 5.3.

Theorem 6.3

Let $\{\mathbf{y}_1(t), \mathbf{y}_2(t), \ldots, \mathbf{y}_n(t)\}$ be a set of n solutions of

$$\mathbf{y}' = P(t)\mathbf{y}, \qquad a < t < b,$$

where the matrix function $P(t)$ is continuous on (a, b). Let $W(t)$ denote the Wronskian of this set of solutions. If there is a point t_0 in $a < t < b$ such that $W(t_0) \neq 0$, then $\{\mathbf{y}_1(t), \mathbf{y}_2(t), \ldots, \mathbf{y}_n(t)\}$ is a fundamental set of solutions.

PROOF: Let $\mathbf{u}(t)$ be any solution of $\mathbf{y}' = P(t)\mathbf{y}$, $a < t < b$. We must show there are constants $c_1, c_2, \ldots, c_n$ such that

$$\mathbf{u}(t) = c_1\mathbf{y}_1(t) + c_2\mathbf{y}_2(t) + \cdots + c_n\mathbf{y}_n(t), \qquad a < t < b. \tag{7}$$

By (5), the linear combination $c_1\mathbf{y}_1(t) + c_2\mathbf{y}_2(t) + \cdots + c_n\mathbf{y}_n(t)$ can be rewritten as

$$c_1\mathbf{y}_1(t) + c_2\mathbf{y}_2(t) + \cdots + c_n\mathbf{y}_n(t) = \begin{bmatrix} y_{1,1}(t) & y_{1,2}(t) & \cdots & y_{1,n}(t) \\ y_{2,1}(t) & y_{2,2}(t) & & y_{2,n}(t) \\ \vdots & & & \vdots \\ y_{n,1}(t) & y_{n,2}(t) & \cdots & y_{n,n}(t) \end{bmatrix} \begin{bmatrix} c_1 \\ c_2 \\ \vdots \\ c_n \end{bmatrix} \tag{8}$$

$$= \Psi(t)\mathbf{c}.$$

The problem of showing $\mathbf{u}(t)$ can be expressed as a linear combination of the form (7) reduces to showing we can find a vector $\mathbf{c}$ such that

$$\mathbf{u}(t) = \Psi(t)\mathbf{c}, \qquad a < t < b.$$

Note that the determinant of $\Psi(t)$ is the Wronskian, $W(t)$, and we have assumed that $W(t_0) \neq 0$. Therefore, $\Psi^{-1}(t_0)$ exists. Let us solve the matrix equation $\mathbf{u}(t_0) = \Psi(t_0)\mathbf{c}$ for $\mathbf{c} = [c_1, c_2, \ldots, c_n]^T$ and then define the vector function $\overline{\mathbf{y}}(t)$ by

$$\overline{\mathbf{y}}(t) = c_1\mathbf{y}_1(t) + c_2\mathbf{y}_2(t) + \cdots + c_n\mathbf{y}_n(t),$$

using the components of $\mathbf{c}$. Since $\overline{\mathbf{y}}(t)$ is a superposition of solutions, it follows that $\overline{\mathbf{y}}(t)$ is also a solution of $\mathbf{y}' = P(t)\mathbf{y}$.

Therefore, $\overline{\mathbf{y}}(t)$ and $\mathbf{u}(t)$ are two solutions of $\mathbf{y}' = P(t)\mathbf{y}, a < t < b$, that satisfy the same initial condition at t_0 [by construction we have $\overline{\mathbf{y}}(t_0) = \mathbf{u}(t_0)$]. However, Theorem 6.1 states that there is one and only one solution and therefore

$$\mathbf{u}(t) = \overline{\mathbf{y}}(t) = c_1\mathbf{y}_1(t) + c_2\mathbf{y}_2(t) + \cdots + c_n\mathbf{y}_n(t), \qquad a < t < b.$$

This identity establishes Theorem 6.3. ▮

EXAMPLE 3

Example 1 listed three solutions, $\mathbf{y}_1(t), \mathbf{y}_2(t)$, and $\mathbf{y}_3(t)$, of a linear system $\mathbf{y}' = A\mathbf{y}$. By Theorem 6.3 and Example 2, $\{\mathbf{y}_1(t), \mathbf{y}_2(t), \mathbf{y}_3(t)\}$ is a fundamental set of solutions for $\mathbf{y}' = A\mathbf{y}$. Use this fact to solve the initial value problem $\mathbf{y}' = A\mathbf{y}$, $\mathbf{y}(0) = \mathbf{y}_0$, where

$$\mathbf{y}_0 = \begin{bmatrix} 3 \\ 7 \\ 4 \end{bmatrix}.$$

Solution: Since the given functions form a fundamental set, the general solution is

$$\mathbf{y}(t) = c_1\mathbf{y}_1(t) + c_2\mathbf{y}_2(t) + c_3\mathbf{y}_3(t) = \Psi(t)\mathbf{c}$$

where

$$\Psi(t) = \begin{bmatrix} e^{2t} & e^{3t} & 0 \\ 2e^{2t} & 3e^{3t} & 0 \\ 0 & 0 & e^{t} \end{bmatrix}.$$

Imposing the initial condition, $\mathbf{y}(0) = \Psi(0)\mathbf{c} = \mathbf{y}_0$, we obtain

$$\begin{bmatrix} 1 & 1 & 0 \\ 2 & 3 & 0 \\ 0 & 0 & 1 \end{bmatrix} \begin{bmatrix} c_1 \\ c_2 \\ c_3 \end{bmatrix} = \begin{bmatrix} 3 \\ 7 \\ 4 \end{bmatrix}.$$

Therefore, the vector $\mathbf{c}$ is given by

$$\mathbf{c} = \Psi(0)^{-1}\mathbf{y}_0 = \begin{bmatrix} 3 & -1 & 0 \\ -2 & 1 & 0 \\ 0 & 0 & 1 \end{bmatrix} \begin{bmatrix} 3 \\ 7 \\ 4 \end{bmatrix} = \begin{bmatrix} 2 \\ 1 \\ 4 \end{bmatrix},$$

and the solution of the initial value problem is

$$\mathbf{y}(t) = \Psi(t)\mathbf{c} = \begin{bmatrix} 2e^{2t} + e^{3t} \\ 4e^{2t} + 3e^{3t} \\ 4e^{t} \end{bmatrix}. \quad \blacktriangle$$

Abel's Theorem

In the next theorem, we present (without proof) a final generalization of Abel's theorem. This generalization states, as did the previous versions in Chapters 4 and 5, that the Wronskian of a set of solutions either vanishes nowhere or it vanishes everywhere on the t-interval of interest.

Before stating this latest generalization of Abel's theorem, we need to define a new quantity. Let A be an $(n \times n)$ matrix,

$$A = \begin{bmatrix} a_{11} & a_{12} & \cdots & a_{1n} \\ a_{21} & a_{22} & & a_{2n} \\ \vdots & & & \vdots \\ a_{n1} & a_{n2} & \cdots & a_{nn} \end{bmatrix}.$$

The **trace of A**, denoted by tr[A], is defined to be the sum of the diagonal elements of A,

$$\text{tr}[A] = a_{11} + a_{22} + a_{33} + \cdots + a_{nn}.$$

For example, the (3×3) matrix A from Example 1,

$$A = \begin{bmatrix} 0 & 1 & 0 \\ -6 & 5 & 0 \\ 0 & 0 & 1 \end{bmatrix},$$

has tr[A] = 6. For the matrix function $P(t)$ in equation (2),

$$\text{tr}[P(t)] = p_{11}(t) + p_{22}(t) + p_{33}(t) + \cdots + p_{nn}(t).$$

Having this preliminary definition, we can state Abel's theorem.

Theorem 6.4

Let $\{\mathbf{y}_1(t), \mathbf{y}_2(t), \ldots, \mathbf{y}_n(t)\}$ be a set of solutions of

$$\mathbf{y}' = P(t)\mathbf{y}, \qquad a < t < b,$$

and let $W(t)$ be the Wronskian of these solutions. Then, $W(t)$ satisfies the scalar differential equation

$$W'(t) = \text{tr}[P(t)]W(t).$$

Moreover, if t_0 is any point in $a < t < b$, then

$$W(t) = W(t_0)e^{\int_{t_0}^{t} \text{tr}[P(s)]\,ds}, \qquad a < t < b. \tag{9}$$

As we see from equation (9), if $W(t_0) = 0$, then the Wronskian vanishes identically on the t-interval of interest. On the other hand, if $W(t_0) \neq 0$, then the Wronskian is never zero in (a, b).

In the special case where an nth order linear scalar differential equation is recast as a first order linear system, the definition of the Wronskian and the conclusion of Abel's theorem stated for systems reduce precisely to their counterparts in Chapters 4 and 5. In particular, if the scalar equation is

$$y^{(n)} + p_{n-1}(t)y^{(n-1)} + \cdots + p_1(t)y' + p_0(t)y = 0,$$

then the corresponding first order linear homogeneous system is $\mathbf{Y}' = P(t)\mathbf{Y}$, where [see equations (5) and (6) in Section 6.2],

$$\mathbf{Y}(t) = \begin{bmatrix} Y_1(t) \\ Y_2(t) \\ \vdots \\ Y_n(t) \end{bmatrix} = \begin{bmatrix} y(t) \\ y'(t) \\ \vdots \\ y^{(n-1)}(t) \end{bmatrix} \tag{10}$$

and

$$P(t) = \begin{bmatrix} 0 & 1 & 0 & \cdots & 0 \\ 0 & 0 & 1 & & 0 \\ \vdots & & & & \vdots \\ 0 & 0 & 0 & & 1 \\ -p_0(t) & -p_1(t) & -p_2(t) & \cdots & -p_{n-1}(t) \end{bmatrix}. \tag{11}$$

As we see from equation (10), each solution vector $\mathbf{Y}_i(t)$ has a solution of the scalar problem, $Y_{1,i}(t) = y(t)$, as its first component while successive derivatives of this solution form the remaining components, $Y_{2,i}(t), Y_{3,i}(t), \ldots Y_{n,i}(t)$. Therefore, the Wronskian determinant has the same structure as that defined in Chapters 4 and 5.

Finally, we see from (11) that $\text{tr}[P(t)] = -p_{n-1}(t)$. Therefore, equation (9) in Theorem 6.4 reduces precisely to equation (6) in Theorem 5.4 (and to a similar equation in Theorem 4.4 when $n = 2$).

EXERCISES

Exercises 1–6:

(a) Rewrite the given system of linear homogeneous differential equations as a homogeneous linear system of the form $\mathbf{y}' = P(t)\mathbf{y}$.

(b) Verify that the given function $\mathbf{y}(t)$ is a solution of $\mathbf{y}' = P(t)\mathbf{y}$.

1. $\begin{aligned} y_1' &= 9y_1 - 4y_2 \\ y_2' &= 15y_1 - 7y_2 \end{aligned}$ $\quad \mathbf{y}(t) = \begin{bmatrix} 2e^{3t} \\ 3e^{3t} \end{bmatrix}$

2. $\begin{aligned} y_1' &= -3y_1 - 2y_2 \\ y_2' &= 4y_1 + 3y_2 \end{aligned}$ $\quad \mathbf{y}(t) = \begin{bmatrix} e^t + e^{-t} \\ -2e^t - e^{-t} \end{bmatrix}$

3. $\begin{aligned} y_1' &= y_1 + 4y_2 \\ y_2' &= -y_1 + y_2 \end{aligned}$ $\quad \mathbf{y}(t) = \begin{bmatrix} 2e^t \cos 2t \\ -e^t \sin 2t \end{bmatrix}$

4. $\begin{aligned} y_1' &= y_2 \\ y_2' &= \frac{2}{t^2}y_1 - \frac{2}{t}y_2 \end{aligned}$, $\quad t > 0, \quad \mathbf{y}(t) = \begin{bmatrix} -t^2 + 3t \\ -2t + 3 \end{bmatrix}$

5. $\begin{aligned} y_1' &= y_2 + y_3 \\ y_2' &= -6y_1 - 3y_2 + y_3 \\ y_3' &= -8y_1 - 2y_2 + 4y_3 \end{aligned}$ $\quad \mathbf{y}(t) = \begin{bmatrix} e^t \\ -e^t \\ 2e^t \end{bmatrix}$

6. $\begin{aligned} y_1' &= 2y_1 + y_2 + y_3 \\ y_2' &= y_1 + y_2 + 2y_3 \\ y_3' &= y_1 + 2y_2 + y_3 \end{aligned}$ $\quad \mathbf{y}(t) = \begin{bmatrix} 2e^t + e^{4t} \\ -e^t + e^{4t} \\ -e^t + e^{4t} \end{bmatrix}$

Exercises 7–15:

(a) Verify the given functions are solutions of the homogeneous linear system.

(b) Compute the Wronskian of the solution set. On the basis of this calculation, can you assert that the set of solutions forms a fundamental set?

(c) If the given solutions are shown in part (b) to form a fundamental set, state the general solution of the linear homogeneous system. Express the general solution as the product $\mathbf{y}(t) = \Psi(t)\mathbf{c}$, where $\Psi(t)$ is a square matrix whose columns are the solutions forming the fundamental set and $\mathbf{c}$ is a column vector of arbitrary constants.

(d) If the solutions are shown in part (b) to form a fundamental set, impose the given initial condition and find the unique solution of the initial value problem.

7. $\mathbf{y}' = \begin{bmatrix} 9 & -4 \\ 15 & -7 \end{bmatrix}\mathbf{y}, \quad \mathbf{y}(0) = \begin{bmatrix} 1 \\ 1 \end{bmatrix}; \quad \mathbf{y}_1(t) = \begin{bmatrix} 2e^{3t} \\ 3e^{3t} \end{bmatrix}, \quad \mathbf{y}_2(t) = \begin{bmatrix} 2e^{-t} \\ 5e^{-t} \end{bmatrix}$

8. $\mathbf{y}' = \begin{bmatrix} 9 & -4 \\ 15 & -7 \end{bmatrix}\mathbf{y}, \quad \mathbf{y}(0) = \begin{bmatrix} 0 \\ 1 \end{bmatrix}; \quad \mathbf{y}_1(t) = \begin{bmatrix} 2e^{3t} - 4e^{-t} \\ 3e^{3t} - 10e^{-t} \end{bmatrix}, \quad \mathbf{y}_2(t) = \begin{bmatrix} 4e^{3t} + 2e^{-t} \\ 6e^{3t} + 5e^{-t} \end{bmatrix}$

9. $\mathbf{y}' = \begin{bmatrix} 3 & 2 \\ -4 & -3 \end{bmatrix}\mathbf{y}, \quad \mathbf{y}(0) = \begin{bmatrix} 1 \\ 1 \end{bmatrix}; \quad \mathbf{y}_1(t) = \begin{bmatrix} e^{-t} \\ -2e^{-t} \end{bmatrix}, \quad \mathbf{y}_2(t) = \begin{bmatrix} -3e^{-t} \\ 6e^{-t} \end{bmatrix}$

10. $\mathbf{y}' = \begin{bmatrix} -3 & -5 \\ 2 & -1 \end{bmatrix}\mathbf{y}, \quad \mathbf{y}(0) = \begin{bmatrix} 5 \\ 2 \end{bmatrix}; \quad \mathbf{y}_1(t) = \begin{bmatrix} -5e^{-2t}\cos 3t \\ e^{-2t}(\cos 3t - 3\sin 3t) \end{bmatrix},$

$\mathbf{y}_2(t) = \begin{bmatrix} -5e^{-2t}\sin 3t \\ e^{-2t}(3\cos 3t + \sin 3t) \end{bmatrix}$

11. $\mathbf{y}' = \begin{bmatrix} -3 & -2 \\ 4 & 3 \end{bmatrix}\mathbf{y}, \quad \mathbf{y}(1) = \begin{bmatrix} 1 \\ -3 \end{bmatrix}; \quad \mathbf{y}_1(t) = \begin{bmatrix} e^{t} \\ -2e^{t} \end{bmatrix}, \quad \mathbf{y}_2(t) = \begin{bmatrix} e^{-t} \\ -e^{-t} \end{bmatrix}$

12. $\mathbf{y}' = \begin{bmatrix} 1 & -1 \\ -2 & 2 \end{bmatrix}\mathbf{y}, \quad \mathbf{y}(-1) = \begin{bmatrix} -2 \\ 4 \end{bmatrix}; \quad \mathbf{y}_1(t) = \begin{bmatrix} 1 \\ 1 \end{bmatrix}, \quad \mathbf{y}_2(t) = \begin{bmatrix} e^{3t} \\ -2e^{3t} \end{bmatrix}$

13. $\mathbf{y}' = \begin{bmatrix} 2t^{-2} & 1 - 2t^{-1} + 2t^{-2} \\ -2t^{-2} & 2t^{-1} - 2t^{-2} \end{bmatrix}\mathbf{y}, \quad \mathbf{y}(2) = \begin{bmatrix} -2 \\ 2 \end{bmatrix}, \quad t > 0; \quad \mathbf{y}_1(t) = \begin{bmatrix} t^2 - 2t \\ 2t \end{bmatrix},$

$\mathbf{y}_2(t) = \begin{bmatrix} t - 1 \\ 1 \end{bmatrix}$

14. $\mathbf{y}' = \begin{bmatrix} -2 & 0 & 0 \\ 0 & 1 & 4 \\ 0 & -1 & 1 \end{bmatrix}\mathbf{y}, \quad \mathbf{y}(0) = \begin{bmatrix} 3 \\ 4 \\ -2 \end{bmatrix}; \quad \mathbf{y}_1(t) = \begin{bmatrix} e^{-2t} \\ 0 \\ 0 \end{bmatrix}, \quad \mathbf{y}_2(t) = \begin{bmatrix} 0 \\ 2e^{t}\cos 2t \\ -e^{t}\sin 2t \end{bmatrix},$

$\mathbf{y}_3(t) = \begin{bmatrix} 0 \\ 2e^{t}\sin 2t \\ e^{t}\cos 2t \end{bmatrix}$

15. $\mathbf{y}' = \begin{bmatrix} -21 & -10 & 2 \\ 22 & 11 & -2 \\ -110 & -50 & 11 \end{bmatrix}\mathbf{y}, \quad \mathbf{y}(0) = \begin{bmatrix} 3 \\ -10 \\ -16 \end{bmatrix}; \quad \mathbf{y}_1(t) = \begin{bmatrix} 5e^{t} \\ -11e^{t} \\ 0 \end{bmatrix}, \quad \mathbf{y}_2(t) = \begin{bmatrix} e^{t} \\ 0 \\ 11e^{t} \end{bmatrix},$

$\mathbf{y}_3(t) = \begin{bmatrix} e^{-t} \\ -e^{-t} \\ 5e^{-t} \end{bmatrix}$

Exercises 16–19:

The given functions are solutions of the homogeneous linear system.

(a) Compute the Wronskian of the solution set and verify the solution set is a fundamental set of solutions.

(b) Compute the trace of the coefficient matrix.

(c) Verify Abel's theorem by showing that, for the given point t_0, $W(t) = W(t_0)e^{\int_{t_0}^{t} \mathrm{tr}[P(s)]\,ds}$.

16. $\mathbf{y}' = \begin{bmatrix} 6 & 5 \\ -7 & -6 \end{bmatrix}\mathbf{y},\ \mathbf{y}_1(t) = \begin{bmatrix} 5e^{-t} \\ -7e^{-t} \end{bmatrix},\ \mathbf{y}_2(t) = \begin{bmatrix} e^{t} \\ -e^{t} \end{bmatrix},\ t_0 = -1, \qquad -\infty < t < \infty$

17. $\mathbf{y}' = \begin{bmatrix} 9 & 5 \\ -7 & -3 \end{bmatrix}\mathbf{y},\ \mathbf{y}_1(t) = \begin{bmatrix} 5e^{2t} \\ -7e^{2t} \end{bmatrix},\ \mathbf{y}_2(t) = \begin{bmatrix} e^{4t} \\ -e^{4t} \end{bmatrix},\ t_0 = 0, \qquad -\infty < t < \infty$

18. $\mathbf{y}' = \begin{bmatrix} 1 & t \\ 0 & -t^{-1} \end{bmatrix}\mathbf{y},\ t \neq 0,\ \mathbf{y}_1(t) = \begin{bmatrix} -1 \\ t^{-1} \end{bmatrix},\ \mathbf{y}_2(t) = \begin{bmatrix} e^{t} \\ 0 \end{bmatrix},\ t_0 = 1, \qquad 0 < t < \infty$

19. $\mathbf{y}' = \begin{bmatrix} 2 & 1 & 1 \\ 1 & 1 & 2 \\ 1 & 2 & 1 \end{bmatrix}\mathbf{y},\ \mathbf{y}_1(t) = \begin{bmatrix} 2e^{t} \\ -e^{t} \\ -e^{t} \end{bmatrix},\ \mathbf{y}_2(t) = \begin{bmatrix} 0 \\ -e^{-t} \\ e^{-t} \end{bmatrix},\ \mathbf{y}_3(t) = \begin{bmatrix} e^{4t} \\ e^{4t} \\ e^{4t} \end{bmatrix},$

$t_0 = 0,\ -\infty < t < \infty$

20. The functions

$$\mathbf{y}_1(t) = \begin{bmatrix} 5 \\ 1 \end{bmatrix} \quad \text{and} \quad \mathbf{y}_2(t) = \begin{bmatrix} 2e^{3t} \\ e^{3t} \end{bmatrix}$$

are known to be solutions of the homogeneous linear system $\mathbf{y}' = A\mathbf{y}$, where A is a real (2×2) constant matrix.

(a) Verify that the two solutions form a fundamental set of solutions.

(b) What is tr[A]? [Hint: Use Abel's theorem and the fundamental theorem of calculus.]

(c) Show that $\Psi(t)$ satisfies the homogeneous differential equation $\Psi' = A\Psi$, where

$$\Psi(t) = [\mathbf{y}_1(t), \mathbf{y}_2(t)] = \begin{bmatrix} 5 & 2e^{3t} \\ 1 & e^{3t} \end{bmatrix}.$$

(d) Use the observation of part (c) to determine the matrix A. [Hint: Compute the matrix product $\Psi'(t)\Psi^{-1}(t)$. It follows from part (a) that $\Psi^{-1}(t)$ exists.] Are the results of parts (b) and (d) consistent?

21. The homogeneous linear system

$$\mathbf{y}' = \begin{bmatrix} 3 & 1 \\ -2 & \alpha \end{bmatrix}\mathbf{y}$$

has a fundamental set of solutions whose Wronskian is constant, $W(t) = 4$, $-\infty < t < \infty$. What is the value α?

6.4 Fundamental Sets and Linear Independence

The results presented in this section for first order homogeneous linear systems (and the arguments used to establish these results) closely parallel analogous results and arguments presented in Sections 4.3 and 5.3 for nth order linear homogeneous scalar differential equations.

Fundamental Sets of Solutions Always Exist

We first show, as in Theorems 4.5 and 5.5, that fundamental sets of solutions always exist.

Theorem 6.5

Consider the first order homogeneous linear system

$$\mathbf{y}' = P(t)\mathbf{y}, \qquad a < t < b,$$

where the $(n \times n)$ matrix function $P(t)$ is continuous on $a < t < b$. A fundamental set of solutions exists for this differential equation.

PROOF: Choose some point t_0 lying in (a, b). Let the vectors $\mathbf{e}_j$, $1 \le j \le n$, denote the columns of the $(n \times n)$ identity matrix I. Thus, in column form, $I = [\mathbf{e}_1, \mathbf{e}_2, \ldots, \mathbf{e}_n]$. Equivalently,

$$\mathbf{e}_1 = \begin{bmatrix} 1 \\ 0 \\ 0 \\ \vdots \\ 0 \end{bmatrix}, \quad \mathbf{e}_2 = \begin{bmatrix} 0 \\ 1 \\ 0 \\ \vdots \\ 0 \end{bmatrix}, \quad \ldots, \quad \mathbf{e}_n = \begin{bmatrix} 0 \\ 0 \\ 0 \\ \vdots \\ 1 \end{bmatrix}.$$

According to Theorem 6.1, each of the following initial value problems has a unique solution on the interval $a < t < b$:

$$\mathbf{y}_j' = P(t)\mathbf{y}_j, \qquad \mathbf{y}_j(t_0) = \mathbf{e}_j, \qquad j = 1, \ldots, n. \tag{1}$$

Let $\Psi(t) = [\mathbf{y}_1(t), \mathbf{y}_2(t), \ldots, \mathbf{y}_n(t)]$ be the matrix of solutions. Because of the initial conditions prescribed in (1), we see that $\Psi(t_0) = I$. Therefore, the Wronskian, $W(t) = \det[\Psi(t)]$, is nonzero at t_0. In fact,

$$W(t_0) = \det[\Psi(t_0)] = \det[I] = 1.$$

By Theorem 6.3, it follows that $\{\mathbf{y}_1(t), \mathbf{y}_2(t), \ldots, \mathbf{y}_n(t)\}$ is a fundamental set of solutions. ❚

The Wronskian of a Fundamental Set Is Never Zero

We now show, as in Theorems 4.6 and 5.6, that the Wronskian of a fundamental set of solutions is never zero.

Theorem 6.6

Let $\{\mathbf{y}_1(t), \mathbf{y}_2(t), \ldots, \mathbf{y}_n(t)\}$ be a fundamental set of solutions of

$$\mathbf{y}' = P(t)\mathbf{y}, \qquad a < t < b,$$

where the $(n \times n)$ matrix function $P(t)$ is continuous on the interval (a, b). Let $W(t)$ denote the Wronskian of the solution set and let t_0 be any point in the interval (a, b). Then $W(t_0)$ is nonzero.

PROOF: Let $\Psi(t) = [\mathbf{y}_1(t), \mathbf{y}_2(t), \ldots, \mathbf{y}_n(t)]$. The general solution of $\mathbf{y}' = P(t)\mathbf{y}$ can be written as

$$\begin{aligned}\mathbf{y}(t) &= c_1\mathbf{y}_1(t) + c_2\mathbf{y}_2(t) + \cdots + c_n\mathbf{y}_n(t)\\ &= \Psi(t)\mathbf{c},\end{aligned} \tag{2}$$

where

$$\mathbf{c} = \begin{bmatrix} c_1 \\ c_2 \\ \vdots \\ c_n \end{bmatrix}.$$

Expression (2) is the general solution and so, for any point t_0 in (a, b) and for any $(n \times 1)$ vector $\mathbf{y}_0$, the matrix equation

$$\Psi(t_0)\mathbf{c} = \mathbf{y}_0$$

has a unique solution for $\mathbf{c}$. We now use a result from matrix theory that states this system of equations has a unique solution for all possible right-hand sides if and only if

$$\det[\Psi(t_0)] = W(t_0) \neq 0.$$

Combining Theorems 6.3 and 6.6, we obtain the following corollary, which characterizes a fundamental set of solutions.

Corollary

Let $\{\mathbf{y}_1(t), \mathbf{y}_2(t), \ldots, \mathbf{y}_n(t)\}$ be a set of solutions of

$$\mathbf{y}' = P(t)\mathbf{y}, \qquad a < t < b,$$

where the $(n \times n)$ matrix function $P(t)$ is continuous on the interval $a < t < b$. Let $W(t)$ denote the corresponding Wronskian for this set of solutions. Then $\{\mathbf{y}_1(t), \mathbf{y}_2(t), \ldots, \mathbf{y}_n(t)\}$ is a fundamental set of solutions if and only if $W(t)$ is never zero in (a, b).

Fundamental Sets of Solutions Are Linearly Independent Sets of Functions

We next extend the definition of linear dependence and independence to vector functions and show that a fundamental set of solutions is a linearly independent set of functions on the t-interval of existence. The proof is left to Exercise 19.

Let $\mathbf{f}_1(t), \mathbf{f}_2(t), \ldots, \mathbf{f}_r(t)$ be a collection of $(n \times 1)$ vector functions defined on $a < t < b$. The set of functions $\{\mathbf{f}_1, \mathbf{f}_2, \ldots, \mathbf{f}_r\}$ is said to be a **linearly dependent set of functions on (a, b)** if there exist constants $k_1, k_2, \ldots, k_r$, not all zero, such that

$$k_1\mathbf{f}_1(t) + k_2\mathbf{f}_2(t) + \cdots + k_r\mathbf{f}_r(t) = \mathbf{0}, \qquad a < t < b.$$

If the set of functions $\{\mathbf{f}_1, \mathbf{f}_2, \ldots, \mathbf{f}_r\}$ is not linearly dependent, then it said to be a **linearly independent set of functions on (a, b)**.

Theorem 6.7

Let $\{\mathbf{y}_1(t), \mathbf{y}_2(t), \ldots, \mathbf{y}_n(t)\}$ be a fundamental set of solutions of

$$\mathbf{y}' = P(t)\mathbf{y}, \qquad a < t < b,$$

where the $(n \times n)$ matrix function $P(t)$ is continuous on the interval $a < t < b$. Then $\{\mathbf{y}_1(t), \mathbf{y}_2(t), \ldots, \mathbf{y}_n(t)\}$ is a linearly independent set of functions on (a, b).

How Fundamental Sets of Solutions Are Related

Even though fundamental sets of solutions are not unique, they are related to one another. Theorem 6.8 describes the relationship between different fundamental sets of solutions; as such, Theorem 6.8, which we present without proof, is similar to Theorems 4.8 and 5.8.

Theorem 6.8

Let $\{\mathbf{y}_1(t), \mathbf{y}_2(t), \ldots, \mathbf{y}_n(t)\}$ be a fundamental set of solutions of

$$\mathbf{y}' = P(t)\mathbf{y}, \qquad a < t < b,$$

where the $(n \times n)$ matrix function $P(t)$ is continuous on the interval (a, b). Let

$$\Psi(t) = [\mathbf{y}_1(t), \mathbf{y}_2(t), \ldots, \mathbf{y}_n(t)]$$

denote the $(n \times n)$ matrix function formed from the fundamental set. Let $\{\hat{\mathbf{y}}_1(t), \hat{\mathbf{y}}_2(t), \ldots, \hat{\mathbf{y}}_n(t)\}$ be any other set of n solutions of the differential equation, and let

$$\hat{\Psi}(t) = [\hat{\mathbf{y}}_1(t), \hat{\mathbf{y}}_2(t), \ldots, \hat{\mathbf{y}}_n(t)]$$

denote the $(n \times n)$ matrix formed from this other set of solutions. Then

(a) There is a unique $(n \times n)$ constant matrix C such that

$$\hat{\Psi}(t) = \Psi(t)C, \qquad a < t < b.$$

(b) Moreover, $\{\hat{\mathbf{y}}_1(t), \hat{\mathbf{y}}_2(t), \ldots, \hat{\mathbf{y}}_n(t)\}$ is also a fundamental set of solutions if and only if the determinant of C is nonzero.

The Fundamental Matrix

As we have seen in the last two sections, it is often convenient to introduce an $(n \times n)$ matrix function of the form

$$\Psi(t) = [\mathbf{y}_1(t), \mathbf{y}_2(t), \ldots, \mathbf{y}_n(t)],$$

where $\Psi(t)$ is created using either a set of solutions or a fundamental set of solutions as its columns. If the columns of $\Psi(t)$ are solutions of $\mathbf{y}' = P(t)\mathbf{y}$, we refer to $\Psi(t)$ as a **solution matrix** of $\mathbf{y}' = P(t)\mathbf{y}$. In addition, if the set of solutions forms a fundamental set of solutions, we refer to the corresponding solution matrix $\Psi(t)$ as a **fundamental matrix** of $\mathbf{y}' = P(t)\mathbf{y}$.

Part (a) of Theorem 6.8 states that any solution matrix can be expressed as any given fundamental matrix multiplied on the right by an $(n \times n)$ constant matrix C. Part (b) of Theorem 6.8 states that any other fundamental matrix can be expressed as the given fundamental matrix multiplied on the right by an invertible $(n \times n)$ constant matrix C.

If solutions of $\mathbf{y}' = P(t)\mathbf{y}$ are used to form a solution matrix $\Psi(t)$, then

$$\begin{aligned}\Psi'(t) &= [\mathbf{y}_1'(t), \mathbf{y}_2'(t), \ldots, \mathbf{y}_n'(t)] \\ &= [P(t)\mathbf{y}_1(t), P(t)\mathbf{y}_2(t), \ldots, P(t)\mathbf{y}_n(t)] \\ &= P(t)[\mathbf{y}_1(t), \mathbf{y}_2(t), \ldots, \mathbf{y}_n(t)] \\ &= P(t)\Psi(t).\end{aligned}$$

Therefore, $\Psi(t)$ is itself a solution of the matrix differential equation

$$\Psi'(t) = P(t)\Psi(t).$$

EXAMPLE 1

The matrix

$$\Psi(t) = \begin{bmatrix} e^t & 2e^{-t} \\ 2e^t & 3e^{-t} \end{bmatrix}$$

is a fundamental matrix for the homogeneous linear system $\mathbf{y}' = A\mathbf{y}$, where

$$A = \begin{bmatrix} -7 & 4 \\ -12 & 7 \end{bmatrix}.$$

Use the fundamental matrix $\Psi(t)$ to create another fundamental matrix $\hat{\Psi}(t)$ that satisfies the condition $\hat{\Psi}(0) = I$. [Thus, $\hat{\Psi}(t)$ is the solution of the matrix initial value problem $\hat{\Psi}' = A\Psi$, $\Psi(0) = I$.]

Solution: According to Theorem 6.8, every fundamental matrix $\hat{\Psi}(t)$ has the form $\hat{\Psi}(t) = \Psi(t)C$, where C is invertible and $\Psi(t)$ is a given fundamental matrix. Imposing the condition $\hat{\Psi}(0) = I$, we obtain $I = \Psi(0)C$, or $C = \Psi(0)^{-1}$. Now,

$$\Psi(0)^{-1} = \begin{bmatrix} 1 & 2 \\ 2 & 3 \end{bmatrix}^{-1} = \begin{bmatrix} -3 & 2 \\ 2 & -1 \end{bmatrix}.$$

(continued)

(continued)

Thus, $\hat{\Psi}(t) = \Psi(t)C$, or

$$\hat{\Psi}(t) = \begin{bmatrix} e^t & 2e^{-t} \\ 2e^t & 3e^{-t} \end{bmatrix} \begin{bmatrix} -3 & 2 \\ 2 & -1 \end{bmatrix} = \begin{bmatrix} -3e^t + 4e^{-t} & 2e^t - 2e^{-t} \\ -6e^t + 6e^{-t} & 4e^t - 3e^{-t} \end{bmatrix}. \blacktriangle$$

EXERCISES

Exercises 1–3:

In each exercise, vector functions $\mathbf{f}_1(t)$ and $\mathbf{f}_2(t)$, defined on the interval $-\infty < t < \infty$, are given. Recall that the linear combination $k_1\mathbf{f}_1(t) + k_2\mathbf{f}_2(t)$ can be written as the matrix-vector product $[\mathbf{f}_1(t), \mathbf{f}_2(t)]\mathbf{k}$, where $[\mathbf{f}_1(t), \mathbf{f}_2(t)]$ is the (2×2) matrix formed using the vector functions as columns.

(a) Evaluate the determinant of the (2×2) matrix $[\mathbf{f}_1(t), \mathbf{f}_2(t)]$.

(b) At $t = 0$, the determinant calculated in part (a) is zero. Therefore we can find a vector $\mathbf{k} \neq \mathbf{0}$ such that $[\mathbf{f}_1(0), \mathbf{f}_2(0)]\mathbf{k} = \mathbf{0}$; see the appendix on matrix theory. Does this fact prove that the given vector functions are linearly dependent on $-\infty < t < \infty$? Explain.

(c) At $t = 1$, the determinant calculated in part (b) is nonzero. Does this fact prove that the given vector functions are linearly independent on $-\infty < t < \infty$? Explain.

1. $\mathbf{f}_1(t) = \begin{bmatrix} t \\ 1 \end{bmatrix}, \quad \mathbf{f}_2(t) = \begin{bmatrix} t^2 \\ 2 \end{bmatrix}$

2. $\mathbf{f}_1(t) = \begin{bmatrix} e^t \\ t \end{bmatrix}, \quad \mathbf{f}_2(t) = \begin{bmatrix} \sin t \\ t^2 \end{bmatrix}$

3. $\mathbf{f}_1(t) = \begin{bmatrix} te^t \\ t-1 \end{bmatrix}, \quad \mathbf{f}_2(t) = \begin{bmatrix} \sin^2 t \\ 2 \end{bmatrix}$

Exercises 4–10:

In each exercise, determine whether the given functions are linearly dependent or linearly independent on the interval $-\infty < t < \infty$.

4. $\begin{bmatrix} t \\ 1 \end{bmatrix}, \begin{bmatrix} t^2 \\ 1 \end{bmatrix}$

5. $\begin{bmatrix} e^t \\ 1 \end{bmatrix}, \begin{bmatrix} e^{-t} \\ 0 \end{bmatrix}$

6. $\begin{bmatrix} e^t \\ 1 \end{bmatrix}, \begin{bmatrix} e^{-t} \\ 1 \end{bmatrix}, \begin{bmatrix} \sinh t \\ 0 \end{bmatrix}$

7. $\begin{bmatrix} 1 \\ t \\ 0 \end{bmatrix}, \begin{bmatrix} 0 \\ 1 \\ t^2 \end{bmatrix}$

8. $\begin{bmatrix} 1 \\ t \\ 0 \end{bmatrix}, \begin{bmatrix} 0 \\ 1 \\ t^2 \end{bmatrix}, \begin{bmatrix} 0 \\ 0 \\ 0 \end{bmatrix}$

9. $\begin{bmatrix} 1 \\ t \\ 0 \end{bmatrix}, \begin{bmatrix} 0 \\ 1 \\ t^2 \end{bmatrix}, \begin{bmatrix} 0 \\ 0 \\ 1 \end{bmatrix}$

10. $\begin{bmatrix} 1 \\ \sin^2 t \\ 0 \end{bmatrix}, \begin{bmatrix} 0 \\ 2 - 2\cos^2 t \\ -2 \end{bmatrix}, \begin{bmatrix} 1 \\ 0 \\ 1 \end{bmatrix}$

Exercises 11–12:

Form a (2×2) matrix using the given functions as columns.

(a) Compute the determinant of the resulting matrix.

(b) Is it possible that the given functions form a fundamental set of solutions for a linear system $\mathbf{y}' = P(t)\mathbf{y}$ where $P(t)$ is continuous on a t-interval containing the point $t = 0$? Explain.

(c) Determine a matrix $P(t)$ such that the given vector functions form a fundamental set for $\mathbf{y}' = P(t)\mathbf{y}$. On what t-interval(s) is the coefficient matrix $P(t)$ continuous? [Hint: The (2×2) matrix formed in part (a), call it $\Psi(t)$, must satisfy the equation $\Psi' = P(t)\Psi$, and it must also be invertible on these t-interval(s).]

11. $\begin{bmatrix} e^t \\ 0 \end{bmatrix}, \quad \begin{bmatrix} t^2 \\ t \end{bmatrix}.$ **12.** $\begin{bmatrix} t^2 \\ 0 \end{bmatrix}, \quad \begin{bmatrix} 2t \\ 1 \end{bmatrix}$

Exercises 13–16:

In each exercise,

(a) Verify that the matrix $\Psi(t)$ is a fundamental matrix of the given linear system.

(b) Determine a constant matrix C such that the given matrix $\hat{\Psi}(t)$ can be represented as $\hat{\Psi}(t) = \Psi(t)C$.

(c) Use your knowledge of the matrix C and assertion (b) of Theorem 6.8 to determine whether $\hat{\Psi}(t)$ is also a fundamental matrix, or simply a solution matrix.

13. $\mathbf{y}' = \begin{bmatrix} 0 & 1 \\ 1 & 0 \end{bmatrix}\mathbf{y}, \quad \Psi(t) = \begin{bmatrix} e^t & e^{-t} \\ e^t & -e^{-t} \end{bmatrix}, \quad \hat{\Psi}(t) = \begin{bmatrix} \sinh t & \cosh t \\ \cosh t & \sinh t \end{bmatrix}.$

14. $\mathbf{y}' = \begin{bmatrix} 0 & 1 \\ 1 & 0 \end{bmatrix}\mathbf{y}, \quad \Psi(t) = \begin{bmatrix} e^t & e^{-t} \\ e^t & -e^{-t} \end{bmatrix}, \quad \hat{\Psi}(t) = \begin{bmatrix} 2e^t - e^{-t} & e^t + 3e^{-t} \\ 2e^t + e^{-t} & e^t - 3e^{-t} \end{bmatrix}.$

15. $\mathbf{y}' = \begin{bmatrix} 1 & 1 \\ 0 & -2 \end{bmatrix}\mathbf{y}, \quad \Psi(t) = \begin{bmatrix} e^t & e^{-2t} \\ 0 & -3e^{-2t} \end{bmatrix}, \quad \hat{\Psi}(t) = \begin{bmatrix} 2e^{-2t} & 0 \\ -6e^{-2t} & 0 \end{bmatrix}.$

16. $\mathbf{y}' = \begin{bmatrix} 1 & 1 & 1 \\ 0 & -1 & 1 \\ 0 & 0 & 2 \end{bmatrix}\mathbf{y}, \quad \Psi(t) = \begin{bmatrix} e^t & e^{-t} & 4e^{2t} \\ 0 & -2e^{-t} & e^{2t} \\ 0 & 0 & 3e^{2t} \end{bmatrix}, \quad \hat{\Psi}(t) = \begin{bmatrix} e^t + e^{-t} & 4e^{2t} & e^t + 4e^{2t} \\ -2e^{-t} & e^{2t} & e^{2t} \\ 0 & 3e^{2t} & 3e^{2t} \end{bmatrix}.$

Exercises 17–18:

The matrix $\Psi(t)$ is a fundamental matrix of the given homogeneous linear system. Find a constant matrix C such that $\hat{\Psi}(t) = \Psi(t)C$ is a fundamental matrix satisfying $\hat{\Psi}(0) = I$, where I is the (2×2) identity matrix.

17. $\mathbf{y}' = \begin{bmatrix} 0 & 1 \\ 1 & 0 \end{bmatrix}\mathbf{y}, \quad \Psi(t) = \begin{bmatrix} e^t & e^{-t} \\ e^t & -e^{-t} \end{bmatrix}.$ **18.** $\mathbf{y}' = \begin{bmatrix} 1 & 1 \\ 0 & -2 \end{bmatrix}\mathbf{y}, \quad \Psi(t) = \begin{bmatrix} e^t & e^{-2t} \\ 0 & -3e^{-2t} \end{bmatrix}.$

19. Prove Theorem 6.7.

6.5 Constant Coefficient Homogeneous Systems

In this section we consider the first order homogeneous equation

$$\mathbf{y}' = A\mathbf{y}, \qquad -\infty < t < \infty,$$

where $\mathbf{y}(t)$ is an $(n \times 1)$ vector function and A is an $(n \times n)$ matrix of real-valued constants. The general solution of $\mathbf{y}' = A\mathbf{y}$ has the form

$$\mathbf{y}(t) = c_1\mathbf{y}_1(t) + c_2\mathbf{y}_2(t) + \cdots + c_n\mathbf{y}_n(t), \tag{1}$$

where $\{\mathbf{y}_1(t), \mathbf{y}_2(t), \ldots, \mathbf{y}_n(t)\}$ is a fundamental set of solutions. We now address the problem of finding a fundamental set of solutions for $\mathbf{y}' = A\mathbf{y}$.

The Eigenvalue Problem

In Chapters 4 and 5, we found solutions of the linear homogeneous constant coefficient scalar equation by looking for solutions of the form $y(t) = e^{\lambda t}$. For the present case, $\mathbf{y}' = A\mathbf{y}$, we take a similar approach. This time, however, we must find solutions that are vector functions. Therefore, we look for solutions of the form

$$\mathbf{y}(t) = e^{\lambda t}\mathbf{x}, \tag{2}$$

where λ is a constant (possibly complex) and $\mathbf{x}$ is an $(n \times 1)$ constant vector. To ensure that $\mathbf{y}(t)$ is a nonzero solution, we require that $\mathbf{x}$ be a nonzero vector.

Substituting the trial form $\mathbf{y}(t) = e^{\lambda t}\mathbf{x}$ into the left-hand side of $\mathbf{y}' = A\mathbf{y}$ leads to

$$\mathbf{y}' = (e^{\lambda t}\mathbf{x})' = (e^{\lambda t})'\mathbf{x} = \lambda e^{\lambda t}\mathbf{x} = e^{\lambda t}(\lambda\mathbf{x}). \tag{3}$$

Substituting $\mathbf{y}(t) = e^{\lambda t}\mathbf{x}$ into the right-hand side of $\mathbf{y}' = A\mathbf{y}$ yields

$$A\mathbf{y} = A(e^{\lambda t}\mathbf{x}) = e^{\lambda t}(A\mathbf{x}). \tag{4}$$

Equating expressions (3) and (4),

$$e^{\lambda t}(\lambda\mathbf{x}) = e^{\lambda t}(A\mathbf{x}).$$

Canceling the nonzero factor $e^{\lambda t}$ and rearranging

$$A\mathbf{x} = \lambda\mathbf{x}, \qquad \mathbf{x} \neq \mathbf{0}. \tag{5}$$

Equation (5) is a problem, known as an **eigenvalue problem**, which is important in mathematics, science, and engineering. The problem posed by equation (5) is that of finding constants λ, called **eigenvalues**, and corresponding nonzero vectors $\mathbf{x}$, called **eigenvectors**, such that $A\mathbf{x} = \lambda\mathbf{x}$.

The combination of an eigenvalue λ and a corresponding eigenvector $\mathbf{x}$ is referred to as an **eigenpair** and denoted by $(\lambda, \mathbf{x})$. For every eigenpair $(\lambda, \mathbf{x})$ of the matrix A, we find the associated vector function

$$\mathbf{y}(t) = e^{\lambda t}\mathbf{x} \tag{6}$$

is a solution of $\mathbf{y}' = A\mathbf{y}$.

If $\mathbf{x}$ is an eigenvector of A corresponding to an eigenvalue λ, then so is the vector $a\mathbf{x}$, where a is any nonzero constant; to verify this fact, you need only multiply equation (5) by the nonzero scalar a. Hence, if $(\lambda, \mathbf{x})$ is an eigenpair for A, then so is $(\lambda, a\mathbf{x})$, $a \neq 0$. Eigenvectors are not unique. In view of equation (6), however, lack of uniqueness is not surprising. That is, if $\mathbf{y}(t) = e^{\lambda t}\mathbf{x}$ is a solution of the homogeneous linear equation $\mathbf{y}' = A\mathbf{y}$, then so is $\tilde{\mathbf{y}}(t) = a\mathbf{y}(t)$ (see Theorem 6.2).

EXAMPLE 1

Consider the homogeneous first order system $\mathbf{y}' = A\mathbf{y}$, where

$$\mathbf{y}(t) = \begin{bmatrix} y_1(t) \\ y_2(t) \end{bmatrix} \quad \text{and} \quad A = \begin{bmatrix} 1 & 2 \\ 2 & 1 \end{bmatrix}.$$

Let

$$\mathbf{x}_1 = \begin{bmatrix} 1 \\ -1 \end{bmatrix} \quad \text{and} \quad \mathbf{x}_2 = \begin{bmatrix} 1 \\ 1 \end{bmatrix}.$$

(a) Use equation (5) to show that $\mathbf{x}_1$ and $\mathbf{x}_2$ are eigenvectors of A. Determine the corresponding eigenvalues λ_1 and λ_2.

(b) Use equation (6) to determine two solutions, $\mathbf{y}_1(t)$ and $\mathbf{y}_2(t)$, of $\mathbf{y}' = A\mathbf{y}$.

(c) Calculate the Wronskian and decide if $\{\mathbf{y}_1(t), \mathbf{y}_2(t)\}$ is a fundamental set of solutions.

Solution:

(a) Calculating $A\mathbf{x}_1$, we find

$$A\mathbf{x}_1 = \begin{bmatrix} 1 & 2 \\ 2 & 1 \end{bmatrix} \begin{bmatrix} 1 \\ -1 \end{bmatrix} = \begin{bmatrix} -1 \\ 1 \end{bmatrix} = (-1)\,\mathbf{x}_1.$$

Therefore, $(\lambda_1, \mathbf{x}_1) = (-1, \mathbf{x}_1)$ is an eigenpair of A. Similarly, calculating $A\mathbf{x}_2$ we obtain

$$A\mathbf{x}_2 = \begin{bmatrix} 1 & 2 \\ 2 & 1 \end{bmatrix} \begin{bmatrix} 1 \\ 1 \end{bmatrix} = \begin{bmatrix} 3 \\ 3 \end{bmatrix} = 3\mathbf{x}_2.$$

Therefore, $(\lambda_2, \mathbf{x}_2) = (3, \mathbf{x}_2)$ is a second eigenpair of A.

(b) From part (a), we have eigenpairs $(\lambda_1, \mathbf{x}_1)$ and $(\lambda_2, \mathbf{x}_2)$. Using equation (6), we can form two solutions, $\mathbf{y}_1(t) = e^{\lambda_1 t}\mathbf{x}_1$ and $\mathbf{y}_2(t) = e^{\lambda_2 t}\mathbf{x}_2$,

$$\mathbf{y}_1(t) = e^{-t}\begin{bmatrix} 1 \\ -1 \end{bmatrix} = \begin{bmatrix} e^{-t} \\ -e^{-t} \end{bmatrix} \quad \text{and} \quad \mathbf{y}_2(t) = e^{3t}\begin{bmatrix} 1 \\ 1 \end{bmatrix} = \begin{bmatrix} e^{3t} \\ e^{3t} \end{bmatrix}.$$

(c) To determine if these two solutions form a fundamental set of solutions, we calculate the Wronskian, $W(t)$, of $\mathbf{y}_1$ and $\mathbf{y}_2$. From part (b), our solution matrix $\Psi(t)$ is

$$\Psi(t) = [\mathbf{y}_1(t), \mathbf{y}_2(t)] = \begin{bmatrix} e^{-t} & e^{3t} \\ -e^{-t} & e^{3t} \end{bmatrix}.$$

The Wronskian is

$$W(t) = \det[\Psi(t)] = 2e^{2t}.$$

Since the Wronskian is never zero, we know by Theorem 6.3 that $\{\mathbf{y}_1(t), \mathbf{y}_2(t)\}$ is a fundamental set of solutions of $\mathbf{y}' = A\mathbf{y}$. ▲

Finding Eigenpairs

Example 1 suggests a procedure to find the general solution of $\mathbf{y}' = A\mathbf{y}$ when A is an $(n \times n)$ constant matrix. Each eigenpair $(\lambda, \mathbf{x})$ gives rise to a solution of the form $\mathbf{y}(t) = e^{\lambda t}\mathbf{x}$. Moreover, the general solution is a linear combination of n different solutions,

$$\mathbf{y}(t) = c_1\mathbf{y}_1(t) + c_2\mathbf{y}_2(t) + \cdots + c_n\mathbf{y}_n(t).$$

Some obvious questions, therefore, are

(a) Given an $(n \times n)$ constant matrix A, do there always exist eigenpairs $(\lambda, \mathbf{x})$? Is it possible to find n different eigenpairs and thereby form n different solutions?

(b) How do we find these eigenpairs?

We now address these questions.

The eigenvalue problem (5) consists of finding scalars λ and nonzero vectors $\mathbf{x}$ such that $A\mathbf{x} = \lambda\mathbf{x}$ or, equivalently, $A\mathbf{x} - \lambda\mathbf{x} = \mathbf{0}$. We can rewrite the equation $A\mathbf{x} - \lambda\mathbf{x} = \mathbf{0}$ as

$$(A - \lambda I)\mathbf{x} = \mathbf{0}, \qquad \mathbf{x} \neq \mathbf{0}, \tag{7}$$

where I denotes the $(n \times n)$ identity matrix.

To solve (7), we use a result from linear algebra stating that the matrix equation $(A - \lambda I)\mathbf{x} = \mathbf{0}$ has a nonzero solution $\mathbf{x}$ if and only if the determinant of $A - \lambda I$ is zero. Therefore, we conclude:

1. λ is an eigenvalue of A if and only if $\det[A - \lambda I] = 0$; that is,

$$\begin{vmatrix} a_{11} - \lambda & a_{12} & \cdots & a_{1n} \\ a_{21} & a_{22} - \lambda & \cdots & a_{2n} \\ \vdots & & & \vdots \\ a_{n1} & a_{n2} & \cdots & a_{nn} - \lambda \end{vmatrix} = 0. \tag{8}$$

   A cofactor expansion of this determinant shows that (8) is a polynomial equation of the form

$$p(\lambda) = 0, \tag{9}$$

   where $p(\lambda)$ is a polynomial of degree n in the variable λ. The polynomial $p(\lambda)$ is called the **characteristic polynomial** and equation (8) is called the **characteristic equation**. The eigenvalues of A are, therefore, the roots of the characteristic equation. Since an nth degree polynomial has n zeros (counting multiplicity), we conclude that an $(n \times n)$ matrix A always has n eigenvalues. The eigenvalues may be zero or nonzero, real or complex, and some of them may be repeated (that is, they may have multiplicity greater than one).

2. For each eigenvalue λ, we know the homogeneous system of equations $(A - \lambda I)\mathbf{x} = \mathbf{0}$ has a nontrivial solution. Therefore, an eigenvector is obtained by forming the homogeneous system $(A - \lambda I)\mathbf{x} = \mathbf{0}$ and finding a nontrivial solution.

EXAMPLE 2

Find the eigenpairs of

$$A = \begin{bmatrix} 4 & -2 \\ 1 & 1 \end{bmatrix}.$$

Solution: We first find the eigenvalues of A by finding the roots of $p(\lambda) = 0$,

$$p(\lambda) = \det(A - \lambda I) = \begin{vmatrix} 4-\lambda & -2 \\ 1 & 1-\lambda \end{vmatrix}$$
$$= (4-\lambda)(1-\lambda) + 2$$
$$= \lambda^2 - 5\lambda + 6$$
$$= (\lambda - 3)(\lambda - 2).$$

The matrix therefore has two real distinct eigenvalues, $\lambda_1 = 3$ and $\lambda_2 = 2$.

For each eigenvalue, we solve the homogeneous system of equations $(A - \lambda I)\mathbf{x} = \mathbf{0}$ and choose a nontrivial solution to serve as the eigenvector. For $\lambda_1 = 3$, the system is $(A - 3I)\mathbf{x} = \mathbf{0}$, or

$$\begin{bmatrix} 1 & -2 \\ 1 & -2 \end{bmatrix} \begin{bmatrix} x_1 \\ x_2 \end{bmatrix} = \begin{bmatrix} 0 \\ 0 \end{bmatrix}.$$

Because the coefficient matrix has zero determinant, the homogeneous system $(A - 3I)\mathbf{x} = \mathbf{0}$ has nontrivial solutions. One such nontrivial solution is

$$\mathbf{x}_1 = \begin{bmatrix} 2 \\ 1 \end{bmatrix}.$$

We next obtain an eigenvector $\mathbf{x}_2$ corresponding to $\lambda_2 = 2$ by solving $(A - 2I)\mathbf{x} = \mathbf{0}$,

$$\begin{bmatrix} 2 & -2 \\ 1 & -1 \end{bmatrix} \begin{bmatrix} x_1 \\ x_2 \end{bmatrix} = \begin{bmatrix} 0 \\ 0 \end{bmatrix}.$$

A convenient nontrivial solution is

$$\mathbf{x}_2 = \begin{bmatrix} 1 \\ 1 \end{bmatrix}.$$

Eigenpairs of A are, therefore, $(\lambda_1, \mathbf{x}_1)$ and $(\lambda_2, \mathbf{x}_2)$, where

$$\lambda_1 = 3, \mathbf{x}_1 = \begin{bmatrix} 2 \\ 1 \end{bmatrix} \quad \text{and} \quad \lambda_2 = 2, \mathbf{x}_2 = \begin{bmatrix} 1 \\ 1 \end{bmatrix}. \; ▲$$

EXAMPLE 3

Find the eigenpairs of the (3×3) matrix

$$A = \begin{bmatrix} 1 & -7 & 3 \\ -1 & -1 & 1 \\ 4 & -4 & 0 \end{bmatrix}.$$

Solution: We find the eigenvalues by solving the characteristic equation $p(\lambda) = 0$. To calculate the characteristic polynomial $p(\lambda) = \det[A - \lambda I]$, we use

(continued)

(continued)

a cofactor expansion along the first row:

$$\begin{aligned} p(\lambda) &= \begin{vmatrix} 1-\lambda & -7 & 3 \\ -1 & -1-\lambda & 1 \\ 4 & -4 & -\lambda \end{vmatrix} \\ &= (1-\lambda)\begin{vmatrix} -1-\lambda & 1 \\ -4 & -\lambda \end{vmatrix} + 7\begin{vmatrix} -1 & 1 \\ 4 & -\lambda \end{vmatrix} + 3\begin{vmatrix} -1 & -1-\lambda \\ 4 & -4 \end{vmatrix} \\ &= (1-\lambda)(\lambda^2+\lambda+4) + 7(\lambda-4) + 3(4\lambda+8) \\ &= -\lambda^3 + 16\lambda \\ &= -\lambda(\lambda-4)(\lambda+4). \end{aligned}$$

The eigenvalues of A are therefore $\lambda_1 = 0$, $\lambda_2 = 4$, and $\lambda_3 = -4$. (In this example, $\lambda_1 = 0$ is an eigenvalue. Although eigenvectors must be nonzero, eigenvalues *can* be zero. In fact, see equation (7), $\lambda = 0$ is an eigenvalue whenever $\det[A] = 0$.)

We now compute the eigenvectors. An eigenvector corresponding to $\lambda_1 = 0$ is a nonzero solution of $(A - \lambda_1 I)\mathbf{x} = A\mathbf{x} = \mathbf{0}$,

$$\begin{bmatrix} 1 & -7 & 3 \\ -1 & -1 & 1 \\ 4 & -4 & 0 \end{bmatrix} \begin{bmatrix} x_1 \\ x_2 \\ x_3 \end{bmatrix} = \begin{bmatrix} 0 \\ 0 \\ 0 \end{bmatrix}. \tag{10}$$

In this case, unlike the situation in Example 2, we cannot obtain a solution by inspection. Instead, we solve system (10) using Gaussian elimination. Elementary row operations can be used to row reduce system (10) into the following equivalent system:

$$\begin{bmatrix} 1 & 0 & -\frac{1}{2} \\ 0 & 1 & -\frac{1}{2} \\ 0 & 0 & 0 \end{bmatrix} \begin{bmatrix} x_1 \\ x_2 \\ x_3 \end{bmatrix} = \begin{bmatrix} 0 \\ 0 \\ 0 \end{bmatrix}.$$

The solution of this system is $x_1 = (1/2)x_3$, $x_2 = (1/2)x_3$. For convenience, we set $x_3 = 2$, obtaining the eigenvector

$$\mathbf{x}_1 = \begin{bmatrix} 1 \\ 1 \\ 2 \end{bmatrix}.$$

Similarly, an eigenvector corresponding to $\lambda_2 = 4$ is a nonzero solution of $(A - \lambda_2 I)\mathbf{x} = (A - 4I)\mathbf{x} = \mathbf{0}$,

$$\begin{bmatrix} -3 & -7 & 3 \\ -1 & -5 & 1 \\ 4 & -4 & -4 \end{bmatrix} \begin{bmatrix} x_1 \\ x_2 \\ x_3 \end{bmatrix} = \begin{bmatrix} 0 \\ 0 \\ 0 \end{bmatrix}.$$

Using Gaussian elimination, we find the equivalent system

$$\begin{bmatrix} 1 & 0 & -1 \\ 0 & 1 & 0 \\ 0 & 0 & 0 \end{bmatrix} \begin{bmatrix} x_1 \\ x_2 \\ x_3 \end{bmatrix} = \begin{bmatrix} 0 \\ 0 \\ 0 \end{bmatrix}.$$

The solution of the system $(A - 4I)\mathbf{x} = \mathbf{0}$ is, therefore, $x_1 = x_3, x_2 = 0$. A convenient choice for x_3 is $x_3 = 1$, leading to the eigenvector

$$\mathbf{x}_2 = \begin{bmatrix} 1 \\ 0 \\ 1 \end{bmatrix}.$$

Lastly, we leave it as an exercise to show that

$$\mathbf{x}_3 = \begin{bmatrix} 2 \\ 1 \\ -1 \end{bmatrix}$$

is an eigenvector corresponding to $\lambda_3 = -4$.

In summary, we have obtained eigenpairs $(\lambda_1, \mathbf{x}_1)$, $(\lambda_2, \mathbf{x}_2)$, $(\lambda_3, \mathbf{x}_3)$, where

$$\lambda_1 = 0, \quad \mathbf{x}_1 = \begin{bmatrix} 1 \\ 1 \\ 2 \end{bmatrix}, \qquad \lambda_2 = 4, \quad \mathbf{x}_2 = \begin{bmatrix} 1 \\ 0 \\ 1 \end{bmatrix}, \qquad \lambda_3 = -4, \quad \mathbf{x}_3 = \begin{bmatrix} 2 \\ 1 \\ -1 \end{bmatrix}. \quad \blacktriangle$$

Eigenpair computations, such as those of the previous two examples, have built-in checks available that should be exploited. When computing eigenvectors, the Gaussian elimination process that transforms the coefficient matrix $A - \lambda I$ to echelon form must produce at least one row of zeros. If that does not occur, you should realize you've made a mistake. Likewise, a final check is simply to compute the product $A\mathbf{x}$ and verify that it equals $\lambda\mathbf{x}$.

Some Properties of Eigenpairs

We now list some relevant properties of eigenpairs. Additional properties are dealt with in the exercises.

Eigenvectors are not unique.
As we noted earlier, if $\mathbf{x}$ is an eigenvector of a matrix A with corresponding eigenvalue λ, then any nonzero constant multiple of this eigenvector is also an eigenvector. In other words, if $A\mathbf{x} = \lambda\mathbf{x}$, $\mathbf{x} \neq \mathbf{0}$, and if $c \neq 0$ is a (real or complex) constant, then $c\mathbf{x}$ is also an eigenvector, since $A(c\mathbf{x}) = \lambda(c\mathbf{x})$, $c\mathbf{x} \neq \mathbf{0}$.

A matrix can have a zero eigenvalue.
Eigenvectors must be nonzero, but eigenvalues can be zero. Example 3 illustrated this fact. If $(0, \mathbf{x})$ is an eigenpair for the matrix A, then the homogeneous linear system $\mathbf{y}' = A\mathbf{y}$ has a constant solution,

$$\mathbf{y}(t) = e^{0t}\mathbf{x} = \mathbf{x}.$$

Note that the matrix A has $\lambda = 0$ as an eigenvalue if and only if $\det[A] = 0$.

A real matrix may have one or more complex eigenvalues and eigenvectors.
A real matrix A can have complex eigenvalues. For example, consider the matrix

$$A = \begin{bmatrix} 1 & 1 \\ -1 & 1 \end{bmatrix}. \tag{11}$$

The characteristic polynomial is

$$p(\lambda) = \begin{vmatrix} 1-\lambda & 1 \\ -1 & 1-\lambda \end{vmatrix} = (1-\lambda)^2 + 1 = \lambda^2 - 2\lambda + 2.$$

Therefore, A has complex conjugate eigenvalues $\lambda_1 = 1 + i$ and $\lambda_2 = \overline{\lambda}_1 = 1 - i$. What about eigenvectors? To obtain an eigenvector corresponding to $\lambda_1 = 1 + i$, we must find a nontrivial solution of $(A - (1+i)I)\mathbf{x} = \mathbf{0}$,

$$\begin{bmatrix} -i & 1 \\ -1 & -i \end{bmatrix} \begin{bmatrix} x_1 \\ x_2 \end{bmatrix} = \begin{bmatrix} 0 \\ 0 \end{bmatrix}.$$

A convenient choice is

$$\mathbf{x}_1 = \begin{bmatrix} 1 \\ i \end{bmatrix}.$$

Similarly, an eigenvector corresponding to the eigenvalue $\lambda_2 = 1 - i$ is

$$\mathbf{x}_2 = \overline{\mathbf{x}}_1 = \begin{bmatrix} 1 \\ -i \end{bmatrix}.$$

In Section 6.6 we examine in detail the complex eigenvalue case, showing that complex eigenvalues and eigenvectors of a real matrix A occur in conjugate pairs. We likewise show how to convert complex-valued solutions of $\mathbf{y}' = A\mathbf{y}$ into real solutions.

Eigenvectors corresponding to distinct eigenvalues are linearly independent.
The proof of this assertion, presented in most linear algebra texts, will not be given here. (We do, however, ask you to prove it for the special case of $n = 2$; see Exercise 36.) We use this assertion to prove the corollary to Theorem 6.9.

Linearly Independent Eigenvectors and Fundamental Sets of Solutions

Let A be an $(n \times n)$ matrix with eigenpairs $(\lambda_1, \mathbf{x}_1), (\lambda_2, \mathbf{x}_2), \ldots, (\lambda_n, \mathbf{x}_n)$. Theorem 6.9 states: if this set of eigenvectors $\{\mathbf{x}_1, \mathbf{x}_2, \ldots, \mathbf{x}_n\}$ is linearly independent, then the set of n solutions $\{e^{\lambda_1 t}\mathbf{x}_1, e^{\lambda_2 t}\mathbf{x}_2, \ldots, e^{\lambda_n t}\mathbf{x}_n\}$ is a fundamental set of solutions of the system $\mathbf{y}' = A\mathbf{y}$.

Theorem 6.9

Consider the homogeneous linear system $\mathbf{y}' = A\mathbf{y}$, $-\infty < t < \infty$. Let the constant $(n \times n)$ matrix A have eigenpairs $(\lambda_1, \mathbf{x}_1), (\lambda_2, \mathbf{x}_2), \ldots, (\lambda_n, \mathbf{x}_n)$ where the eigenvectors are linearly independent. Then, the set of solutions

$$\{e^{\lambda_1 t}\mathbf{x}_1, e^{\lambda_2 t}\mathbf{x}_2, \ldots, e^{\lambda_n t}\mathbf{x}_n\}$$

is a fundamental set of solutions.

PROOF: Let $\Psi(t)$ denote the solution matrix,

$$\Psi(t) = [e^{\lambda_1 t}\mathbf{x}_1, e^{\lambda_2 t}\mathbf{x}_2, \ldots, e^{\lambda_n t}\mathbf{x}_n].$$

We must show that the Wronskian, $W(t) = \det[\Psi(t)]$, is nonzero on $-\infty < t < \infty$.

From matrix theory, we know that scalar multiplication of a column of a square matrix multiplies the determinant by this same factor. Therefore,

$$W(t) = \det[\Psi(t)] = e^{\lambda_1 t}e^{\lambda_2 t}\cdots e^{\lambda_n t}\det([\mathbf{x}_1, \mathbf{x}_2, \ldots, \mathbf{x}_n]). \quad (12)$$

The product of exponentials is nonzero and, therefore, $W(t)$ is zero if and only if $\det([\mathbf{x}_1, \mathbf{x}_2, \ldots, \mathbf{x}_n])$ is zero.

However, the set $\{\mathbf{x}_1, \mathbf{x}_2, \ldots, \mathbf{x}_n\}$ is linearly independent if and only if $\det([\mathbf{x}_1, \mathbf{x}_2, \ldots, \mathbf{x}_n]) \neq 0$. Since the eigenvectors are assumed to form a linearly independent set, we know by (12) that $W(t) \neq 0$, $-\infty < t < \infty$. Thus, the set of solutions is a fundamental set.

We noted previously that eigenvectors corresponding to distinct eigenvalues are linearly independent. Therefore, if an $(n \times n)$ matrix A has n distinct eigenvalues, then the corresponding set of eigenvectors $\{\mathbf{x}_1, \mathbf{x}_2, \ldots, \mathbf{x}_n\}$ is linearly independent. Hence, the following corollary of Theorem 6.9 is immediate.

Corollary

Consider the homogeneous linear system $\mathbf{y}' = A\mathbf{y}$, $-\infty < t < \infty$. Let the constant $(n \times n)$ matrix A have eigenpairs $(\lambda_1, \mathbf{x}_1), (\lambda_2, \mathbf{x}_2), \ldots, (\lambda_n, \mathbf{x}_n)$, where the eigenvalues are distinct. Then, the set of solutions

$$\{e^{\lambda_1 t}\mathbf{x}_1,\ e^{\lambda_2 t}\mathbf{x}_2, \ldots,\ e^{\lambda_n t}\mathbf{x}_n\}$$

is a fundamental set of solutions.

REMARK: In Examples 1–3, the eigenvalues of the given matrix A were distinct. Therefore, we can use the Corollary to conclude directly:

For Example 1: If

$$A = \begin{bmatrix} 1 & 2 \\ 2 & 1 \end{bmatrix},$$

the general solution of $\mathbf{y}' = A\mathbf{y}$, is

$$\mathbf{y}(t) = c_1 \begin{bmatrix} e^{-t} \\ -e^{-t} \end{bmatrix} + c_2 \begin{bmatrix} e^{3t} \\ e^{3t} \end{bmatrix} = \begin{bmatrix} e^{-t} & e^{3t} \\ -e^{-t} & e^{3t} \end{bmatrix} \begin{bmatrix} c_1 \\ c_2 \end{bmatrix}.$$

For Example 2: If

$$A = \begin{bmatrix} 4 & -2 \\ 1 & 1 \end{bmatrix},$$

the general solution of $\mathbf{y}' = A\mathbf{y}$ is

$$\mathbf{y}(t) = c_1 \begin{bmatrix} 2e^{3t} \\ e^{3t} \end{bmatrix} + c_2 \begin{bmatrix} e^{2t} \\ e^{2t} \end{bmatrix} = \begin{bmatrix} 2e^{3t} & e^{2t} \\ e^{3t} & e^{2t} \end{bmatrix} \begin{bmatrix} c_1 \\ c_2 \end{bmatrix}.$$

For Example 3: If

$$A = \begin{bmatrix} 1 & -7 & 3 \\ -1 & -1 & 1 \\ 4 & -4 & 0 \end{bmatrix},$$

the general solution of $\mathbf{y}' = A\mathbf{y}$ is

$$\mathbf{y}(t) = c_1 e^{-4t} \begin{bmatrix} 2 \\ 1 \\ -1 \end{bmatrix} + c_2 \begin{bmatrix} 1 \\ 1 \\ 2 \end{bmatrix} + c_3 e^{4t} \begin{bmatrix} 1 \\ 0 \\ 1 \end{bmatrix} = \begin{bmatrix} 2e^{-4t} & 1 & e^{4t} \\ e^{-4t} & 1 & 0 \\ -e^{-4t} & 2 & e^{4t} \end{bmatrix} \begin{bmatrix} c_1 \\ c_2 \\ c_3 \end{bmatrix}.$$

EXERCISES

Exercises 1–6:

In each exercise, a (2×2) matrix A and vectors $\mathbf{x}_1$ and $\mathbf{x}_2$ are given.

(a) Decide which, if any, of the given vectors is an eigenvector of A, and determine the corresponding eigenvalue λ_k.

(b) For each eigenpair found in part (a), form a solution $\mathbf{y}_k(t)$ of the first order system $\mathbf{y}' = A\mathbf{y}$.

(c) If two solutions were found in part (b), do they form a fundamental set of solutions for $\mathbf{y}' = A\mathbf{y}$?

1. $A = \begin{bmatrix} 4 & 2 \\ -1 & 1 \end{bmatrix}, \quad \mathbf{x}_1 = \begin{bmatrix} 1 \\ -1 \end{bmatrix}, \quad \mathbf{x}_2 = \begin{bmatrix} -2 \\ 1 \end{bmatrix}$

2. $A = \begin{bmatrix} 7 & -3 \\ 16 & -7 \end{bmatrix}, \quad \mathbf{x}_1 = \begin{bmatrix} 3 \\ 8 \end{bmatrix}, \quad \mathbf{x}_2 = \begin{bmatrix} 1 \\ 2 \end{bmatrix}$

3. $A = \begin{bmatrix} 11 & 5 \\ -22 & -10 \end{bmatrix}, \quad \mathbf{x}_1 = \begin{bmatrix} 0 \\ 0 \end{bmatrix}, \quad \mathbf{x}_2 = \begin{bmatrix} 1 \\ -2 \end{bmatrix}$

4. $A = \begin{bmatrix} -5 & 2 \\ -18 & 7 \end{bmatrix}, \quad \mathbf{x}_1 = \begin{bmatrix} 1 \\ 3 \end{bmatrix}, \quad \mathbf{x}_2 = \begin{bmatrix} 1 \\ 2 \end{bmatrix}$

5. $A = \begin{bmatrix} 0 & 1 \\ 1 & 0 \end{bmatrix}, \quad \mathbf{x}_1 = \begin{bmatrix} 1 \\ -1 \end{bmatrix}, \quad \mathbf{x}_2 = \begin{bmatrix} 2 \\ 2 \end{bmatrix}$

6. $A = \begin{bmatrix} 2 & -1 \\ -4 & 2 \end{bmatrix}, \quad \mathbf{x}_1 = \begin{bmatrix} 1 \\ -2 \end{bmatrix}, \quad \mathbf{x}_2 = \begin{bmatrix} 1 \\ 2 \end{bmatrix}$

Exercises 7–13:

In each exercise, λ is an eigenvalue of the given matrix A. Determine an eigenvector corresponding to λ.

7. $A = \begin{bmatrix} -4 & 3 \\ -4 & 4 \end{bmatrix}, \quad \lambda = 2$

8. $A = \begin{bmatrix} 5 & 3 \\ -4 & -3 \end{bmatrix}, \quad \lambda = -1$

9. $A = \begin{bmatrix} 1 & 1 \\ -4 & 6 \end{bmatrix}, \quad \lambda = 5$

10. $A = \begin{bmatrix} 1 & -7 & 3 \\ -1 & -1 & 1 \\ 4 & -4 & 0 \end{bmatrix}, \quad \lambda = -4$

11. $A = \begin{bmatrix} 3 & 1 & 1 \\ -1 & 1 & -1 \\ 2 & 1 & 2 \end{bmatrix}, \quad \lambda = 2$

12. $A = \begin{bmatrix} 1 & 3 & 1 \\ 2 & 1 & 2 \\ 4 & 3 & -2 \end{bmatrix}, \quad \lambda = 5$

13. $A = \begin{bmatrix} -2 & 3 & 1 \\ -8 & 13 & 5 \\ 11 & -17 & -6 \end{bmatrix}, \quad \lambda = 0$

Exercises 14–20:

For the given matrix A, calculate the characteristic equation and find the eigenvalues of A. [Hint: Every characteristic equation has integer roots.]

14. $A = \begin{bmatrix} -5 & 1 \\ 0 & 4 \end{bmatrix}$

15. $A = \begin{bmatrix} 8 & 0 \\ 3 & 2 \end{bmatrix}$

16. $A = \begin{bmatrix} 3 & -3 \\ -6 & 6 \end{bmatrix}$

17. $A = \begin{bmatrix} 5 & 2 \\ 4 & 3 \end{bmatrix}$

18. $A = \begin{bmatrix} 5 & 0 & 0 \\ 0 & 1 & 3 \\ 0 & 2 & 2 \end{bmatrix}$

19. $A = \begin{bmatrix} -2 & 3 & 1 \\ -8 & 13 & 5 \\ 11 & -17 & -6 \end{bmatrix}$

20. $A = \begin{bmatrix} 1 & -7 & 3 \\ -1 & -1 & 1 \\ 4 & -4 & 0 \end{bmatrix}$

Exercises 21–30:

In each exercise, the given matrix A has distinct eigenvalues. Using the corollary to Theorem 6.9, determine a fundamental set of solutions of the first order linear system $\mathbf{y}' = A\mathbf{y}$. Exercises 21–22 are partly done in Exercises 7–8. If an initial condition $\mathbf{y}(t_0) = \mathbf{y}_0$ is given, form the general solution and solve the corresponding initial value problem.

21. $A = \begin{bmatrix} -4 & 3 \\ -4 & 4 \end{bmatrix}, \quad \mathbf{y}(0) = \begin{bmatrix} 4 \\ 4 \end{bmatrix}$

22. $A = \begin{bmatrix} 5 & 3 \\ -4 & -3 \end{bmatrix}, \quad \mathbf{y}(1) = \begin{bmatrix} 2 \\ 0 \end{bmatrix}$

23. $A = \begin{bmatrix} 1 & 2 \\ -4 & 7 \end{bmatrix}, \quad \mathbf{y}(0) = \begin{bmatrix} 7 \\ 11 \end{bmatrix}$

24. $A = \begin{bmatrix} -0.09 & 0.02 \\ 0.04 & -0.07 \end{bmatrix}$

25. $A = \begin{bmatrix} 2 & 1 & 2 \\ 0 & 3 & 2 \\ 0 & 0 & 1 \end{bmatrix}, \quad \mathbf{y}(0) = \begin{bmatrix} 4 \\ 3 \\ -1 \end{bmatrix}$

26. $A = \begin{bmatrix} 4 & 2 & 0 \\ 0 & 1 & 3 \\ 0 & 0 & -2 \end{bmatrix}, \quad \mathbf{y}(0) = \begin{bmatrix} -1 \\ 0 \\ 3 \end{bmatrix}$

27. $A = \begin{bmatrix} 2 & 1 & 2 \\ 0 & 2 & 2 \\ 0 & 4 & 0 \end{bmatrix}$

28. $A = \begin{bmatrix} 1 & 2 & 0 \\ -4 & 7 & 0 \\ 0 & 0 & 1 \end{bmatrix}$

29. $A = \begin{bmatrix} 3 & 1 & 0 \\ -6 & -5 & 2 \\ -7 & -8 & 4 \end{bmatrix}$

30. $A = \begin{bmatrix} 3 & 1 & 0 \\ -8 & -6 & 2 \\ -9 & -9 & 4 \end{bmatrix}$

31. Find x so that $\mathbf{u}$ is an eigenvector of A. What is the corresponding eigenvalue?

$$A = \begin{bmatrix} 2 & x \\ 1 & -5 \end{bmatrix}, \qquad \mathbf{u} = \begin{bmatrix} 1 \\ -1 \end{bmatrix}$$

32. Find x and y so that $\mathbf{u}$ is an eigenvector of A corresponding to the eigenvalue $\lambda = 1$

$$A = \begin{bmatrix} x & y \\ 2x & -y \end{bmatrix}, \qquad \mathbf{u} = \begin{bmatrix} -1 \\ 1 \end{bmatrix}$$

33. Consider the (2×2) matrix

$$A = \begin{bmatrix} a & b \\ b & d \end{bmatrix},$$

where A has all real entries. Show that A has only real eigenvalues. [Hint: Calculate the characteristic polynomial and use the quadratic formula.] The matrix A is a **symmetric matrix** since $A = A^T$. In general, A^T denotes the **transpose of A**, where A^T is obtained by interchanging the rows and columns of A. For example, if

$$A = \begin{bmatrix} x & y \\ u & v \end{bmatrix} \quad \text{then} \quad A^T = \begin{bmatrix} x & u \\ y & v \end{bmatrix}.$$

34. Consider the (2×2) matrix

$$A = \begin{bmatrix} a & b \\ -b & a \end{bmatrix}$$

where a and b are real numbers and b is nonzero. Show that every eigenvalue of A is complex.

35. Let A be a (2×2) matrix. Show that A and A^T have the same characteristic polynomial, and hence the same eigenvalues.

36. Let A be a (2×2) matrix with eigenvalues λ_1 and λ_2 where $\lambda_1 \neq \lambda_2$. Let $\mathbf{x}_1$ and $\mathbf{x}_2$ be corresponding eigenvectors. Show that $\{\mathbf{x}_1, \mathbf{x}_2\}$ is a linearly independent set. [Hint: Suppose $k_1\mathbf{x}_1 + k_2\mathbf{x}_2 = \mathbf{0}$. Multiply this equation by A and obtain $k_1\lambda_1\mathbf{x}_1 + k_2\lambda_2\mathbf{x}_2 = \mathbf{0}$. Next, multiply $k_1\mathbf{x}_1 + k_2\mathbf{x}_2 = \mathbf{0}$ by λ_1 and obtain $k_1\lambda_1\mathbf{x}_1 + k_2\lambda_1\mathbf{x}_2 = \mathbf{0}$. Show these two resulting equations imply that $k_2 = 0$. If $k_2 = 0$, what does that say about the linear independence condition $k_1\mathbf{x}_1 + k_2\mathbf{x}_2 = \mathbf{0}$?]

37. Let A be an $(n \times n)$ matrix with eigenvalue λ and eigenvector $\mathbf{x}$. Let α be any constant. Use the definition, $A\mathbf{x} = \lambda\mathbf{x}$, $\mathbf{x} \neq \mathbf{0}$, to show that $\lambda + \alpha$ is an eigenvalue of the matrix $A + \alpha I$ and $\alpha\lambda$ is an eigenvalue of αA. Similarly, if A is invertible, show that $\lambda \neq 0$ and that $1/\lambda$ is an eigenvalue of A^{-1}.

38. Let A be an $(n \times n)$ matrix with eigenvalue λ and eigenvector $\mathbf{x}$.

(a) Use the definition, $A\mathbf{x} = \lambda\mathbf{x}$, $\mathbf{x} \neq \mathbf{0}$, to show that λ^2 is an eigenvalue of A^2.

(b) Let A be a (2×2) matrix such that $A\mathbf{x} = \lambda\mathbf{x}$ where $\lambda = -2$ and

$$\mathbf{x} = \begin{bmatrix} 3 \\ -1 \end{bmatrix}.$$

Determine the vector $A^3\mathbf{x}$. [This is a special case of the general result: λ^k is an eigenvalue of A^k.]

Tank-Flushing Problems Consider the flow systems schematically shown in the figures for Exercises 39 and 40. In each case, a flushing out of the system is initiated at time $t = 0$. Fresh water is pumped into each tank and well-stirred mixtures flow out. Note that each flow rate is equal to r gal/min and we let $Q_j(t)$ represent the amount of solute (in pounds) in the jth tank at time t. The flow configuration is such that the volumes in each tank remain constant; we designate the volume as V. Note that all tanks in the system have an identical flow environment.

39. Consider the two-tank flow system shown in the figure. Apply the "conservation of salt" principle to each tank and derive the homogeneous linear system for $Q_j(t)$,

$$\frac{d}{dt}\begin{bmatrix} Q_1 \\ Q_2 \end{bmatrix} = \frac{r}{V}\begin{bmatrix} -2 & 1 \\ 1 & -2 \end{bmatrix}\begin{bmatrix} Q_1 \\ Q_2 \end{bmatrix}.$$

Making the change of independent variable $\tau = (r/V)t$, this system can be rewritten as

$$\frac{d}{d\tau}\begin{bmatrix} Q_1 \\ Q_2 \end{bmatrix} = \begin{bmatrix} -2 & 1 \\ 1 & -2 \end{bmatrix}\begin{bmatrix} Q_1 \\ Q_2 \end{bmatrix}.$$

(a) The fact that the two tanks have identical capacity and experience the same environment reflects itself in the fact that the coefficient matrix is a (real) symmetric matrix. Determine the eigenvalues and corresponding eigenvectors of this matrix and form the general solution of the homogeneous linear system.

(b) Assume that the initial amount of solute in each tank is $Q_1(0) = Q_0$ and $Q_2(0) = 2Q_0$, where Q_0 is a positive constant. Assume that $r/V = 0.02\,\text{sec}^{-1}$. Determine the amount of flushing time required to reduce the amount of salt in each of the two tanks to $0.01Q_0$ or less.

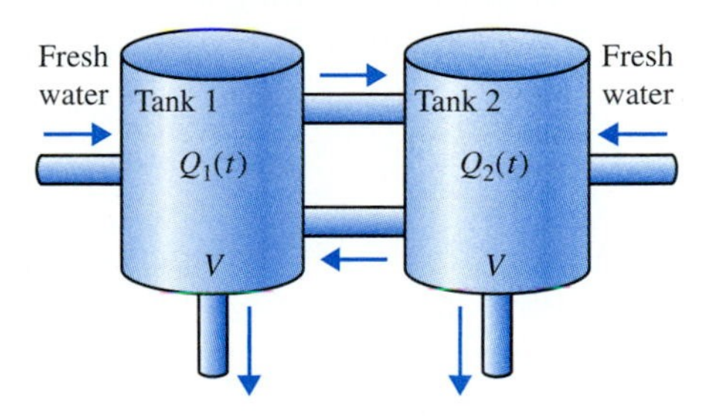

Figure for Exercise 39

40. Consider the three-tank flow system shown in the figure on the next page. As in Exercise 39, derive a homogeneous linear system for $Q_j(t)$,

$$\frac{d}{dt}\begin{bmatrix} Q_1 \\ Q_2 \\ Q_3 \end{bmatrix} = \frac{r}{V}\begin{bmatrix} -3 & 1 & 1 \\ 1 & -3 & 1 \\ 1 & 1 & -3 \end{bmatrix}\begin{bmatrix} Q_1 \\ Q_2 \\ Q_3 \end{bmatrix}.$$

(a) Determine the eigenvalues and eigenvectors of the coefficient matrix. You will find that one eigenvalue appears as a root of multiplicity two of the characteristic equation. You should, however, be able to find two linearly independent eigenvectors corresponding to this repeated eigenvalue. Form the general solution of the homogeneous linear system.

(b) If the initial amounts of salt in each tank are $Q_1(0) = Q_0$, $Q_2(0) = 2Q_0$, and $Q_3(0) = 3Q_0$, respectively, determine the solution of the resulting initial value problem. (Your answer will involve the constants r/V and Q_0 as well as time t.)

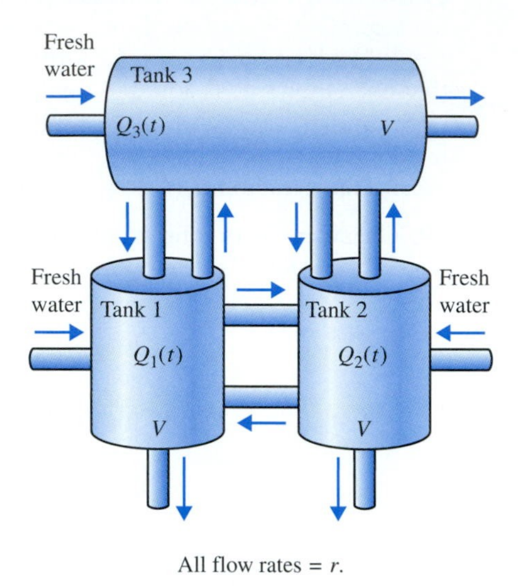

Figure for Exercise 40

6.6 Complex Eigenvalues

In this section we study the differential equation $\mathbf{y}' = A\mathbf{y}$, where A is a real constant $(n \times n)$ matrix possessing complex conjugate eigenvalues. Equation (11) of Section 6.5, provided an example,

$$A = \begin{bmatrix} 1 & 1 \\ -1 & 1 \end{bmatrix}.$$

As was shown, A has eigenvalues $\lambda_1 = 1 + i$ and $\lambda_2 = 1 - i$ and corresponding eigenvectors

$$\mathbf{x}_1 = \begin{bmatrix} 1 \\ i \end{bmatrix} \quad \text{and} \quad \mathbf{x}_2 = \begin{bmatrix} 1 \\ -i \end{bmatrix}.$$

For this matrix, the eigenvalues and eigenvectors occur in conjugate pairs. In fact, this property is common to every real matrix with complex eigenvalues. If A is a real matrix, then

1. Complex eigenvalues always occur in conjugate pairs.
2. If λ is a complex eigenvalue with a corresponding eigenvector $\mathbf{x}$, then $\overline{\mathbf{x}}$ is an eigenvector for the complex conjugate eigenvalue, $\overline{\lambda}$.

Therefore, if $(\lambda, \mathbf{x})$ is an eigenpair of a real matrix A, then so is $(\overline{\lambda}, \overline{\mathbf{x}})$. Note the computational implications: Once an eigenvector $\mathbf{x}$ corresponding to a complex eigenvalue λ has been determined, we need only form the complex conjugate $\overline{\mathbf{x}}$ to obtain an eigenvector corresponding to $\overline{\lambda}$.

We now describe how to convert the complex solutions of $\mathbf{y}' = A\mathbf{y}$ that arise from complex eigenpairs into real-valued solutions, which are generally more meaningful in applications.

The Real and Imaginary Parts of a Complex-Valued Solution Are Also Solutions

As noted earlier, the matrix

$$A = \begin{bmatrix} 1 & 1 \\ -1 & 1 \end{bmatrix}$$

has eigenvalues $\lambda_1 = 1 + i$ and $\lambda_2 = \overline{\lambda}_1 = 1 - i$. Corresponding eigenvectors are

$$\mathbf{x}_1 = \begin{bmatrix} 1 \\ i \end{bmatrix} \quad \text{and} \quad \mathbf{x}_2 = \overline{\mathbf{x}}_1 = \begin{bmatrix} 1 \\ -i \end{bmatrix}.$$

Since the eigenvalues are distinct, we conclude from the corollary of Theorem 6.9 that

$$\{\mathbf{y}_1(t), \mathbf{y}_2(t)\} = \left\{ e^{(1+i)t} \begin{bmatrix} 1 \\ i \end{bmatrix}, e^{(1-i)t} \begin{bmatrix} 1 \\ -i \end{bmatrix} \right\}, \tag{1}$$

is a fundamental set of solutions for $\mathbf{y}' = A\mathbf{y}$.

As given in equation (1), these two solutions are somewhat unsatisfactory since we generally want real-valued solutions. Sections 4.6 and 5.3 showed how to convert complex-valued solutions of scalar equations into real-valued solutions. In this subsection, we generalize those results to first order systems.

As in Section 4.6 and 5.3, the key result is that both the real and imaginary parts of a complex-valued solution are also solutions.

Theorem 6.10

Consider the differential equation $\mathbf{y}' = A\mathbf{y}$, $-\infty < t < \infty$, where A is an $(n \times n)$ real matrix. Let $\mathbf{y}(t) = \mathbf{u}(t) + i\mathbf{v}(t)$ be a complex-valued solution of this differential equation, where $\mathbf{u}(t)$ and $\mathbf{v}(t)$ are each real-valued $(n \times 1)$ vector functions representing the real and imaginary parts of $\mathbf{y}(t)$, respectively. Then, $\mathbf{u}(t)$ and $\mathbf{v}(t)$ are each solutions of $\mathbf{y}' = A\mathbf{y}$, $-\infty < t < \infty$.

PROOF: Substitute $\mathbf{y}(t) = \mathbf{u}(t) + i\mathbf{v}(t)$ into the left-hand side of the differential equation $\mathbf{y}' = A\mathbf{y}$ to obtain

$$(\mathbf{u}(t) + i\mathbf{v}(t))' = \begin{bmatrix} (u_1(t) + iv_1(t))' \\ (u_2(t) + iv_2(t))' \\ \vdots \\ (u_n(t) + iv_n(t))' \end{bmatrix} = \begin{bmatrix} u_1'(t) + iv_1'(t) \\ u_2'(t) + iv_2'(t) \\ \vdots \\ u_n'(t) + iv_n'(t) \end{bmatrix} = \mathbf{u}'(t) + i\mathbf{v}'(t). \tag{2}$$

Substituting $\mathbf{y}(t) = \mathbf{u}(t) + i\mathbf{v}(t)$ into the right-hand side of the differential equation $\mathbf{y}' = A\mathbf{y}$, we obtain

$$A(\mathbf{u}(t) + i\mathbf{v}(t)) = A\mathbf{u}(t) + A(i\mathbf{v}(t)) = A\mathbf{u}(t) + iA\mathbf{v}(t). \tag{3}$$

Two complex quantities are equal if and only if their corresponding real and imaginary parts are equal. Therefore, equating expressions (2) and (3), we see that

$$\mathbf{u}'(t) = A\mathbf{u}(t), \qquad \mathbf{v}'(t) = A\mathbf{v}(t), \qquad -\infty < t < \infty.$$ ❚

The following example illustrates the use of Theorem 6.10 to convert a set of complex-valued solutions into a set of real-valued solutions.

EXAMPLE 1

Find a real-valued fundamental set of solutions of $\mathbf{y}' = A\mathbf{y}$, $-\infty < t < \infty$, where

$$A = \begin{bmatrix} 1 & 1 \\ -1 & 1 \end{bmatrix}.$$

Solution: We know, from equation (1) that

$$\mathbf{y}(t) = e^{(1+i)t}\begin{bmatrix} 1 \\ i \end{bmatrix} \tag{4}$$

is a complex-valued solution of $\mathbf{y}' = A\mathbf{y}$. From Euler's formula [see equation (9) in Section 4.6],

$$e^{(1+i)t} = e^t e^{it} = e^t(\cos t + i\sin t) = e^t\cos t + ie^t\sin t.$$

Therefore, we can write solution (4) as

$$\begin{aligned} \mathbf{y}(t) &= (e^t\cos t + ie^t\sin t)\begin{bmatrix} 1 \\ i \end{bmatrix} \\ &= \begin{bmatrix} e^t\cos t \\ -e^t\sin t \end{bmatrix} + i\begin{bmatrix} e^t\sin t \\ e^t\cos t \end{bmatrix} \\ &= \mathbf{u}(t) + i\mathbf{v}(t). \end{aligned}$$

It follows from Theorem 6.10 that the two real functions,

$$\mathbf{u}(t) = \begin{bmatrix} e^t\cos t \\ -e^t\sin t \end{bmatrix} \quad \text{and} \quad \mathbf{v}(t) = \begin{bmatrix} e^t\sin t \\ e^t\cos t \end{bmatrix}$$

are also solutions of the differential equation. (You can also verify this claim by direct substitution into the differential equation.)

To show that they form a fundamental set of solutions on $-\infty < t < \infty$, we calculate the Wronskian of these two real-valued solutions and find

$$W(t) = \begin{vmatrix} e^t\cos t & e^t\sin t \\ -e^t\sin t & e^t\cos t \end{vmatrix} = e^{2t}(\cos^2 t + \sin^2 t) = e^{2t} \neq 0.$$

Therefore, the general solution of the differential equation is $\mathbf{y}(t) = c_1\mathbf{u}(t) + c_2\mathbf{v}(t)$:

$$\mathbf{y}(t) = c_1\begin{bmatrix} e^t\cos t \\ -e^t\sin t \end{bmatrix} + c_2\begin{bmatrix} e^t\sin t \\ e^t\cos t \end{bmatrix} = \begin{bmatrix} e^t\cos t & e^t\sin t \\ -e^t\sin t & e^t\cos t \end{bmatrix}\begin{bmatrix} c_1 \\ c_2 \end{bmatrix}.$$ ▲

❚ **REMARK:** According to Theorem 6.8 in Section 6.4, the two fundamental sets [the original set of complex-valued solutions (1) and the real-valued set constructed in Example 1] are related via multiplication on the right by a constant nonsingular matrix. We can illustrate Theorem 6.8 here by noting:

$$\begin{bmatrix} e^t \cos t & e^t \sin t \\ -e^t \sin t & e^t \cos t \end{bmatrix} = \begin{bmatrix} e^{(1+i)t} & e^{(1-i)t} \\ ie^{(1+i)t} & -ie^{(1-i)t} \end{bmatrix} \begin{bmatrix} \frac{1}{2} & \frac{-i}{2} \\ \frac{1}{2} & \frac{i}{2} \end{bmatrix},$$

where the constant matrix on the right-hand side is nonsingular since

$$\begin{vmatrix} \frac{1}{2} & \frac{-i}{2} \\ \frac{1}{2} & \frac{i}{2} \end{vmatrix} = \frac{i}{2} \neq 0. ▮$$

EXAMPLE 2

Solve the initial value problem $\mathbf{y}' = A\mathbf{y}$, $-\infty < t < \infty$,

$$A = \begin{bmatrix} 1 & 2 & -2 \\ 2 & 5 & -2 \\ 4 & 12 & -5 \end{bmatrix}, \qquad \mathbf{y}(0) = \begin{bmatrix} 1 \\ 1 \\ 0 \end{bmatrix}.$$

Solution: Since the characteristic polynomial is a cubic polynomial with real coefficients, there will be at least one real eigenvalue. Using a cofactor expansion along the first row, we obtain

$$p(\lambda) = \det(A - \lambda I) = \begin{vmatrix} 1-\lambda & 2 & -2 \\ 2 & 5-\lambda & -2 \\ 4 & 12 & -5-\lambda \end{vmatrix}$$

$$= (1-\lambda)\begin{vmatrix} 5-\lambda & -2 \\ 12 & -5-\lambda \end{vmatrix} - 2\begin{vmatrix} 2 & -2 \\ 4 & -5-\lambda \end{vmatrix} - 2\begin{vmatrix} 2 & 5-\lambda \\ 4 & 12 \end{vmatrix}$$

$$= -(\lambda^3 - \lambda^2 + 3\lambda + 5).$$

Computer software can certainly be used to determine the roots of $p(\lambda) = 0$. In this particular case, however, we see by inspection that $\lambda = -1$ is a root. Therefore, synthetic division and subsequent use of the quadratic formula can be used to obtain the factorization

$$p(\lambda) = -(\lambda + 1)(\lambda^2 - 2\lambda + 5) = -(\lambda + 1)(\lambda - 1 - 2i)(\lambda - 1 + 2i).$$

The eigenvalues are $\lambda_1 = -1$, $\lambda_2 = 1 + 2i$, $\lambda_3 = \overline{\lambda}_2 = 1 - 2i$. We now compute the eigenvectors.

(a) For $\lambda_1 = -1$, we solve $(A - \lambda_1 I)\mathbf{x}_1 = (A + I)\mathbf{x}_1 = \mathbf{0}$, or

$$\begin{bmatrix} 2 & 2 & -2 \\ 2 & 6 & -2 \\ 4 & 12 & -4 \end{bmatrix} \begin{bmatrix} x_1 \\ x_2 \\ x_3 \end{bmatrix} = \begin{bmatrix} 0 \\ 0 \\ 0 \end{bmatrix}.$$

Using elementary row operations, we obtain an equivalent homogeneous system

$$\begin{bmatrix} 1 & 0 & -1 \\ 0 & 1 & 0 \\ 0 & 0 & 0 \end{bmatrix} \begin{bmatrix} x_1 \\ x_2 \\ x_3 \end{bmatrix} = \begin{bmatrix} 0 \\ 0 \\ 0 \end{bmatrix}.$$

(continued)

(continued)

A nontrivial solution is

$$\mathbf{x}_1 = \begin{bmatrix} 1 \\ 0 \\ 1 \end{bmatrix}.$$

(b) For $\lambda_2 = 1 + 2i$, we solve $(A - \lambda_2 I)\mathbf{x} = (A - (1 + 2i)I)\mathbf{x} = \mathbf{0}$, or

$$\begin{bmatrix} -2i & 2 & -2 \\ 2 & 4 - 2i & -2 \\ 4 & 12 & -6 - 2i \end{bmatrix} \begin{bmatrix} x_1 \\ x_2 \\ x_3 \end{bmatrix} = \begin{bmatrix} 0 \\ 0 \\ 0 \end{bmatrix}.$$

In this case, elementary row operations lead to an equivalent system

$$\begin{bmatrix} 1 & 0 & -\frac{i}{2} \\ 0 & 1 & -\frac{1}{2} \\ 0 & 0 & 0 \end{bmatrix} \begin{bmatrix} x_1 \\ x_2 \\ x_3 \end{bmatrix} = \begin{bmatrix} 0 \\ 0 \\ 0 \end{bmatrix}.$$

A nontrivial solution is

$$\mathbf{x}_2 = \begin{bmatrix} i \\ 1 \\ 2 \end{bmatrix}.$$

Although we do not need it to solve the given initial value problem, we know an eigenvector $\mathbf{x}_3$ corresponding to the eigenvalue $\lambda_3 = \overline{\lambda}_2 = 1 - 2i$ is given by

$$\mathbf{x}_3 = \overline{\mathbf{x}}_2 = \begin{bmatrix} -i \\ 1 \\ 2 \end{bmatrix}.$$

We now develop a real-valued fundamental set of solutions. One solution [corresponding to the eigenpair $(\lambda_1, \mathbf{x}_1)$ is

$$\mathbf{y}_1(t) = e^{-t} \begin{bmatrix} 1 \\ 0 \\ 1 \end{bmatrix} = \begin{bmatrix} e^{-t} \\ 0 \\ e^{-t} \end{bmatrix}.$$

Since $\mathbf{y}' = A\mathbf{y}$ is a system of three first order equations, we know that a fundamental set must consist of three solutions. To obtain the two other solutions needed, we take the complex-valued solution determined by the eigenpair $(\lambda_2, \mathbf{x}_2)$,

$$\mathbf{y}(t) = e^{(1+2i)t} \begin{bmatrix} i \\ 1 \\ 2 \end{bmatrix},$$

and decompose it into the form $\mathbf{u}(t) + i\mathbf{v}(t)$. [The other complex-valued solution, determined by the eigenpair $(\lambda_3, \mathbf{x}_3)$, decomposes into $\mathbf{u}(t) - i\mathbf{v}(t)$, yielding (essentially) the same pair of real-valued solutions.] Decomposing $\mathbf{y}(t)$, we obtain

$$\mathbf{y}(t) = e^{(1+2i)t}\begin{bmatrix} i \\ 1 \\ 2 \end{bmatrix} = e^t(\cos 2t + i\sin 2t)\begin{bmatrix} i \\ 1 \\ 2 \end{bmatrix}$$
$$= \begin{bmatrix} -e^t\sin 2t \\ e^t\cos 2t \\ 2e^t\cos 2t \end{bmatrix} + i\begin{bmatrix} e^t\cos 2t \\ e^t\sin 2t \\ 2e^t\sin 2t \end{bmatrix}.$$

We complete the fundamental set by setting

$$\mathbf{y}_2(t) = \begin{bmatrix} -e^t\sin 2t \\ e^t\cos 2t \\ 2e^t\cos 2t \end{bmatrix} \quad \text{and} \quad \mathbf{y}_3(t) = \begin{bmatrix} e^t\cos 2t \\ e^t\sin 2t \\ 2e^t\sin 2t \end{bmatrix}.$$

The general solution, $\mathbf{y}(t) = c_1\mathbf{y}_1(t) + c_2\mathbf{y}_2(t) + c_3\mathbf{y}_3(t)$, is therefore

$$\mathbf{y}(t) = c_1\begin{bmatrix} e^{-t} \\ 0 \\ e^{-t} \end{bmatrix} + c_2\begin{bmatrix} -e^t\sin 2t \\ e^t\cos 2t \\ 2e^t\cos 2t \end{bmatrix} + c_3\begin{bmatrix} e^t\cos 2t \\ e^t\sin 2t \\ 2e^t\sin 2t \end{bmatrix}$$
$$= \begin{bmatrix} e^{-t} & -e^t\sin 2t & e^t\cos 2t \\ 0 & e^t\cos 2t & e^t\sin 2t \\ e^{-t} & 2e^t\cos 2t & 2e^t\sin 2t \end{bmatrix}\begin{bmatrix} c_1 \\ c_2 \\ c_3 \end{bmatrix}. \tag{5}$$

The fact that our three solutions form a fundamental set can be verified directly by noting that the Wronskian, evaluated at $t = 0$, is

$$W(0) = \begin{vmatrix} 1 & 0 & 1 \\ 0 & 1 & 0 \\ 1 & 2 & 0 \end{vmatrix} = -1 \neq 0.$$

To solve the given initial value problem, we impose the initial condition in equation (5), finding

$$\begin{bmatrix} 1 & 0 & 1 \\ 0 & 1 & 0 \\ 1 & 2 & 0 \end{bmatrix}\begin{bmatrix} c_1 \\ c_2 \\ c_3 \end{bmatrix} = \begin{bmatrix} 1 \\ 1 \\ 0 \end{bmatrix}.$$

The solution is

$$\begin{bmatrix} c_1 \\ c_2 \\ c_3 \end{bmatrix} = \begin{bmatrix} -2 \\ 1 \\ 3 \end{bmatrix}$$

and, therefore, the unique solution of the initial value problem is

$$\mathbf{y}(t) = -2\begin{bmatrix} e^{-t} \\ 0 \\ e^{-t} \end{bmatrix} + \begin{bmatrix} -e^t\sin 2t \\ e^t\cos 2t \\ 2e^t\cos 2t \end{bmatrix} + 3\begin{bmatrix} e^t\cos 2t \\ e^t\sin 2t \\ 2e^t\sin 2t \end{bmatrix} = \begin{bmatrix} -2e^{-t} - e^t\sin 2t + 3e^t\cos 2t \\ e^t\cos 2t + 3e^t\sin 2t \\ -2e^{-t} + 2e^t\cos 2t + 6e^t\sin 2t \end{bmatrix}.$$

▲

We now apply the results developed in this section to a problem arising in charged particle ballistics.

Motion of a Charged Particle in a Magnetic Field

Consider a particle, having mass m and electric charge q, moving in a magnetic field. The magnetic field is a vector field $\mathbf{B}$. The motion of the particle in this magnetic field is most conveniently described in terms of vectors.

Let $\mathbf{i}, \mathbf{j}$, and $\mathbf{k}$ represent unit vectors in the x, y, and z directions, respectively. The position of the particle is defined by the position vector,

$$\mathbf{r}(t) = x(t)\mathbf{i} + y(t)\mathbf{j} + z(t)\mathbf{k},$$

and its velocity by the corresponding velocity vector,

$$\mathbf{v}(t) = v_x(t)\mathbf{i} + v_y(t)\mathbf{j} + v_z(t)\mathbf{k}.$$

Since $\mathbf{v}(t) = d\mathbf{r}/dt$, the velocity components are

$$v_x(t) = \frac{d}{dt}x(t), \qquad v_y(t) = \frac{d}{dt}y(t), \qquad v_z(t) = \frac{d}{dt}z(t).$$

Figure 6.2 illustrates the problem.

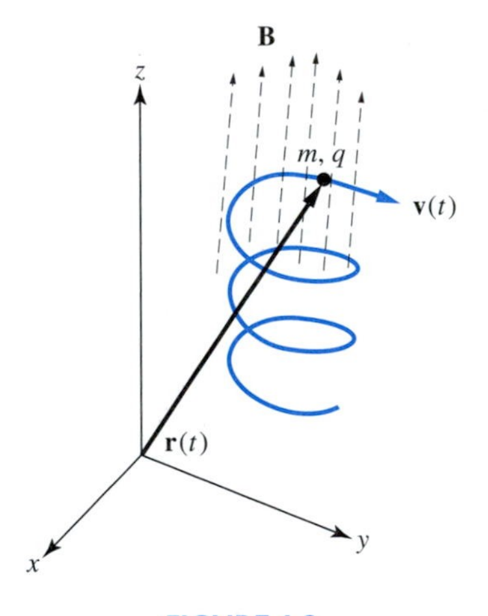

FIGURE 6.2

A charged particle having mass m and electric charge q moves in the magnetic field $\mathbf{B}$. Its motion is described by its position vector $\mathbf{r}(t)$ and velocity vector $\mathbf{v}(t)$.

If the weight of the charged particle is neglected, the only force acting on it is the Lorentz[1] force (the force the magnetic field exerts on the moving charge). The Lorentz force, described using vector notation, is $q\mathbf{v}(t) \times \mathbf{B}$. Thus,

[1] Hendrik Lorentz (1853–1928) was professor of mathematical physics at Leiden University from 1878 until 1912; he thereafter directed research at the Teyler Institute in Haarlem. Lorentz is noted for his studies of atomic structure and of the mathematical transformations (called Lorentz transformations) which form the basis of Einstein's theory of special relativity. Along with his student Pieter Zeeman, Lorentz was awarded the Nobel Prize in 1902.

an application of Newton's second law of motion leads to the vector equation

$$m \frac{d}{dt}\mathbf{v}(t) = q\mathbf{v}(t) \times \mathbf{B}. \tag{6}$$

This vector formulation of the equations of motion provides physical insight. Since a vector cross product is perpendicular to both of its constituent vectors, the Lorentz force acting on the particle is perpendicular to both the instantaneous velocity and to the magnetic field. The instantaneous acceleration, therefore, has no tangential component. In other words, the Lorentz force does not alter the speed of the particle. Kinetic energy is conserved and the Lorentz force acts only to change the direction of the particle's path. To verify, mathematically, that kinetic energy is conserved, note that

$$\frac{d}{dt}\left(\frac{1}{2}m\mathbf{v}\cdot\mathbf{v}\right) = m\mathbf{v}\cdot\frac{d\mathbf{v}}{dt} = q\mathbf{v}\cdot(\mathbf{v}\times\mathbf{B}) = q\mathbf{B}\cdot(\mathbf{v}\times\mathbf{v}) = 0.$$

While the vector formulation (6) is a concise and insightful summary of the relevant physics, we must rewrite equation (6) as a first order linear system in order to determine $\mathbf{v}(t)$. An illustration is given in the next example.

EXAMPLE 3

Assume that a constant magnetic field, given by $\mathbf{B} = B_0\mathbf{k}$, occupies all of space. At time $t = 0$ a particle having mass m and electric charge q leaves the origin with initial velocity $\mathbf{v}(0) = v_1\mathbf{i} + v_2\mathbf{j} + v_3\mathbf{k}$. Where will the particle be at later times, $t > 0$?

Solution: For brevity, let $\omega_C = qB_0/m$. With this, equation (6) can be rewritten as

$$\frac{d}{dt}\mathbf{v}(t) = \omega_C\mathbf{v}(t) \times \mathbf{k},$$

or, in terms of components,

$$\frac{dv_x}{dt}\mathbf{i} + \frac{dv_y}{dt}\mathbf{j} + \frac{dv_z}{dt}\mathbf{k} = \omega_C(v_x\mathbf{i} + v_y\mathbf{j} + v_z\mathbf{k}) \times \mathbf{k} = \omega_C v_y\mathbf{i} - \omega_C v_x\mathbf{j}.$$

Since vectors are equal if and only if their corresponding components are equal, we obtain the following system of equations:

$$\begin{aligned} v_x' &= \omega_C v_y \\ v_y' &= -\omega_C v_x \\ v_z' &= 0 \end{aligned} \tag{7}$$

Define the (3×1) vector function $\mathbf{v}(t)$ by

$$\mathbf{v}(t) = \begin{bmatrix} v_x(t) \\ v_y(t) \\ v_z(t) \end{bmatrix}.$$

Equation (7), together with the given initial conditions, leads to the initial value problem

$$\mathbf{v}'(t) = \begin{bmatrix} 0 & \omega_C & 0 \\ -\omega_C & 0 & 0 \\ 0 & 0 & 0 \end{bmatrix}\mathbf{v}(t), \qquad \mathbf{v}(0) = \begin{bmatrix} v_1 \\ v_2 \\ v_3 \end{bmatrix}. \tag{8}$$

(continued)

(continued)

The eigenvalues of the coefficient matrix are $\lambda_1 = 0$, $\lambda_2 = i\omega_C$, and $\lambda_3 = -i\omega_C$. Corresponding eigenvectors are

$$\mathbf{x}_1 = \begin{bmatrix} 0 \\ 0 \\ 1 \end{bmatrix}, \qquad \mathbf{x}_2 = \begin{bmatrix} 1 \\ i \\ 0 \end{bmatrix}, \qquad \mathbf{x}_3 = \begin{bmatrix} 1 \\ -i \\ 0 \end{bmatrix}.$$

To obtain a real fundamental set, we decompose the complex solution corresponding to the eigenpair $(\lambda_2, \mathbf{x}_2)$ into two real solutions,

$$e^{i\omega_C t}\begin{bmatrix} 1 \\ i \\ 0 \end{bmatrix} = (\cos\omega_C t + i\sin\omega_C t)\begin{bmatrix} 1 \\ i \\ 0 \end{bmatrix} = \begin{bmatrix} \cos\omega_C t \\ -\sin\omega_C t \\ 0 \end{bmatrix} + i\begin{bmatrix} \sin\omega_C t \\ \cos\omega_C t \\ 0 \end{bmatrix}.$$

The general solution is therefore

$$\mathbf{v}(t) = c_1\begin{bmatrix} 0 \\ 0 \\ 1 \end{bmatrix} + c_2\begin{bmatrix} \cos\omega_C t \\ -\sin\omega_C t \\ 0 \end{bmatrix} + c_3\begin{bmatrix} \sin\omega_C t \\ \cos\omega_C t \\ 0 \end{bmatrix} = \begin{bmatrix} 0 & \cos\omega_C t & \sin\omega_C t \\ 0 & -\sin\omega_C t & \cos\omega_C t \\ 1 & 0 & 0 \end{bmatrix}\begin{bmatrix} c_1 \\ c_2 \\ c_3 \end{bmatrix}.$$

Imposing the initial condition leads to the solution of the given initial value problem,

$$\mathbf{v}(t) = \begin{bmatrix} v_1\cos\omega_C t + v_2\sin\omega_C t \\ -v_1\sin\omega_C t + v_2\cos\omega_C t \\ v_3 \end{bmatrix}.$$

We next define a (3×1) position vector,

$$\mathbf{r}(t) = \begin{bmatrix} x(t) \\ y(t) \\ z(t) \end{bmatrix}.$$

Since $\mathbf{v}(t) = \mathbf{r}'(t)$,

$$\mathbf{r}(t) = \mathbf{r}(0) + \int_0^t \mathbf{v}(s)\,ds = \int_0^t \mathbf{v}(s)\,ds.$$

(The last equality follows because the particle starts at the origin.) Therefore,

$$\mathbf{r}(t) = \begin{bmatrix} \omega_C^{-1}(v_1\sin\omega_C t + v_2(1-\cos\omega_C t)) \\ \omega_C^{-1}(v_2\sin\omega_C t - v_1(1-\cos\omega_C t)) \\ v_3 t \end{bmatrix}.$$

At this point, the motivation for the notation $\omega_c = qB_0/m$ is perhaps clear. This constant is a radian frequency (see Section 4.7) called the **cyclotron frequency**. The particle executes circular motion (at the cyclotron frequency) in the xy-plane while simultaneously moving with constant speed in the z-direction; we ask you to verify the form of the particle's motion in Exercise 21 by showing that

$$(x(t) - \omega_C^{-1}v_2)^2 + (y(t) + \omega_C^{-1}v_1)^2 = \omega_C^{-2}(v_1^2 + v_2^2). \qquad \textbf{(9)}$$

The trajectory of the charged particle is a helix if $v_3 \neq 0$ and a circle if $v_3 = 0$. This type of behavior is the basis for using magnetic fields to spatially confine clouds of charged particles, called plasmas. In terms of more wide-ranging applications, we note that the Lorentz force is (ultimately) the basis for the operation of electric motors and generators. ▲

EXERCISES

Exercises 1–6:

Find the eigenvalues and eigenvectors of the given matrix A. [Hint: In Exercise 6, $\lambda = 2$ is an eigenvalue.]

1. $A = \begin{bmatrix} 2 & 1 \\ -1 & 2 \end{bmatrix}$ **2.** $A = \begin{bmatrix} 0 & -9 \\ 1 & 0 \end{bmatrix}$ **3.** $A = \begin{bmatrix} 6 & -13 \\ 1 & 0 \end{bmatrix}$

4. $A = \begin{bmatrix} 3 & 1 \\ -2 & 1 \end{bmatrix}$ **5.** $A = \begin{bmatrix} -2 & -2 & -9 \\ -1 & 1 & -3 \\ 1 & 1 & 4 \end{bmatrix}$ **6.** $A = \begin{bmatrix} 1 & -4 & -1 \\ 3 & 2 & 3 \\ 1 & 1 & 3 \end{bmatrix}$

Exercises 7–12:

In each exercise, one or more eigenvalues and corresponding eigenvectors are given for a real matrix A. Determine a fundamental set of solutions for $\mathbf{y}' = A\mathbf{y}$, where the fundamental set consists entirely of *real* solutions.

7. A is (2×2) with an eigenvalue $\lambda = 4 + 2i$ and corresponding eigenvector

$$\mathbf{x} = \begin{bmatrix} 4 \\ -1 + i \end{bmatrix}.$$

8. A is (2×2) with an eigenvalue $\lambda = i$ and corresponding eigenvector

$$\mathbf{x} = \begin{bmatrix} -2 + i \\ 5 \end{bmatrix}.$$

9. A is (2×2) with an eigenvalue $\lambda = 2i$ and corresponding eigenvector

$$\mathbf{x} = \begin{bmatrix} -1 - i \\ 1 \end{bmatrix}.$$

10. A is (2×2) with an eigenvalue $\lambda = 1 + i$ and corresponding eigenvector

$$\mathbf{x} = \begin{bmatrix} -1 + i \\ i \end{bmatrix}$$

11. A is (3×3) with a complex eigenvalue $\lambda = 2 + 3i$ and corresponding eigenvector

$$\mathbf{x} = \begin{bmatrix} -5 + 3i \\ 3 + 3i \\ 2 \end{bmatrix},$$

and a real eigenvalue $\lambda = 2$ with corresponding eigenvector

$$\mathbf{x} = \begin{bmatrix} 1 \\ 0 \\ -1 \end{bmatrix}.$$

12. A is (4×4) and has two different complex eigenvalues: $\lambda = 1 + 5i$ with corresponding eigenvector

$$\mathbf{x} = \begin{bmatrix} i \\ 1 \\ 0 \\ 0 \end{bmatrix},$$

and $\lambda = 1 + 2i$ with corresponding eigenvector

$$\mathbf{x} = \begin{bmatrix} 0 \\ 0 \\ i \\ 1 \end{bmatrix}.$$

Exercises 13–18:

Solve the initial value problem $\mathbf{y}' = A\mathbf{y}$, $\mathbf{y}(0) = \mathbf{y}_0$. Eigenpairs were determined in Exercises 1–6.

13. $A = \begin{bmatrix} 2 & 1 \\ -1 & 2 \end{bmatrix}, \quad \mathbf{y}_0 = \begin{bmatrix} 4 \\ 7 \end{bmatrix}$

14. $A = \begin{bmatrix} 0 & -9 \\ 1 & 0 \end{bmatrix}, \quad \mathbf{y}_0 = \begin{bmatrix} 6 \\ 2 \end{bmatrix}$

15. $A = \begin{bmatrix} 6 & -13 \\ 1 & 0 \end{bmatrix}, \quad \mathbf{y}_0 = \begin{bmatrix} 1 \\ 3 \end{bmatrix}$

16. $A = \begin{bmatrix} 3 & 1 \\ -2 & 1 \end{bmatrix}, \quad \mathbf{y}_0 = \begin{bmatrix} 8 \\ 6 \end{bmatrix}$

17. $A = \begin{bmatrix} -2 & -2 & -9 \\ -1 & 1 & -3 \\ 1 & 1 & 4 \end{bmatrix}, \quad \mathbf{y}_0 = \begin{bmatrix} 6 \\ 1 \\ 2 \end{bmatrix}$

18. $A = \begin{bmatrix} 1 & -4 & -1 \\ 3 & 2 & 3 \\ 1 & 1 & 3 \end{bmatrix}, \quad \mathbf{y}_0 = \begin{bmatrix} -1 \\ 9 \\ 4 \end{bmatrix}$

19. Let A be a real (2×2) matrix having $\lambda = \alpha + i\beta$ as a complex eigenvalue with β nonzero. Show that any eigenvector $\mathbf{x}$ corresponding to λ must have at least one complex component. [Hint: Assume that $\mathbf{x}$ is a real vector and arrive at a contradiction.]

20. Let A be a real (2×2) matrix with a complex eigenvalue $\lambda = \alpha + i\beta$, where $\beta \neq 0$. As we know from Exercise 19, $\mathbf{x}$ must have at least one complex component. Show that $\mathbf{x}$ must have two nonzero components. [Hint: Show that assuming

$$\mathbf{x} = \begin{bmatrix} \mu + i\delta \\ 0 \end{bmatrix} \quad \text{or} \quad \mathbf{x} = \begin{bmatrix} 0 \\ \mu + i\delta \end{bmatrix},$$

where $\delta \neq 0$ leads to a contradiction.]

21. Verify equation (9), showing that the charged particle in Example 3 follows a circular path when $v_3 = 0$ and a helical path when $v_3 \neq 0$.

Exercises 22–25:

Consider the initial value problem $\mathbf{y}' = A\mathbf{y}$, $\mathbf{y}(0) = \mathbf{y}_0$. In each exercise, the given matrix A contains a real parameter μ. Determine all values of μ, if any, such that A has distinct eigenvalues and such that: $\sqrt{y_1(t)^2 + y_2(t)^2} \to 0$ as $t \to \infty$ for every initial vector $\mathbf{y}_0$.

22. $A = \begin{bmatrix} 1 & 3 \\ \mu & -2 \end{bmatrix}$

23. $A = \begin{bmatrix} -2 & \mu \\ 1 & -3 \end{bmatrix}$

24. $A = \begin{bmatrix} -3 & -\mu \\ \mu & 1 \end{bmatrix}$

25. $A = \begin{bmatrix} -3 & \mu \\ \mu & 1 \end{bmatrix}$

Vector Differential Equations We now consider differential equations of the type arising in the study of charged particle motion in a magnetic field. The differential equations are conveniently described in vector form; the exercises ask you to solve the equivalent homogeneous linear systems.

26. Let $\mathbf{r}(t) = x(t)\mathbf{i} + y(t)\mathbf{j}$ represent the position vector of a particle moving in the xy-plane and let $\mathbf{v}(t) = v_x(t)\mathbf{i} + v_y(t)\mathbf{j}$ be the corresponding velocity vector. Assume that Newton's law of motion leads to the following vector equation for the velocity,

$$\frac{d}{dt}\mathbf{v}(t) = 2\mathbf{k} \times \mathbf{v}(t). \tag{10}$$

(a) Write the component differential equations arising from (10) as a homogeneous linear system of the form

$$\frac{d}{dt}\begin{bmatrix} v_x \\ v_y \end{bmatrix} = \begin{bmatrix} a_{11} & a_{12} \\ a_{21} & a_{22} \end{bmatrix}\begin{bmatrix} v_x \\ v_y \end{bmatrix} \tag{11}$$

and obtain the general solution of (11).

(b) Assume that at time $t = 0$, the particle is located at $\mathbf{r}(0) = 2\mathbf{i} + \mathbf{j}$ and has initial velocity $\mathbf{v}(0) = \mathbf{i} + 2\mathbf{j}$. What will be the velocity $\mathbf{v}(t)$ and the position $\mathbf{r}(t)$ of the particle at time $t = 3\pi/2$?

27. Let $\mathbf{r}(t) = x(t)\mathbf{i} + y(t)\mathbf{j} + z(t)\mathbf{k}$ represent the position vector of a particle moving in space, and let $\mathbf{v}(t) = v_x(t)\mathbf{i} + v_y(t)\mathbf{j} + v_z(t)\mathbf{k}$ be the corresponding velocity vector. Assume that Newton's law of motion leads to the following vector equation for the velocity,

$$\frac{d}{dt}\mathbf{v}(t) = (\mathbf{i} + \mathbf{j} + \mathbf{k}) \times \mathbf{v}(t). \tag{12}$$

(a) Write the component differential equations arising from (12) as a homogeneous linear system of the form

$$\frac{d}{dt}\begin{bmatrix} v_x \\ v_y \\ v_z \end{bmatrix} = \begin{bmatrix} a_{11} & a_{12} & a_{13} \\ a_{21} & a_{22} & a_{23} \\ a_{31} & a_{32} & a_{33} \end{bmatrix}\begin{bmatrix} v_x \\ v_y \\ v_z \end{bmatrix} \tag{13}$$

(b) Show that the differential equation (13) that arises from (12) has a nontrivial equilibrium solution. Determine this equilibrium solution.

(c) An eigenpair of the (3×3) coefficient matrix in (13) is

$$\lambda = i\sqrt{3}, \qquad \mathbf{x} = \begin{bmatrix} 2 \\ -1 - i\sqrt{3} \\ -1 + i\sqrt{3} \end{bmatrix}.$$

Determine the general solution of homogeneous linear system (13).

(d) Solve the initial value problem consisting of equation (13) and initial condition $v_x(0) = v_0$, $v_y(0) = v_0$, $v_z(0) = v_0$. How does the initial condition, rewritten as a vector, relate to the vector $\mathbf{i} + \mathbf{j} + \mathbf{k}$ appearing on the right-hand side of differential equation (12)? If the vector $\mathbf{i} + \mathbf{j} + \mathbf{k}$ represents the vector direction of a magnetic field, explain the physical significance of the solution obtained.

28. Consider the vector differential equation $m\mathbf{v}'(t) = -\gamma\mathbf{v}(t) + q\mathbf{v}(t) \times \mathbf{B}$. Such an equation arises in modeling the motion of a charged particle when the particle is subjected to both a Lorentz force and a drag force proportional to velocity.

(a) Let $\mathbf{v}(t) = v_x(t)\mathbf{i} + v_y(t)\mathbf{j} + v_z(t)\mathbf{k}$ and $\mathbf{B} = B_x\mathbf{i} + B_y\mathbf{j} + B_z\mathbf{k}$. Show that the vector differential equation is equivalent to the homogeneous linear system

$$\frac{d}{dt}\begin{bmatrix} v_x \\ v_y \\ v_z \end{bmatrix} = -\frac{\gamma}{m}\begin{bmatrix} 1 & 0 & 0 \\ 0 & 1 & 0 \\ 0 & 0 & 1 \end{bmatrix}\begin{bmatrix} v_x \\ v_y \\ v_z \end{bmatrix} + \frac{q}{m}\begin{bmatrix} 0 & B_z & -B_y \\ -B_z & 0 & B_x \\ B_y & -B_x & 0 \end{bmatrix}\begin{bmatrix} v_x \\ v_y \\ v_z \end{bmatrix}. \quad (14)$$

(b) Let

$$A = \frac{q}{m}\begin{bmatrix} 0 & B_z & -B_y \\ -B_z & 0 & B_x \\ B_y & -B_x & 0 \end{bmatrix}.$$

If the eigenpairs of A are represented as λ_1, $\mathbf{x}_1$, λ_2, $\mathbf{x}_2$ and λ_3, $\mathbf{x}_3$, what are the corresponding eigenpairs of $-(\gamma/m)I + A$? If $\Psi(t)$ represents a fundamental matrix for the homogeneous linear system $\mathbf{y}' = A\mathbf{y}$, what is a corresponding fundamental matrix for the homogeneous linear system $\mathbf{y}' = (-(\gamma/m)I + A)\mathbf{y}$?

(c) For simplicity, let $B_x = B_y = 0$, $(q/m)B_z = \omega_C$ and $v_x(0) = v_{x_0}$, $v_y(0) = v_{y_0}$, $v_z(0) = v_{z_0}$. The speed of the particle at time t is defined as

$$v(t) = \sqrt{v_x^2(t) + v_y^2(t) + v_z^2(t)}.$$

Show from the solution of the initial value problem that

$$v(t) = e^{-(\gamma/m)t}\sqrt{v_{x_0}^2 + v_{y_0}^2 + v_{z_0}^2} = e^{-(\gamma/m)t}v(0).$$

Thus, drag causes the particle to slow down as time evolves. [This result also follows from the vector form of the differential equation if we take the dot product of both sides of the differential equation with the velocity vector $\mathbf{v}(t)$.]

29. In Chapter 4 we saw that studying the motion of a mass suspended from the end of a hanging spring leads to the initial value problem

$$my'' + ky = 0, \qquad y(0) = y_0, \qquad y'(0) = y_0'. \quad (15)$$

In (15), $y(t)$ represents the displacement of the mass from its equilibrium rest position (positive in the downward direction); the constants y_0 and y_0' represent the initial displacement and velocity, respectively. Although the techniques of Chapter 4 are most conveniently used to obtain the solution, it is instructive to also solve this problem by recasting it as an initial value problem for a first order system.

(a) Let $z_1(t) = y(t)$, $z_2(t) = y'(t)$ and

$$\mathbf{z}(t) = \begin{bmatrix} z_1(t) \\ z_2(t) \end{bmatrix}.$$

Rewrite initial value problem (15) as an initial value problem of the form $\mathbf{z}' = A\mathbf{z}$, $\mathbf{z}(0) = \mathbf{z}_0$.

(b) Determine the eigenpairs of the coefficient matrix A.

(c) Represent the general solution of the initial value problem as $\mathbf{z}(t) = \Psi(t)\mathbf{c}$,here $\Psi(t)$ is a real-valued fundamental matrix and $\mathbf{c}$ is a (2×1) vector of arbitrary constants.

(d) Impose the initial conditions to determine $\mathbf{c}$ and obtain the unique solution of the initial value problem. As a check, does the second component, $z_2(t)$, equal the derivative of the first component, $z_1(t)$?

6.7 Repeated Eigenvalues

In Section 4.5 we discussed the second order homogeneous scalar equation

$$y'' - 2\alpha y' + \alpha^2 y = 0. \quad (1)$$

Looking for solutions of the form $y(t) = e^{\lambda t}$ led us to a characteristic polynomial with repeated roots,

$$\lambda^2 - 2\alpha\lambda + \alpha^2 = (\lambda - \alpha)^2.$$

One solution of equation (1) is $y_1(t) = e^{\alpha t}$. A second solution (needed to form a fundamental set of solutions) was found to be $y_2(t) = te^{\alpha t}$.

In our present study of the homogeneous first order linear system $\mathbf{y}' = A\mathbf{y}$, an analogous situation arises when the constant coefficient matrix A has repeated eigenvalues. (We say that A has **repeated eigenvalues** whenever the characteristic equation, $\det[A - \lambda I] = 0$, has repeated roots.) The process of finding a fundamental set of solutions when A has repeated eigenvalues is more complicated than in the repeated root scalar case. We give an example that illustrates the complications.

Complications from Repeated Eigenvalues

If A has repeated eigenvalues, are there enough linearly independent eigenvectors to form a fundamental set of solutions for $\mathbf{y}' = A\mathbf{y}$? To illustrate the complications that may arise from repeated eigenvalues, consider the linear system $\mathbf{y}' = A\mathbf{y}$, where

$$A = \begin{bmatrix} \alpha & \beta \\ 0 & \alpha \end{bmatrix}. \quad (2)$$

The characteristic polynomial for A is

$$p(\lambda) = (\lambda - \alpha)^2.$$

Thus, $\lambda = \alpha$ is a repeated eigenvalue of A.

If $\beta = 0$ in equation (2), then

$$\mathbf{x}_1 = \begin{bmatrix} 1 \\ 0 \end{bmatrix} \quad \text{and} \quad \mathbf{x}_2 = \begin{bmatrix} 0 \\ 1 \end{bmatrix}$$

are eigenvectors corresponding to the eigenvalue $\lambda = \alpha$. [In fact, the matrix $A - \alpha I$ is the zero matrix and thus any nonzero (2×1) vector is an eigenvector.] Since the set $\{\mathbf{x}_1, \mathbf{x}_2\}$ is linearly independent, we know from Theorem 6.9 that the functions

$$\mathbf{y}_1(t) = e^{\alpha t}\mathbf{x}_1 \quad \text{and} \quad \mathbf{y}_2(t) = e^{\alpha t}\mathbf{x}_2 \quad (3)$$

form a fundamental set of solutions for $\mathbf{y}' = A\mathbf{y}$. However, what happens if $\beta \neq 0$ in equation (2)? In this event, as shown below, there is (essentially) only *one* eigenvector $\mathbf{x}_1$. Hence, we need a further analysis of $\mathbf{y}' = A\mathbf{y}$ in order to find the second solution needed for a fundamental set of solutions.

When the value β in equation (2) is nonzero, A does not have a set of two linearly independent eigenvectors. The eigenvector equation is $(A - \alpha I)\mathbf{x} = \mathbf{0}$, or

$$\begin{bmatrix} 0 & \beta \\ 0 & 0 \end{bmatrix} \begin{bmatrix} x_1 \\ x_2 \end{bmatrix} = \begin{bmatrix} 0 \\ 0 \end{bmatrix}.$$

Hence, the equation $(A - \alpha I)\mathbf{x} = \mathbf{0}$ reduces to $\beta x_2 = 0$. Since $\beta \neq 0$, it follows that $x_2 = 0$. Thus, every eigenvector corresponding to $\lambda = \alpha$ has the form

$$\mathbf{x} = \begin{bmatrix} x_1 \\ x_2 \end{bmatrix} = \begin{bmatrix} c \\ 0 \end{bmatrix} = c \begin{bmatrix} 1 \\ 0 \end{bmatrix} = c\mathbf{x}_1. \tag{4}$$

We now have one member of a fundamental set of solutions, namely

$$\mathbf{y}_1(t) = e^{\alpha t}\mathbf{x}_1 = \begin{bmatrix} e^{\alpha t} \\ 0 \end{bmatrix}.$$

How do we find a second solution, $\mathbf{y}_2(t)$?

Finding a Second Solution When the Value β in Equation (2) Is Nonzero

For the simple example (2), we can find a second solution $\mathbf{y}_2(t)$ by sequentially solving the component equations. Let

$$\mathbf{y}(t) = \begin{bmatrix} y_1(t) \\ y_2(t) \end{bmatrix}.$$

In component form, the differential equation $\mathbf{y}' = A\mathbf{y}$ is

$$\begin{aligned} y_1' &= \alpha y_1 + \beta y_2 \\ y_2' &= \alpha y_2. \end{aligned} \tag{5}$$

We first solve the second equation, finding $y_2(t) = c_2 e^{\alpha t}$. Next, we substitute this expression for $y_2(t)$ into the first equation, obtaining

$$y_1' = \alpha y_1 + \beta c_2 e^{\alpha t}.$$

Solving this first order linear equation we arrive at $y_1(t) = c_1 e^{\alpha t} + c_2 \beta t e^{\alpha t}$. Therefore, the general solution of $\mathbf{y}' = A\mathbf{y}$ is

$$\begin{aligned} \mathbf{y}(t) = \begin{bmatrix} y_1(t) \\ y_2(t) \end{bmatrix} &= \begin{bmatrix} c_1 e^{\alpha t} + c_2 \beta t e^{\alpha t} \\ c_2 e^{\alpha t} \end{bmatrix} = \begin{bmatrix} c_1 e^{\alpha t} \\ 0 \end{bmatrix} + \begin{bmatrix} c_2 \beta t e^{\alpha t} \\ c_2 e^{\alpha t} \end{bmatrix} \\ &= c_1 e^{\alpha t} \begin{bmatrix} 1 \\ 0 \end{bmatrix} + c_2 \left\{ t e^{\alpha t} \begin{bmatrix} \beta \\ 0 \end{bmatrix} + e^{\alpha t} \begin{bmatrix} 0 \\ 1 \end{bmatrix} \right\} \\ &= c_1 \mathbf{y}_1(t) + c_2 \mathbf{y}_2(t). \end{aligned}$$

As in the repeated-root scalar case, we see that the function $te^{\alpha t}$ enters into the second solution,

$$\mathbf{y}_2(t) = t e^{\alpha t} \begin{bmatrix} \beta \\ 0 \end{bmatrix} + e^{\alpha t} \begin{bmatrix} 0 \\ 1 \end{bmatrix}.$$

We also see, however, that the second solution is not simply of the form $te^{\alpha t}\mathbf{x}_1$; rather, it has the form $\mathbf{y}_2(t) = te^{\alpha t}\mathbf{v}_1 + e^{\alpha t}\mathbf{v}_2$, where $\mathbf{v}_1$ and $\mathbf{v}_2$ are nonzero constant vectors. [Note that the vector we are calling $\mathbf{v}_1$ is actually an eigenvector since $\mathbf{v}_1 = \beta\mathbf{x}_1$ (see equation (4)).]

To verify, for this example, that $\{\mathbf{y}_1(t), \mathbf{y}_2(t)\}$ is a fundamental set of solutions, let $\Psi(t) = [\mathbf{y}_1(t), \mathbf{y}_2(t)]$. At $t = 0$, the Wronskian is nonzero since

$$W(0) = \det[\Psi(0)] = \det[\mathbf{y}_1(0), \mathbf{y}_2(0)] = \begin{vmatrix} 1 & 0 \\ 0 & 1 \end{vmatrix} = 1.$$

How do we find a second solution when we cannot solve the component equations sequentially as we did in equation (5)? In the next subsection we will use this example as a guide to develop a procedure for finding a second solution of $\mathbf{y}' = A\mathbf{y}$ in the case where A is a (2×2) constant matrix with a repeated eigenvalue. Later in this section we comment on the general case where A is an $(n \times n)$ matrix.

Finding a Fundamental Set of Solutions When an Eigenvalue Is Repeated

Consider the differential equation $\mathbf{y}' = A\mathbf{y}$, $-\infty < t < \infty$, where A is a real constant (2×2) matrix and where the characteristic polynomial is $p(\lambda) = (\lambda - \alpha)^2$.

If $\{\mathbf{x}_1, \mathbf{x}_2\}$ is a set of two linearly independent eigenvectors corresponding to the repeated eigenvalue $\lambda = \alpha$, then

$$\mathbf{y}_1(t) = e^{\alpha t}\mathbf{x}_1 \quad \text{and} \quad \mathbf{y}_2(t) = e^{\alpha t}\mathbf{x}_2$$

form a fundamental set of solutions of $\mathbf{y}' = A\mathbf{y}$. Suppose however, that a set of two linearly independent eigenvectors corresponding to the eigenvalue $\lambda = \alpha$ does not exist. In such a case, $\mathbf{y}_1(t) = e^{\alpha t}\mathbf{x}_1$ is one solution. How do we find a second solution, $\mathbf{y}_2(t)$?

Motivated by the previous example, we look for a second solution of the form

$$\mathbf{y}_2(t) = te^{\alpha t}\mathbf{v}_1 + e^{\alpha t}\mathbf{v}_2, \tag{6}$$

where $\mathbf{v}_1$ and $\mathbf{v}_2$ are nonzero constant (2×1) vectors to be determined. Substituting this representation into the differential equation we obtain

$$(e^{\alpha t} + \alpha te^{\alpha t})\mathbf{v}_1 + \alpha e^{\alpha t}\mathbf{v}_2 = A(te^{\alpha t}\mathbf{v}_1 + e^{\alpha t}\mathbf{v}_2).$$

We can rewrite this equation as

$$te^{\alpha t}(A\mathbf{v}_1 - \alpha\mathbf{v}_1) + e^{\alpha t}(A\mathbf{v}_2 - \alpha\mathbf{v}_2 - \mathbf{v}_1) = \mathbf{0}, \qquad -\infty < t < \infty. \tag{7}$$

The set $\{te^{\alpha t}, e^{\alpha t}\}$ is a linearly independent set of functions on any t-interval of interest. Therefore, if equation (7) is to hold, each of the constant matrix coefficients must vanish. We obtain, therefore, a pair of matrix equations that the nonzero vectors $\mathbf{v}_1$ and $\mathbf{v}_2$ must satisfy:

$$\begin{aligned} (A - \alpha I)\mathbf{v}_1 &= \mathbf{0}, \\ (A - \alpha I)\mathbf{v}_2 &= \mathbf{v}_1. \end{aligned} \tag{8}$$

The first of these equations is simply the eigenvector equation, and we take $\mathbf{v}_1$ to be an eigenvector corresponding to $\lambda = \alpha$. Consider now the second equation,

$$(A - \alpha I)\mathbf{v}_2 = \mathbf{v}_1. \tag{9}$$

At first glance, equation (9) should give us cause for concern. The coefficient matrix $A - \alpha I$ is singular (that is, noninvertible) since α is an eigenvalue of A. We recall that a nonhomogeneous system of equations having a singular coefficient matrix, such as system (9), either has no solution or infinitely many solutions. In the present case, however, it can be shown that equation (9) always has infinitely many solutions. Selecting one particular solution of (9) determines $\mathbf{v}_2$; having $\mathbf{v}_2$, we can form a second solution, $\mathbf{y}_2(t)$. It can be shown that the pair of solutions obtained, $\{\mathbf{y}_1(t), \mathbf{y}_2(t)\}$, is a fundamental set of solutions. The vector $\mathbf{v}_2$ in equation (8) is called a "generalized eigenvector of order 2." See Exercises 30–35.

EXAMPLE 1

Solve the initial value problem

$$\mathbf{y}' = \begin{bmatrix} 2 & -1 \\ 1 & 4 \end{bmatrix}\mathbf{y}, \qquad \mathbf{y}(1) = \begin{bmatrix} 1 \\ 3 \end{bmatrix}.$$

Solution: The characteristic polynomial is

$$p(\lambda) = \det[A - \lambda I] = \begin{vmatrix} 2-\lambda & -1 \\ 1 & 4-\lambda \end{vmatrix} = \lambda^2 - 6\lambda + 9 = (\lambda - 3)^2.$$

The coefficient matrix, therefore, has $\lambda = 3$ as a repeated eigenvalue. The corresponding eigenvector equation is $(A - 3I)\mathbf{x} = \mathbf{0}$, or

$$\begin{bmatrix} -1 & -1 \\ 1 & 1 \end{bmatrix}\begin{bmatrix} x_1 \\ x_2 \end{bmatrix} = \begin{bmatrix} 0 \\ 0 \end{bmatrix}.$$

This equation reduces to $x_1 + x_2 = 0$ and hence the eigenvectors have the form

$$\mathbf{x} = \begin{bmatrix} x_1 \\ x_2 \end{bmatrix} = \begin{bmatrix} -x_2 \\ x_2 \end{bmatrix} = x_2\begin{bmatrix} -1 \\ 1 \end{bmatrix}, \qquad x_2 \neq 0.$$

A convenient choice for an eigenvector is

$$\mathbf{v}_1 = \begin{bmatrix} -1 \\ 1 \end{bmatrix}.$$

Therefore, one solution of $\mathbf{y}' = A\mathbf{y}$ is

$$\mathbf{y}_1(t) = e^{3t}\mathbf{v}_1.$$

The eigenvalue $\lambda = 3$ does not have a set of two linearly independent eigenvectors; every eigenvector is a nonzero multiple of $\mathbf{v}_1$. Therefore, we now look for a second solution having the form $\mathbf{y}_2(t) = te^{3t}\mathbf{v}_1 + e^{3t}\mathbf{v}_2$, where [see equation (9)] $\mathbf{v}_2$ satisfies the equation $(A - 3I)\mathbf{x} = \mathbf{v}_1$ or

$$\begin{bmatrix} -1 & -1 \\ 1 & 1 \end{bmatrix}\begin{bmatrix} x_1 \\ x_2 \end{bmatrix} = \begin{bmatrix} -1 \\ 1 \end{bmatrix}.$$

This equation reduces to $x_1 + x_2 = 1$ and hence the solution is

$$\mathbf{v}_2 = \begin{bmatrix} 1 - x_2 \\ x_2 \end{bmatrix} = \begin{bmatrix} 1 \\ 0 \end{bmatrix} + x_2 \begin{bmatrix} -1 \\ 1 \end{bmatrix}.$$

Since the constant x_2 is arbitrary, there are infinitely many choices for x_2. We only need one solution, however, and choosing $x_2 = 0$ for convenience, we obtain

$$\mathbf{v}_2 = \begin{bmatrix} 1 \\ 0 \end{bmatrix}.$$

These calculations give us a second solution, $\mathbf{y}_2(t) = te^{3t}\mathbf{v}_1 + e^{3t}\mathbf{v}_2$:

$$\mathbf{y}_2(t) = te^{3t} \begin{bmatrix} -1 \\ 1 \end{bmatrix} + e^{3t} \begin{bmatrix} 1 \\ 0 \end{bmatrix} = \begin{bmatrix} -te^{3t} + e^{3t} \\ te^{3t} \end{bmatrix}.$$

Computing the Wronskian of the two solutions, we find

$$W(t) = \begin{vmatrix} -e^{3t} & -te^{3t} + e^{3t} \\ e^{3t} & te^{3t} \end{vmatrix} = -e^{6t} \neq 0.$$

Therefore, these two solutions form a fundamental set and the general solution is given by $\mathbf{y}(t) = c_1\mathbf{y}_1(t) + c_2\mathbf{y}_2(t)$:

$$\mathbf{y}(t) = c_1 \begin{bmatrix} -e^{3t} \\ e^{3t} \end{bmatrix} + c_2 \begin{bmatrix} -te^{3t} + e^{3t} \\ te^{3t} \end{bmatrix} = \begin{bmatrix} -e^{3t} & -te^{3t} + e^{3t} \\ e^{3t} & te^{3t} \end{bmatrix} \begin{bmatrix} c_1 \\ c_2 \end{bmatrix}.$$

Imposing the initial condition,

$$\mathbf{y}(1) = \begin{bmatrix} 1 \\ 3 \end{bmatrix} = \begin{bmatrix} -e^3 & 0 \\ e^3 & e^3 \end{bmatrix} \begin{bmatrix} c_1 \\ c_2 \end{bmatrix},$$

leads to $c_1 = -e^{-3}$ and $c_2 = 4e^{-3}$. The solution of the initial value problem is, therefore,

$$\mathbf{y}(t) = \begin{bmatrix} -4te^{3(t-1)} + 5e^{3(t-1)} \\ 4te^{3(t-1)} - e^{3(t-1)} \end{bmatrix}. \quad \blacktriangle$$

Algebraic Multiplicity and Geometric Multiplicity

So far in this section we have concentrated on the equation $\mathbf{y}' = A\mathbf{y}$ in the case where A is a constant (2×2) matrix with a repeated eigenvalue. The case where A is an $(n \times n)$ matrix with repeated eigenvalues is more complicated, and a comprehensive treatment is beyond the scope of our present discussion. We do, however, give some indications of the underlying ideas in the Exercises. In addition, some observations about the $(n \times n)$ case can be made at this point.

When looking for a fundamental set of solutions for $\mathbf{y}' = A\mathbf{y}$, there is more to consider than simply whether A has repeated eigenvalues. If an eigenvalue is repeated, the question then arises as to whether there exist enough linearly independent eigenvectors. These considerations lead to the following definitions.

Let α be an eigenvalue of an $(n \times n)$ matrix A. The **algebraic multiplicity** of α is the order of α as a root of the characteristic equation, $p(\lambda) = 0$. If

$$p(\lambda) = (\lambda - \alpha)^r q(\lambda),$$

where $q(\lambda)$ is a polynomial of degree $n - r$ and $q(\alpha) \neq 0$, then α is an eigenvalue of algebraic multiplicity r. The **geometric multiplicity** of the eigenvalue α is the number of linearly independent eigenvectors that can be found corresponding to this eigenvalue.

Since the characteristic polynomial has degree n, the algebraic multiplicity of α is an integer r, where $1 \leq r \leq n$. Similarly, a set of $(n \times 1)$ vectors cannot be linearly independent unless the set contains n or fewer vectors. Thus, the geometric multiplicity of α is an integer s, where $1 \leq s \leq n$. As we will note later in equation (12), the inequality $s \leq r$ holds for every matrix A.

EXAMPLE 2

In each of the following cases, determine the algebraic and geometric multiplicity of the eigenvalue $\lambda = 2$.

$$\text{(a)}\quad A_1 = \begin{bmatrix} 2 & 1 & 0 \\ 0 & 2 & 1 \\ 0 & 0 & 2 \end{bmatrix}, \qquad \text{(b)}\quad A_2 = \begin{bmatrix} 2 & 1 & 0 \\ 0 & 2 & 0 \\ 0 & 0 & 2 \end{bmatrix}, \qquad \text{(c)}\quad A_3 = \begin{bmatrix} 2 & 0 & 0 \\ 0 & 2 & 0 \\ 0 & 0 & 2 \end{bmatrix}.$$

Solution: In each case, the characteristic polynomial is $p(\lambda) = -(\lambda - 2)^3$. Therefore, in each case, the eigenvalue $\lambda = 2$ has algebraic multiplicity 3.

(a) All solutions (see Exercise 9) of the eigenvector equation $(A_1 - 2I)\mathbf{x} = \mathbf{0}$ have the form

$$\mathbf{x} = \begin{bmatrix} a \\ 0 \\ 0 \end{bmatrix} = a \begin{bmatrix} 1 \\ 0 \\ 0 \end{bmatrix} = a\mathbf{x}_1.$$

Therefore, for the matrix A_1, the geometric multiplicity of the eigenvalue $\lambda = 2$ is 1.

(b) All solutions of the eigenvector equation $(A_2 - 2I)\mathbf{x} = \mathbf{0}$ have the following form (see Exercise 9):

$$\mathbf{x} = \begin{bmatrix} b \\ 0 \\ c \end{bmatrix} = b \begin{bmatrix} 1 \\ 0 \\ 0 \end{bmatrix} + c \begin{bmatrix} 0 \\ 0 \\ 1 \end{bmatrix} = b\mathbf{x}_1 + c\mathbf{x}_2,$$

where $\mathbf{x}_1$ and $\mathbf{x}_2$ are linearly independent. Therefore, for the matrix A_2, the geometric multiplicity of the eigenvalue $\lambda = 2$ is 2.

(c) Finally, since the matrix $A_3 - 2I$ is the (3×3) zero matrix, every (3×1) vector is an eigenvector of A_3. Therefore, for the matrix A_3, the geometric multiplicity of $\lambda = 2$ is 3. ▲

If we want to solve the equation $\mathbf{y}' = A_1\mathbf{y}$ in Example 2, we see that we know one solution, $\mathbf{y}_1(t) = e^{2t}\mathbf{x}_1$. We need to find two more solutions to form a fundamental set of solutions. By contrast, if we want to solve $\mathbf{y}' = A_2\mathbf{y}$, we know two linearly independent solutions, $\mathbf{y}_1(t) = e^{2t}\mathbf{x}_1$ and $\mathbf{y}_2(t) = e^{2t}\mathbf{x}_2$. Thus, to form a fundamental set of solutions, we need find only one additional solution. Finally, if we want to solve $\mathbf{y}' = A_3\mathbf{y}$, then (since every three-dimensional vector is an eigenvector of A_3) we need only select three linearly independent vectors $\mathbf{u}$, $\mathbf{v}$, $\mathbf{w}$ and we are assured (by Theorem 6.9) that the solutions $\mathbf{y}_1(t) = e^{2t}\mathbf{u}$, $\mathbf{y}_2(t) = e^{2t}\mathbf{v}$, and $\mathbf{y}_3(t) = e^{2t}\mathbf{w}$ form a fundamental set of solutions.

Defective Matrices

With regard to multiplicity, the following inequality is established in advanced texts:

$$\begin{bmatrix} \text{The geometric multiplicity} \\ \text{of an eigenvalue} \end{bmatrix} \le \begin{bmatrix} \text{The algebraic multiplicity} \\ \text{of an eigenvalue.} \end{bmatrix} \tag{10}$$

In determining the structure of a fundamental set of solutions of $\mathbf{y}' = A\mathbf{y}$, the key question is "How does the geometric multiplicity of each eigenvalue relate to its algebraic multiplicity?" If the two multiplicities for a given eigenvalue λ are equal, the corresponding solutions entering into the fundamental set will all be of the form $e^{\lambda t}\mathbf{x}$ (whether the eigenvalue is distinct or repeated). If the geometric multiplicity of an eigenvalue is strictly less than its algebraic multiplicity, we have a "deficiency of eigenvectors" and solutions of a more complicated form become part of the fundamental set. A matrix that has at least one eigenvalue with a geometric multiplicity that is strictly less than its algebraic multiplicity is called a **defective matrix**. Thus, in Example 2, the matrices A_1 and A_2 are defective, whereas the matrix A_3 is not defective. When the matrix A is not defective, we say that A has a **full set of eigenvectors**.

There are important classes of $(n \times n)$ matrices that always have a full set of eigenvectors even if some of the eigenvalues are repeated. We consider one such class of matrices now.

Symmetric Matrices and Hermitian Matrices

If A is an $(m \times n)$ matrix, the **transpose** of A, denoted by A^T, is the $(n \times m)$ matrix obtained by interchanging the rows and columns of A; if a_{ij} denotes the (i, j)-th entry of A, then the (i, j)-th entry of A^T is a_{ji}. An example of a matrix and its transpose is

$$A = \begin{bmatrix} 1 & 2 & 3 \\ 4 & 5 & 6 \\ 7 & 8 & 9 \end{bmatrix}, \qquad A^T = \begin{bmatrix} 1 & 4 & 7 \\ 2 & 5 & 8 \\ 3 & 6 & 9 \end{bmatrix}.$$

A real $(n \times n)$ matrix A satisfying the equation $A = A^T$ is called a **real symmetric matrix**. For such matrices, $a_{ij} = a_{ji}$, $1 \le i, j \le n$. For instance,

$$A = \begin{bmatrix} -1 & 2 & -3 \\ 2 & 5 & 7 \\ -3 & 7 & 2 \end{bmatrix},$$

is a real symmetric (3×3) matrix.

Real symmetric matrices are important in applications—they often arise because of an underlying reciprocity principle present in the physical system being modeled. Real symmetric matrices are a special case of a larger class of matrices, known as Hermitian matrices. An $(n \times n)$ matrix A is called a **Hermitian matrix** if it equals its complex conjugate transpose; that is, if $A = \overline{A}^T$. For example, the matrix

$$A = \begin{bmatrix} -1 & 2i & 1-i \\ -2i & 0 & -3 \\ 1+i & -3 & 1 \end{bmatrix}$$

is a (3×3) Hermitian matrix. Obviously, real Hermitian matrices are real symmetric matrices.

If A is a real symmetric matrix or a Hermitian matrix, then all the eigenvalues of A are real (see Exercises 28 and 29). In more advanced texts it is shown that real symmetric matrices (and Hermitian matrices) always have a full set of eigenvectors; that is, each eigenvalue has geometric multiplicity equal to its algebraic multiplicity. Therefore, when we study the homogeneous linear first order system $\mathbf{y}' = A\mathbf{y}$, where A is an $(n \times n)$ real symmetric matrix, we know that all solutions forming a fundamental set are of the form $e^{\lambda t}\mathbf{x}$, where λ is a real eigenvalue and $\mathbf{x}$ is an associated eigenvector. In Exercise 27 we ask you to establish a special case, showing that a (2×2) real symmetric matrix cannot be defective. Also see Exercise 36.

EXAMPLE 3

Find the general solution of $\mathbf{y}' = A\mathbf{y}$ where

$$A = \begin{bmatrix} 1 & -1 & -1 & -1 \\ -1 & 1 & -1 & -1 \\ -1 & -1 & 1 & -1 \\ -1 & -1 & -1 & 1 \end{bmatrix}.$$

Solution: Since the matrix is real and symmetric, we know all the eigenvalues are real. A calculation shows that the characteristic polynomial is

$$p(\lambda) = \det[A - \lambda I] = (\lambda - 2)^3(\lambda + 2).$$

Therefore, the eigenvalues are $\lambda = 2$ (an eigenvalue having algebraic multiplicity 3) and $\lambda = -2$ (an eigenvalue having algebraic multiplicity 1).

We noted above that every eigenvalue of a real symmetric matrix has geometric multiplicity equal to its algebraic multiplicity. Therefore, we know that we can find three linearly independent eigenvectors corresponding to $\lambda = 2$. For $\lambda = 2$, the eigenvector equation $(A - 2I)\mathbf{x} = \mathbf{0}$ is

$$\begin{bmatrix} -1 & -1 & -1 & -1 \\ -1 & -1 & -1 & -1 \\ -1 & -1 & -1 & -1 \\ -1 & -1 & -1 & -1 \end{bmatrix} \begin{bmatrix} x_1 \\ x_2 \\ x_3 \\ x_4 \end{bmatrix} = \begin{bmatrix} 0 \\ 0 \\ 0 \\ 0 \end{bmatrix}.$$

This system reduces to $x_1 + x_2 + x_3 + x_4 = 0$. Hence, the solution of $(A - 2I)\mathbf{x} = \mathbf{0}$ is

$$\mathbf{x} = \begin{bmatrix} x_1 \\ x_2 \\ x_3 \\ x_4 \end{bmatrix} = \begin{bmatrix} -x_2 - x_3 - x_4 \\ x_2 \\ x_3 \\ x_4 \end{bmatrix} = x_2 \begin{bmatrix} -1 \\ 1 \\ 0 \\ 0 \end{bmatrix} + x_3 \begin{bmatrix} -1 \\ 0 \\ 1 \\ 0 \end{bmatrix} + x_4 \begin{bmatrix} -1 \\ 0 \\ 0 \\ 1 \end{bmatrix}.$$

There are three linearly independent eigenvectors corresponding to $\lambda = 2$.

Finally, an eigenvector corresponding to $\lambda = -2$ is

$$\mathbf{x} = \begin{bmatrix} 1 \\ 1 \\ 1 \\ 1 \end{bmatrix}.$$

If we designate the eigenvectors, in the order found, as $\mathbf{x}_1$, $\mathbf{x}_2$, $\mathbf{x}_3$, $\mathbf{x}_4$, then the general solution is

$$\mathbf{y}(t) = c_1 e^{2t}\mathbf{x}_1 + c_2 e^{2t}\mathbf{x}_2 + c_3 e^{2t}\mathbf{x}_3 + c_4 e^{-2t}\mathbf{x}_4,$$

or

$$\mathbf{y}(t) = c_1 e^{2t} \begin{bmatrix} -1 \\ 1 \\ 0 \\ 0 \end{bmatrix} + c_2 e^{2t} \begin{bmatrix} -1 \\ 0 \\ 1 \\ 0 \end{bmatrix} + c_3 e^{2t} \begin{bmatrix} -1 \\ 0 \\ 0 \\ 1 \end{bmatrix} + c_4 e^{-2t} \begin{bmatrix} 1 \\ 1 \\ 1 \\ 1 \end{bmatrix}$$

$$= \begin{bmatrix} -e^{2t} & -e^{2t} & -e^{2t} & e^{-2t} \\ e^{2t} & 0 & 0 & e^{-2t} \\ 0 & e^{2t} & 0 & e^{-2t} \\ 0 & 0 & e^{2t} & e^{-2t} \end{bmatrix} \begin{bmatrix} c_1 \\ c_2 \\ c_3 \\ c_4 \end{bmatrix}. \blacktriangle$$

EXERCISES

Exercises 1–8:

Consider the initial value problem $\mathbf{y}' = A\mathbf{y}$, $\mathbf{y}(0) = \mathbf{y}_0$, where A and the initial condition $\mathbf{y}_0$ are given.

(a) Compute the eigenvalues and eigenvectors of the matrix A. If the eigenvalue is repeated, how many linearly independent eigenvectors can be found for this eigenvalue? (In other words, if the algebraic multiplicity of the eigenvalue is 2, is the corresponding geometric multiplicity 1 or 2?)

(b) Construct a fundamental set of solutions for the given differential equation. Use this fundamental set of solutions to construct a fundamental matrix, $\Psi(t)$.

(c) Impose the initial condition to obtain the unique solution of the initial value problem.

1. $A = \begin{bmatrix} 2 & 1 \\ 0 & 2 \end{bmatrix}$, $\mathbf{y}_0 = \begin{bmatrix} 1 \\ -1 \end{bmatrix}$

2. $A = \begin{bmatrix} 3 & 2 \\ 0 & 3 \end{bmatrix}$, $\mathbf{y}_0 = \begin{bmatrix} 4 \\ 1 \end{bmatrix}$

3. $A = \begin{bmatrix} 6 & 0 \\ 2 & 6 \end{bmatrix}$, $\mathbf{y}_0 = \begin{bmatrix} -2 \\ 0 \end{bmatrix}$

4. $A = \begin{bmatrix} 3 & 0 \\ 1 & 3 \end{bmatrix}$, $\mathbf{y}_0 = \begin{bmatrix} 2 \\ -3 \end{bmatrix}$

5. $A = \begin{bmatrix} 5 & -1 \\ 4 & 1 \end{bmatrix}$, $\mathbf{y}_0 = \begin{bmatrix} 1 \\ 1 \end{bmatrix}$

6. $A = \begin{bmatrix} -3 & -36 \\ 1 & 9 \end{bmatrix}$, $\mathbf{y}_0 = \begin{bmatrix} 0 \\ 2 \end{bmatrix}$

7. $A = \begin{bmatrix} 1 & -1 \\ 1 & 3 \end{bmatrix}$, $\mathbf{y}_0 = \begin{bmatrix} 4 \\ -1 \end{bmatrix}$

8. $A = \begin{bmatrix} 6 & 1 \\ -1 & 4 \end{bmatrix}$, $\mathbf{y}_0 = \begin{bmatrix} 4 \\ -4 \end{bmatrix}$

9. Recalling Example 2, find the eigenvalues and eigenvectors of

$$\text{(a) } A_1 = \begin{bmatrix} 2 & 1 & 0 \\ 0 & 2 & 1 \\ 0 & 0 & 2 \end{bmatrix} \text{ and (b) } A_2 = \begin{bmatrix} 2 & 1 & 0 \\ 0 & 2 & 0 \\ 0 & 0 & 2 \end{bmatrix}.$$

10. Consider the homogeneous linear system $\mathbf{y}' = A_1\mathbf{y}$, where A_1 is given in Exercise 9.

(a) Write the three component differential equations of $\mathbf{y}' = A_1\mathbf{y}$ and solve these equations sequentially, first finding $y_3(t)$, then $y_2(t)$, and then $y_1(t)$. [For example, the third component equation is $y_3' = 2y_3$. Therefore, $y_3(t) = c_3e^{2t}$.]

(b) Rewrite the component solutions obtained in part (a) as a single matrix equation of the form $\mathbf{y}(t) = \Psi(t)\mathbf{c}$, where $\Psi(t)$ is a (3×3) solution matrix and

$$\mathbf{c} = \begin{bmatrix} c_1 \\ c_2 \\ c_3 \end{bmatrix}.$$

Show that $\Psi(t)$ is, in fact, a fundamental matrix. [Note that this observation is consistent with the fact that the component solutions obtained in part (a) form the general solution.]

11. Repeat Exercise 10 for the homogeneous linear system $\mathbf{y}' = A_2\mathbf{y}$.

12. **The Scalar Repeated-Root Equation Revisited** Consider the homogeneous scalar equation $y'' - 2\alpha y' + \alpha^2 y = 0$, where α is a real constant. Recall from Section 4.5 that the general solution is $y(t) = c_1e^{\alpha t} + c_2te^{\alpha t}$.

(a) Recast $y'' - 2\alpha y' + \alpha^2 y = 0$ as a first order linear system $\mathbf{z}' = A\mathbf{z}$.

(b) Show that the (2×2) matrix A has eigenvalue α with algebraic multiplicity 2 and geometric multiplicity 1.

(c) Obtain the general solution of $\mathbf{z}' = A\mathbf{z}$. As a check, does $z_1(t)$ equal the general solution $y(t) = c_1e^{\alpha t} + c_2te^{\alpha t}$? Is $z_2(t)$ equal to $z_1'(t)$?

Exercises 13–19:

For each matrix A, find the eigenvalues and eigenvectors. Give the geometric and algebraic multiplicity of each eigenvalue. Does A have a full set of eigenvectors?

13. $A = \begin{bmatrix} 5 & 0 & 0 \\ 1 & 5 & 0 \\ 1 & 1 & 5 \end{bmatrix}$

14. $A = \begin{bmatrix} 5 & 0 & 0 \\ 1 & 5 & 0 \\ 1 & 0 & 5 \end{bmatrix}$

15. $A = \begin{bmatrix} 5 & 0 & 1 \\ 0 & 5 & 0 \\ 0 & 0 & 5 \end{bmatrix}$

16. $A = \begin{bmatrix} 5 & 0 & 0 \\ 0 & 5 & 0 \\ 0 & 0 & 5 \end{bmatrix}$

17. $A = \begin{bmatrix} 2 & 0 & 0 & 0 \\ 1 & 2 & 0 & 0 \\ 0 & 0 & 3 & 0 \\ 0 & 0 & 1 & 3 \end{bmatrix}$

18. $A = \begin{bmatrix} 2 & 0 & 0 & 0 \\ 0 & 2 & 0 & 0 \\ 0 & 0 & 2 & 0 \\ 0 & 0 & 1 & 2 \end{bmatrix}$

19. $A = \begin{bmatrix} 2 & 0 & 0 & 0 \\ 0 & 2 & 0 & 0 \\ 0 & 0 & 2 & 0 \\ 0 & 0 & 1 & 3 \end{bmatrix}$

20. Let A be a real (2×2) matrix with an eigenvalue $\lambda_1 = a + ib$ where $b \neq 0$. Can A have a repeated eigenvalue? Can A be defective? Explain.

Exercises 21–23:

The matrix A is known to be real and symmetric. What are the possible vaues for the constants x and y?

21. $A = \begin{bmatrix} 2 & x & 7 \\ 9 & 5 & y \\ 7 & 4 & 3 \end{bmatrix}$ **22.** $A = \begin{bmatrix} 0 & 1 & x \\ y & 2 & 2 \\ 6 & 2 & 7 \end{bmatrix}$ **23.** $A = \begin{bmatrix} 9 & 1 & x^2 - 1 \\ 2/y & 3 & 8 \\ 0 & 8 & 1 \end{bmatrix}$

Exercises 24–25:

The matrix A is known to be Hermitian. What are the possible values for the real constants x and y?

24. $A = \begin{bmatrix} 2 & x + 3i & 7 \\ 9 - 3i & 5 & 2 + yi \\ 7 & 2 + 5i & 3 \end{bmatrix}$ **25.** $A = \begin{bmatrix} 2 + xi & 1 + 2i & 7 \\ 1 + yi & x^2 & y^2 + x - i \\ 7 & 4 + i & 3 \end{bmatrix}$

26. (a) Give an example of a (2×2) matrix A that is not invertible but which has a full set of eigenvectors.

(b) Give an example of a (2×2) matrix A that is invertible but does not have a full set of eigenvectors.

27. Consider the (2×2) real symmetric matrix

$$A = \begin{bmatrix} a & b \\ b & c \end{bmatrix}.$$

Show: if A has a repeated eigenvalue then A must have the form

$$A = \begin{bmatrix} r & 0 \\ 0 & r \end{bmatrix}.$$

[Hint: The characteristic polynomial is $p(\lambda) = \lambda^2 - (a + c)\lambda + ac - b^2$. What does the quadratic formula become when A has a repeated eigenvalue?]

28. **Real Symmetric Matrices Have Only Real Eigenvalues** Exercise 33 in Section 6.5 shows that a (2×2) real symmetric matrix has only real eigenvalues. This exercise shows that every $(n \times n)$ real symmetric matrix has only real eigenvalues. Let A be an $(n \times n)$ real symmetric matrix. Suppose $A\mathbf{x} = \lambda\mathbf{x}$, where $\mathbf{x} \neq \mathbf{0}$. Multiply both sides of $A\mathbf{x} = \lambda\mathbf{x}$ by $\overline{\mathbf{x}}^T$, producing an equality between the two scalars, $\overline{\mathbf{x}}^T A\mathbf{x}$ and $\lambda\overline{\mathbf{x}}^T\mathbf{x}$. Take the transpose of each side [recall that $(PQ)^T = Q^T P^T$] and show that $\overline{\lambda} = \lambda$.

29. **Hermitian Matrices Have Only Real Eigenvalues** Let A be an $(n \times n)$ Hermitian matrix. Modify your proof in Exercise 28 to show that A has only real eigenvalues.

Generalized Eigenvectors Let A be an $(n \times n)$ matrix. The ideas introduced in equation (8) can be extended. Let $\mathbf{v}_1 \neq \mathbf{0}$ be an eigenvector of A corresponding to the eigenvalue λ and suppose we can generate the following "chain" of nonzero vectors:

$$\begin{aligned} (A - \lambda I)\mathbf{v}_1 &= \mathbf{0} \\ (A - \lambda I)\mathbf{v}_2 &= \mathbf{v}_1 \\ (A - \lambda I)\mathbf{v}_3 &= \mathbf{v}_2 \\ &\vdots \\ (A - \lambda I)\mathbf{v}_r &= \mathbf{v}_{r-1} \end{aligned} \tag{11}$$

In (11), the vector $\mathbf{v}_j$ is called a **generalized eigenvector of order j**. Define

$$\mathbf{y}_k(t) = e^{\lambda t}\left(\mathbf{v}_k + t\mathbf{v}_{k-1} + \cdots + \frac{t^{k-1}}{(k-1)!}\mathbf{v}_1\right). \tag{12}$$

Exercise 34 asks you to show, for $r = 3$ and $k = 1, 2, 3$ that $\mathbf{y}_k(t)$ is a solution of $\mathbf{y}' = A\mathbf{y}$. If λ has algebraic multiplicity m and geometric multiplicity 1, then it can be shown that there is a chain (11) consisting of m different generalized eigenvectors and that these generalized eigenvectors are linearly independent (see Exercise 35). Thus, equation (12) defines a set of m linearly independent solutions of $\mathbf{y}' = A\mathbf{y}$—as many solutions as the multiplicity of the eigenvalue. (If λ has geometric multiplicity 2 or larger, the situation is more complicated.)

Exercises 30–33:

Using equations (11) and (12), find a fundamental set of solutions for the linear system $\mathbf{y}' = A\mathbf{y}$. In each exercise, you are given an eigenvalue λ, where λ has algebraic multiplicity 3 and geometric multiplicity 1 and an eigenvector $\mathbf{v}_1$. In Exercise 30, compare the fundamental set of solutions found with those obtained in Exercise 10.

30. $A = \begin{bmatrix} 2 & 1 & 0 \\ 0 & 2 & 1 \\ 0 & 0 & 2 \end{bmatrix}$, $\lambda = 2$, $\mathbf{v}_1 = \begin{bmatrix} 1 \\ 0 \\ 0 \end{bmatrix}$

31. $A = \begin{bmatrix} 4 & 0 & 0 \\ 2 & 4 & 0 \\ 1 & 3 & 4 \end{bmatrix}$, $\lambda = 4$, $\mathbf{v}_1 = \begin{bmatrix} 0 \\ 0 \\ 1 \end{bmatrix}$

32. $A = \begin{bmatrix} 3 & -8 & -10 \\ -2 & 7 & 8 \\ 2 & -6 & -7 \end{bmatrix}$, $\lambda = 1$, $\mathbf{v}_1 = \begin{bmatrix} 1 \\ -1 \\ 1 \end{bmatrix}$

33. $A = \begin{bmatrix} -6 & -8 & 22 \\ 2 & 4 & -4 \\ -2 & -2 & 8 \end{bmatrix}$, $\lambda = 2$, $\mathbf{v}_1 = \begin{bmatrix} 1 \\ -1 \\ 0 \end{bmatrix}$

34. Let $\mathbf{v}_1$, $\mathbf{v}_2$, and $\mathbf{v}_3$ be a chain of nonzero vectors, as in equation (11). Show that the vector function $\mathbf{y}_3(t)$ defined in equation (12) is a solution of $\mathbf{y}' = A\mathbf{y}$.

35. Let $\mathbf{v}_1$, $\mathbf{v}_2$, and $\mathbf{v}_3$ be a chain of nonzero vectors, as in equation (11). Show that these vectors form a linearly independent set of vectors. [Hint: Begin with the dependence equation $c_1\mathbf{v}_1 + c_2\mathbf{v}_2 + c_3\mathbf{v}_3 = \mathbf{0}$ and multiply both sides by $A - \lambda I$.]

36. In Section 6.5 we treated the flushing of interconnected systems of tanks with fresh water (see Exercises 39–40). The tanks had identical capacity and were interconnected so that they all experienced the same flow environment. We now obtain the general solution of this problem for the case of three tanks. Let $Q_j(t)$, $j = 1, 2, 3$ denote the amount of salt in the jth tank at time t. Let V denote the common tank capacity and r the common flow rate. As in Exercise 40 of Section 6.5, the homogeneous linear system describing the flushing process is $\mathbf{Q}' = (r/V)A\mathbf{Q}$, where

$$A = \begin{bmatrix} -3 & 1 & 1 \\ 1 & -3 & 1 \\ 1 & 1 & -3 \end{bmatrix}.$$

Since A is a real symmetric matrix, it has a full set of eigenvectors. Therefore, we can find three linearly independent eigenvectors, and the solutions forming a fundamental set will all have the form $e^{\lambda t}\mathbf{x}$, where λ is an eigenvalue and $\mathbf{x}$ is a corresponding eigenvector.

(a) Show that $\lambda_1 = -4$ is an eigenvalue of A having geometric multiplicity 2 by obtaining two linearly independent solutions of the matrix equation $(A + 4I)\mathbf{x} = \mathbf{0}$.

(b) Show that $\lambda_2 = -1$ is also an eigenvalue of A by finding a nontrivial solution of the matrix equation $(A + I)\mathbf{x} = \mathbf{0}$.

(c) Using the results of parts (a) and (b), form the general solution of the homogeneous linear system.

6.8 Nonhomogeneous Linear Systems

We now address the problem of finding the general solution of a nonhomogeneous first order linear system,

$$\mathbf{y}' = P(t)\mathbf{y} + \mathbf{g}(t), \qquad a < t < b. \tag{1}$$

In (1), $\mathbf{y}(t)$ is an $(n \times 1)$ vector function, $P(t)$ is an $(n \times n)$ matrix function, and the nonhomogeneous term, $\mathbf{g}(t)$, is an $(n \times 1)$ vector function. The component functions of $P(t)$ and $\mathbf{g}(t)$ are assumed to be continuous on $a < t < b$.

The Structure of the General Solution

In analyzing the structure of the general solution of (1), we return once more to a theme that has permeated our entire discussion of nonhomogeneous linear equations. If $\mathbf{y}_1(t)$ and $\mathbf{y}_2(t)$ represent any two solutions of $\mathbf{y}' = P(t)\mathbf{y} + \mathbf{g}(t)$, we ask: "How do they differ?"

To answer this question, we form the difference function, $\mathbf{w}(t) = \mathbf{y}_1(t) - \mathbf{y}_2(t)$. Differentiating $\mathbf{w}(t)$, we find

$$\begin{aligned}
\mathbf{w}' &= (\mathbf{y}_1 - \mathbf{y}_2)' \\
&= \mathbf{y}_1' - \mathbf{y}_2' \\
&= [P(t)\mathbf{y}_1 + \mathbf{g}(t)] - [P(t)\mathbf{y}_2 + \mathbf{g}(t)] \\
&= P(t)(\mathbf{y}_1 - \mathbf{y}_2) \\
&= P(t)\mathbf{w}.
\end{aligned}$$

Thus, the difference between any two solutions of the nonhomogeneous linear equation is a solution of the homogeneous linear equation. This leads to the familiar decomposition

The general solution of the nonhomogeneous linear system $\mathbf{y}' = P(t)\mathbf{y} + \mathbf{g}(t)$	=	The general solution of the homogeneous linear system $\mathbf{y}' = P(t)\mathbf{y}$	+	A particular solution of the nonhomogeneous linear system $\mathbf{y}' = P(t)\mathbf{y} + \mathbf{g}(t)$.

As before, we refer to the general solution of the homogeneous system, $\mathbf{y}' = P(t)\mathbf{y}$, as the *complementary solution* and denote it by $\mathbf{y}_C(t)$. A solution of the nonhomogeneous system that we have somehow found is called a *particular solution* and is denoted by $\mathbf{y}_P(t)$.

The following theorem, an analog of the superposition principle given in Theorems 4.9 and 5.9, holds for nonhomogeneous linear systems. We leave the proof as an exercise.

Theorem 6.11

Let $\mathbf{u}(t)$ be a solution of

$$\mathbf{y}' = P(t)\mathbf{y} + \mathbf{g}_1(t), \qquad a < t < b,$$

and let $\mathbf{v}(t)$ be a solution of

$$\mathbf{y}' = P(t)\mathbf{y} + \mathbf{g}_2(t), \qquad a < t < b.$$

Let a_1 and a_2 be any constants. Then the vector function $\mathbf{y}_P(t) = a_1\mathbf{u}(t) + a_2\mathbf{v}(t)$ is a particular solution of

$$\mathbf{y}' = P(t)\mathbf{y} + a_1\mathbf{g}_1(t) + a_2\mathbf{g}_2(t), \qquad a < t < b.$$

The following example illustrates Theorem 6.11.

EXAMPLE 1

Find the general solution of

$$\mathbf{y}' = \begin{bmatrix} 1 & 2 \\ 2 & 1 \end{bmatrix}\mathbf{y} + \begin{bmatrix} e^{2t} \\ -2t \end{bmatrix}, \qquad -\infty < t < \infty.$$

Solution: We saw earlier (in Example 1 of Section 6.5) that the general solution of the homogeneous equation,

$$\mathbf{y}' = \begin{bmatrix} 1 & 2 \\ 2 & 1 \end{bmatrix}\mathbf{y}, \qquad -\infty < t < \infty,$$

is

$$\mathbf{y}_C(t) = c_1\begin{bmatrix} e^{-t} \\ -e^{-t} \end{bmatrix} + c_2\begin{bmatrix} e^{3t} \\ e^{3t} \end{bmatrix} = \begin{bmatrix} e^{-t} & e^{3t} \\ -e^{-t} & e^{3t} \end{bmatrix}\begin{bmatrix} c_1 \\ c_2 \end{bmatrix}.$$

Having the complementary solution, $\mathbf{y}_C(t)$, we turn our attention to the task of somehow finding a particular solution, $\mathbf{y}_P(t)$, of the nonhomogeneous equation. Note that the nonhomogeneous term,

$$\mathbf{g}(t) = \begin{bmatrix} e^{2t} \\ -2t \end{bmatrix},$$

can be decomposed as follows:

$$\begin{bmatrix} e^{2t} \\ -2t \end{bmatrix} = e^{2t}\begin{bmatrix} 1 \\ 0 \end{bmatrix} + t\begin{bmatrix} 0 \\ -2 \end{bmatrix}.$$

Using the superposition principle in Theorem 6.11, we decompose the differential equation and separately find particular solutions of

$$\mathbf{u}' = \begin{bmatrix} 1 & 2 \\ 2 & 1 \end{bmatrix}\mathbf{u} + e^{2t}\begin{bmatrix} 1 \\ 0 \end{bmatrix} \quad \text{and} \quad \mathbf{v}' = \begin{bmatrix} 1 & 2 \\ 2 & 1 \end{bmatrix}\mathbf{v} + t\begin{bmatrix} 0 \\ -2 \end{bmatrix}. \tag{2}$$

Consider the first equation in (2). Remembering the method of undetermined coefficients, we look for a solution of the form $\mathbf{u}_P(t) = e^{2t}\mathbf{a}$, where $\mathbf{a}$ is a constant (2×1) vector to be determined. Substituting $\mathbf{u}_P(t) = e^{2t}\mathbf{a}$ into the first differential equation in (2) leads to

$$2e^{2t}\mathbf{a} = \begin{bmatrix} 1 & 2 \\ 2 & 1 \end{bmatrix}(e^{2t}\mathbf{a}) + e^{2t}\begin{bmatrix} 1 \\ 0 \end{bmatrix}, \qquad -\infty < t < \infty.$$

Canceling the common e^{2t} factor and rearranging terms, we see that **a** must satisfy the condition

$$\begin{bmatrix} 1 & -2 \\ -2 & 1 \end{bmatrix}\mathbf{a} = \begin{bmatrix} 1 \\ 0 \end{bmatrix} \quad \text{or} \quad \mathbf{a} = \begin{bmatrix} -\dfrac{1}{3} \\ -\dfrac{2}{3} \end{bmatrix}.$$

Thus, a particular solution is

$$\mathbf{u}_P(t) = e^{2t}\mathbf{a} = \begin{bmatrix} -\dfrac{1}{3}e^{2t} \\ -\dfrac{2}{3}e^{2t} \end{bmatrix}.$$

To find a particular solution of the second equation in (2), we look for a solution having the form $\mathbf{v}_P(t) = t\mathbf{b} + \mathbf{c}$, where **b** and **c** are constant (2×1) vectors to be determined. Substituting this guess into the differential equation leads to

$$\mathbf{b} = \begin{bmatrix} 1 & 2 \\ 2 & 1 \end{bmatrix}(t\mathbf{b} + \mathbf{c}) + t\begin{bmatrix} 0 \\ -2 \end{bmatrix}$$

or, after collecting like powers of t,

$$t\left(\begin{bmatrix} 1 & 2 \\ 2 & 1 \end{bmatrix}\mathbf{b} + \begin{bmatrix} 0 \\ -2 \end{bmatrix}\right) + \left(\begin{bmatrix} 1 & 2 \\ 2 & 1 \end{bmatrix}\mathbf{c} - \mathbf{b}\right) = \mathbf{0}, \qquad -\infty < t < \infty.$$

Since the set of functions $\{t, 1\}$ is linearly independent on any t-interval, this equation holds only if the coefficients of t and 1 are $\mathbf{0}$; that is,

$$\begin{bmatrix} 1 & 2 \\ 2 & 1 \end{bmatrix}\begin{bmatrix} b_1 \\ b_2 \end{bmatrix} = \begin{bmatrix} 0 \\ 2 \end{bmatrix} \quad \text{and} \quad \begin{bmatrix} 1 & 2 \\ 2 & 1 \end{bmatrix}\begin{bmatrix} c_1 \\ c_2 \end{bmatrix} = \begin{bmatrix} b_1 \\ b_2 \end{bmatrix}. \tag{3}$$

The two matrix equations in (3) can be solved sequentially, yielding

$$\mathbf{b} = \begin{bmatrix} \dfrac{4}{3} \\ -\dfrac{2}{3} \end{bmatrix} \quad \text{and} \quad \mathbf{c} = \begin{bmatrix} -\dfrac{8}{9} \\ \dfrac{10}{9} \end{bmatrix}.$$

Therefore, the solution $\mathbf{v}(t) = t\mathbf{b} + \mathbf{c}$ is

$$\mathbf{v}(t) = \begin{bmatrix} \dfrac{4}{3}t - \dfrac{8}{9} \\ -\dfrac{2}{3}t + \dfrac{10}{9} \end{bmatrix}.$$

(continued)

(continued)

Applying Theorem 6.11, a particular solution of the given differential equation is

$$\mathbf{y}_P(t) = \mathbf{u}(t) + \mathbf{v}(t) = \begin{bmatrix} -\frac{1}{3}e^{2t} + \frac{4}{3}t - \frac{8}{9} \\ -\frac{2}{3}e^{2t} - \frac{2}{3}t + \frac{10}{9} \end{bmatrix}.$$

The general solution is therefore

$$\mathbf{y}(t) = \mathbf{y}_C(t) + \mathbf{y}_P(t) = c_1 \begin{bmatrix} e^{-t} \\ -e^{-t} \end{bmatrix} + c_2 \begin{bmatrix} e^{3t} \\ e^{3t} \end{bmatrix} + \begin{bmatrix} -\frac{1}{3}e^{2t} + \frac{4}{3}t - \frac{8}{9} \\ -\frac{2}{3}e^{2t} - \frac{2}{3}t + \frac{10}{9} \end{bmatrix}. \quad \blacktriangle$$

Comparing Solution Methods

In Chapters 4 and 5, the method of undetermined coefficients was shown to be an effective way to find particular solutions when the differential equation had constant coefficients and when the nonhomogeneous term was of a certain form (see Tables 4.1 and 5.1). Example 1 provided a simple illustration of how these ideas can be extended and applied to the constant coefficient linear system $\mathbf{y}' = A\mathbf{y} + \mathbf{g}(t)$ when the nonhomogeneous vector function $\mathbf{g}(t)$ has components of a certain form. The exercises give additional illustrations.

However, in contrast to the scalar problem, the complexity of the matrix problem makes the "educated guesswork" at the core of this method difficult to implement systematically. We shall not discuss the method of undetermined coefficients any further.

The method of variation of parameters (considered in Sections 4.10 and 5.5 for scalar linear equations) extends very naturally to linear systems. Therefore, we shall concentrate on the method of variation of parameters. In Chapter 7, we show how Laplace transforms also can be used to solve constant coefficient nonhomogeneous linear first order systems.

As the first step in studying the method of variation of parameters as applied to systems of differential equations, we revisit the concept of a fundamental matrix.

Fundamental Matrices

In Section 6.4, we introduced the concepts of a solution matrix and a fundamental matrix. Recall that a *solution matrix,* $\Psi(t)$, is an $(n \times n)$ matrix whose n columns are each solutions of the homogeneous linear first order system $\mathbf{y}' = P(t)\mathbf{y}$, $a < t < b$. Thus, if $\mathbf{y}_1(t), \mathbf{y}_2(t), \ldots, \mathbf{y}_n(t)$ are solutions of $\mathbf{y}' = P(t)\mathbf{y}$, then

$$\Psi(t) = [\mathbf{y}_1(t), \mathbf{y}_2(t), \ldots, \mathbf{y}_n(t)]$$

is a solution matrix. In addition, if these solutions form a fundamental set of solutions, then the solution matrix $\Psi(t)$ is called a *fundamental matrix* (see Section 6.4).

When we introduced the concepts of solution matrix and fundamental matrix in Section 6.4, our primary focus was on the solutions $\{\mathbf{y}_1, \mathbf{y}_2, \ldots, \mathbf{y}_n\}$. We used solution matrices and fundamental matrices as a way to organize solutions into an array. Initial conditions are conveniently imposed using such arrays. We also use solution matrices to define the Wronskian of the solution set.

At this point we begin a subtle but important shift of emphasis. We now view solution matrices and fundamental matrices as $(n \times n)$ matrix functions that are mathematical entities in their own right. In particular, solution matrices and fundamental matrices for $\mathbf{y}' = P(t)\mathbf{y}$ can themselves be viewed as solutions of the *matrix* differential equation $\Psi' = P(t)\Psi$. Some important properties of solution matrices and fundamental matrices are summarized in Theorem 6.12.

Theorem 6.12

Consider the homogeneous linear first order system

$$\mathbf{y}' = P(t)\mathbf{y}, \qquad a < t < b,$$

where $\mathbf{y}(t)$ is an $(n \times 1)$ vector function and $P(t)$ is an $(n \times n)$ coefficient matrix, continuous on (a, b).

(a) Let $\Psi(t)$ be any solution matrix of $\mathbf{y}' = P(t)\mathbf{y}$, $a < t < b$. Then $\Psi(t)$ satisfies the matrix differential equation

$$\Psi' = P(t)\Psi, \qquad a < t < b.$$

(b) Let Ψ_0 represent any given constant $(n \times n)$ matrix, and let t_0 be any fixed point in the interval $a < t < b$. Then there is a unique $(n \times n)$ matrix $\Psi(t)$ that solves the initial value problem

$$\Psi' = P(t)\Psi, \qquad \Psi(t_0) = \Psi_0, \qquad a < t < b.$$

Moreover, if the constant matrix Ψ_0 is invertible, then the matrix $\Psi(t)$ is a fundamental matrix of $\mathbf{y}' = P(t)\mathbf{y}$, $a < t < b$.

(c) If $\Psi(t)$ is any fundamental matrix and $\hat{\Psi}(t)$ is any solution matrix of $\mathbf{y}' = P(t)\mathbf{y}$, $a < t < b$, then there exists an $(n \times n)$ constant matrix C such that

$$\hat{\Psi}(t) = \Psi(t)C, \qquad a < t < b.$$

Moreover, the matrix $\hat{\Psi}(t)$ is also a fundamental matrix if and only if $\det[C] \neq 0$.

PROOF:

(a) Express the solution matrix in column form as

$$\Psi(t) = [\mathbf{y}_1(t), \mathbf{y}_2(t), \ldots, \mathbf{y}_n(t)].$$

Recall, from Section 6.1, that $\Psi'(t) = [\mathbf{y}_1'(t), \mathbf{y}_2'(t), \ldots, \mathbf{y}_n'(t)]$. We now use the fact that each column of $\Psi(t)$ is a solution of the given homogeneous linear system to obtain the desired result:

$$\begin{aligned}\Psi'(t) &= [\mathbf{y}_1'(t), \mathbf{y}_2'(t), \ldots, \mathbf{y}_n'(t)] \\ &= [P(t)\mathbf{y}_1(t), P(t)\mathbf{y}_2(t), \ldots, P(t)\mathbf{y}_n(t)] \\ &= P(t)[\mathbf{y}_1(t), \mathbf{y}_2(t), \ldots, \mathbf{y}_n(t)] \\ &= P(t)\Psi(t), \quad a < t < b.\end{aligned}$$

(b) We build upon the argument of part (a). Again represent $\Psi(t)$ in terms of its columns as $\Psi(t) = [\mathbf{y}_1(t), \mathbf{y}_2(t), \ldots, \mathbf{y}_n(t)]$. Likewise represent the constant matrix Ψ_0 in terms of its columns as $\Psi_0 = [\boldsymbol{\Psi}_1, \boldsymbol{\Psi}_2, \ldots, \boldsymbol{\Psi}_n]$. Observe that the initial value problem

$$\Psi' = P(t)\Psi, \qquad \Psi(t_0) = \Psi_0, \qquad a < t < b$$

is equivalent to the n separate initial value problems

$$\mathbf{y}_j' = P(t)\mathbf{y}_j, \qquad \mathbf{y}_j(t_0) = \boldsymbol{\Psi}_j, \qquad 1 \le j \le n, \qquad a < t < b.$$

The uniqueness of the solution of each of these initial value problems is established in Theorem 6.1. Hence, the solution matrix $\Psi(t)$ is also unique. Moreover, if Ψ_0 is invertible, then $\det[\Psi(t_0)] = \det[\Psi_0] \neq 0$. Thus, see Theorem 6.3, $\Psi(t)$ is a fundamental matrix since the Wronskian, $W(t_0) = \det[\Psi(t_0)]$, is nonzero at t_0.

(c) This result is simply a restatement of Theorem 6.8 and the proof is given there. ∎

Example 2 illustrates several parts of Theorem 6.12.

EXAMPLE 2

Find the unique matrix solution of the initial value problem $\Psi' = A\Psi$, $\Psi(0) = \Psi_0$, $-\infty < t < \infty$, where

$$A = \begin{bmatrix} 1 & 2 \\ 2 & 1 \end{bmatrix}, \qquad \Psi_0 = \begin{bmatrix} 3 & 2 \\ 1 & -4 \end{bmatrix}.$$

Solution: We saw earlier, in Example 1, that the two vector functions

$$\mathbf{y}_1(t) = \begin{bmatrix} e^{-t} \\ -e^{-t} \end{bmatrix} \quad \text{and} \quad \mathbf{y}_2(t) = \begin{bmatrix} e^{3t} \\ e^{3t} \end{bmatrix}$$

form a fundamental set of solutions of $\mathbf{y}' = A\mathbf{y}$, $-\infty < t < \infty$. Therefore, the (2×2) matrix function

$$\Psi(t) = \begin{bmatrix} e^{-t} & e^{3t} \\ -e^{-t} & e^{3t} \end{bmatrix}$$

is a fundamental matrix for $\mathbf{y}' = A\mathbf{y}$. By part (a) of Theorem 6.12, the matrix $\Psi(t)$ satisfies the given differential equation, $\Psi' = A\Psi$. Note that $\Psi(t)$ does not satisfy the initial condition since

$$\Psi(0) = \begin{bmatrix} 1 & 1 \\ -1 & 1 \end{bmatrix} \neq \Psi_0.$$

By part (c) of Theorem 6.12, however, we know we can represent the desired solution, $\hat{\Psi}(t)$, as

$$\hat{\Psi}(t) = \Psi(t)C,$$

where C is a constant nonsingular (2×2) matrix. Imposing the initial conditions, we see that C must satisfy the equation $\hat{\Psi}(0) = \Psi(0)C = \Psi_0$. Solving for C we find

$$C = [\Psi(0)]^{-1}\Psi_0 \quad \text{or} \quad C = \begin{bmatrix} \frac{1}{2} & -\frac{1}{2} \\ \frac{1}{2} & \frac{1}{2} \end{bmatrix} \begin{bmatrix} 3 & 2 \\ 1 & -4 \end{bmatrix} = \begin{bmatrix} 1 & 3 \\ 2 & -1 \end{bmatrix}.$$

Having C, we know the solution of the initial value problem is $\hat{\Psi}(t) = \Psi(t)C$ or

$$\hat{\Psi}(t) = \begin{bmatrix} e^{-t} & e^{3t} \\ -e^{-t} & e^{3t} \end{bmatrix} \begin{bmatrix} 1 & 3 \\ 2 & -1 \end{bmatrix} = \begin{bmatrix} e^{-t} + 2e^{3t} & 3e^{-t} - e^{3t} \\ -e^{-t} + 2e^{3t} & -3e^{-t} - e^{3t} \end{bmatrix}. \quad \blacktriangle$$

The Variation of Parameters Formula

We now use fundamental matrices in the method of variation of parameters to obtain a representation for the solution of the nonhomogeneous initial value problem

$$\mathbf{y}' = P(t)\mathbf{y} + \mathbf{g}(t), \qquad \mathbf{y}(t_0) = \mathbf{y}_0, \qquad a < t < b. \tag{4}$$

In (4), we assume that the $(n \times n)$ coefficient matrix $P(t)$ and the $(n \times 1)$ vector function $\mathbf{g}(t)$ are continuous on (a, b) and that t_0 is some point lying in this interval.

The method of variation of parameters, as developed in Sections 4.10 and 5.5, is based on an assumed knowledge of the complementary solution. This knowledge is then used to facilitate finding a particular solution. For the linear system (4), the analogous assumption is a knowledge of a fundamental set of solutions of the homogeneous problem; that is, a knowledge of a fundamental matrix. Therefore, we assume that we know a fundamental matrix, $\Psi(t)$, where

$$\Psi' = P(t)\Psi, \qquad a < t < b. \tag{5}$$

The complementary solution of (4) has the form $\mathbf{y}_C(t) = \Psi(t)\mathbf{c}$, where $\mathbf{c}$ is an arbitrary constant $(n \times 1)$ vector. Therefore, we "vary the parameter" and look for a solution of the nonhomogeneous equation (4) of the form $\mathbf{y}(t) = \Psi(t)\mathbf{u}(t)$, where $\mathbf{u}(t)$ is an unknown $(n \times 1)$ matrix function to be determined. Substituting this representation into equation (4) leads to

$$(\Psi(t)\mathbf{u}(t))' = P(t)(\Psi(t)\mathbf{u}(t)) + \mathbf{g}(t), \tag{6}$$

or, after differentiating the product on the left-hand side of (6),

$$\Psi'(t)\mathbf{u}(t) + \Psi(t)\mathbf{u}'(t) = P(t)(\Psi(t)\mathbf{u}(t)) + \mathbf{g}(t). \tag{7}$$

Using the fact that $\Psi'(t) = P(t)\Psi(t)$, equation (7) reduces to $\Psi(t)\mathbf{u}'(t) = \mathbf{g}(t)$, or

$$\mathbf{u}'(t) = \Psi^{-1}(t)\mathbf{g}(t). \tag{8}$$

[Note that $\Psi(t)$, being a fundamental matrix, is invertible.]

We next solve for the unknown matrix function $\mathbf{u}(t)$ in equation (8) by antidifferentiation:

$$\mathbf{u}(t) = \mathbf{u}_0 + \int_{t_0}^{t} \Psi^{-1}(s)\mathbf{g}(s)\,ds, \tag{9}$$

where $\mathbf{u}(t_0) = \mathbf{u}_0$ is an arbitrary constant $(n \times 1)$ vector. By including an arbitrary vector $\mathbf{u}_0$ in (9), we have found a representation $\mathbf{y}(t) = \Psi(t)\mathbf{u}(t)$ for the *general solution* of $\mathbf{y}' = P(t)\mathbf{y} + \mathbf{g}(t)$, namely:

$$\mathbf{y}(t) = \Psi(t)\mathbf{u}_0 + \Psi(t)\int_{t_0}^{t} \Psi^{-1}(s)\mathbf{g}(s)\,ds. \tag{10}$$

Note that equation (10) has the structure mentioned at the beginning of this section; that is, $\mathbf{y}(t)$ is the sum of the complementary solution, $\mathbf{y}_C(t) = \Psi(t)\mathbf{u}_0$, and a particular solution

$$\mathbf{y}_P(t) = \Psi(t)\int_{t_0}^{t} \Psi^{-1}(s)\mathbf{g}(s)\,ds.$$

(This particular solution is the one that vanishes at $t = t_0$.)

We can solve initial value problem (4) by imposing the initial condition in equation (10). Since $\mathbf{y}(t_0) = \Psi(t_0)\mathbf{u}_0$, we see that $\mathbf{u}_0$ is given by

$$\mathbf{u}_0 = \Psi^{-1}(t_0)\mathbf{y}_0.$$

Thus, the solution of initial value problem (4) is

$$\mathbf{y}(t) = \Psi(t)\Psi^{-1}(t_0)\mathbf{y}_0 + \Psi(t)\int_{t_0}^{t} \Psi^{-1}(s)\mathbf{g}(s)\,ds. \tag{11}$$

We refer to equation (11) as the **variation of parameters** formula for the solution of the initial value problem. Note, in equation (11), that the matrix product $\Psi(t)\Psi^{-1}(t_0)$ is the fundamental matrix that reduces to the identity matrix when $t = t_0$. Likewise, the matrix product $\Psi(t)\Psi^{-1}(s)$ is the fundamental matrix that reduces to the identity when $t = s$.

A Control Problem in Charged Particle Ballistics

The variation of parameters formula can be used to analyze a control problem that builds on a problem considered in Example 3 of Section 6.6. We pose the problem here and develop its solution in the exercises.

Assume that a constant uniform magnetic field $\mathbf{B}$ and a constant uniform electric field $\mathbf{E}$ exist throughout all of space. A particle having charge q and mass m is launched from the origin at time $t = 0$. As in Section 6.6, let the position of the particle at time t be denoted by the position vector

$$\mathbf{r}(t) = x(t)\mathbf{i} + y(t)\mathbf{j} + z(t)\mathbf{k}$$

and let its velocity vector be given by

$$\mathbf{v}(t) = \frac{d}{dt}\mathbf{r}(t) = v_x(t)\mathbf{i} + v_y(t)\mathbf{j} + v_z(t)\mathbf{k}. \tag{12}$$

If we include the force of gravity, then Newton's second law of motion leads to the following vector equation for the motion of the particle:

$$m\mathbf{v}' = q\mathbf{v} \times \mathbf{B} + q\mathbf{E} + m\mathbf{g}, \qquad \mathbf{v}(0) = \mathbf{v}_0. \tag{13}$$

We saw in Section 6.6 that the Lorentz force exerted upon the particle by the magnetic field is given by the term $q\mathbf{v} \times \mathbf{B}$; this force tends to move the charged particle in a helical path. The forces $q\mathbf{E}$ and $m\mathbf{g}$ are the additional forces exerted upon the particle by the electric and gravitational fields, respectively. For simplicity, we assume that the magnetic field $\mathbf{B}$ is directed in the positive z-direction and gravitational acceleration is directed in the negative z-direction; thus

$$\mathbf{B} = B\mathbf{k} \quad \text{and} \quad \mathbf{g} = -g\mathbf{k}.$$

We place no restriction on the orientation of the electric field, $\mathbf{E} = E_x\mathbf{i} + E_y\mathbf{j} + E_z\mathbf{k}$.

The problem we want to consider is the following. Suppose we have total control over the launching of the charged particle from the origin at time $t = 0$. In other words, we are free to specify $\mathbf{v}(0) = \mathbf{v}_0$. Is it always possible to select an appropriate initial velocity $\mathbf{v}_0$ so that the charged particle will arrive at some arbitrarily specified target location in space, say $\tilde{\mathbf{r}} = \tilde{x}\mathbf{i} + \tilde{y}\mathbf{j} + \tilde{z}\mathbf{k}$, at some specified later time $t = \tau > 0$? Figure 6.3 illustrates the problem.

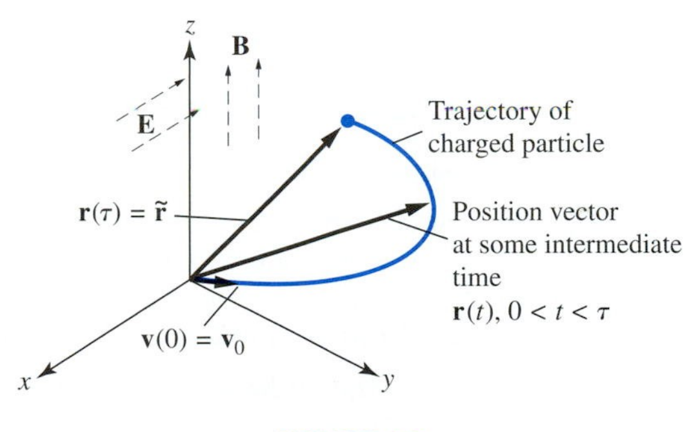

FIGURE 6.3

A charged particle is launched from the origin at time $t = 0$. The position, $\mathbf{r}(t)$, and the velocity, $\mathbf{v}(t)$, of the particle are governed by equation (13). For a given target $\tilde{\mathbf{r}}$ in space and for a given time of arrival $\tau > 0$, is it always possible to choose the initial condition $\mathbf{v}(0) = \mathbf{v}_0$ so that the particle reaches the target at time τ? That is, can we choose $\mathbf{v}_0$ in (13) so that $\mathbf{r}(\tau) = \tilde{\mathbf{r}}$?

EXERCISES

Exercises 1–5:

Consider the initial value problem $\mathbf{y}' = A\mathbf{y} + \mathbf{g}(t)$, $\mathbf{y}(0) = \mathbf{y}_0$.

(a) Find the eigenpairs of the matrix A and form the complementary solution of the differential equation.

(b) Construct a particular solution by assuming a solution of the form suggested and solving for the undetermined constant vectors, $\mathbf{a}$, $\mathbf{b}$, and $\mathbf{c}$.

(c) Form the general solution of the nonhomogeneous differential equation.

(d) Impose the initial condition to obtain the unique solution of the initial value problem.

1. $\mathbf{y}' = \begin{bmatrix} -2 & 1 \\ 1 & -2 \end{bmatrix} \mathbf{y} + \begin{bmatrix} 1 \\ 1 \end{bmatrix}, \quad \mathbf{y}_0 = \begin{bmatrix} 3 \\ 1 \end{bmatrix}.$ Try $\mathbf{y}_P(t) = \mathbf{a}$.

2. $\mathbf{y}' = \begin{bmatrix} 2 & 1 \\ 1 & 2 \end{bmatrix} \mathbf{y} + \begin{bmatrix} e^{-t} \\ 0 \end{bmatrix}, \quad \mathbf{y}_0 = \begin{bmatrix} 0 \\ 0 \end{bmatrix}.$ Try $\mathbf{y}_P(t) = e^{-t}\mathbf{a}$.

3. $\mathbf{y}' = \begin{bmatrix} 0 & 1 \\ 1 & 0 \end{bmatrix} \mathbf{y} + \begin{bmatrix} t \\ -1 \end{bmatrix}, \quad \mathbf{y}_0 = \begin{bmatrix} 2 \\ -1 \end{bmatrix}.$ Try $\mathbf{y}_P(t) = t\mathbf{a} + \mathbf{b}$.

4. $\mathbf{y}' = \begin{bmatrix} 0 & -1 \\ -1 & 0 \end{bmatrix} \mathbf{y} + \begin{bmatrix} t \\ e^{2t} \end{bmatrix}, \quad \mathbf{y}_0 = \begin{bmatrix} 0 \\ 1 \end{bmatrix}.$ Try $\mathbf{y}_P(t) = e^{2t}\mathbf{a} + \mathbf{b}t + \mathbf{c}$.

5. $\mathbf{y}' = \begin{bmatrix} -3 & -2 \\ 4 & 3 \end{bmatrix} \mathbf{y} + \begin{bmatrix} \sin t \\ 0 \end{bmatrix}, \quad \mathbf{y}_0 = \begin{bmatrix} 0 \\ 0 \end{bmatrix}.$ Try $\mathbf{y}_P(t) = (\sin t)\mathbf{a} + (\cos t)\mathbf{b}$.

6. As an illustration of the difficulties that might arise in using the method of undetermined coefficients, consider

$$\mathbf{y}' = \begin{bmatrix} 1 & 1 \\ 1 & 1 \end{bmatrix} \mathbf{y} + \begin{bmatrix} e^{2t} \\ 0 \end{bmatrix}.$$

(a) Determine the complementary solution.

(b) Show that seeking a particular solution of the form $\mathbf{y}_P(t) = e^{2t}\mathbf{a}$ does not work.

(c) Since a member of the fundamental set of solutions comprising the complementary solution is a constant vector multiplied by e^{2t}, we might consider $\mathbf{y}_P(t) = te^{2t}\mathbf{a}$ to be a reasonable guess. Show that this guess does not work either. (We obtain the general solution in Exercise 21, using the method of variation of parameters.)

7. Consider the initial value problem

$$\mathbf{y}' = \begin{bmatrix} 0 & 2 \\ -2 & 0 \end{bmatrix} \mathbf{y} + \mathbf{g}(t), \qquad \mathbf{y}\left(\frac{\pi}{2}\right) = \mathbf{y}_0.$$

Suppose we know that

$$\mathbf{y}(t) = \begin{bmatrix} 1 + \sin 2t \\ e^t + \cos 2t \end{bmatrix}$$

is the unique solution. Determine $\mathbf{g}(t)$ and $\mathbf{y}_0$.

8. Consider the initial value problem

$$\mathbf{y}' = \begin{bmatrix} 1 & t \\ t^2 & 1 \end{bmatrix} \mathbf{y} + \mathbf{g}(t), \qquad \mathbf{y}(1) = \begin{bmatrix} 2 \\ -1 \end{bmatrix}.$$

Suppose we know that

$$\mathbf{y}(t) = \begin{bmatrix} t + \alpha \\ t^2 + \beta \end{bmatrix}$$

is the unique solution. Find $\mathbf{g}(t)$ and the constants α and β.

9. Let $P(t)$ be a (2×2) matrix with continuous entries. Consider the differential equation $\mathbf{y}' = P(t)\mathbf{y} + \mathbf{g}(t)$. Suppose we know the solution is $\mathbf{y}_1(t)$ when $\mathbf{g}(t) = \mathbf{g}_1(t)$ and $\mathbf{y}_2(t)$ when $\mathbf{g}(t) = \mathbf{g}_2(t)$. Determine $P(t)$ if

$$\mathbf{g}_1(t) = \begin{bmatrix} -2 \\ 0 \end{bmatrix} \text{ and } \mathbf{y}_1(t) = \begin{bmatrix} 1 \\ e^{-t} \end{bmatrix}; \quad \mathbf{g}_2(t) = \begin{bmatrix} e^t \\ -1 \end{bmatrix} \text{ and } \mathbf{y}_2(t) = \begin{bmatrix} e^t \\ -1 \end{bmatrix}.$$

[Hint: Form the matrix equation $[\mathbf{y}_1', \mathbf{y}_2'] = P(t)[\mathbf{y}_1, \mathbf{y}_2] + [\mathbf{g}_1(t), \mathbf{g}_2(t)]$.]

Equilibrium Solutions Consider the linear system $\mathbf{y}' = A\mathbf{y} + \mathbf{b}$ where A is a constant $(n \times n)$ matrix and $\mathbf{b}$ is a constant $(n \times 1)$ vector. An *equilibrium solution,* $\mathbf{y}(t)$, is a constant solution of the differential equation.

10. If the matrix A is invertible, show that $\mathbf{y}' = A\mathbf{y} + \mathbf{b}$ has a unique equilibrium solution. If the matrix A is not invertible, must the differential equation $\mathbf{y}' = A\mathbf{y} + \mathbf{b}$ possess an equilibrium solution? If an equilibrium solution does exist in this case, is it unique?

Exercises 11–15:

In each exercise, determine all equilibrium solutions (if any).

11. $\mathbf{y}' = \begin{bmatrix} 1 & 4 \\ -1 & -3 \end{bmatrix}\mathbf{y} + \begin{bmatrix} 2 \\ 1 \end{bmatrix}$

12. $\mathbf{y}' = \begin{bmatrix} 2 & -1 \\ -1 & 1 \end{bmatrix}\mathbf{y} + \begin{bmatrix} 2 \\ -1 \end{bmatrix}$

13. $\mathbf{y}' = \begin{bmatrix} 1 & -1 \\ -1 & 1 \end{bmatrix}\mathbf{y} + \begin{bmatrix} 2 \\ -2 \end{bmatrix}$

14. $\mathbf{y}' = \begin{bmatrix} 1 & 1 & 0 \\ 0 & -1 & 2 \\ 0 & 0 & 1 \end{bmatrix}\mathbf{y} + \begin{bmatrix} 2 \\ 3 \\ 2 \end{bmatrix}$

15. $\mathbf{y}' = \begin{bmatrix} 1 & 0 & 0 \\ 0 & 1 & 1 \\ 0 & 1 & 1 \end{bmatrix}\mathbf{y} + \begin{bmatrix} -2 \\ 0 \\ 0 \end{bmatrix}$

Exercises 16–20:

Consider the homogeneous linear system $\mathbf{y}' = A\mathbf{y}$. Recall that any associated fundamental matrix satisfies the matrix differential equation $\Psi' = A\Psi$. In each exercise, construct a fundamental matrix that solves the matrix initial value problem $\Psi' = A\Psi$, $\Psi(t_0) = \Psi_0$.

16. $\Psi' = \begin{bmatrix} 1 & -1 \\ -1 & 1 \end{bmatrix}\Psi, \quad \Psi(1) = \begin{bmatrix} 1 & 0 \\ 0 & 1 \end{bmatrix}$

17. $\Psi' = \begin{bmatrix} 0 & 2 \\ -2 & 0 \end{bmatrix}\Psi, \quad \Psi\left(\frac{\pi}{4}\right) = \begin{bmatrix} 1 & -1 \\ 0 & 1 \end{bmatrix}$

18. $\Psi' = \begin{bmatrix} 1 & -1 \\ -1 & 1 \end{bmatrix}\Psi, \quad \Psi(0) = \begin{bmatrix} 1 & 0 \\ 2 & 1 \end{bmatrix}$

19. $\Psi' = \begin{bmatrix} 3 & -4 \\ 2 & -3 \end{bmatrix}\Psi, \quad \Psi(0) = \begin{bmatrix} 1 & 0 \\ 0 & 1 \end{bmatrix}$

20. $\Psi' = \begin{bmatrix} 1 & 4 \\ -1 & 1 \end{bmatrix}\Psi, \quad \Psi\left(\frac{\pi}{4}\right) = \begin{bmatrix} 1 & 0 \\ 0 & 1 \end{bmatrix}$

Exercises 21–24:

Use the method of variation of parameters to solve the given initial value problem.

21. $\mathbf{y}' = \begin{bmatrix} 1 & 1 \\ 1 & 1 \end{bmatrix}\mathbf{y} + \begin{bmatrix} e^{2t} \\ 0 \end{bmatrix}, \quad \mathbf{y}(0) = \begin{bmatrix} 0 \\ 0 \end{bmatrix}$

22. $\mathbf{y}' = \begin{bmatrix} 9 & -4 \\ 15 & -7 \end{bmatrix} \mathbf{y} + \begin{bmatrix} e^t \\ 0 \end{bmatrix}, \quad \mathbf{y}(0) = \begin{bmatrix} 2 \\ 5 \end{bmatrix}$

23. $\mathbf{y}' = \begin{bmatrix} 0 & 1 \\ -1 & 0 \end{bmatrix} \mathbf{y} + \begin{bmatrix} 2 \\ 1 \end{bmatrix}, \quad \mathbf{y}(0) = \begin{bmatrix} 0 \\ 1 \end{bmatrix}$

24. $\mathbf{y}' = \begin{bmatrix} 1 & 1 \\ 0 & 1 \end{bmatrix} \mathbf{y} + \begin{bmatrix} 1 \\ 1 \end{bmatrix}, \quad \mathbf{y}(0) = \begin{bmatrix} 0 \\ 0 \end{bmatrix}$

25. Consider the two-tank flow system shown in the figure. The volume of fluid in each tank is V gallons, and all flow rates are r gallons per minute. Initially both tanks contain fresh water. At time $t = 0$, a salt solution having concentration c pounds per gallon begins to flow into Tank 1 while fresh water flows into Tank 2. Well-stirred mixtures are exchanged between the two tanks and also exit the system from both tanks as shown. If the "conservation of salt" principle is applied to both tanks, the following initial value problem is obtained,

$$\frac{d}{dt} \begin{bmatrix} Q_1 \\ Q_2 \end{bmatrix} = \frac{r}{V} \begin{bmatrix} -2 & 1 \\ 1 & -2 \end{bmatrix} \begin{bmatrix} Q_1 \\ Q_2 \end{bmatrix} + \begin{bmatrix} cr \\ 0 \end{bmatrix}, \qquad \begin{bmatrix} Q_1(0) \\ Q_2(0) \end{bmatrix} = \begin{bmatrix} 0 \\ 0 \end{bmatrix},$$

where $Q_1(t)$ and $Q_2(t)$ denote the amount of salt in the two tanks at time t.

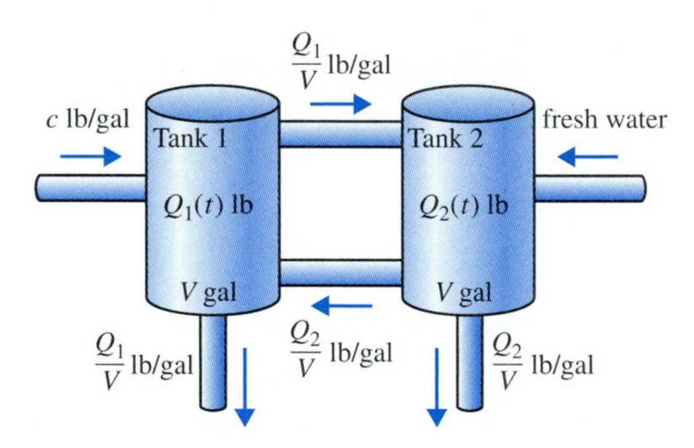

Figure for Exercise 25

(a) Does this nonhomogeneous system possess an equilibrium solution? If so, determine the equilibrium solution and determine whether it is physically meaningful. (Since Q_1 and Q_2 represent amounts of salt, their equilibrium values must be nonnegative.)

(b) Find the eigenpairs of the coefficient matrix

$$\frac{r}{V} \begin{bmatrix} -2 & 1 \\ 1 & -2 \end{bmatrix}$$

and form the complementary solution $\mathbf{Q}_C(t)$.

(c) Look for a particular solution of the form $\mathbf{Q}_P(t) = \mathbf{a}$. How does this calculation relate to looking for an equilibrium solution?

(d) Form the general solution of the nonhomogeneous system and impose the initial condition to obtain the unique solution of the initial value problem.

(e) Does

$$\lim_{t\to\infty}\begin{bmatrix}\dfrac{Q_1(t)}{V}\\[2mm] \dfrac{Q_2(t)}{V}\end{bmatrix}$$

exist? If so, what is the limit?

26. Consider the RL network shown in the figure. Assume that the loop currents I_1 and I_2 are zero until a voltage source, $V_s(t)$, having the polarity shown, is turned on at time $t = 0$. Applying Kirchhoff's voltage law to each loop, we obtain the equations

$$-V_S(t) + L_1\frac{dI_1}{dt} + R_1I_1 + R_3(I_1 - I_2) = 0,$$

$$R_3(I_2 - I_1) + R_2I_2 + L_2\frac{dI_2}{dt} = 0.$$

(a) Formulate the initial value problem for the loop currents,

$$\begin{bmatrix}I_1(t)\\ I_2(t)\end{bmatrix}, \quad \text{assuming that } L_1 = L_2 = 0.5\,H, \qquad R_1 = R_2 = 1\,k\Omega, \quad \text{and} \quad R_3 = 2\,k\Omega.$$

(b) Determine a fundamental matrix for the associated linear homogeneous system.

(c) Use the method of variation of parameters to solve the initial value problem for the case where $V_S(t) = 1$ for $t > 0$.

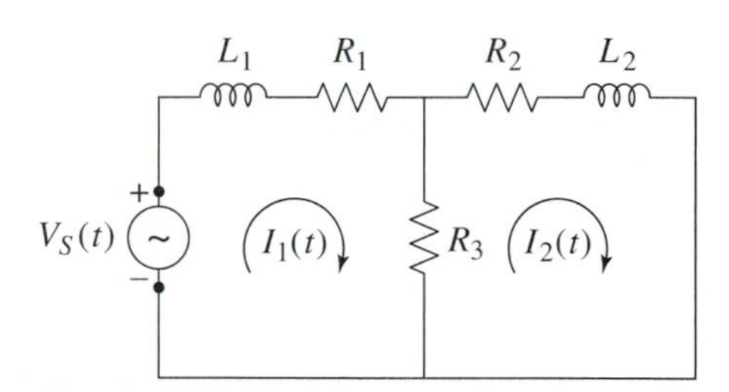

Figure for Exercise 26

27. A particle moves in the xy-plane with velocity vector $\mathbf{v}(t) = v_x(t)\mathbf{i} + v_y(t)\mathbf{j}$. Assume that the motion is governed by the differential equation

$$\mathbf{v}'(t) = v_x'(t)\mathbf{i} + v_y'(t)\mathbf{j} = -\mathbf{v}(t) + \mathbf{v}(t) \times \mathbf{k} + \mathbf{f}, \tag{14}$$

where $\mathbf{f} = f_x\mathbf{i} + f_y\mathbf{j}$ is a constant vector. Differential equation (14) is the type of equation that describes the motion of a charged particle moving under the combined influence of a drag force, a Lorentz force, and a constant applied force.

(a) Form the vector

$$\begin{bmatrix}v_x(t)\\ v_y(t)\end{bmatrix}$$

and rewrite equation (14) as a nonhomogeneous linear system of the form

$$\frac{d}{dt}\begin{bmatrix}v_x\\ v_y\end{bmatrix} = \begin{bmatrix}a_{11} & a_{12}\\ a_{21} & a_{22}\end{bmatrix}\begin{bmatrix}v_x\\ v_y\end{bmatrix} + \begin{bmatrix}f_x\\ f_y\end{bmatrix}.$$

(b) Assume that $\mathbf{f} = (\mathbf{i} + \sqrt{3}\,\mathbf{j})/2$. Is it possible to launch the particle with an initial velocity $\mathbf{v}(0) = \mathbf{v}_0$ in such a way that the particle will continue to move with this same constant velocity? If so, determine this initial velocity.

28. Charged Particle Control Problem In this exercise, we ask you to develop the solution of the problem described in Figure 6.3. The equation of motion for the particle of mass m and charge q, moving in constant magnetic field $\mathbf{B} = B\mathbf{k}$ and electric field $\mathbf{E}$ is

$$m\mathbf{v}' = q\mathbf{v} \times (B\mathbf{k}) + q\mathbf{E} - mg\mathbf{k},$$

where the weight of the particle is directed in the negative $\mathbf{k}$ (vertically downward) direction. Is it possible to select an initial velocity $\mathbf{v}(0) = \mathbf{v}_0$ so that the particle, launched from the origin at time $t = 0$, will arrive at some desired location $\tilde{\mathbf{r}}$ at some prescribed time $t = \tau$?

(a) Let $qB/m = \omega_C$. As in Section 6.6, ω_C is called the (radian) cyclotron frequency. Show that the given vector equation is equivalent to the following nonhomogeneous linear system.

$$\frac{d}{dt}\begin{bmatrix} v_x \\ v_y \\ v_z \end{bmatrix} = \begin{bmatrix} 0 & \omega_C & 0 \\ -\omega_C & 0 & 0 \\ 0 & 0 & 0 \end{bmatrix}\begin{bmatrix} v_x \\ v_y \\ v_z \end{bmatrix} + \begin{bmatrix} \frac{q}{m}E_x \\ \frac{q}{m}E_y \\ \frac{q}{m}E_z - g \end{bmatrix}. \tag{15}$$

(b) Find the eigenvalues and eigenvectors of the coefficient matrix in (15). Construct a fundamental matrix.

(c) Let $\Phi(t)$ denote the fundamental matrix that reduces to the identity matrix at time $t = 0$. Show that

$$\Phi(t) = \begin{bmatrix} \cos(\omega_C t) & \sin(\omega_C t) & 0 \\ -\sin(\omega_C t) & \cos(\omega_C t) & 0 \\ 0 & 0 & 1 \end{bmatrix}.$$

(d) Use the variation of parameters formula to show that the solution of (15) can be written as

$$\mathbf{v}(t) = \Phi(t)\mathbf{v}_0 + \mathbf{f}(t),$$

where

$$\mathbf{v}(t) = \begin{bmatrix} v_x(t) \\ v_y(t) \\ v_z(t) \end{bmatrix}, \qquad \mathbf{v}(0) = \mathbf{v}_0, \quad \text{and} \quad \mathbf{f}(t) = \Phi(t)\int_0^t \Phi^{-1}(s)\begin{bmatrix} \frac{q}{m}E_x \\ \frac{q}{m}E_y \\ \frac{q}{m}E_z - g \end{bmatrix} ds.$$

The initial condition $\mathbf{v}(0) = \mathbf{v}_0$ is to be determined. However, as you will see, the precise nature of $\mathbf{f}(t)$ does not matter in answering the question we have posed.

(e) Since the particle is initially at the origin, its position at the specified time $t = \tau$ is given by

$$\mathbf{r}(\tau) = \int_0^\tau \mathbf{v}(t)\,dt = \left[\int_0^\tau \Phi(t)\,dt\right]\mathbf{v}_0 + \int_0^\tau \mathbf{f}(t)\,dt.$$

Since we want this position to be the specified location $\tilde{\mathbf{r}}$, the initial velocity must be a solution of the following linear system of equations,

$$\left[\int_0^\tau \Phi(t)\,dt\right]\mathbf{v}_0 = \tilde{\mathbf{r}} - \int_0^\tau \mathbf{f}(t)\,dt.$$

This system of equations has a unique solution if and only if $D = \det\left[\int_0^\tau \Phi(t)\,dt\right] \neq 0$. Compute D and show that

$$D = \frac{4\tau}{\omega_C^2}\sin^2\left(\frac{\omega_C \tau}{2}\right).$$

Therefore, we can, in principle, choose the initial velocity $\mathbf{v}_0$ so that the particle arrives at position $\tilde{\mathbf{r}}$ at time τ as long as $\tau \neq 2n\pi/\omega_C$, $n = 1, 2, 3, \ldots$. Since $2\pi/\omega_C$ is the cyclotron period, time τ cannot be an integral number of cyclotron periods.

6.9 Euler's Method for Systems of Differential Equations

We introduced Euler's method in Section 3.8 as a simple numerical algorithm for approximating the solution of a first order scalar initial value problem,

$$y' = f(t, y), \qquad y(t_0) = y_0, \qquad a \le t \le b. \tag{1}$$

In this section we extend Euler's method to the first order linear system

$$\mathbf{y}' = P(t)\mathbf{y} + \mathbf{g}(t), \qquad \mathbf{y}(a) = \mathbf{y}_0, \qquad a \le t \le b.$$

Later, in Chapter 9, we will extend Euler's method (and the other numerical methods developed in Chapter 9) to systems of first order nonlinear equations, $\mathbf{y}' = \mathbf{f}(t, \mathbf{y})$.

Before beginning, we recall (from Section 3.8) the form of Euler's method for a scalar equation. For initial value problem (1), we often partition the interval $[a, b]$ into N subintervals, each having length h, where

$$h = \frac{b - a}{N}.$$

For a given step size h, we define grid points $t_0, t_1, \ldots, t_N$ by the formula

$$t_k = t_0 + kh, \qquad 0 \le k \le N, \quad \text{where} \quad t_0 = a.$$

Note that $a = t_0 < t_1 < t_2 < \cdots < t_N = b$ and that $t_{k+1} = t_k + h$, $0 \le k \le N - 1$.

The numerical solution of initial value problem (1) consists of the points (t_k, y_k), $0 \le k \le N$, where the values, y_k, are numerical approximations to the exact solution values $y(t_k)$; that is $y_k \approx y(t_k)$.

Euler's Method Is a Finite Difference Method

In Section 3.8, we developed Euler's method from geometrical considerations, using the direction field for $y' = f(t, y)$ as a starting point. We now give a different derivation, one that generalizes to systems of first order differential equations.

Let $y(t)$ be the solution of the initial value problem (1), and let $h > 0$ be a given step size. From calculus we know the derivative, $y'(t)$, is defined by

$$\lim_{\Delta t \to 0} \frac{y(t + \Delta t) - y(t)}{\Delta t} = y'(t).$$

Therefore, if $y(t)$ is the solution of initial value problem (1) and if the step size h is small, then we expect the following approximation to be good:

$$\frac{y(t+h) - y(t)}{h} \approx y'(t) = f(t, y(t)). \tag{2}$$

(In approximation (2), we assume that t and $t+h$ are in the interval $[a, b]$.) Evaluating approximation (2) at a grid point $t = t_k$, we obtain

$$y(t_k + h) \approx y(t_k) + hf(t_k, y(t_k)), \qquad 0 \le k \le N - 1. \tag{3}$$

Therefore, once we have an estimate y_k of $y(t_k)$, approximation (3) leads us to an estimate y_{k+1} of $y(t_{k+1})$:

$$y_{k+1} = y_k + hf(t_k, y_k), \qquad 0 \le k \le N - 1, \qquad y_0 = y(t_0). \tag{4}$$

Equation (4) is Euler's method applied to the scalar initial value problem

$$y' = f(t, y), \qquad y(t_0) = y_0.$$

The approximation (2) is called a **finite difference approximation** to $y'(t)$. Methods derived from finite difference approximations, such as Euler's method, are known as **finite difference methods**. As we see in the next subsection, such finite difference methods can be generalized in a natural way to systems of first order differential equations.

Extending Euler's Method to First Order Linear Systems

Consider the initial value problem

$$\mathbf{y}' = P(t)\mathbf{y} + \mathbf{g}(t), \qquad \mathbf{y}(t_0) = \mathbf{y}_0, \qquad a \le t \le b, \tag{5}$$

where the $(n \times n)$ matrix function $P(t)$ and the $(n \times 1)$ vector function $\mathbf{g}(t)$ are continuous on $[a, b]$. Euler's method for initial value problem (5) begins with a generalization of the finite difference approximation (2). In particular, let $\mathbf{y}(t)$ be the unique solution of initial value problem (5), where

$$\mathbf{y}(t) = \begin{bmatrix} y_1(t) \\ \vdots \\ y_n(t) \end{bmatrix}.$$

As we know from Section 6.1,

$$\mathbf{y}'(t) = \begin{bmatrix} y_1'(t) \\ \vdots \\ y_n'(t) \end{bmatrix} = \begin{bmatrix} \lim\limits_{\Delta t \to 0} \dfrac{y_1(t+\Delta t) - y_1(t)}{\Delta t} \\ \vdots \\ \lim\limits_{\Delta t \to 0} \dfrac{y_n(t+\Delta t) - y_n(t)}{\Delta t} \end{bmatrix}$$

$$= \lim_{\Delta t \to 0} \frac{1}{\Delta t} \begin{bmatrix} y_1(t+\Delta t) - y_1(t) \\ \vdots \\ y_n(t+\Delta t) - y_n(t) \end{bmatrix}$$

$$= \lim_{\Delta t \to 0} \frac{1}{\Delta t} [\mathbf{y}(t+\Delta t) - \mathbf{y}(t)].$$

As before, let $h > 0$ be a given step size where $h = (b-a)/N$ and let

$$t_k = t_0 + kh, \qquad 0 \le k \le N, \qquad t_0 = a$$

be a set of grid points for $[a, b]$. Since $\mathbf{y}'(t) = P(t)\mathbf{y}(t) + \mathbf{g}(t)$, if the step size h is small,

$$\frac{1}{h}[\mathbf{y}(t+h) - \mathbf{y}(t)] \approx P(t)\mathbf{y}(t) + \mathbf{g}(t).$$

Evaluating this approximation at the grid point $t = t_k$, we obtain

$$\mathbf{y}(t_k + h) \approx \mathbf{y}(t_k) + h[P(t_k)\mathbf{y}(t_k) + \mathbf{g}(t_k)].$$

Therefore, once we have an estimate $\mathbf{y}_k$ of $\mathbf{y}(t_k)$, this approximation gives us an estimate $\mathbf{y}_{k+1}$ of $\mathbf{y}(t_{k+1})$; we define

$$\mathbf{y}_{k+1} = \mathbf{y}_k + h[P(t_k)\mathbf{y}_k + \mathbf{g}(t_k)], \qquad 0 \le k \le N-1, \qquad \mathbf{y}_0 = \mathbf{y}(t_0). \tag{6}$$

Iteration (6) is Euler's method for the initial value problem (5).

There are obvious mathematical questions about the algorithm (similar to those raised in Section 3.8 for the scalar problem) that need to be answered. These will be addressed in Chapter 9. For now, as in Section 3.8, we assume that the use of a sufficiently small step size h will yield a reasonable approximation of the exact solution. In the rest of this section, we focus on showing how the algorithm is applied.

EXAMPLE 1

Consider the two-tank mixing problem formulated in the Introduction to this chapter. The flow schematic is shown in Figure 6.4 on the next page. The corresponding initial value problem is

$$\frac{d}{dt}\begin{bmatrix} Q_1 \\ Q_2 \end{bmatrix} = \begin{bmatrix} -\dfrac{6}{200+10t} & \dfrac{6}{300} \\ \dfrac{2}{200+10t} & -\dfrac{14}{300} \end{bmatrix} \begin{bmatrix} Q_1 \\ Q_2 \end{bmatrix} + \begin{bmatrix} 5 \\ 0 \end{bmatrix}, \qquad 0 \le t \le 30, \qquad \begin{bmatrix} Q_1(0) \\ Q_2(0) \end{bmatrix} = \begin{bmatrix} 0 \\ 50 \end{bmatrix}. \tag{7}$$

(continued)

(continued)

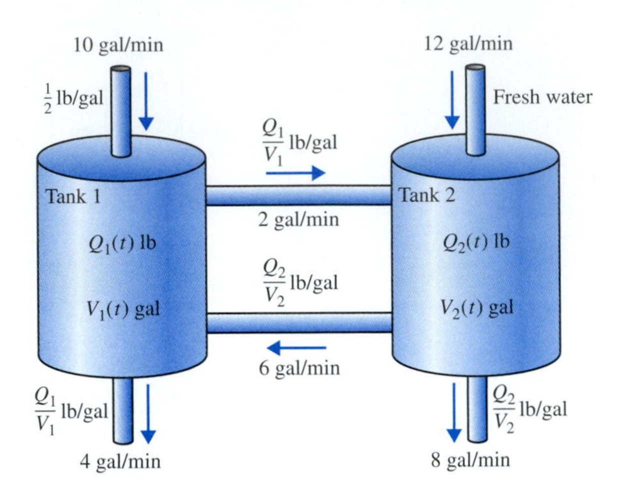

FIGURE 6.4

The two-tank mixing problem discussed in Example 1.

In (7), $Q_1(t)$ and $Q_2(t)$ represent the amount of salt (pounds) in Tanks 1 and 2, respectively, at time t (minutes). Recall that salt solutions enter and leave Tank 1 at different rates, leading to the variable coefficient matrix in (7). At time $t = 30$ min, Tank 1 is filled to capacity and the flow stops. Use Euler's method (6) to estimate $Q_1(t)$ and $Q_2(t)$ on this time interval; use a step size of $h = 0.01$. Plot $Q_1(t)$ and $Q_2(t)$ as functions of t and also plot the concentration of salt in each tank as a function of t.

Solution: The first order system has the form

$$\mathbf{Q}' = P(t)\mathbf{Q} + \mathbf{g}, \qquad 0 \le t \le 30,$$

where, see equation (7), $P(t)$ is a (2×2) matrix function and $\mathbf{g}$ is a (2×1) constant vector. Euler's method, applied to this problem, is the iteration

$$\mathbf{Q}_{k+1} = \mathbf{Q}_k + h[P(t_k)\mathbf{Q}_k + \mathbf{g}], \qquad 0 \le k \le N-1, \qquad \mathbf{Q}_0 = \begin{bmatrix} 0 \\ 50 \end{bmatrix},$$

where $t_k = kh, h = 30/N$. In component form, the algorithm is

$$\begin{aligned} Q_{1,k+1} &= Q_{1,k} + h\left[-\frac{6}{200 + 10t_k}Q_{1,k} + \frac{6}{300}Q_{2,k} + 5\right] \\ Q_{2,k+1} &= Q_{2,k} + h\left[\frac{2}{200 + 10t_k}Q_{1,k} - \frac{14}{300}Q_{2,k}\right], \qquad 0 \le k \le N-1, \end{aligned} \tag{8}$$

where $Q_{1,0} = 0, Q_{2,0} = 50$.

The vector $\mathbf{Q}_j$ is an approximation to the exact solution, $\mathbf{Q}(t_j)$, at time $t_j = jh$. Using $h = 0.01$ corresponds to the value $N = 3000$.

Figure 6.5(a) shows the result of implementing Euler's method with $h = 0.01$. In Figure 6.5(b) we display the Euler's method approximations to

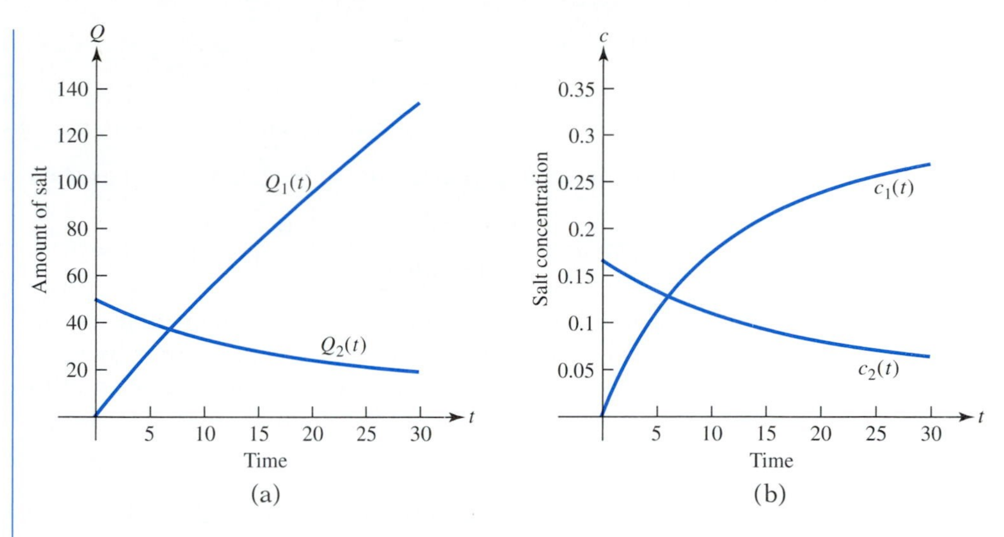

FIGURE 6.5

The results of applying Euler's method to the initial value problem in Example 1.

the concentrations,

$$c_m(t) = \frac{Q_m(t)}{V_m(t)}, \qquad m = 1, 2.$$

As graphed in Figure 6.5, the answers seem reasonable. We would expect the salt concentration in Tank 1 to increase, but we would not expect it to exceed the maximum inflow concentration of 0.5 lb/gal. Likewise, the 2 gal/min inflow from Tank 1 into Tank 2 mitigates the "flushing out" of Tank 2 that would otherwise occur. The concentration in Tank 2 therefore decreases somewhat slowly over the 30-min interval. ▲

Solving Variable-Coefficient Scalar Equations

Euler's method (and the numerical methods we introduce in Chapter 9) can also be used to solve initial value problems involving variable-coefficient scalar equations. Consider, for example, an initial value problem of the type considered in Chapter 4,

$$y'' + p(t)y' + q(t)y = g(t), \qquad y(t_0) = y_0, \qquad y'(t_0) = y_0', \qquad a < t < b, \tag{9}$$

where $p(t)$, $q(t)$, and $g(t)$ are continuous on (a, b) and t_0 lies in (a, b).

Suppose we want to solve this problem on a subinterval $[c, d]$, where $[c, d]$ is contained within (a, b). We have seen that the second order scalar equation (9) can be recast as a first order system. Let

$$z_1(t) = y(t), \qquad z_2(t) = y'(t) \quad \text{and} \quad \mathbf{z}(t) = \begin{bmatrix} z_1(t) \\ z_2(t) \end{bmatrix}.$$

Then,

$$z_1' = y' = z_2$$
$$z_2' = y'' = -p(t)y' - q(t)y + g(t) = -q(t)z_1 - p(t)z_2 + g(t).$$

Thus, we can rewrite initial value problem (9) as the first order linear system,

$$\mathbf{z}' = \begin{bmatrix} 0 & 1 \\ -q(t) & -p(t) \end{bmatrix} \mathbf{z} + \begin{bmatrix} 0 \\ g(t) \end{bmatrix}, \qquad a < t < b, \qquad \mathbf{z}(t_0) = \begin{bmatrix} y_0 \\ y_0' \end{bmatrix}. \tag{10}$$

Equation (10) is an initial value problem that we can solve numerically with any algorithm that is suitable for first order linear systems. In the following example we illustrate the idea by using Euler's method to study the vibrations of a spring-mass system in which the spring undergoes "real-time" stiffening.

EXAMPLE 2

In Section 4.7 we considered unforced mechanical vibrations, specifically a vibrating spring-mass system. In the case where no damping is present, Newton's second law of motion, together with Hooke's law, leads to a differential equation of the following form [see Section 4.7]:

$$my'' + ky = 0. \tag{11}$$

In equation (11), m is the mass of the body, $k > 0$ is the spring constant and $y(t)$ is the displacement of the mass from its equilibrium position at time t. A small value of k corresponds to a "soft" spring—one in which a relatively small force is all that is necessary to achieve a given stretching. By contrast, a large value of k corresponds to a "hard" spring—where a greater force is required to achieve the same amount of stretching.

Suppose we have some mechanism by which we can control the stiffness of the spring in real time—that is, while the mass-spring system is actually vibrating. If we implement such real-time modifications of spring stiffness, how will the system behave?

As a way to gain some insight into the behavior of a stiffening spring, investigate the following model problem using Euler's method:

$$my'' + k(t)y = 0, \qquad 0 \le t \le 3, \qquad y(0) = 0, \qquad y'(0) = 2\pi, \tag{12}$$

where $k(t) = 4m\pi^2(1 + t)$.

Solution: Figure 6.6 displays the variation of $k(t)/m$ with time t. As is shown in the figure, the spring constant increases linearly from an initial value of $k = 4m\pi^2$ at time $t = 0$ to a final value of $k = 16m\pi^2$ at time $t = 3$.

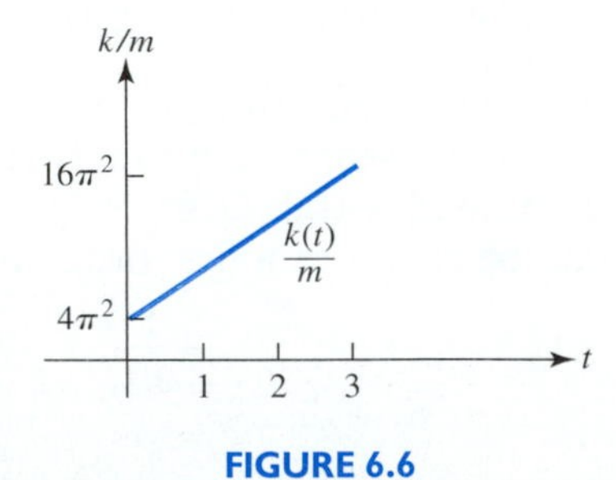

FIGURE 6.6

The graph of $k(t)/m = 4\pi^2(1 + t)$ increases linearly with time.

What should we expect? In Section 4.7 we saw, when k/m is constant, that the vibrations occur with constant amplitude and with a radian frequency given by

$$\omega = \sqrt{\frac{k}{m}}.$$

Therefore, as k/m increases linearly with time, it seems reasonable to expect that the vibrations will become increasingly rapid since ω is growing larger. What will happen to vibration amplitude as time increases, however, is not quite so obvious. To gain some overall insight, we numerically solve equation (12).

The first step in solving initial value problem (12) numerically, is to recast it as a first order linear system. Setting $z_1(t) = y(t)$, $z_2(t) = y'(t)$ and

$$\mathbf{z}(t) = \begin{bmatrix} z_1(t) \\ z_2(t) \end{bmatrix},$$

we have

$$\mathbf{z}' = \begin{bmatrix} 0 & 1 \\ -4\pi^2(1+t) & 0 \end{bmatrix}\mathbf{z}, \qquad \mathbf{z}(0) = \begin{bmatrix} 0 \\ 2\pi \end{bmatrix}, \qquad 0 \le t \le 3.$$

As applied to this initial value problem, the Euler's method algorithm is

$$\mathbf{z}_{k+1} = \mathbf{z}_k + h\begin{bmatrix} 0 & 1 \\ -4\pi^2(1+t_k) & 0 \end{bmatrix}\mathbf{z}_k, \qquad \mathbf{z}_0 = \begin{bmatrix} 0 \\ 2\pi \end{bmatrix}, \qquad k = 0, 1, \ldots, N-1,$$

where

$$h = \frac{3}{N} \quad \text{and} \quad t_k = kh, \qquad 0 \le k \le N.$$

Euler's method was used with a step size $h = 0.005 (N = 600)$. Figure 6.7 displays the results, showing the behavior of both displacement $z_1(t) = y(t)$ and velocity $z_2(t) = y'(t)$ as functions of time. As we conjectured, the vibrations tend

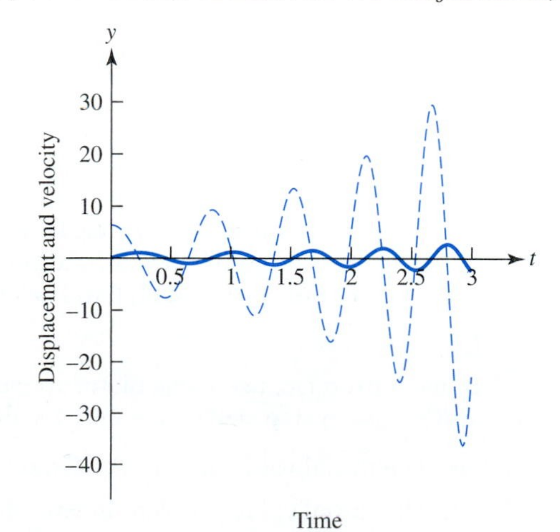

FIGURE 6.7

A graph of displacement, $y(t)$, and velocity, $y'(t)$, for the spring-mass system in Example 2.

(continued)

(continued)

to become more rapid as the spring stiffens; the frequency roughly doubles over the time interval considered. The amplitude of these displacement vibrations also increases. Both of these effects, in turn, contribute to increasing the peak values of the velocity $y'(t)$. ▲

EXERCISES

Exercises 1–5:

In each exercise, assume that a numerical solution is desired on the interval $t_0 \le t \le t_0 + T$ using a uniform step size h.

(a) As in equation (8), write the Euler's method algorithm in explicit form for the given initial value problem. Specify the starting values t_0 and y_0.

(b) Give a formula for the kth t-value, t_k. What is the range of the index k if we choose $h = 0.01$?

1. $\mathbf{y}' = \begin{bmatrix} 1 & 2 \\ 2 & 3 \end{bmatrix} \mathbf{y} + \begin{bmatrix} 1 \\ 1 \end{bmatrix}, \quad \mathbf{y}(0) = \begin{bmatrix} -1 \\ 1 \end{bmatrix}, \qquad 0 \le t \le 1$

2. $\mathbf{y}' = \begin{bmatrix} 1 & t \\ 2+t & 2 \end{bmatrix} \mathbf{y} + \begin{bmatrix} 1 \\ t \end{bmatrix}, \quad \mathbf{y}(1) = \begin{bmatrix} 2 \\ 1 \end{bmatrix}, \qquad 1 \le t \le 1.5$

3. $\mathbf{y}' = \begin{bmatrix} -t^2 & t \\ 2-t & 0 \end{bmatrix} \mathbf{y} + \begin{bmatrix} 1 \\ t \end{bmatrix}, \quad \mathbf{y}(1) = \begin{bmatrix} 2 \\ 0 \end{bmatrix}, \qquad 1 \le t \le 4$

4. $\mathbf{y}' = \begin{bmatrix} 1 & 0 & 1 \\ 3 & 2 & 1 \\ 1 & 2 & 0 \end{bmatrix} \mathbf{y} + \begin{bmatrix} 0 \\ 2 \\ t \end{bmatrix}, \quad \mathbf{y}(-1) = \begin{bmatrix} 0 \\ 0 \\ 1 \end{bmatrix}, \qquad -1 \le t \le 0$

5. $\mathbf{y}' = \begin{bmatrix} \frac{1}{t} & \sin t \\ 1-t & 1 \end{bmatrix} \mathbf{y} + \begin{bmatrix} 0 \\ t^2 \end{bmatrix}, \quad \mathbf{y}(1) = \begin{bmatrix} 0 \\ 0 \end{bmatrix}, \qquad 1 \le t \le 6$

Checking a Program We do not normally carry out Euler's method (or other numerical algorithms) by hand. The iteration is usually programmed and run on a computer or some other programmable device. However, to guard against coding errors, we generally need to perform some hand calculations as a check on our program. Exercises 6–10 ask you to carry out several such hand calculations.

Exercises 6–10:

In each exercise, use a calculator to carry out two steps of Euler's method, finding $\mathbf{y}_1$ and $\mathbf{y}_2$. Use a step size of $h = 0.01$ for the given initial value problem.

6. The initial value problem in Exercise 1

7. The initial value problem in Exercise 2

8. The initial value problem in Exercise 3

9. The initial value problem in Exercise 4

10. The initial value problem in Exercise 5

Exercises 11–14:

In each exercise,

(a) As in equation (9), rewrite the given scalar initial value problem as an equivalent initial value problem for a first order system, defining $z_1 = y, z_2 = y', \ldots, z_n = y^{(n-1)}$.

(b) Write the Euler's method algorithm, $\mathbf{z}_{k+1} = \mathbf{z}_k + h[P(t_k)\mathbf{z}_k + \mathbf{g}(t_k)]$, in explicit form for the problem being considered. Specify the starting values t_0 and $\mathbf{z}_0$.

(c) Using a calculator and a uniform step size of $h = 0.01$, carry out two steps of Euler's method, finding $\mathbf{z}_1$ and $\mathbf{z}_2$. What are the corresponding numerical approximations to the solution $y(t)$ at the times $t = 0.01$ and $t = 0.02$?

11. $y'' + y = t^{3/2}, \quad y(0) = 1, \quad y'(0) = 0$

12. $y'' + y' + t^2 y = 2, \quad y(1) = 1, \quad y'(1) = 1$

13. $y''' + 2y' + ty = t + 1, \quad y(0) = 1, \quad y'(0) = -1, \quad y''(0) = 0$

14. $\dfrac{d}{dt}\left[e^t \dfrac{dy}{dt}\right] + y = 2e^t, \quad y(0) = -1, \quad y'(0) = 1$

15. Select one of the problems in Exercises 1–5.

(a) Write a program that carries out Euler's method for the problem selected. Use a step size of $h = 0.01$.

(b) Run your program on the interval given.

(c) Check your numerical solution by comparing the first two values, $\mathbf{y}_1$ and $\mathbf{y}_2$, with the hand calculations made in Exercises 6–10.

(d) Plot the components of the numerical solution on a common graph over the time interval of interest.

16. Repeat the steps of Exercise 15 selecting one of the problems in Exercises 11–14.

Estimating the Numerical Error of Euler's Method When solving problems, we should apply all available tests or checks before accepting an answer. In addition to the checks provided by common sense, the physics of the problem being modeled, and the mathematical theory of differential equations, there are additional checks available for testing the accuracy of numerical algorithms. We now describe such a check.

Suppose we apply Euler's method to the initial value problem $\mathbf{y}' = P(t)\mathbf{y} + \mathbf{g}(t)$, $\mathbf{y}(t_0) = \mathbf{y}_0$, $a \le t \le b$. We observed in Section 3.9 that the error in Euler's method is reduced approximately in half when the step size h is halved (this result will be justified in Chapter 9). This approximate halving of the error leads to a process for estimating the error. In particular, let t^* be a point in $[a, b]$ and choose a step size h by $h = (t^* - t_0)/n$ where n is a positive integer. Let $\mathbf{y}_n$ denote the Euler's method estimate to $\mathbf{y}(t^*)$, obtained using a step size h. Let $\overline{\mathbf{y}}_{2n}$ denote the Euler's method estimate to $\mathbf{y}(t^*)$, obtained using a step size of $h/2$. We anticipate, by halving the step size, that the error will be (approximately) halved as well:

$$\mathbf{y}(t^*) - \overline{\mathbf{y}}_{2n} \approx \frac{\mathbf{y}(t^*) - \mathbf{y}_n}{2}.$$

Some algebraic manipulation leads to the following estimate of the error, $\mathbf{y}(t^*) - \overline{\mathbf{y}}_{2n}$,

$$\mathbf{y}(t^*) - \overline{\mathbf{y}}_{2n} \approx \overline{\mathbf{y}}_{2n} - \mathbf{y}_n. \tag{13}$$

Exercises 17–20:

Compute the error estimate (13) by applying your Euler's method program on the given initial value problem. In each case, let $t^* = 1$. Use $h = 0.01$ and $h = 0.005$. Compare the actual error $\mathbf{y}(t^*) - \overline{\mathbf{y}}_{2n}$ with the estimate of the error $\overline{\mathbf{y}}_{2n} - \mathbf{y}_n$. [Note: Estimate (13) is also applicable at any of the intermediate points $0.01, 0.02, \ldots, 0.99$.]

17. $y'' - y = t, \quad y(0) = 2, y'(0) = -1$

18. $y'' + 4y = 3\cos t + 3\sin t, \quad y(0) = 4, y'(0) = 5$

19. $\mathbf{y}' = \begin{bmatrix} 2 & -1 \\ 1 & 2 \end{bmatrix}\mathbf{y}, \quad \mathbf{y}(0) = \begin{bmatrix} 0 \\ 2 \end{bmatrix}$

20. $\mathbf{y}' = \begin{bmatrix} -1 & 1 \\ 1 & -1 \end{bmatrix}\mathbf{y}, \quad \mathbf{y}(0) = \begin{bmatrix} 3 \\ -1 \end{bmatrix}$

21. Draining a Two-Tank System Consider the flow system shown in the figure. Tank 1 initially contains 40 lb of salt dissolved in 200 gal of water while Tank 2 initially contains 40 lb of salt dissolved in 500 gal of water. At time $t = 0$ the draining process is initiated, with well-stirred mixtures flowing at the rates and directions shown. The volumes of fluid in each tank change with time. Note, in particular, that Tank 1 empties in 20 min. Therefore, the flow processes shown in the figure cease to be valid after 20 min. (Tank 2 will still contain 100 gal of fluid after 20 min.)

(a) Let $Q_j(t)$ represent the amount of salt in tank j at time t, $j = 1, 2$. Formulate the initial value problem for

$$\mathbf{Q}(t) = \begin{bmatrix} Q_1(t) \\ Q_2(t) \end{bmatrix},$$

valid for $0 < t < 20$ min.

(b) Solve the initial value problem on the interval $0 \le t \le 19.9$ min using Euler's method. Use a uniform step size with $h = 0.01$.

(c) Plot the amounts of salt (pounds) in each tank, $Q_1(t)$ and $Q_2(t)$, on the same graph for $0 \le t \le 19.9$ min.

(d) On what positive t-interval does Theorem 6.1 guarantee a unique solution of the initial value problem formulated in (a)? Should we be on the alert for possible numerical problems as time t increases to 20 min? Explain.

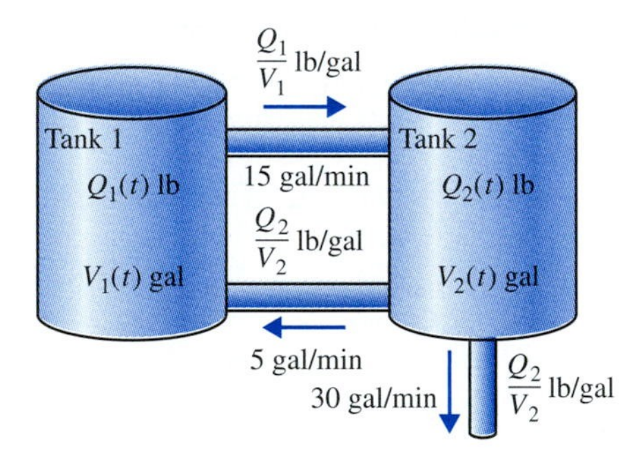

Figure for Exercise 21

22. A Spring-Mass-Dashpot System with Variable Damping As we saw in Section 4.7, the differential equation modeling unforced damped motion of a mass suspended from a spring is $my'' + \gamma y' + ky = 0$, where $y(t)$ represents the downward displacement of the mass from its equilibrium position. Assume a mass $m = 1$ kg and a spring constant $k = 4\pi^2$ N/m. Also assume the damping coefficient γ varying with time:

$$\gamma(t) = 2te^{-t/2} \text{ kg/sec.}$$

Assume, at time $t = 0$, the mass is pulled down 20 cm and released.

(a) Formulate the appropriate initial value problem for the second order scalar differential equation, and rewrite it as an equivalent initial value problem for a first order linear system.

(b) Applying Euler's method, numerically solve this problem on the interval $0 \le t \le 10$ min. Use a step size of $h = 0.005$.

(c) Plot the numerical solution on the time interval $0 \le t \le 10$ min. Explain, in qualitative terms, the effect of the variable damping upon the solution.

23. **The Cloudy Apple Juice Problem Revisited** In the exercises of Section 4.4, we derived a differential equation describing the radial motion of a particle in a simple centrifuge, a tube rotating in the horizontal plane about a fixed pivot. We noted, if drag forces are included, that this equation is essentially the one used by scientists employing centrifuges to study particle size distributions within cloudy apple juice. The differential equation modeling the radial distance of the particle from the pivot was found to be

$$r'' + \gamma r' - (\theta')^2 r = 0,$$

where $r(t)$ denotes the radial distance of the particle from the pivot, γ is a drag coefficient per unit mass, and $\theta'(t)$ is the angular velocity of the rotating tube. Up to now, we have assumed constant angular velocities and used the theory of Chapter 4 to solve such problems. A more realistic scenario is to assume the centrifuge starts from rest with a given constant angular acceleration, α. In this case, the differential equation becomes

$$r'' + \gamma r' - \alpha^2 t^2 r = 0.$$

As a model problem, assume at time $t = 0$ that a particle is at rest, located 2 cm from the pivot. At that time, the tube begins to rotate with a constant acceleration of $\pi/4$ rad/s^2. Assume the particle, as it moves outward, experiences a drag force characterized by a coefficient $\gamma = 0.5$ s^{-1}. What is the distance of the particle from the pivot at time $t = 3$ s?

(a) Formulate the initial value problem for the second order scalar equation, and rewrite it as an equivalent initial value problem for a first order linear system.

(b) Using Euler's method with step size $h = 0.01$, obtain a numerical solution on the interval $0 \le t \le 3$ sec.

(c) Plot the graph of $r(t)$ versus t for $0 \le t \le 3$ sec. What is your estimate of $r(3)$?

6.10 Diagonalization

In this section we introduce the concepts of similar matrices and matrix diagonalization. These ideas are central to a change of dependent variable transformation that is useful for solving constant coefficient linear systems. We use this transformation to rederive the solution of the first order constant coefficient linear system discussed in Section 6.8. We also apply the technique to solve a second order constant coefficient linear system that arises in studying the motion of a coupled spring-mass system.

Similar Matrices

We say that an $(n \times n)$ matrix A is **similar** to an $(n \times n)$ matrix B if there exists an $(n \times n)$ invertible matrix T such that

$$T^{-1}AT = B.$$

If A is similar to B, it follows that B is similar to A since $T^{-1}AT = B$ implies

$$A = TBT^{-1} = (T^{-1})^{-1}B(T^{-1}).$$

Therefore, it is appropriate to refer to the matrices A and B as a pair of **similar matrices**. The act of forming the matrix product $T^{-1}AT$ is often referred to as a **similarity transformation**. Among other things, the concept of similarity is important because:

(a) If A and B are similar $(n \times n)$ matrices, then A and B have the same characteristic polynomial and hence, the same eigenvalues (see Theorem 6.13).

(b) If A and B are similar $(n \times n)$ matrices, then solutions of $\mathbf{w}' = B\mathbf{w}$ are related to solutions of $\mathbf{y}' = A\mathbf{y}$ by the transformation $\mathbf{y}(t) = T\mathbf{w}(t)$.

Theorem 6.13

Let A and B be similar $(n \times n)$ matrices. Then A and B have the same characteristic polynomial.

PROOF: Since A and B are similar, there is an invertible $(n \times n)$ matrix T such that $T^{-1}AT = B$. Let $p_A(\lambda)$ and $p_B(\lambda)$ denote the characteristic polynomials of matrices A and B, respectively. Observe that

$$\begin{aligned} p_B(\lambda) &= \det[B - \lambda I] = \det[T^{-1}AT - \lambda I] = \det[T^{-1}AT - \lambda T^{-1}T] \\ &= \det[T^{-1}(A - \lambda I)T] = \det[T^{-1}]\det[A - \lambda I]\det[T] \\ &= p_A(\lambda). \end{aligned}$$

(To obtain the last equality, we used the fact that $\det[T^{-1}]\det[T] = \det[I] = 1$.)

Diagonalization

Consider a similarity transformation of the form $T^{-1}AT = D$, where the matrix D is a diagonal matrix,

$$D = \begin{bmatrix} d_{11} & 0 & 0 & \cdots & 0 \\ 0 & d_{22} & 0 & \cdots & 0 \\ 0 & 0 & d_{33} & \cdots & 0 \\ \vdots & & & & \vdots \\ 0 & 0 & 0 & \cdots & d_{nn} \end{bmatrix}.$$

In such cases we say that matrix A is **similar to a diagonal matrix**, or that A is **diagonalizable**. Two questions arise with regard to diagonalization:

(i) When is it possible to diagonalize a square matrix A?

(ii) If it is possible to diagonalize a given matrix A, how do we find the matrix T that accomplishes the diagonalization?

Theorem 6.14 and its corollary address these two questions.

Theorem 6.14

Let A be an $(n \times n)$ matrix similar to a diagonal matrix D. Let T be an invertible matrix such that $T^{-1}AT = D$. Then the diagonal elements of matrix D are the eigenvalues of matrix A and the columns of matrix T are corresponding eigenvectors of matrix A.

PROOF: It follows from Theorem 6.13 that matrices A and D have the same eigenvalues. The eigenvalues of diagonal matrix D, however, are its diagonal elements. Therefore, the diagonal elements of D are the eigenvalues of A.

To finish the proof, let D be the diagonal matrix

$$D = \begin{bmatrix} \lambda_1 & 0 & 0 & \cdots & 0 \\ 0 & \lambda_2 & 0 & \cdots & 0 \\ 0 & 0 & \lambda_3 & \cdots & 0 \\ \vdots & & & & \vdots \\ 0 & 0 & 0 & \cdots & \lambda_n \end{bmatrix}$$

and let $\mathbf{t}_1, \mathbf{t}_2, \ldots, \mathbf{t}_n$ denote the n columns of the matrix T so that $T = [\mathbf{t}_1, \mathbf{t}_2, \ldots, \mathbf{t}_n]$. Because $T^{-1}AT = D$, it follows that $AT = TD$:

$$A[\mathbf{t}_1, \mathbf{t}_2, \ldots, \mathbf{t}_n] = [\mathbf{t}_1, \mathbf{t}_2, \ldots, \mathbf{t}_n] \begin{bmatrix} \lambda_1 & 0 & 0 & \cdots & 0 \\ 0 & \lambda_2 & 0 & \cdots & 0 \\ 0 & 0 & \lambda_3 & \cdots & 0 \\ \vdots & & & & \vdots \\ 0 & 0 & 0 & \cdots & \lambda_n \end{bmatrix}. \qquad (1)$$

The matrix product on the left-hand side of equation (1) can be rewritten as $[A\mathbf{t}_1, A\mathbf{t}_2, \ldots, A\mathbf{t}_n]$. Similarly, the matrix product on the right-hand side of equation (1) can be rewritten as $[\lambda_1\mathbf{t}_1, \lambda_2\mathbf{t}_2, \ldots, \lambda_n\mathbf{t}_n]$. Since matrix equality implies that corresponding columns are equal, we obtain

$$A\mathbf{t}_j = \lambda_j\mathbf{t}_j, \qquad j = 1, 2, \ldots, n. \qquad (2)$$

To complete the argument, note that invertibility of the matrix T implies that none of its columns can be the $(n \times 1)$ zero vector. Therefore, $\mathbf{t}_j \neq \mathbf{0}$, $1 \leq j \leq n$, and this fact, in conjunction with (2), shows that the columns of T are eigenvectors of A. ∎

A corollary of Theorem 6.14 characterizes diagonalizable matrices.

Corollary

An $(n \times n)$ matrix A is diagonalizable if and only if it has a set of n linearly independent eigenvectors.

From what we know already, matrices with distinct eigenvalues as well as real symmetric matrices and Hermitian matrices are diagonalizable. In general, A is diagonalizable if and only if A has a full set of eigenvectors.

EXAMPLE 1

As noted, real symmetric matrices are diagonalizable. Let A be the matrix

$$A = \begin{bmatrix} 1 & 2 \\ 2 & 1 \end{bmatrix}.$$

Find an invertible (2×2) matrix T such that $T^{-1}AT = D$.

Solution: We saw in Example 1 of Section 6.5 that eigenpairs of A are

$$\lambda_1 = 3, \quad \mathbf{x}_1 = \begin{bmatrix} 1 \\ 1 \end{bmatrix} \quad \text{and} \quad \lambda_2 = -1, \quad \mathbf{x}_2 = \begin{bmatrix} 1 \\ -1 \end{bmatrix}.$$

Therefore, a matrix T made from the eigenvectors will diagonalize A. So, let T be

$$T = \begin{bmatrix} 1 & 1 \\ 1 & -1 \end{bmatrix}.$$

A direct calculation shows,

$$T^{-1}AT = \begin{bmatrix} \frac{1}{2} & \frac{1}{2} \\ \frac{1}{2} & -\frac{1}{2} \end{bmatrix} \begin{bmatrix} 1 & 2 \\ 2 & 1 \end{bmatrix} \begin{bmatrix} 1 & 1 \\ 1 & -1 \end{bmatrix} = \begin{bmatrix} 3 & 0 \\ 0 & -1 \end{bmatrix}. \quad \blacktriangle$$

Solving a First Order Constant Coefficient Linear System

We now describe a method based on diagonalization for solving the initial value problem

$$\mathbf{y}' = A\mathbf{y} + \mathbf{g}(t), \qquad \mathbf{y}(t_0) = \mathbf{y}_0, \qquad a < t < b. \tag{3}$$

In (3), the $(n \times 1)$ vector function $\mathbf{g}(t)$ is assumed to be continuous on (a, b), t_0 is an arbitrary point in (a, b), and A is an $(n \times n)$ constant diagonalizable matrix.

We already know how to solve initial value problem (3); the variation of parameters formula [see equation (11) in Section 6.8] provides an explicit formula for the solution. We now take a different solution approach, one involving diagonalization. We take this different approach because the underlying ideas are useful and can be adapted to solve second order linear systems.

Since we are assuming the matrix A in (3) is diagonalizable, there is an $(n \times n)$ invertible matrix T (composed of eigenvectors of A) such that

$$T^{-1}AT = D = \begin{bmatrix} \lambda_1 & 0 & 0 & \cdots & 0 \\ 0 & \lambda_2 & 0 & \cdots & 0 \\ 0 & 0 & \lambda_3 & \cdots & 0 \\ \vdots & & & & \vdots \\ 0 & 0 & 0 & \cdots & \lambda_n \end{bmatrix}.$$

We use the constant matrix T to introduce a change of dependent variable:

$$\mathbf{y}(t) = T\mathbf{z}(t) \qquad \text{or} \qquad \mathbf{z}(t) = T^{-1}\mathbf{y}(t). \tag{4}$$

Substituting this change of variables into differential equation (3) leads to

$$T\mathbf{z}'(t) = AT\mathbf{z}(t) + \mathbf{g}(t).$$

Multiplying on the left by T^{-1} and using the fact that $T^{-1}AT = D$, we obtain

$$\mathbf{z}'(t) = D\mathbf{z}(t) + T^{-1}\mathbf{g}(t), \qquad \mathbf{z}(t_0) = T^{-1}\mathbf{y}_0, \qquad a < t < b. \tag{5}$$

Since coefficient matrix D is diagonal, the transformation (4) uncouples the components of $\mathbf{z}(t)$. Each component satisfies a first order scalar differential equation of the type we studied in Chapter 2. In particular, let us define $\mathbf{h}(t) = T^{-1}\mathbf{g}(t)$, $a < t < b$, and set $\mathbf{z}_0 = T^{-1}\mathbf{y}_0$. Then initial value problem (3) is transformed into the following set of uncoupled first order scalar initial value problems:

$$\begin{aligned} z_1' &= \lambda_1 z_1 + h_1(t), \qquad & z_1(t_0) &= z_1^0 \\ z_2' &= \lambda_2 z_2 + h_2(t), \qquad & z_2(t_0) &= z_2^0 \\ &\vdots \\ z_n' &= \lambda_n z_n + h_n(t), \qquad & z_n(t_0) &= z_n^0. \end{aligned} \tag{6}$$

In (6), $h_j(t)$ is the jth component of the vector function $\mathbf{h}(t) = T^{-1}\mathbf{g}(t)$ and z_j^0 is the jth component of the $(n \times 1)$ constant vector $\mathbf{z}_0 = T^{-1}\mathbf{y}_0$. We can solve each of these first order scalar initial value problems using the integrating factor approach developed in Section 2.3, obtaining

$$z_j(t) = e^{\lambda_j(t-t_0)}z_j^0 + \int_{t_0}^{t} e^{\lambda_j(t-s)}h_j(s)\,ds, \qquad j = 1, 2, \ldots, n. \tag{7}$$

Having the vector function $\mathbf{z}(t)$ that solves equation (5), we then find the solution of the original initial value problem [equation (3)] by forming the matrix product $\mathbf{y}(t) = T\mathbf{z}(t)$.

The diagonalization approach yields the same solution for first order systems as was obtained in Section 6.8. In particular (see Exercise 26), the n scalar solutions given in equation (7) can be used to reconstruct the matrix variation of parameters formula given in Section 6.8.

EXAMPLE 2

Obtain the general solution of

$$\mathbf{y}' = \begin{bmatrix} 1 & 2 \\ 2 & 1 \end{bmatrix}\mathbf{y} + \begin{bmatrix} e^{2t} \\ -2t \end{bmatrix}, \qquad -\infty < t < \infty. \tag{8}$$

Solution: This problem was solved by the method of undetermined coefficients in Section 6.8 (see Example 1). In the present example, we use a different approach based on diagonalizing the coefficient matrix

$$A = \begin{bmatrix} 1 & 2 \\ 2 & 1 \end{bmatrix}. \tag{9}$$

As we know from Example 1 in this section, a diagonalizing matrix T for A is

$$T = \begin{bmatrix} 1 & 1 \\ 1 & -1 \end{bmatrix}.$$

(continued)

(continued)

Let $\mathbf{y}(t) = T\mathbf{z}(t)$. Substituting this change of variables into the given differential equation leads to

$$\frac{d}{dt}\begin{bmatrix} z_1 \\ z_2 \end{bmatrix} = \begin{bmatrix} 3 & 0 \\ 0 & -1 \end{bmatrix}\begin{bmatrix} z_1 \\ z_2 \end{bmatrix} + \begin{bmatrix} \dfrac{e^{2t}}{2} - t \\ \dfrac{e^{2t}}{2} + t \end{bmatrix}. \tag{10}$$

Thus, the given first order system transforms into two uncoupled scalar equations

$$\begin{aligned} z_1' &= 3z_1 + \frac{e^{2t}}{2} - t, \\ z_2' &= -z_2 + \frac{e^{2t}}{2} + t. \end{aligned} \tag{11}$$

The general solutions of these first order linear scalar equations are

$$z_1(t) = -\frac{e^{2t}}{2} + \frac{1}{3}\left(t + \frac{1}{3}\right) + c_1 e^{3t}$$

and

$$z_2(t) = \frac{e^{2t}}{6} + t - 1 + c_2 e^{-t}.$$

Since $\mathbf{y}(t) = T\mathbf{z}(t)$, it follows that

$$\begin{aligned} \mathbf{y}(t) &= \begin{bmatrix} 1 & 1 \\ 1 & -1 \end{bmatrix} \begin{bmatrix} -\dfrac{e^{2t}}{2} + \dfrac{1}{3}\left(t + \dfrac{1}{3}\right) + c_1 e^{3t} \\ \dfrac{e^{2t}}{6} + t - 1 + c_2 e^{-t} \end{bmatrix} \\ &= c_1 \begin{bmatrix} e^{3t} \\ e^{3t} \end{bmatrix} + c_2 \begin{bmatrix} e^{-t} \\ -e^{-t} \end{bmatrix} + \begin{bmatrix} -\dfrac{e^{2t}}{3} + \dfrac{4t}{3} - \dfrac{8}{9} \\ -\dfrac{2e^{2t}}{3} - \dfrac{2t}{3} + \dfrac{10}{9} \end{bmatrix}. \end{aligned}$$

Except for a renaming of the two arbitrary constants, this is the general solution previously obtained. ▲

A Coupled Spring-Mass System

Diagonalization techniques can be used to solve second order linear constant coefficient systems such as those that arise in the study of coupled spring-mass systems.

Consider the configuration shown in Figure 6.8. The three identical springs of unstretched length l are assumed to be weightless, and the two identical masses are assumed to slide on a frictionless surface. In the vertical direction, the surface exerts a normal force upon each mass equal and opposite to its weight. Therefore, we need only consider motion in the horizontal direction.

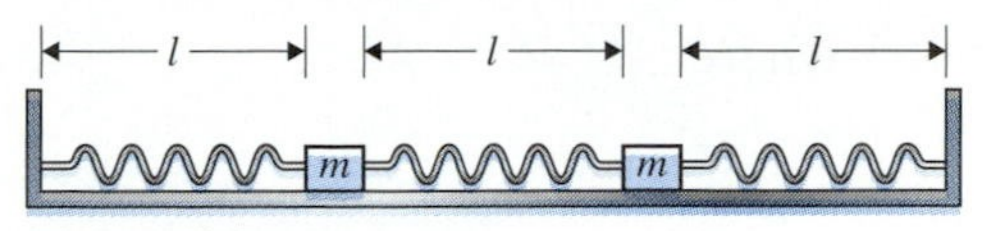

(a) Equilibrium state

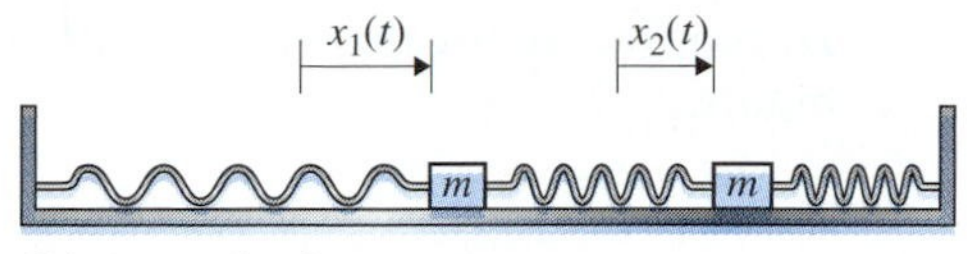

(b) Perturbed state

FIGURE 6.8

A coupled spring-mass system. (a) The equilibrium state. (b) The perturbed state.

Assume the system is set in motion at time $t = 0$ and there are no externally applied forces. Let $x_1(t)$ and $x_2(t)$ represent the respective horizontal displacements of the two masses from their equilibrium positions, measured positive to the right as shown. The application of Newton's second law of motion to each mass leads to the system of equations:

$$\begin{aligned} mx_1'' &= k(x_2 - x_1) - kx_1, \\ mx_2'' &= -k(x_2 - x_1) - kx_2, \qquad t > 0. \end{aligned}$$

These equations can be rewritten as a second order linear system

$$m\mathbf{x}'' = k\begin{bmatrix} -2 & 1 \\ 1 & -2 \end{bmatrix}\mathbf{x}, \qquad t > 0 \quad \text{where} \quad \mathbf{x}(t) = \begin{bmatrix} x_1(t) \\ x_2(t) \end{bmatrix}. \tag{12}$$

In addition to the equations of motion (12), we specify the initial position and velocity of each mass by

$$\mathbf{x}(0) = \mathbf{x}_0, \qquad \mathbf{x}'(0) = \mathbf{x}_0' \tag{13}$$

where $\mathbf{x}_0$ and $\mathbf{x}_0'$ are given (2×1) constant vectors. (We assume that the initial conditions are such that the two masses do not collide with each other or with the confining walls.)

Equations (12) and (13) form the initial value problem of interest. The new feature is that we have a second order constant coefficient linear system of differential equations. How do we go about solving this problem?

One approach would be to rewrite the second order system for the (2×1) vector function $\mathbf{x}(t)$ as a first order linear system for a new (4×1) vector function $\mathbf{z}(t)$ where $z_1(t) = x_1(t), z_2(t) = x_1'(t), z_3(t) = x_2(t), z_4(t) = x_2'(t)$. The techniques of Section 6.5 could be used to solve this first order system initial value problem. This approach requires us to find the eigenpairs of a (4×4) coefficient matrix.

We will use another approach to solving the initial value problem, an approach based on the idea of diagonalization. Note that (12) can be rewritten as the second order linear system

$$\mathbf{x}'' + A\mathbf{x} = \mathbf{0}, \tag{14}$$

where

$$A = \frac{k}{m}\begin{bmatrix} 2 & -1 \\ -1 & 2 \end{bmatrix}. \tag{15}$$

The (2×2) coefficient matrix A is real and symmetric; therefore, it is diagonalizable. Its eigenvalues are $\lambda_1 = k/m$ and $\lambda_2 = 3k/m$. A corresponding matrix of eigenvectors is

$$T = \begin{bmatrix} 1 & 1 \\ 1 & -1 \end{bmatrix}.$$

We can transform equation (14) into a pair of uncoupled scalar second order equations by introducing [as in equation (4)] the change of dependent variable $\mathbf{x}(t) = T\mathbf{w}(t)$. Substituting this change of variables into equation (14), we obtain $T\mathbf{w}'' + AT\mathbf{w} = \mathbf{0}$ or

$$\mathbf{w}'' + D\mathbf{w} = \mathbf{0} \tag{16a}$$

where

$$D = \begin{bmatrix} \dfrac{k}{m} & 0 \\ 0 & 3\dfrac{k}{m} \end{bmatrix}.$$

As initial conditions for (16a), we have

$$\mathbf{w}(0) = T^{-1}\mathbf{x}(0) = T^{-1}\mathbf{x}_0 \quad \text{and} \quad \mathbf{w}'(0) = T^{-1}\mathbf{x}'(0) = T^{-1}\mathbf{x}'_0. \tag{16b}$$

The components of differential equation (16a) uncouple and can be individually solved using the techniques of Chapter 4. In particular, equation (16a) reduces to

$$\begin{aligned} w_1'' + \frac{k}{m}w_1 &= 0, \\ w_2'' + 3\frac{k}{m}w_2 &= 0. \end{aligned} \tag{17a}$$

Solving the uncoupled equations in (17a), we obtain

$$\begin{aligned} w_1(t) &= c_1 \cos\sqrt{\frac{k}{m}}\,t + d_1 \sin\sqrt{\frac{k}{m}}\,t, \\ w_2(t) &= c_2 \cos\sqrt{3\frac{k}{m}}\,t + d_2 \sin\sqrt{3\frac{k}{m}}\,t. \end{aligned} \tag{17b}$$

The four arbitrary constants are determined by imposing the initial conditions. That is,

$$\mathbf{x}(0) = \mathbf{x}_0 = T\mathbf{w}(0) = \begin{bmatrix} 1 & 1 \\ 1 & -1 \end{bmatrix}\begin{bmatrix} c_1 \\ c_2 \end{bmatrix} \tag{18a}$$

and

$$\mathbf{x}'(0) = \mathbf{x}'_0 = T\mathbf{w}'(0) = \begin{bmatrix} 1 & 1 \\ 1 & -1 \end{bmatrix} \begin{bmatrix} \sqrt{\frac{k}{m}}\, d_1 \\ \sqrt{3\frac{k}{m}}\, d_2 \end{bmatrix}. \tag{18b}$$

To gain some insight into the behavior of this spring-mass system, we consider two particular types of excitation.

Case 1: Let

$$\mathbf{x}_0 = \begin{bmatrix} -\frac{l}{4} \\ \frac{l}{4} \end{bmatrix}, \qquad \mathbf{x}'_0 = \mathbf{0}. \tag{19}$$

In this case, the two masses are symmetrically stretched apart and then released from rest. From equations (18a) and (18b), we obtain $c_1 = 0$, $c_2 = -l/4$, $d_1 = d_2 = 0$. Hence,

$$\mathbf{x}(t) = T\mathbf{w}(t) = \frac{l}{4} \begin{bmatrix} -\cos\left(\sqrt{3\frac{k}{m}}\, t\right) \\ \cos\left(\sqrt{3\frac{k}{m}}\, t\right) \end{bmatrix}.$$

The two masses therefore pulsate back and forth, executing symmetric motion about the centerline. The motion occurs at radian frequency $\sqrt{3(k/m)}$.

Case 2: Let

$$\mathbf{x}_0 = \begin{bmatrix} \frac{l}{4} \\ \frac{l}{4} \end{bmatrix}, \qquad \mathbf{x}'_0 = \mathbf{0}. \tag{20}$$

In this case, the masses are both shifted an equal amount to the right and then released from rest. From equations (18a) and (18b) we obtain $c_1 = l/4$, $c_2 = 0$, $d_1 = d_2 = 0$. Hence,

$$\mathbf{x}(t) = T\mathbf{w}(t) = \frac{l}{4} \begin{bmatrix} \cos\left(\sqrt{\frac{k}{m}}\, t\right) \\ \cos\left(\sqrt{\frac{k}{m}}\, t\right) \end{bmatrix}.$$

The two masses therefore move back and forth maintaining a constant separation—the two masses and center spring move as if they formed a rigid body. In this case, the radian frequency is $\sqrt{k/m}$. Since this radian frequency is less than that of Case 1, the vibrations will be correspondingly slower; see Figure 6.9. The motion of the masses in the two cases considered is illustrated in Figure 6.9 for parameter values $l = 1$ and $k = m$.

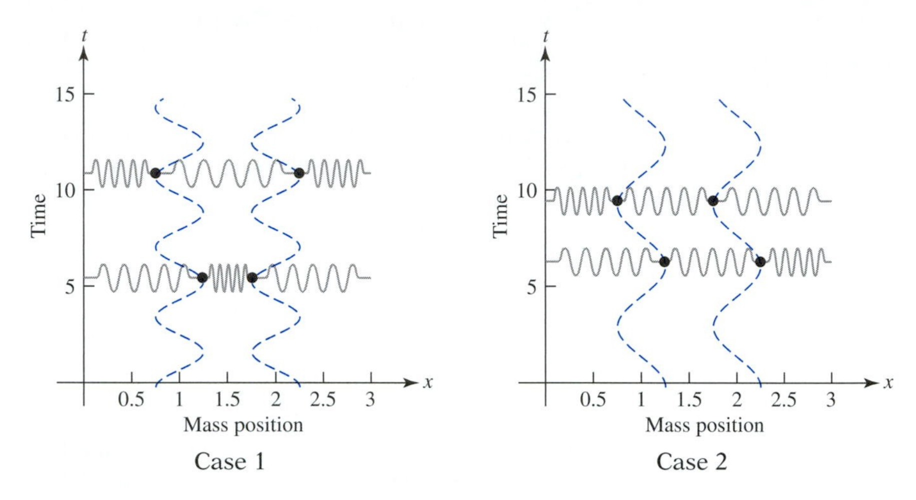

FIGURE 6.9

The motion of the coupled spring-mass system shown in Figure 6.8, for the two cases considered in equations (19) and (20). For these two cases, $l = 1$ and $k = m$.

EXERCISES

Exercises 1–8:

In each exercise, the given matrix is diagonalizable. Find matrices T and D such that $T^{-1}AT = D$.

1. $A = \begin{bmatrix} 5 & -6 \\ 3 & -4 \end{bmatrix}$ **2.** $A = \begin{bmatrix} 3 & 4 \\ -2 & -3 \end{bmatrix}$ **3.** $A = \begin{bmatrix} 1 & 1 \\ 1 & 1 \end{bmatrix}$ **4.** $A = \begin{bmatrix} 2 & 3 \\ 2 & 3 \end{bmatrix}$

5. $A = \begin{bmatrix} 2 & 3 \\ 3 & 2 \end{bmatrix}$ **6.** $A = \begin{bmatrix} 1 & 2 \\ 2 & 1 \end{bmatrix}$ **7.** $A = \begin{bmatrix} 2 & 0 \\ 1 & 1 \end{bmatrix}$ **8.** $A = \begin{bmatrix} -2 & 2 \\ 0 & 3 \end{bmatrix}$

Exercises 9–13:

In each exercise, you are given the characteristic polynomial for the matrix A. Determine the algebraic and geometric multiplicities of each eigenvalue. If the matrix A is diagonalizable, find matrices T and D such that $T^{-1}AT = D$.

9. $A = \begin{bmatrix} 25 & -8 & 30 \\ 24 & -7 & 30 \\ -12 & 4 & -14 \end{bmatrix}, \quad p(\lambda) = (\lambda - 1)^2(\lambda - 2)$

10. $A = \begin{bmatrix} 7 & -2 & 2 \\ 8 & -1 & 4 \\ -8 & 4 & -1 \end{bmatrix}, \quad p(\lambda) = (\lambda - 3)^2(\lambda + 1)$

11. $A = \begin{bmatrix} 1 & 0 & 1 \\ 2 & 2 & -3 \\ 0 & 0 & 1 \end{bmatrix}, \quad p(\lambda) = (\lambda - 1)^2(\lambda - 2)$

12. $A = \begin{bmatrix} 5 & -1 & 1 \\ 14 & -3 & 6 \\ 5 & -2 & 5 \end{bmatrix}, \quad p(\lambda) = (\lambda - 2)^2(\lambda - 3)$

13. $A = \begin{bmatrix} 4 & -1 & 1 \\ 10 & -2 & 3 \\ 1 & 0 & 1 \end{bmatrix}, \quad p(\lambda) = (\lambda - 1)^3$

14. At least two (and possibly more) of the four matrices are diagonalizable. You should be able to recognize two by inspection. Choose them and give a reason for your choice.

(a) $A = \begin{bmatrix} 5 & 6 \\ 3 & 4 \end{bmatrix}$ (b) $A = \begin{bmatrix} 3 & 6 \\ 6 & 9 \end{bmatrix}$ (c) $A = \begin{bmatrix} 3 & 0 \\ 3 & -4 \end{bmatrix}$ (d) $A = \begin{bmatrix} 1 & 3 \\ -1 & 4 \end{bmatrix}$

Exercises 15–18:

Consider the given system $\mathbf{y}' = A\mathbf{y} + \mathbf{g}(t)$. In each case, the coefficient matrix A is diagonalizable. As in Example 2, obtain the general solution by making the change of variables $\mathbf{y} = T\mathbf{z}$.

15. $\mathbf{y}' = \begin{bmatrix} 6 & -6 \\ 2 & -1 \end{bmatrix}\mathbf{y} + \begin{bmatrix} 4 + 3e^t \\ 2 + 2e^t \end{bmatrix}$

16. $\mathbf{y}' = \begin{bmatrix} -4 & -6 \\ 3 & 5 \end{bmatrix}\mathbf{y} + \begin{bmatrix} e^{2t} - 2e^t \\ -e^{2t} + e^t \end{bmatrix}$

17. $\mathbf{y}' = \begin{bmatrix} 1 & 1 \\ 2 & 2 \end{bmatrix}\mathbf{y} + \begin{bmatrix} t \\ -t + 3 \end{bmatrix}$

18. $\mathbf{y}' = \begin{bmatrix} 3 & 2 \\ 1 & 4 \end{bmatrix}\mathbf{y} + \begin{bmatrix} 4t + 4 \\ -2t + 1 \end{bmatrix}$

Exercises 19–22:

In each exercise, the given matrix A is diagonalizable. Use the diagonalization technique illustrated in equations (16)–(18) to find the general solution of $\mathbf{x}'' = A\mathbf{x}$.

19. $A = \begin{bmatrix} -9 & -5 \\ 8 & 4 \end{bmatrix}$

20. $A = \begin{bmatrix} 6 & 7 \\ -15 & -16 \end{bmatrix}$

21. $A = \begin{bmatrix} -2 & -1 \\ 3 & 2 \end{bmatrix}$

22. $A = \begin{bmatrix} 4 & 2 \\ 2 & 1 \end{bmatrix}$

23. We know that similar matrices have the same eigenvalues (in fact, they have the same characteristic polynomial). There are many examples that show the converse is not true; that is, there are examples of matrices A and B that have the same characteristic polynomial but are not similar. Show that the matrices A and B cannot be similar:

$$A = \begin{bmatrix} 1 & 0 \\ 3 & 1 \end{bmatrix} \quad \text{and} \quad B = \begin{bmatrix} 1 & 0 \\ 0 & 1 \end{bmatrix}.$$

24. Drawing on the ideas involved in working Exercise 23, show that if an $(n \times n)$ real matrix A is similar to the $(n \times n)$ identity I, then $A = I$.

25. Give an example that shows, while similar matrices have the same eigenvalues, they may not have the same eigenvectors.

26. Recovering the Variation of Parameters Formula Show that equation (7) is equivalent to the variation of parameters formula,

$$\mathbf{y}(t) = \Psi(t)\Psi^{-1}(t_0)\mathbf{y}_0 + \Psi(t)\int_{t_0}^{t} \Psi^{-1}(s)\mathbf{g}(s)\,ds,$$

where $\Psi(t)$ is a fundamental matrix for $\mathbf{y}' = A\mathbf{y}$. Proceed as follows:

(a) Define the $(n \times n)$ diagonal matrix

$$\Lambda(t) = \begin{bmatrix} e^{\lambda_1 t} & 0 & 0 & \cdots & 0 \\ 0 & e^{\lambda_2 t} & 0 & \cdots & 0 \\ 0 & 0 & e^{\lambda_3 t} & \cdots & 0 \\ \vdots & \vdots & \vdots & & \vdots \\ 0 & 0 & 0 & \cdots & e^{\lambda_n t} \end{bmatrix},$$

and show that equation (7) can be written as the matrix equation,

$$\mathbf{z}(t) = \Lambda(t - t_0)\mathbf{z}_0 + \int_{t_0}^{t} \Lambda(t - s)\mathbf{h}(s)\,ds.$$

(b) Using $\mathbf{y}(t) = T\mathbf{z}(t)$, $\mathbf{y}_0 = T\mathbf{z}_0$ and $\mathbf{g}(t) = T\mathbf{h}(t)$, show that

$$\mathbf{y}(t) = T\Lambda(t - t_0)T^{-1}\mathbf{y}_0 + \int_{t_0}^{t} T\Lambda(t - s)T^{-1}\mathbf{g}(s)\,ds$$

(c) Show that the matrix $\Lambda(t)$ satisfies the differential equation $\Lambda'(t) = D\Lambda(t)$.

(d) Use part (c) to show that the matrix product $T\Lambda(t)$ satisfies the differential equation $(T\Lambda(t))' = A(T\Lambda(t))$. Likewise, show that $T\Lambda(t)$ is invertible. Therefore, $T\Lambda(t)$ is a fundamental matrix for $\mathbf{y}' = A\mathbf{y}$. Define $T\Lambda(t) = \Psi(t)$.

(e) Show that $T\Lambda(t) = \Psi(t)$ implies $T\Lambda(t - t_0)T^{-1} = \Psi(t)\Psi^{-1}(t_0)$ and likewise that $T\Lambda(t - s)T^{-1} = \Psi(t)\Psi^{-1}(s)$. This completes the recovery of the variation of parameters formula.

27. Coupled Spring-Mass Systems Consider the spring-mass system shown in the figure. The figure shows two springs of negligible weight. The springs have unstretched lengths l_1 and l_2 and have spring constants k_1 and k_2. Masses m_1 and m_2 are attached and the springs stretch appropriately to achieve the rest configuration shown. The amount of stretching done by each spring is determined by imposing conditions of static equilibrium on each mass. Since the sum of the vertical forces acting on each mass must vanish, we obtain $k_1Y_1 = m_1g + k_2Y_2$ and $m_2g = k_2Y_2$. Therefore, $k_1Y_1 = m_1g + m_2g$, where g represents acceleration due to gravity. When the system is disturbed from its equilibrium state, both masses move vertically. Let $y_1(t)$ and $y_2(t)$ represent the displacements of the masses from their equilibrium positions at time t, as shown in the figure. Newton's second law of motion leads to the following system of second order differential equations:

$$\begin{aligned} m_1 y_1'' &= -(k_1 + k_2)y_1 + k_2 y_2 \\ m_2 y_2'' &= k_2 y_1 - k_2 y_2 \end{aligned}$$

We assume that the system is set into motion at time $t = 0$ with initial displacements and velocities

$$y_1(0) = y_{1,0}, \qquad y_2(0) = y_{2,0}, \qquad y_1'(0) = y_{1,0}', \qquad y_2'(0) = y_{2,0}'.$$

Assume that $m_1 = m_2 = 2$ kg, $k_1 = 600$ N/m and $k_2 = 400$ N/m. Mass 2 is pulled downward by applying a force of 20 N and the system is then released. The 60 N

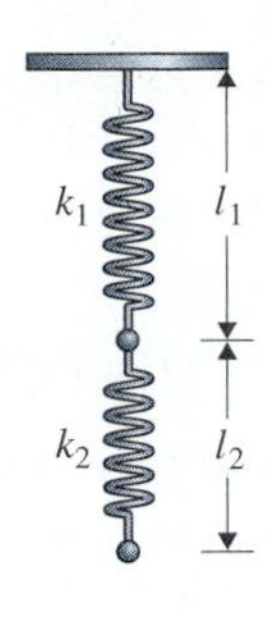

(a) Unstretched springs

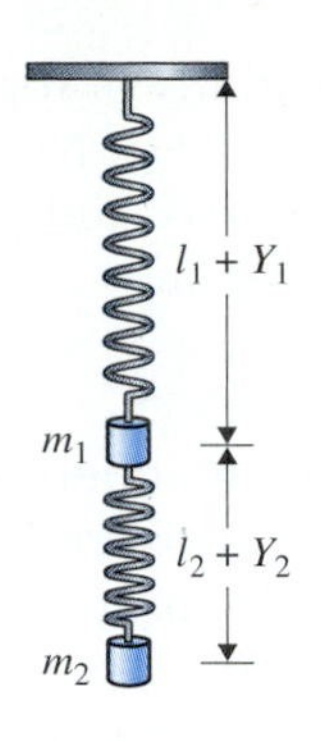

(b) System at rest with masses attached

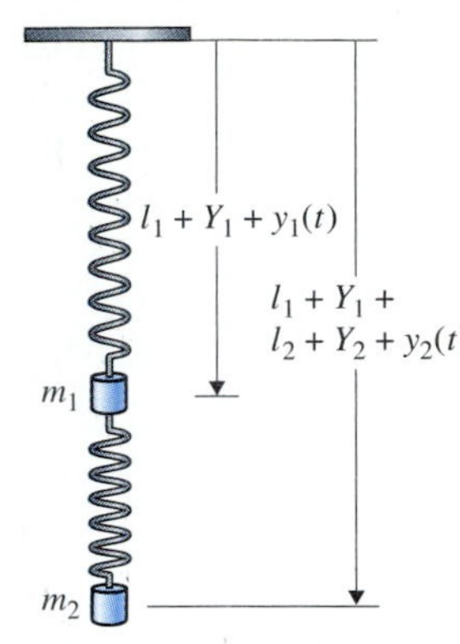

(c) System disturbed from rest state

Figure for Exercise 27

force stretches the upper spring 10 cm and the lower spring 15 cm downward from their equilibrium rest positions; the initial value problem to be solved becomes

$$\mathbf{y}'' + \begin{bmatrix} 500 & -200 \\ -200 & 200 \end{bmatrix} \mathbf{y} = \mathbf{0}, \qquad \mathbf{y}(0) = \begin{bmatrix} 0.10 \\ 0.15 \end{bmatrix}, \qquad \mathbf{y}'(0) = \mathbf{0}.$$

(a) Find the eigenpairs of the coefficient matrix

$$A = \begin{bmatrix} 500 & -200 \\ -200 & 200 \end{bmatrix}.$$

(b) Form the (2×2) matrix of eigenvectors, T, and define the new dependent variable $\mathbf{z}(t)$ by setting $\mathbf{y}(t) = T\mathbf{z}(t)$. Determine the two scalar initial value problems satisfied by the components of $\mathbf{z}(t)$.

(c) Solve the problems for the two components of $\mathbf{z}(t)$ using the techniques of Chapter 4. Form the desired solution $\mathbf{y}(t)$.

6.11 Propagator Matrices, Functions of a Matrix, and the Exponential Matrix

We focus first on a special type of fundamental matrix, called a "propagator matrix." Consider the differential equation $\mathbf{y}' = P(t)\mathbf{y}$, $a < t < b$, and let s be an arbitrary but fixed point in the interval (a, b). We use the notation $\Phi(t, s)$ to denote the unique solution of the initial value problem

$$\Phi' = P(t)\Phi, \qquad \Phi(s, s) = I, \qquad a < t < b,$$

where I denotes the $(n \times n)$ identity matrix. By part (b) of Theorem 6.12, $\Phi(t, s)$ is a fundamental matrix of $\mathbf{y}' = P(t)\mathbf{y}$ because I is nonsingular. In particular, $\Phi(t, s)$ is the fundamental matrix that reduces to the identity matrix when $t = s$.

We now show that we can relate a solution $\mathbf{y}(t)$ of $\mathbf{y}' = P(t)\mathbf{y}$ at time t to its value at time s by the simple formula

$$\mathbf{y}(t) = \Phi(t, s)\mathbf{y}(s), \qquad a < t < b. \tag{1}$$

To see the validity of formula (1), let $\mathbf{z}(t) = \Phi(t, s)\mathbf{y}(s)$ represent the right-hand side of formula (1). Note that $\mathbf{z}(t)$ is a solution of $\mathbf{y}' = P(t)\mathbf{y}$ because $\Phi(t, s)$ is a solution matrix and $\mathbf{y}(s)$ is a constant $(n \times 1)$ vector. Evaluating $\mathbf{z}(t)$ at the point $t = s$, we find

$$\mathbf{z}(s) = \Phi(s, s)\mathbf{y}(s) = I\mathbf{y}(s) = \mathbf{y}(s).$$

Therefore, by uniqueness, $\mathbf{z}(t)$ is equal to $\mathbf{y}(t)$ for all t in (a, b).

Because of formula (1), the matrix $\Phi(t, s)$ is referred to as a **propagator matrix**. It "propagates" any solution of $\mathbf{y}' = P(t)\mathbf{y}$ from time s to time t. Since propagator matrices are fundamental matrices, they possess all the properties listed in Theorem 6.12 as well as the additional properties which we state without proof in Theorem 6.15.

Theorem 6.15

Consider the homogeneous first order linear system

$$\mathbf{y}' = P(t)\mathbf{y}, \qquad a < t < b,$$

where $P(t)$ is an $(n \times n)$ coefficient matrix, continuous on the interval $a < t < b$.

(a) Let $\Psi(t)$ be any fundamental matrix, and let s be any point in the interval $a < t < b$. The propagator matrix $\Phi(t, s)$ is given by

$$\Phi(t, s) = \Psi(t)\Psi^{-1}(s).$$

(b) Let t, s, and τ be any three points in (a, b). Then

$$\Phi(t, \tau) = \Phi(t, s)\Phi(s, \tau).$$

(c) Let t and s be any two points in (a, b). Then

$$\Phi^{-1}(t, s) = \Phi(s, t).$$

(d) Let $\Phi(t, s)$ be a fundamental matrix solution of $\Phi' = A\Phi$, $-\infty < t < \infty$, where A is an $(n \times n)$ constant matrix and where $\Phi(s, s) = I$. Then

$$\Phi(t, s) = \Phi(t - s, 0).$$

REMARKS: We make two remarks about Theorem 6.15.

1. The formula in part (a) shows us how to create a desired propagator matrix from any given fundamental matrix.
2. In part (d), the equation $\mathbf{y}' = A\mathbf{y}$ is autonomous, and hence the fundamental matrix Φ depends upon the variables t and s in the specific combination $t - s$. For this special autonomous case, therefore, we can dispense with the second argument when writing the propagator matrix Φ and simply refer to it as $\Phi(t - s)$ with the understanding that $\Phi(0) = I$.

EXAMPLE 1

Find the propagator matrix $\Phi(t,s)$ for the first order system

$$\mathbf{y}' = \begin{bmatrix} 1 & 2 \\ 2 & 1 \end{bmatrix} \mathbf{y}, \qquad -\infty < t < \infty.$$

Solution: We saw in Example 1 in Section 6.8 that the matrix

$$\Psi(t) = \begin{bmatrix} e^{-t} & e^{3t} \\ -e^{-t} & e^{3t} \end{bmatrix}$$

is a fundamental matrix. The inverse, expressed as a function of s, is

$$\Psi^{-1}(s) = \frac{1}{2}\begin{bmatrix} e^{s} & -e^{s} \\ e^{-3s} & e^{-3s} \end{bmatrix}.$$

From part (a) of Theorem 6.15

$$\begin{aligned}
\Phi(t,s) &= \Psi(t)\Psi^{-1}(s) \\
&= \begin{bmatrix} e^{-t} & e^{3t} \\ -e^{-t} & e^{3t} \end{bmatrix} \begin{bmatrix} \dfrac{e^{s}}{2} & -\dfrac{e^{s}}{2} \\ \dfrac{e^{-3s}}{2} & \dfrac{e^{-3s}}{2} \end{bmatrix} \\
&= \begin{bmatrix} \dfrac{e^{-(t-s)} + e^{3(t-s)}}{2} & \dfrac{-e^{-(t-s)} + e^{3(t-s)}}{2} \\ \dfrac{-e^{-(t-s)} + e^{3(t-s)}}{2} & \dfrac{e^{-(t-s)} + e^{3(t-s)}}{2} \end{bmatrix}.
\end{aligned}$$

Note that the matrix is a function of $t-s$; we have agreed to use the notation $\Phi(t-s)$. ▲

The solution of the scalar initial value problem

$$y' = ay, \qquad y(0) = 1 \tag{2}$$

is $y(t) = e^{at}$. Given the matrix problem

$$\Phi' = A\Phi, \qquad \Phi(0) = I, \tag{3}$$

we know that the unique solution of this problem is the propagator matrix, $\Phi(t)$. We consider the following question: Is it possible to represent this solution in the form

$$\Phi(t) = e^{tA}?$$

There are two important issues to be resolved. First, how do we give meaning to a function of a square matrix? In particular, how do we define the exponential of a matrix, e^{tA}? Second, if we can somehow give meaning to e^{tA}, is it the unique solution of initial value problem (3)? In this section, we will show that both issues can be answered affirmatively.

Functions of a Matrix

We have three basic matrix operations at our disposal: scalar multiplication, matrix addition, and matrix multiplication. Given a square matrix, A, we can form powers of A. For example, $A^2 = AA, A^3 = AAA = AA^2 = A^2A$ and, in general,

$$A^{m+1} = AA^m = A^m A, \qquad m = 1, 2, \ldots.$$

Since we can form powers, we can form polynomials of a square matrix. For example, if $p(z) = z^3 - 2z + 4$ and

$$A = \begin{bmatrix} -1 & 1 \\ 2 & 0 \end{bmatrix},$$

then we form the **matrix polynomial** $p(A)$:

$$p(A) = A^3 - 2A + 4I = \begin{bmatrix} -5 & 3 \\ 6 & -2 \end{bmatrix} - 2\begin{bmatrix} -1 & 1 \\ 2 & 0 \end{bmatrix} + 4\begin{bmatrix} 1 & 0 \\ 0 & 1 \end{bmatrix} = \begin{bmatrix} 1 & 1 \\ 2 & 2 \end{bmatrix}.$$

To see how we can proceed from polynomials of a matrix to a function of a matrix such as e^A, recall how the scalar function e^z is represented in terms of its Maclaurin series:

$$e^z = 1 + z + \frac{z^2}{2!} + \frac{z^3}{3!} + \frac{z^4}{4!} + \cdots = \sum_{m=0}^{\infty} \frac{z^m}{m!}.$$

For a given scalar z, this power series is given meaning in terms of the limit of the sequence of partial sums. That is, we form the partial sums

$$S_M(z) = 1 + z + \frac{z^2}{2!} + \cdots + \frac{z^M}{M!}, \qquad M = 1, 2, 3, \ldots$$

and define e^z by

$$e^z = \lim_{M\to\infty} S_M(z).$$

Can we define e^A simply by replacing the scalar z by the matrix A in the Maclaurin series for e^z?

Dealing with power series requires two things. We must be able to form polynomials, and we must also have a means for measuring distance, a yardstick to determine whether the partial sums are convergent or not. For square matrices we have both of these ingredients. As the preceding discussion shows, we can form matrix polynomials and hence we can form partial sums $S_M(A)$. Moreover, there exist yardsticks, called matrix norms, for measuring the size of a matrix and the distance between matrices. In principle, therefore, we can determine whether or not the matrix partial sums are converging to some limit matrix as $M \to \infty$.

One commonly used matrix norm is

$$\|A\| = \sqrt{\sum_{i=1}^{n}\sum_{j=1}^{n}(a_{ij})^2} \quad \text{where} \quad A = \begin{bmatrix} a_{11} & \cdots & a_{1n} \\ \vdots & & \vdots \\ a_{n1} & \cdots & a_{nn} \end{bmatrix}.$$

Every scalar Maclaurin series (and, in general, every scalar power series) has associated with it a **radius of convergence**, R. If $0 < R < \infty$, then the Maclaurin series converges for all $|z| < R$ and diverges for all $|z| > R$. If $R = 0$, the Maclaurin series converges only for the value $z = 0$. If the Maclaurin series converges for all values z, we define $R = \infty$. The mathematical arguments used to establish these results for scalar power series can be adapted to prove the following theorem.

Theorem 6.16

Let $f(z)$ possess a Maclaurin series expansion

$$f(z) = f(0) + f'(0)z + \frac{f''(0)}{2!}z^2 + \cdots = \sum_{n=0}^{\infty} \frac{f^{(n)}(0)}{n!} z^n$$

with radius of convergence $R > 0$. Let A be a square matrix such that $\|A\| < R$. Then the matrix power series

$$f(0)I + f'(0)A + \frac{f''(0)}{2!}A^2 + \cdots = \sum_{m=0}^{\infty} \frac{f^{(m)}(0)}{m!} A^m$$

converges to a limiting matrix, which we call $f(A)$.

We will use Theorem 6.16 to define e^{tA}.

The Exponential Matrix

Recall that the radius of convergence of the Maclaurin series for e^z is infinite; that is, the power series

$$e^z = 1 + z + \frac{z^2}{2!} + \frac{z^3}{3!} + \frac{z^4}{4!} + \cdots = \sum_{m=0}^{\infty} \frac{z^m}{m!}$$

converges for all z. It therefore follows from Theorem 6.16 that the matrix power series

$$\begin{aligned} e^{tA} &= I + tA + \frac{1}{2!}(tA)^2 + \frac{1}{3!}(tA)^3 + \cdots \\ &= I + tA + \frac{t^2}{2!}A^2 + \frac{t^3}{3!}A^3 + \cdots \qquad (4) \\ &= \sum_{m=0}^{\infty} \frac{t^m}{m!} A^m \end{aligned}$$

converges for any square matrix A and for any scalar t. We refer to this limiting matrix as the **exponential matrix**.

It can be shown that the matrix function of t defined by (4) is differentiable and that its derivative can be calculated by termwise differentiation. Assuming the validity of term-by-term differentiation, it follows easily that

$$\frac{d}{dt}e^{tA} = Ae^{tA}. \tag{5}$$

Because of differentiation formula (5), the fact that e^{tA} reduces to I when $t = 0$, and because the solution of initial value problem (3) is unique, we see that the exponential matrix e^{tA} is in fact identical to the propagator matrix $\Phi(t)$.

The Exponential of a Diagonalizable Matrix

In the special case where A is diagonalizable, we obtain a simple closed form expression for the exponential matrix. In addition, we can use this form to show, explicitly, that $e^{tA} = \Phi(t)$. Let $\lambda_1, \lambda_2, \ldots, \lambda_n$ denote the eigenvalues of A and let T be the invertible matrix whose columns are corresponding eigenvectors of A. Then (see Theorem 6.14) it follows that $T^{-1}AT = D$, where

$$D = \begin{bmatrix} \lambda_1 & 0 & 0 & \cdots & 0 \\ 0 & \lambda_2 & 0 & \cdots & 0 \\ 0 & 0 & \lambda_3 & \cdots & 0 \\ \vdots & & & & \vdots \\ 0 & 0 & 0 & \cdots & \lambda_n \end{bmatrix}.$$

Since $T^{-1}AT = D$, we also have $A = TDT^{-1}$. This expression provides a convenient method for computing powers of A:

$$\begin{aligned} A^2 &= (TDT^{-1})(TDT^{-1}) = TD(T^{-1}T)DT^{-1} = TDDT^{-1} = TD^2T^{-1}, \\ A^3 &= (TDT^{-1})(TDT^{-1})(TDT^{-1}) = TD(T^{-1}T)D(T^{-1}T)DT^{-1} = TD^3T^{-1}, \end{aligned}$$

and, in general,

$$A^m = TD^mT^{-1}, \qquad m = 1, 2, 3, \ldots. \tag{6}$$

Equation (6) is useful because powers of a diagonal matrix are easy to compute. That is, $A^m = TD^mT^{-1}$, where

$$D^m = \begin{bmatrix} \lambda_1^m & 0 & 0 & \cdots & 0 \\ 0 & \lambda_2^m & 0 & \cdots & 0 \\ 0 & 0 & \lambda_3^m & \cdots & 0 \\ \vdots & & & & \vdots \\ 0 & 0 & 0 & \cdots & \lambda_n^m \end{bmatrix}, \qquad m = 1, 2, 3, \ldots.$$

We can also use equation (6) to simplify the matrix power series for e^{tA} given in equation (5). In particular, the partial sum

$$S_M(tA) = I + tA + \frac{t^2}{2!}A^2 + \cdots + \frac{t^M}{M!}$$

can be written as

$$S_M(tA) = T\left[I + tD + \frac{t^2}{2!}D^2 + \cdots \frac{t^M}{M!}D^M\right]T^{-1}.$$

It follows that

$$\begin{aligned} e^{tA} &= \lim_{M\to\infty} S_M(tA) \\ &= T\begin{bmatrix} e^{\lambda_1 t} & 0 & 0 & \cdots & 0 \\ 0 & e^{\lambda_2 t} & 0 & \cdots & 0 \\ 0 & 0 & e^{\lambda_3 t} & \cdots & 0 \\ \vdots & & & & \vdots \\ 0 & 0 & 0 & \cdots & e^{\lambda_n t} \end{bmatrix} T^{-1} \\ &= T\Lambda(t)T^{-1}. \end{aligned} \tag{7}$$

EXAMPLE 2

Find e^{tA} for the matrix

$$A = \begin{bmatrix} 1 & 2 \\ 2 & 1 \end{bmatrix}.$$

Solution: The matrix A is diagonalizable since it is real and symmetric. We have seen this matrix in previous examples and we know that $T^{-1}AT = D$, where

$$T = \begin{bmatrix} 1 & 1 \\ 1 & -1 \end{bmatrix} \quad \text{and} \quad D = \begin{bmatrix} 3 & 0 \\ 0 & -1 \end{bmatrix}.$$

Thus, by equation (7),

$$\begin{aligned} e^{tA} &= T\Lambda(t)T^{-1} \\ &= \begin{bmatrix} 1 & 1 \\ 1 & -1 \end{bmatrix}\begin{bmatrix} e^{3t} & 0 \\ 0 & e^{-t} \end{bmatrix}\begin{bmatrix} \frac{1}{2} & \frac{1}{2} \\ \frac{1}{2} & -\frac{1}{2} \end{bmatrix} \\ &= \frac{1}{2}\begin{bmatrix} e^{3t} + e^{-t} & e^{3t} - e^{-t} \\ e^{3t} - e^{-t} & e^{3t} + e^{-t} \end{bmatrix}. \end{aligned}$$

As a check, you can verify that $de^{tA}/dt = Ae^{tA}$. ▲

Although we have explicitly shown $e^{tA} = \Phi(t)$ when A is diagonalizable, it follows from equations (4) and (5) that this equality holds in general—even when A does not have a full set of eigenvectors. In that case, however, the matrix exponential e^{tA} has a more complicated form than the one shown in (7).

Other Functions of a Matrix

Among the functions that possess a Maclaurin series convergent for all z are

$$\cos z = 1 - \frac{z^2}{2!} + \frac{z^4}{4!} - \frac{z^6}{6!} + \cdots = \sum_{m=0}^{\infty} \frac{(-1)^m z^{2m}}{(2m)!}$$

and

$$\sin z = z - \frac{z^3}{3!} + \frac{z^5}{5!} - \frac{z^7}{7!} + \cdots = \sum_{m=0}^{\infty} \frac{(-1)^m z^{2m+1}}{(2m+1)!}.$$

Therefore, given a square matrix A, we can define the matrix functions

$$\cos(tA) = I - \frac{t^2}{2!}A^2 + \frac{t^4}{4!}A^4 - \frac{t^6}{6!}A^6 + \cdots = \sum_{m=0}^{\infty} \frac{(-1)^m t^{2m}}{(2m)!}A^{2m}$$

and

$$\sin(tA) = tA - \frac{t^3}{3!}A^3 + \frac{t^5}{5!}A^5 - \frac{t^7}{7!}A^7 + \cdots = \sum_{m=0}^{\infty} \frac{(-1)^m t^{2m+1}}{(2m+1)!}A^{2m+1}.$$

By Theorem 6.16, we are assured that the two matrix power series converge for all square matrices A and all values t. Moreover, if A is an $(n \times n)$ diagonalizable matrix, we obtain the closed-form representations

$$\cos(tA) = T \begin{bmatrix} \cos\lambda_1 t & 0 & 0 & \cdots & 0 \\ 0 & \cos\lambda_2 t & 0 & \cdots & 0 \\ 0 & 0 & \cos\lambda_3 t & \cdots & 0 \\ \vdots & & & & \vdots \\ 0 & 0 & 0 & \cdots & \cos\lambda_n t \end{bmatrix} T^{-1}$$

and

$$\sin(tA) = T \begin{bmatrix} \sin\lambda_1 t & 0 & 0 & \cdots & 0 \\ 0 & \sin\lambda_2 t & 0 & \cdots & 0 \\ 0 & 0 & \sin\lambda_3 t & \cdots & 0 \\ \vdots & & & & \vdots \\ 0 & 0 & 0 & \cdots & \sin\lambda_n t \end{bmatrix} T^{-1}.$$

We will use these two matrix functions (as well as the following ideas involving the roots of a diagonalizable matrix) when we revisit the coupled spring-mass system studied in Section 6.10.

Roots of a Diagonalizable Matrix

Given an $(n \times n)$ diagonalizable matrix A, is it possible to find an $(n \times n)$ matrix B such that $A = B^2$? If so, we call such a matrix B a **square root** of A and adopt the notation $B = \sqrt{A}$. As we will see, the generalization of this idea to nth roots of a matrix is straightforward.

We first observe that the problem of finding a square root is easy if the matrix is diagonal. For example, if D is the diagonal matrix

$$D = \begin{bmatrix} \lambda_1 & 0 & 0 & \cdots & 0 \\ 0 & \lambda_2 & 0 & \cdots & 0 \\ 0 & 0 & \lambda_3 & \cdots & 0 \\ \vdots & & & & \vdots \\ 0 & 0 & 0 & \cdots & \lambda_n \end{bmatrix},$$

then (as we can check by direct multiplication)

$$\sqrt{D} = \begin{bmatrix} \sqrt{\lambda_1} & 0 & 0 & \cdots & 0 \\ 0 & \sqrt{\lambda_2} & 0 & \cdots & 0 \\ 0 & 0 & \sqrt{\lambda_3} & \cdots & 0 \\ \vdots & \vdots & \vdots & \ddots & \vdots \\ 0 & 0 & 0 & \cdots & \sqrt{\lambda_n} \end{bmatrix}.$$

[Note: If some of the diagonal entries of D are negative or complex, then $\sqrt{D}$ may have complex entries.]

How do we find $\sqrt{A}$ if A is not a diagonal matrix? Suppose A is diagonalizable and that $T^{-1}AT = D$, where D is diagonal. Recalling equation (4) and the definitions of e^A, $\cos A$ and $\sin A$, we might suspect that

$$\sqrt{A} = T\sqrt{D}\,T^{-1}. \tag{8}$$

In fact, equation (6) does define a square root of A, as we can verify by squaring the right side of (6):

$$\left(T\sqrt{D}\,T^{-1}\right)\left(T\sqrt{D}\,T^{-1}\right) = T\sqrt{D}\left(T^{-1}T\right)\sqrt{D}\,T^{-1} = TDT^{-1} = A.$$

When A is diagonalizable, we can define kth roots in a similar fashion:

$$\sqrt[k]{A} = T\sqrt[k]{D}\,T^{-1},$$

where

$$\sqrt[k]{D} = \begin{bmatrix} \sqrt[k]{\lambda_1} & 0 & \cdots & 0 \\ 0 & \sqrt[k]{\lambda_2} & \cdots & 0 \\ \vdots & & & \vdots \\ 0 & 0 & \cdots & \sqrt[k]{\lambda_n} \end{bmatrix}.$$

EXAMPLE 3

Find $\sqrt{A}$, where

$$A = \frac{k}{m}\begin{bmatrix} 2 & -1 \\ -1 & 2 \end{bmatrix}.$$

Note that A is the coefficient matrix for the coupled spring-mass system studied in Section 6.10; see equation (14) in Section 6.10.

(continued)

(continued)

Solution: Since A is real and symmetric, A is diagonalizable. As we saw in Section 6.10, $T^{-1}AT = D$, where

$$T = \begin{bmatrix} 1 & 1 \\ 1 & -1 \end{bmatrix} \quad \text{and} \quad D = \begin{bmatrix} \frac{k}{m} & 0 \\ 0 & 3\frac{k}{m} \end{bmatrix}.$$

Therefore,

$$\sqrt{A} = T\sqrt{D}\,T^{-1} = \begin{bmatrix} 1 & 1 \\ 1 & -1 \end{bmatrix} \begin{bmatrix} \sqrt{\frac{k}{m}} & 0 \\ 0 & \sqrt{\frac{3k}{m}} \end{bmatrix} \begin{bmatrix} \frac{1}{2} & \frac{1}{2} \\ \frac{1}{2} & -\frac{1}{2} \end{bmatrix}$$

$$= \frac{1}{2}\sqrt{\frac{k}{m}} \begin{bmatrix} \sqrt{3}+1 & -\sqrt{3}+1 \\ -\sqrt{3}+1 & \sqrt{3}+1 \end{bmatrix}.$$

As an exercise, you should check by direct multiplication that $\sqrt{A}\sqrt{A} = A$. ▲

The Coupled Spring-Mass Problem Revisited

The second order linear system modeling the coupled spring-mass system in Section 6.10 has the form

$$\mathbf{x}'' + A\mathbf{x} = \mathbf{0}, \tag{9}$$

where the matrix A in equation (9) is

$$A = \frac{k}{m}\begin{bmatrix} 2 & -1 \\ -1 & 2 \end{bmatrix}.$$

Equation (9) resembles the scalar second order equation $x'' + ax = 0$, which has general solution

$$x(t) = c\cos\sqrt{a}\,t + d\sin\sqrt{a}\,t,$$

where c and d are arbitrary constants. Is it possible to write the general solution of $\mathbf{x}'' + A\mathbf{x} = \mathbf{0}$ in an analogous form as

$$\mathbf{x}(t) = \left[\cos\left(t\sqrt{A}\right)\right]\mathbf{c} + \left[\sin\left(t\sqrt{A}\right)\right]\mathbf{d}, \tag{10}$$

where $\mathbf{c}$ and $\mathbf{d}$ are arbitrary (2×1) constant vectors? [Since $\sqrt{A}$ is a (2×2) matrix, it follows that $\cos(t\sqrt{A})$ and $\sin(t\sqrt{A})$ are also (2×2) matrices.]

We computed the general solution of $\mathbf{x}'' + A\mathbf{x} = \mathbf{0}$ in Section 6.10, so we need only examine it and see if it can be put into the suggested form (10). The general solution, given in equation (17b) of Section 6.10 is

$$\mathbf{x}(t) = T\mathbf{w}(t) = T \begin{bmatrix} c_1 \cos\sqrt{\frac{k}{m}}\,t + d_1 \sin\sqrt{\frac{k}{m}}\,t \\ c_2 \cos\sqrt{3\frac{k}{m}}\,t + d_2 \sin\sqrt{3\frac{k}{m}}\,t \end{bmatrix}$$

$$= T \begin{bmatrix} \cos\sqrt{\frac{k}{m}}\,t & 0 \\ 0 & \cos\sqrt{3\frac{k}{m}}\,t \end{bmatrix} \begin{bmatrix} c_1 \\ c_2 \end{bmatrix} + T \begin{bmatrix} \sin\sqrt{\frac{k}{m}}\,t & 0 \\ 0 & \sin\sqrt{3\frac{k}{m}}\,t \end{bmatrix} \begin{bmatrix} d_1 \\ d_2 \end{bmatrix}.$$

From Example 2,

$$\cos\left(t\sqrt{A}\right) = T \begin{bmatrix} \cos\sqrt{\frac{k}{m}}\,t & 0 \\ 0 & \cos\sqrt{3\frac{k}{m}}\,t \end{bmatrix} T^{-1}$$

and

$$\sin\left(t\sqrt{A}\right) = T \begin{bmatrix} \sin\sqrt{\frac{k}{m}}\,t & 0 \\ 0 & \sin\sqrt{3\frac{k}{m}}\,t \end{bmatrix} T^{-1}.$$

Therefore, if we redefine the (2×1) vectors of arbitrary constants to be

$$\begin{bmatrix} c_1 \\ c_2 \end{bmatrix} = T^{-1}\mathbf{c} \quad \text{and} \quad \begin{bmatrix} d_1 \\ d_2 \end{bmatrix} = T^{-1}\mathbf{d},$$

we see that the general solution does indeed have the conjectured form (10). The exercises contain some additional examples of this type.

A Warning

The ability to formulate matrix-valued analogs of familiar functions and to express solutions of systems of differential equations in terms of these functions is intrinsically elegant. We must be careful, however, not to push such analogies beyond their legitimate bounds.

A case in point is given by the following example. We know from Chapter 2 that the solution of the initial value problem

$$y' = p(t)y, \qquad y(0) = 1$$

is

$$y(t) = e^{\int_0^t p(s)\,ds}. \tag{11a}$$

Given our success with the constant coefficient matrix problem $\Phi' = A\Phi$, $\Phi(0) = I$, we are tempted to ask if we can write the solution of the variable coefficient matrix problem

$$\Phi' = P(t)\Phi, \qquad \Phi(0) = I$$

as the matrix exponential

$$\Phi(t) = e^{\int_0^t P(s)\,ds}. \tag{11b}$$

Despite the obvious resemblance between the right-hand sides of equations (11a) and (11b), the matrix $\Phi(t)$ in (11b) is *not* generally a solution of $\Phi' = P(t)\Phi$. The reason, ultimately, stems from the noncommutative nature of matrix multiplication.

We can certainly define the matrix antiderivative

$$Q(t) = \int_0^t P(s)\,ds$$

and then form the matrix exponential $e^{Q(t)}$ using the Maclaurin series representation. However, $e^{Q(t)}$ is generally *not* a solution of the differential equation $\Phi' = P(t)\Phi$. To explain why, suppose we attempt to duplicate our prior argument, the one that led us successfully from equation (4) to equation (5). As before, we begin with the series for $e^{Q(t)}$,

$$e^{Q(t)} = I + Q(t) + \frac{1}{2!}Q^2(t) + \frac{1}{3!}Q^3(t) + \cdots.$$

When we differentiate this series termwise, we need to calculate derivatives of various powers of the matrix $Q(t)$. As an example of "what goes wrong," consider the derivative of $Q^2(t)$:

$$\frac{d}{dt}Q^2(t) = \frac{d}{dt}[Q(t)Q(t)] = Q'(t)Q(t) + Q(t)Q'(t) = P(t)Q(t) + Q(t)P(t) \neq 2P(t)Q(t).$$

The matrix $Q(t)$ generally does not commute with its derivative, $P(t)$. In general, when $P(t)$ is a matrix, we usually find that

$$\frac{d}{dt}\,e^{\int_0^t P(s)\,ds} \neq P(t)\,e^{\int_0^t P(s)\,ds}.$$

EXAMPLE 4

To illustrate the fact that a matrix function does not generally commute with its derivative, consider

$$Q(t) = \begin{bmatrix} t & \sin t \\ t^2 & 0 \end{bmatrix}.$$

Form the products $Q(t)Q'(t)$ and $Q'(t)Q(t)$. Are they equal?

Solution: We find

$$Q'(t) = \begin{bmatrix} 1 & \cos t \\ 2t & 0 \end{bmatrix}.$$

Therefore, as we see below, $Q(t)Q'(t) \neq Q'(t)Q(t)$:

$$Q(t)Q'(t) = \begin{bmatrix} t & \sin t \\ t^2 & 0 \end{bmatrix} \begin{bmatrix} 1 & \cos t \\ 2t & 0 \end{bmatrix} = \begin{bmatrix} t + 2t\sin t & t\cos t \\ t^2 & t^2\cos t \end{bmatrix}$$

$$Q'(t)Q(t) = \begin{bmatrix} 1 & \cos t \\ 2t & 0 \end{bmatrix} \begin{bmatrix} t & \sin t \\ t^2 & 0 \end{bmatrix} = \begin{bmatrix} t + t^2\cos t & \sin t \\ 2t^2 & 2t\sin t \end{bmatrix}. \quad ▲$$

EXERCISES

Exercises 1–4:

For the given autonomous system $\mathbf{y}' = A\mathbf{y}$,

(a) Determine the propagator matrix, $\Phi(t) = e^{tA}$.

(b) For the given values of s and $\mathbf{y}(s)$, use the propagator matrix to determine $\mathbf{y}(t)$ at the specified value of t.

1. $\mathbf{y}' = \begin{bmatrix} 5 & -4 \\ 5 & -4 \end{bmatrix}\mathbf{y}, \quad \mathbf{y}(-1) = \begin{bmatrix} 1 \\ 0 \end{bmatrix}, \quad \mathbf{y}(2) = ?$

2. $\mathbf{y}' = \begin{bmatrix} 2 & 1 \\ 0 & 2 \end{bmatrix}\mathbf{y}, \quad \mathbf{y}(1) = \begin{bmatrix} 1 \\ 2 \end{bmatrix}, \quad \mathbf{y}(2) = ?$

3. $\mathbf{y}' = \begin{bmatrix} 6 & 5 \\ 1 & 2 \end{bmatrix}\mathbf{y}, \quad \mathbf{y}(0) = \begin{bmatrix} 1 \\ 1 \end{bmatrix}, \quad \mathbf{y}(-1) = ?$

4. $\mathbf{y}' = \begin{bmatrix} 1 & 1 & 1 \\ 0 & 2 & 1 \\ 0 & 0 & -1 \end{bmatrix}\mathbf{y}, \quad \mathbf{y}(0) = \begin{bmatrix} 1 \\ 1 \\ 0 \end{bmatrix}, \quad \mathbf{y}(1) = ?$

5. The nonautonomous linear system

$$\mathbf{y}' = \begin{bmatrix} 0 & 1 \\ -2t^{-2} & 2t^{-1} \end{bmatrix}\mathbf{y}, \qquad t \neq 0$$

has fundamental matrix

$$\Psi(t) = \begin{bmatrix} t & t^2 \\ 1 & 2t \end{bmatrix}.$$

(a) Determine the propagator matrix $\Phi(t, s)$ where t and s are positive. Is $\Phi(t, s)$ a function of the difference variable $t - s$?

(b) Suppose

$$\mathbf{y}(1) = \begin{bmatrix} 1 \\ -1 \end{bmatrix}.$$

What is $\mathbf{y}(3)$?

6. Let A be a diagonalizable (2×2) matrix where

$$T^{-1}AT = \begin{bmatrix} \lambda_1 & 0 \\ 0 & \lambda_2 \end{bmatrix}.$$

Consider the matrix polynomial $p(A) = 2A^3 - A + 3I$. Find a matrix B such that $p(A) = TBT^{-1}$.

7. Let A be a diagonalizable (3×3) matrix where

$$T^{-1}AT = \begin{bmatrix} \lambda_1 & 0 & 0 \\ 0 & \lambda_2 & 0 \\ 0 & 0 & \lambda_3 \end{bmatrix}.$$

Let $p(\lambda) = \det(A - \lambda I)$ denote the characteristic polynomial of A. Show that

$$p(A) = T \begin{bmatrix} p(\lambda_1) & 0 & 0 \\ 0 & p(\lambda_2) & 0 \\ 0 & 0 & p(\lambda_3) \end{bmatrix} T^{-1}$$

and conclude, therefore, that $p(A)$ is the (3×3) zero matrix. (This is a special case of a result known as the Cayley-Hamilton theorem.[2] The Cayley-Hamilton theorem states that every $(n \times n)$ matrix satisfies its own characteristic equation. Exercise 7 illustrates the Cayley-Hamilton theorem when $n = 3$.)

8. Let A be an invertible and diagonalizable $(n \times n)$ matrix, where $T^{-1}AT = D$ with D diagonal.

(a) Show that D is invertible.

(b) Show that $T^{-1}A^{-1}T = D^{-1}$; that is, A^{-1} is also diagonalizable and by the same similarity transformation. [Hint: Recall that $(PQ)^{-1} = Q^{-1}P^{-1}$.]

Exercises 9–12:

Consider the (2×2) matrix A and the eigenpairs of A listed:

$$A = \begin{bmatrix} 17 & -12 \\ 24 & -17 \end{bmatrix}, \qquad \lambda_1 = 1, \mathbf{x}_1 = \begin{bmatrix} 3 \\ 4 \end{bmatrix}, \qquad \lambda_2 = -1, \mathbf{x}_2 = \begin{bmatrix} 2 \\ 3 \end{bmatrix}.$$

9. (a) What is the power A^n when n is an even positive integer? [Hint: Begin with $T^{-1}AT = D$.]

(b) What is the power A^n when n is an odd positive integer?

(c) What is the power A^{-n} when n is an even positive integer? An odd positive integer?

10. Find four different diagonal (2×2) matrices D such that $A^{1/2} = TDT^{-1}$.

11. Find a (2×2) matrix B such that $B^2 = A^{-1}$.

12. Let T be an invertible matrix whose columns are eigenvectors of A. Find a (2×2) matrix B such that $A^2 + A^{1/2} = TBT^{-1}$.

13. Determine A^3 if A is a real (2×2) matrix such that $T^{-1}AT = D$ where

$$T = \begin{bmatrix} 1 & 1 \\ 2 & 1 \end{bmatrix} \quad \text{and} \quad D = \begin{bmatrix} -2 & 0 \\ 0 & 2 \end{bmatrix}.$$

---

[2]Arthur Cayley (1821–1895) graduated from Trinity College, Cambridge in 1842 and spent the next four years teaching at Cambridge. Thereafter he became a lawyer to support himself. During his 14 years as a lawyer he maintained an active involvement in mathematics, publishing more than 200 mathematical papers. In 1863 Cayley received a professorial appointment at Cambridge. In that capacity, he published over 900 papers and notes on virtually every aspect of modern mathematics. For details on Sir William Rowan Hamilton (1805–1865), see Section 8.3.

14. Let A be a diagonalizable (2×2) matrix with eigenpairs

$$\lambda_1 = \frac{1}{4}, \quad \mathbf{x}_1 = \begin{bmatrix} 2 \\ 5 \end{bmatrix}, \quad \lambda_2 = \frac{1}{2}, \quad \mathbf{x}_2 = \begin{bmatrix} 1 \\ 3 \end{bmatrix}.$$

Compute the matrix functions $f_1(A) = \cos(\pi A)$ and $f_2(A) = \sin(\pi A)$.

15. Let A be a diagonalizable (2×2) matrix with eigenvalues $\lambda_1 = 1/3$ and $\lambda_2 = 7/3$. Without knowing the corresponding eigenvectors, compute the matrix functions $f_1(A) = \cos(\pi A)$ and $f_2(A) = \sin(\pi A)$.

Exercises 16–19:

The matrix

$$A = \begin{bmatrix} 3 & 2 \\ -4 & -3 \end{bmatrix}$$

has eigenpairs

$$\lambda_1 = -1, \quad \mathbf{x}_1 = \begin{bmatrix} 1 \\ -2 \end{bmatrix}, \quad \lambda_2 = 1, \quad \mathbf{x}_2 = \begin{bmatrix} 1 \\ -1 \end{bmatrix}.$$

Obtain the general solution of the given linear system.

16. $\mathbf{y}'' + A\mathbf{y} = \begin{bmatrix} 1 \\ 0 \end{bmatrix}$

17. $A\mathbf{y}' + \mathbf{y} = \begin{bmatrix} 1 \\ 1 \end{bmatrix}$

18. $\mathbf{y}'' + \mathbf{y}' + A\mathbf{y} = \mathbf{0}$

19. $\mathbf{y}'' + 2A\mathbf{y}' = \mathbf{0}$

20. Consider the spring-mass system shown in the figure. The system can execute one-dimensional motion on the frictionless horizontal surface. The unperturbed and perturbed systems are labeled (a) and (b) respectively.

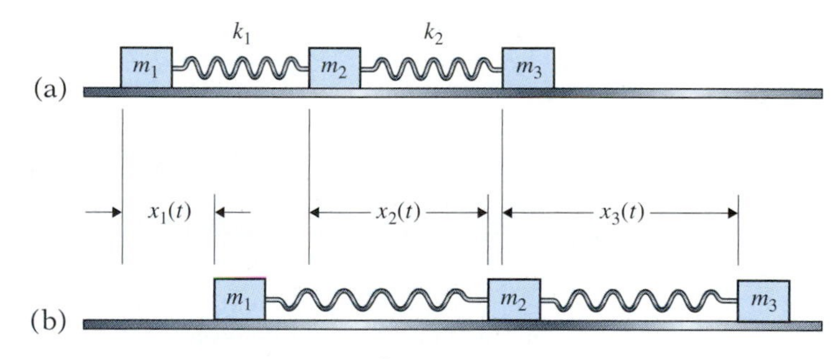

Figure for Exercise 20

(a) Show that an application of Newton's second law of motion leads the second order system $M\mathbf{x}'' + K\mathbf{x} = \mathbf{0}$, where

$$M = \begin{bmatrix} m_1 & 0 & 0 \\ 0 & m_2 & 0 \\ 0 & 0 & m_3 \end{bmatrix}, \quad K = \begin{bmatrix} k_1 & -k_1 & 0 \\ -k_1 & k_1 + k_2 & -k_2 \\ 0 & -k_2 & k_2 \end{bmatrix}, \quad \mathbf{x}(t) = \begin{bmatrix} x_1(t) \\ x_2(t) \\ x_3(t) \end{bmatrix}.$$

(b) Show that the matrix K has an eigenvalue $\lambda = 0$. Determine a corresponding eigenvector and denote it as $\mathbf{v}_0$.

(c) Explain the physical significance of the eigenpair $(0, \mathbf{v}_0)$. In particular, what motion does the system execute if the initial conditions are $\mathbf{x}(0) = \mathbf{0}$, $\mathbf{x}'(0) = \mathbf{v}_0$? [Hint: Look for a solution of the form $\mathbf{x}(t) = f(t)\mathbf{v}_0$ where $f(t)$ is a scalar function to be determined.] Describe, in words, how the system is behaving.

21. Consider the spring-mass system described in Exercise 20. Let $m_1 = m_2 = m_3 = m$ and $k_1 = k_2 = k$.

(a) Determine the eigenpairs of $A = M^{-1}K$.

(b) Obtain the general solution of $\mathbf{x}'' + A\mathbf{x} = \mathbf{0}$.

EXTENDED PROBLEM: THREE CHEMICAL REACTANTS

We consider a system of first order chemical reactions studied by Aris.[3] The triangular reaction scheme involving three reactants, A, B, and C, is shown in the figure. The six nonnegative constants $k_1, k_2, k_3, k_1', k_2', k_3'$ represent reaction rates per unit reactant. These rates are associated with arrows indicating the direction in which the reactions are occurring. To interpret this diagram, consider reactant A for example. In the differential time segment dt, reactant A is decreased by an amount $k_1 A(t)\,dt$, which is converted to reactant B and by an amount $k_3' A(t)\,dt$, which is converted to reactant C. During the same differential time segment, however, reactant A is increased by amounts $k_1' B(t)\,dt$ and $k_3 C(t)\,dt$. Therefore, the net change in the amount of reactant A over the differential time segment dt is

$$dA = -k_1 A\,dt - k_3' A\,dt + k_1' B\,dt + k_3 C\,dt. \tag{1}$$

Dividing by dt and letting $dt \to 0$, we obtain a differential equation for the rate of change of reactant $A(t)$.

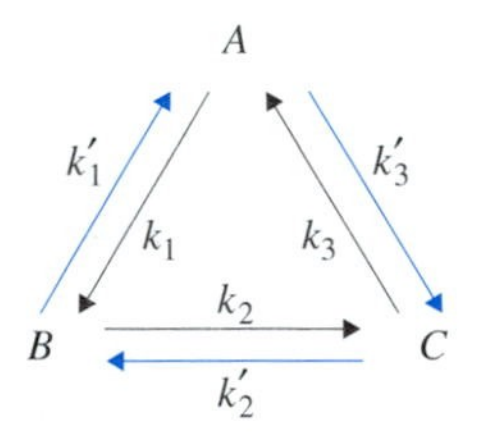

A chemical reaction scheme involving three reactants

1. Apply the same conservation principle to reactants B and C to obtain a linear homogeneous first order system of differential equations describing the reaction dynamics. Your answer should have the form $\mathbf{y}' = K\mathbf{y}$, where

$$\mathbf{y}(t) = \begin{bmatrix} A(t) \\ B(t) \\ C(t) \end{bmatrix}$$

and K is a constant (3×3) matrix. Assume the reaction begins at $t = 0$ with $A(0) = A_0$, $B(0) = B_0$, and $C(0) = C_0$ representing the initial amounts of reactants present. The initial value problem of interest therefore has the form

$$\mathbf{y}' = K\mathbf{y}, \qquad t > 0, \qquad \mathbf{y}(0) = \mathbf{y}_0. \tag{2}$$

What to Expect: As an initial step, we will examine the structure of problem (2) and try to determine qualitatively what we should expect. In particular, we'll

[3] Rutherford Aris, *Elementary Chemical Reactor Analysis* (Englewood Cliffs, NJ, Prentice-Hall, 1969).

consider the eigenpairs of K and the related question of equilibrium solutions of equation (2).

2. Is the coefficient matrix K real and symmetric?
3. Show that $\lambda_1 = 0$ is an eigenvalue of K. [Hint: You can, but need not, compute the characteristic polynomial of K. An alternative approach is to show that $\det(K) = 0$ or, equivalently, find a nonzero vector $\mathbf{x}$ such that $K\mathbf{x} = \mathbf{0}$.]
4. Why does the fact that K has 0 as an eigenvalue imply that the differential equation $\mathbf{y}' = K\mathbf{y}$ possesses a nontrivial equilibrium solution? What property must the corresponding eigenvector possess in order that this equilibrium solution be physically relevant?
5. Consider the particular case $k_1 = 1, k_2 = 2, k_3 = 3, k_1' = 3, k_2' = 1, k_3' = 2$. Compute the eigenpairs for the corresponding matrix K. Your computations should show that K has a repeated negative real eigenvalue with geometric multiplicity 2 and that the eigenvector corresponding to the eigenvalue 0 can be chosen to have real components.

Numerical Solution

6. Consider the initial conditions $A(0) = 2$, $B(0) = 0$, $C(0) = 0$. Let K be as in Problem 5. Using the computations of Problem 5, determine the solution $\mathbf{y}(t)$. What is $\lim_{t\to\infty} \mathbf{y}(t)$?
7. Plot the three components of $\mathbf{y}(t)$, namely, $A(t)$, $B(t)$, and $C(t)$, as functions of time t. Select a time interval T large enough so that at time T all three reactants are within 1% of their respective limiting values.
8. Use Euler's method to solve the initial value problem numerically on the interval $0 \le t \le T$, where T is the time determined in Problem 7. Use step size $h = 0.01$. Compare your numerically generated approximation of $\mathbf{y}(T)$ with that determined in Problem 6.

7

Laplace Transforms

CHAPTER OVERVIEW

Introduction

When you begin to study a new topic such as the Laplace transform, two questions arise: "What is it?" and "Why is it important?" A scientist often uses Laplace[1] transforms to solve a mathematical problem for the same reason that a motorist leaves a congested highway and travels a network of back roads to reach his or her destination. The most easily traveled path between two points is not always the most direct one.

The philosophy underlying the use of Laplace transforms is illustrated in Figure 7.1. We have a problem to solve—for example, determining the behavior of some mechanical or electrical system. Instead of attacking the problem directly, we transform (or map) the original problem into a new problem. This mapping is accomplished by means of the mathematical operation known as

[1] Pierre-Simon Laplace (1749–1827) was a French scientist renowned for his work in mathematics, celestial mechanics, probability theory, and physics. The Laplace transform, the Laplace probability density function, and Laplace's equation (arising in the study of potential theory) are some mathematical entities named in his honor.

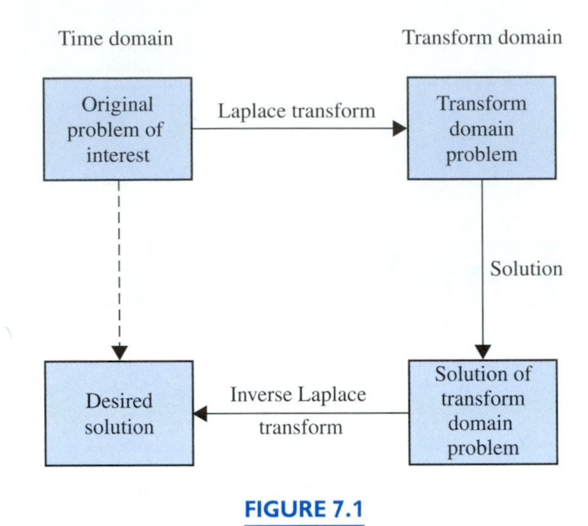

FIGURE 7.1

The philosophy underlying the use of Laplace transforms.

the Laplace transform. The original problem is often referred to as the "time domain problem" since the independent variable for the original problem is usually time. The new problem, resulting from the Laplace transformation, is referred to as the "transform domain problem." After obtaining this new transform domain problem, we solve it and then transform the solution back into the time domain by performing another mapping, known as the inverse Laplace transform. The inverse mapping thus gives us the solution of the original time domain problem, the problem of interest.

In order for the problem-solving approach diagrammed in Figure 7.1 to be attractive, the following three steps must be easier to implement than the direct solution approach:

1. Calculating the Laplace transform
2. Solving the transformed problem
3. Calculating the inverse Laplace transform

For many of the problems we treat, this will be the case. Constant coefficient linear *differential* equations will be transformed into *algebraic* equations. The resulting transform domain problem typically entails solving a single linear algebraic equation or a system of linear algebraic equations.

In Section 7.1, we define the Laplace transform and give some examples showing how to compute the Laplace transform. Theorem 7.1 gives conditions that guarantee a function has a Laplace transform. Section 7.2 derives a number of Laplace transform pairs and shows how the Laplace transform of derivatives and antiderivatives of $f(t)$ relate to the Laplace transform of $f(t)$. We also describe how to use Laplace transforms to solve certain initial value problems. Section 7.3 provides a review of partial fraction expansions.

Section 7.4 derives an expression for the Laplace transform of a periodic function and introduces the concept of a system transfer function. Laplace transforms of vector functions are defined in Section 7.5, where we show how to use Laplace transforms to solve initial value problems of the form $\mathbf{y}' = A\mathbf{y} + \mathbf{g}(t)$, $\mathbf{y}(0) = \mathbf{y}_0$. In Section 7.6 we consider a mathematical operation known as convolution and discuss its Laplace transform. Section 7.7 introduces the delta function and its use in modeling impulsive forces applied to linear systems.

We will consider a variety of constant coefficient linear differential equations (both scalar equations and systems) and show how these problems can be solved using Laplace transforms. We will also consider several problems, such as the parameter identification problem in the following example, which are not so straightforwardly solved using the techniques developed so far.

EXAMPLE 1

Consider a vibrating mechanical system that exists in a "black box," as in Figure 7.2. Assume you are confident that the vibrating system can be modeled as the spring-mass-dashpot arrangement shown, but you do not have the internal access needed to directly measure the spring constant k, the mass m, or the damping constant γ. You can only apply a force at the external access point and measure the resulting displacement.

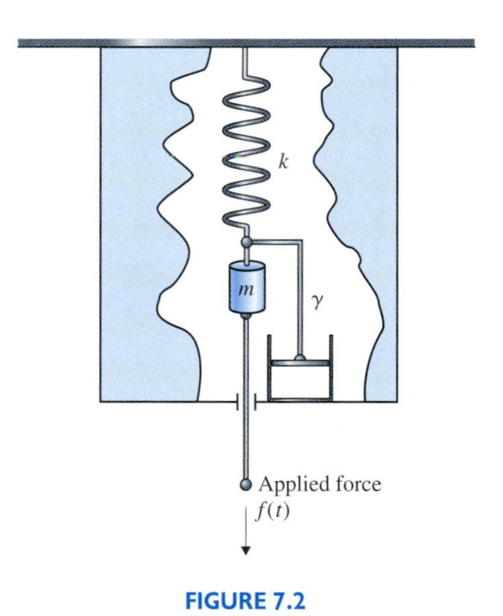

FIGURE 7.2

A cutaway schematic of a "black box" vibrating system.

Mathematically (as we saw in Section 4.11), the mechanical system in Figure 7.2 is described by the initial value problem,

$$my'' + \gamma y' + ky = f(t), \qquad t > 0, \qquad y(0) = 0, \qquad y'(0) = 0. \tag{1}$$

The system is at rest at time $t = 0$ when force $f(t)$ is applied. The applied force $f(t)$ is known for $t \geq 0$; the displacement $y(t)$ is monitored and is thus known for $t \geq 0$. The parameters m, γ, and k, however, are unknown.

(continued)

(continued)

Assuming we know the input-output relation [that is, $f(t)$ and $y(t)$] for one given applied force and the corresponding measured displacement, we ask the following two questions:

1. Is it possible to predict what the output will be if another input is applied to the system? In other words, if we apply a different external force, $\tilde{f}(t)$, is it possible to predict the resulting displacement, $\tilde{y}(t)$?
2. Is it possible to determine the constants m, γ, and k from a knowledge of the single input-output history given by $f(t)$ and $y(t)$?

Suppose we decide to attempt a direct "time domain approach." Using the method of variation of parameters (Section 4.10), we can derive a representation for displacement $y(t)$ in terms of the applied force and the unknown constants m, γ, and k. Suppose we are able to infer (from our observation of the output) that the mechanical system is underdamped. In this case, the variation of parameters solution is

$$y(t) = \int_0^t \varphi(t-\lambda) f(\lambda)\, d\lambda, \tag{2}$$

where $\varphi(u)$ is given by

$$\varphi(u) = \frac{e^{-(\gamma/2m)u} \sin\left(\sqrt{\dfrac{k}{m} - \dfrac{\gamma^2}{4m^2}}\, u\right)}{m\sqrt{\dfrac{k}{m} - \dfrac{\gamma^2}{4m^2}}}, \tag{3}$$

With respect to equations (2) and (3), we assume we know both $f(t)$ and $y(t)$ for all $t \geq 0$. All of the information about the mechanical system is contained in the integrand factor $\varphi(u)$.

In order to answer the first question affirmatively, we would have to devise a strategy to determine $\varphi(t), t \geq 0$, from a knowledge of $y(t)$ and $f(t)$. In this way we could then use (2) to predict the displacement $\tilde{y}(t)$ arising from a different force, $\tilde{f}(t)$. To answer the second question affirmatively, we would have to be able to extract the constants m, k, and γ from a knowledge of $\varphi(t)$ for all $t \geq 0$. The "time domain approach," therefore, does not appear to be very simple or straightforward. You will see in Section 7.4, however, that the use of Laplace transforms provides a relatively easy way to answer both questions affirmatively. ▲

7.1 The Laplace Transform

The first use of Laplace transforms as an operational tool for solving constant coefficient linear differential equations is often attributed to the British physicist Oliver Heaviside.[2] As we mentioned in the Introduction, the Laplace transform maps a function of t, say $f(t)$, into a new function, $F(s)$, of a new "transform variable" s. [In terms of notation, we generally use lowercase letters to designate time domain functions, such as $f(t)$, and capital letters to denote corresponding transform domain functions, $F(s)$.]

The Definition of the Laplace Transform

Let $f(t)$ be a function defined on the interval $0 \le t < \infty$. The **Laplace transform of $f(t)$** is defined by the improper integral

$$\mathcal{L}\{f(t)\} = F(s) = \int_0^\infty f(t)e^{-st}\,dt. \tag{1}$$

As the notation of equation (1) indicates, we often denote $F(s)$, the Laplace transform of $f(t)$, by the symbol $\mathcal{L}\{f(t)\}$. The new transform variable s is assumed to be a real variable. (In more advanced treatments of the Laplace transform, the variable s is allowed to be a complex variable.)

Looking at equation (1), the first issue we must settle is that of identifying the properties $f(t)$ must possess in order for its Laplace transform to exist; that is, in order for the improper integral in equation (1) to converge. Recall from calculus that the improper integral in (1) is defined by

$$\int_0^\infty f(t)e^{-st}\,dt = \lim_{T\to\infty} \int_0^T f(t)e^{-st}\,dt, \tag{2}$$

provided the limit exists. When the limit (2) exists, we say that the improper integral **converges** and we define the improper integral to be this limit value. If the limit in (2) does not exist, we say that the improper integral **diverges**.

Whether or not limit (2) exists depends on the nature of $f(t)$ and on the value of the transform variable s; note that s plays the role of a parameter in the integral. In the present section we identify a large class of functions that possess Laplace transforms. It is important to realize, however, that not every function has a Laplace transform. The third function considered in Example 1 illustrates this fact.

EXAMPLE 1

Find the Laplace transform, if it exists, of

(a) $f(t) = e^{at}$ (b) $f(t) = t$ (c) $f(t) = e^{t^2}$

Solution:

(a) Applying the definition, we see that

$$\mathcal{L}\{e^{at}\} = \int_0^\infty e^{at}e^{-st}\,dt = \lim_{T\to\infty} \int_0^T e^{-(s-a)t}\,dt, \tag{3}$$

(continued)

[2]Oliver Heaviside (1850–1925) studied electricity and magnetism while employed as a telegrapher. He is remembered for his great simplification of Maxwell's equations, his contributions to vector analysis, and his development of operational calculus. In 1902, Heaviside conjectured that a conducting layer exists in the atmosphere which allows radio waves to follow the curvature of Earth. This layer, now called the Heaviside layer, was detected in 1923.

(continued)

provided the limit exists. Since

$$\int_0^T e^{-(s-a)t}\,dt = \begin{cases} T, & s = a \\ \dfrac{1 - e^{-(s-a)T}}{s-a}, & s \neq a, \end{cases}$$

the improper integral defined by the limit (3) exists if and only if $s > a$. Therefore,

$$F(s) = \mathcal{L}\{e^{at}\} = \frac{1}{s-a}, \qquad s > a. \tag{4}$$

(b) Similarly,

$$\mathcal{L}\{t\} = \int_0^\infty te^{-st}\,dt = \lim_{T\to\infty}\int_0^T te^{-st}\,dt, \tag{5}$$

provided the limit exists. Since

$$\int_0^T te^{-st}\,dt = \begin{cases} \dfrac{T^2}{2}, & s = 0 \\ -\dfrac{e^{-sT}}{s^2}(1 + sT) + \dfrac{1}{s^2}, & s \neq 0, \end{cases}$$

the improper integral defined by the limit (5) exists if and only if $s > 0$. Therefore,

$$F(s) = \mathcal{L}\{t\} = \frac{1}{s^2}, \qquad s > 0. \tag{6}$$

(c) Applying the definition,

$$\mathcal{L}\{e^{t^2}\} = \int_0^\infty e^{t^2}e^{-st}\,dt = \lim_{T\to\infty}\int_0^T e^{t(t-s)}\,dt,$$

provided the limit exists. For any fixed value of s, however, the integrand, $e^{t(t-s)}$, is greater than 1 whenever $t \geq s$. Therefore, the limit does not exist for any value s and the function $f(t) = e^{t^2}$ does not possess a Laplace transform. ▲

Piecewise Continuous Functions and Exponentially Bounded Functions

We now identify a class of functions that possess Laplace transforms. As you will see, if $f(t)$ is a member of this class, then its Laplace transform, $F(s)$, exists for all $s > a$, where a is a constant that depends upon the particular function f.

We begin with two definitions. A function $f(t)$ is called **piecewise continuous on the interval** $0 \leq t \leq T$ if

(i) There are at most finitely many points, $0 \le t_1 < t_2 < \cdots < t_{N_T} \le T$, at which $f(t)$ is not continuous, and

(ii) At any point of discontinuity, t_j, the one-sided limits

$$\lim_{t \to t_j^-} f(t) \quad \text{and} \quad \lim_{t \to t_j^+} f(t)$$

both exist. (If a discontinuity occurs at an endpoint, 0 or T, then only the interior one-sided limits need exist.) These discontinuities are called **jump discontinuities**.

Condition (ii) says the only discontinuities allowed for a piecewise continuous function are jump discontinuities. Condition (i) says a function that is piecewise continuous on the interval $[0, T]$ never has more than a finite number of jump discontinuities in $[0, T]$. Note that the number of discontinuity points is a nondecreasing function of the interval length T; that is, if $T_2 > T_1$, then $N_{T_2} \ge N_{T_1}$.

A function defined on $0 \le t < \infty$ is called **piecewise continuous on the interval** $0 \le t < \infty$ if it is piecewise continuous on $0 \le t \le T$ for all $T > 0$.

A typical example of a function $f(t)$ that is piecewise continuous on $0 \le t < \infty$ is the "square wave" shown in Figure 7.3, where

$$f(t) = \begin{cases} 1, & 0 \le t \le 1 \\ 0, & 1 < t < 2 \end{cases} \qquad f(t+2) = f(t).$$

Note that $f(t)$ is a periodic function with period 2. Every discontinuity of $f(t)$ is a jump discontinuity. While $f(t)$ has infinitely many discontinuities in $0 \le t < \infty$, the function never has more than a finite number of discontinuities in a finite interval $0 \le t \le T$.

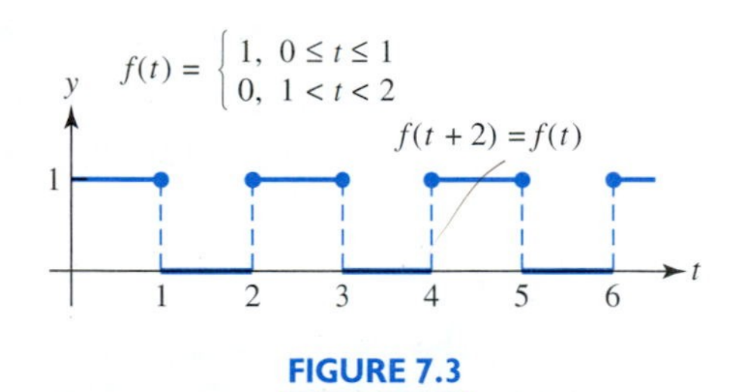

FIGURE 7.3

The function $f(t)$ is called a "square wave." While $f(t)$ has infinitely many discontinuities in $0 \le t < \infty$, they are all jump discontinuities and there are never more than a finite number in any finite subinterval of $0 \le t < \infty$. Therefore, $f(t)$ is piecewise continuous on $0 \le t < \infty$.

Our next definition is concerned with measuring how fast $|f(t)|$ grows as $t \to \infty$. A function $f(t)$ defined on $0 \le t < \infty$ is called **exponentially bounded** on $0 \le t < \infty$ if there are constants M and a, with $M \ge 0$, such that

$$|f(t)| \le Me^{at}, \qquad 0 \le t < \infty.$$

Figure 7.4 illustrates the nature of this definition. A function $f(t)$ is exponentially bounded if we can find constants M and a such that the graph of $|f(t)|$ is contained in the region R, where R is bounded above by $y = Me^{at}$ and below by $y = -Me^{at}$.

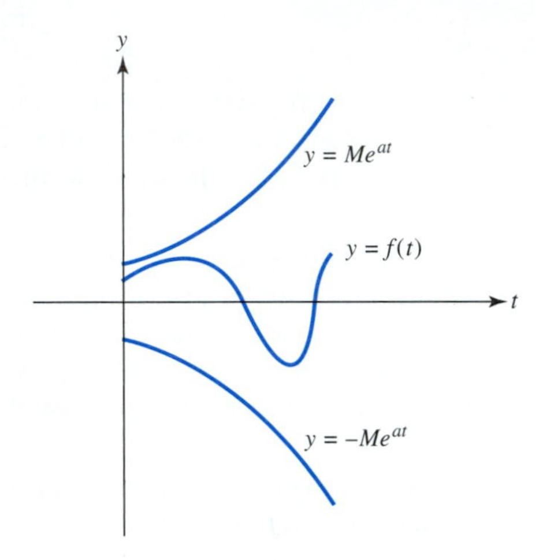

FIGURE 7.4

The function $f(t)$ is exponentially bounded because its graph is bounded above by $y = Me^{at}$ and below by $y = -Me^{at}$. (In the event that $a < 0, f(t) \to 0$ as $t \to \infty$.) Hence, $|f(t)| \le Me^{at}, 0 \le t < \infty$.

If a function $f(t)$ is bounded on $0 \le t < \infty$, then it is also exponentially bounded; that is, if $|f(t)| \le M, 0 \le t < \infty$, then $|f(t)| \le Me^{at}, 0 \le t < \infty$, with $a = 0$.

Existence of the Laplace Transform

Theorem 7.1 establishes the existence of the Laplace transform for all functions that are piecewise continuous and exponentially bounded on $0 \le t < \infty$. The proof is given in advanced texts.

Theorem 7.1

Let $f(t)$ be piecewise continuous and exponentially bounded on $0 \le t < \infty$, where $|f(t)| \le Me^{at}, 0 \le t < \infty$. Then the Laplace transform,

$$F(s) = \int_0^\infty f(t)e^{-st}\,dt,$$

exists for all $s > a$.

In this chapter, we restrict our attention to functions that are piecewise continuous and exponentially bounded on $0 \le t < \infty$. The next theorem, stated without proof, gives some closure properties for this special class of functions, asserting that we can form linear combinations and products of functions in the class and any new functions produced will also belong to the class (and thus have Laplace transforms).

Theorem 7.2

Let $f_1(t)$ and $f_2(t)$ be piecewise continuous and exponentially bounded on $0 \le t < \infty$, where

$$|f_1(t)| \le M_1 e^{a_1 t} \quad \text{and} \quad |f_2(t)| \le M_2 e^{a_2 t}.$$

(i) Let $f(t) = c_1 f_1(t) + c_2 f_2(t)$, where c_1 and c_2 are arbitrary constants. Then, $f(t)$ is also piecewise continuous and exponentially bounded on $0 \le t < \infty$. In fact, $|f(t)| \le Me^{at}$, where $M = |c_1|M_1 + |c_2|M_2$ and $a = \max\{a_1, a_2\}$. Moreover, $F(s) = \mathcal{L}\{f(t)\}$ is given by

$$\begin{aligned} F(s) &= \mathcal{L}\{c_1 f_1(t) + c_2 f_2(t)\} = c_1 \mathcal{L}\{f_1(t)\} + c_2 \mathcal{L}\{f_2(t)\} \\ &= c_1 F_1(s) + c_2 F_2(s), \qquad s > a. \end{aligned}$$

(ii) Let $g(t) = f_1(t) f_2(t)$. Then $g(t)$ is also piecewise continuous and exponentially bounded on $0 \le t < \infty$. In fact, $|g(t)| \le Me^{at}$ where $M = M_1 M_2$ and $a = a_1 + a_2$. It follows that $G(s) = \mathcal{L}\{g(t)\}$ exists for $s > a$.

Since the Laplace transform satisfies the formula

$$\mathcal{L}\{c_1 f_1(t) + c_2 f_2(t)\} = c_1 \mathcal{L}\{f_1(t)\} + c_2 \mathcal{L}\{f_2(t)\}, \tag{7}$$

we say that the Laplace transform is a **linear transformation** on the set of piecewise continuous, exponentially bounded functions.

EXAMPLE 2

Determine whether the functions are exponentially bounded and piecewise continuous on $0 \le t < \infty$:

(a) $f(t) = \begin{cases} 1, & 0 \le t \le 1, \\ 0, & 1 < t < 2, \end{cases} \qquad f(t) = f(t-2) \quad \text{for} \quad t \ge 2,$

(b) $g(t) = te^t, \qquad 0 \le t < \infty,$

(c) $k(t) = e^{t^2}, \qquad 0 \le t < \infty.$

Solution:

(a) The function $f(t)$ is defined on $0 \le t < \infty$ as a periodic function having period 2. This function, whose graph was shown earlier in Figure 7.3, has jump discontinuities at the positive integers. As was noted in Figure 7.3, $f(t)$ is piecewise continuous on $0 \le t < \infty$. Since the function is bounded it is also exponentially bounded; in fact, we have

$$|f(t)| \le Me^{at} \quad \text{with} \quad M = 1 \quad \text{and} \quad a = 0.$$

We will calculate the Laplace transform of $f(t)$ in Section 7.4 when we discuss Laplace transforms of periodic functions.

(continued)

(continued)

(b) Since $g(t)$ is continuous, it is certainly piecewise continuous on $0 \le t < \infty$. It remains to show that $g(t)$ is exponentially bounded. Let $\alpha > 1$ and consider the function

$$\varphi(t) = g(t)e^{-\alpha t} = te^{-(\alpha-1)t}, \qquad 0 \le t < \infty.$$

It can be shown that

$$0 \le \varphi(t) \le \frac{1}{(\alpha - 1)e}, \qquad 0 \le t < \infty.$$

This inequality implies that

$$0 \le g(t) \le \frac{1}{(\alpha - 1)e} e^{\alpha t}, \qquad 0 \le t < \infty,$$

and we conclude that $g(t)$ is exponentially bounded on $0 \le t < \infty$, with $M = 1/[(\alpha - 1)e]$ and $a = \alpha > 1$.

(c) If $k(t)$ were exponentially bounded, then there would be constants M and a such that $e^{t^2} \le Me^{at}$ for all nonnegative values t. In turn, this inequality would imply

$$e^{t(t-a)} \le M, \qquad 0 \le t < \infty.$$

But, as t grows, the inequality has to fail eventually. Thus, $k(t) = e^{t^2}$ is not exponentially bounded. The function is, however, piecewise continuous since it is continuous. ▲

The Inverse Laplace Transform and Uniqueness

Using the Laplace transform to solve problems involves three separate steps: (1) apply the transform to obtain a new "transform domain problem," (2) solve the new transform domain problem, and (3) apply the inverse transform that maps the transform domain solution back to the time domain, resulting in the solution of the problem of interest.

In order to define the inverse mapping (that is, the inverse Laplace transform), we need to know that the Laplace transform operation, when applied to functions that are piecewise continuous and exponentially bounded, possesses an underlying uniqueness property. In particular, given a transform domain function $F(s)$, we want unambiguously to identify a function $f(t)$ that has $F(s)$ as its transform. The following theorem, which we present without proof, addresses the uniqueness question.

Theorem 7.3

Let $f_1(t)$ and $f_2(t)$ be piecewise continuous and exponentially bounded on $0 \le t < \infty$. Let $F_1(s)$ and $F_2(s)$ represent their respective Laplace transforms. Suppose, for some constant a, that

$$F_1(s) = F_2(s), \qquad s > a.$$

Then $f_1(t) = f_2(t)$ at all points $t \ge 0$ where both functions are continuous.

This theorem gives about the best result we can hope for. As an illustration, consider the function $f_1(t) = e^{-t}, t \geq 0$. We saw in Example 1 that

$$\mathcal{L}\{e^{at}\} = \frac{1}{s-a}, \qquad s > a.$$

Therefore,

$$\mathcal{L}\{e^{-t}\} = F(s) = \frac{1}{s+1}, \qquad s > -1.$$

Suppose we create a new function $f_2(t)$ by simply redefining $f_1(t)$ to be zero at each of the positive integers:

$$f_2(t) = \begin{cases} e^{-t}, & t \text{ not an integer} \\ 0, & t \text{ an integer.} \end{cases} \tag{8}$$

The graph of the function $f_2(t)$ is shown in Figure 7.5. Observe, even though $f_1(t)$ and $f_2(t)$ are different functions, that

$$\int_0^T f_1(t)e^{-st}\,dt = \int_0^T f_2(t)e^{-st}\,dt \tag{9}$$

for every $T > 0$. Therefore, $\mathcal{L}\{f_1(t)\} = \mathcal{L}\{f_2(t)\} = F(s), s > -1$.

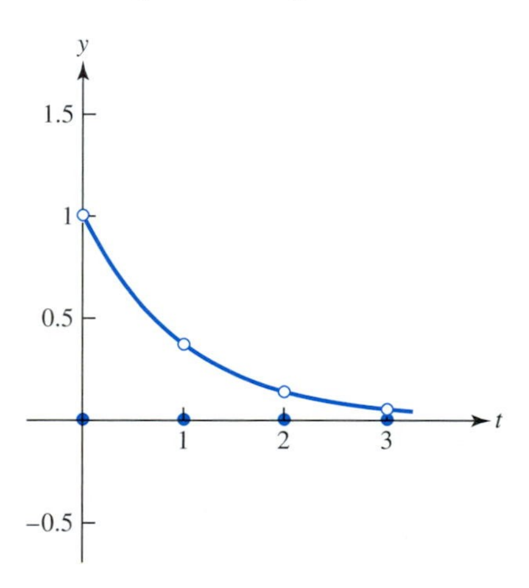

FIGURE 7.5

The graph of the function $f_2(t)$ defined by equation (8). Note that the graph of $f_2(t)$ is identical to the graph of $f_1(t) = e^{-t}$ except at $t = 0, 1, 2, \ldots$. Even though the functions $f_1(t)$ and $f_2(t)$ are different, their Laplace transforms are the same (see equation (9)).

As we see from equation (9), the improper integral defining the Laplace transform is insensitive to changes in the value of a function at a finite number of points in $0 \leq t \leq T$. This insensitivity, however, does not pose a serious practical problem since we are interested in physically relevant functions. For example, in defining the inverse Laplace transform of

$$F(s) = \frac{1}{s-a},$$

we will choose it to be the continuous function $f(t) = e^{at}, t \geq 0$.

Our approach to determining inverse Laplace transforms will be a tabular one. In the next several sections we will compute Laplace transforms of functions and build up a library of **Laplace transform pairs**, such as the pair

$$f(t) = e^{at}, \qquad t \geq 0 \quad \text{and} \quad F(s) = \frac{1}{s-a}, \qquad s > a. \tag{10}$$

Determining an inverse Laplace transform will essentially consist of a simple "table look-up" process. That is, we find the appropriate transform domain function $F(s)$ in the table and then take the corresponding time domain function $f(t)$ in the table to be the inverse transform of $F(s)$. The next example illustrates this approach.

[Note: We will use the symbol $\mathcal{L}^{-1}\{\ \}$ to denote the operation of taking the inverse Laplace transform.]

EXAMPLE 3

What is the inverse Laplace transform of

$$F(s) = \frac{2s}{s^2-1}, \qquad s > 1?$$

That is, for what function $f(t)$, do we have $\mathcal{L}\{f(t)\} = F(s) = 2s/(s^2-1)$?

Solution: We first observe that the rational function $F(s)$ has the following partial fraction expansion

$$\frac{2s}{s^2-1} = \frac{1}{s-1} + \frac{1}{s+1}.$$

(The topic of partial fractions is reviewed in Section 7.3. For now you can verify this claim by simply recombining the right-hand side.) Since the Laplace transform is a linear transformation, the inverse Laplace transform is likewise a linear transformation. In particular,

$$\mathcal{L}^{-1}\left\{\frac{2s}{s^2-1}\right\} = \mathcal{L}^{-1}\left\{\frac{1}{s-1} + \frac{1}{s+1}\right\} = \mathcal{L}^{-1}\left\{\frac{1}{s-1}\right\} + \mathcal{L}^{-1}\left\{\frac{1}{s+1}\right\}.$$

Recalling the Laplace transform pair listed earlier in equation (10), we obtain

$$\mathcal{L}^{-1}\left\{\frac{1}{s-1}\right\} = e^{t} \quad \text{and} \quad \mathcal{L}^{-1}\left\{\frac{1}{s+1}\right\} = e^{-t}.$$

Therefore,

$$\mathcal{L}^{-1}\left\{\frac{2s}{s^2-1}\right\} = e^{t} + e^{-t} = 2\cosh t, \qquad t \geq 0. \ \blacktriangle$$

EXERCISES

Exercises 1–8:

As in Example 1, use the definition to find the Laplace transform for $f(t)$, if it exists. In each exercise, the given function $f(t)$ is defined on the interval $0 \leq t < \infty$. If the Laplace transform exists, give the domain of $F(s)$. In Exercises 9–12, also sketch the graph of $f(t)$.

1. $f(t) = 1$ **2.** $f(t) = e^{3t}$ **3.** $f(t) = te^{-t}$ **4.** $f(t) = t - 5$

5. $f(t) = te^{t\sqrt{t}}$ **6.** $f(t) = e^{(t-1)^2}$ **7.** $f(t) = |t - 1|$ **8.** $f(t) = (t - 2)^2$

9. $f(t) = \begin{cases} 0, & 0 \le t < 1 \\ 1, & 1 \le t \end{cases}$ **10.** $f(t) = \begin{cases} 0, & 0 \le t < 1 \\ t-1, & 1 \le t \end{cases}$

11. $f(t) = \begin{cases} 0, & 0 \le t < 1 \\ 1, & 1 \le t < 2 \\ 0, & 2 \le t \end{cases}$ **12.** $f(t) = \begin{cases} 0, & 0 \le t < 1 \\ t-1, & 1 \le t < 2 \\ 0, & 2 \le t \end{cases}$

13. Let n be a positive integer. Using integration by parts, establish the reduction formula

$$\int t^n e^{-st}\,dt = -\frac{t^n e^{-st}}{s} + \frac{n}{s}\int t^{n-1}e^{-st}\,dt, \qquad s > 0.$$

14. For $s > 0$ and n a positive integer, evaluate the limits

$$\text{(a)} \lim_{t\to 0} t^n e^{-st} \quad \text{and} \quad \text{(b)} \lim_{t\to\infty} t^n e^{-st}.$$

[Hint: Use L'Hôpital's rule to establish limit (b).]

15. (a) Use Exercises 13 and 14 to derive a reduction formula for the Laplace transform of $f(t) = t^n$,

$$\mathcal{L}\{t^n\} = \frac{n}{s}\mathcal{L}\{t^{n-1}\}, \qquad s > 0. \tag{11}$$

(b) From Example 1, we have $\mathcal{L}\{t\} = 1/s^2$, $s > 0$. Use this fact, together with reduction formula (11), to calculate $\mathcal{L}\{t^k\}$, for $k = 2, 3, \ldots, 5$.

(c) Formulate a conjecture as to the Laplace transform of $f(t) = t^m$, where m is an arbitrary positive integer.

Exercises 16–21:

From a table of integrals,

$$\int e^{\alpha u}\sin\beta u\,du = e^{\alpha u}\frac{\alpha\sin\beta u - \beta\cos\beta u}{\alpha^2+\beta^2}$$

$$\int e^{\alpha u}\cos\beta u\,du = e^{\alpha u}\frac{\alpha\cos\beta u + \beta\sin\beta u}{\alpha^2+\beta^2}.$$

Use these integrals to find the Laplace transform of $f(t)$, if it exists. If the Laplace transform exists, give the domain of $F(s)$.

16. $f(t) = \cos\omega t$ **17.** $f(t) = \sin\omega t$ **18.** $f(t) = \cos[\omega(t-2)]$

19. $f(t) = \sin[\omega(t-2)]$ **20.** $f(t) = e^{3t}\sin t$ **21.** $f(t) = e^{-2t}\cos 4t$

Exercises 22–23:

Use the linearity property (7) along with the transforms found in Example 1,

$$\mathcal{L}\{e^{at}\} = \frac{1}{s-a}, \qquad s > a \quad \text{and} \quad \mathcal{L}\{t\} = \frac{1}{s^2}, \qquad s > 0,$$

to calculate the Laplace transform $R(s) = \mathcal{L}\{r(t)\}$ of the given function $r(t)$. For what values s does the Laplace transform exist?

22. $r(t) = 2e^{-5t} + 6t$ **23.** $r(t) = 5e^{-7t} + t + 2e^{2t}$

Exercises 24–31:

In each exercise, a function $f(t)$ is given. In Exercises 28 and 29, the symbol $[\![u]\!]$ denotes the **greatest integer function**, $[\![u]\!] = n$ when $n \le u < n+1$, n an integer, $n = \ldots, -2, -1, 0, 1, 2, \ldots$.

(a) Is $f(t)$ continuous on $0 \le t < \infty$, discontinuous but piecewise continuous on $0 \le t < \infty$, or neither?

(b) Is $f(t)$ exponentially bounded on $0 \le t < \infty$? If so, determine values of M and a such that $|f(t)| \le Me^{at}$, $0 \le t < \infty$.

24. $f(t) = \tan t$ **25.** $f(t) = e^t \sin t$ **26.** $f(t) = t^2 e^{-t}$ **27.** $f(t) = \cosh 2t$

28. $f(t) = [\![t]\!]$ **29.** $f(t) = [\![e^{2t}]\!]$ **30.** $f(t) = \dfrac{e^{t^2}}{e^{2t}+1}$ **31.** $f(t) = \dfrac{1}{t}$

Exercises 32–35:

Determine whether the given improper integral converges. If the integral converges, give its value.

32. $\displaystyle\int_0^\infty \frac{1}{1+t^2}\,dt$ **33.** $\displaystyle\int_0^\infty \frac{t}{1+t^2}\,dt$

34. $\displaystyle\int_0^\infty e^{-t}\cos(e^{-t})\,dt$ **35.** $\displaystyle\int_0^\infty te^{-t^2}\,dt$

Exercises 36–39:

Suppose that $\mathcal{L}\{f_1(t)\} = F_1(s)$ and $\mathcal{L}\{f_2(t)\} = F_2(s)$, $s > a$. Use the fact that

$$\mathcal{L}^{-1}\{c_1F_1(s) + c_2F_2(s)\} = c_1\mathcal{L}^{-1}\{F_1(s)\} + c_2\mathcal{L}^{-1}\{F_2(s)\}, \qquad a < s$$

to determine the inverse Laplace transform of the given function. Refer to the examples in this section and equation (11) in Exercise 15.

36. $F(s) = \dfrac{3}{s-2}$ **37.** $F(s) = -\dfrac{2}{s^2} + \dfrac{1}{s+1}$

38. $F(s) = \dfrac{4s}{s^2-4} = \dfrac{2}{s+2} + \dfrac{2}{s-2}$ **39.** $F(s) = \dfrac{2}{s^2-1} = \dfrac{1}{s-1} - \dfrac{1}{s+1}$

7.2 Laplace Transform Pairs

This section develops a library of Laplace transform pairs that we will use to solve problems. We begin by defining a function known as the "unit step function" or the "Heaviside step function."

The Unit Step Function

The **unit step function** or **Heaviside step function**, $h(t)$, is the piecewise continuous function defined by

$$h(t) = \begin{cases} 1, & t \ge 0 \\ 0, & t < 0. \end{cases}$$

Figure 7.6 displays graphs of $h(t)$ and its "shifted argument" counterpart, $h(t-\alpha)$, $\alpha > 0$.

The Laplace transform of the unit step function $h(t)$ is given by

$$\mathcal{L}\{h(t)\} = \int_0^\infty h(t)e^{-st}\,dt = \int_0^\infty e^{-st}\,dt = -\frac{e^{-st}}{s}\bigg|_0^\infty = \frac{1}{s}, \qquad s > 0. \qquad \textbf{(1a)}$$

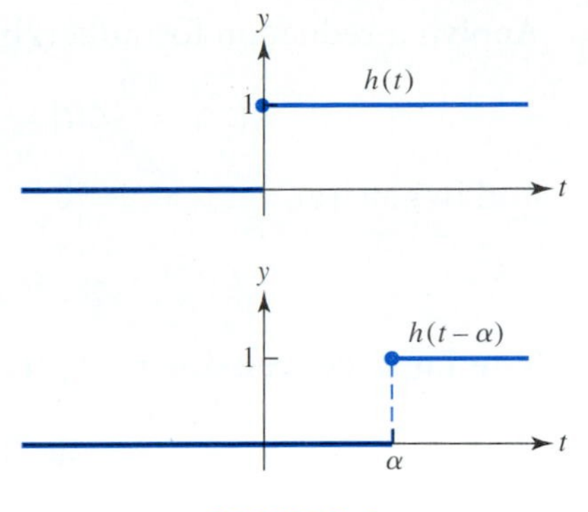

FIGURE 7.6

The graphs of the unit step function, $h(t)$, and the shifted step function, $h(t-\alpha)$.

In equation (1a), we use a common notation:

$$f(t)\Big|_a^\infty = \lim_{t\to\infty} f(t) - f(a).$$

For the shifted step function, $h(t-\alpha)$, we obtain the Laplace transform

$$\mathcal{L}\{h(t-\alpha)\} = \int_0^\infty h(t-\alpha)e^{-st}\,dt = \int_\alpha^\infty e^{-st}\,dt = -\frac{e^{-st}}{s}\Big|_\alpha^\infty = \frac{e^{-s\alpha}}{s}, \qquad \textbf{(1b)}$$

$$s > 0 \quad \text{and} \quad \alpha \ge 0.$$

Note that the unit step function $h(t)$ and the constant function $f(t) = 1$ are identical on $0 \le t < \infty$. Therefore, they have the same Laplace transform,

$$\mathcal{L}\{1\} = \frac{1}{s}, \qquad s > 0. \qquad \textbf{(1c)}$$

Transforms of Polynomial, Exponential, and Trigonometric Functions

In this subsection we develop some common transform pairs, starting with the polynomial function, $f(t) = t^n$. Also see Exercise 15 in Section 7.1.

The Laplace transform of $f(t) = t^n$: For $n = 1$, we use integration by parts to obtain

$$\mathcal{L}\{t\} = \int_0^\infty te^{-st}\,dt = \left[-\frac{te^{-st}}{s} - \frac{e^{-st}}{s^2}\right]\Bigg|_0^\infty = \frac{1}{s^2}, \qquad s > 0. \qquad \textbf{(2a)}$$

In general, for any positive integer n, integration by parts yields

$$\mathcal{L}\{t^n\} = \int_0^\infty t^n e^{-st}\,dt = -\frac{t^n e^{-st}}{s}\Big|_0^\infty + \frac{n}{s}\int_0^\infty t^{n-1}e^{-st}\,dt.$$

You can use L'Hôpital's rule to show that $\lim_{t\to\infty} t^n e^{-st} = 0, s > 0$. Therefore, we obtain the following reduction formula for $\mathcal{L}\{t^n\}$:

$$\mathcal{L}\{t^n\} = \frac{n}{s}\,\mathcal{L}\{t^{n-1}\}, \qquad s > 0. \qquad \textbf{(2b)}$$

Applying reduction formula (2b) recursively, we find for $s > 0$,

$$\mathcal{L}\{t^2\} = \frac{2}{s}\mathcal{L}\{t\} = \frac{2}{s^3}, \qquad \mathcal{L}\{t^3\} = \frac{3}{s}\mathcal{L}\{t^2\} = \frac{3\cdot 2}{s^4},$$

and in general,

$$\mathcal{L}\{t^n\} = \frac{n!}{s^{n+1}}, \qquad n = 1, 2, 3, \ldots, \qquad s > 0. \tag{3}$$

The Laplace transform of $f(t) = e^{\alpha t}$: We saw in Section 7.1 that

$$\mathcal{L}\{e^{\alpha t}\} = \frac{1}{s-\alpha}, \qquad s > \alpha. \tag{4}$$

The Laplace transforms of $f(t) = \sin\omega t$ and $f(t) = \cos\omega t$: Using integration by parts twice,

$$\begin{aligned}\mathcal{L}\{\sin\omega t\} &= \int_0^\infty e^{-st}\sin\omega t\,dt \\ &= \left[-\frac{e^{-st}\sin\omega t}{s} - \frac{\omega e^{-st}\cos\omega t}{s^2}\right]\Bigg|_0^\infty - \frac{\omega^2}{s^2}\int_0^\infty e^{-st}\sin\omega t\,dt \\ &= \frac{\omega}{s^2} - \frac{\omega^2}{s^2}\mathcal{L}\{\sin\omega t\}, \qquad s > 0.\end{aligned}$$

Solving for $\mathcal{L}\{\sin\omega t\}$, we find

$$\mathcal{L}\{\sin\omega t\} = \frac{\omega}{s^2+\omega^2}, \qquad s > 0. \tag{5a}$$

Similarly, you can show that

$$\mathcal{L}\{\cos\omega t\} = \frac{s}{s^2+\omega^2}, \qquad s > 0. \tag{5b}$$

We know from Section 7.1 that the Laplace transform defines a linear transformation on the set of piecewise continuous and exponentially bounded functions; that is, if $f(t)$ and $g(t)$ are piecewise continuous and exponentially bounded, then

$$\mathcal{L}\{c_1 f(t) + c_2 g(t)\} = c_1\mathcal{L}\{f(t)\} + c_2\mathcal{L}\{g(t)\}.$$

We can use this linearity property to extend our library of transforms. For example, combining linearity with the transform for $\mathcal{L}\{t^n\}$ listed in equation (3), we obtain the Laplace transform of any polynomial. The next example illustrates this point.

EXAMPLE 1

Use the transform pairs developed above to find the Laplace transform of

(a) $p(t) = 2t^3 + 5t - 3, \quad t \ge 0$ (b) $f(t) = 4\cos^2 3t, \quad t \ge 0$

Solution:

(a) Using linearity and equation (3),

$$\begin{aligned}\mathcal{L}\{p(t)\} &= \mathcal{L}\{2t^3 + 5t - 3\} = 2\mathcal{L}\{t^3\} + 5\mathcal{L}\{t\} - 3\mathcal{L}\{1\} \\ &= 2\frac{3!}{s^4} + 5\frac{1}{s^2} - 3\frac{1}{s} = \frac{12 + 5s^2 - 3s^3}{s^4}, \qquad s > 0.\end{aligned}$$

(b) We have no transform pair directly involving $\cos^2 \omega t$. However, we can use a trigonometric identity to rewrite $f(t) = 4\cos^2 3t$ as $f(t) = 2 + 2\cos 6t$. Using linearity and equation (5b),

$$\begin{aligned}\mathcal{L}\{f(t)\} &= 2\mathcal{L}\{1\} + 2\mathcal{L}\{\cos 6t\} \\ &= 2\frac{1}{s} + 2\frac{s}{s^2+36} = \frac{4s^2+72}{s(s^2+36)}, \qquad s > 0.\end{aligned}$$ ▲

Two Shift Theorems

The next two results, established in Theorem 7.4, are often referred to as the **first and second shift theorems**. As with the linearity property just illustrated in Example 1, the shift theorems increase the number of functions for which we can easily find Laplace transforms.

Theorem 7.4

Let $f(t)$ be piecewise continuous and exponentially bounded on $0 \le t < \infty$ where $|f(t)| \le Me^{at}$, $0 \le t < \infty$. Let $F(s) = \mathcal{L}\{f(t)\}$ and let $h(t)$ denote the unit step function. Then

(a) $\mathcal{L}\{e^{\alpha t} f(t)\} = F(s - \alpha), \qquad s > a + \alpha$

(b) $\mathcal{L}\{f(t-\alpha)h(t-\alpha)\} = e^{-\alpha s}F(s), \qquad \alpha \ge 0, \qquad s > a.$

Since $h(t - \alpha) = 0$ when $t < \alpha$ (see Figure 7.6),

$$f(t-\alpha)h(t-\alpha) = 0, \qquad t < \alpha.$$

The graph of $f(t-\alpha)h(t-\alpha)$ looks just like the graph of $f(t)$, except for the fact that it is shifted to the right and remains zero until $t = \alpha$. Figure 7.7 provides an example.

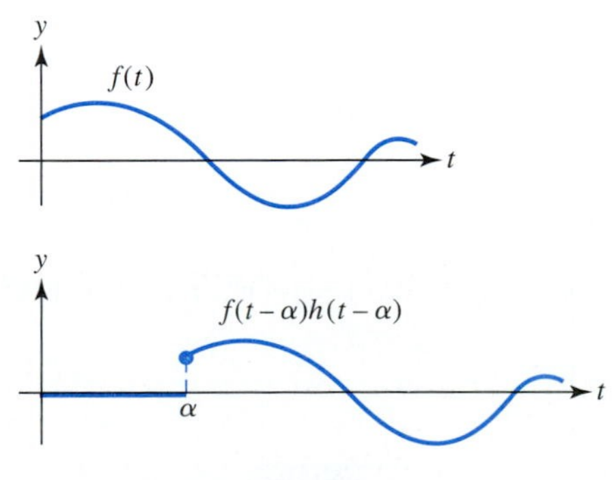

FIGURE 7.7

The graphs of $f(t)$ and $f(t-\alpha)h(t-\alpha)$. Note that the graph of $f(t-\alpha)h(t-\alpha)$ looks like the graph of $f(t)$, except that it is shifted α units to the right.

PROOF OF THEOREM 7.4:

(a) The following calculation establishes part (a),

$$\mathcal{L}\{e^{\alpha t}f(t)\} = \int_0^\infty e^{\alpha t}f(t)e^{-st}\,dt = \int_0^\infty f(t)e^{-(s-\alpha)t}\,dt = F(s-\alpha),$$
$$s > a + \alpha.$$

(b) To establish the second shift theorem, we begin with

$$\mathcal{L}\{f(t-\alpha)h(t-\alpha)\} = \int_0^\infty f(t-\alpha)h(t-\alpha)e^{-st}\,dt = \int_\alpha^\infty f(t-\alpha)e^{-st}\,dt.$$

Making the change of variable $\tau = t - \alpha$, we have

$$\mathcal{L}\{f(t-\alpha)h(t-\alpha)\} = \int_\alpha^\infty f(t-\alpha)e^{-st}\,dt = \int_0^\infty f(\tau)e^{-s(\tau+\alpha)}\,d\tau$$
$$= e^{-s\alpha}\int_0^\infty f(\tau)e^{-s\tau}\,d\tau = e^{-s\alpha}F(s), \qquad s > a.$$

Note that parts (a) and (b) of Theorem 7.4 possess a certain duality. Roughly speaking, multiplying a function by an exponential function in the time domain shifts the argument of its Laplace transform. Likewise, shifting the argument in the time domain leads to an exponential multiplicative factor in the transform domain.

EXAMPLE 2

Find

(a) $\mathcal{L}\{e^{2t}t^4\}$ (b) $\mathcal{L}\{e^{\alpha t}\cos\omega t\}$ (c) $\mathcal{L}^{-1}\left\{\dfrac{e^{-5s}}{s^2}\right\}$ (d) $\mathcal{L}^{-1}\left\{\dfrac{e^{-\alpha s}}{s^2+1}\right\}$

Solution:

(a) By Theorem 7.4, multiplying $f(t)$ by $e^{\alpha t}$ shifts the argument of its transform, $F(s)$. That is,

$$\mathcal{L}\{e^{2t}t^4\} = \mathcal{L}\{t^4\}\Big|_{s\to s-2} = \frac{4!}{(s-2)^5}, \qquad s > 2.$$

(b) As in part (a),

$$\mathcal{L}\{e^{\alpha t}\cos\omega t\} = \mathcal{L}\{\cos\omega t\}|_{s\to s-\alpha} = \frac{s-\alpha}{(s-\alpha)^2+\omega^2}, \qquad s > \alpha$$

(c) We know that $\mathcal{L}\{t\} = 1/s^2, t \ge 0$. Thus, by the second shift theorem,

$$\frac{e^{-5s}}{s^2} = \mathcal{L}\{(t-5)h(t-5)\}.$$

Therefore,

$$\mathcal{L}^{-1}\left\{\frac{e^{-5s}}{s^2}\right\} = (t-5)h(t-5) = \begin{cases} 0, & 0 \le t < 5 \\ t-5, & 5 \le t < \infty. \end{cases}$$

The graph of this inverse transform is shown in Figure 7.8.

(d) We know that $\mathcal{L}\{\sin t\} = 1/(s^2+1), t \geq 0$. Using the second shift theorem,

$$\frac{e^{-\alpha s}}{s^2+1} = \mathcal{L}\{[\sin(t-\alpha)]h(t-\alpha)\}.$$

Therefore,

$$\mathcal{L}^{-1}\left\{\frac{e^{-\alpha s}}{s^2+1}\right\} = [\sin(t-\alpha)]h(t-\alpha) = \begin{cases} 0, & 0 \leq t < \alpha \\ \sin(t-\alpha), & \alpha \leq t < \infty. \end{cases}$$ ▲

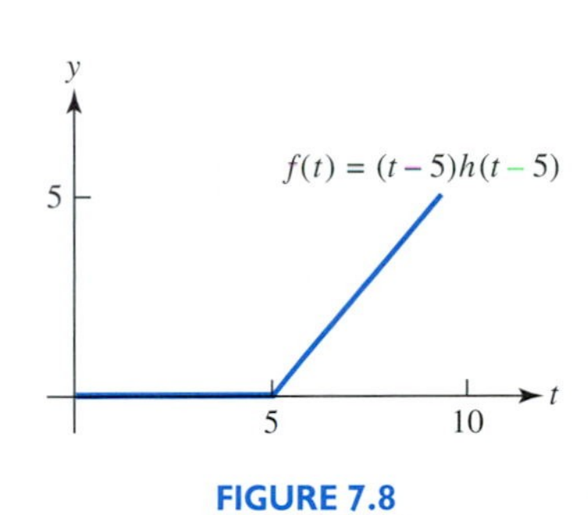

FIGURE 7.8

The graph of $f(t) = (t-5)h(t-5)$.

The Laplace Transform of Derivatives and Antiderivatives

The utility of Laplace transforms as a tool to solve problems involving constant coefficient linear differential equations is due in large part to the transform pairs in the next theorem, Theorem 7.5, that relate the Laplace transform of derivatives and integrals of a function to the Laplace transform of the function itself. We will make extensive use of these results in Sections 7.4–7.7.

Theorem 7.5

(a) Let $f(t)$ be continuous on $0 \leq t < \infty$, and let $f'(t)$ exist as a piecewise continuous, exponentially bounded function on $0 \leq t < \infty$, where $|f'(t)| \leq Me^{at}, 0 \leq t < \infty$. Then

$$\mathcal{L}\{f'(t)\} = s\mathcal{L}\{f(t)\} - f(0) = sF(s) - f(0), \qquad s > \max\{a, 0\}.$$

(b) Let $f'(t)$ be continuous on $0 \leq t < \infty$, and let $f''(t)$ exist as a piecewise continuous, exponentially bounded function on $0 \leq t < \infty$, where $|f''(t)| \leq Me^{at}, 0 \leq t < \infty$. Then

$$\begin{aligned}\mathcal{L}\{f''(t)\} &= s\mathcal{L}\{f'(t)\} - f'(0) = s\,[sF(s) - f(0)] - f'(0)\\ &= s^2F(s) - sf(0) - f'(0), \qquad s > \max\{a, 0\}.\end{aligned}$$

(c) Let $f(t)$ be piecewise continuous and exponentially bounded on $0 \leq t < \infty$, where $|f(t)| \leq Me^{at}, 0 \leq t < \infty$. Then

$$\mathcal{L}\left\{\int_0^t f(u)\,du\right\} = \frac{\mathcal{L}\{f(t)\}}{s} = \frac{F(s)}{s}, \qquad s > \max\{a, 0\}.$$

PROOF: The proof of part (a) is presented to illustrate the relevant ideas. By hypothesis, the function $f(t)$ is continuous. We now show it is also exponentially bounded and thus $f(t)$ has a Laplace transform. We have

$$|f(t)| = \left| f(0) + \int_0^t f'(u)\,du \right| \le |f(0)| + \left| \int_0^t f'(u)\,du \right| \le |f(0)| + \int_0^t Me^{au}\,du, \qquad t \ge 0.$$

Therefore, for $0 \le t < \infty$,

$$|f(t)| \le \begin{cases} |f(0)| + \dfrac{M}{a}(e^{at} - 1) \le \left[|f(0)| + \dfrac{M}{a} \right] e^{at}, & a > 0 \\ |f(0)| + Mt, \quad a = 0 \\ |f(0)| + \dfrac{M}{|a|}(1 - e^{at}) \le |f(0)| + \dfrac{M}{|a|}, & a < 0. \end{cases} \tag{6}$$

From these inequalities, we are able to conclude that $\mathcal{L}\{f(t)\} = F(s)$ exists for $s > \max\{a, 0\}$.

To obtain (a), consider the interval $0 \le t \le T$ for some arbitrary $T > 0$. Let $t_1 < t_2 < \cdots < t_N$ represent the points of discontinuity of $f'(t)$ on this interval. Then

$$\int_0^T f'(t)e^{-st}\,dt = \int_0^{t_1} f'(t)e^{-st}\,dt + \int_{t_1}^{t_2} f'(t)e^{-st}\,dt + \cdots + \int_{t_{N-1}}^{t_N} f'(t)e^{-st}\,dt + \int_{t_N}^{T} f'(t)e^{-st}\,dt.$$

Performing integration by parts on each of these integrals yields

$$\int_0^T f'(t)e^{-st}\,dt = f(t)e^{-st}\Big|_0^{t_1} + f(t)e^{-st}\Big|_{t_1}^{t_2} + \cdots + f(t)e^{-st}\Big|_{t_{N-1}}^{t_N} + f(t)e^{-st}\Big|_{t_N}^{T} + s\left[\int_0^{t_1} f(t)e^{-st}\,dt + \int_{t_1}^{t_2} f(t)e^{-st}\,dt + \cdots + \int_{t_{N-1}}^{t_N} f(t)e^{-st}\,dt + \int_{t_N}^{T} f(t)e^{-st}\,dt \right].$$

Since $f(t)$ is continuous, the sum of the endpoint evaluations reduces to $f(T)e^{-sT} - f(0)$. Similarly, the sum of integrals on the right-hand side can be expressed as a single integral from 0 to T. Therefore, we obtain

$$\int_0^T f'(t)e^{-st}\,dt = f(T)e^{-sT} - f(0) + s\int_0^T f(t)e^{-st}\,dt.$$

Now let $T \to \infty$, while assuming that $s > \max\{a, 0\}$. For these values of s, $\lim_{T\to\infty} f(T)e^{-sT} = 0$ and $\lim_{T\to\infty} \int_0^T f(t)e^{-st}\,dt = \mathcal{L}\{f(t)\} = F(s)$. Therefore, the result follows.

Note that differentiation in the time domain corresponds, roughly speaking, to multiplication by s in the transform domain, while antidifferentiation in the time domain corresponds to division by s in the transform domain.

Solving Initial Value Problems

Consider the second order linear initial value problem

$$y'' + ay' + by = g(t), \qquad y(0) = y_0, \qquad y'(0) = y_0',$$

where a and b are constants and where $g(t)$ is piecewise continuous and exponentially bounded on the interval $0 \le t < \infty$. Let $y(t)$ denote the solution of this problem, and let $Y(s)$ denote the Laplace transform of $y(t)$. We can use linearity of the Laplace transform to obtain an algebraic equation for $Y(s)$. Then, if we can find the inverse transform of $Y(s)$, we will have the solution, $y(t)$.

In particular, let $y(t)$ be the solution of the initial value problem and assume that $y(t), y'(t), y''(t)$, and $g(t)$ have Laplace transforms that exist on some common domain. Applying the Laplace transform to both sides of the equation $y''(t) + ay'(t) + by(t) = g(t)$, we obtain

$$\mathcal{L}\{y''(t)\} + a\mathcal{L}\{y'(t)\} + b\mathcal{L}\{y(t)\} = \mathcal{L}\{g(t)\}, \qquad 0 \le t < \infty.$$

Let $G(s)$ denote the Laplace transform of $g(t)$ and $Y(s)$ the Laplace transform of $y(t)$. Using parts (a) and (b) of Theorem 7.5, we have

$$[s^2Y(s) - sy(0) - y'(0)] + a[sY(s) - y(0)] + bY(s) = G(s).$$

Grouping terms,

$$(s^2 + as + b)Y(s) - (s + a)y(0) - y'(0) = G(s).$$

Solving for $Y(s)$, we obtain

$$Y(s) = \frac{G(s)}{(s^2 + as + b)} + \frac{(s + a)y_0}{(s^2 + as + b)} + \frac{y_0'}{(s^2 + as + b)}.$$

If we can find the inverse transform of $Y(s)$, we will have the solution of the initial value problem.

The next example, while quite simple, illustrates the process. Following a review of the method of partial fractions in Section 7.3, we give a more detailed discussion of using Laplace transforms to solve initial value problems.

EXAMPLE 3

Consider the initial value problem

$$y' - 3y = g(t), \qquad y(0) = 1,$$

where $g(t)$ is the step function given by

$$g(t) = \begin{cases} 0, & 0 \le t < 2 \\ 6, & 2 \le t < \infty. \end{cases}$$

Let $Y(s)$ denote the Laplace transform of $y(t)$, where $y(t)$ is the unique solution of this initial value problem. Using Theorem 7.5, derive an equation for $Y(s)$ and, taking the inverse Laplace transform, find $y(t)$.

Solution: The nonhomogeneous term $g(t)$ can be represented as $g(t) = 6h(t - 2)$, where $h(t - 2)$ denotes the shifted Heaviside step function. (See Figure 7.6.) Taking Laplace transforms of

$$y'(t) - 3y(t) = 6h(t - 2), \qquad 0 \le t < \infty,$$

we have

$$\mathcal{L}\{y'(t)\} - 3\mathcal{L}\{y(t)\} = 6\mathcal{L}\{h(t - 2)\}.$$

(continued)

(continued)

Using part (a) of Theorem 7.5 to evaluate $\mathcal{L}\{y'(t)\}$ and part (b) of Theorem 7.4 to evaluate $\mathcal{L}\{h(t-2)\}$, we find

$$[sY(s) - y(0)] - 3Y(s) = \frac{6e^{-2s}}{s}.$$

Solving for $Y(s)$ and using the fact that $y(0) = 1$, we obtain

$$Y(s) = \frac{1}{s-3} + \frac{6e^{-2s}}{s(s-3)}, \qquad s > 3.$$

Using a partial fraction expansion for the second term on the right-hand side,

$$Y(s) = \frac{1}{s-3} + e^{-2s}\left(\frac{2}{s-3} - \frac{2}{s}\right), \qquad s > 3.$$

Therefore,

$$y(t) = \mathcal{L}^{-1}\left\{\frac{1}{s-3}\right\} + 2\mathcal{L}^{-1}\left\{e^{-2s}\frac{1}{s-3}\right\} - 2\mathcal{L}^{-1}\left\{e^{-2s}\frac{1}{s}\right\}.$$

Using the second shifting theorem, we see that $y(t) = e^{3t} + 2h(t-2)[e^{3t-6} - 1]$, $t \geq 0$, is the unique solution of the initial value problem. In particular, $y(t)$ is the piecewise-defined function

$$y(t) = \begin{cases} e^{3t}, & 0 \leq t < 2 \\ e^{3t} + 2[e^{3t-6} - 1], & 2 \leq t < \infty. \end{cases}$$

Note that $y(t)$ is continuous at $t = 2$, but it is not differentiable at $t = 2$. ▲

We have restricted our attention in Theorem 7.5 to transform relations for first and second derivatives since these derivatives appear most frequently in applications. It should be clear that, with appropriate hypotheses, the arguments establishing Theorem 7.5 can be extended and we can obtain similar formulas for higher derivatives. In general,

$$\begin{aligned} \mathcal{L}\{f^{(n)}(t)\} &= s\mathcal{L}\{f^{(n-1)}(t)\} - f^{(n-1)}(0) = s[s\mathcal{L}\{f^{(n-2)}(t)\} - f^{(n-2)}(0)] - f^{(n-1)}(0) \\ &= s^2[s\mathcal{L}\{f^{(n-3)}(t)\} - f^{(n-3)}(0)] - sf^{(n-2)}(0) - f^{(n-1)}(0) = \cdots \\ &= s^nF(s) - s^{n-1}f(0) - s^{n-2}f'(0) - \cdots - sf^{(n-2)}(0) - f^{(n-1)}(0), \\ &\qquad n = 1, 2, 3, \ldots \end{aligned}$$

The Laplace transform pairs and relations that we have developed so far are summarized in Table 7.1 at the end of this section.

The importance of Theorem 7.5 cannot be overemphasized. As we have seen, derivatives and antiderivatives of a function $f(t)$ transform into algebraic expressions involving the function's transform $F(s)$. This is the key to the problem simplification achieved by working in the transform domain. Example 4, solving for the response of a simple *RLC* network, illustrates how Laplace transforms achieve such simplifications. The example also illustrates the important role that partial fraction expansions play in the use of Laplace transforms. Partial fraction expansions will be reviewed in Section 7.3.

Changing Our Point of View

Until now, we have carefully stated the values of s for which the improper integral

$$\mathcal{L}\{f(t)\} = \int_0^\infty f(t)e^{-st}\,dt$$

converges and thus defines the domain of the Laplace transform $F(s)$. Each entry of our table of transform pairs, Table 7.1 at the end of this section, includes the domain of $F(s)$.

It is important to understand and appreciate the underlying mathematical issues. However, when we begin to use Laplace transforms to actually solve initial value problems, we will no longer be so attentive to these details. Part of the reason for this change is that we will be "computing" the Laplace transform, $Y(s)$, of the *unknown* solution, $y(t)$. Since the solution $y(t)$ is unknown, we cannot easily determine the domain of its transform, $Y(s)$.

Instead, we are going to use the Laplace transform as an operational tool. We will simply assume that the unknown solution is a piecewise continuous and exponentially bounded function whose Laplace transform exists for $s > a$ for some value a. After we formally execute the three steps:

1. Laplace transformation,
2. solution of the transformed problem,
3. inverse Laplace transformation,

we will obtain what we can regard as a *candidate* for the solution of our initial value problem. If we can directly verify that the "candidate solution" obtained by the use of transforms is, in fact, the unique solution of the original problem of interest, then we are done. (In most examples we will leave this final task of verification as an exercise.) Example 4 illustrates these points and the three-step Laplace transform solution procedure, using it to analyze a network problem.

EXAMPLE 4

The series RLC network shown in Figure 7.9 is assumed to be initially quiescent; that is, the current and the charge on the capacitor are both zero for $t \leq 0$. At time $t = 0$ a voltage source, $v(t) = v_0 te^{-\alpha t}$ having the polarity shown, is turned on. Determine the current $i(t)$ for $t \geq 0$.

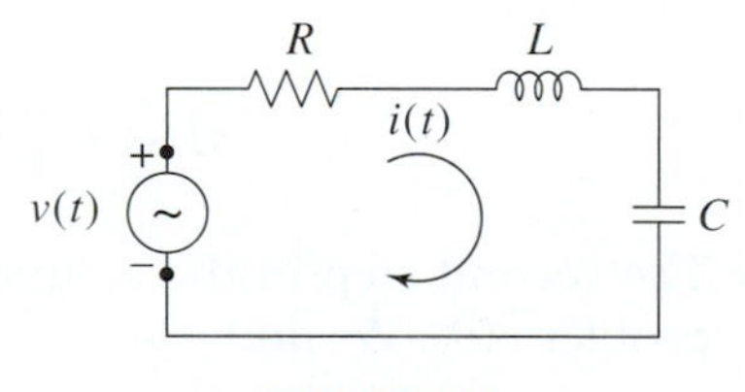

FIGURE 7.9

The RLC network analyzed in Example 4.

Solution: Recall that the underlying principle for describing our problem mathematically is Kirchhoff's voltage law (see Section 4.11). As we traverse the

(continued)

(continued)

network in a clockwise manner, the voltage rise through the source must equal the sum of the voltage drops through the resistor R, inductor L, and capacitor C. These voltage drops are

$$Ri(t), \qquad L\,\frac{di(t)}{dt}, \quad \text{and} \quad \frac{1}{C}\int_0^t i(u)\,du,$$

respectively. The resulting equation, along with the accompanying supplementary conditions, is

$$v(t) = Ri(t) + L\,\frac{di(t)}{dt} + \frac{1}{C}\int_0^t i(u)\,du, \qquad i(0) = 0, \qquad t \geq 0$$

or

$$\frac{di(t)}{dt} + \frac{R}{L}i(t) + \frac{1}{LC}\int_0^t i(u)\,du = \frac{1}{L}v(t), \qquad i(0) = 0, \qquad t \geq 0. \tag{7}$$

When we considered this problem in Section 4.11, we differentiated equation (7) to obtain a second order differential equation for the current $i(t)$. [We also observed that equation (7) could be viewed as a second order equation for the charge $q(t)$.] Now, however, we will work directly with equation (7), which is an integro-differential equation for the unknown current, $i(t)$.

The first step in the solution process is to compute the Laplace transform of both sides of equation (7), obtaining

$$\mathcal{L}\left\{\frac{di}{dt}\right\} + \frac{R}{L}\mathcal{L}\{i\} + \frac{1}{LC}\mathcal{L}\left\{\int_0^t i(u)\,du\right\} = \frac{1}{L}\mathcal{L}\{v(t)\}.$$

Using Theorem 7.5, this equation can be written as

$$sI(s) - i(0) + \frac{R}{L}I(s) + \frac{1}{LC}\frac{I(s)}{s} = \frac{1}{L}V(s). \tag{8}$$

Notice that the supplementary condition involving $i(0)$ enters directly into the transformed equation (8). In our case, $i(0) = 0$. Since $v(t) = v_0 te^{-\alpha t}$, we have

$$V(s) = v_0\mathcal{L}\{te^{-\alpha t}\} = v_0\mathcal{L}\{t\}\big|_{s\to s+\alpha} = v_0\frac{1}{(s+\alpha)^2}.$$

The transform domain problem is therefore entirely defined by the algebraic equation:

$$sI(s) + \frac{R}{L}I(s) + \frac{1}{LC}\frac{I(s)}{s} = \frac{v_0}{L}\frac{1}{(s+\alpha)^2}. \tag{9}$$

The second step in the solution process is that of solving transform domain problem (9). We find

$$I(s) = \frac{v_0}{L}\frac{s}{(s+\alpha)^2\left(s^2 + \dfrac{R}{L}s + \dfrac{1}{LC}\right)}. \tag{10}$$

The third step in the solution process is finding the inverse Laplace transform of $I(s)$. To accomplish this, we use a partial fraction expansion (see Section 7.3) to decompose rational function (10) into a sum of terms, each of whose inverse Laplace transform is known.

In order to illustrate this third step with a specific case, suppose that $I(s)$ in equation (10) is given by

$$I(s) = \frac{50s}{(s+1)^2(s^2+4s+13)}. \tag{11}$$

Expression (11) has the partial fraction expansion

$$\begin{aligned} I(s) &= \frac{6}{s+1} - \frac{5}{(s+1)^2} - \frac{6s+13}{(s+2)^2+9} \\ &= \frac{6}{s+1} - \frac{5}{(s+1)^2} - 6\frac{s+2}{(s+2)^2+9} - \frac{1}{3}\frac{3}{(s+2)^2+9}, \end{aligned} \tag{12}$$

where we have used the fact that $s^2 + 4s + 13 = (s+2)^2 + 9$. The algebraic manipulations leading to the last expression in (12) were done in anticipation of the inverse Laplace transform computation.

The third and final step in the solution process amounts to determining the current $i(t)$ by applying the inverse Laplace transform to $I(s)$:

$$\begin{aligned} i(t) &= \mathcal{L}^{-1}\{I(s)\} \\ &= 6\mathcal{L}^{-1}\left\{\frac{1}{s+1}\right\} - 5\mathcal{L}^{-1}\left\{\frac{1}{(s+1)^2}\right\} \\ &\quad - 6\mathcal{L}^{-1}\left\{\frac{s+2}{(s+2)^2+9}\right\} - \frac{1}{3}\mathcal{L}^{-1}\left\{\frac{3}{(s+2)^2+9}\right\}. \end{aligned} \tag{13}$$

The inverse required transforms can be obtained from Table 7.1 at the end of this section.

$$\mathcal{L}^{-1}\left\{\frac{1}{s+1}\right\} = e^{-t}$$

(See transform pair 3, with $\alpha = -1$.)

$$\mathcal{L}^{-1}\left\{\frac{1}{(s+1)^2}\right\} = te^{-t}$$

(See transform pair 10, with $n = 1$ and $\alpha = -1$.)

$$\mathcal{L}^{-1}\left\{\frac{s+2}{(s+2)^2+9}\right\} = e^{-2t}\cos 3t$$

(See transform pair 12, with $\alpha = -2$ and $\omega = 3$.)

$$\mathcal{L}^{-1}\left\{\frac{3}{(s+2)^2+9}\right\} = e^{-2t}\sin 3t$$

(See transform pair 11, with $\alpha = -2$ and $\omega = 3$.)

Using these inverse transforms in equation (13), it follows that

$$i(t) = 6e^{-t} - 5te^{-t} - 6e^{-2t}\cos 3t - \frac{1}{3}e^{-2t}\sin 3t, \qquad t \geq 0.$$

As a final check, one should verify that this expression for the network current is, in fact, the desired solution. ▲

The network current is plotted in Figure 7.10; its behavior seems reasonable. Since the source voltage is proportional to te^{-t}, one would expect the current to exhibit a transient behavior followed by an approach to zero as $t \to \infty$.

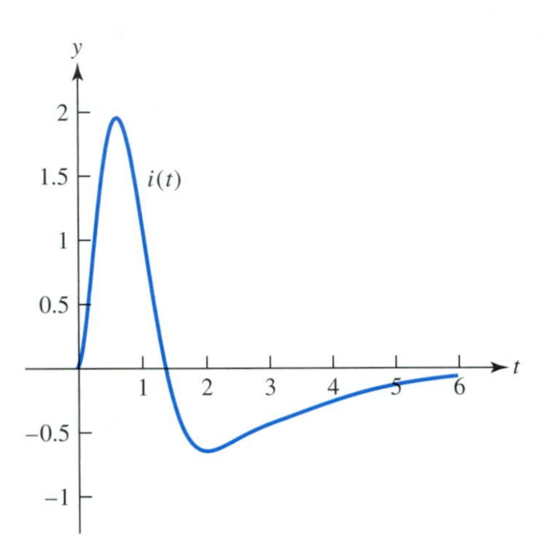

FIGURE 7.10

The network current found in Example 4 for the *RLC* network of Figure 7.9. Since the source voltage is proportional to te^{-t}, we expect the current to consist of an initial transient variation followed by an approach to zero as time increases.

TABLE 7.1

A Table of Laplace Transforms Pairs

Time Domain Function $f(t)$, $t \geq 0$	Laplace Transform $F(s)$
1. $h(t) = \begin{cases} 1, & t \geq 0 \\ 0, & t < 0 \end{cases}$	$\frac{1}{s}, \quad s > 0$
2. $t^n, \quad n = 1, 2, 3, \ldots$	$\frac{n!}{s^{n+1}}, \quad s > 0$
3. $e^{\alpha t}$	$\frac{1}{s - \alpha}, \quad s > \alpha$
4. $\sin \omega t$	$\frac{\omega}{s^2 + \omega^2}, \quad s > 0$
5. $\cos \omega t$	$\frac{s}{s^2 + \omega^2}, \quad s > 0$
6. $\sinh \alpha t$	$\frac{\alpha}{s^2 - \alpha^2}, \quad s > \|\alpha\|$
7. $\cosh \alpha t$	$\frac{s}{s^2 - \alpha^2}, \quad s > \|\alpha\|$
8. $e^{\alpha t} f(t)$, with $\|f(t)\| \leq M e^{at}$	$F(s - \alpha), \quad s > \alpha + a$

TABLE 7.1 (continued)

$f(t)$	$F(s)$
Four special cases of (8) include	
9. $e^{\alpha t}h(t)$	$\dfrac{1}{s-\alpha}, \quad s > \alpha$
10. $e^{\alpha t}t^n, \quad n = 1, 2, 3, \ldots$	$\dfrac{n!}{(s-\alpha)^{n+1}}, \quad s > \alpha$
11. $e^{\alpha t} \sin \omega t$	$\dfrac{\omega}{(s-\alpha)^2 + \omega^2}, \quad s > \alpha$
12. $e^{\alpha t} \cos \omega t$	$\dfrac{(s-\alpha)}{(s-\alpha)^2 + \omega^2}, \quad s > \alpha$
13. $f(t-\alpha)h(t-\alpha), \quad \alpha \geq 0$ with $\lvert f(t) \rvert \leq Me^{at}$	$e^{-\alpha s}F(s), \quad s > a$
A special case of (13) is	
14. $h(t-\alpha), \quad \alpha \geq 0$	$\dfrac{e^{-\alpha s}}{s}, \quad s > 0$
15. $f'(t)$, with $f(t)$ continuous and $\lvert f'(t) \rvert \leq Me^{at}$	$sF(s) - f(0)$, $s > \max\{a, 0\}$
16. $f''(t)$, with $f'(t)$ continuous and $\lvert f''(t) \rvert \leq Me^{at}$	$s^2F(s) - sf(0) - f'(0)$, $s > \max\{a, 0\}$
17. $f^{(n)}(t)$, with $f^{(n-1)}(t)$ continuous and $\lvert f^{(n)}(t) \rvert \leq Me^{at}$	$s^nF(s) - s^{n-1}f(0) - \cdots - sf^{(n-2)}(0) - f^{(n-1)}(0)$, $s > \max\{a, 0\}, \quad n = 1, 2, 3, \ldots$
18. $\displaystyle\int_0^t f(u)\,du$, with $\lvert f(t) \rvert \leq Me^{at}$	$\dfrac{F(s)}{s}, \quad s > \max\{a, 0\}$
19. $\dfrac{1}{2\omega^3}[\sin \omega t - \omega t \cos \omega t]$	$\dfrac{1}{(s^2 + \omega^2)^2}, \quad s > 0$
20. $\dfrac{t}{2\omega} \sin \omega t$	$\dfrac{s}{(s^2 + \omega^2)^2}, \quad s > 0$
21. $tf(t)$	$-F'(s)$

EXERCISES

Exercises 1–12:

Use Table 7.1 to find $\mathcal{L}\{f(t)\}$ for the given function $f(t)$ defined on the interval $t \geq 0$.

1. $f(t) = 3t^2 + 2t + 1$
2. $f(t) = 2e^t + 5$
3. $f(t) = 1 + \sin 3t$
4. $f(t) = e^{3t-3}h(t-1)$
5. $f(t) = (t-1)^2h(t-1)$
6. $f(t) = \sin^2 \omega t$
7. $f(t) = 2te^{-2t}$
8. $f(t) = \sin 3t \cos 3t$
9. $f(t) = 2th(t-2)$

10. $f(t) = e^{2t}\cos 3t$ **11.** $f(t) = e^{3t}h(t-1)$ **12.** $f(t) = e^{4t}(t^2 + 3t + 5)$

Exercises 13–21:

Use Table 7.1 to find $\mathcal{L}^{-1}\{F(s)\}$ for the given $F(s)$.

13. $F(s) = \dfrac{3}{s} + \dfrac{24}{s^4}$

14. $F(s) = \dfrac{10}{s^2+25} + \dfrac{4}{s-3}$

15. $F(s) = \dfrac{2s-4}{(s-2)^2+9}$

16. $F(s) = \dfrac{5}{(s-3)^4}$

17. $F(s) = e^{-2s}\dfrac{3}{s^2+9}$

18. $F(s) = \dfrac{e^{-2s}}{s-9}$

19. $F(s) = \dfrac{4s-6}{s^2-2s+10}$

20. $F(s) = \dfrac{e^{-3s}(2s+7)}{s^2+16}$

21. $F(s) = \dfrac{48(e^{-3s}+2e^{-5s})}{s^5}$

Combinations of Shifted Heaviside Step Functions Exercises 22–33 deal with combinations of Heaviside step functions. We can use combinations of shifted Heaviside step functions to represent pulses such as those graphed in (a) and (b).

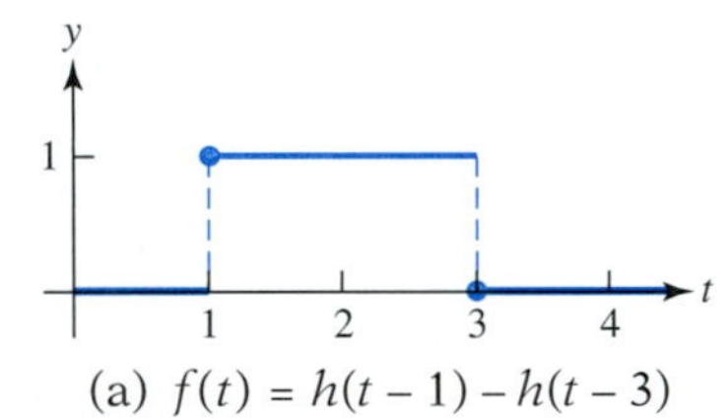

(a) $f(t) = h(t-1) - h(t-3)$

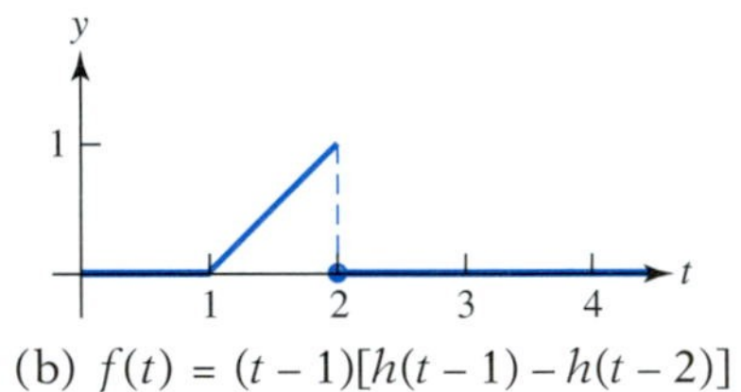

(b) $f(t) = (t-1)[h(t-1) - h(t-2)]$

Figure for Exercises 22–33

Exercises 22–33:

In each exercise, graph the function $f(t)$ for $0 \le t < \infty$, and use Table 7.1 to find the Laplace transform of $f(t)$.

22. $f(t) = h(t-1) + h(t-3)$

23. $f(t) = \sin(t-2\pi)h(t-2\pi)$

24. $f(t) = t[h(t-1) - h(t-3)]$

25. $f(t) = h(t) - h(t-3)$

26. $f(t) = 3[h(t-1) - h(t-4)]$

27. $f(t) = (2-t)[h(t-1) - h(t-3)]$

28. $f(t) = |2-t|[h(t-1) - h(t-3)]$

29. $f(t) = [h(t-1) - h(t-2)] - [h(t-2) - h(t-3)]$

30. $h(2-t)$

31. $e^{-2t}h(1-t)$

32. $h(t-1) + h(4-t)$

33. $h(t-2) - h(3-t)$

Exercises 34–37:

In each exercise, the graph of $f(t)$ is given. Represent $f(t)$ as a combination of Heaviside step functions, and use Table 7.1 to calculate the Laplace transform of $f(t)$.

34.

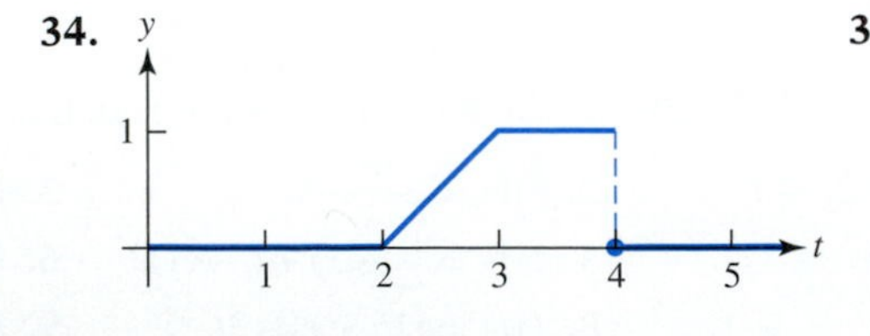

35.

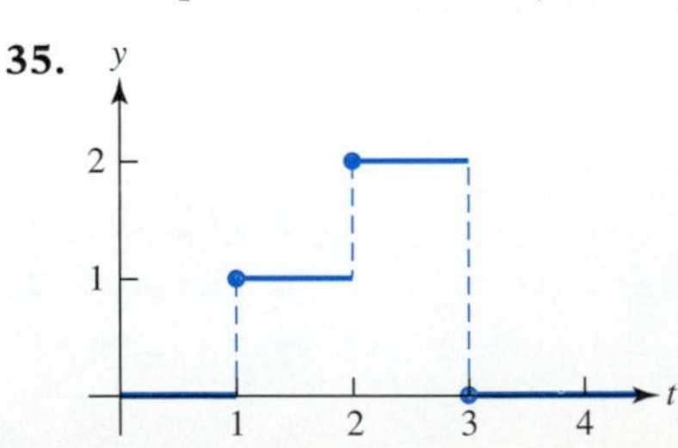

36.

37.

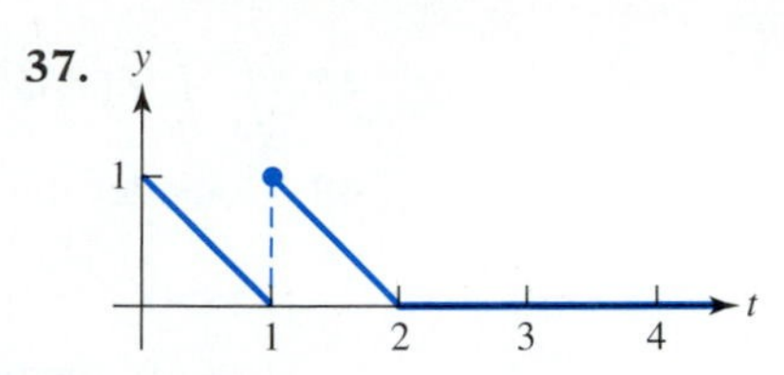

An Introduction to the Method of Partial Fractions We will present a review of the method of partial fractions in Section 7.3. By way of introduction, however, we consider here a special case of the method and use it to solve some initial value problems, as in Examples 3 and 4. Suppose $F(s) = 1/Q(s)$, where $Q(s) = (s - r_1)(s - r_2)\cdots(s - r_n)$ and where the roots $r_1, r_2, \ldots, r_n$ are real and distinct. In this case, there are constants $A_1, A_2, \ldots, A_n$ such that

$$F(s) = \frac{1}{(s - r_1)(s - r_2)\cdots(s - r_n)} = \frac{A_1}{s - r_1} + \frac{A_2}{s - r_2} + \cdots + \frac{A_n}{s - r_n}. \tag{14}$$

One way to determine the constants $A_1, A_2, \ldots, A_n$ is to recombine the right-hand side into a single rational function and equate the resulting numerator to 1.

Exercises 38–41:

Using a partial fraction expansion, find $\mathcal{L}^{-1}\{F(s)\}$. In Exercise 40, compare your answer with (6) in Table 7.1.

38. $F(s) = \dfrac{12}{(s - 3)(s + 1)}$

39. $F(s) = \dfrac{4}{s(s + 2)}$

40. $F(s) = \dfrac{24e^{-5s}}{s^2 - 9}$

41. $F(s) = \dfrac{10e^{-s}}{s^2 - 5s + 6}$

Exercises 42–45:

As in Example 3, use Laplace transform techniques to solve the initial value problem.

42. $y' + 4y = g(t),\ \ y(0) = 2,\ \ g(t) = \begin{cases} 0, & 0 \le t < 1 \\ 12, & 1 \le t < 3 \\ 0, & 3 \le t < \infty \end{cases}$

43. $y' - y = g(t),\ \ y(0) = 1,\ \ g(t) = \begin{cases} 0, & 0 \le t < 4 \\ e^{3t}, & 4 \le t < \infty \end{cases}$

44. $y'' - 4y = e^{3t},\ \ y(0) = 0,\ \ y'(0) = 0$

45. $y'' - 2y' - 8y = e^{t},\ \ y(0) = 0,\ \ y'(0) = 1$

46. Let $f(t)$ be piecewise continuous and exponentially bounded on the interval $0 \le t < \infty$ and let $F(s)$ denote the Laplace transform of $f(t)$. It is shown in advanced calculus[3] that it is possible to differentiate under the integral sign with respect to the parameter s. That is,

$$\frac{d}{ds}F(s) = \frac{d}{ds}\int_0^\infty e^{-st}f(t)\,dt = \int_0^\infty \frac{d}{ds}[e^{-st}f(t)]\,dt.$$

(a) Use this result to show that $\mathcal{L}\{tf(t)\} = -F'(s)$.

(b) Use the result of part (a) to establish formula (20) in Table 7.1.

Exercises 47–48:

Obtain the Laplace transform of the given function in terms of $\mathcal{L}\{f(t)\} = F(s)$. For Exercise 48, note that $\int_a^t f(\lambda)\,d\lambda = \int_0^t f(\lambda)\,d\lambda - \int_0^a f(\lambda)\,d\lambda$.

[3]David V. Widder, *Advanced Calculus*, 2nd edition (Englewood Cliffs, NJ: Prentice Hall, 1961).

47. $\displaystyle\int_0^t \int_0^{\lambda} f(\sigma)\,d\sigma\,d\lambda$ **48.** $\displaystyle\int_2^t f(\lambda)\,d\lambda$, given that $\displaystyle\int_0^2 f(\lambda)\,d\lambda = 3$.

49. Consider the functions f and g defined on $0 \le t < \infty$,

$$f(t) = h(t)h(3-t) \quad \text{and} \quad g(t) = h(t) - h(t-3).$$

(a) Are the two functions identical?

(b) Determine $F(s) = \mathcal{L}\{f(t)\}$ and $G(s) = \mathcal{L}\{g(t)\}$. Is $F(s) = G(s)$?

7.3 The Method of Partial Fractions

When we solve a problem in the transform domain, the solution is often a rational function of s; this was the case in Examples 3 and 4 of Section 7.2. The function $F(s)$ is called a **rational function** if it has the form

$$F(s) = \frac{N(s)}{D(s)}, \tag{1}$$

where $N(s)$ and $D(s)$ are polynomials.

In order to find $\mathcal{L}^{-1}\{F(s)\}$, we proceed as in Examples 3 and 4 of Section 7.2, using the method of partial fractions to decompose the rational function (1) into a sum of simpler expressions whose inverse transform can be recognized from a table of transform pairs.

The method of partial fractions is usually studied in calculus when computing antiderivatives of rational functions. You should refer back to your calculus text for a comprehensive discussion of the technique. The goal of this section is simply to remind you of the underlying ideas.

The Method of Partial Fractions

The starting point for the method of partial fractions is a rational function where *the degree of the numerator polynomial is strictly less than the degree of the denominator polynomial*. The rational functions we will encounter in the transform domain will have this form.

Let $F(s) = N(s)/D(s)$, where $N(s)$ and $D(s)$ are polynomials having real coefficients and where the degree of $N(s)$ is less than the degree of $D(s)$. Recall that the form of the partial fraction expansion is totally determined by the factorization of the denominator polynomial, $D(s)$. Table 7.2 lists the possible factors of this denominator polynomial. For each of the factors in the left-hand column, we need to include the terms in the right-hand column as part of the partial fraction expansion. The complete partial fraction expansion is the sum of the contributions from all of the denominator factors. This expansion contains constants that must subsequently be determined.

Since the denominator polynomial, $D(s)$, has real coefficients, any complex zeros will occur in complex conjugate pairs. Therefore, irreducible quadratic factors (which correspond to complex conjugate pairs of zeros) will have the forms listed in Table 7.2. Cases 3 and 4 are special versions of cases 5 and 6 that correspond to $\alpha = 0$. Since the numerator polynomial also has real coefficients, the constants in the partial fraction expansion will likewise be real valued.

TABLE 7.2

Denominator Polynomial Factors and Their Corresponding Terms in the Partial Fraction Expansion

Denominator Factor	Partial Fraction Expansion Term
1. Simple real root $s-\alpha$	$\dfrac{A}{s-\alpha}$
2. Repeated real root $(s-\alpha)^n$	$\dfrac{A_n}{(s-\alpha)^n}+\dfrac{A_{n-1}}{(s-\alpha)^{n-1}}+\cdots+\dfrac{A_1}{s-\alpha}$
3. Irreducible quadratic factor $s^2+\omega^2$	$\dfrac{Bs+C}{s^2+\omega^2}$
4. Repeated irreducible quadratic factor $(s^2+\omega^2)^n$	$\dfrac{B_ns+C_n}{(s^2+\omega^2)^n}+\dfrac{B_{n-1}s+C_{n-1}}{(s^2+\omega^2)^{n-1}}+\cdots+\dfrac{B_1s+C_1}{s^2+\omega^2}$
5. Irreducible quadratic factor $s^2+2\alpha s+\beta^2,\ \ \beta^2>\alpha^2$	$\dfrac{Bs+C}{s^2+2\alpha s+\beta^2}$
6. Repeated irreducible quadratic factor $(s^2+2\alpha s+\beta^2)^n,\ \ \beta^2>\alpha^2$	$\dfrac{B_ns+C_n}{(s^2+2\alpha s+\beta^2)^n}+\cdots+\dfrac{B_1s+C_1}{s^2+2\alpha s+\beta^2}$

These constants, denoted by capital letters on the right-hand side of Table 7.2, must be determined.

When looking for the inverse transform of a term that arises in case 5 or 6, we usually rewrite the term first. In particular, by completing the square, an irreducible quadratic factor of the form $s^2+2\alpha s+\beta^2$, $\beta^2>\alpha^2$, can be rewritten as

$$(s^2+2\alpha s+\alpha^2)+(\beta^2-\alpha^2)=(s+\alpha)^2+\omega^2.$$

This form is associated with the Laplace transforms of $e^{-\alpha t}\sin\omega t$ and $e^{-\alpha t}\cos\omega t$; see the transform pairs (11) and (12) in Table 7.1.

The next few examples illustrate partial fraction expansions and their use in computing inverse Laplace transforms. These examples also illustrate several different techniques for evaluating the expansion constants.

EXAMPLE 1

Find $\mathcal{L}^{-1}\{F(s)\}$, where

$$F(s)=\frac{s^2+4}{s^4-s^2}.$$

(continued)

(continued)

Solution: The function $F(s)$ is a rational function with numerator $N(s) = s^2 + 4$ having degree less than the denominator $D(s) = s^4 - s^2$. The denominator factors into

$$D(s) = s^2(s^2 - 1) = s^2(s - 1)(s + 1).$$

Therefore, the denominator has $s = 0$ as a repeated real root and $s = \pm 1$ as simple real roots. According to Table 7.2, $F(s)$ has a partial fraction expansion of the form

$$F(s) = \frac{s^2 + 4}{s^4 - s^2} = \frac{A_2}{s^2} + \frac{A_1}{s} + \frac{B}{s - 1} + \frac{C}{s + 1}, \tag{2}$$

where the four constants A_1, A_2, B, and C must be determined.

One way to evaluate the unknown constants is to recombine the right-hand side of (2), obtaining

$$\frac{s^2 + 4}{s^4 - s^2} = \frac{(A_1 + B + C)s^3 + (A_2 + B - C)s^2 - A_1 s - A_2}{s^4 - s^2}. \tag{3}$$

Since the two expressions in (3) must be equal, the two numerator polynomials must be identical. We therefore obtain four equations for the four unknown constants,

$$\begin{aligned} A_1 \qquad + B + C &= 0 \\ A_2 + B - C &= 1 \\ -A_1 \qquad\qquad &= 0 \\ -A_2 \qquad\quad &= 4. \end{aligned}$$

This system is easily solved and we obtain $A_1 = 0, A_2 = -4, B = 5/2$, and $C = -5/2$. Thus, we have the partial fraction expansion

$$F(s) = -\frac{4}{s^2} + \frac{\frac{5}{2}}{s - 1} - \frac{\frac{5}{2}}{s + 1}. \tag{4}$$

From Table 7.1 in the previous section,

$$\begin{aligned} \mathcal{L}^{-1}\{F(s)\} &= -4\mathcal{L}^{-1}\left\{\frac{1}{s^2}\right\} + \frac{5}{2}\mathcal{L}^{-1}\left\{\frac{1}{s - 1}\right\} - \frac{5}{2}\mathcal{L}^{-1}\left\{\frac{1}{s + 1}\right\} \\ &= -4t + \frac{5}{2}e^t - \frac{5}{2}e^{-t} = -4t + 5\sinh t, \qquad t \ge 0. \end{aligned}$$ ▲

Alternative Approaches for Determining the Constants in a Partial Fraction Expansion

We now consider two alternative approaches for determining the constants in a partial fraction expansion. We explain the first approach below by reworking the expansion in Example 1. We explain the second approach later, in Example 2.

Consider equation (2) in Example 1,

$$\frac{s^2+4}{s^2(s-1)(s+1)} = \frac{A_2}{s^2} + \frac{A_1}{s} + \frac{B}{s-1} + \frac{C}{s+1}.$$

If we multiply both sides by $(s-1)$ and cancel common factors, we obtain:

$$\frac{s^2+4}{s^2(s+1)} = A_2\frac{(s-1)}{s^2} + A_1\frac{(s-1)}{s} + B + C\frac{(s-1)}{(s+1)}. \tag{5}$$

We now determine B by setting $s=1$ on both sides of expression (5), finding

$$B = \left.\frac{s^2+4}{s^2(s+1)}\right|_{s=1} = \frac{5}{2}.$$

Similarly, multiplying both sides by $(s+1)$ leads to

$$\frac{s^2+4}{s^2(s-1)} = A_2\frac{(s+1)}{s^2} + A_1\frac{(s+1)}{s} + B\frac{(s+1)}{(s-1)} + C,$$

and we find

$$C = \left.\frac{s^2+4}{s^2(s-1)}\right|_{s=-1} = -\frac{5}{2}.$$

So far, we have determined the constants B and C in the expansion

$$\frac{s^2+4}{s^4-s^2} = \frac{A_2}{s^2} + \frac{A_1}{s} + \frac{B}{s-1} + \frac{C}{s+1}.$$

We next multiply both sides by s^2, obtaining

$$\frac{s^2+4}{s^2-1} = A_2 + A_1 s + B\frac{s^2}{s-1} + C\frac{s^2}{s+1}. \tag{6}$$

We now evaluate the constant A_2 by setting $s=0$,

$$A_2 = \left.\frac{s^2+4}{s^2-1}\right|_{s=0} = -4.$$

To determine A_1, we differentiate both sides of equation (6), finding

$$\frac{-10s}{(s^2-1)^2} = A_1 + B\frac{s^2-2s}{(s-1)^2} + C\frac{s^2+2s}{(s+1)^2}.$$

Setting $s=0$ leads to $A_1=0$. In this way, we once more obtain expansion (4) for $F(s)$.

EXAMPLE 2

Find $\mathcal{L}^{-1}\{F(s)\}$, where

$$F(s) = \frac{s^2+s-1}{(s^2+4s+4)(s^2+2s+5)}.$$

Solution: Since the numerator degree is less than the denominator degree, we proceed with the partial fraction expansion. The first quadratic factor in

(continued)

(continued)

the denominator can be factored as $s^2 + 4s + 4 = (s+2)^2$. The second quadratic factor, $s^2 + 2s + 5$, is irreducible. Therefore, the partial fraction expansion of $F(s)$ has the form:

$$\frac{s^2 + s - 1}{(s+2)^2(s^2 + 2s + 5)} = \frac{A_2}{(s+2)^2} + \frac{A_1}{s+2} + \frac{Bs + C}{s^2 + 2s + 5}. \tag{7}$$

We can certainly recombine the right-hand side of (7), obtaining four equations for the four unknown constants. However, as an alternative strategy, we first determine A_1 and A_2 as explained above and then use another approach to determine B and C. Multiplying equation (7) by $(s+2)^2$, we obtain A_2:

$$A_2 = \left. \frac{s^2 + s - 1}{s^2 + 2s + 5} \right|_{s=-2} = \frac{1}{5}.$$

Next, we find A_1:

$$A_1 = \frac{d}{ds}\left[\frac{s^2 + s - 1}{s^2 + 2s + 5}\right]\Bigg|_{s=-2} = \left. \frac{s^2 + 12s + 7}{(s^2 + 2s + 5)^2} \right|_{s=-2} = -\frac{13}{25}.$$

Using these values in (7), we have

$$\frac{s^2 + s - 1}{(s+2)^2(s^2 + 2s + 5)} = \frac{\frac{1}{5}}{(s+2)^2} - \frac{\frac{13}{25}}{s+2} + \frac{Bs + C}{s^2 + 2s + 5}. \tag{8}$$

We now determine the constants B and C by selecting two convenient values of s and evaluating (8) at these values. Setting $s = 0$ in (8) leads to

$$-\frac{1}{20} = \frac{1}{20} - \frac{13}{50} + \frac{C}{5} \quad \text{or} \quad C = \frac{4}{5}.$$

Similarly, setting $s = -1$ in (8) gives

$$-\frac{1}{4} = \frac{1}{5} - \frac{13}{25} + \frac{-B + \frac{4}{5}}{4} \quad \text{or} \quad B = \frac{13}{25}.$$

Therefore, the partial fraction expansion is

$$F(s) = \frac{\frac{1}{5}}{(s+2)^2} - \frac{\frac{13}{25}}{(s+2)} + \frac{\frac{13}{25}s + \frac{4}{5}}{(s+1)^2 + 4}. \tag{9}$$

We use Table 7.1 to compute the inverse transform. From formulas 10 and 3 of Table 7.1:

$$\mathcal{L}^{-1}\left\{\frac{\frac{1}{5}}{(s+2)^2}\right\} = \frac{1}{5}te^{-2t} \quad \text{and} \quad \mathcal{L}^{-1}\left\{-\frac{\frac{13}{25}}{(s+2)}\right\} = -\frac{13}{25}e^{-2t}, \qquad t \geq 0.$$

To obtain the inverse transform of the third expression on the right-hand side of (9), we recall formulas 11 and 12 of Table 7.1. To apply these formulas we first rewrite this third term as

$$\frac{\frac{13}{25}s+\frac{4}{5}}{(s+1)^2+4}=\frac{\frac{13}{25}(s+1)+\frac{7}{25}}{(s+1)^2+4}=\frac{13}{25}\frac{(s+1)}{(s+1)^2+4}+\frac{7}{50}\frac{2}{(s+1)^2+4}.$$

Then, from formulas 11 and 12 we conclude that

$$\mathcal{L}^{-1}\left\{\frac{\frac{13}{25}s+\frac{4}{5}}{(s+1)^2+4}\right\}=\frac{13}{25}\mathcal{L}^{-1}\left\{\frac{(s+1)}{(s+1)^2+4}\right\}+\frac{7}{50}\mathcal{L}^{-1}\left\{\frac{2}{(s+1)^2+4}\right\}$$

$$=\frac{13}{25}e^{-t}\cos 2t+\frac{7}{50}e^{-t}\sin 2t,\qquad t\ge 0.$$

Combining these results, we obtain the final answer:

$$f(t)=\mathcal{L}^{-1}\left\{\frac{s^2+s-1}{(s^2+4s+4)(s^2+2s+5)}\right\}$$

$$=\frac{1}{5}te^{-2t}-\frac{13}{25}e^{-2t}+\frac{13}{25}e^{-t}\cos 2t+\frac{7}{50}e^{-t}\sin 2t,\qquad t\ge 0. \blacktriangle$$

EXAMPLE 3

Use Laplace transforms to solve the initial value problem

$$y''+4y=4t+8,\qquad y(0)=4,\qquad y'(0)=-1.$$

Solution: Let $y(t)$ denote the solution, and let $Y(s)=\mathcal{L}\{y(t)\}$. Taking Laplace transforms of both sides of $y''+4y=4t+8$, we obtain

$$\mathcal{L}\{y''\}+4\mathcal{L}\{y\}=\frac{4}{s^2}+\frac{8}{s}. \tag{10}$$

From Theorem 7.5,

$$\mathcal{L}\{y''\}=s^2Y(s)-sy(0)-y'(0),$$

and therefore

$$s^2Y(s)-4s+1+4Y(s)=\frac{4}{s^2}+\frac{8}{s}. \tag{11}$$

Note that both initial conditions enter into equation (11). The solution of the problem in the transform domain is

$$Y(s)=\frac{4s^3-s^2+8s+4}{s^2(s^2+4)},$$

which has partial fraction expansion

$$\frac{4s^3-s^2+8s+4}{s^2(s^2+4)}=\frac{A}{s^2}+\frac{B}{s}+\frac{Cs+D}{s^2+4}$$

$$=\frac{A(s^2+4)+Bs(s^2+4)+Cs^3+Ds^2}{s^2(s^2+4)}.$$

(continued)

(continued)

Since the numerator polynomials in (7) must be identical, we equate coefficients of like powers of s and obtain the system of equations

$$B + C = 4, \qquad A + D = -1, \qquad 4B = 8, \qquad 4A = 4,$$

which has solution $A = 1, B = 2, C = 2$, and $D = -2$. Therefore,

$$Y(s) = \frac{1}{s^2} + \frac{2}{s} + \frac{2s}{s^2+4} - \frac{2}{s^2+4}.$$

We obtain the solution of the original initial value problem by taking the inverse Laplace transform,

$$\begin{aligned} y(t) &= \mathcal{L}^{-1}\left\{\frac{1}{s^2}\right\} + 2\mathcal{L}^{-1}\left\{\frac{1}{s}\right\} + 2\mathcal{L}^{-1}\left\{\frac{s}{s^2+4}\right\} - \mathcal{L}^{-1}\left\{\frac{2}{s^2+4}\right\} \\ &= t + 2 + 2\cos 2t - \sin 2t, \qquad t \geq 0. \; \blacktriangle \end{aligned}$$

In Example 3, we chose to combine the transform domain solution

$$Y(s) = \frac{4s}{s^2+4} - \frac{1}{s^2+4} + \frac{4}{s^2(s^2+4)} + \frac{8}{s(s^2+4)}$$

into a single rational function whose partial fraction expansion was then considered. The resulting system of equations for the unknown coefficients was easily solved. In some problems, however, it may be simpler to deal with individual terms, rather than a single combined term, when computing the inverse Laplace transform. Each problem should be scrutinized to determine the best solution strategy.

EXERCISES

Exercises 1–8:

Give the *form* of the partial fraction expansion for the given rational function $F(s)$. You need not evaluate the constants in the expansion. However, if the denominator of $F(s)$ contains irreducible quadratic factors of the form $s^2 + 2\alpha s + \beta^2$, $\beta^2 > \alpha^2$, complete the square and rewrite this factor in the form $(s+\alpha)^2 + \omega^2$.

1. $F(s) = \dfrac{2s+3}{(s-1)(s-2)^2}$

2. $F(s) = \dfrac{s^3+3s+1}{(s-1)^3(s-2)^2}$

3. $F(s) = \dfrac{s^2+1}{s^2(s^2+2s+10)}$

4. $F(s) = \dfrac{s^2+5s-3}{(s^2+16)(s-2)}$

5. $F(s) = \dfrac{s^2-1}{(s^2-9)^2}$

6. $F(s) = \dfrac{s^3-1}{(s^2+1)^2(s+4)^2}$

7. $F(s) = \dfrac{s^2+s+2}{(s^2+8s+17)(s^2+6s+13)}$

8. $F(s) = \dfrac{s^4+5s^2+2s-9}{(s^2+8s+17)^2(s-2)^2}$

Exercises 9–17:

Find the inverse Laplace transform.

9. $F(s) = \dfrac{2}{s-3}$

10. $F(s) = \dfrac{1}{(s+1)^3}$

11. $F(s) = \dfrac{4s+5}{s^2+9}$

12. $F(s) = \dfrac{2s - 3}{s^2 - 3s + 2}$ **13.** $F(s) = \dfrac{3s + 7}{s^2 + 4s + 3}$ **14.** $F(s) = \dfrac{4s^2 + s + 1}{s^3 + s}$

15. $F(s) = \dfrac{3s^2 + s + 8}{s^3 + 4s}$ **16.** $F(s) = \dfrac{s^2 + 6s + 8}{s^4 + 8s^2 + 16}$ **17.** $F(s) = \dfrac{s}{s^3 - 3s^2 + 3s - 1}$

Exercises 18–29:

Use the Laplace transform to solve the initial value problem.

18. $y' + 2y = 26 \sin 3t, \;\; y(0) = 3$ **19.** $y' - 3y = 13 \cos 2t, \;\; y(0) = 1$

20. $y' + 2y = 4t, \;\; y(0) = 3$ **21.** $y' - 3y = e^{3t}, \;\; y(0) = 1$

22. $y'' + 3y' + 2y = 6e^{-t}, \;\; y(0) = 1, \;\; y'(0) = 2$

23. $y'' + 4y = 8t, \;\; y(0) = 2, \;\; y'(0) = 6$

24. $y'' + 4y = \cos 2t, \;\; y(0) = 1, \;\; y'(0) = 1$

25. $y'' + 4y = \sin 2t, \;\; y(0) = 1, \;\; y'(0) = 0$

26. $y'' - 2y' + y = e^{2t}, \;\; y(0) = 0, \;\; y'(0) = 0$

27. $y'' + 2y' + y = e^{-t}, \;\; y(0) = 0, \;\; y'(0) = 1$

28. $y'' + 9y = g(t), \;\; y(0) = 1, \;\; y'(0) = 3, \qquad g(t) = \begin{cases} 6, & 0 \le t < \pi \\ 0, & \pi \le t < \infty \end{cases}$

29. $y'' + y = g(t), \;\; y(0) = 1, \;\; y'(0) = 0, \qquad g(t) = \begin{cases} t, & 0 \le t < 2 \\ 0, & 2 \le t < \infty \end{cases}$

Exercises 30–32:

Consider the initial value problem $y'' + \alpha y' + \beta y = 0, y(0) = y_0, y'(0) = y'_0$. The Laplace transform of the solution, $Y(s) = \mathcal{L}\{y(t)\}$, is given. Determine the constants α, β, y_0, and y'_0.

30. $Y(s) = \dfrac{2s - 1}{s^2 + s + 2}$ **31.** $Y(s) = \dfrac{3}{s^2 - 4}$ **32.** $Y(s) = \dfrac{s}{(s + 1)^2}$

7.4 Laplace Transforms of Periodic Functions and System Transfer Functions

In many applications, the nonhomogeneous term in a linear differential equation is a periodic function. We now derive a formula for the Laplace transform of such periodic functions.

Theorem 7.6

Let $f(t)$ be a piecewise continuous periodic function defined on $0 \le t < \infty$, where $f(t)$ has period T. Then

$$\mathcal{L}\{f(t)\} = \frac{\displaystyle\int_0^T e^{-st} f(t)\, dt}{1 - e^{-sT}}, \qquad s > 0.$$

PROOF: Since $f(t)$ is piecewise continuous, it is bounded on the interval $0 \le t \le T$. Since the function is also periodic, it follows that $f(t)$ is bounded on

$0 \le t < \infty$. Therefore, by Theorem 7.1, its Laplace transform exists for $s > 0$. Computing the transform, we find

$$F(s) = \mathcal{L}\{f(t)\} = \int_0^\infty f(t)e^{-st}\,dt = \sum_{n=0}^{\infty} \int_{nT}^{(n+1)T} f(t)e^{-st}\,dt. \tag{1}$$

[In the last step of equation (1), we have decomposed the improper integral into a sum of integrals over the constituent periods.] Consider a representative integral in (1),

$$\int_{nT}^{(n+1)T} f(t)e^{-st}\,dt,$$

where n is an arbitrary but fixed integer. Making the change of variables $\tau = t - nT$, we obtain

$$\int_{nT}^{(n+1)T} f(t)e^{-st}\,dt = \int_0^T f(\tau + nT)e^{-s(\tau+nT)}\,d\tau = e^{-snT}\int_0^T f(\tau)e^{-s\tau}\,d\tau.$$

Thus, equation (1) reduces to

$$F(s) = \sum_{n=0}^{\infty} e^{-snT}\int_0^T f(\tau)e^{-s\tau}\,d\tau = \left[\int_0^T f(\tau)e^{-s\tau}\,d\tau\right]\sum_{n=0}^{\infty} e^{-snT}. \tag{2}$$

Since $s > 0$, it follows that $0 < e^{-sT} < 1$. Therefore, the infinite series in equation (2) is a convergent geometric series,

$$\sum_{n=0}^{\infty} e^{-snT} = \sum_{n=0}^{\infty} (e^{-sT})^n = \frac{1}{1 - e^{-sT}},$$

and Theorem 7.6 follows. ▮

EXAMPLE 1

Let T be a positive constant and consider the square wave,

$$f(t) = \begin{cases} 1, & 0 \le t \le \dfrac{T}{2}, \\ 0, & \dfrac{T}{2} < t < T, \end{cases} \qquad f(t+T) = f(t), \qquad t \ge 0.$$

The graph of $f(t)$ is shown in Figure 7.11. Use Theorem 7.6 to determine the Laplace transform of $f(t)$.

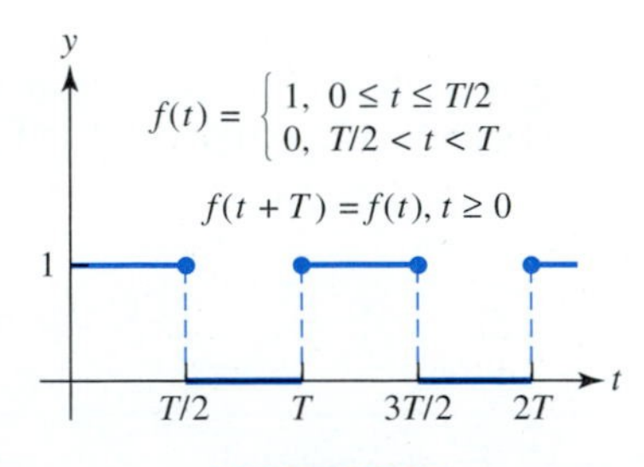

FIGURE 7.11

The graph of the square wave, $f(t)$, treated in Example 1. Note that $f(t)$ is periodic with period T and is piecewise continuous on $0 \le t < \infty$.

Solution: By Theorem 7.6,

$$\mathcal{L}\{f(t)\} = \frac{\int_0^T e^{-st} f(t)\, dt}{1 - e^{-sT}} = \frac{\int_0^{T/2} e^{-st}\, dt}{1 - e^{-sT}}, \qquad s > 0.$$

Evaluating this last integral, we find

$$\mathcal{L}\{f(t)\} = \frac{\left(1 - e^{-sT/2}\right)/s}{1 - e^{-sT}} = \frac{1}{s\left(1 + e^{-sT/2}\right)}. \quad ▲$$

EXAMPLE 2

Find the inverse transform of

$$F(s) = \frac{1}{s^2} - \frac{e^{-s}}{s(1 - e^{-s})}, \qquad s > 0.$$

Solution: Applying the inverse transform,

$$\mathcal{L}^{-1}\{F(s)\} = \mathcal{L}^{-1}\left\{\frac{1}{s^2}\right\} - \mathcal{L}^{-1}\left\{\frac{e^{-s}}{s(1 - e^{-s})}\right\} = t - \mathcal{L}^{-1}\left\{\frac{e^{-s}}{s(1 - e^{-s})}\right\}.$$

Expanding the factor $(1 - e^{-s})^{-1}$ as a geometric series and assuming for the moment that we can apply the inverse operation termwise to the convergent series, we obtain

$$\begin{aligned}
\mathcal{L}^{-1}\left\{\frac{e^{-s}}{s(1 - e^{-s})}\right\} &= \mathcal{L}^{-1}\left\{\frac{e^{-s}}{s}[1 + e^{-s} + e^{-2s} + e^{-3s} + \cdots]\right\} \\
&= \mathcal{L}^{-1}\left\{\frac{1}{s}[e^{-s} + e^{-2s} + e^{-3s} + \cdots]\right\} \\
&= \mathcal{L}^{-1}\left\{\frac{e^{-s}}{s}\right\} + \mathcal{L}^{-1}\left\{\frac{e^{-2s}}{s}\right\} + \mathcal{L}^{-1}\left\{\frac{e^{-3s}}{s}\right\} + \cdots \\
&= h(t-1) + h(t-2) + h(t-3) + \cdots .
\end{aligned}$$

The function $g(t) = h(t-1) + h(t-2) + h(t-3) + \cdots$ has the staircase-like graph shown in Figure 7.12(a). Thus, the inverse transform of $F(s)$ is the sawtooth wave function $f(t) = t - g(t)$ whose graph is shown in Figure 7.12(b). [Note: For any fixed value of t, $g(t) = h(t-1) + h(t-2) + h(t-3) + \cdots$ is actually a finite sum since $h(t - \alpha) = 0$ when $t < \alpha$. For instance, if $t = 2.5$, then $g(2.5) = 2$.] ▲

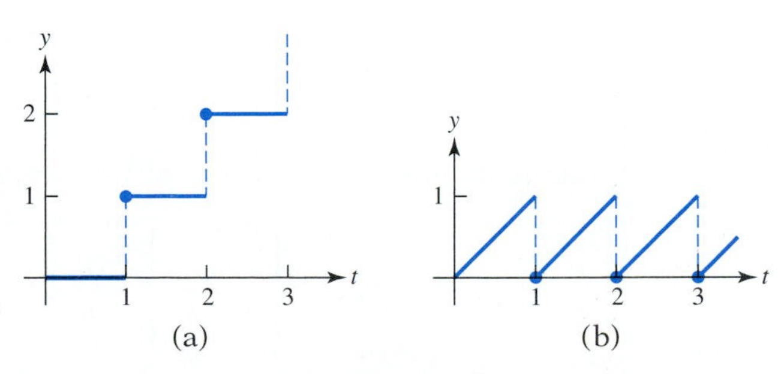

FIGURE 7.12

(a) The graph of $g(t) = h(t-1) + h(t-2) + h(t-3) + \cdots$ resembles a staircase. (b) The inverse transform of $F(s)$ in Example 2 is $f(t) = t - g(t)$; this graph is often called a sawtooth wave.

One-Dimensional Motion with Drag and Periodic Thrust

In Example 3, we use Laplace transform methods to solve a nonhomogeneous linear differential equation, where the nonhomogeneous term is a square wave.

EXAMPLE 3

Assume a projectile of mass m moves in a straight line on a horizontal surface with velocity $v(t)$, where $v(0) = v_0$. The projectile is subjected to a frictional force proportional to $v(t)$ (see Section 3.6). The projectile has a propulsion engine that provides the thrust needed to sustain motion through the medium. Instead of firing continuously, the engine has a periodic on/off cycle of period T. The engine therefore provides a constant thrust for half the cycle and then shuts off for the other half. The thrust is a square wave function such as that shown in Figure 7.11. Determine the velocity $v(t), t \geq 0$.

Solution: The relevant mathematical problem, arising from Newton's laws of motion, is the first order linear nonhomogeneous initial value problem

$$\begin{aligned} &mv' + kv = f(t), \qquad t \geq 0 \\ &v(0) = v_0 \\ &f(t) = \begin{cases} f_0, & 0 \leq t \leq \dfrac{T}{2}, \\ 0, & \dfrac{T}{2} < t < T, \end{cases} \qquad f(t+T) = f(t), \qquad t \geq 0. \end{aligned} \tag{3}$$

In (3), k is a positive friction coefficient and f_0 denotes the magnitude of the thrust when the engine is firing.

Although the ideas of Chapter 2 could be used to solve this problem, Laplace transform methods provide a convenient alternative solution strategy. Taking Laplace transforms of (3) leads to

$$m(sV(s) - v_0) + kV(s) = F(s),$$

where, by Example 1,

$$F(s) = \frac{f_0}{s\left(1 + e^{-sT/2}\right)}, \qquad s > 0.$$

The transform domain solution is

$$V(s) = \frac{v_0}{s + \dfrac{k}{m}} + \frac{f_0}{m} \frac{1}{\left(s + \dfrac{k}{m}\right) s\left(1 + e^{-sT/2}\right)}.$$

Taking the inverse transform of $V(s)$ we obtain the time domain solution

$$\begin{aligned} v(t) &= v_0 \mathcal{L}^{-1}\left\{\frac{1}{s + \dfrac{k}{m}}\right\} + \frac{f_0}{m}\mathcal{L}^{-1}\left\{\frac{1}{\left(s + \dfrac{k}{m}\right) s\left(1 + e^{-sT/2}\right)}\right\} \\ &= v_0 e^{-(k/m)t} + \frac{f_0}{m}\mathcal{L}^{-1}\left\{\frac{1}{\left(s + \dfrac{k}{m}\right) s\left(1 + e^{-sT/2}\right)}\right\}. \end{aligned} \tag{4}$$

To evaluate the last term on the right-hand side of (4), we expand $1/(1+e^{-sT/2})$ in a geometric series and use the second shifting theorem (Theorem 7.4 (b)). Note first that

$$\frac{1}{1+e^{-sT/2}} = \sum_{n=0}^{\infty}(-1)^n\left(e^{-sT/2}\right)^n = \sum_{n=0}^{\infty}(-1)^n e^{-nsT/2}.$$

Therefore,

$$\begin{aligned}\mathcal{L}^{-1}\left\{\frac{1}{\left(s+\frac{k}{m}\right)s\left(1+e^{-sT/2}\right)}\right\} &= \mathcal{L}^{-1}\left\{\frac{1}{\left(s+\frac{k}{m}\right)s}\sum_{n=0}^{\infty}(-1)^n e^{-nsT/2}\right\}\\ &= \mathcal{L}^{-1}\left\{\sum_{n=0}^{\infty}(-1)^n\frac{e^{-nsT/2}}{\left(s+\frac{k}{m}\right)s}\right\}\\ &= \sum_{n=0}^{\infty}(-1)^n\mathcal{L}^{-1}\left\{\frac{e^{-nsT/2}}{\left(s+\frac{k}{m}\right)s}\right\}.\end{aligned}$$

(We have assumed that interchanging the operations of inverse Laplace transformation and infinite summation is permissible.) We now examine the nth term in the sum above and apply the second shifting theorem, finding

$$\mathcal{L}^{-1}\left\{e^{-nsT/2}\frac{1}{\left(s+\frac{k}{m}\right)s}\right\} = h\left(t-\frac{nT}{2}\right)g\left(t-\frac{nT}{2}\right),$$

where $h(t)$ is the Heaviside step function and where

$$g(t) = \mathcal{L}^{-1}\left\{\frac{1}{\left(s+\frac{k}{m}\right)s}\right\} = \mathcal{L}^{-1}\left\{\frac{m}{k}\left(\frac{1}{s}-\frac{1}{s+\frac{k}{m}}\right)\right\} = \frac{m}{k}\left(1-e^{-(k/m)t}\right).$$

Therefore,

$$\begin{aligned}\frac{f_0}{m}\mathcal{L}^{-1}\left\{\frac{1}{(s+\frac{k}{m})s\left(1+e^{-sT/2}\right)}\right\} &= \frac{f_0}{m}\sum_{n=0}^{\infty}(-1)^n\mathcal{L}^{-1}\left\{e^{-nsT/2}\left(\frac{1}{s\left(s+\frac{k}{m}\right)}\right)\right\}\\ &= \frac{f_0}{k}\sum_{n=0}^{\infty}(-1)^n\left(1-e^{-(k/m)(t-nT/2)}\right)h\left(t-\frac{nT}{2}\right).\end{aligned} \tag{5}$$

Expression (5) looks more formidable than it actually is since the step function $h(t)$ vanishes for negative values of its argument. In particular, for any fixed positive time t, the infinite series in (5) reduces to a finite sum. The solution of initial value problem (3) is, therefore,

$$v(t) = v_0 e^{-(k/m)t} + \frac{f_0}{k}\sum_{n=0}^{M}(-1)^n\left(1-e^{-(k/m)(t-nT/2)}\right), \tag{6}$$

where M denotes the largest integer less than or equal to the number $2t/T$. ▲

We have chosen to consider one-dimensional projectile motion in a medium with friction. The same *mathematical* problem arises in other contexts as well. For instance, recall the models studied in Chapter 2. Radioactive decay with periodic material replenishment would lead to a similar problem, as would Newton's law of cooling applied to a body immersed in colder surroundings and subjected to a periodic heating cycle. Some of these applications are considered in the exercises.

Solution (6) would be relatively useless if it could not be conveniently evaluated in order to reveal its physical content. As you will see below, we can plot and study the solution using the capabilities of modern computational software.

A Special Case To illustrate the general behavior of solution (6), we set $T = 2$ and set all the other constants equal to one. This leads to the model problem

$$\begin{aligned} &y' + y = f(t), \qquad t \geq 0, \\ &y(0) = 1, \\ &f(t) = \begin{cases} 1, & 0 \leq t \leq 1, \\ 0, & 1 < t < 2, \end{cases} \qquad f(t+2) = f(t). \end{aligned} \tag{7}$$

A plot of the solution, for $0 \leq t \leq 20$, is shown in Figure 7.13. Your initial response to it should probably be "Can this be correct?" On one hand, it seems reasonable to expect the solution to "settle into" a periodic pattern such as the one shown—the solution increases during the portion of the cycle in which the nonhomogeneous term $f(t)$ is equal to 1 and decays exponentially when $f(t)$ is equal to 0. At points where $f(t)$ is discontinuous, the solution is continuous but its derivative is not. On the other hand, the constant behavior of the solution for $0 \leq t \leq 1$ might seem somewhat strange and unexpected.

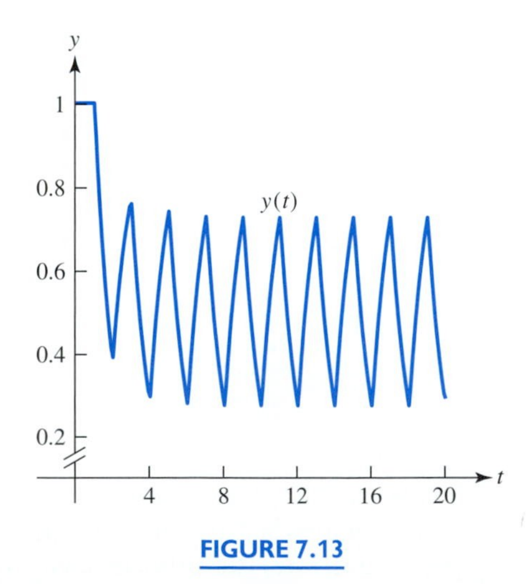

FIGURE 7.13

The graph of the solution of initial value problem (7).

A closer examination, however, shows that the initial portion of the graph is, in fact, correct. To see why, observe that on the interval $0 \leq t \leq 1$ our model problem reduces to

$$\begin{aligned} y' + y &= 1, \qquad 0 \le t \le 1 \\ y(0) &= 1. \end{aligned} \tag{8}$$

The unique solution of this initial value problem is the equilibrium solution $y(t) = 1, 0 \le t \le 1$. Therefore, on the interval $[0, 1]$, the solution remains constant (as depicted in Figure 7.13). At time $t = 1$, the source "turns off" and the solution is displaced from its equilibrium value.

Solution of Parameter Identification Problems and the System Transfer Function

In the Introduction to this chapter, we posed the problem of studying a "black box" that housed a spring-mass-dashpot mechanical system (see Figure 7.2). Two specific questions posed were

(a) If we subject the initially quiescent system to a known force $f(t)$, starting at $t = 0$, and measure the subsequent displacement $y(t)$ for $t \ge 0$, can we use our measurements to predict what the displacement $\tilde{y}(t)$ would be if a different force $\tilde{f}(t)$ were applied?

(b) Can we use our knowledge of the input-output relation [that is, our knowledge of $f(t)$ and $y(t)$] to determine the mass m, the spring constant k and the damping coefficient γ of the unknown mechanical system?

The relevant mathematical problem is

$$\begin{aligned} my'' + \gamma y' + ky &= f(t), \qquad t > 0 \\ y(0) = 0, &\qquad y'(0) = 0. \end{aligned} \tag{9}$$

We now use Laplace transforms to provide affirmative answers to these two questions.

Taking Laplace transforms of both sides of equation (9) and noting the zero initial conditions, we have

$$ms^2 Y(s) + \gamma s Y(s) + kY(s) = F(s)$$

or

$$Y(s) = \left[\frac{1}{ms^2 + \gamma s + k}\right] F(s). \tag{10}$$

Although the computations in (10) are simple, the result is very important. In the time domain, we obtain the output $y(t)$ from the input $f(t)$ by solving an initial value problem. In the transform domain, however, we obtain the output $Y(s)$ from the input $F(s)$ by multiplying $F(s)$ by the function

$$\Phi(s) = \frac{1}{ms^2 + \gamma s + k}. \tag{11}$$

Note that the function $\Phi(s)$ in (11) depends only upon the mechanical system; it is sometimes referred to as the **system transfer function**. If we know $\Phi(s)$, we can determine the output $Y(s)$ arising from a given input $F(s)$ by multiplication (see equation (10) and Example 4 below). Conversely, if we know some input-output pair, $F(s)$ and $Y(s)$, then we can determine the system transfer function

$\Phi(s)$ by forming the quotient

$$\Phi(s) = \frac{Y(s)}{F(s)}.$$

The role of the system transfer function is shown schematically in Figure 7.14.

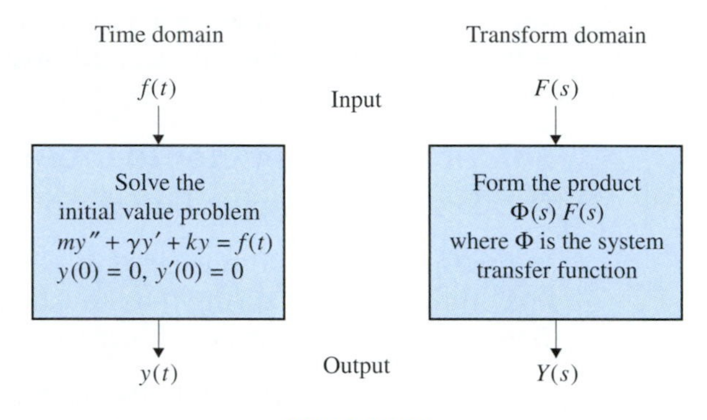

FIGURE 7.14

There are two ways to analyze the mechanical spring-mass-dashpot system. In the time domain, solve the initial value problem given by equation (9). In the transform domain, form the product of the system transfer function $\Phi(s)$ and the input $F(s)$, as in equation (10).

Example 4 illustrates how these transform domain ideas can be used to answer the two questions posed above.

EXAMPLE 4

Suppose we somehow know that the response of an initially quiescent mechanical system to an applied force can be modeled as the solution of the spring-mass-dashpot initial value problem

$$my'' + \gamma y' + y = f(t), \qquad t > 0$$
$$y(0) = 0, \qquad y'(0) = 0.$$

As in the discussion in the introduction to this chapter, assume that we know a certain applied force $f(t)$ and that we can measure the resulting displacement $y(t)$ for $t \geq 0$. We are, however, unable to directly determine the parameters m, γ, and k.

In particular, suppose that when we apply a unit step force, $f(t) = h(t)$, the displacement is measured to be

$$y(t) = -\frac{1}{2}e^{-t}\cos t - \frac{1}{2}e^{-t}\sin t + \frac{1}{2}, \qquad t \geq 0.$$

Use this information to:

(a) Predict the displacement should the force $\tilde{f}(t) = e^{-2t}, t \geq 0$ be applied.

(b) Determine the parameters m, γ, and k.

Solution: To solve the problem, we first compute Laplace transforms of the applied force $f(t) = h(t)$ and the ensuing response $y(t) = -(1/2)e^{-t}\cos t - (1/2)e^{-t}\sin t + (1/2)$. We obtain

$$F(s) = \frac{1}{s},$$

$$Y(s) = -\frac{1}{2}\frac{s+1}{(s+1)^2+1} - \frac{1}{2}\frac{1}{(s+1)^2+1} + \frac{1}{2s} = \frac{1}{s(s^2+2s+2)}.$$

In the transform domain, $Y(s) = \Phi(s)F(s)$, where $\Phi(s)$ is the system transfer function. Therefore, the system transfer function is given by

$$\Phi(s) = \frac{Y(s)}{F(s)} = \frac{1}{s^2+2s+2}. \tag{12}$$

Once we know the system transfer function, we can readily predict the output corresponding to any input.

(a) Suppose the applied force is $\tilde{f}(t) = e^{-2t}$. In this event, the Laplace transform of the applied force is $\tilde{F}(s) = 1/(s+2)$. We can now find the transform of the corresponding displacement, $\tilde{y}(t)$, from the relationship $\tilde{Y}(s) = \Phi(s)\tilde{F}(s)$,

$$\tilde{Y}(s) = \left[\frac{1}{s^2+2s+2}\right]\frac{1}{(s+2)} = \frac{1}{(s^2+2s+2)(s+2)}$$
$$= \frac{1}{2}\frac{1}{s+2} - \frac{1}{2}\frac{s+1}{(s+1)^2+1} + \frac{1}{2}\frac{1}{(s+1)^2+1}.$$

The corresponding time domain output is thus

$$\tilde{y}(t) = \mathcal{L}^{-1}\{\tilde{Y}(s)\} = \frac{1}{2}e^{-2t} - \frac{1}{2}e^{-t}\cos t + \frac{1}{2}e^{-t}\sin t, \qquad t \geq 0.$$

(b) The problem posed in question (b) can be solved by comparing the transfer function

$$\Phi(s) = \frac{1}{s^2+2s+2},$$

with the previously determined form of the transfer function in equation (11),

$$\Phi(s) = \frac{1}{ms^2+\gamma s+k}.$$

Comparing coefficients, we conclude that

$$m = 1, \qquad \gamma = 2, \qquad k = 2. \ \blacktriangle$$

In the preceding discussion, we assumed that the mechanical system was initially at rest and an applied force then activated the system at time $t = 0$. If the initial conditions are nonzero but known, however, the same general approach can be used. These ideas are developed in the exercises.

EXERCISES

Exercises 1–8:

Find the Laplace transform of the periodic function whose graph is shown.

1.

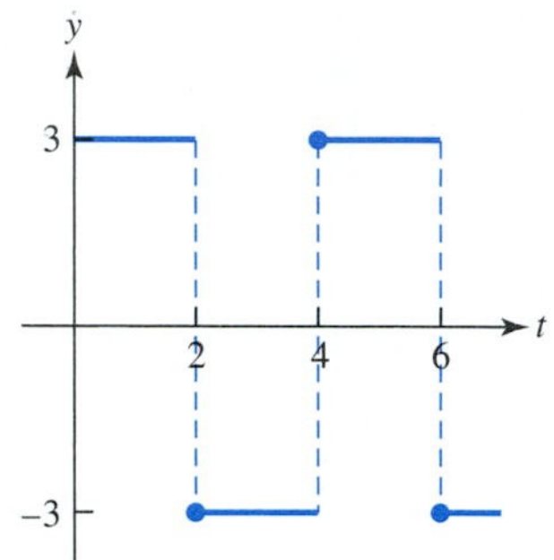

2.

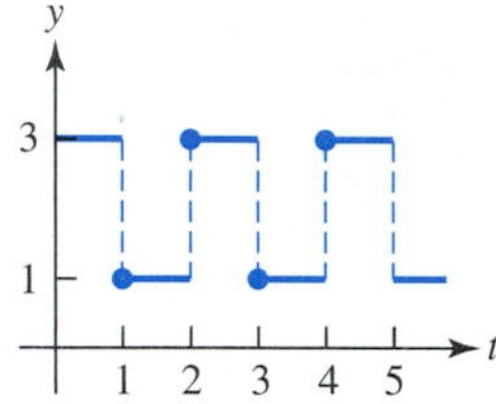

3.

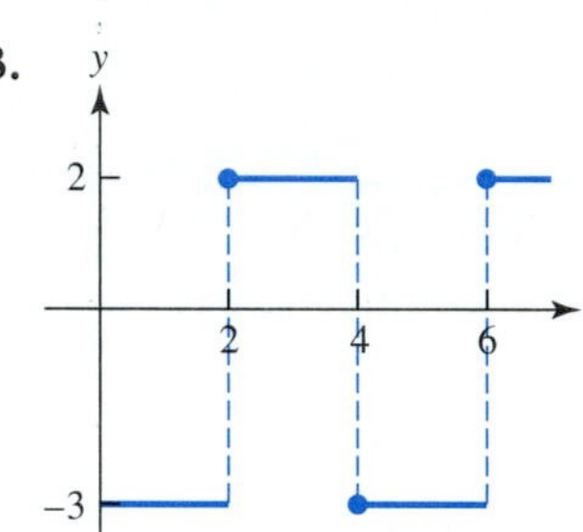

4.

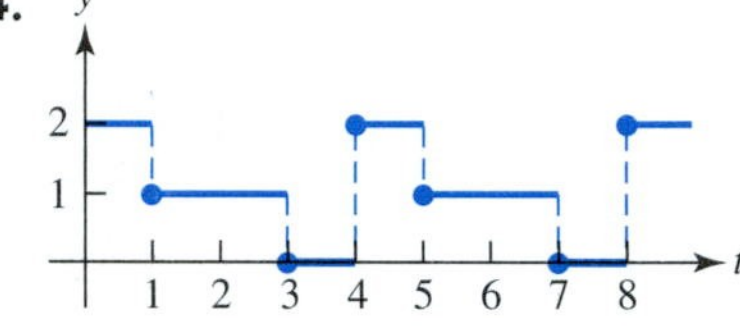

5.

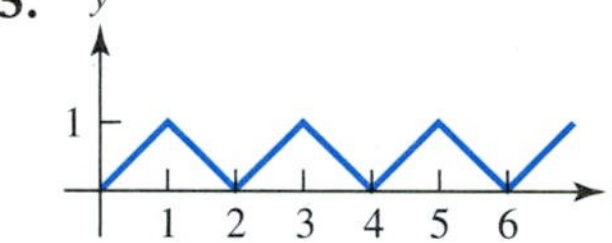

6.

7.

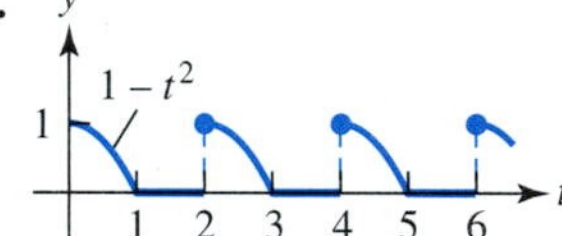

8.

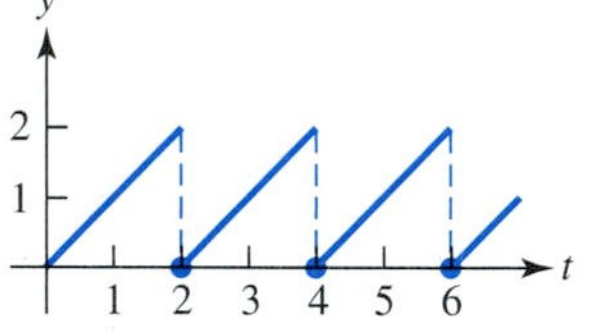

Exercises 9–12:

Sketch the graph of $f(t)$, state the period of $f(t)$, and find $\mathcal{L}\{f(t)\}$.

9. $f(t) = |\sin 2t|$

10. $f(t) = \begin{cases} \sin t, & 0 \le t < \pi, \\ 0, & \pi \le t < 2\pi, \end{cases} \qquad f(t + 2\pi) = f(t)$

11. $f(t) = e^{-t}, \;\; 0 \le t < 1, \;\; f(t+1) = f(t)$

12. $f(t) = 1 - e^{-t}, \;\; 0 \le t < 2, \;\; f(t+2) = f(t)$

13. Let α be a positive constant. As in Example 2, show that

$$\mathcal{L}^{-1}\left\{\frac{e^{-\alpha s}}{s(1 - e^{-\alpha s})}\right\} = h(t - \alpha) + h(t - 2\alpha) + h(t - 3\alpha) + \cdots.$$

Sketch the graph of $g(t) = h(t - \alpha) + h(t - 2\alpha) + h(t - 3\alpha) + \cdots$ for $\alpha = 1$ and $0 \le t < 5$.

Exercises 14–15:

In each exercise, use linearity of the inverse transformation and Exercise 13 to find $f(t) = \mathcal{L}^{-1}\{F(s)\}$ for the given transform $F(s)$. Sketch the graph of $f(t)$ for $0 \le t < 5$ in Exercise 14 and $0 \le t < 10$ in Exercise 15.

14. $F(s) = \dfrac{s^2 - s}{s^3} + \dfrac{e^{-s}}{s(1 - e^{-s})}$

15. $F(s) = \dfrac{3}{s^2} - \dfrac{3e^{-2s}}{s(1 - e^{-2s})}$

16. As in Example 2, find $f(t) = \mathcal{L}^{-1}\{F(s)\}$ for $F(s) = 1/2s^2 - (1/s^2)(e^{-2s}/(1 + e^{-2s}))$. Sketch the graph of $f(t)$ for $0 \le t < 12$.

17. A lake containing 50 million gallons of fresh water has a stream flowing through it. Water enters the lake at a constant rate of 5 million gal/day and leaves at the same rate. At some initial time, an upstream manufacturer begins to discharge pollutants into the feeder stream. Each day, during the hours of 8 A.M. to 8 P.M., the stream has a pollutant concentration of 1 mg/gal (10^{-6} kg/gal); at other times the stream feeds in fresh water. Assume that a well-stirred mixture leaves the lake and that the manufacturer operates seven days per week.

(a) Let $t = 0$ denote the instant that pollutants first enter the lake. Let $q(t)$ denote the amount of pollutant (kg) present in the lake at time t (days). Use a "conservation of pollutant" principle (rate of change = rate in − rate out) to formulate the initial value problem satisfied by $q(t)$.

(b) Apply Laplace transforms to the problem formulated in (a) and determine $Q(s) = \mathcal{L}\{q(t)\}$.

(c) Determine $q(t) = \mathcal{L}^{-1}\{Q(s)\}$, using the ideas of Example 2. In particular, what is $q(t)$ for $1 \le t < 2$, the second day of manfacturing?

18. An object having mass m is initially at rest on a frictionless horizontal surface. At time $t = 0$, a periodic force is applied horizontally to the object, causing it to move in the positive x-direction. The force, in newtons, is given by

$$f(t) = \begin{cases} f_0, & 0 \le t \le T/2, \\ 0, & T/2 < t < T, \end{cases} \qquad f(t + T) = f(t).$$

The initial value problem for the horizontal position, $x(t)$, of the object is $mx''(t) = f(t), x(0) = 0, x'(0) = 0$.

(a) Use Laplace transforms to determine the velocity, $v(t) = x'(t)$, and the position, $x(t)$, of the object.

(b) Let $m = 1$ kg, $f_0 = 1$ N and $T = 1$ s. What is the velocity, v, and position, x, of the object at $t = 1.25$ s?

Transfer Function Problems Consider the initial value problem

$$\begin{aligned} &ay'' + by' + cy = f(t), \qquad 0 < t < \infty \\ &y(0) = 0, \qquad y'(0) = 0 \end{aligned} \tag{13}$$

where a, b, and c are constants and $f(t)$ is a known function. We can view problem (13) as defining a linear system, as schematically shown in the figure on the next page, where $f(t)$ is a known input and the corresponding solution $y(t)$ is the output. As we have seen, Laplace transforms of the input and output functions satisfy the multiplicative relation, $Y(s) = \Phi(s)F(s)$, where $\Phi(s)$ is the system transfer function.

19. Show that the term "linear system" is appropriate. In particular, show that if an input $f_1(t)$ produces an output $y_1(t)$ and an input $f_2(t)$ produces an output $y_2(t)$, then the input $f(t) = c_1f_1(t) + c_2f_2(t)$ produces the output $y(t) = c_1y_1(t) + c_2y_2(t)$. [Hint: Use the superposition principle discussed in Section 4.8.]

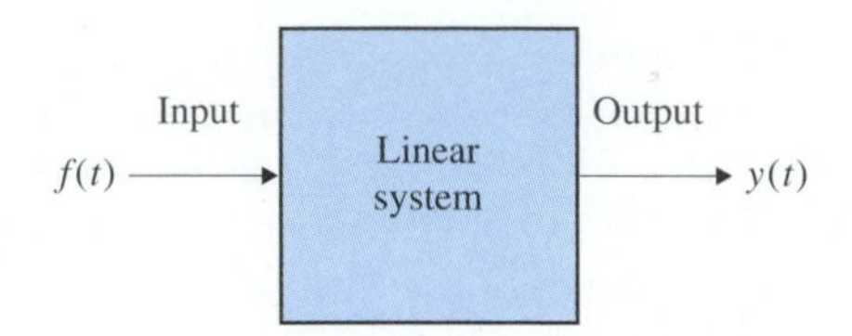

Figure for Exercises 19–29

20. Suppose that the transfer function for linear system (13) is $\Phi(s) = 1/(2s^2 + 5s + 2)$.

(a) What are the constants a, b, and c?

(b) If $f(t) = e^{-t}$, determine $F(s)$, $Y(s)$, and $y(t)$.

21. Suppose an input $f(t) = t$, when applied to linear system (13), produces the output $y(t) = 2(e^{-t} - 1) + t(e^{-t} + 1), t \geq 0$. What is the system transfer function $\Phi(s)$?

22. Suppose an input $f(t) = t$, when applied to linear system (13), produces the output $y(t) = 2(e^{-t} - 1) + t(e^{-t} + 1), t \geq 0$. What will be the output if a Heaviside unit step input $f(t) = h(t)$ is applied to the system?

Exercises 23–27:

For the linear system defined by the given initial value problem,

(a) Determine the system transfer function, $\Phi(s)$.

(b) Determine the Laplace transform of the output, $Y(s)$, corresponding to the specified input, $f(t)$.

23. $y'' + 4y = f(t), \quad y(0) = 0, \quad y'(0) = 0; \quad f(t) = t^2$

24. $y'' + y' + y = f(t), \quad y(0) = 0, \quad y'(0) = 0; \quad f(t) = \begin{cases} 1, & 0 \leq t \leq 1, \\ -1, & 1 < t < 2, \end{cases} \quad f(t+2) = f(t)$

25. $y'' + 4y' + 4y = f(t), \quad y(0) = 0, \quad y'(0) = 0; \quad f(t) = t, \quad 0 \leq t < 1, \quad f(t+1) = f(t)$

26. $y''' - 4y = f(t), \quad y(0) = 0, \quad y'(0) = 0, \quad y''(0) = 0; \quad f(t) = e^t + t$

27. $y''' + 4y' = f(t), \quad y(0) = 0, \quad y'(0) = 0, \quad y''(0) = 0; \quad f(t) = \cos 2t$

Exercises 28–29:

We now allow the initial values to be nonzero. Consider the initial value problem

$$y'' + by' + cy = f(t), \qquad 0 < t < \infty,$$
$$y(0) = y_0, \qquad y'(0) = y_0'.$$

The input function $f(t)$ and the Laplace transform of the output function, $Y(s)$, are given. Determine the constants b, c, y_0, and y_0'.

28. $f(t) = h(t)$, the Heaviside unit step function, $Y(s) = (s^2 + 2s + 1)/(s^3 + 3s^2 + 2s)$.

29. $f(t) = e^{-t}, Y(s) = (s^2 + s + 1)/((s + 1)(s^2 + 4))$

7.5 Solving Systems of Differential Equations

In this section we extend the definition of the Laplace transform to matrix-valued functions and take note of some simple consequences of the extension. We then see how to use Laplace transforms to solve problems involving systems of differential equations.

Laplace Transforms of Matrix-Valued Functions

As we saw in Section 6.1, the integral of a matrix-valued function is simply the matrix of integrals. Similarly, the Laplace transform of a matrix-valued function is the matrix of Laplace transforms. Consider the vector-valued function

$$\mathbf{y}(t) = \begin{bmatrix} y_1(t) \\ y_2(t) \\ \vdots \\ y_n(t) \end{bmatrix}, \tag{1}$$

where each of the component functions is piecewise continuous and exponentially bounded on $0 \le t < \infty$. The Laplace transform, $\mathcal{L}\{\mathbf{y}(t)\}$, is

$$\mathcal{L}\{\mathbf{y}(t)\} = \int_0^\infty \mathbf{y}(t)e^{-st}\,dt$$

$$= \int_0^\infty \begin{bmatrix} y_1(t) \\ y_2(t) \\ \vdots \\ y_n(t) \end{bmatrix} e^{-st}\,dt = \begin{bmatrix} \int_0^\infty y_1(t)e^{-st}\,dt \\ \int_0^\infty y_2(t)e^{-st}\,dt \\ \vdots \\ \int_0^\infty y_n(t)e^{-st}\,dt \end{bmatrix} = \begin{bmatrix} Y_1(s) \\ Y_2(s) \\ \vdots \\ Y_n(s) \end{bmatrix} = \mathbf{Y}(s). \tag{2}$$

We will use uppercase bold letters to denote the Laplace transform of a vector-valued function.

Similarly, the Laplace transform of an $(m \times n)$ matrix-valued function is the $(m \times n)$ matrix consisting of the Laplace transforms of the component functions. In general, if each component function of a matrix-valued function is Laplace transformable, we say that the matrix function itself is **Laplace transformable**.

EXAMPLE 1

Compute $\mathcal{L}\{\mathbf{y}(t)\}$, where

$$\mathbf{y}(t) = \begin{bmatrix} t \\ -1 \\ e^t \end{bmatrix}.$$

Solution: Using Table 7.1,

$$\mathbf{Y}(s) = \begin{bmatrix} \mathcal{L}\{t\} \\ \mathcal{L}\{-1\} \\ \mathcal{L}\{e^t\} \end{bmatrix} = \begin{bmatrix} \dfrac{1}{s^2} \\ -\dfrac{1}{s} \\ \dfrac{1}{s-1} \end{bmatrix}, \qquad s > 1.$$

Note that the domain of $\mathbf{Y}(s)$ is the intersection of the domains of the component functions. ▲

Some Useful Matrix Formulas

The following results can be established by taking Laplace transforms of each component function and then reassembling the components into a single expression.

1. Let A be a constant $(n \times n)$ matrix and let $\mathbf{y}(t)$ be an $(n \times p)$ Laplace transformable matrix function. Then

$$\mathcal{L}\{A\mathbf{y}(t)\} = A\mathcal{L}\{\mathbf{y}(t)\} = A\mathbf{Y}(s). \tag{3}$$

2. If each component function satisfies the appropriate hypotheses of Theorem 7.5, then

$$\begin{aligned} \mathcal{L}\{\mathbf{y}'(t)\} &= s\mathbf{Y}(s) - \mathbf{y}(0), \\ \mathcal{L}\{\mathbf{y}''(t)\} &= s^2\mathbf{Y}(s) - s\mathbf{y}(0) - \mathbf{y}'(0), \\ \mathcal{L}\left\{\int_0^t \mathbf{y}(u)\,du\right\} &= \frac{1}{s}\mathbf{Y}(s). \end{aligned} \tag{4}$$

Solution of the Initial Value Problem for a Nonhomogeneous System

Consider the initial value problem:

$$\mathbf{y}' = A\mathbf{y} + \mathbf{g}(t), \qquad t > 0, \qquad \mathbf{y}(0) = \mathbf{y}_0, \tag{5}$$

where $\mathbf{y}(t)$ is the $(n \times 1)$ vector of unknowns and A is a real-valued $(n \times n)$ constant matrix. We also assume the nonhomogeneous term,

$$\mathbf{g}(t) = \begin{bmatrix} g_1(t) \\ g_2(t) \\ \vdots \\ g_n(t) \end{bmatrix},$$

is a Laplace transformable vector function.

Using formulas (3) and (4), we can take the Laplace transform of system (5) and work directly with the matrices rather than dealing with the component equations. We obtain

$$s\mathbf{Y}(s) - \mathbf{y}(0) = A\mathbf{Y}(s) + \mathbf{G}(s),$$

or

$$(sI - A)\mathbf{Y}(s) = \mathbf{y}_0 + \mathbf{G}(s),$$

where $\mathbf{G}(s) = \mathcal{L}\{\mathbf{g}(t)\}$. The solution of the transform domain problem is therefore

$$\mathbf{Y}(s) = (sI - A)^{-1}[\mathbf{y}_0 + \mathbf{G}(s)]. \tag{6}$$

To compute the desired time domain solution, $\mathbf{y}(t) = \mathcal{L}^{-1}\{\mathbf{Y}(s)\}$, we compute the inverse Laplace transform of each component of $\mathbf{Y}(s)$. [Note that $(sI - A)^{-1}$ does not exist when s is an eigenvalue of A.]

EXAMPLE 2

Solve the initial value problem

$$\mathbf{y}' = A\mathbf{y} + \mathbf{g}(t), \qquad 0 \le t < \infty, \qquad \mathbf{y}(0) = \mathbf{y}_0,$$

where

$$A = \begin{bmatrix} 1 & 2 \\ 2 & 1 \end{bmatrix}, \qquad \mathbf{g}(t) = \begin{bmatrix} e^{2t} \\ -2t \end{bmatrix}, \qquad \mathbf{y}_0 = \begin{bmatrix} 1 \\ -2 \end{bmatrix}.$$

(We computed the general solution of this nonhomogeneous linear first order system earlier, in Example 1 of Section 6.8.)

Solution: Taking Laplace transforms and using equations (3) and (4), we obtain [as in equation (6)]:

$$\mathbf{Y}(s) = (sI - A)^{-1}[\mathbf{y}_0 + \mathbf{G}(s)]$$

where

$$\mathbf{G}(s) = \begin{bmatrix} \dfrac{1}{s-2} \\ -\dfrac{2}{s^2} \end{bmatrix}.$$

Note that

$$(sI - A)^{-1} = \begin{bmatrix} s-1 & -2 \\ -2 & s-1 \end{bmatrix}^{-1} = \frac{1}{(s+1)(s-3)} \begin{bmatrix} s-1 & 2 \\ 2 & s-1 \end{bmatrix}.$$

Therefore, the transform domain solution is

$$\mathbf{Y}(s) = \frac{1}{(s+1)(s-3)} \begin{bmatrix} s-1 & 2 \\ 2 & s-1 \end{bmatrix} \begin{bmatrix} 1 + \dfrac{1}{s-2} \\ -2 - \dfrac{2}{s^2} \end{bmatrix} = \begin{bmatrix} \dfrac{s^4 - 6s^3 + 9s^2 - 4s + 8}{s^2(s+1)(s-2)(s-3)} \\ \dfrac{-2s^4 + 8s^3 - 8s^2 + 6s - 4}{s^2(s+1)(s-2)(s-3)} \end{bmatrix}.$$

To obtain the time domain solution, we need to determine the inverse Laplace transform of each component of $\mathbf{Y}(s)$. Using a partial fraction expansion, we write $Y_1(s)$ and $Y_2(s)$ as

$$Y_1(s) = \frac{s^4 - 6s^3 + 9s^2 - 4s + 8}{s^2(s+1)(s-2)(s-3)} = \frac{4}{3}\frac{1}{s^2} - \frac{8}{9}\frac{1}{s} + \frac{7}{3}\frac{1}{s+1} - \frac{1}{3}\frac{1}{s-2} - \frac{1}{9}\frac{1}{s-3},$$

$$Y_2(s) = \frac{-2s^4 + 8s^3 - 8s^2 + 6s - 4}{s^2(s+1)(s-2)(s-3)} = -\frac{2}{3}\frac{1}{s^2} + \frac{10}{9}\frac{1}{s} - \frac{7}{3}\frac{1}{s+1} - \frac{2}{3}\frac{1}{s-2} - \frac{1}{9}\frac{1}{s-3}.$$

Therefore,

$$y_1(t) = \mathcal{L}^{-1}\{Y_1(s)\} = \frac{4}{3}t - \frac{8}{9} + \frac{7}{3}e^{-t} - \frac{1}{3}e^{2t} - \frac{1}{9}e^{3t},$$

$$y_2(t) = \mathcal{L}^{-1}\{Y_2(s)\} = -\frac{2}{3}t + \frac{10}{9} - \frac{7}{3}e^{-t} - \frac{2}{3}e^{2t} - \frac{1}{9}e^{3t}, \qquad t \ge 0.$$

(continued)

(continued)

We can regroup these terms into the following vector solution,

$$\mathbf{y}(t) = t\begin{bmatrix} \frac{4}{3} \\ -\frac{2}{3} \end{bmatrix} + \begin{bmatrix} -\frac{8}{9} \\ \frac{10}{9} \end{bmatrix} + e^{-t}\begin{bmatrix} \frac{7}{3} \\ -\frac{7}{3} \end{bmatrix} + e^{2t}\begin{bmatrix} -\frac{1}{3} \\ -\frac{2}{3} \end{bmatrix} + e^{3t}\begin{bmatrix} -\frac{1}{9} \\ -\frac{1}{9} \end{bmatrix}, \qquad t \ge 0. \tag{7}$$

As a check, you can compare solution (7) with the general solution obtained in Section 6.8, Example 1. What values of c_1 and c_2 are needed in Example 1 of Section 6.8 in order to replicate solution (7)? ▲

The System Transfer Function

The preceding discussion indicates that we can identify a system transfer function for a linear constant coefficient system. Consider, in particular, the transform domain solution given by equation (6),

$$\mathbf{Y}(s) = (sI - A)^{-1}[\mathbf{y}_0 + \mathbf{G}(s)].$$

The vector $\mathbf{y}_0 + \mathbf{G}(s)$ is the sum of the initial condition and the transformed nonhomogeneous term; this sum represents the system input. The system output, $\mathbf{Y}(s)$, is obtained by premultiplying the input by the square matrix $(sI - A)^{-1}$. Therefore, the matrix $(sI - A)^{-1}$ is the system transfer function (also called the system transfer matrix). Note that the system transfer matrix for $\mathbf{y}' = A\mathbf{y} + \mathbf{g}(t)$ depends only on the coefficient matrix A.

We now show that the system transfer matrix, $(sI - A)^{-1}$, is actually the Laplace transform of the exponential matrix, e^{tA}. To see why, consider the matrix initial value problem

$$\Phi' = A\Phi, \qquad \Phi(0) = I, \tag{8}$$

where A is a constant $(n \times n)$ matrix and I is the $(n \times n)$ identity matrix. As we saw in Section 6.11, the solution of initial value problem (8) is

$$\Phi(t) = e^{tA}.$$

However, when we take the Laplace transform of equation (8) we obtain

$$s\mathcal{L}\{\Phi\} - I = A\mathcal{L}\{\Phi\}.$$

Solving for $\mathcal{L}\{\Phi\}$, we find $\mathcal{L}\{\Phi\} = (sI - A)^{-1}$. But, since $\Phi(t) = e^{tA}$, we are led to

$$\mathcal{L}\{e^{tA}\} = (sI - A)^{-1}.$$

This equation is an elegant generalization of the familiar formula

$$\mathcal{L}\{e^{\alpha t}\} = (s - \alpha)^{-1}.$$

A Network Example

Laplace transforms provide a convenient tool for analyzing networks having a more complicated structure than the single loop or single node networks we have studied thus far. As an example, consider the two-loop network shown in Figure 7.15. We assume that the network is initially quiescent; that is, both

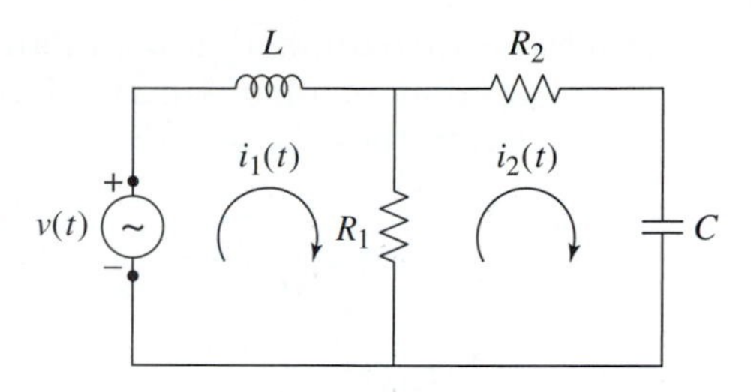

FIGURE 7.15

A two-loop network. The loop currents, $i_1(t)$ and $i_2(t)$, are found by solving the linear system (9).

loop currents are zero at time $t = 0$ and the capacitor has no initial charge. At time $t = 0$, the voltage source $v(t)$ is turned on.

The mathematical description of this network's behavior is obtained by applying Kirchhoff's voltage law to each loop; as we traverse a loop in the clockwise direction, the sum of the voltage rises must equal the sum of the voltage drops. Thus, we obtain the following system of equations:

$$\begin{aligned} v(t) &= L\frac{di_1}{dt} + R_1 i_1 - R_1 i_2 \\ 0 &= R_1 i_2 - R_1 i_1 + R_2 i_2 + \frac{1}{C}\int_0^t i_2(\lambda)\, d\lambda \\ i_1(0) &= i_2(0) = 0. \end{aligned} \tag{9}$$

Taking the Laplace transform in equation (9), we obtain the following system of two linear equations for $I_1(s)$ and $I_2(s)$:

$$\begin{aligned} V(s) &= sLI_1(s) + R_1 I_1(s) - R_1 I_2(s) \\ 0 &= R_1 I_2(s) - R_1 I_1(s) + R_2 I_2(s) + \frac{1}{Cs} I_2(s). \end{aligned}$$

This system can be written in matrix form as

$$\begin{bmatrix} R_1 + sL & -R_1 \\ -R_1 & R_1 + R_2 + \dfrac{1}{Cs} \end{bmatrix} \begin{bmatrix} I_1(s) \\ I_2(s) \end{bmatrix} = \begin{bmatrix} V(s) \\ 0 \end{bmatrix}, \tag{10}$$

where $V(s)$ is the Laplace transform of the known voltage $v(t)$. Note that equation (10) incorporates the initial conditions $i_1(0) = i_2(0) = 0$. We obtain the transform domain solution,

$$\begin{bmatrix} I_1(s) \\ I_2(s) \end{bmatrix} = \frac{sV(s)}{(R_1 + R_2)Ls^2 + \left(R_1 R_2 + \dfrac{L}{C}\right)s + \dfrac{R_1}{C}} \begin{bmatrix} R_1 + R_2 + \dfrac{1}{Cs} \\ R_1 \end{bmatrix}.$$

As a convenient particular case, we'll assume the following network element values

$$R_1 = R_2 = 1\ \text{k}\Omega, \qquad L = 0.5\ \text{H}, \qquad C = 0.5\ \mu\text{F}.$$

Likewise we assume that the input voltage is $v(t) = h(t)$, where $h(t)$ is the unit step function; in other words, a 1-volt DC voltage source is switched on at time $t = 0$. Given these values, the transform domain solutions in (10) become

$$I_1(s) = \frac{2(s+1)}{s(s^2+2s+2)} = \frac{1}{s} - \left[\frac{s+1}{(s+1)^2+1} - \frac{1}{(s+1)^2+1}\right]$$
$$I_2(s) = \frac{1}{s^2+2s+2} = \frac{1}{(s+1)^2+1}.$$

Therefore, the resulting time domain network loop currents are

$$\begin{aligned} i_1(t) &= 1 - e^{-t}[\cos t - \sin t], \\ i_2(t) &= e^{-t}\sin t, \qquad t \geq 0 \end{aligned} \tag{11}$$

where the units of current and time are milliamperes and milliseconds, respectively.

The loop currents behave qualitatively as one would expect. In particular, as $t \to \infty$ the current in Loop 1 approaches a constant unit value and the current in Loop 2 tends to zero. In the limit, the inductor voltage tends to zero and the unit current produces a voltage drop across resistor R_1 equal to the source voltage of one volt. In Loop 2, the capacitor voltage,

$$\frac{1}{C}\int_0^t i_2(\lambda)\,d\lambda = 2\int_0^t e^{-\lambda}\sin\lambda\,d\lambda = 1 - e^{-t}[\sin t + \cos t],$$

tends to unity as $t \to \infty$. In this loop, the voltage across resistor R_2 tends to zero; the buildup of charge and accompanying voltage drop across the capacitor ultimately balances the voltage across resistor R_1. Graphs of the loop currents, $i_1(t)$ and $i_2(t)$, are displayed in Figure 7.16.

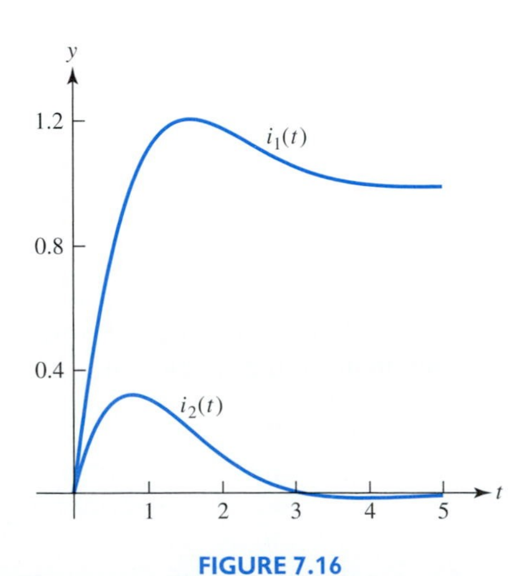

FIGURE 7.16

Graphs of the time domain loop currents given in equation (11).

EXERCISES

Exercises 1–5:

Compute the Laplace transform of the given matrix-valued function $\mathbf{y}(t)$.

1. $\mathbf{y}(t) = \begin{bmatrix} \cos t \\ t \\ te^t \end{bmatrix}$

2. $\mathbf{y}(t) = \dfrac{d}{dt}\begin{bmatrix} e^{-t}\cos 2t \\ 0 \\ t + e^t \end{bmatrix}$

3. $\mathbf{y}(t) = \begin{bmatrix} 1 & -1 \\ 0 & 2 \end{bmatrix}\begin{bmatrix} 2t \\ h(t-2) \end{bmatrix}$

4. $\mathbf{y}(t) = \displaystyle\int_0^t \begin{bmatrix} 1 \\ \lambda \\ e^{-\lambda} \end{bmatrix} d\lambda$

5. $\mathbf{y}(t) = \begin{bmatrix} h(t-1)\sin(t-1) & 0 \\ e^{t-1} & t \end{bmatrix}\begin{bmatrix} 1 \\ -2 \end{bmatrix}$

Exercises 6–8:

Compute the inverse Laplace transform of the given matrix function $\mathbf{Y}(s)$.

6. $\mathbf{Y}(s) = \begin{bmatrix} \dfrac{1}{s} \\ \dfrac{2}{s^2+2s+2} \\ \dfrac{1}{s^2+s} \end{bmatrix}$

7. $\mathbf{Y}(s) = e^{-s}\begin{bmatrix} 1 & -1 \\ 0 & 2 \end{bmatrix}\begin{bmatrix} \dfrac{1}{s} \\ \dfrac{1}{s^2+1} \end{bmatrix}$

8. $\mathbf{Y}(s) = \begin{bmatrix} 1 & -1 & 2 \\ 2 & 0 & 3 \\ 1 & -2 & 1 \end{bmatrix}\begin{bmatrix} \mathcal{L}\{t^3\} \\ \mathcal{L}\{e^{2t}\} \\ \mathcal{L}\{\sin t\} \end{bmatrix}$

Exercises 9–20:

Use Laplace transforms to solve the given initial value problem.

9. $\mathbf{y}' = \begin{bmatrix} 5 & -4 \\ 5 & -4 \end{bmatrix}\mathbf{y}, \quad \mathbf{y}(0) = \begin{bmatrix} 5 \\ 6 \end{bmatrix}$

10. $\mathbf{y}' = \begin{bmatrix} 5 & -4 \\ 5 & -4 \end{bmatrix}\mathbf{y} + \begin{bmatrix} 0 \\ 1 \end{bmatrix}, \quad \mathbf{y}(0) = \begin{bmatrix} 0 \\ 0 \end{bmatrix}$

11. $\mathbf{y}' = \begin{bmatrix} 5 & -4 \\ 3 & -2 \end{bmatrix}\mathbf{y} + \begin{bmatrix} t \\ 1 \end{bmatrix}, \quad \mathbf{y}(0) = \begin{bmatrix} 0 \\ 0 \end{bmatrix}$

12. $\mathbf{y}' = \begin{bmatrix} 5 & -4 \\ 3 & -2 \end{bmatrix}\mathbf{y}, \quad \mathbf{y}(0) = \begin{bmatrix} 3 \\ 2 \end{bmatrix}$

13. $\mathbf{y}' = \begin{bmatrix} 1 & 4 \\ -1 & 1 \end{bmatrix}\mathbf{y}, \quad \mathbf{y}(0) = \begin{bmatrix} 2 \\ 0 \end{bmatrix}$

14. $\mathbf{y}' = \begin{bmatrix} 1 & 4 \\ -1 & 1 \end{bmatrix}\mathbf{y} + \begin{bmatrix} 0 \\ 3e^t \end{bmatrix}, \quad \mathbf{y}(0) = \begin{bmatrix} 3 \\ 0 \end{bmatrix}$

15. $\mathbf{y}' = \begin{bmatrix} 6 & -3 \\ 8 & -5 \end{bmatrix}\mathbf{y}, \quad \mathbf{y}(1) = \begin{bmatrix} 5 \\ 10 \end{bmatrix}$ [Hint: Make the change of variable $\tau = t - 1$.]

16. $\mathbf{y}'' = \begin{bmatrix} -3 & -2 \\ 4 & 3 \end{bmatrix}\mathbf{y}, \quad \mathbf{y}(0) = \begin{bmatrix} 1 \\ 0 \end{bmatrix}, \quad \mathbf{y}'(0) = \begin{bmatrix} 0 \\ 1 \end{bmatrix}$

17. $\mathbf{y}'' = \begin{bmatrix} 1 & -1 \\ 1 & -1 \end{bmatrix}\mathbf{y} + \begin{bmatrix} t \\ 1 \end{bmatrix}, \quad \mathbf{y}(0) = \begin{bmatrix} 0 \\ 0 \end{bmatrix}, \quad \mathbf{y}'(0) = \begin{bmatrix} 0 \\ 0 \end{bmatrix}$

18. $\mathbf{y}'' = \begin{bmatrix} 1 & -1 \\ 1 & -1 \end{bmatrix}\mathbf{y} + \begin{bmatrix} 2 \\ 1 \end{bmatrix}, \quad \mathbf{y}(0) = \begin{bmatrix} 0 \\ 1 \end{bmatrix}, \quad \mathbf{y}'(0) = \begin{bmatrix} 0 \\ 0 \end{bmatrix}$

19. $\mathbf{y}' = \begin{bmatrix} 6 & 5 & 0 \\ -7 & -6 & 0 \\ 0 & 0 & -2 \end{bmatrix} \mathbf{y}, \quad \mathbf{y}(0) = \begin{bmatrix} 2 \\ -4 \\ -1 \end{bmatrix}$

20. $\mathbf{y}' = \begin{bmatrix} 1 & 0 & 0 \\ 0 & -1 & 1 \\ 0 & 0 & 2 \end{bmatrix} \mathbf{y} + \begin{bmatrix} e^t \\ 1 \\ -2t \end{bmatrix}, \quad \mathbf{y}(0) = \begin{bmatrix} 0 \\ 0 \\ 0 \end{bmatrix}$

21. The Laplace transform was applied to the initial value problem $\mathbf{y}' = A\mathbf{y}, \mathbf{y}(0) = \mathbf{y}_0$, where $\mathbf{y}(t) = \begin{bmatrix} y_1(t) \\ y_2(t) \end{bmatrix}$, A is a (2×2) constant matrix, and $\mathbf{y}_0 = \begin{bmatrix} y_{1,0} \\ y_{2,0} \end{bmatrix}$. The following transform domain solution was obtained:

$$\mathcal{L}\{\mathbf{y}(t)\} = \mathbf{Y}(s) = \frac{1}{s^2 - 9s + 18} \begin{bmatrix} s-2 & -1 \\ 4 & s-7 \end{bmatrix} \begin{bmatrix} y_{1,0} \\ y_{2,0} \end{bmatrix}$$

(a) What are the eigenvalues of the coefficient matrix A?

(b) What is the coefficient matrix A?

22. A System Cascade Consider the linear system defined as follows:

$$\mathbf{y}_1' = A\mathbf{y}_1 + \mathbf{g}(t), \qquad \mathbf{y}_1(0) = \mathbf{0},$$
$$\mathbf{y}_2' = A\mathbf{y}_2 + \mathbf{y}_1(t), \qquad \mathbf{y}_2(0) = \mathbf{0},$$

where $\mathbf{y}_1(t), \mathbf{y}_2(t)$, and $\mathbf{g}(t)$ are (2×1) vector functions and A is a (2×2) constant matrix. A schematic of the system is shown in the figure. It consists of two identical stages connected in cascade. The input $\mathbf{g}(t)$ is applied to the first stage, producing an output $\mathbf{y}_1(t)$. This output is then used as input to the second stage, producing an output $\mathbf{y}_2(t)$. We can view this cascade connection as forming an overall linear system determined by input $\mathbf{g}(t)$ and output $\mathbf{y}_2(t)$. Let $\mathbf{Y}_1(s), \mathbf{Y}_2(s)$, and $\mathbf{G}(s)$ denote the Laplace transforms of $\mathbf{y}_1(t), \mathbf{y}_2(t)$, and $\mathbf{g}(t)$, respectively.

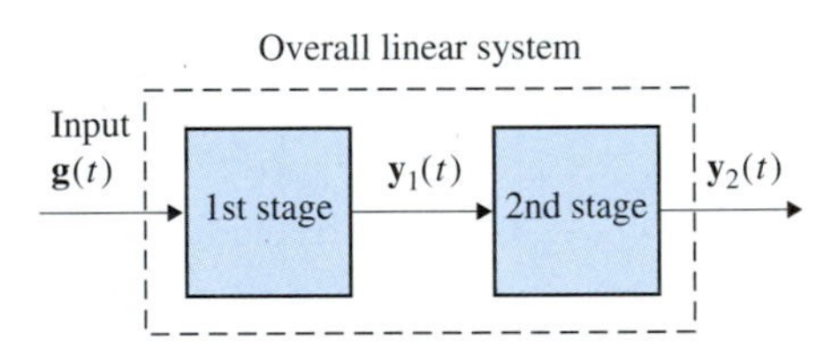

Figure for Exercise 22

(a) Show that $\mathbf{Y}_2(s)$ and $\mathbf{G}(s)$ are related by an equation of the form $\mathbf{Y}_2(s) = \Omega(s)\mathbf{G}(s)$, where $\Omega(s)$ is a (2×2) matrix transfer function for the cascade system. How is $\Omega(s)$ related to the coefficient matrix A?

(b) Suppose

$$A = \begin{bmatrix} 1 & -1 \\ 1 & -1 \end{bmatrix} \quad \text{and} \quad g(t) = \begin{bmatrix} 1 \\ t \end{bmatrix}.$$

Determine $\Omega(s)$ and $y_2(t), t \geq 0$.

Exercises 23–24:

System Identification We consider a system analog of the parameter identification problem studied in Section 7.4. Assume that a linear system can be modeled by the initial value problem $\mathbf{y}' = A\mathbf{y} + \mathbf{g}(t), \mathbf{y}(0) = \mathbf{y}_0$.

Assume we can select the input $\mathbf{g}(t)$ and the initial state $\mathbf{y}_0$ and can measure the output $\mathbf{y}(t)$, but we have no direct way of measuring the coefficient matrix A. The task is to determine A by exciting the system with an appropriate selection of inputs and/or initial states and measuring the corresponding outputs. Exercises 23–24 treat particular two-dimensional cases.

In each exercise, use the given input-output information to determine the coefficient matrix A. One approach is to use Laplace transforms. Let $\mathbf{Y}(s)$ and $\mathbf{G}(s)$ represent the Laplace transforms of $\mathbf{y}(t)$ and $\mathbf{g}(t)$, respectively. Then, we know that $\mathbf{Y}(s) = (sI - A)^{-1}(\mathbf{y}_0 + \mathbf{G}(s))$. Form (2×2) matrices

$$[\mathbf{Y}_1(s), \mathbf{Y}_2(s)] \quad \text{and} \quad [\mathbf{y}_{1,0} + \mathbf{G}_1(s), \mathbf{y}_{2,0} + \mathbf{G}_2(s)],$$

using the transformed information as columns. We can obtain an equation relating these two (2×2) matrices and use this equation to determine A.

23. When $\mathbf{g}(t) = \mathbf{0}$ and $\mathbf{y}_0 = \begin{bmatrix} 1 \\ 1 \end{bmatrix}$, the observed output is $\mathbf{y}(t) = \begin{bmatrix} e^{3t} \\ e^{3t} \end{bmatrix}$. When $\mathbf{g}(t) = \mathbf{0}$ and $\mathbf{y}_0 = \begin{bmatrix} 0 \\ 5 \end{bmatrix}$, the observed output is $\mathbf{y}(t) = \begin{bmatrix} 3e^{-2t} - 3e^{3t} \\ 8e^{-2t} - 3e^{3t} \end{bmatrix}$. Determine coefficient matrix A.

24. When $\mathbf{g}(t) = \mathbf{0}$ and $\mathbf{y}_0 = \begin{bmatrix} 0 \\ 1 \end{bmatrix}$, the observed output is $\mathbf{y}(t) = \begin{bmatrix} te^{-2t} \\ e^{-2t} \end{bmatrix}$. When $\mathbf{g}(t) = \begin{bmatrix} 2 \\ 0 \end{bmatrix}$ and $\mathbf{y}_0 = \begin{bmatrix} 1 \\ 0 \end{bmatrix}$, the observed output is $\mathbf{y}(t) = \begin{bmatrix} 1 \\ 0 \end{bmatrix}$. Determine the coefficient matrix A.

25. For the network shown, both loop currents are initially zero and no charge is present on the capacitor. At time $t = 0$, both voltage sources are turned on. An application of Kirchhoff's voltage law, equating the algebraic sum of the voltage drops in a clockwise traversal of each loop to zero, leads to the system of equations

$$-v_1(t) + R_1 i_1 + L\frac{di_1}{dt} + R_2(i_1 - i_2) = 0, \qquad i_1(0) = 0$$

$$R_2(i_2 - i_1) + \frac{1}{C}\int_0^t i_2(\lambda)\,d\lambda + R_3 i_2 + v_2(t) = 0, \qquad i_2(0) = 0.$$

(a) Apply the Laplace transform to this system of equations. Solve the transformed system of equations for the (2×1) vector of transformed loop currents,

$$\begin{bmatrix} I_1(s) \\ I_2(s) \end{bmatrix}.$$

(b) For simplicity, let $R_1 = R_2 = R_3 = 1\,\text{k}\Omega$, $L = 1$ H, and $C = 1\ \mu\text{F}$; let $v_1(t) = v_2(t) = te^{-t}$ volts. Solve for the currents, $i_1(t)$ and $i_2(t)$, $t > 0$, (the units being mA).

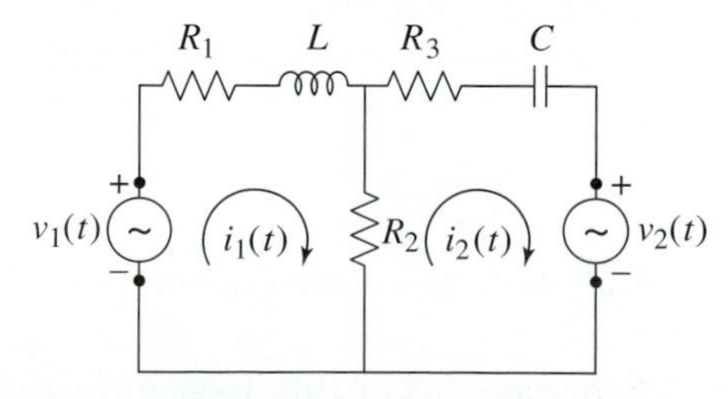

Figure for Exercise 25

7.6 Convolution

When we use Laplace transforms, we often need to find the inverse transform of a product,

$$\mathcal{L}^{-1}\{F(s)G(s)\}.$$

For example, we have seen that the Laplace transform of a system output is the product of the system transfer function and the Laplace transform of the system input. To obtain the time domain output, we must determine the inverse Laplace transform of this product. It is clear from the integral definition of the Laplace transform that the inverse transform of a product of transforms is *not* the product of the inverse transforms. What, then, is it?

In this section we introduce a mathematical operation known as "convolution." The convolution operation, denoted by the symbol $*$, starts with two functions $f(t)$ and $g(t)$ defined on $0 \le t < \infty$ and creates a new function, $f * g$, also defined on $0 \le t < \infty$. After we define the convolution operation, we will state the convolution theorem; this theorem shows that the Laplace transform of the newly created function $f * g$ is, in fact, the product of the Laplace transforms of the two original functions. Thus, by the convolution theorem,

$$\mathcal{L}^{-1}\{F(s)G(s)\} = (f * g)(t),$$

where $\mathcal{L}\{f(t)\} = F(s)$ and $\mathcal{L}\{g(t)\} = G(s)$.

Although the terminology we introduce is new, we will see that convolution is an operation we have already encountered several times in our study of linear constant coefficient differential equations.

The Convolution Integral

Let $f(t)$ and $g(t)$ be two functions defined on $0 \le t < \infty$. The **convolution of $f(t)$ and $g(t)$**, denoted $f * g$, is the function defined by

$$(f * g)(t) = \int_0^t f(t-\lambda)g(\lambda)\,d\lambda, \qquad 0 \le t < \infty, \tag{1}$$

provided the integral exists. It can be shown that integral (1) exists whenever $f(t)$ and $g(t)$ are piecewise continuous on $0 \le t < \infty$. Moreover, the function $(f * g)(t)$ is piecewise continuous and exponentially bounded on $0 \le t < \infty$ if both $f(t)$ and $g(t)$ possess these properties.

As equation (1) indicates, we use the notation $(f * g)(t)$ to denote the newly created function of t. When we want to designate the convolution of specific functions such as $f(t) = e^{-t}$ and $g(t) = \sin 2t$, we may simply write

$$e^{-t} * \sin 2t.$$

EXAMPLE 1

Calculate the convolution $f * g$ where $f(t) = t$ and $g(t) = e^{-t}$.

Solution: According to definition (1),

$$t * e^{-t} = \int_0^t (t-\lambda)e^{-\lambda}\,d\lambda = t\int_0^t e^{-\lambda}\,d\lambda - \int_0^t \lambda e^{-\lambda}\,d\lambda.$$

Evaluating these integrals, we find

$$t * e^{-t} = t\left[-e^{-\lambda}\right]_0^t - \left[e^{-\lambda}(-\lambda - 1)\right]_0^t = t + e^{-t} - 1. \ \blacktriangle$$

The next example illustrates convolution from a geometric point of view.

EXAMPLE 2

Calculate the convolution $f * g$ where

$$f(t) = \begin{cases} t, & 0 \le t < 1 \\ 0, & 1 \le t < \infty, \end{cases} \qquad g(t) = \begin{cases} 0, & 0 \le t < 2 \\ 1, & 2 \le t < 3 \\ 0, & 3 \le t < \infty. \end{cases}$$

Solution: The piecewise definition of these two functions provides an opportunity to illustrate the graphical aspects of the convolution operation,

$$(f * g)(t) = \int_0^t f(t - \lambda)g(\lambda)\, d\lambda.$$

The functions in the integrand, $f(\lambda)$ and $g(\lambda)$, are shown in Figure 7.17(a) on the next page. Forming $f(t - \lambda)$ reverses the orientation of the right triangle and translates (or slides) the triangle so that the intersection point of its hypotenuse with the λ-axis occurs at $\lambda = t$, see Figure 7.17(b). As t increases, we can envision this triangle as translating to the right and passing through the region $2 \le \lambda < 3$, where $g(\lambda) \ne 0$ (see Figures 7.17(c) and 7.17(d)). At each value of t, the integrand is nonzero only in the overlap region of the right triangle [the graph of $f(t - \lambda)$] and the rectangle [the graph of $g(\lambda)$]. For t in the interval $[2, 4]$, the value $(f * g)(t)$ is equal to the area of the overlap region. Therefore, the convolution integral $\int_0^t f(t - \lambda)g(\lambda)\, d\lambda$ can be evaluated graphically, as is shown in Figures 7.17(b) through 7.17(d). We find,

$$(f * g)(t) = \begin{cases} 0, & 0 \le t < 2 \\ \dfrac{1}{2}(t-2)^2, & 2 \le t < 3 \\ \dfrac{1}{2}(4-t)(t-2), & 3 \le t < 4 \\ 0, & 4 \le t < \infty. \end{cases}$$

The graph of the resulting function, $(f * g)(t)$, is given in Figure 7.17(f). ▲

Algebraic Properties of the Convolution Operation

Let f, g, and k be three scalar functions defined on $0 \le t < \infty$ and let c_1 and c_2 represent arbitrary constants. It can be shown that

$$f * (c_1 g + c_2 k) = c_1(f * g) + c_2(f * k), \tag{2a}$$

$$f * g = g * f, \tag{2b}$$

$$(f * g) * k = f * (g * k). \tag{2c}$$

The distributive property, equation (2a), says that the convolution of a function f with a linear combination of functions equals the linear combination of the

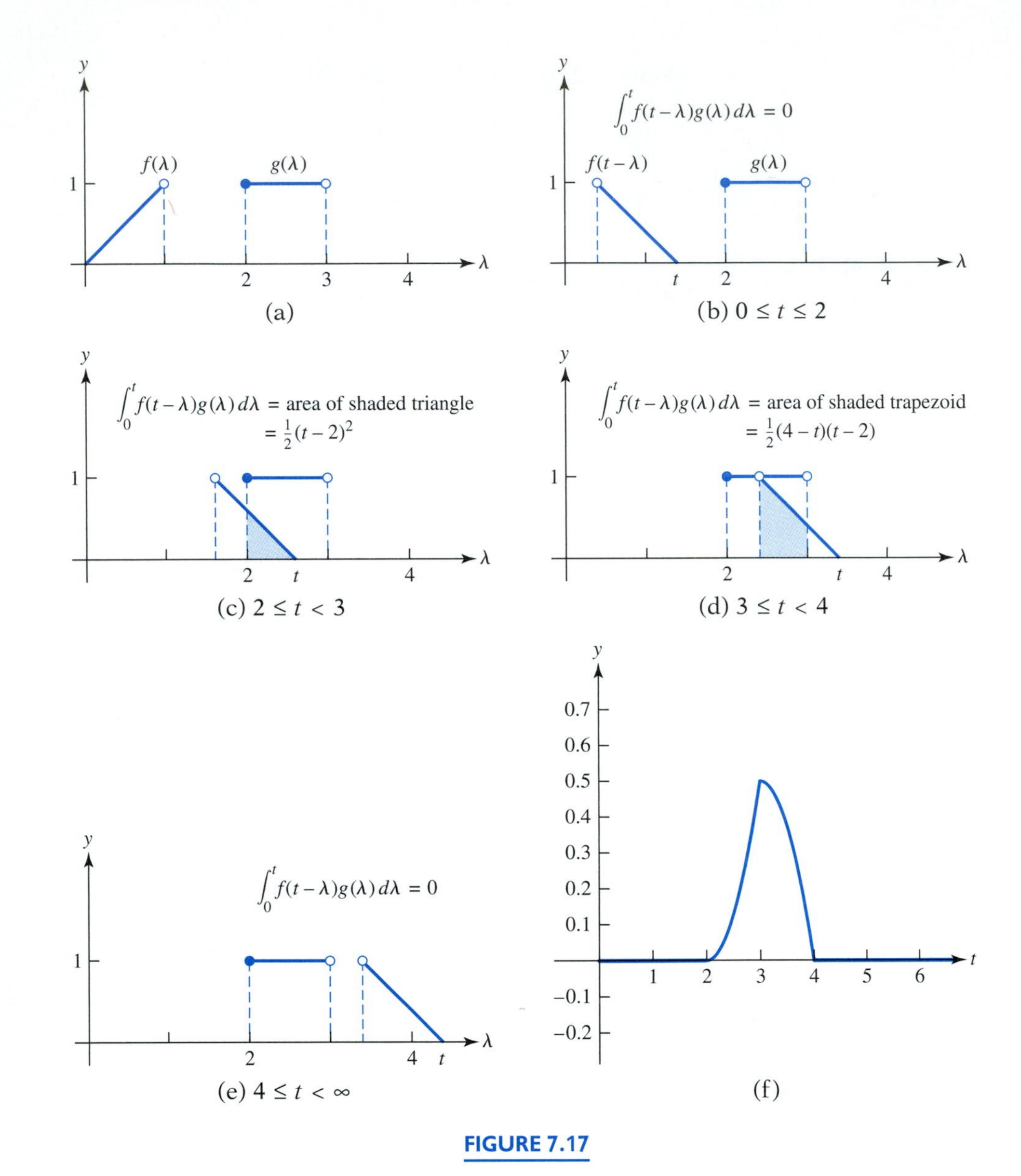

FIGURE 7.17

A graphical interpretation of the calculation $(f * g)(t)$, where $f(t)$ and $g(t)$ are the functions shown in (a); the details are given in Example 2. A graph of the function $(f * g)(t)$ is shown in (f). Note that $(f * g)(t)$ is continuous even though $f(t)$ and $g(t)$ have jump discontinuities.

convolutions. Property (2b) asserts that the convolution operation is commutative; that is, the order in which we choose the two functions doesn't matter; see Exercise 1. The associative property, equation (2c), says the convolution of three functions can be done in any order. Therefore, parentheses are unnecessary in (2c) and we can simply write $f * g * k$.

Some Remarks about Convolution

While we have defined the convolution integral only for scalar functions $f(t)$ and $g(t)$, it should be clear that the definition can be extended to compatibly dimensioned matrix-valued functions. For example, if $\mathbf{f}(t)$ is an $(m \times n)$ matrix function and $\mathbf{g}(t)$ is an $(n \times p)$ matrix function, then $(\mathbf{f} * \mathbf{g})(t)$ is the $(m \times p)$ matrix function defined by

$$(\mathbf{f} * \mathbf{g})(t) = \int_0^t \mathbf{f}(t - \lambda)\mathbf{g}(\lambda)\, d\lambda, \qquad 0 \le t < \infty. \tag{3}$$

We also note that we have encountered the convolution integral in previous chapters even though we did not use the term convolution there. For example:

(a) In Chapter 2, we saw that the solution of the initial value problem $y' + \alpha y = g(t), y(0) = y_0$ is given by

$$y(t) = e^{-\alpha t}y_0 + \int_0^t e^{-\alpha(t-\lambda)}g(\lambda)\, d\lambda.$$

The integral term is a convolution integral; therefore we can interpret the solution as

$$y(t) = e^{-\alpha t}y_0 + e^{-\alpha t} * g(t). \tag{4a}$$

(b) In the discussion of first order constant coefficient linear systems in Section 6.8, we developed the variation of parameters formula for the solution of the initial value problem

$$\mathbf{y}' = A\mathbf{y} + \mathbf{g}(t), \qquad \mathbf{y}(0) = \mathbf{y}_0.$$

The solution was found to be

$$\mathbf{y}(t) = \Phi(t)\mathbf{y}_0 + \int_0^t \Phi(t - \lambda)\mathbf{g}(\lambda)\, d\lambda,$$

where $\Phi(t)$ is the fundamental matrix that reduces to the identity matrix at $t = 0$. Therefore,

$$\mathbf{y}(t) = \Phi(t)\mathbf{y}_0 + \Phi(t) * \mathbf{g}(t). \tag{4b}$$

In Section 6.11, we saw that $\Phi(t) = e^{tA}$. Therefore, we can also write the solution as

$$\mathbf{y}(t) = e^{tA}\mathbf{y}_0 + e^{tA} * \mathbf{g}(t).$$

(c) In our discussion of the scalar parameter identification problem in the Introduction to this chapter, the solution of the initial value problem

$$my'' + \gamma y' + ky = g(t), \qquad y(0) = 0, \qquad y'(0) = 0$$

was found to be

$$y(t) = \int_0^t \frac{e^{-(\gamma/2m)(t-s)} \sin\left(\sqrt{\dfrac{k}{m} - \dfrac{\gamma^2}{4m^2}}\,(t-s)\right)}{m\sqrt{\dfrac{k}{m} - \dfrac{\gamma^2}{4m^2}}}\, g(s)\, ds, \qquad t \ge 0.$$

The solution is thus given by the convolution

$$y(t) = \left[\frac{e^{-(\gamma/2m)t} \sin \sqrt{\dfrac{k}{m} - \dfrac{\gamma^2}{4m^2}}\, t}{m\sqrt{\dfrac{k}{m} - \dfrac{\gamma^2}{4m^2}}} \right] * g(t). \tag{4c}$$

The Convolution Theorem

Equations (4a)–(4c) show three instances where the solution of an initial value problem can be related to convolution of functions in the time domain. Theorem 7.7 (the convolution theorem) establishes the connection between convolution of functions in the time domain and multiplication of Laplace transforms in the transform domain.

Theorem 7.7

Let $f(t)$ and $g(t)$ be piecewise continuous and exponentially bounded functions defined on $0 \le t < \infty$. Let $F(s)$ and $G(s)$ denote their respective Laplace transforms. Then $(f * g)(t)$ is a Laplace transformable function and its Laplace transform equals the product of $F(s)$ and $G(s)$; that is,

$$\mathcal{L}\{f * g\} = F(s)G(s). \tag{5}$$

PROOF: We first show that $f * g$ is Laplace transformable. Then we establish the result in equation (5).

The function $f * g$ is piecewise continuous on $0 \le t < \infty$ whenever $f(t)$ and $g(t)$ are piecewise continuous on $0 \le t < \infty$. Thus (see Theorem 7.1), to show $f * g$ is Laplace transformable we need only show that $f * g$ is exponentially bounded on $0 \le t < \infty$. From our hypotheses, we know that $|f(t)| \le M_1 e^{a_1 t}$ and $|g(t)| \le M_2 e^{a_2 t}$. Therefore,

$$\begin{aligned} |(f * g)(t)| &= \left| \int_0^t f(t-\lambda) g(\lambda)\, d\lambda \right| \le \int_0^t \left| f(t-\lambda) g(\lambda) \right| d\lambda \\ &\le \int_0^t M_1 e^{a_1 (t-\lambda)} M_2 e^{a_2 \lambda}\, d\lambda = \begin{cases} M_1 M_2 t e^{a_1 t}, & a_1 = a_2 \\ M_1 M_2 \dfrac{e^{a_2 t} - e^{a_1 t}}{a_2 - a_1}, & a_1 \ne a_2 \end{cases} \end{aligned}$$

and $f * g$ is exponentially bounded.

To complete the argument, we need to establish relation (5). From the definition,

$$\begin{aligned} \mathcal{L}\{(f * g)(t)\} &= \int_0^\infty \left[\int_0^t f(t-\lambda) g(\lambda)\, d\lambda \right] e^{-st}\, dt \\ &= \int_0^\infty \int_0^t f(t-\lambda) g(\lambda) e^{-st}\, d\lambda\, dt, \end{aligned} \tag{6}$$

where we view the integral in (6) as a double integral over the portion of the λt-plane shown in Figure 7.18.

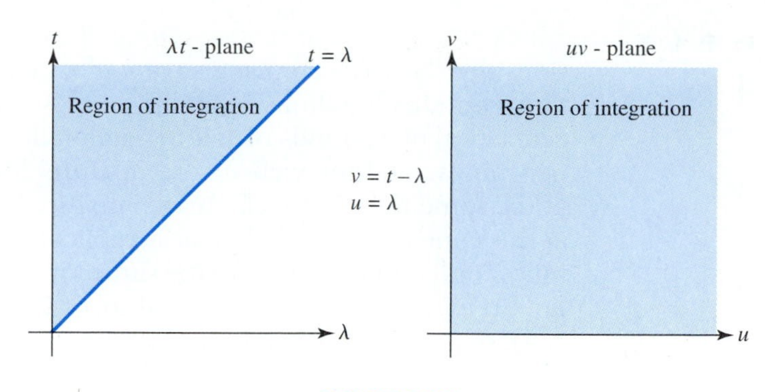

FIGURE 7.18

The regions of integration for the double integrals in equations (6) and (7).

We now introduce the change of variables $u = \lambda, v = t - \lambda$. The boundary lines $\lambda = 0$ and $\lambda = t$ transform into the lines $u = 0$ and $v = 0$, respectively. Note that the Jacobian determinant of this transformation is equal to 1. Therefore, we can rewrite the integral in equation (6) as

$$\begin{aligned}\mathcal{L}\{f * g\} &= \int_0^\infty \int_0^\infty f(v)g(u)e^{-s(u+v)}\, du\, dv \\ &= \left(\int_0^\infty f(v)e^{-sv}\, dv\right)\left(\int_0^\infty g(u)e^{-su}\, du\right) = F(s)G(s). \end{aligned} \tag{7}$$

As noted earlier, Theorem 7.7 can be used to find $\mathcal{L}^{-1}\{F(s)G(s)\}$:

$$\mathcal{L}^{-1}\{F(s)G(s)\} = \int_0^t f(t-\lambda)g(\lambda)\, d\lambda. \tag{8}$$

EXAMPLE 3

Use equation (8) to find

$$\mathcal{L}^{-1}\left\{\frac{1}{s^2(s+1)}\right\}.$$

Solution: Applying equation (8) $F(s) = 1/s^2$ and $G(s) = 1/(s+1)$, we have

$$\mathcal{L}^{-1}\left\{\frac{1}{s^2(s+1)}\right\} = \mathcal{L}^{-1}\left\{\frac{1}{s^2}\right\} * \mathcal{L}^{-1}\left\{\frac{1}{s+1}\right\} = t * e^{-t} = t + e^{-t} - 1.$$

(Recall that the convolution $t * e^{-t}$ was computed earlier, in Example 1.) ▲

Multiple Convolutions

In some applications, such as a cascade connection of linear systems, the solution of the problem of interest is a "multiple convolution." Suppose $f_1(t), f_2(t), \ldots, f_n(t)$ are Laplace transformable functions with Laplace transforms $F_1(s), F_2(s), \ldots, F_n(s)$, respectively. From a repeated application of the convolution theorem, it follows that

$$\mathcal{L}\{f_1 * f_2 * \cdots * f_n\} = F_1(s)F_2(s)\cdots F_n(s).$$

Our next example treats such an application.

EXAMPLE 4

Consider the serial connection of n identical tanks shown in Figure 7.19. Each tank contains V gallons of fresh water. At time $t = 0$, a solution having a concentration of c pounds of salt per gallon flows into Tank 1 at a rate of r gallons per minute and the well-stirred mixture flows out of Tank 1 and into Tank 2 at the same rate. The well-stirred mixture in Tank 2, in turn, flows into Tank 3 at the same rate. This behavior is replicated throughout the cascade. Since the inflow and outflow rates are the same for each tank, the volume of fluid in each tank remains constant and equal to V. Determine the outflow concentration, $c_n(t)$, of Tank n as a function of time.

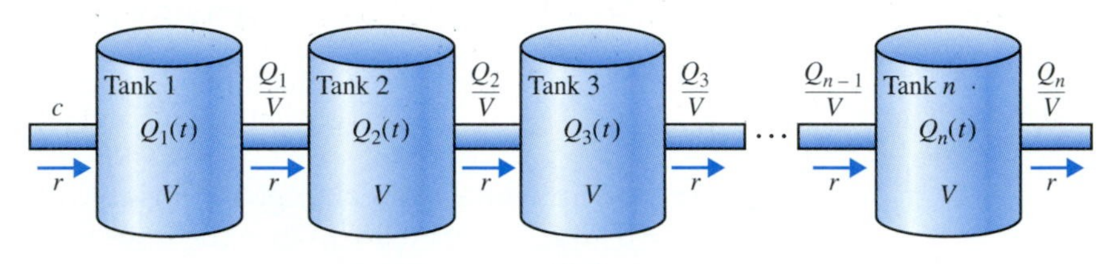

FIGURE 7.19

The n-tank cascade described in Example 4.

Solution: As in Section 2.4, we apply the "conservation of salt" principle to each tank. Let $Q_j(t), j = 1, 2, \dots, n$ represent the amount of salt (in pounds) in the jth tank at time t (in minutes). The following system of initial value problems models this process:

$$Q_1' = rc - r\frac{Q_1}{V}, \qquad Q_1(0) = 0$$

$$Q_j' = r\frac{Q_{j-1}}{V} - r\frac{Q_j}{V}, \qquad Q_j(0) = 0, \qquad j = 2, 3, \dots, n.$$

The solutions of these differential equations can be obtained recursively. We use a convolution representation for each of the solutions. From equation (4a),

$$Q_1(t) = e^{-(r/V)t} * rc,$$
$$Q_j(t) = e^{-(r/V)t} * \frac{r}{V}Q_{j-1}(t), \qquad j = 2, 3, \dots, n.$$

Therefore, the outflow concentration of the nth tank can be represented as the following multiple convolution

$$\begin{aligned}
c_n(t) &= \frac{1}{V}Q_n(t) \\
&= \frac{1}{V}e^{-(r/V)t} * \frac{r}{V}Q_{n-1}(t) \\
&= \frac{1}{V}e^{-(r/V)t} * \frac{r}{V}e^{-(r/V)t} * \frac{r}{V}Q_{n-2}(t) = \cdots \\
&= \frac{1}{V}e^{-(r/V)t} * \overbrace{\frac{r}{V}e^{-(r/V)t} * \frac{r}{V}e^{-(r/V)t} * \cdots * \frac{r}{V}e^{-(r/V)t}}^{n-1 \text{ functions}} * rc \\
&= c\left(\frac{r}{V}\right)^n \overbrace{e^{-(r/V)t} * e^{-(r/V)t} * \cdots * e^{-(r/V)t}}^{n \text{ functions}} * 1.
\end{aligned}$$

By the convolution theorem,

$$\mathcal{L}\{c_n(t)\} = c\left(\frac{r}{V}\right)^n \left[\mathcal{L}\left\{e^{-(r/V)t}\right\}\right]^n \mathcal{L}\{1\}$$

$$= c\left(\frac{r}{V}\right)^n \left[\frac{1}{\left(s+\frac{r}{V}\right)^n}\right]\frac{1}{s}.$$

We can recover $c_n(t)$ by taking the inverse transform. Recall, from equations (10) and (18) in Table 7.1,

$$\mathcal{L}^{-1}\left\{\frac{1}{s}F(s)\right\} = \int_0^t f(u)\,du \quad \text{and} \quad \mathcal{L}^{-1}\left\{\frac{1}{\left(s+\frac{r}{V}\right)^n}\right\} = \frac{t^{n-1}}{(n-1)!}e^{-(r/V)t}.$$

Therefore,

$$c_n(t) = c\left(\frac{r}{V}\right)^n \int_0^t \frac{u^{n-1}}{(n-1)!}e^{-(r/V)u}\,du.$$

We can simplify this integral by making the change of variable $w = (r/V)u$, obtaining

$$c_n(t) = \frac{c}{(n-1)!}\int_0^{rt/V} w^{n-1}e^{-w}\,dw. \tag{9}$$

This final expression can be evaluated for modestly large n using integration-by-parts or by using computer software. Figure 7.20 shows a plot of normalized concentration c_n/c vs. rt/V for $n = 3$ and $n = 10$. As we would expect, both normalized concentrations approach a horizontal asymptote of unity. As time evolves, the concentration in all tanks in the cascade builds up to the inflow concentration, c. Figure 7.20 shows, as one would expect, that the concentration in the last tank of the three-tank cascade builds up to this limiting value more rapidly than does the concentration of the last tank in the ten-tank cascade. ▲

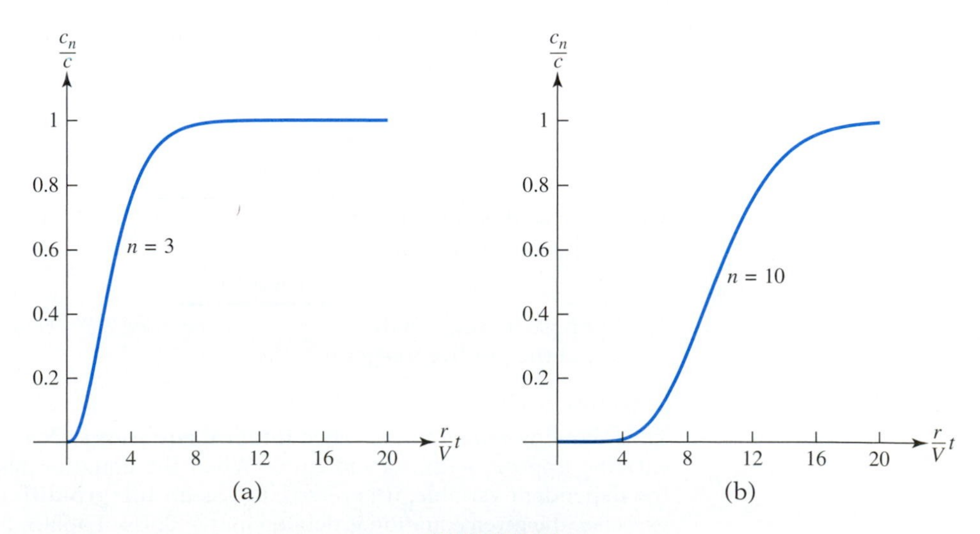

FIGURE 7.20

Graphs of c_n/c vs. rt/V for $n = 3$ and $n = 10$. (See equation (9) in Example 4.)

EXERCISES

1. Show that $f * g = g * f$. That is, show $\int_0^t f(t-\lambda)g(\lambda)\,d\lambda = \int_0^t g(t-\sigma)f(\sigma)\,d\sigma$. [Hint: Use the change of integration variable $\sigma = t - \lambda$. This exercise shows that the convolution operation is commutative.]

Exercises 2–7:

For the given functions $f(t)$ and $g(t)$ defined on $0 \le t < \infty$, compute $f * g$ in two different ways,

(a) by directly evaluating the integral

(b) by computing $\mathcal{L}^{-1}\{F(s)G(s)\}$, where $F(s) = \mathcal{L}\{f(t)\}$ and $G(s) = \mathcal{L}\{g(t)\}$.

2. $f(t) = g(t) = h(t)$

3. $f(t) = t, \quad g(t) = t^2$

4. $f(t) = e^t, \quad g(t) = e^{-2t}$

5. $f(t) = t, \quad g(t) = \sin t$

6. $f(t) = \sin t, \quad g(t) = \cos t$

7. $f(t) = t, \quad g(t) = h(t) - h(t-1)$

Exercises 8–9:

In each exercise, use Laplace transforms to compute the convolution.

8. $P * \mathbf{y}$, where $P(t) = \begin{bmatrix} h(t) & e^t \\ 0 & t \end{bmatrix}$ and $\mathbf{y}(t) = \begin{bmatrix} h(t) \\ e^{-t} \end{bmatrix}$

9. $t * \begin{bmatrix} t \\ \cos t \end{bmatrix}$

Exercises 10–12:

Compute and graph $f * g$.

10. $f(t) = h(t), \quad g(t) = t[h(t) - h(t-2)]$

11. $f(t) = g(t) = h(t-1) - h(t-2)$

12. $f(t) = h(t) - h(t-1), \quad g(t) = h(t-1) - 2h(t-2)$

Exercises 13–15:

Compute the given multiple convolution. (Convolution operations, particularly multiple convolutions, have important applications in probability theory, for example, when computing the probability density function for a sum of independent random variables.[4])

13. $t * t * t$

14. $h(t) * e^{-t} * e^{-2t}$

15. $t * e^{-t} * e^t$

16. Suppose it is known that $\overbrace{h(t) * h(t) * \cdots * h(t)}^{n \text{ functions}} = Ct^8$. Determine the constant C and the positive integer n.

17. Suppose it is known that $\overbrace{e^{-t} * e^{-t} * \cdots * e^{-t}}^{n \text{ functions}} = Ct^4 e^{\alpha t}$. Determine the constants C and α and the positive integer n.

Exercises 18–24:

The following equations are called **integral equations** because the unknown dependent variable appears within an integral. When the equation also contains derivatives of the dependent variable, it is referred to as an **integro-differential equation**. In each exercise, the given equation is defined for $t \ge 0$. Use Laplace transforms to solve for $y(t)$.

[4] Walter C. Giffin, *Transform Techniques for Probability Modeling* (New York: Academic Press, 1975).

18. $\int_0^t \sin(t-\lambda)y(\lambda)\,d\lambda = t^2$

19. $t^2e^{-t} = \int_0^t \cos(t-\lambda)y(\lambda)\,d\lambda$

20. $y(t) - \int_0^t e^{t-\lambda}y(\lambda)\,d\lambda = t$

21. $\int_0^t y(t-\lambda)y(\lambda)\,d\lambda = 6t^3$. Is the solution $y(t)$ unique? If not, find all possible solutions.

22. $t * y(t) = t^2(1-e^{-t})$

23. $\frac{dy}{dt} + \int_0^t y(t-\lambda)e^{-2\lambda}\,d\lambda = 1,\ \ y(0)=0$

24. $\mathbf{y}' = h(t) * \mathbf{y},\ \ \mathbf{y}(0) = \begin{bmatrix} 1 \\ 2 \end{bmatrix}$

Exercises 25–26:

Solve the given initial value problem.

25. $\frac{dy}{dt} = t * t,\ \ y(0) = 1$

26. $y' - y = \int_0^t (t-\lambda)e^{\lambda}\,d\lambda,\ \ y(0) = -1$

7.7 The Delta Function and Impulse Response

In applications we often need to determine the behavior of a linear system, initially at rest, that is suddenly subjected to an input of short duration and large amplitude. In an electrical network, such an input might be a large applied voltage spike. In a mechanical system, the input might be a very sharp applied force.

System excitations of this sort cause a system response that approximates what is known as the "impulse response" of the linear system. In this section, we discuss the concept of an impulse response and show that it is equal to the inverse Laplace transform of the system transfer function.

An Example of Impulse Response

To introduce the idea of impulse response, we begin with a mass-spring-dashpot system. An example of the "short duration/large amplitude" scenario we want to examine is the following initial value problem:

$$\begin{aligned} &my'' + \gamma y' + ky = p_\varepsilon(t), \qquad t > 0 \\ &y(0) = 0, \qquad y'(0) = 0. \end{aligned} \tag{1a}$$

In (1a), we assume ε is a small positive parameter and that

$$p_\varepsilon(t) = \begin{cases} \dfrac{1}{\varepsilon}, & 0 \le t \le \varepsilon \\ 0, & \text{otherwise.} \end{cases} \tag{1b}$$

Since ε is small, the applied force p_ε is a pulse of short duration, ε, and large amplitude, $1/\varepsilon$; Figure 7.21 on the next page shows the graph of a typical pulse. Note that the applied force p_ε has "unit strength" in the sense that the area under the graph in Figure 7.21 is equal to 1 for any choice of ε.[5] By choosing ε smaller

[5]In physics, the linear impulse produced by a constant force is the product of the force times the duration of its application. In that context, the applied force p_ε has a linear impulse of unity for all ε.

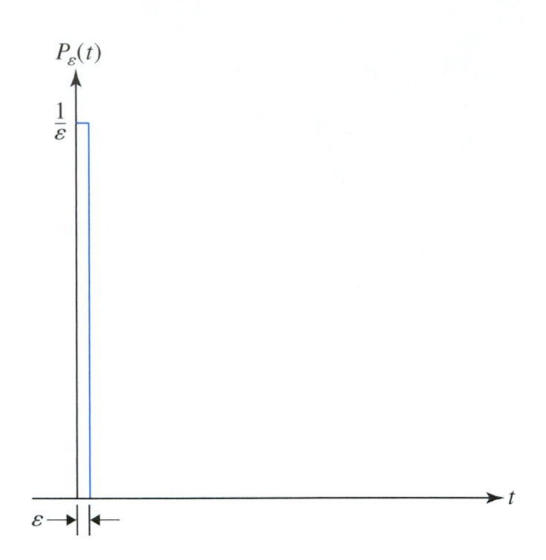

FIGURE 7.21

The function $p_\varepsilon(t)$ is a pulse; see equation (1b). Note that $\int_{-\infty}^{\infty} p_\varepsilon(t)\,dt = 1$.

and smaller, the pulse p_ε can model applied forces having larger and larger amplitudes over shorter and shorter periods. Therefore, it is natural to ask the question:

> What happens to the system behavior as we make the applied force progressively "sharper" and "stronger?"

In other words, what happens to the solution of initial value problem (1a) as we let $\varepsilon \to 0$?

We saw in Section 7.1 that the solution of initial value problem (1a) can be expressed as the convolution

$$y_\varepsilon(t) = \int_0^t \phi(t-\lambda) p_\varepsilon(\lambda)\, d\lambda, \tag{2a}$$

where, if the system is underdamped,

$$\phi(t) = \frac{e^{-(\gamma/2m)t} \sin \sqrt{\dfrac{k}{m} - \dfrac{\gamma^2}{4m^2}}\, t}{m\sqrt{\dfrac{k}{m} - \dfrac{\gamma^2}{4m^2}}}. \tag{2b}$$

We use the subscript ε in equation (2a) to denote the fact that the solution $y_\varepsilon(t)$ depends upon the parameter ε. For $t \geq \varepsilon$, we see from equation (2a) that

$$y_\varepsilon(t) = \frac{1}{\varepsilon}\int_0^\varepsilon \phi(t-\lambda)\, d\lambda, \qquad t \geq \varepsilon. \tag{3}$$

Since $\phi(t)$ is continuous for all t, we can apply the mean value theorem for integrals in equation (3), obtaining

$$y_\varepsilon(t) = \phi(t-\xi), \tag{4}$$

where ξ is some value in the interval $0 \leq \lambda \leq \varepsilon$. As is typical with mean value theorems, the value ξ is known to be sandwiched between 0 and ε but is otherwise unknown. Because of this sandwiching, ξ must approach 0 as $\varepsilon \to 0^+$. Thus, since ϕ is continuous,

$$\lim_{\varepsilon \to 0^+} y_\varepsilon(t) = \phi(t).$$

Therefore, as we make ε progressively smaller (that is, as we make the applied force both shorter in duration and correspondingly larger in amplitude), the system response approaches $\phi(t)$, where $\phi(t)$ is the function given in equation (2b). This limiting response is called the **impulse response** of the linear system.

Figure 7.22 shows the impulse response $\phi(t)$ of an underdamped spring-mass-dashpot system with parameters $m = 1$, $\gamma = 2$, and $k = 5$. For these parameters, the function $\phi(t)$ is given by

$$\phi(t) = 0.5e^{-t} \sin 2t. \tag{5}$$

From a heuristic point of view, $\phi(t)$ represents the response of the mechanical system to an "impulsive force," a force having essentially zero duration and infinite amplitude, but unit area.

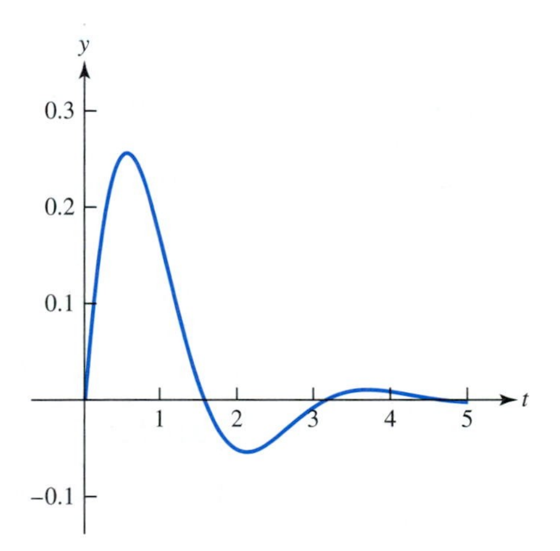

FIGURE 7.22

The impulse response function $\phi(t)$ in equation (5).

The Delta Function

From the point of view of applications, it would be nice to have a function $\delta(t)$ that we could use to model an impulsive force. That is, we would like to be able to write

$$\begin{aligned} \lim_{\varepsilon \to 0^+} y_\varepsilon(t) &= \lim_{\varepsilon \to 0} \int_0^t \phi(t - \lambda) p_\varepsilon(\lambda)\, d\lambda \\ &= \int_0^t \phi(t - \lambda)\delta(\lambda)\, d\lambda \\ &= \phi(t). \end{aligned} \tag{6}$$

The role of the function $\delta(\lambda)$ in (6) would be to evaluate the integrand at $\lambda = 0$. It is important to appreciate, however, that we cannot simply obtain $\delta(\lambda)$ as a limit of $p_\varepsilon(\lambda)$ as $\varepsilon \to 0^+$; that is, we cannot interchange the operations of limit and integration in the first line of (6) because (see Figure 7.21),

$$\lim_{\varepsilon \to 0^+} p_\varepsilon(\lambda) = \begin{cases} 0, & \lambda \neq 0 \\ \infty, & \lambda = 0. \end{cases}$$

The **delta function**, denoted by $\delta(t)$, is actually given precise mathematical meaning as a "generalized function" within a branch of mathematics known as the theory of distributions. For our purposes we will define the delta function, $\delta(t)$, by the limit

$$\int_a^b f(t)\delta(t-t_0)\,dt = \lim_{\varepsilon \to 0^+} \int_a^b f(t)p_\varepsilon(t-t_0)\,dt, \tag{7a}$$

whenever $f(t)$ is a function defined and continuous on $[a, b]$. That is, $\delta(t)$ has the property that

$$\int_a^b f(t)\delta(t-t_0)\,dt = \begin{cases} f(t_0), & a \le t_0 < b \\ 0, & \text{otherwise.} \end{cases} \tag{7b}$$

REMARK: The delta function is sometimes referred to as the **Dirac delta function**.[6] Our definition of the delta function in equation (7a) follows directly from our original definition of the pulse function, p_ε. We therefore obtain the value $f(a)$ when $t_0 = a$ and 0 when $t_0 = b$. Other references use a pulse function that is an even function (having value $1/\varepsilon$ in the interval $-\varepsilon/2 \le t \le \varepsilon/2$) as the basis for their definition. In that case, the values in (7b) obtained when $t_0 = a$ and $t_0 = b$ will differ from ours. The reader should always check the definition used by the reference being consulted.

The Laplace Transform of the Delta Function

Equation (7b) can be used as the basis for defining the Laplace transform of $\delta(t)$. We obtain

$$\mathcal{L}\{\delta(t-t_0)\} = \int_0^\infty e^{-st}\delta(t-t_0)\,dt = e^{-st_0}, \qquad t_0 \ge 0. \tag{8}$$

As a special case, when $t_0 = 0$, we have

$$\mathcal{L}\{\delta(t)\} = 1.$$

The Delta Function as a Formal Modeling Tool

It is important to be aware that the delta function is different from the typical function encountered in calculus. Nevertheless, in many applications people have found it convenient to ignore this distinction; the delta function is often viewed and formally treated as an ordinary function, usually modeling an impulsive input. The solution of the problem of interest typically is given as a convolution integral involving the delta function and so the answer obtained

[6] Paul Adrien Maurice Dirac (1902–1984) was an English mathematical physicist who held the Lucasian Professorship of Mathematics at Cambridge University from 1932 until 1969. After retiring, he moved to Florida, where he continued his research. Dirac is known for his many contributions to quantum theory, particularly the unification theories of quantum mechanics and special relativity.

makes physical sense and can be interpreted as the system response to an idealized impulsive input. The following example illustrates such a formal use of the delta function.

EXAMPLE 1

A body of mass m is at the origin at time $t = 0$, moving in the positive x-direction with velocity v_0. Assume that a frictional force, proportional to the velocity with proportionality constant k, acts to retard the motion. At a time $t_0 > 0$, an impulsive force of strength F_0 acts upon the moving body in the direction of the motion. Find the velocity and position of the body as a function of time t.

Solution: We can use the delta function to formally model the impulsive force as

$$F_0\delta(t - t_0), \qquad t > 0.$$

Given this model of the impulsive force, Newton's laws of motion lead to the following initial value problem

$$\begin{aligned} mv' + kv &= F_0\delta(t - t_0), \qquad t > 0, \\ v(0) &= v_0. \end{aligned} \tag{9}$$

Once we know $v(t)$, the position of the body is given by

$$x(t) = \int_0^t v(\lambda)\, d\lambda. \tag{10}$$

We will use Laplace transforms to solve the problem. Let

$$V(s) = \mathcal{L}\{v(t)\} \quad \text{and} \quad X(s) = \mathcal{L}\{x(t)\}.$$

Noting equation (8), the Laplace transform of equation (9) is

$$m[sV(s) - v_0] + kV(s) = F_0 e^{-st_0}.$$

Therefore,

$$V(s) = \frac{v_0}{s + \dfrac{k}{m}} + \frac{F_0}{m}\frac{e^{-st_0}}{\left(s + \dfrac{k}{m}\right)},$$

and hence

$$v(t) = v_0 e^{-(k/m)t} + \frac{F_0}{m} e^{-(k/m)(t-t_0)} h(t - t_0), \qquad t \geq 0. \tag{11}$$

We can find position $x(t)$ by computing the antiderivative of velocity $v(t)$, as in equation (10). Alternatively, we can use the fact that

$$X(s) = \frac{1}{s}V(s)$$

to obtain

$$X(s) = v_0\frac{m}{k}\left[\frac{1}{s} - \frac{1}{s + \dfrac{k}{m}}\right] + \frac{F_0}{k}e^{-st_0}\left[\frac{1}{s} - \frac{1}{s + \dfrac{k}{m}}\right].$$

Taking inverse transforms, we find

$$x(t) = v_0\frac{m}{k}\left[1 - e^{-(k/m)t}\right] + \frac{F_0}{k}\left[1 - e^{-(k/m)(t-t_0)}\right] h(t - t_0), \qquad t \geq 0. \ \blacktriangle \tag{12}$$

The solid curves in Figure 7.23 are graphs of velocity and position of the body for the parameter values

$$m = 5 \text{ kg}, \quad k = 0.5 \text{ kg/s}, \quad v_0 = 20 \text{ m/s}, \quad F_0 = 500 \text{ N}, \quad t_0 = 3 \text{ s}. \tag{13}$$

As the graph illustrates, application of the impulsive force creates a jump discontinuity in the velocity. This jump is the idealization of the very rapid velocity transition that would occur if the applied force were a very narrow pulse of integrated strength 500 newton-seconds.

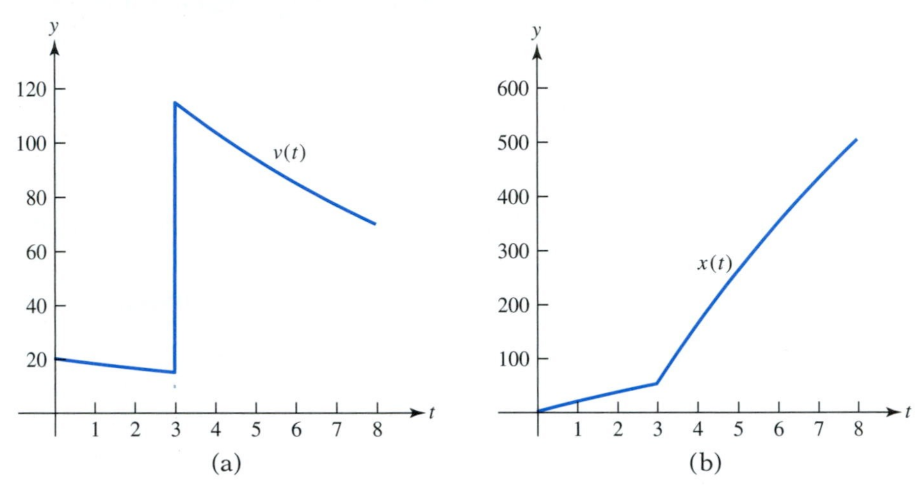

FIGURE 7.23

The results from Example 1. (a) The graph of velocity, $v(t)$, as given in equation (11). The discontinuity is the result of an impulsive force applied at $t = 3$. (b) The graph of position, $x(t)$, as given in equation (12).

To illustrate this point, the dashed curves in Figure 7.24 show the velocity and position that would result from a force of 5000 N being applied during the interval $3 \le t \le 3.1$ sec. In other words, the dashed curves arise from solving the initial value problem

$$\begin{aligned} &mv' + kv = F_0 p_\varepsilon(t - t_0), \qquad t > 0, \\ &v(0) = v_0 \end{aligned} \tag{14}$$

with $\varepsilon = 0.1$ sec and all other parameter values as given by (13). The comparison of these graphs illustrates the "idealizing nature" of using the delta function in modeling applications. [The solution of problem (14) is outlined in the exercises.]

The Impulse Response and the System Transfer Function

The formal use of the delta function as an impulsive source leads to the fact that the impulse response and the system transfer function form a Laplace transform pair. For example, consider the initial value problem

$$\begin{aligned} &a_n \frac{d^n y}{dt^n} + a_{n-1} \frac{d^{n-1} y}{dt^{n-1}} + a_{n-2} \frac{d^{n-2} y}{dt^{n-2}} + \cdots + a_1 \frac{dy}{dt} + a_0 y = \delta(t), \\ &y^{(n-1)}(0) = 0, \qquad y^{(n-2)}(0) = 0, \ldots, \qquad y'(0) = 0, \qquad y(0) = 0, \end{aligned} \tag{15}$$

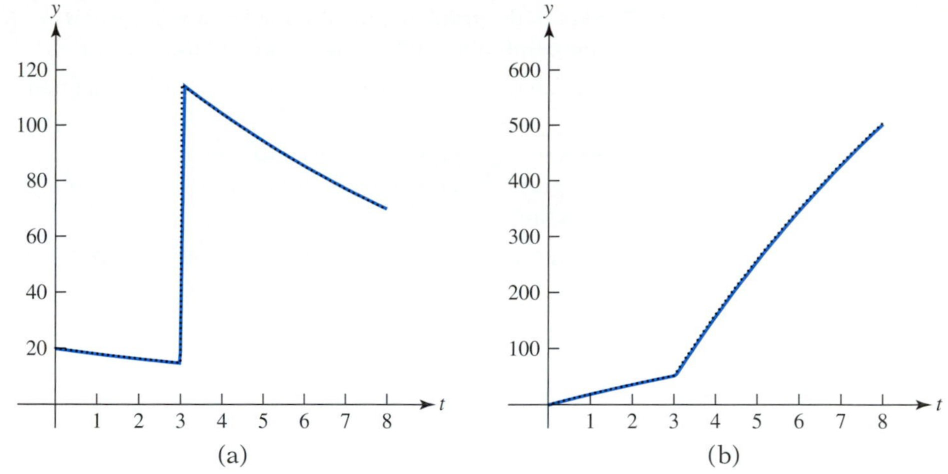

FIGURE 7.24

The graphs of (a) velocity and (b) position for the problems described by equations (9) and (14). Equation (9) models the idealized problem, using the delta function. Equation (14) models the problem using a large (but finite) pulse applied over a t-interval of small (but nonzero) duration. As you can see, the graphs are qualitatively similar.

where we use the delta function to model an impulseive nonhomogeneous term. The solution of (15) is the impulse response of an nth order linear system.

Taking Laplace transforms of (15) and using equation (8) we find

$$(a_n s^n + a_{n-1}s^{n-1} + a_{n-2}s^{n-2} + \cdots + a_1 s + a_0)Y(s) = 1,$$

and therefore

$$Y(s) = \frac{1}{(a_n s^n + a_{n-1}s^{n-1} + a_{n-2}s^{n-2} + \cdots + a_1 s + a_0)}. \qquad (16)$$

The right-hand side of equation (16) is the Laplace transform of the impulse response and is equal to the system transfer function.

EXERCISES

1. Evaluate

(a) $\displaystyle\int_0^3 (1 + e^{-t})\delta(t-2)\,dt$

(b) $\displaystyle\int_{-2}^1 (1 + e^{-t})\delta(t-2)\,dt$

(c) $\displaystyle\int_{-1}^2 \begin{bmatrix} \cos 2t \\ te^{-t} \end{bmatrix} \delta(t)\,dt$

(d) $\displaystyle\int_{-3}^2 (e^{2t} + t) \begin{bmatrix} \delta(t+2) \\ \delta(t-1) \\ \delta(t-3) \end{bmatrix} dt$

2. Let $f(t)$ be a function defined and continuous on $0 \le t < \infty$. Determine

$$f * \delta = \int_0^t f(t-\lambda)\delta(\lambda)\,d\lambda.$$

3. Determine a value of the constant t_0 such that $\int_0^1 \sin^2[\pi(t-t_0)]\delta(t-(1/2))\,dt = 3/4$.

4. If $\int_1^5 t^n \delta(t-2)\,dt = 8$, what is the exponent n?

5. Sketch the graph of the function $f(t)$ defined by $f(t) = \int_0^t \delta(\lambda - 1)\, d\lambda$, $0 \le t < \infty$. Can the graph obtained be characterized in terms of a Heaviside step function?

6. Sketch the graph of the function $g(t)$ which is defined by $g(t) = \int_0^t \int_0^\lambda \delta(\sigma - 1)\, d\sigma\, d\lambda$, $0 \le t < \infty$.

7. Sketch the graph of the function $k(t) = \int_0^t [\delta(\lambda - 1) - \delta(\lambda - 2)]\, d\lambda$, $0 \le t < \infty$. Can the graph be characterized in terms of a Heaviside step function or Heaviside step functions?

8. The graph of the function $g(t) = \int_0^t e^{\alpha t} \delta(t - t_0)\, dt$, $0 \le t < \infty$ is shown. Determine the constants α and t_0.

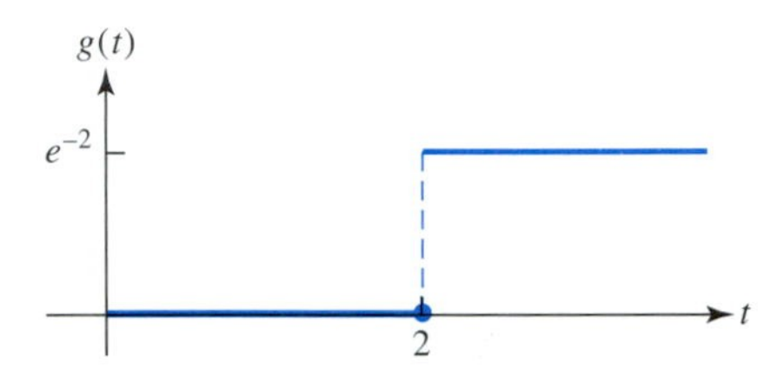

Figure for Exercise 8

Exercises 9–11:

In each exercise, a function $g(t)$ is given.

(a) Solve the initial value problem $y' - y = g(t)$, $y(0) = 0$ using the techniques developed in Chapter 2.

(b) Use Laplace transforms to determine the transfer function $\phi(t)$,

$$\phi' - \phi = \delta(t), \qquad \phi(0) = 0.$$

(c) Evaluate the convolution integral $\phi * g = \int_0^t \phi(t - \lambda) g(\lambda)\, d\lambda$ and compare the resulting function with the solution obtained in part (a).

9. $g(t) = h(t)$ 10. $g(t) = e^t$ 11. $g(t) = t$

Exercises 12–20:

Solve the given initial value problem, in which inputs of large amplitude and short duration have been idealized as delta functions. Graph the solution that you obtain on the indicated interval. (In Exercises 19 and 20, plot the two components of the solution on the same graph.)

12. $y' + y = 2 + \delta(t - 1), \quad y(0) = 0, \quad 0 \le t \le 6$

13. $y' + y = \delta(t - 1) - \delta(t - 2), \quad y(0) = 0, \quad 0 \le t \le 6$

14. $y'' = \delta(t - 1) - \delta(t - 3), \quad y(0) = 0, \quad y'(0) = 0, \quad 0 \le t \le 6$

15. $y'' + 4\pi^2 y = 2\pi\delta(t - 2), \quad y(0) = 0, \quad y'(0) = 0, \quad 0 \le t \le 6$

16. $y'' - 2y' = \delta(t - 1), \quad y(0) = 1, \quad y'(0) = 0, \quad 0 \le t \le 2$

17. $y'' + 2y' + 2y = \delta(t - 1), \quad y(0) = 0, \quad y'(0) = 0, \quad 0 \le t \le 6$

18. $y'' + 2y' + y = \delta(t - 2), \quad y(0) = 0, \quad y'(0) = 1, \quad 0 \le t \le 6$

19. $\dfrac{d}{dt}\begin{bmatrix} y_1 \\ y_2 \end{bmatrix} = \begin{bmatrix} 1 & 1 \\ 1 & 1 \end{bmatrix}\begin{bmatrix} y_1 \\ y_2 \end{bmatrix} + \delta(t - 1)\begin{bmatrix} 1 \\ 0 \end{bmatrix}, \quad \begin{bmatrix} y_1(0) \\ y_2(0) \end{bmatrix} = \begin{bmatrix} 0 \\ 0 \end{bmatrix}, \quad 0 \le t \le 2$

20. $\dfrac{d}{dt}\begin{bmatrix} y_1 \\ y_2 \end{bmatrix} = \begin{bmatrix} 2 & 1 \\ 0 & 1 \end{bmatrix}\begin{bmatrix} y_1 \\ y_2 \end{bmatrix} + \begin{bmatrix} 0 \\ 1 \end{bmatrix} - \delta(t - 1)\begin{bmatrix} 1 \\ 0 \end{bmatrix}, \quad \begin{bmatrix} y_1(0) \\ y_2(0) \end{bmatrix} = \begin{bmatrix} 0 \\ 0 \end{bmatrix}, \quad 0 \le t \le 2$

EXTENDED PROBLEM: LOCATING A TRANSMISSION LINE FAULT

Laplace transformation is an operational tool that can be used to map a given problem into a simpler "transformed problem." We have seen in this chapter how problems involving ordinary differential equations can be transformed into problems involving simpler algebraic equations. The purpose of this extended problem is to illustrate a variation on this same theme.

We consider here an instance where Laplace transforms are used to transform a problem involving partial differential equations into a simpler problem involving ordinary differential equations. The steps outlined in Figure 7.1 remain the same; we first solve this simpler problem and then use the inverse Laplace transform to find the desired solution.

The problem considered is a simple application of the idea of echo location. For example, knowing how fast sound travels in air, we can determine the distance to a reflection point by measuring the time separation between when we emit a sound and when its echo is heard. This basic idea can be used to determine where a transmission line fault or disruption is located.

Problem Formulation

A transmission line is an example of a distributed network. In contrast to the networks considered earlier, the voltage and current on a transmission line are functions of both space and time. Consider Figure 7.25 where the transmission line is represented by the two parallel cables. The variable x measures distance along the line, with a voltage source or generator positioned at $x = 0$ and the fault (assumed to be an open circuit) located at $x = l$. We assume that the location of the fault is unknown; our goal is to locate it by sending a short pulse down the line and measuring the two-way transit time, the time it takes for the pulse to travel to the fault, be reflected by it, and return to the source.

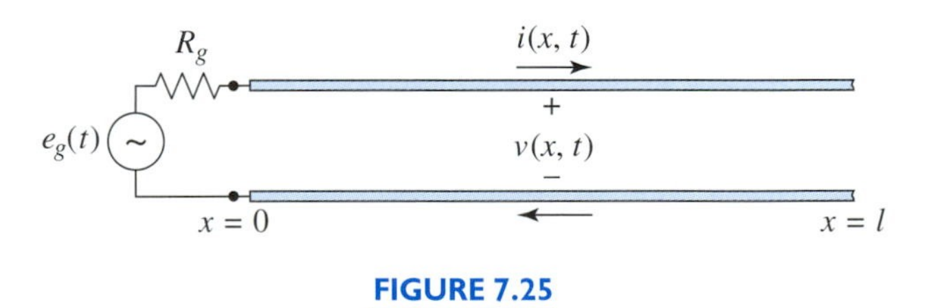

FIGURE 7.25

A transmission line network. A voltage generator is connected at $x = 0$ and an open circuit is assumed to exist at the unknown fault location, $x = l$.

As shown in Figure 7.25, we represent the voltage across the line and the current along the line at position x and time t by $v(x, t)$ and $i(x, t)$, respectively. The voltage generator is assumed to have an internal resistance R_g. Figure 7.26 on the next page depicts a snapshot of a differential transmission line segment taken at some time t. As the figure indicates, the transmission line itself is characterized by a series inductance per unit length, L, and a shunt capacitance per unit length, C. To determine how the transmission line voltage and current behave as functions of space and time, we apply Kirchhoff's voltage and current laws to this differential segment of line. The voltage drop across the inductance is

$$L\frac{\partial i(x,t)}{\partial t}\,dx$$

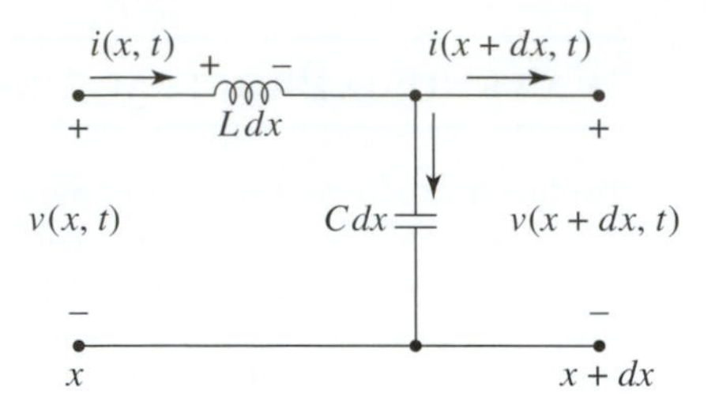

FIGURE 7.26

Differential transmission line segment equivalent circuit.

while the current flow through the capacitance is

$$C\frac{\partial v(x,t)}{\partial t}\,dx.$$

If we apply Kirchhoff's voltage law to the circuit in Figure 7.26, equating the sum of the voltage rises in a clockwise traversal of the network to zero, we obtain

$$v(x,t) - L\frac{\partial i(x,t)}{\partial t}\,dx - v(x+dx,t) = 0.$$

Similarly, applying Kirchhoff's current law (equating the sum of the currents entering the upper node to zero) leads to

$$i(x,t) - C\frac{\partial v(x,t)}{\partial t}\,dx - i(x+dx,t) = 0.$$

If we divide by dx and let $dx \to 0$, we obtain a pair of partial differential equations

$$\frac{\partial v(x,t)}{\partial x} = -L\frac{\partial i(x,t)}{\partial t},$$
$$\frac{\partial i(x,t)}{\partial x} = -C\frac{\partial v(x,t)}{\partial t}. \quad \textbf{(1)}$$

We assume that the transmission line is quiescent for $t \le 0$. That is, we assume

$$i(x,0) = 0, \qquad v(x,0) = 0, \qquad 0 \le x \le l. \quad \textbf{(2)}$$

At time $t = 0$ the voltage generator is turned on, emitting a signal $e_g(t)$, $t > 0$. Applying Kirchhoff's voltage law at the generator leads to

$$e_g(t) - i(0,t)R_g - v(0,t) = 0, \qquad t > 0. \quad \textbf{(3)}$$

Lastly, the assumption that an open circuit exists at fault location $x = l$ leads us to the constraint

$$i(l,t) = 0, \qquad t > 0. \quad \textbf{(4)}$$

Equations (1)–(4) constitute the mathematical problem of interest. We are free to select the generator voltage $e_g(t)$. Our goal is to determine a formula for $v(0,t)$, $t > 0$. As we will see, this formula will contain l as a parameter. Since $v(0,t)$ is a quantity that can be measured, we can use a voltage measurement to determine the distance l to the fault. Knowing the location of the fault, appropriate repairs can be made.

Solution Using Laplace Transforms

Since $0 < t < \infty$ is the time interval of interest, we can define the Laplace transforms,

$$V(x,s) \equiv \int_0^\infty v(x,t)e^{-st}\,dt, \qquad I(x,s) \equiv \int_0^\infty i(x,t)e^{-st}\,dt. \quad \textbf{(5)}$$

In (5), the variable x is treated as a parameter.

1. Apply the Laplace transform (5) to both sides of equations (1). Assume that the order of operations can be interchanged. For example,

$$\int_0^\infty \frac{\partial v(x,t)}{\partial x} e^{-st}\,dt = \frac{\partial}{\partial x}\left[\int_0^\infty v(x,t)e^{-st}\,dt\right] = \frac{\partial V(x,s)}{\partial x}. \tag{6}$$

Likewise, use the formula for the transform of a derivative and condition (2). For example

$$\int_0^\infty \frac{\partial v(x,t)}{\partial t} e^{-st}\,dt = sV(x,s) - v(x,0) = sV(x,s). \tag{7}$$

Show that an application of the Laplace transform operation to (1) leads to the system of differential equations

$$\begin{aligned}\frac{\partial V(x,s)}{\partial x} &= -sLI(x,s),\\ \frac{\partial I(x,s)}{\partial x} &= -sCV(x,s).\end{aligned} \tag{8}$$

2. Apply the Laplace transform to equations (3) and (4), obtaining

$$\begin{aligned}E_g(s) - I(0,s)R_g &= V(0,s),\\ I(l,s) &= 0,\end{aligned} \tag{9}$$

where $E_g(s)$ denotes the Laplace transform of $e_g(t)$.

Equations (8) and (9) constitute the transformed problem. Note that problem is in fact simpler. The only differentiation performed in (8) is with respect to the spatial variable x. If we view the transform variable s as a parameter, then equation (8) is basically a linear system of ordinary differential equations. A problem involving partial differential equations has been transformed into one that *de facto* involves only ordinary differential equations. Note that problems (8) and (9) do not constitute an initial value problem. It is a two-point boundary value problem; the spatial domain is $0 \le x \le l$ and supplementary conditions (9) involve constraints at the two endpoints.

3. Obtain the general solution of equation (8), viewed as a linear system of ordinary differential equations. Note that since transform variable s is being viewed as a parameter, the two arbitrary constants appearing in the general solution can be functions of s.

4. The quantity $\sqrt{L/C}$ has the dimensions of resistance; it is called the **characteristic impedance** of the transmission line and is often denoted by the symbol Z_0. Assume that $R_g = Z_0$. When this condition holds, the voltage generator is said to be "matched to the transmission line." Impose constraint (9) on the general solution obtained in step 3, assuming that $R_g = Z_0$. This, then, represents the solution of the transformed problem. In particular, show that

$$V(0,s) = \frac{E_g(s)}{2}\left[1 + e^{-s(2l\sqrt{LC})}\right]. \tag{10}$$

5. Determine $v(0,t), t > 0$ by computing the inverse Laplace transform of (10). The product $l\sqrt{LC}$ has the dimensions of time. To illustrate the underlying ideas, assume that generator voltage $e_g(t)$ is a pulse, say

$$e_g(t) = \begin{cases} 10, & 0 < t \le 0.1 \\ 0, & 0.1 < t < \infty, \end{cases} \qquad \text{and that} \quad l\sqrt{LC} = 5.$$

Graph $v(0,t)$ as a function of time t for $t > 0$.

6. Explain the physical significance of the two terms comprising $v(0,t)$. Suppose, we know the properties of the transmission line; specifically, suppose we know L, C, and therefore $\sqrt{LC}$. Explain how your solution obtained in step (5) can be used to determine unknown distance l.

7. The fact that we modeled the fault as an open circuit is not particularly important. Suppose, for example, the disruption was a short circuit. In that event, boundary condition (4) would be replaced by the constraint
$$v(l, t) = 0, \qquad 0 < t < \infty. \tag{11}$$
Solve the problem in this case, obtaining $v(0, t), t > 0$. How does the "echo" measured at $x = 0$ differ from that measured in the case of an open circuit fault?
8. As a final exercise, reconsider the partial differential equations (1). Compute the partial derivative of the first equation with respect to x and the partial derivative of the second equation with respect to t. Assume that the order of mixed second partial derivatives can be interchanged; that is, $\partial^2 i(x, t)/\partial x\, \partial t = \partial^2 i(x, t)/\partial t\, \partial x$. Show that the voltage $v(x, t)$ is a solution of the partial differential equation
$$\frac{\partial^2 v(x, t)}{\partial x^2} = LC\frac{\partial^2 v(x, t)}{\partial t^2}. \tag{12}$$
Equation (12) is an important differential equation in mathematical physics known as the **wave equation**. The positive constant $1/\sqrt{LC}$ is the speed with which waves propagate. Is this statement consistent with your observations in (6)?
9. Show that the current $i(x, t)$ also satisfies wave equation (12).

8

Nonlinear Systems

CHAPTER OVERVIEW

Introduction

In this chapter we consider systems of nonlinear differential equations,

$$\begin{aligned} y_1' &= f_1(t, y_1, y_2, \ldots, y_n), \\ y_2' &= f_2(t, y_1, y_2, \ldots, y_n), \\ &\vdots \\ y_n' &= f_n(t, y_1, y_2, \ldots, y_n), \qquad a < t < b. \end{aligned} \tag{1}$$

System (1) consists of n differential equations for n dependent variables $y_1(t), y_2(t), \ldots, y_n(t)$, where $a < t < b$. An initial value problem is formulated once we specify n initial conditions,

$$y_1(t_0) = y_1^0, \quad y_2(t_0) = y_2^0, \quad \ldots, \quad y_n(t_0) = y_n^0, \tag{2}$$

where t_0 is some point belonging to the interval $a < t < b$. The special case where $n = 1$ reduces to the scalar nonlinear problem, $y' = f(t, y)$, treated in Chapter 3.

An existence and uniqueness theorem for nonlinear systems is given in Section 8.1. Section 8.2 focuses on two-dimensional autonomous systems of the form

$$\begin{aligned} x' &= f(x, y), \\ y' &= g(x, y), \end{aligned}$$

and discusses geometric issues such as direction fields and the phase plane. Section 8.2 shows how these geometric concepts can be used to make qualitative predictions about the behavior of solutions of two-dimensional autonomous systems. In Section 8.3 we consider second order scalar equations that satisfy a conservation law and discuss the related idea of a Hamiltonian system.

Section 8.4 considers the stability properties of equilibrium points of autonomous systems. Section 8.5 discusses linearization as a way to analyze the stability properties of an equilibrium point. Section 8.5 also introduces the concept of an almost linear system. In Section 8.6 we study phase-plane behavior of the linear system $\mathbf{y}' = A\mathbf{y}$, where A is a constant (2×2) invertible matrix; we classify the types of behavior possible for solutions near the equilibrium point $\mathbf{y} = \mathbf{0}$. Section 8.7 applies the ideas of Sections 8.1–8.6 to predator-prey population models.

The Vector Form for a Nonlinear System

We can express the nonlinear system (1) in a compact fashion using vector notation. In particular, define the vector functions

$$\mathbf{y}(t) = \begin{bmatrix} y_1(t) \\ y_2(t) \\ \vdots \\ y_n(t) \end{bmatrix}, \qquad \mathbf{f}(t, \mathbf{y}) = \begin{bmatrix} f_1(t, y_1, y_2, \ldots, y_n) \\ f_2(t, y_1, y_2, \ldots, y_n) \\ \vdots \\ f_n(t, y_1, y_2, \ldots, y_n) \end{bmatrix}, \qquad \mathbf{y}_0 = \begin{bmatrix} y_1^0 \\ y_2^0 \\ \vdots \\ y_n^0 \end{bmatrix}.$$

With this notation, we can write the initial value problem succinctly as

$$\begin{aligned} \mathbf{y}'(t) &= \mathbf{f}(t, \mathbf{y}(t)), \qquad a < t < b \\ \mathbf{y}(t_0) &= \mathbf{y}_0. \end{aligned} \tag{3}$$

The linear systems considered in Chapter 6 correspond to a special case of equation (3) where $\mathbf{f}(t, \mathbf{y}) = A(t)\mathbf{y} + \mathbf{g}(t)$.

Autonomous Systems

An important special case occurs when none of the n functions appearing on the right-hand side of system (1) is an explicit function of the independent variable t. In this case, system (1) has the form

$$\begin{aligned} y_1' &= f_1(y_1, y_2, \ldots, y_n), \\ y_2' &= f_2(y_1, y_2, \ldots, y_n), \\ &\vdots \\ y_n' &= f_n(y_1, y_2, \ldots, y_n). \end{aligned} \tag{4}$$

System (4) is called an **autonomous** system. When $n = 1$, the system reduces to an autonomous scalar differential equation, discussed in Section 3.1. In the autonomous case, initial value problems have the form

$$\begin{aligned}\mathbf{y}' &= \mathbf{f}(\mathbf{y}),\\ \mathbf{y}(t_0) &= \mathbf{y}_0.\end{aligned} \tag{5}$$

Autonomous systems arise in many applications and we give two examples shortly.

An important feature of solutions of autonomous systems is the nature of their dependence on the independent variable t and the initial value t_0. In Section 3.1 we argued that the solution of a scalar autonomous equation is a function of the *time difference* $t - t_0$; what matters is the value of time t measured relative to the starting time t_0. The same basic argument can be applied to the autonomous system (5), showing that solutions are functions of the difference variable $t - t_0$.

Much of our attention in this chapter will be directed toward the study of autonomous systems because they arise so often in applications. To provide some motivation, we introduce two examples:

(a) A two-species population model
(b) A pendulum

These two applications are discussed in more detail throughout this chapter.

Two-Species Population Models

Modeling the interaction of different species of organisms is important in biological and ecological studies. In particular, suppose we consider two species coexisting in some confined environment, say a lake or an island. In some cases, the two species may interact benignly with each other except for the fact that they both compete for the same limited food supply. In other cases, one species may act as a predator and depend on the second species (the prey) as its food supply. Not surprisingly, these two models are referred to as the competing species model and the predator-prey model, respectively.

The ideas underlying the Verhulst population model discussed in Section 3.5 can be extended to describe two-species interactions. We now have two dependent variables, the populations $P_1(t)$ and $P_2(t)$, and their interaction is often modeled by the autonomous nonlinear system,

$$\begin{aligned}P_1' &= r_1(1 - \alpha_1 P_1 - \beta_1 P_2)P_1\\ P_2' &= r_2(1 - \beta_2 P_1 - \alpha_2 P_2)P_2,\end{aligned} \tag{6}$$

where the constants $r_1, r_2, \alpha_1, \alpha_2, \beta_1$, and β_2 are all positive. These equations are a natural generalization of the logistic equation. The relative birth rates per unit population are now $r_1(1 - \alpha_1 P_1 - \beta_1 P_2)$ and $r_2(1 - \beta_2 P_1 - \alpha_2 P_2)$, respectively.

The nonlinear terms having β_1 and β_2 as coefficients are the interaction terms that couple population dynamics. When β_1 and β_2 are positive, an increase in either population decreases the relative birth rate of both populations since any population increase puts additional stress upon the available resources needed by both. If β_1 and β_2 are both zero in equation (6), then the two populations evolve independently of each other; in fact, the two differential equations uncouple and each population satisfies a separate logistic equation of the type discussed in Section 3.5.

We treat the predator-prey model in Section 8.7. For now, we leave it as an exercise for you to decide how the competing species model (6) should be modified if one population, say P_1, were a population of predators that depends on the second population, P_2, for its food supply.

The Pendulum

Some problems, such as the motion of the pendulum in Figure 8.1, give rise to second order scalar nonlinear differential equations. Such equations can be recast as first order nonlinear systems and studied as such; this will be our approach in the present chapter.

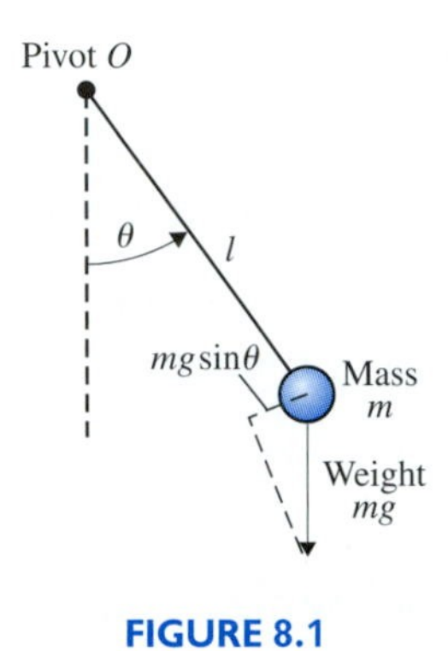

FIGURE 8.1

The pendulum.

Consider the pendulum shown in Figure 8.1. A mass m is attached to the end of a rigid rod of length l. We neglect the weight of the rod and assume the pivot is frictionless. Because of the constraining action of the rod, the (assumed planar) motion of the pendulum mass occurs on the circumference of a circle of radius l centered at the pivot.

The equation of motion for the pendulum can be obtained by equating the sum of the moments about pivot O to the product of the pendulum's moment of inertia and angular acceleration. The resulting formula,

$$\sum M_O = I_O \alpha,$$

can be viewed as a rotational analog of Newton's second law of motion, $F = ma$.

The moment of inertia of the pendulum about pivot O is ml^2. Taking the counterclockwise direction as positive, the moment sum is

$$\sum M_O = -mgl \sin\theta,$$

while the angular acceleration is $\alpha = \theta''$. Therefore, we obtain $-mgl\sin\theta = ml^2\theta''$, or

$$\theta'' + \frac{g}{l}\sin\theta = 0. \tag{7}$$

To study pendulum motion, we typically specify pendulum position and angular velocity at some initial time, say $t = 0$. Nonlinear differential equation (7), together with the initial conditions $\theta(0) = \theta_0$, $\theta'(0) = \theta_0'$, forms the initial value problem of interest.

We can recast this initial value problem as an initial value problem for a first order nonlinear system using the ideas introduced in Section 6.2 for linear problems. In particular, let $y_1(t) = \theta(t)$, $y_2(t) = \theta'(t)$, and let

$$\mathbf{y}(t) = \begin{bmatrix} y_1(t) \\ y_2(t) \end{bmatrix}.$$

Under this change of variables we have

$$y_1' = \theta' = y_2 \quad \text{and} \quad y_2' = \theta'' = -\frac{g}{l}\sin\theta = -\frac{g}{l}\sin y_1.$$

Therefore, equation (7) can be rewritten as the first order nonlinear system

$$\begin{aligned} y_1' &= y_2, \\ y_2' &= -\frac{g}{l}\sin y_1. \end{aligned}$$

Note that this first order system is autonomous. In vector form, the associated initial value problem is

$$\mathbf{y}' = \mathbf{f}(\mathbf{y}), \qquad \mathbf{y}(0) = \mathbf{y}_0,$$

where

$$\mathbf{f}(\mathbf{y}) = \begin{bmatrix} y_2 \\ -\frac{g}{l}\sin y_1 \end{bmatrix}, \qquad \mathbf{y}_0 = \begin{bmatrix} \theta_0 \\ \theta_0' \end{bmatrix}.$$

Writing Higher Order Scalar Equations as First Order Systems

Being able to rewrite an initial value problem involving a higher order scalar differential equation such as equation (7) as an initial value problem for a first order system has several important consequences. For example, in Section 8.1 we will discuss existence-uniqueness theory for first order nonlinear systems; this theory generalizes the scalar results of Chapter 3 in much the same way as the scalar results of Chapter 2 are generalized to linear systems by Theorem 6.1. Since systems of higher order nonlinear differential equations can be recast as first order nonlinear systems, the theory presented in Section 8.1 will accommodate initial value problems for these scalar higher order equations as well.

In many cases it is not possible to explicitly solve a nonlinear differential equation and numerical methods are needed to obtain quantitative information about the solutions. Chapter 9 is devoted to a discussion of these numerical algorithms, building upon and improving Euler's method, described in Sections 3.8 and 6.9. As we shall see in Chapter 9, these improved algorithms are typically developed for first order scalar equations and then extended to first order systems. Therefore, the fact that we can reformulate a system of higher order nonlinear scalar problems as a first order system enables us to apply these algorithms and obtain numerical solutions.

8.1 Existence and Uniqueness

We first discussed existence and uniqueness for nonlinear differential equations in Section 3.1, where we treated scalar nonlinear problems of the form

$$y' = f(t, y), \qquad y(t_0) = y_0. \tag{1}$$

In Section 3.1 we stated Theorem 3.1, an existence and uniqueness result for the scalar equation (1). According to Theorem 3.1, if the initial condition point (t_0, y_0) lies in an open rectangle R in the ty-plane defined by $a < t < b, \alpha < y < \beta$, and if the functions

$$f(t, y) \quad \text{and} \quad \frac{\partial f(t, y)}{\partial y}$$

are continuous on the open rectangle R, then there is a unique solution $y(t)$ of initial value problem (1) that exists on some interval $c < t < d$ containing t_0. Theorem 3.1 says nothing, however, about the size of the interval $c < t < d$.

We present (without proof) a generalization of Theorem 3.1 to systems of nonlinear differential equations. Consider the initial value problem

$$\mathbf{y}' = \mathbf{f}(t, \mathbf{y}), \qquad \mathbf{y}(t_0) = \mathbf{y}_0, \tag{2}$$

where

$$\mathbf{y}(t) = \begin{bmatrix} y_1(t) \\ y_2(t) \\ \vdots \\ y_n(t) \end{bmatrix} \quad \text{and} \quad \mathbf{f}(t, \mathbf{y}) = \begin{bmatrix} f_1(t, y_1, y_2, \ldots, y_n) \\ f_2(t, y_1, y_2, \ldots, y_n) \\ \vdots \\ f_n(t, y_1, y_2, \ldots, y_n) \end{bmatrix}. \tag{3}$$

Because there are now n dependent variables, we consider initial value problem (2) in the $(n + 1)$-dimensional open rectangular region R defined by the inequalities

$$a < t < b, \quad \alpha_1 < y_1 < \beta_1, \quad \alpha_2 < y_2 < \beta_2, \quad \ldots, \quad \alpha_n < y_n < \beta_n. \tag{4}$$

Assume the initial condition point $(t_0, \mathbf{y}_0)$ lies in the region R. Theorem 8.1 asserts that continuity of the n component functions of $\mathbf{f}(t, \mathbf{y})$ in equation (3) along with continuity of the n^2 partial derivatives

$$\begin{array}{cccc} \dfrac{\partial f_1(t, y_1, y_2, \ldots, y_n)}{\partial y_1}, & \dfrac{\partial f_1(t, y_1, y_2, \ldots, y_n)}{\partial y_2}, & \ldots, & \dfrac{\partial f_1(t, y_1, y_2, \ldots, y_n)}{\partial y_n} \\ \dfrac{\partial f_2(t, y_1, y_2, \ldots, y_n)}{\partial y_1}, & \dfrac{\partial f_2(t, y_1, y_2, \ldots, y_n)}{\partial y_2}, & \ldots, & \dfrac{\partial f_2(t, y_1, y_2, \ldots, y_n)}{\partial y_n} \\ \vdots & & & \vdots \\ \dfrac{\partial f_n(t, y_1, y_2, \ldots, y_n)}{\partial y_1}, & \dfrac{\partial f_n(t, y_1, y_2, \ldots, y_n)}{\partial y_2}, & \ldots, & \dfrac{\partial f_n(t, y_1, y_2, \ldots, y_n)}{\partial y_n} \end{array} \tag{5}$$

is sufficient to ensure the existence of a unique solution of the initial value problem on some interval $c < t < d$ containing t_0. As in the scalar case, however, Theorem 8.1 gives no insight into the size of the interval $c < t < d$.

Theorem 8.1

Consider the initial value problem

$$\mathbf{y}' = \mathbf{f}(t, \mathbf{y}), \qquad \mathbf{y}(t_0) = \mathbf{y}_0,$$

where the initial value point $(t_0, \mathbf{y}_0)$ lies in the region R defined by the inequalities in (4). Let $\mathbf{f}(t, \mathbf{y})$ and the partial derivatives in (5) be continuous in R. Then the initial value problem has a unique solution $\mathbf{y}(t)$ that exists on some t-interval (c, d) containing t_0.

The following example illustrates the application of Theorem 8.1 to a pendulum problem similar to the one shown in Figure 8.1.

EXAMPLE 1

Consider the following initial value problem for a forced pendulum,

$$\begin{aligned} ml\theta'' + mg\sin\theta &= F_0 \sin\omega t \\ \theta(0) = 0, \qquad \theta'(0) &= 0. \end{aligned} \tag{6}$$

This equation describes a sinusoidal tangential force having amplitude F_0 and radian frequency ω applied to the pendulum. At time $t = 0$ the pendulum is in the vertically downward position with no initial angular velocity. What can we conclude about solutions of (6) from Theorem 8.1?

Solution: In order to apply Theorem 8.1, we need to write the second order differential equation as a first order system. As in the introduction to this chapter, let $y_1(t) = \theta(t)$, $y_2(t) = \theta'(t)$ and let

$$\mathbf{y}(t) = \begin{bmatrix} y_1(t) \\ y_2(t) \end{bmatrix}.$$

With this change of variables, we can recast the given differential equation as a first order system $\mathbf{y}' = \mathbf{f}(t, \mathbf{y})$, where

$$\mathbf{f}(t, \mathbf{y}) = \begin{bmatrix} y_2 \\ -\dfrac{g}{l}\sin y_1 + \dfrac{F_0}{ml}\sin\omega t \end{bmatrix}.$$

According to Theorem 8.1, we need to examine continuity of the functions

$$f_1(t, y_1, y_2) = y_2 \quad \text{and} \quad f_2(t, y_1, y_2) = -\frac{g}{l}\sin y_1 + \frac{F_0}{ml}\sin\omega t$$

and the four partial derivatives

$$\begin{aligned} \frac{\partial f_1(t, y_1, y_2)}{\partial y_1} &= 0, & \frac{\partial f_1(t, y_1, y_2)}{\partial y_2} &= 1, \\ \frac{\partial f_2(t, y_1, y_2)}{\partial y_1} &= -\frac{g}{l}\cos y_1, & \frac{\partial f_2(t, y_1, y_2)}{\partial y_2} &= 0. \end{aligned}$$

(continued)

(continued)

The functions f_1 and f_2 along with the four partial derivatives are continuous for all values (t, y_1, y_2) in $(t, \mathbf{y})$-space. Therefore, applying Theorem 8.1, we can take R to be any open three-dimensional rectangular region in $(t, \mathbf{y})$-space that contains the initial condition point $(t_0, \mathbf{y}_0) = (0, 0, 0)$. Theorem 8.1 concludes that a unique solution of the initial value problem (6) exists on *some* t-interval containing $t = 0$. ▲

Example 1 not only illustrates the application of the existence-uniqueness theorem; it also highlights its shortcomings. The theorem concludes that there is some t-interval of existence-uniqueness but gives no insight into how large this interval might be. On the one hand, a theorem such as Theorem 8.1 that deals with a very general class of nonlinear systems cannot be expected to do more; it cannot give precise results for particular cases. As shown in Chapter 3, nonlinear initial value problems can have solutions exhibiting a wide variety of behavior. On the other hand, our everyday experience with pendulums would suggest that the particular initial value problem considered in Example 1 should have a unique solution on an arbitrarily large t-interval; that is, we don't expect such a mechanical system to behave catastrophically.

For initial value problems involving nonlinear systems, there are virtually no techniques for finding explicit or implicit representations of solutions; we must look for other ways to understand the behavior of these solutions. Our attention, therefore, will be focused in two directions: on determining qualitative information by graphical means and on obtaining quantitative information from numerical methods.

EXERCISES

Exercises 1–5:

In each exercise,

(a) Rewrite the given nth order scalar initial value problem as $\mathbf{y}' = \mathbf{f}(t, \mathbf{y}), \mathbf{y}(t_0) = \mathbf{y}_0$, by defining $y_1(t) = y(t), y_2(t) = y'(t), \ldots, y_n(t) = y^{(n-1)}(t)$ and

$$\mathbf{y}(t) = \begin{bmatrix} y_1(t) \\ y_2(t) \\ \vdots \\ y_n(t) \end{bmatrix}.$$

(b) Compute the n^2 partial derivatives $\partial f_i(t, y_1, \ldots, y_n)/\partial y_j$, $i, j = 1, \ldots, n$.

(c) For the system obtained in part (a), determine where in $(n+1)$-dimensional $t\mathbf{y}$-space the hypotheses of Theorem 8.1 are *not* satisfied. In other words, at what points $(t, y_1, \ldots, y_n)$, if any, does at least one component function $f_i(t, y_1, \ldots, y_n)$ and/or at least one partial derivative function $\partial f_i(t, y_1, \ldots, y_n)/\partial y_j$, $i, j = 1, \ldots, n$, fail to be continuous?

1. $y'' + ty = \sin y', \quad y(0) = 0, \quad y'(0) = 1$

2. $y'' + (y')^3 + y^{1/3} = \tan\left(\dfrac{t}{2}\right), \quad y(1) = 1, \quad y'(1) = -2$

3. $ty'' + \dfrac{1}{1 + y + 2y'} = e^{-t}, \quad y(2) = 2, \quad y'(2) = 1$

4. $y''' + \cos(ty') = t(y'')^2, \quad y(0) = 1, \quad y'(0) = 1, \quad y''(0) = -2$

5. $y''' + \dfrac{2t^{1/3}}{(y-2)(y''+2)} = 0, \quad y(0) = 0, \quad y'(0) = 2, \quad y''(0) = 2$

Exercises 6–9:

In each exercise, an initial value problem for a first order nonlinear system is given. Rewrite the problem as an equivalent initial value problem for a higher order nonlinear scalar differential equation.

6. $\dfrac{d}{dt}\begin{bmatrix} y_1 \\ y_2 \end{bmatrix} = \begin{bmatrix} y_2 \\ t\cos^2(y_2) - 3y_1 + t^4 \end{bmatrix}, \quad \begin{bmatrix} y_1(2) \\ y_2(2) \end{bmatrix} = \begin{bmatrix} 1 \\ -1 \end{bmatrix}$

7. $\dfrac{d}{dt}\begin{bmatrix} y_1 \\ y_2 \end{bmatrix} = \begin{bmatrix} y_2 \\ y_2\tan(y_1) + e^{y_2} \end{bmatrix}, \quad \begin{bmatrix} y_1(0) \\ y_2(0) \end{bmatrix} = \begin{bmatrix} 0 \\ 1 \end{bmatrix}$

8. $\dfrac{d}{dt}\begin{bmatrix} y_1 \\ y_2 \\ y_3 \end{bmatrix} = \begin{bmatrix} y_2 \\ y_3 \\ y_1y_2 + y_3^2 \end{bmatrix}, \quad \begin{bmatrix} y_1(-1) \\ y_2(-1) \\ y_3(-1) \end{bmatrix} = \begin{bmatrix} -1 \\ 2 \\ -4 \end{bmatrix}$

9. $\dfrac{d}{dt}\begin{bmatrix} y_1 \\ y_2 \\ y_3 \end{bmatrix} = \begin{bmatrix} y_2 \\ y_3 \\ \sqrt{y_2y_3 + t^2} \end{bmatrix}, \quad \begin{bmatrix} y_1(1) \\ y_2(1) \\ y_3(1) \end{bmatrix} = \begin{bmatrix} 1 \\ \frac{1}{2} \\ 3 \end{bmatrix}$

10. Consider the initial value problem

$$\frac{d}{dt}\begin{bmatrix} y_1 \\ y_2 \end{bmatrix} = \begin{bmatrix} \frac{5}{4}y_1^{1/5} + y_2^2 \\ 3y_1y_2 \end{bmatrix}, \quad \begin{bmatrix} y_1(0) \\ y_2(0) \end{bmatrix} = \begin{bmatrix} 0 \\ 0 \end{bmatrix}.$$

For the given autonomous system, the two functions, $f_1(y_1, y_2) = (5/4)y_1^{1/5} + y_2^2$ and $f_2(y_1, y_2) = 3y_1y_2$ are continuous functions for all (y_1, y_2).

(a) Show by direct substitution that

$$y_1(t) = \begin{cases} 0, & -\infty < t \le c, \\ (t-c)^{5/4}, & c < t < \infty, \end{cases} \qquad y_2(t) = 0$$

is a solution of this initial value problem on $-\infty < t < \infty$ for any positive constant c.

(b) Since c is an arbitrary positive constant, the solution of the given initial value problem is clearly not unique. Does this example contradict Theorem 8.1? Explain your answer.

11. Consider the initial value problem $y'' + y^2 = f(t), y(0) = y_0, y'(0) = y_0'$, where $f(t)$ is a continuous function defined on $0 \le t < \infty$ and where y_0 and y_0' are given constants. Can Laplace transforms be used to solve this initial value problem? Explain your answer.

Exercises 12–13:

Give an example of a two-dimensional nonlinear first order system for which the hypotheses of Theorem 8.1 are not satisfied at precisely the specified points in (t, y_1, y_2)-space.

12. The points satisfying $1 + t + y_1 + 3y_2 = 0$

13. The points $(t, y_1, y_2) = (1, n\pi, 2), n = 0, \pm 1, \pm 2, \ldots$

Applications We now introduce a number of applications that give rise to initial value problems for nonlinear differential equations. We will return to these problems at different times throughout this chapter.

14. Nonlinear Spring-Mass Systems Hooke's law assumes the restoring force exerted by a spring under tension or compression is proportional to the displacement (the distance stretched or foreshortened). This assumption cannot be valid for large displacements since there are limits to the amount a spring can be stretched or compressed. Suppose we assume that the restoring force $F_R(x)$ is related to spring displacement x by

$$F_R(x) = -\frac{2k\delta}{\pi}\tan\left(\frac{\pi x}{2\delta}\right),$$

In this model the restoring force has vertical asymptotes at $x = \pm\delta$; the value δ represents the maximum amount the spring can be stretched or compressed. Consider the figure illustrating a mass m attached to such a spring. Assume the mass moves on a frictionless horizontal surface and that the spring has unstretched length l. Newton's second law of motion leads to the nonlinear differential equation

$$mx'' + \frac{2k\delta}{\pi}\tan\left(\frac{\pi x}{2\delta}\right) = 0. \tag{7}$$

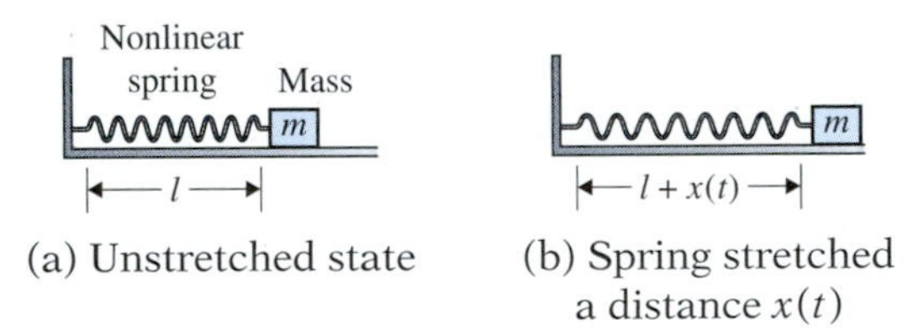

Figure for Exercise 14

(a) Consider $\tan(\pi x/2\delta)$ as a function of x defined on $-\delta < x < \delta$. Expand this function in a Maclaurin series. Show that if we assume $|\pi x/2\delta|$ is small and approximate $\tan(\pi x/2\delta)$ by the first nonvanishing term in this series, we obtain the linear differential equation found previously when we assumed Hooke's law to be valid.

(b) Show that if the first two nonvanishing terms of the Maclaurin expansion are retained, we obtain the differential equation

$$mx'' + k\left[x + \frac{1}{3}\left(\frac{\pi}{2\delta}\right)^2 x^3\right] = 0. \tag{8}$$

Equation (8) is often used to model the onset of nonlinear effects and is referred to as modeling a spring-mass system with cubic nonlinearity.

(c) Rewrite differential equations (7) and (8) as equivalent first order systems.

(d) For each nonlinear system obtained in part (c), determine the points, if any, where the hypotheses of Theorem 8.1 are not satisfied.

15. Chemical Reactions Nonlinear systems often arise when chemical reactions are modeled. One example is described in the reaction diagram in the figure. In the reaction shown, substance A interacts reversibly with enzyme E to form complex C. Complex C, in turn, decomposes irreversibly into the reaction product B and the original enzyme E. The reaction rates k_1, k_1' and k_2 (assumed to be constant) are shown in the figure. Using lowercase symbols to designate concentrations, the governing differential equations are

$$\begin{aligned}\frac{da}{dt} &= -k_1 ae + k_1' c,\\ \frac{db}{dt} &= k_2 c,\\ \frac{dc}{dt} &= k_1 ae - (k_1' + k_2)c,\\ \frac{de}{dt} &= -k_1 ae + (k_1' + k_2)c.\end{aligned} \tag{9}$$

The equations forming this autonomous system arise from the basic principle: Rate of change = rate of gain − rate of loss. Note that the rate at which the complex is formed is proportional to the product of concentrations a and e. Typical initial conditions are $a(0) = a_0$, $b(0) = 0$, $c(0) = 0$, $e(0) = e_0$.

$$A + E \underset{k_1'}{\overset{k_1}{\rightleftarrows}} C \qquad C \xrightarrow{k_2} B + E$$

Figure for Exercise 15

(a) Show that the differential equations (9) imply that $d(c(t) + e(t))/dt = 0$, which implies that $c(t) + e(t) = c(0) + e(0) = e_0$.

(b) Use the observation made in part (a) to eliminate $e(t)$ in (9) and obtain a two-dimensional nonlinear system for the dependent variables $a(t)$ and $c(t)$.

(c) For the two-dimensional system obtained in part (b), at what points in (t, a, c)-space are the hypotheses of Theorem 8.1 satisfied?

16. **A Bobbing Sphere** In Chapter 4 we considered the vertical bobbing motion of floating cylindrical objects that arises when they are displaced from their equilibrium positions. The Extended Problem in Chapter 4 dealt with two noncylindrical objects. Models describing the bobbing motion of such objects are more complex and nonlinear second order differential equations typically arise. Here we consider another case of a floating noncylindrical object, the sphere.

Consider the figure shown. Assume a sphere of radius R weighs half as much as an equivalent volume of water. In its equilibrium state, the sphere floats half-submerged as shown. The sphere is disturbed from equilibrium at some instant; its position is as shown, with displacement $y(t)$ measured positive downward.

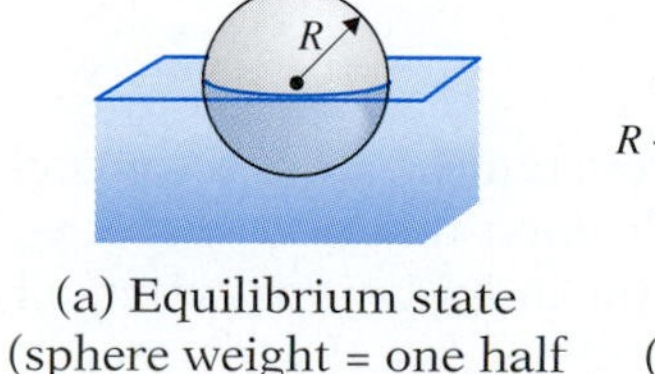

(a) Equilibrium state (sphere weight = one half the weight of an equal volume of water)

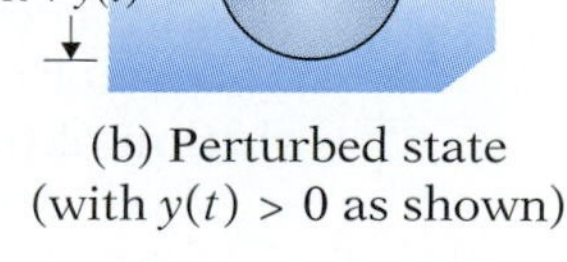

(b) Perturbed state (with $y(t) > 0$ as shown)

Figure for Exercise 16

(a) Compute the volume of the submerged portion of the sphere at the instant when the displacement from equilibrium is $y(t)$. (This is the volume of the water displaced at that instant.) Archimedes' law of buoyancy states that the upward force acting upon the sphere at the given instant is the weight of the water displaced at that instant.

(b) Apply Newton's second law of motion to obtain the differential equation governing the bobbing motion of the sphere. Considering only the weight and buoyant force, equate my'' (where m is the mass of the sphere) to the net downward force (which is the sphere weight minus the upward buoyant force). Show that the resulting equation can be written as

$$y'' + \frac{g}{2}\left(\frac{3y}{R} - \frac{y^3}{R^3}\right) = 0, \tag{10}$$

where g represents the acceleration due to gravity. For what range of values of $y(t)$ is differential equation (10) physically relevant? Is equation (10) applicable when $|y(t)| > R$?

(c) Rewrite differential equation (10) as an equivalent two-dimensional first order system.

8.2 Equilibrium Solutions and Direction Fields

In this section we extend the concepts of direction fields and equilibrium solutions to systems of autonomous equations. As you will see, this extension provides a large-scale overview of the qualitative behavior of solutions of autonomous systems.

Equilibrium Solutions

We begin by extending the definition of an equilibrium solution. Consider a system of n autonomous differential equations:

$$\begin{aligned} y_1' &= f_1(y_1, y_2, \ldots, y_n), \\ y_2' &= f_2(y_1, y_2, \ldots, y_n), \\ &\vdots \\ y_n' &= f_n(y_1, y_2, \ldots, y_n), \end{aligned}$$

or, in vector terms,

$$\mathbf{y}' = \mathbf{f}(\mathbf{y}). \tag{1}$$

Let $\mathbf{y}_e$ be a constant $(n \times 1)$ vector such that $\mathbf{f}(\mathbf{y}_e) = \mathbf{0}$. The corresponding constant vector-valued function, $\mathbf{y}(t) = \mathbf{y}_e$, $-\infty < t < \infty$, is called an **equilibrium solution** of the autonomous system (1).

EXAMPLE 1

Find the equilibrium solutions for the pendulum equation

$$\begin{aligned} y_1' &= y_2, \\ y_2' &= -\frac{g}{l}\sin y_1. \end{aligned}$$

Solution: In vector form, the equation is $\mathbf{y}' = \mathbf{f}(\mathbf{y})$, where

$$\mathbf{y} = \begin{bmatrix} y_1 \\ y_2 \end{bmatrix} \quad \text{and} \quad \mathbf{f}(\mathbf{y}) = \begin{bmatrix} y_2 \\ -\frac{g}{l}\sin y_1 \end{bmatrix}. \tag{2}$$

From (2), the equation $\mathbf{f}(\mathbf{y}) = \mathbf{0}$ requires $y_2 = 0$ and $y_1 = m\pi$, $m = 0, \pm 1, \pm 2, \ldots$. Therefore, the equilibrium solutions are

$$\mathbf{y}_m(t) = \begin{bmatrix} m\pi \\ 0 \end{bmatrix}, \qquad m = 0, \pm 1, \pm 2, \ldots.$$

These constant solutions of the pendulum equation have a simple physical interpretation. Recall that $y_1 = \theta$ and $y_2 = \theta'$ (see Figure 8.1). Thus, for an even value of m, the pendulum is at rest (since it has zero angular velocity) and is positioned so that it hangs downward. For an odd value of m, the pendulum is also at rest, but it is positioned in the vertically upward position. ▲

EXAMPLE 2

Find the equilibrium solutions of the competing species model

$$P_1' = r_1(1 - \alpha_1 P_1 - \beta_1 P_2)P_1,$$
$$P_2' = r_2(1 - \beta_2 P_1 - \alpha_2 P_2)P_2.$$

Solution: Setting both right-hand sides simultaneously equal to zero leads to four equilibrium solutions. One of the equilibrium solutions is the trivial one,

$$\text{(i)}\ P_1 = P_2 = 0 \quad \text{or} \quad \mathbf{P}_1^{(e)} = \mathbf{0}.$$

[This uninteresting equilibrium solution corresponds to the absence of both species from the colony.]

Two additional equilibrium solutions are

$$\text{(ii)}\ P_1 = 0, \qquad P_2 = \frac{1}{\alpha_2}, \quad \text{or} \quad \mathbf{P}_2^{(e)} = \begin{bmatrix} 0 \\ \frac{1}{\alpha_2} \end{bmatrix}.$$

$$\text{(iii)}\ P_1 = \frac{1}{\alpha_1}, \qquad P_2 = 0, \quad \text{or} \quad \mathbf{P}_3^{(e)} = \begin{bmatrix} \frac{1}{\alpha_1} \\ 0 \end{bmatrix}.$$

[Equilibrium solutions (ii) and (iii) correspond to the absence of one species. The remaining species has the equilibrium value of the corresponding scalar logistic equation (see Section 3.5).]

If neither P_1 nor P_2 is zero, we obtain a fourth equilibrium solution,

$$\text{(iv)}\ P_1 = \frac{\alpha_2 - \beta_1}{\alpha_1\alpha_2 - \beta_1\beta_2}, \qquad P_2 = \frac{\alpha_1 - \beta_2}{\alpha_1\alpha_2 - \beta_1\beta_2}, \quad \text{or} \quad \mathbf{P}_4^{(e)} = \begin{bmatrix} \frac{\alpha_2 - \beta_1}{\alpha_1\alpha_2 - \beta_1\beta_2} \\ \frac{\alpha_1 - \beta_2}{\alpha_1\alpha_2 - \beta_1\beta_2} \end{bmatrix}.$$

(continued)

(continued)

In (iv) we tacitly assume that $\alpha_1\alpha_2 - \beta_1\beta_2 \neq 0$. Since populations are nonnegative quantities, equilibrium solution (iv) is physically meaningful only if the constants α_1, α_2, β_1, and β_2 are such that each component of $\mathbf{P}_4^{(e)}$ is positive. Equilibrium solution (iv) corresponds to a state where both populations are present and coexist at constant levels within the colony. ▲

Two-Dimensional Autonomous Systems and the Phase Plane

We now consider a special case—the two-dimensional autonomous system

$$\begin{aligned} y_1' &= f_1(y_1, y_2), \\ y_2' &= f_2(y_1, y_2). \end{aligned} \tag{3}$$

The qualitative behavior of solutions of system (3) can be described and studied graphically. Solution trajectories are plotted in a two- dimensional setting known as the **phase plane**.

By way of introducing the phase plane, recall that when we studied scalar differential equations, it was natural to graph solution curves, $y(t)$, in the two-dimensional ty-plane. If we decide to graph the solution curves $\mathbf{y}(t)$ of system (3) in a similar fashion, the graph requires three-dimensional (t, y_1, y_2)-space. We now consider an alternative way to represent and study solutions of the system. In particular, let $\mathbf{y}(t)$ be a solution of system (3), where

$$\mathbf{y}(t) = \begin{bmatrix} y_1(t) \\ y_2(t) \end{bmatrix}. \tag{4}$$

By viewing the independent variable t as a parameter, we can also represent $\mathbf{y}(t)$ as a parameterized curve in the (y_1, y_2)-plane. In the context of system (3), the (y_1, y_2)-plane is referred to as the *phase plane*. The following example contrasts the two approaches and illustrates the phase plane in a specific case. For simplicity, Example 3 treats a linear autonomous differential equation that can be explicitly solved.

EXAMPLE 3

Consider the initial value problem

$$\begin{aligned} y_1' &= -y_1 - 6y_2, \\ y_2' &= 6y_1 - y_2, \\ y_1(0) &= 1, \qquad y_2(0) = 0. \end{aligned}$$

In matrix terms, this autonomous initial value problem has the form $\mathbf{y}' = A\mathbf{y}$, $\mathbf{y}(0) = \mathbf{y}_0$, where

$$\mathbf{y} = \begin{bmatrix} y_1 \\ y_2 \end{bmatrix}, \qquad A = \begin{bmatrix} -1 & -6 \\ 6 & -1 \end{bmatrix}, \quad \text{and} \quad \mathbf{y}_0 = \begin{bmatrix} 1 \\ 0 \end{bmatrix}. \tag{5}$$

The solution of initial value problem (5) can be found using the techniques of Chapter 6. The solution is

$$\begin{bmatrix} y_1(t) \\ y_2(t) \end{bmatrix} = \begin{bmatrix} e^{-t}\cos 6t \\ e^{-t}\sin 6t \end{bmatrix}. \tag{6}$$

Figure 8.2(a) shows the solution graphed in three-dimensional (t, y_1, y_2)-space. The graph evolves in a "shrinking" helical or screwlike manner as time increases.

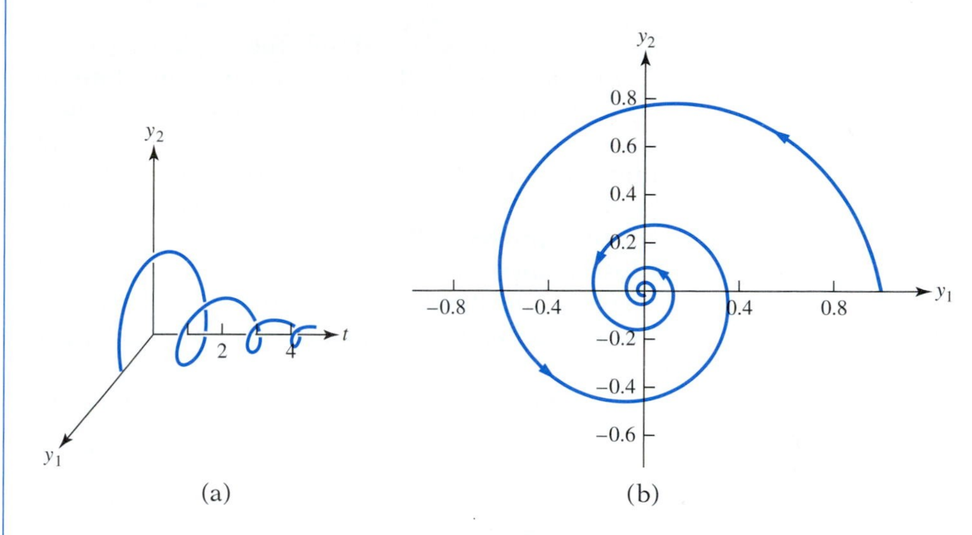

FIGURE 8.2

(a) The solution of equation (5), graphed in (t, y_1, y_2)-space. (b) The solution of equation (5) projected onto the (y_1, y_2)-phase plane. The arrows show how the phase-plane trajectory is traversed as time increases. Figure 8.2(b) corresponds to what an observer would see standing behind the plane $t = 0$ in Figure 8.2(a) and looking in the direction of increasing t.

The two-dimensional phase-plane representation is found by graphing the parameterized curve defined by (6). Figure 8.2(b) shows the graph of (6) for $0 \leq t \leq 3$. The phase-plane graph is a spiral; it is simply the projection of the three-dimensional "helical" trajectory of Figure 8.2(a) upon the two-dimensional (y_1, y_2)-plane. As time increases, the solution point spirals counterclockwise inward toward the phase-plane origin (0, 0). ▲

Using the Phase Plane to Gather Qualitative Information about Solutions

When studying the geometric aspects of two-dimensional systems of the form

$$\begin{aligned} y_1' &= f_1(t, y_1, y_2), \\ y_2' &= f_2(t, y_1, y_2), \end{aligned}$$

it is often desirable to change the notation. In particular, we can drop the subscripts and denote the dependent variables as $x(t)$ and $y(t)$. With this change of notation, the general nonlinear two-dimensional system has the form

$$\begin{aligned} x' &= f(t, x, y), \\ y' &= g(t, x, y), \end{aligned}$$

and a two-dimensional autonomous system has the form

$$\begin{aligned} x' &= f(x, y), \\ y' &= g(x, y). \end{aligned} \tag{7}$$

The phase plane is now simply the xy-plane.

As illustrated by Figure 8.2(b), we can think of the solution components $x(t)$ and $y(t)$ as defining the coordinates of a point, $(x(t), y(t))$, that is moving in the phase plane; we refer to this point as the **solution point**. [For equilibrium solutions, the components $x(t)$ and $y(t)$ are constant for all t. Therefore, the solution point corresponding to an equilibrium solution is often called an **equilibrium point**.] The plane curve traced out by a solution point is called a **solution curve**. The question we now address is "What qualitative information can we obtain about the motion of a solution point without actually solving the system of differential equations?" The answer builds on the direction field ideas of Chapters 1 and 3.

As we know from vector calculus, the vector $\mathbf{v}(t)$ given by

$$\mathbf{v}(t) = x'(t)\mathbf{i} + y'(t)\mathbf{j}$$

is tangent to the solution curve at the point $(x(t), y(t))$ (see Figure 8.3). In particular, if we think of the solution point $(x(t), y(t))$ as moving along the solution curve, then $\mathbf{v}(t)$ is the velocity vector and points in the direction of instantaneous motion at time t.

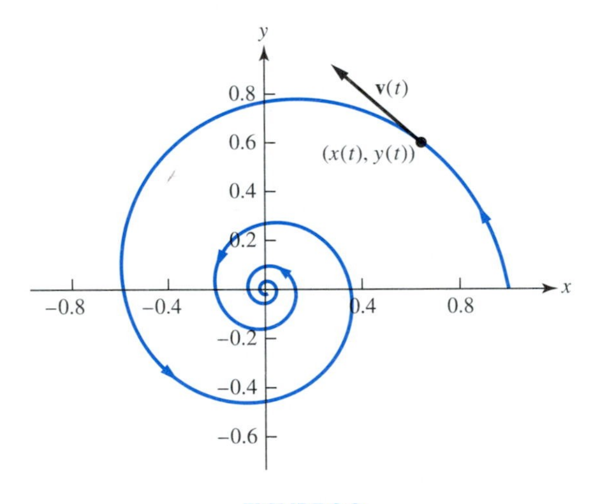

FIGURE 8.3

The solution point $(x(t), y(t))$ lies on a solution curve of system (7). The velocity vector $\mathbf{v}(t) = x'(t)\mathbf{i} + y'(t)\mathbf{j}$ is tangent to the solution curve at $(x(t), y(t))$ and points in the direction the solution point is moving.

Suppose we select an arbitrary point in the phase plane, say $(\bar{x}, \bar{y})$. Assume that a solution curve passes through this point at some time $t = \bar{t}$; that is,

$$(\bar{x}, \bar{y}) = (x(\bar{t}), y(\bar{t})).$$

At the point $(\bar{x}, \bar{y})$, the velocity vector is given by

$$\begin{aligned} \bar{\mathbf{v}} &= x'(\bar{t})\mathbf{i} + y'(\bar{t})\mathbf{j} \\ &= f(\bar{x}, \bar{y})\mathbf{i} + g(\bar{x}, \bar{y})\mathbf{j}. \end{aligned} \tag{8}$$

As illustrated in Figure 8.3, the vector $\bar{\mathbf{v}}$ is tangent to the solution curve at $(\bar{x}, \bar{y})$ and points in the direction of motion. Therefore, by simply evaluating the two right-hand sides of system (7) at a point $(\bar{x}, \bar{y})$ in the phase plane, we can deduce the direction of motion of the solution point at the instant it passes through $(\bar{x}, \bar{y})$. In particular, we see from (8) that the slope m of the line tangent to the solution curve at $(\bar{x}, \bar{y})$ is given by

$$m = \frac{g(\bar{x}, \bar{y})}{f(\bar{x}, \bar{y})}, \qquad f(\bar{x}, \bar{y}) \neq 0. \tag{9}$$

[If $f(\bar{x}, \bar{y}) = 0$, but $g(\bar{x}, \bar{y}) \neq 0$, then the solution curve has a vertical tangent at the point $(\bar{x}, \bar{y})$. If numerator and denominator both vanish, then $(\bar{x}, \bar{y})$ is an equilibrium point.]

EXAMPLE 4

Consider the autonomous system

$$x' = \frac{1}{2}\left(1 - \frac{1}{2}x - \frac{1}{2}y\right)x,$$
$$y' = \frac{1}{4}\left(1 - \frac{1}{3}x - \frac{2}{3}y\right)y.$$

Let $(x(t), y(t))$ denote a solution curve in the phase plane. Determine the velocity vector when the solution curve passes through the given point (x, y).

(a) $(x, y) = (2, 2)$ (b) $(x, y) = \left(\frac{1}{2}, \frac{3}{2}\right)$

Solution: This system is the competing species autonomous system discussed in Example 2, with the dependent variables renamed and specific values assigned to the constants.

(a) Since $f(2, 2) = -1$ and $g(2, 2) = -1/2$, the velocity vector at the point $(2, 2)$ is

$$\mathbf{v} = -\mathbf{i} - \frac{1}{2}\mathbf{j}.$$

[At the instant a solution point passes through $(2, 2)$, it is moving downward and to the left. The slope of the line tangent to the phase-plane trajectory at $(2, 2)$ is $1/2$.]

(b) At this point, $f(1/2, 3/2) = 0$ and $g(1/2, 3/2) = -1/16$. Therefore, the velocity vector at $(1/2, 3/2)$ is

$$\mathbf{v} = -\frac{1}{16}\mathbf{j}.$$

[At the instant a solution point passes through $(1/2, 3/2)$, it is moving vertically downward.] ▲

Phase-Plane Direction Fields

Using equations (8) and (9), we can determine the tangent to a solution curve at a point $(\bar{x}, \bar{y})$ in the phase plane. As in Example 4, we can also determine the instantaneous direction of motion of the solution point when it passes through $(\bar{x}, \bar{y})$. We can use this information to deduce the qualitative behavior of solutions.

A **phase-plane direction field** is constructed by first choosing a suitably dense grid of sampling points in the phase plane. At each grid point we draw a small arrow, directed along the velocity vector at the grid point. In this way, we generate a qualitative picture of how the solution point moves as time increases. Figure 8.4 shows a direction field for the competing species system treated in Example 4. In Figure 8.4, every arrow is drawn with the same length; this is the construction referred to as a *direction field*. When the arrows at each phase-plane point are drawn so that their lengths are proportional to the speed at that point, the construction is referred to as a **vector field**. Since a vector field contains more information, giving both magnitude and direction of velocity, it might seem that such constructions are always more desirable. However, attempts to construct vector fields sometimes lead to information overload because the vector field graphs may have intersecting arrows and be visually confusing.

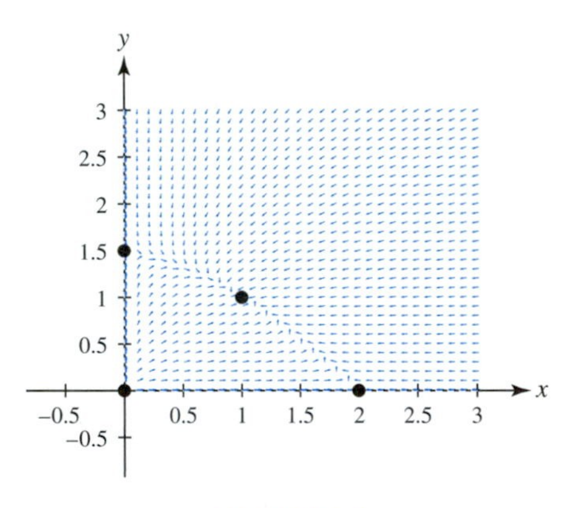

FIGURE 8.4

A portion of the direction field for the autonomous system discussed in Example 4. The arrow at a grid point $(\bar{x}, \bar{y})$ indicates the direction of motion of the solution point as it moves along a solution curve passing through $(\bar{x}, \bar{y})$.

Figure 8.4 provides a good overview of the qualitative behavior of solutions of the autonomous system discussed in Example 4. Recall that the variables x and y correspond to the two populations P_1 and P_2. On the coordinate axes, one or the other populations is zero and the direction field arrows point toward the single-species equilibrium points, $(2, 0)$ and $(0, 3/2)$. Therefore, in the case where only one population is present in the colony, population tends toward the appropriate nonzero equilibrium value, $x = 2$ or $y = 3/2$, as time increases. In the general case, where both populations are initially nonzero, the direction field graphed in Figure 8.4 suggests that the populations tend toward the equilibrium point $(1, 1)$ as time evolves.

Putting the Pieces Together

There are computer packages available that generate detailed direction fields such as the one shown in Figure 8.4. However, it is often possible to combine, by hand, the ideas discussed in this section and develop a less detailed phase-plane picture that nevertheless displays the essential qualitative features of the system's behavior. To illustrate, we once again consider the competing species model discussed in Example 4,

$$\begin{aligned} x' &= \frac{1}{2}\left(1 - \frac{1}{2}x - \frac{1}{2}y\right)x, \\ y' &= \frac{1}{4}\left(1 - \frac{1}{3}x - \frac{2}{3}y\right)y. \end{aligned} \tag{10}$$

First note that

$$f(x, y) = \frac{1}{2}\left(1 - \frac{1}{2}x - \frac{1}{2}y\right)x$$

vanishes on the phase-plane lines $x = 0$ and $x + y = 2$. Likewise,

$$g(x, y) = \frac{1}{4}\left(1 - \frac{1}{3}x - \frac{2}{3}y\right)y$$

vanishes on the lines $y = 0$ and $x + 2y = 3$. Such phase-plane curves, where one of the right-side functions vanishes, are called **nullclines**. Equilibrium points can occur only at a place where two nullclines intersect (although not every intersection point leads to an equilibrium point). For autonomous system (10), there are four equilibrium points:

$$(0, 0), \qquad (2, 0), \qquad \left(0, \frac{3}{2}\right), \qquad \text{and} \quad (1, 1).$$

The nullclines and equilibrium points for system (10) are shown in Figure 8.5(a). The nullclines divide the first quadrant of the phase plane into regions where the functions f and g are either positive or negative. Figure 8.5(b) shows the four phase-plane regions defined by the nullclines of system (10) and the corresponding algebraic signs of f and g. From equation (8), we can see that the signs of f and g determine the orientation of the direction field arrows. In

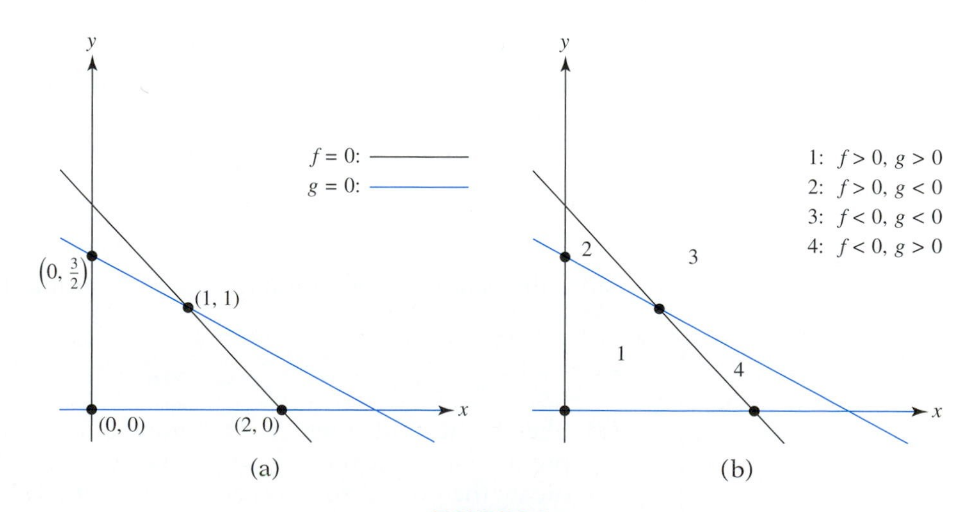

FIGURE 8.5

(a) The lines denote the nullclines for system (10).
(b) The nullclines divide the direction field of Figure 8.4 into regions where arrows all have the same general orientation.

region 1, for instance, $f > 0$ and $g > 0$. Therefore, all the direction field arrows in region 1 point upward and to the right.

Figure 8.6 shows, in schematic form, the general orientation of direction field arrows in each of the four open regions. It also shows the orientation of the direction field arrows on the nullclines; these arrows are either horizontal (if $f \neq 0$ and $g = 0$) or vertical (if $f = 0$ and $g \neq 0$). The information in Figure 8.6 is sufficient to deduce the general qualitative behavior of phase-plane trajectories. For example, if x and y are both initially nonzero, then solutions will tend toward the equilibrium state $x = y = 1$ as time increases.

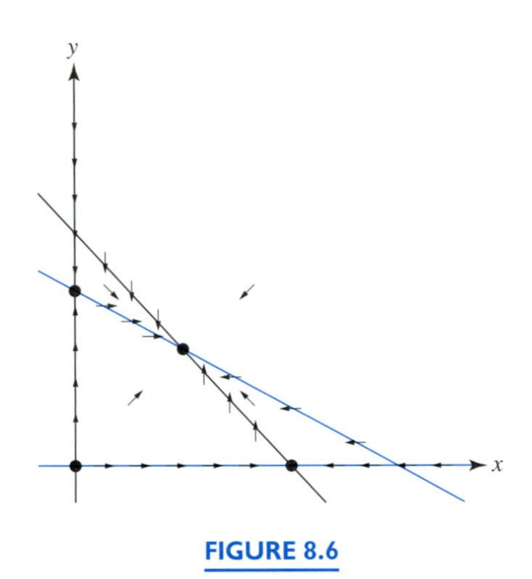

FIGURE 8.6

Along a nullcline, the arrows are either vertical or horizontal. In an open region bounded by nullclines, the single arrow indicates the general orientation of the solution curves in that region. This figure *suggests* that any solution curve starting in one of the open regions will move toward the equilibrium point $(1, 1)$.

While quite useful, a rough graph such as the one in Figure 8.6 may not be detailed enough to yield a good qualitative picture of the phase-plane trajectories. The next example illustrates this point.

EXAMPLE 5

Consider the pendulum equation treated in Example 1, where $g/l = 1$,

$$x' = y,$$
$$y' = -\sin x.$$

(a) Sketch the nullclines, as in Figure 8.5(a), marking the equilibrium points with a heavy dot. Then, as in Figure 8.6, add arrows that indicate the flow of solution curves. Does this sketch have enough detail to predict the qualitative nature of phase-plane trajectories?

(b) Using a computer, sketch a portion of the direction field for $-8 \leq x \leq 8$ and $-6 \leq y \leq 6$. Using your sketch, describe the two different types of phase-plane trajectories and give a physical interpretation for each type. [Recall that x denotes angular displacement from the pendulum's downward-hanging equilibrium position and y denotes angular velocity.]

Solution:

(a) The nullclines consist of the x-axis and the infinite set of vertical lines, $x = m\pi$, $m = 0, \pm 1, \pm 2, \ldots$. A portion of the phase plane is shown in Figure 8.7. This figure also shows the direction-field arrows on the nullclines as well as the general "sense of direction" arrows within the vertical phase-plane strips. The arrows are vertical on the nullcline $f(x, y) = 0$ (the x-axis) and are horizontal on the nullclines $g(x, y) = 0$ (the vertical lines $x = m\pi$, $m = 0, \pm 1, \pm 2, \ldots$). In order to gain a good qualitative picture of the phase plane trajectories, however, this level of description is inadequate.

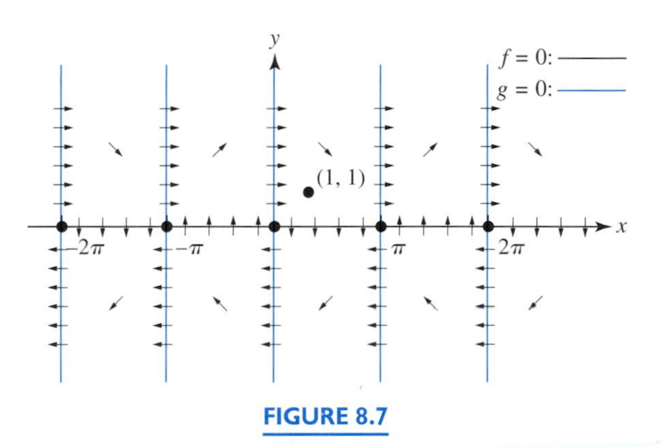

FIGURE 8.7

A sketch showing the main features of the direction field for the pendulum equation in Example 5. Note that we cannot tell from this sketch whether a solution curve passing through the point (1, 1) will continue down until it is below the x-axis or whether it will remain above the x-axis.

For example, consider a solution curve passing through the point $(1, 1)$. As we see from Figure 8.7, the solution point is moving downward and to the right when it passes through $(1, 1)$. But, is the curve falling fast enough so that it crosses the x-axis and continues to move down but now to the left? Or, is its rate of descent slowing enough so that it intersects the line $x = \pi$ in the upper half of the phase plane and then moves up and to the right? We need a more detailed direction field in order to give a reasonable assessment.

(b) Figure 8.8 on the next page presents a more detailed direction field plot. This plot indicates that there are two basic types of trajectories. Near the x-axis, there appear to be closed phase-plane trajectories; as time increases, the solution point seems to make clockwise orbits around these closed curves. Far from the x-axis, the trajectories no longer appear to be closed curves. Rather, they appear to be undulating curves that are basically horizontally oriented and that tend to become flatter as distance from the x-axis increases. ▲

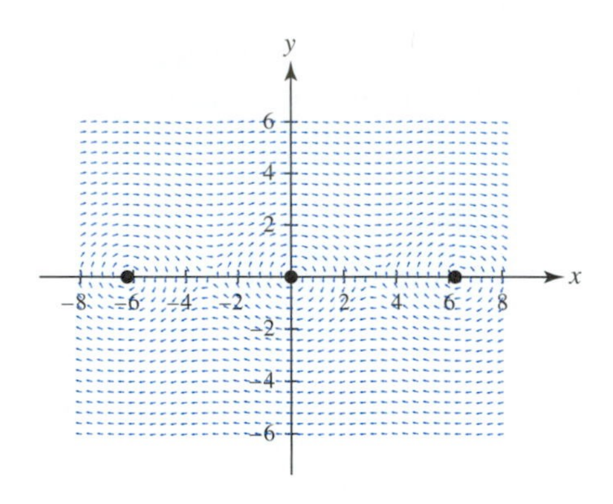

FIGURE 8.8

A portion of the direction field for the pendulum equation in Example 5. Some phase-plane trajectories appear to be closed curves, centered at equilibrium points on the x-axis. Other trajectories appear to be undulating curves where the solution point moves basically in one direction (to the right above the x-axis and to the left below the x-axis). The typical closed trajectory corresponds to the pendulum swinging back and forth, motion in which the pendulum never reaches the vertically upward position. The undulating trajectories correspond to a pendulum that continues to rotate in one direction (counterclockwise if $y = \theta'$ is positive and clockwise if $y = \theta'$ is negative).

We can understand Figure 8.8 in terms of its physical interpretation. Recall that $x = \theta$ represents the angular displacement of the pendulum from its downward hanging equilibrium position and $y = \theta'$ represents the instantaneous angular velocity of the pendulum. The closed orbits close to the x-axis therefore correspond to motion in which the pendulum swings back and forth. For example, consider a closed trajectory centered at the equilibrium point $(0, 0)$. On such a trajectory, the maximum excursion of $x = \theta$ from 0 is less than π. This maximum displacement is reached when the trajectory intersects the x-axis; that is, when angular velocity y is zero. In such a motion, the pendulum never reaches the vertically upward position. It swings up to some maximum angular displacement less than π and then swings back the same amount in the other direction. The continual orbiting of the solution point around a closed trajectory corresponds to this continual back-and-forth swing of the pendulum.

The horizontally configured undulating trajectories, however, correspond to motion in which the pendulum continually rotates about its pivot. In the upper half-plane, $y = \theta'$ is positive and the pendulum is always rotating in the counterclockwise direction. In the lower half-plane, $y = \theta'$ is negative and the pendulum is always rotating in the clockwise direction. Since $y = \theta'$ is never zero, the pendulum never stops and consequently it never changes direction.

The pendulum is an example of a conservative system (that is, a system in which energy is conserved). For such systems, and for a more general two-dimensional class of autonomous systems known as *Hamiltonian systems*, we can derive equations for the phase-plane trajectories. We consider such systems in the next section.

EXERCISES

Exercises 1–10:

Find all equilibrium points of the autonomous system.

1. $x' = -x + xy$
$y' = y - xy$

2. $x' = y(x+3)$
$y' = (x-1)(y-2)$

3. $x' = (x-2)(y+1)$
$y' = x^2 - 4x + 3$

4. $x' = xy - y + x - 1$
$y' = xy - 2y$

5. $x' = x^2 - 2xy$
$y' = 3xy - y^2$

6. $x' = y^2 - xy$
$y' = 2xy + x^2$

7. $x' = x^2 + y^2 - 8$
$y' = x^2 - y^2$

8. $x' = x^2 + 2y^2 - 3$
$y' = 2x^2 + y^2 - 3$

9. $x' = y - 1$
$y' = xy + x^2$
$z' = 2y - yz$

10. $x' = z^2 - 1$
$y' = z - 2xz + yz$
$z' = -(1 - x - y)^2$

Exercises 11–15:

Rewrite the given scalar differential equation as a first order system and find all equilibrium points of the resulting system.

11. $y'' + y + y^3 = 0.$

12. $y'' + e^y y' + \sin^2 \pi y = 1$

13. $y'' + \dfrac{2y'}{1+y^4} + y^2 = 1$

14. $y''' - y'' + 2\sin y = 1$

15. $y''' - (y')^2 + \dfrac{4-y^2}{2+(y')^2} = 0$

Exercises 16–20:

Use the information provided to determine the unspecified constants.

16. The system

$$x' = x + \alpha xy + \beta,$$
$$y' = \gamma y - 3xy + \delta$$

has equilibrium points at $(x, y) = (0, 0)$ and $(2, 1)$. Is $(-2, -2)$ also an equilibrium point?

17. The system

$$x' = \alpha x + \beta xy + 2,$$
$$y' = \gamma x + \delta y^2 - 1$$

has equilibrium points at $(x, y) = (1, 1)$ and $(2, 0)$.

18. Consider the system

$$x' = x + \alpha y^3,$$
$$y' = x + \beta y + y^4.$$

The slopes of the phase-plane trajectories passing through the points $(x, y) = (2, 1)$ and $(1, -1)$ are 1 and 0, respectively.

19. Consider the system

$$x' = \alpha x^2 + \beta y + 1,$$
$$y' = x + \gamma y + y^2.$$

The slopes of the phase-plane trajectories passing through the points $(x, y) = (1, 1)$ and $(1, -1)$ are 0 and 4, respectively. The phase-plane trajectory passing through the point $(x, y) = (0, -1)$ has a vertical tangent.

20. Consider the system

$$x' = x + y^2 - xy^n,$$
$$y' = -x + y^{-1}.$$

The slope of the phase-plane trajectory passing through the point $(x, y) = (1, 2)$ is $1/6$. Determine the exponent n.

21. The scalar differential equation $y'' - y' + 2y^2 = \alpha$, when rewritten as a first order system results in a system having an equilibrium point at $(x, y) = (2, 0)$. Determine the constant α.

22. For the given system, compute the velocity vector $\mathbf{v}(t) = x'(t)\mathbf{i} + y'(t)\mathbf{j}$ at the point $(x, y) = (2, 3)$.

(a) $x' = -x + xy$
$y' = y - xy$

(b) $x' = y(x + 3)$
$y' = (x - 1)(y - 2)$

(c) $x' = (x - 2)(y + 1)$
$y' = x^2 - 4x + 3$

23. Let A be a (2×2) constant matrix and let $(\lambda, \mathbf{u})$ be an eigenpair for A. Assume that λ is real, $\lambda \neq 0$, and

$$\mathbf{u} = \begin{bmatrix} u_1 \\ u_2 \end{bmatrix}.$$

Consider the phase plane for the autonomous linear system $\mathbf{y}' = A\mathbf{y}$. We can define a phase-plane line through the origin by the parametric equations $x = \tau u_1, y = \tau u_2$, $-\infty < \tau < \infty$. Let P be any point on this line, say $P = (\tau_0 u_1, \tau_0 u_2)$ for some $\tau_0 \neq 0$.

(a) Show that at the point P, $x' = \tau_0 \lambda u_1$ and $y' = \tau_0 \lambda u_2$.

(b) How is the velocity vector $\mathbf{v}(t) = x'(t)\mathbf{i} + y'(t)\mathbf{j}$ at point P oriented relative to the line?

Exercises 24–27:

In each exercise, a matrix A is given. For each matrix, the vectors

$$\mathbf{u}_1 = \begin{bmatrix} 1 \\ 1 \end{bmatrix} \quad \text{and} \quad \mathbf{u}_2 = \begin{bmatrix} 1 \\ -1 \end{bmatrix}$$

are eigenvectors of A. As discussed in Exercise 23, these eigenvectors are associated with the phase-plane lines $y = x$ and $y = -x$. Solution points of $\mathbf{y}' = A\mathbf{y}$ originating on these two lines remain on these lines as time evolves. Match the given matrix A to one of the four direction fields shown for $\mathbf{y}' = A\mathbf{y}$.

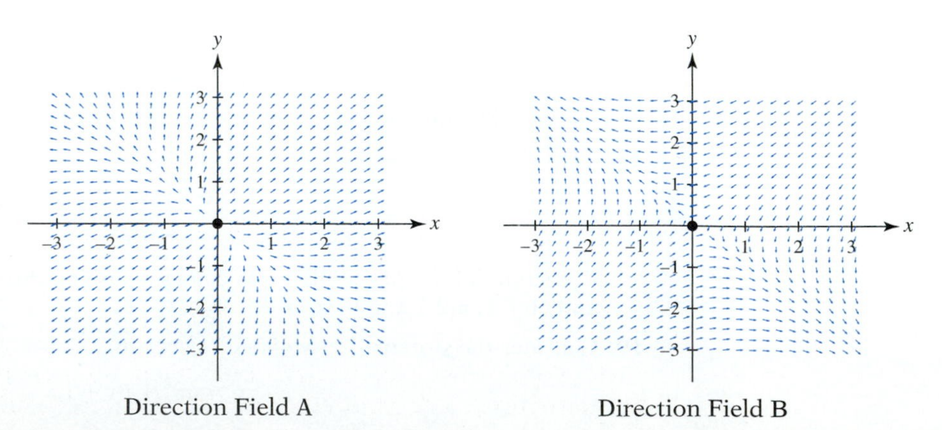

Figure for Exercises 24–27

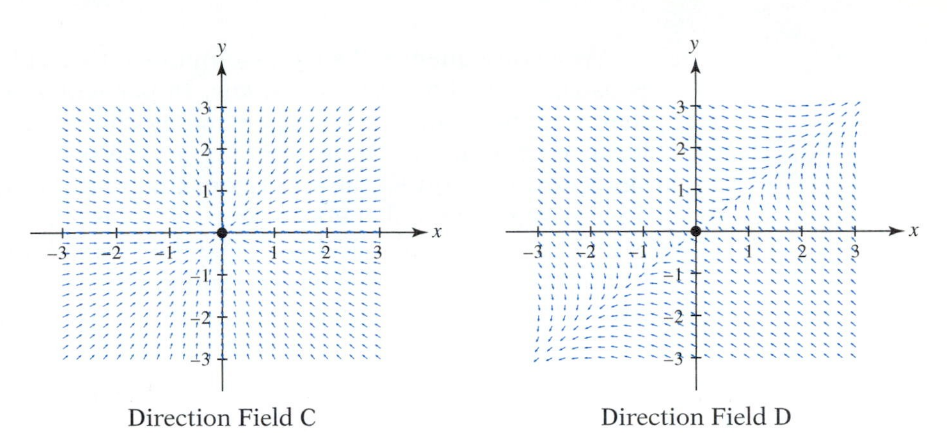

Figure for Exercises 24–27 (continued)

24. $A = \begin{bmatrix} -9 & 1 \\ 1 & -9 \end{bmatrix}$ **25.** $A = \begin{bmatrix} -1 & -3 \\ -3 & -1 \end{bmatrix}$ **26.** $A = \begin{bmatrix} -4 & 6 \\ 6 & -4 \end{bmatrix}$ **27.** $A = \begin{bmatrix} 4 & 2 \\ 2 & 4 \end{bmatrix}$

28. Suppose that the nonlinear second order equation $y'' + f(y) = 0$ is recast as an autonomous first order system. Show that the nullclines for the resulting system are the horizontal line $y = 0$ and vertical lines of the form $x = \alpha$, where α is a root of $f(y) = 0$. For each such root, what is the nature of the phase-plane point $(x, y) = (\alpha, 0)$?

Exercises 29–31:

Rewrite the given second order equation as an equivalent first order system.

(a) Graph the nullclines of the autonomous system and locate all equilibrium points.

(b) As in Figure 8.6, sketch direction field arrows on the nullclines. Also, sketch an arrow in each open region that suggests the direction in which a solution point is moving when it is in that region.

29. $y'' + y + y^3 = 0.$ **30.** $y'' + y(1 - y^2) = 0$ **31.** $y'' + 2\sin^2 y = 1$

Exercises 32–36:

In each exercise

(a) Graph the nullclines of the autonomous system and locate all equilibrium points.

(b) As in Figure 8.6, sketch direction field arrows on the nullclines. Also, sketch an arrow in each open region that suggests the direction in which a solution point is moving when it is in that region.

32. $x' = 3x - y - 2$, $y' = x - y$

33. $x' = -x - y + 2$, $y' = x - y$

34. $x' = (2x - y - 2)(4 - x - y)$, $y' = x - 2y$

35. $x' = (2x - y - 6)(x - y)$, $y' = x + y$

36. $x' = x^2 + y - 1$, $y' = -x^2 + y + 1$

8.3 Conservative Systems

When we formulate a mathematical model for a physical system, we often neglect effects such as friction or electrical resistance if they are small enough. We have already encountered several of these idealized mathematical models in our discussion of spring-mass systems, buoyant bodies, and pendulums.

As a consequence of such assumptions, these idealized models obey what is usually called a *conservation law*. In particular, a conservation law means that a quantity, such as energy, remains constant. For example, consider an idealized pendulum. On its upswing, as the bob elevates and simultaneously slows down, energy is converted from kinetic energy to potential energy. On the downswing, potential energy is, in turn, transformed back into kinetic energy. We will show, for this idealized pendulum model, that total pendulum energy (the sum of kinetic and potential energy) remains constant in time. Thus, total pendulum energy is a conserved quantity in the idealized pendulum model.

In general, consider a second order scalar differential equation

$$y'' = f(t, y, y')$$

and let $y(t)$ be a solution of this differential equation. If there is a function of two variables, $H(u, v)$, such that $H(y(t), y'(t))$ remains constant in time, then we call H a **conserved quantity** and say that the differential equation $y'' = f(t, y, y')$ possesses a **conservation law**. We use the same terms to describe the general case of an n-dimensional system with solution components $y_1(t), y_2(t), \ldots, y_n(t)$ for which some function $H_1(y_1(t), y_2(t), \ldots, y_n(t))$ remains constant.

In this section we are interested in the following questions. Given a mathematical model, how can we determine (from the structure of the differential equation itself) whether or not it satisfies a conservation law? If the model does in fact possess a conserved quantity, how can we explicitly describe this quantity and use its mathematical description to better understand the system's dynamics? In particular, what phase-plane information can be deduced from the conservation law?

An Important Class of Second Order Scalar Equations

Consider the differential equation

$$y'' + f(y) = 0. \tag{1}$$

Such differential equations often arise when we apply Newton's laws to a body in one-dimensional motion. In such applications, y'' corresponds to acceleration and the term $-f(y)$ corresponds to the applied force per unit mass. In fact, three of the mathematical models we have discussed—the undamped mass-spring system, the buoyant body, and the pendulum—all have the structure of equation (1).

We now show that equation (1) possesses a conservation law. To see why, and to obtain a description of this law, we first multiply the equation by y', obtaining

$$y'y'' + f(y)y' = 0. \tag{2}$$

Consider now the terms in (2). Recalling the chain rule of calculus, we see that the first term on the left-hand side can be written as

$$y'(t)y''(t) = \frac{d}{dt}\left[\frac{1}{2}(y'(t))^2\right]. \tag{3}$$

Likewise, if $F(y)$ denotes an antiderivative of $f(y)$,

$$\frac{dF(y)}{dy} = f(y),$$

then the chain rule allows us to express the second term in (2) as

$$f(y(t))y'(t) = \frac{d}{dt} F(y(t)). \tag{4}$$

Using (3) and (4), we can rewrite (2) in the form

$$\frac{d}{dt}\left[\frac{1}{2}(y'(t))^2 + F(y(t))\right] = 0. \tag{5}$$

From (5) we conclude, therefore, that

$$\frac{1}{2}y'(t)^2 + F(y(t)) = C. \tag{6}$$

Equation (6) is the underlying conservation law. For instance, if $y(t)$ represents a displacement, then the term $(1/2)y'(t)^2$ is kinetic energy per unit mass. The other term, $F(y(t))$, is potential energy per unit mass (measured relative to some reference value that depends upon the particular antiderivative F chosen). The constant C can be interpreted as the (constant) total energy per unit mass.

Phase-Plane Interpretation

Differential equation (1) can be recast as the first order autonomous system

$$x' = y,$$
$$y' = -f(x).$$

where x and y now play the roles of $y(t)$ and $y'(t)$, respectively. Thus, the conservation law (6) takes the form

$$\frac{1}{2}(y)^2 + F(x) = C. \tag{7}$$

The family of curves obtained by graphing this equation for different values of C is a set of phase-plane trajectories describing the motion. In the next example, we develop these ideas for the pendulum.

EXAMPLE 1

Consider the pendulum equation [recall Example 5, Section 8.2]

$$x' = y,$$
$$y' = -\sin x.$$

From (7), the corresponding conservation law is

$$\frac{1}{2}y^2 - \cos x = C.$$

If we revert to the original variables, θ and θ', and use E to denote the constant energy of the system, the conservation law has the form[1]

$$\frac{1}{2}(\theta')^2 - \cos\theta = E. \tag{8}$$

(continued)

[1]In Section 3.7, Exercise 12, we derived this conservation law in a different way. A change of independent variable was used to transform $\theta'' + (g/l)\sin\theta = 0$ into a first order separable differential equation. The implicit solution of this equation yielded the conservation law (8).

(continued)

The term $-\cos\theta$ represents the potential energy of the pendulum bob, measured relative to a zero reference at the horizontal position. Equation (8) is graphed in Figure 8.9 for various energy levels. The direction of solution point motion on these trajectories is indicated with arrows. (The direction field arrows in Figure 8.8 of Section 8.2 are tangent to these phase-plane trajectories.) The entire phase plane is simply a periodic repetition of the portion shown; every 2π increment in $x = \theta$ brings the pendulum back to the same physical configuration.

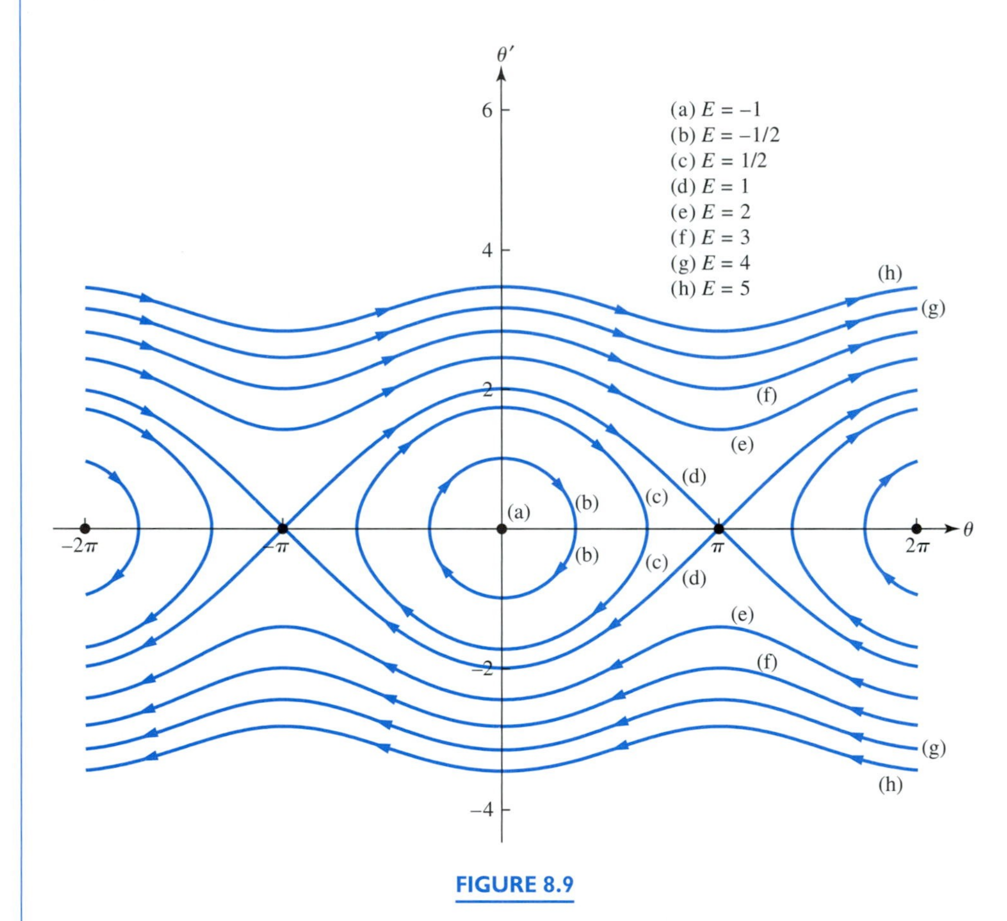

FIGURE 8.9

Some of the phase-plane trajectories for the pendulum equation discussed in Example 1. These curves are graphs of equation (8) for various energy levels E.

To understand Figure 8.9, observe from equation (8) that the equilibrium point $(\theta, \theta') = (0, 0)$ [with the bob at rest hanging vertically downward] corresponds to the energy level $E = -1$. In a similar fashion, the equilibrium points $(\theta, \theta') = (\pi, 0)$ and $(\theta, \theta') = (-\pi, 0)$ [with the bob at rest and positioned vertically upward] correspond to the energy level $E = 1$.

The energy value $E = 1$ is a delineating or "separating" value. Phase-plane trajectories for $-1 < E < 1$ are the closed curves in Figure 8.9 centered at $(0, 0)$.

These trajectories correspond to motion in which the pendulum continuously swings back and forth; it does not have enough energy to reach the vertical upward position. The pendulum swings upward to some $\theta_{max} < \pi$, stops, and then swings downward, achieving the same maximum elevation on its backswing. The two closed curves, labeled b and c in Figure 8.9, correspond to energy levels $E = -1/2$ and $E = 1/2$. The maximum angular displacements achieved by the pendulum in these two cases are $|\theta_{max}| = \pi/3$ and $|\theta_{max}| = 2\pi/3$, respectively. These values correspond to the θ-axis intercepts of the curves.

Energy levels $E > 1$ correspond to motion in which the system possesses enough energy to permit the pendulum to reach the vertical upward position and continue to rotate. Since energy is conserved in this idealized model, the pendulum continues to rotate forever. For each energy level greater than 1, two trajectories are possible. These are not closed trajectories, since total angular displacement increases or decreases monotonically. The trajectories in the upper half-plane (where $\theta' > 0$) correspond to counterclockwise pendulum rotation, while the counterpart trajectories in the lower half-plane (where $\theta' < 0$) represent clockwise pendulum rotation. The eight such trajectories shown in Figure 8.9 correspond to $E = 2$, $E = 3$, $E = 4$, and $E = 5$.

The two trajectories corresponding to $E = 1$ appear to connect the equilibrium points $(-\pi, 0)$ and $(\pi, 0)$. These trajectories are called **separatrices**; they separate the closed trajectories from the open trajectories. On the upper separatrix, the solution point approaches the equilibrium point $(\pi, 0)$ as $t \to \infty$. The pendulum swings upward in the counterclockwise direction, slowing down as it approaches the vertical upward position. The pendulum bob approaches this inverted position in the limit as $t \to \infty$. On the lower separatrix, the solution point approaches the equilibrium point $(-\pi, 0)$ as $t \to \infty$. In this case, the pendulum swings upward in the clockwise direction, again slowing down and approaching the inverted position in the limit as $t \to \infty$. ▲

Hamiltonian Systems

We now discuss a class of autonomous first order systems, called "Hamiltonian systems," that always satisfy a conservation law. We restrict our attention to two-dimensional systems; the exercises show how the underlying principle can be extended to higher dimensional systems. Hamiltonian systems include the second order scalar equation (1) as a special case.

As a first step, recall the following chain rule from calculus. Assume that a function $H(x, y)$, viewed as a function of two independent variables x and y, is continuous and has continuous first and second partial derivatives (it will be apparent later why continuous second partial derivatives are required). We now form a composition, replacing the variables x and y with differentiable functions of t; we refer to these two functions as $x(t)$ and $y(t)$. The resulting composite function,

$$H(x(t), y(t)),$$

is a differentiable function of t and its derivative can be found by the chain rule:

$$\frac{d}{dt}H(x(t), y(t)) = \frac{\partial H(x(t), y(t))}{\partial x}\frac{dx}{dt} + \frac{\partial H(x(t), y(t))}{\partial y}\frac{dy}{dt}. \tag{9}$$

[In Section 3.3, when discussing exact differential equations, we used a special case of the chain rule, (9), in which $x(t) = t$.]

Consider now the two-dimensional autonomous system,

$$\begin{aligned} x'(t) &= f(x(t), y(t)), \\ y'(t) &= g(x(t), y(t)). \end{aligned} \tag{10}$$

System (10) is called a **Hamiltonian system**[2] if there exists a function of two variables, $H(x, y)$, that is continuous, with continuous first and second partial derivatives, and such that

$$\begin{aligned} \frac{\partial H(x,y)}{\partial x} &= -g(x, y), \\ \frac{\partial H(x,y)}{\partial y} &= f(x, y). \end{aligned} \tag{11}$$

The function $H(x, y)$ is called the **Hamiltonian function** (or simply, the **Hamiltonian**) of the system.

If system (10) is a Hamiltonian system, then the composition $H(x(t), y(t))$ is a conserved quantity of the system. To see why, note first that

$$\begin{aligned} \frac{d}{dt} H(x(t), y(t)) &= \frac{\partial H(x(t), y(t))}{\partial x} \frac{dx(t)}{dt} + \frac{\partial H(x(t), y(t))}{\partial y} \frac{dy(t)}{dt} \\ &= [-g(x(t), y(t))] f(x(t), y(t)) + \left[f(x(t), y(t))\right] g(x(t), y(t)) \\ &= 0. \end{aligned}$$

It follows, therefore, that

$$H(x(t), y(t)) = C.$$

Two important questions about Hamiltonian systems are "How can we determine whether a given autonomous system is a Hamiltonian system?" and "If a system is known to be a Hamiltonian system, how do we determine the conserved quantity H?" We will address both of these questions after the next example, which shows that the second order scalar equation (1), when recast as an autonomous first order system, is a Hamiltonian system.

EXAMPLE 2

If the second order scalar equation $y'' + v(y) = 0$ is rewritten as a first order system, we obtain the autonomous system

$$\begin{aligned} x' &= y, \\ y' &= -v(x). \end{aligned}$$

Show that this system is a Hamiltonian system with $H(x, y) = (1/2)y^2 + V(x)$, where $V(x)$ is any antiderivative of $v(x)$.

---

[2] Sir William Rowan Hamilton (1805–1865) was an Irish mathematician noted for his contributions to optics and dynamics and for the development of the theory of quaternions. Shortly before his death he was elected the first foreign member of the United States National Academy of Sciences.

Solution: Using the notation of (10) and (11), $f(x, y) = y$ and $g(x, y) = -v(x)$. Calculating the partial derivatives of $H(x, y) = (1/2)y^2 + V(x)$, we find

$$\frac{\partial H(x,y)}{\partial x} = \frac{dV(x)}{dx} = v(x) = -g(x,y),$$
$$\frac{\partial H(x,y)}{\partial y} = y = f(x,y).$$

Thus, from equation (11), the system is a Hamiltonian system. ▲

Recognizing a Hamiltonian System

The following discussion about identifying Hamiltonian systems and constructing Hamiltonians closely parallels the discussion in Section 3.3 about identifying exact differential equations and constructing solutions of exact equations.

In particular, suppose $H(x, y)$ is a Hamiltonian for system (10). Then, from equation (11),

$$\frac{\partial H(x,y)}{\partial x} = -g(x,y) \quad \text{and} \quad \frac{\partial H(x,y)}{\partial y} = f(x,y). \tag{12}$$

From calculus, if the second partial derivatives of $H(x, y)$ exist and are continuous, then the second mixed partial derivatives are equal; that is,

$$\frac{\partial^2 H(x,y)}{\partial x \partial y} = \frac{\partial^2 H(x,y)}{\partial y \partial x}.$$

Therefore, if system (10) is a Hamiltonian system, it necessarily follows that

$$\frac{\partial f(x,y)}{\partial x} = -\frac{\partial g(x,y)}{\partial y}.$$

The following theorem, stated without proof, asserts that this condition is, in fact, both necessary and sufficient.

Theorem 8.2

Consider the two-dimensional autonomous system

$$x' = f(x,y),$$
$$y' = g(x,y).$$

Assume that $f(x, y)$ and $g(x, y)$ are continuous in the xy-plane. Assume as well that the partial derivatives

$$\frac{\partial f}{\partial x}, \quad \frac{\partial f}{\partial y}, \quad \frac{\partial g}{\partial x}, \quad \frac{\partial g}{\partial y}$$

exist and are continuous in the xy-plane. Then, the system is a Hamiltonian system if and only if

$$\frac{\partial f(x,y)}{\partial x} = -\frac{\partial g(x,y)}{\partial y} \tag{13}$$

for all (x, y).

Constructing Hamiltonians

Once a system is known to be a Hamiltonian system, we can construct the Hamiltonian function by the same process of anti-partial-differentiation we used to solve exact differential equations in Section 3.3. We illustrate the ideas in the next example by constructing a Hamiltonian function for a Hamiltonian system.

EXAMPLE 3

Consider the autonomous system

$$x' = y^2 + \cos x,$$
$$y' = 2x + 1 + y \sin x.$$

(a) Use Theorem 8.2 to show this system is a Hamiltonian system.

(b) Find a Hamiltonian function for the system.

Solution:

(a) Calculating the partial derivatives required by test (13), we find

$$\frac{\partial f}{\partial x} = -\sin x \quad \text{and} \quad \frac{\partial g}{\partial y} = \sin x.$$

Since $\partial f/\partial x = -\partial g/\partial y$, we know the system is a Hamiltonian system.

(b) Since the given system is a Hamiltonian system, there must be a function $H(x, y)$ such that

$$\begin{aligned} \frac{\partial H(x,y)}{\partial x} &= -g(x,y) = -2x - 1 - y\sin x, \\ \frac{\partial H(x,y)}{\partial y} &= f(x,y) = y^2 + \cos x. \end{aligned} \tag{14}$$

We choose one of these equations, say the first, and compute an anti-partial-derivative, obtaining

$$H(x, y) = -x^2 - x + y \cos x + q(y), \tag{15}$$

where $q(y)$ is an arbitrary differentiable function of y. [Note: Since the variable y is held fixed when performing partial differentiation with respect to x, we must allow for this arbitrary function of y when reversing the operation.] From equations (14) and (15), we find

$$\frac{\partial H(x,y)}{\partial y} = y^2 + \cos x = \cos x + \frac{dq(y)}{dy}.$$

Therefore,

$$\frac{dq(y)}{dy} = y^2$$

and $q(y) = y^3/3 + C$. We can drop the additive arbitrary constant and obtain a Hamiltonian

$$H(x, y) = -x^2 - x + y \cos x + \frac{y^3}{3}. \tag{16}$$

Figure 8.10 shows some phase-plane trajectories [that is, graphs of $H(x, y) = C$] for a few representative values of the constant C. ▲

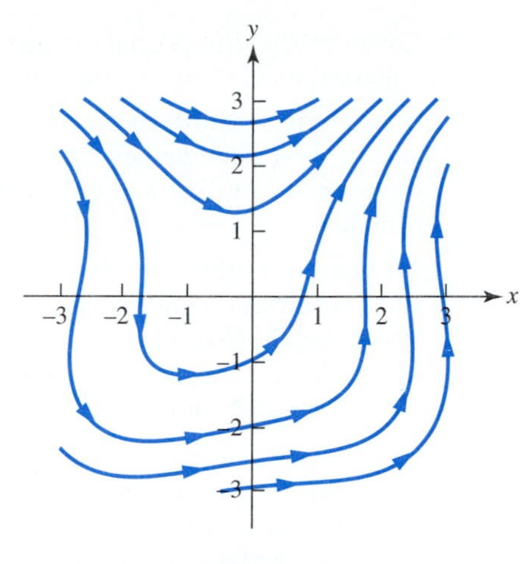

FIGURE 8.10

Some phase-plane trajectories for the autonomous system in Example 3. These curves are graphs of the Hamiltonian (16).

EXERCISES

Exercises 1–6:

In each exercise,

(a) As in Example 1, derive a conservation law for the given autonomous equation $x'' + u(x) = 0$. (Your answer should contain an arbitrary constant and therefore define a one-parameter family of conserved quantities.)

(b) Rewrite the given autonomous equation as a first order system of the form

$$x' = f(x, y),$$
$$y' = g(x, y),$$

by setting $y(t) = x'(t)$. The phase plane is then the xy-plane. Express the family of conserved quantities found in (a) in terms of x and y. Determine the equation of the particular conserved quantity whose graph passes through the phase-plane point $(x, y) = (1, 1)$.

(c) Plot the phase-plane graph of the conserved quantity found in part (b), using a computer if necessary. Determine the velocity vector $\mathbf{v} = f(x, y)\mathbf{i} + g(x, y)\mathbf{j}$ at the phase-plane point $(1, 1)$. Add this vector to your graph with the initial point of the vector at $(1, 1)$. What is the geometric relation of this velocity vector to the graph? What information does the velocity vector give about the direction in which the solution point traverses the graph as time increases?

1. $x'' + 4x = 0$ **2.** $x'' - x = 1$ **3.** $x'' + x^3 = 0$

4. $x'' - x^3 = \pi \sin(\pi x)$ **5.** $x'' + x^2 = 0$ **6.** $x'' + \dfrac{x}{1 + x^2} = 0$

Exercises 7–8:

The conservation law for an autonomous second order scalar differential equation, having the form $x'' + f(x) = 0$, is given (where y corresponds to x'). Determine the differential equation.

7. $y^2 + x^2 \cos x = C$ **8.** $y^2 - e^{-x^2} = C$

9. Consider the differential equation $x'' + x + x^3 = 0$. It has the same structure as the equation used to model the cubically nonlinear spring.

(a) Rewrite the differential equation as a first order system. On the xy-phase plane, sketch the nullclines and locate any equilibrium point(s). Place direction field arrows on the nullclines, indicating the direction the solution point traverses the nullclines.

(b) Compute the velocity vector $\mathbf{v} = x'\mathbf{i} + y'\mathbf{j}$ at the four phase-plane points $(x, y) = (\pm 1, \pm 1)$. Locate these points and draw the velocity vectors on your phase-plane sketch. Use this information, together with the information obtained in part (a), to draw a rough sketch of some phase-plane solution trajectories. Indicate the direction in which the solution point moves on these trajectories.

(c) Determine the conservation law satisfied by solutions of the given differential equation. Determine the equation of the conserved quantity whose graph passes through the phase-plane point $(x, y) = (1, 1)$. Plot the graph of this equation on your phase plane, using computational software as appropriate. Does the graph pass through the other three points, $(-1, 1)$, $(-1, -1)$, and $(1, -1)$ as well? Is the sketch made in part (b) consistent with this graph of the conserved quantity?

10. Each figure shows a phase-plane graph of a conserved quantity for the autonomous differential equation $x'' + \alpha x = 0$, where α is a real constant.

(a) Determine the value of the constant α in each case. What is the equation whose phase-plane graph is shown?

(b) Indicate the direction in which the solution point travels along these phase-plane curves as time increases.

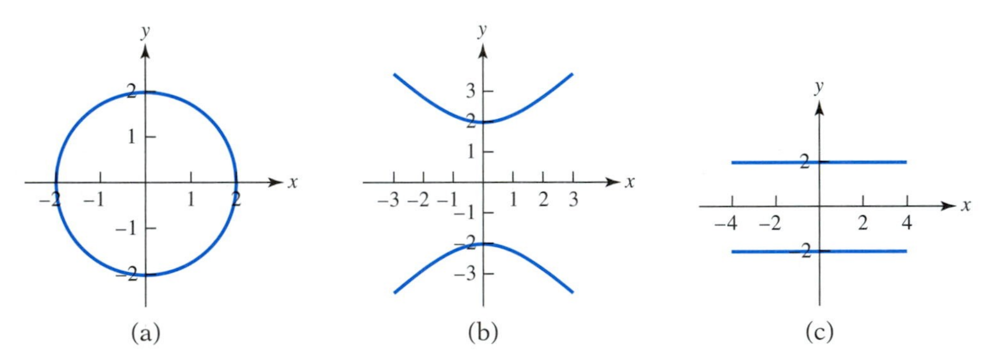

Figure for Exercise 10

11. Consider the autonomous third order scalar equation $y''' + f(y') = 0$, where f is a continuous function. Does this differential equation have a conservation law? If so, obtain the equation of the family of conserved quantities.

12. Consider the equation $mx'' + kx = 0$. We saw in Chapter 4 that this equation models the vibrations of a spring-mass system. The conserved quantity $(1/2)m[x']^2 + (1/2)kx^2 = E$ is the (constant) total energy of the system. The first term, $(1/2)m[x']^2$, is the kinetic energy, while the second term, $(1/2)kx^2$, is the elastic potential energy. Suppose that damping is now added to the system. The differential equation $mx'' + \gamma x' + kx = 0$ now models the motion (with γ a positive constant). Define $E(t) = (1/2)m[x']^2 + (1/2)kx^2$.

(a) Show, in the case of damping, that $E(t)$ is no longer constant. Show, rather, that $dE(t)/dt \leq 0$.

(b) Discuss the physical relevance of the observation made in part (a).

Exercises 13–20:

For the given system,

(a) Use Theorem 8.2 to show the system is a Hamiltonian system.

(b) Find a Hamiltonian function for the system.

(c) Use computational software to graph the phase-plane trajectory passing through $(1, 1)$. Also, indicate the direction of motion for the solution point.

13. $x' = 2x$, $y' = -2y$

14. $x' = 2xy$, $y' = -y^2$

15. $x' = x - x^2 + 1$, $y' = -y + 2xy + 4x$

16. $x' = -8y$, $y' = 2x$

17. $x' = 2y\cos x$, $y' = y^2 \sin x$

18. $x' = 2y - x + 3$, $y' = y + 4x^3 - 2x$

19. $x' = -2y$, $y' = 3x^2$

20. $x' = xe^{xy}$, $y' = -2x - ye^{xy}$

Exercises 21–26:

Use Theorem 8.2 to decide if the given system is a Hamiltonian system. If it is, find a Hamiltonian function for the system.

21. $x' = x^3 + 3\sin(2x + 3y)$, $y' = -3x^2y - 2\sin(2x + 3y)$

22. $x' = e^{xy} + y^3$, $y' = -e^{xy} - x^3$

23. $x' = -\sin(2xy) - x$, $y' = \sin(2xy) + y$

24. $x' = -3x^2 + xe^y$, $y' = 6xy + 3x - e^y$

25. $x' = y$, $y' = x - x^2$

26. $x' = x + 2y$, $y' = x^3 - 2x + y$

Exercises 27–30:

Let $f(u)$ and $g(u)$ be defined and continuously differentiable on the interval $-\infty < u < \infty$, and let $F(u)$ and $G(u)$ be antiderivatives for $f(u)$ and $g(u)$ respectively. In each exercise, use Theorem 8.2 to show that the given system is Hamiltonian. Determine a Hamiltonian function for the system, expressed in terms of F and/or G.

27. $x' = f(y)$, $y' = g(x)$

28. $x' = f(y) + 2y$, $y' = g(x) + 6x$

29. $x' = 3f(y) - 2xy$, $y' = g(x) + y^2 + 1$

30. $x' = f(x - y) + 2y$, $y' = f(x - y)$

31. A Generalized Hamiltonian System Consider the two-dimensional autonomous system

$$x' = f(x, y),$$
$$y' = g(x, y).$$

Suppose there exist two functions $K(x, y)$ and $\mu(x, y)$ satisfying

$$\frac{\partial K(x, y)}{\partial x} = -\mu(x, y)g(x, y),$$
$$\frac{\partial K(x, y)}{\partial y} = \mu(x, y)f(x, y).$$

Does the given autonomous system have a conserved quantity? If so, what is the conserved quantity?

32. Higher-Dimensional Autonomous Systems The ideas underlying Hamiltonian systems extend to higher-dimensional systems. For example, consider the three-dimensional autonomous system

$$\begin{aligned} x' &= f(x, y, z), \\ y' &= g(x, y, z), \\ z' &= h(x, y, z). \end{aligned} \tag{17}$$

(a) Use the chain rule to show that autonomous system (17) has a conserved quantity if there exists a function $H(x, y, z)$ for which

$$\frac{\partial}{\partial x} H(x, y, z) f(x, y, z) + \frac{\partial}{\partial y} H(x, y, z) g(x, y, z) + \frac{\partial}{\partial z} H(x, y, z) h(x, y, z) = 0.$$

(b) Show that $H(x, y, z) = \cos^2(x) + ye^z$ is a conserved quantity for the system

$$\begin{aligned} x' &= ye^z \\ y' &= y \cos x \sin x \\ z' &= \cos x \sin x. \end{aligned}$$

8.4 Stability

It should be evident from the examples considered thus far that differential equations can model different physical behavior at different equilibrium points. For instance, consider the pendulum. The equilibrium points are $(\theta, \theta') = (m\pi, 0)$, where m is an integer. Equilibrium points with m an even integer correspond to the pendulum bob hanging vertically downward while the equilibrium points for m an odd integer correspond to the pendulum bob resting in the inverted position. Consider what happens if a pendulum bob, initially in an equilibrium state, is subjected to a slight perturbation; in other words, it is given a slight displacement and/or a very small angular velocity. If the pendulum is initially hanging downward, we expect the perturbation to remain small—the bob will swing back and forth, making small excursions from the vertical. If the pendulum is initially in the inverted position, however, we expect dramatic changes. The pendulum bob, displaced from its precarious equilibrium state, will fall, ultimately making large departures from its initial equilibrium position.

In everyday language, we might describe the pendulum's downward rest position as a "stable" configuration and the inverted rest position as an "unstable" configuration. Mathematicians have taken these everyday terms and given them precise definitions consistent with our intuitive notion of stable and unstable. In this section we present and discuss these mathematical definitions. In the next section, we introduce the technique of linearization, which, in many cases, enables us to study and characterize equilibrium point stability by analyzing a simpler associated linear system.

The pendulum example illustrates the question of primary concern. If perturbed slightly from an equilibrium state, will a system exhibit a markedly different behavior? In the case of mechanical systems, instability often means vibrations that grow in amplitude, leading to possible system failure.

Stable and Unstable Equilibrium Points

Consider the autonomous system

$$\begin{aligned} y_1' &= f_1(y_1, y_2, \ldots, y_n), \\ y_2' &= f_2(y_1, y_2, \ldots, y_n), \\ &\vdots \\ y_n' &= f_n(y_1, y_2, \ldots, y_n), \end{aligned}$$

which we write in vector form as

$$\mathbf{y}' = \mathbf{f}(\mathbf{y}). \tag{1}$$

Assume that the constant vector function

$$\mathbf{y}(t) = \mathbf{y}_e = \begin{bmatrix} y_{1e} \\ y_{2e} \\ \vdots \\ y_{ne} \end{bmatrix}$$

is an equilibrium solution of the system; that is, $\mathbf{f}(\mathbf{y}_e) = \mathbf{0}$.

In order to define precisely what it means for the equilibrium point $\mathbf{y}_e$ to be stable or unstable, we need to be able to compute the distance between points in n-dimensional space. Let $\mathbf{u}$ and $\mathbf{v}$ denote two points in n-dimensional space,

$$\mathbf{u} = \begin{bmatrix} u_1 \\ u_2 \\ \vdots \\ u_n \end{bmatrix} \quad \text{and} \quad \mathbf{v} = \begin{bmatrix} v_1 \\ v_2 \\ \vdots \\ v_n \end{bmatrix}.$$

We define the **norm** of $\mathbf{u}$, denoted by $\|\mathbf{u}\|$, by

$$\|\mathbf{u}\| = \sqrt{u_1^2 + u_2^2 + \cdots + u_n^2}.$$

The **distance** between $\mathbf{u}$ and $\mathbf{v}$, denoted by $\|\mathbf{u} - \mathbf{v}\|$, is the size (the norm) of the difference $\mathbf{u} - \mathbf{v}$:

$$\|\mathbf{u} - \mathbf{v}\| = \sqrt{(u_1 - v_1)^2 + (u_2 - v_2)^2 + \cdots + (u_n - v_n)^2}. \tag{2}$$

Now, let $\mathbf{y}_e$ be an equilibrium point of the autonomous system $\mathbf{y}' = \mathbf{f}(\mathbf{y})$. We say that the equilibrium point $\mathbf{y}_e$ is **stable** if

> given any $\varepsilon > 0$, there exists a corresponding $\delta > 0$ such that every solution satisfying $\|\mathbf{y}(0) - \mathbf{y}_e\| < \delta$ also satisfies $\|\mathbf{y}(t) - \mathbf{y}_e\| < \varepsilon$ for all $t \geq 0$.

If an equilibrium point of $\mathbf{y}' = \mathbf{f}(\mathbf{y})$ is not stable, it is called **unstable**.

Interpreting Stability in the Phase Plane

When $n = 2$, we can use the phase plane to give a graphical interpretation of stability. Consider the autonomous system

$$x' = f(x, y),$$
$$y' = g(x, y),$$

having equilibrium solution

$$\mathbf{y}_e = \begin{bmatrix} x_e \\ y_e \end{bmatrix}.$$

We have adopted the notation

$$\mathbf{y}(t) = \begin{bmatrix} x(t) \\ y(t) \end{bmatrix} \quad \text{and} \quad \mathbf{f}(\mathbf{y}) = \begin{bmatrix} f(x, y) \\ g(x, y) \end{bmatrix}$$

so that we can speak of the phase plane as the xy-plane.

We can identify $\mathbf{y}_e$ as the point in the phase plane having coordinates (x_e, y_e). The set of all phase-plane points $\mathbf{y}$ satisfying $\|\mathbf{y} - \mathbf{y}_e\| < r$ is the set of all points lying within a circle of radius r centered at (x_e, y_e).

Consider now the definition of stability. It involves two circles centered at (x_e, y_e), one of radius ε and the other of radius δ (see Figure 8.11). The stability criterion requires that all solutions lying within the circle of radius δ at the initial time $t = 0$ remain within the circle of radius ε for all subsequent time. This situation must hold for all possible choices of $\varepsilon > 0$; whenever we are given an $\varepsilon > 0$, we must be able to find a corresponding $\delta > 0$ that "works." The real

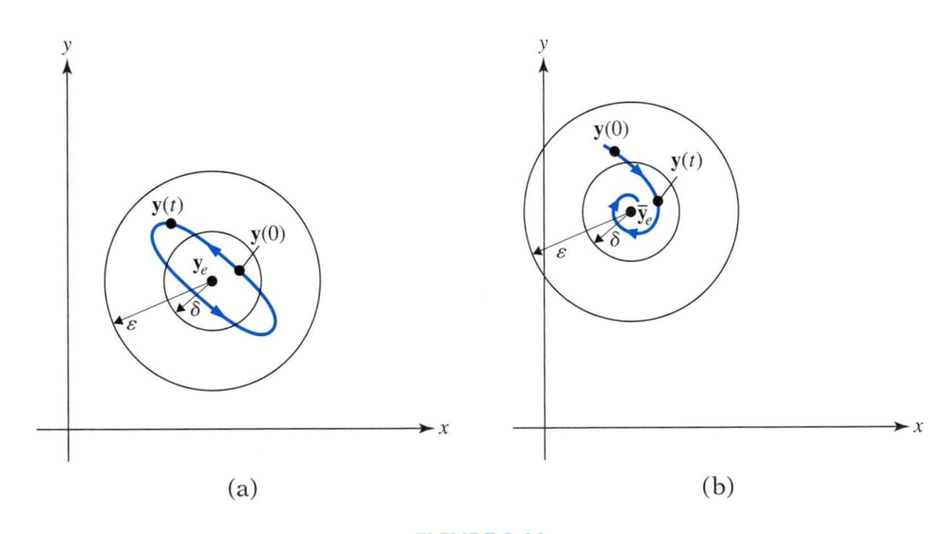

FIGURE 8.11

Two examples of behavior near a stable equilibrium point are shown. In each case, $\mathbf{y}(t) = (x(t), y(t))$ represents a typical solution trajectory near an equilibrium point of the autonomous system $\mathbf{y}' = \mathbf{f}(\mathbf{y})$.
(a) When the initial point $\mathbf{y}(0)$ is sufficiently close to $\mathbf{y}_e$, the solution trajectory is a closed curve surrounding $\mathbf{y}_e$.
(b) When $\mathbf{y}(0)$ is sufficiently close to $\overline{\mathbf{y}}_e$, the solution trajectory spirals in toward $\overline{\mathbf{y}}_e$.

test of the definition occurs as we consider smaller and smaller $\varepsilon > 0$. Can we continue to find corresponding values $\delta > 0$ that work? If the answer is Yes, the equilibrium point is stable; if not, it is unstable.

For higher order systems, the same geometrical ideas hold. However, instead of circles in the phase plane, we must consider n-dimensional spheres. We illustrate the concept of stability with two examples involving autonomous linear systems for which explicit general solutions are known.

EXAMPLE 1

Consider the two-dimensional autonomous linear system $\mathbf{y}' = A\mathbf{y}$, where

$$A = \begin{bmatrix} -2 & 1 \\ 1 & -2 \end{bmatrix}. \tag{3}$$

Show that $\mathbf{y}_e = \mathbf{0}$ is the only equilibrium point, and determine whether it is stable or unstable.

Solution: In this case, $\mathbf{f}(\mathbf{y}) = A\mathbf{y}$. The matrix A is invertible and, therefore, solving $A\mathbf{y} = \mathbf{0}$ leads to a single equilibrium point, $\mathbf{y}_e = \mathbf{0}$.

To determine the stability properties of this equilibrium point, we apply the stability definition directly to the general solution of this first order linear system. Using the methods of Chapter 6, we find the general solution is

$$\mathbf{y}(t) = c_1 e^{-t}\begin{bmatrix} 1 \\ 1 \end{bmatrix} + c_2 e^{-3t}\begin{bmatrix} 1 \\ -1 \end{bmatrix} = \begin{bmatrix} c_1 e^{-t} + c_2 e^{-3t} \\ c_1 e^{-t} - c_2 e^{-3t} \end{bmatrix}. \tag{4}$$

In order to apply the definition of stability, we need to relate the following two quantities:

1. The distance, $\|\mathbf{y}(0) - \mathbf{y}_e\|$, between the initial point and the equilibrium point.
2. The distance, $\|\mathbf{y}(t) - \mathbf{y}_e\|$, between $\mathbf{y}(t)$ and $\mathbf{y}_e$ for $t \ge 0$.

Since $\mathbf{y}_e = \mathbf{0}$, it follows from (4) that

$$\|\mathbf{y}(0) - \mathbf{y}_e\| = \|\mathbf{y}(0)\| = \sqrt{(c_1 + c_2)^2 + (c_1 - c_2)^2} = \sqrt{2(c_1^2 + c_2^2)}.$$

Similarly, for $t \ge 0$,

$$\begin{aligned} \|\mathbf{y}(t) - \mathbf{y}_e\| = \|\mathbf{y}(t)\| &= \sqrt{(c_1 e^{-t} + c_2 e^{-3t})^2 + (c_1 e^{-t} - c_2 e^{-3t})^2} \\ &= \sqrt{2(c_1^2 e^{-2t} + c_2^2 e^{-6t})} = \sqrt{2(c_1^2 + c_2^2 e^{-4t})e^{-2t}} \\ &\le \sqrt{2(c_1^2 + c_2^2)e^{-2t}} = \sqrt{2(c_1^2 + c_2^2)}e^{-t} \\ &= \|\mathbf{y}(0)\|e^{-t}. \end{aligned} \tag{5}$$

Now consider the definition of stability where some value $\varepsilon > 0$ is given. We need to determine a corresponding value $\delta > 0$ such that

If $\mathbf{y}(t)$ is a solution of $\mathbf{y}' = A\mathbf{y}$ and if $\|\mathbf{y}(0)\| < \delta$, then $\|\mathbf{y}(t)\| < \varepsilon$ for all $t \ge 0$.

(continued)

(continued)

By (5), we know that

$$\|\mathbf{y}(t)\| \le \|\mathbf{y}(0)\| e^{-t} \le \|\mathbf{y}(0)\|, \qquad t \ge 0.$$

Therefore, we can guarantee that $\|\mathbf{y}(t)\| < \varepsilon$ by taking $\delta \le \varepsilon$. This shows the equilibrium point $\mathbf{y}_e = \mathbf{0}$ is a stable equilibrium point of the system $\mathbf{y}' = A\mathbf{y}$. ▲

EXAMPLE 2

Consider the two-dimensional autonomous linear system $\mathbf{y}' = A\mathbf{y}$, where

$$A = \begin{bmatrix} -1 & -2 \\ -2 & -1 \end{bmatrix}. \tag{6}$$

Show that $\mathbf{y}_e = \mathbf{0}$ is the only equilibrium point, and determine whether it is stable or unstable.

Solution: As in Example 1, $\mathbf{f}(\mathbf{y}) = A\mathbf{y}$ and we see that the matrix A is invertible. Therefore, solving $A\mathbf{y} = \mathbf{0}$ leads to a single equilibrium point, $\mathbf{y}_e = \mathbf{0}$.

The general solution of $\mathbf{y}' = A\mathbf{y}$ is

$$\mathbf{y}(t) = c_1 e^{t} \begin{bmatrix} 1 \\ -1 \end{bmatrix} + c_2 e^{-3t} \begin{bmatrix} 1 \\ 1 \end{bmatrix} = \begin{bmatrix} c_1 e^{t} + c_2 e^{-3t} \\ -c_1 e^{t} + c_2 e^{-3t} \end{bmatrix}. \tag{7}$$

In Example 1, the coefficient matrix had two negative eigenvalues. In this example, however, the coefficient matrix has a positive eigenvalue, $\lambda = 1$. Since $\lim_{t\to\infty} e^t = \infty$, we anticipate that any solution (7) with $c_1 \ne 0$ will become unboundedly large in norm as t increases. Therefore, we anticipate that $\mathbf{y}_e = \mathbf{0}$ is an unstable equilibrium point of this system.

To prove that $\mathbf{y}_e = \mathbf{0}$ is an unstable equilibrium point, we show that given an $\varepsilon > 0$, there is no δ that works. That is, for every $\delta > 0$, there is at least one solution $\mathbf{y}(t)$ that originates in the circle of radius δ but that eventually gets outside the circle of radius ε. In particular, solutions given by (7) with $c_2 = 0$ and $c_1 \ne 0$ have the form

$$\mathbf{y}(t) = c_1 e^{t} \begin{bmatrix} 1 \\ -1 \end{bmatrix}, \qquad t \ge 0.$$

Moreover,

$$\|\mathbf{y}(0)\| = \sqrt{2}|c_1| \quad \text{and} \quad \|\mathbf{y}(t)\| = \sqrt{2}|c_1|e^{t}.$$

This particular family of solutions has phase-plane trajectories that lie on the line $y = -x$ (see Figure 8.12 and Exercise 23 in Section 8.2). Since $\|\mathbf{y}(t)\| = \sqrt{2}|c_1|e^t$, the solution moves away from the origin along this line, growing in norm as t increases. No matter what value $\delta > 0$ we take, we can always choose $|c_1| \ne 0$ but sufficiently small so that $\mathbf{y}(0)$ is within the circle of radius δ. But, as long as $|c_1| \ne 0$, the solution $\mathbf{y}(t)$ eventually exits the circle of radius ε. Therefore, $\mathbf{y}_e = \mathbf{0}$ is an unstable equilibrium point of the autonomous system $\mathbf{y}' = A\mathbf{y}$. ▲

In Examples 1 and 2 it was relatively straightforward to determine the stability properties of the equilibrium point $\mathbf{y}_e = \mathbf{0}$ because we had an explicit representation of the general solution. The obvious question is "How do we analyze the stability properties of equilibrium points of nonlinear problems,

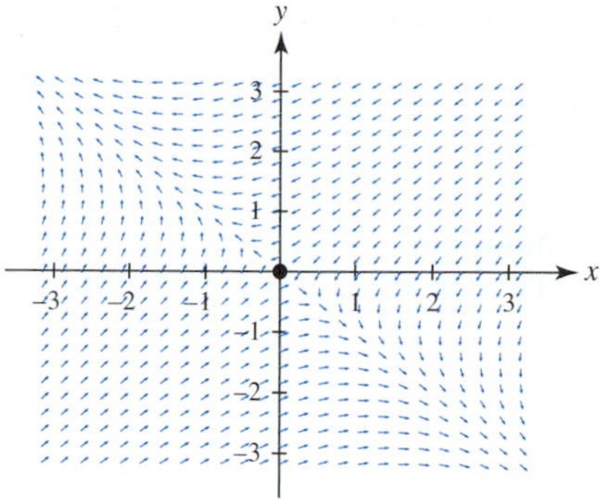

FIGURE 8.12

The direction field for the system in Example 2. The arrows indicate the direction the solution point moves as time increases.

such as the pendulum and competing species problems, when explicit solutions are not attainable?" We address this issue in the next section. In the remainder of this section, we introduce and illustrate the concept of asymptotic stability.

Asymptotic Stability

Example 1 has an interesting aspect. The equilibrium point $\mathbf{y}_e = \mathbf{0}$ is not only stable but it also has the feature that all solutions approach it in the limit as $t \to \infty$. This additional feature, wherein all solutions originating sufficiently close to a stable equilibrium point actually approach the equilibrium point as $t \to \infty$, is important enough to warrant its own definition.

Let $\mathbf{y}_e$ be an equilibrium point of the autonomous system $\mathbf{y}' = \mathbf{f}(\mathbf{y})$. We say that $\mathbf{y}_e$ is an **asymptotically stable equilibrium point** if

(i) it is a stable equilibrium point, and

(ii) there exists a $\delta_0 > 0$ such that $\lim_{t\to\infty} \mathbf{y}(t) = \mathbf{y}_e$ for all solutions initially satisfying $\|\mathbf{y}(0) - \mathbf{y}_e\| < \delta_0$.

Roughly speaking, all solutions starting close enough to a *stable* equilibrium point remain close to it for all subsequent time. All solutions starting sufficiently close to an *asymptotically stable* equilibrium point not only remain close to it for all subsequent time but, in fact, approach it in the limit as $t \to \infty$. Note that asymptotic stability implies stability. However, as the next example shows, an equilibrium point can be stable but not asymptotically stable.

EXAMPLE 3

Consider the two-dimensional autonomous linear system $\mathbf{y}' = A\mathbf{y}$, where

$$A = \begin{bmatrix} 0 & 1 \\ -1 & 0 \end{bmatrix}.$$

(continued)

(continued)

Show that $\mathbf{y}_e = \mathbf{0}$ is the only equilibrium point and determine if it is asymptotically stable.

Solution: As in Examples 1 and 2, $\mathbf{f}(\mathbf{y}) = A\mathbf{y}$, where the matrix A is invertible. Therefore, solving $A\mathbf{y} = \mathbf{0}$ leads to a single equilibrium point, $\mathbf{y}_e = \mathbf{0}$.

Eigenpairs of the matrix A are

$$\lambda_1 = i, \qquad \mathbf{u}_1 = \begin{bmatrix} 1 \\ i \end{bmatrix}, \quad \text{and} \quad \lambda_2 = \overline{\lambda_1} = -i, \qquad \mathbf{u}_2 = \overline{\mathbf{u}}_1 = \begin{bmatrix} 1 \\ -i \end{bmatrix}.$$

Using the ideas of Section 6.6, we find a real-valued general solution of this system to be

$$\mathbf{y}(t) = c_1 \begin{bmatrix} \cos t \\ -\sin t \end{bmatrix} + c_2 \begin{bmatrix} \sin t \\ \cos t \end{bmatrix}. \tag{8}$$

It follows from (8) that:

$$\begin{aligned} \|\mathbf{y}(t) - \mathbf{y}_e\| = \|\mathbf{y}(t)\| &= \sqrt{(c_1 \cos t + c_2 \sin t)^2 + (-c_1 \sin t + c_2 \cos t)^2} \\ &= \sqrt{c_1^2 + c_2^2} = \|\mathbf{y}(0)\|, \qquad t \geq 0. \end{aligned} \tag{9}$$

Therefore, the distance of a solution from the phase-plane origin remains constant in time—the phase-plane trajectories are circles centered at the origin. Thus, the equilibrium point $\mathbf{y}_e = \mathbf{0}$ is stable (given $\varepsilon > 0$, we simply take $\delta = \varepsilon$). The equilibrium point is not asymptotically stable, however. In particular, no nonzero solution approaches the equilibrium point as $t \to \infty$. ▲

Stability Characteristics of $\mathbf{y}' = A\mathbf{y}$

The three examples previously considered illustrate the following general stability result, which we present without proof.

Theorem 8.3

Let A be a real invertible (2×2) matrix. Then, the autonomous linear system $\mathbf{y}' = A\mathbf{y}$ has a unique equilibrium point, $\mathbf{y}_e = \mathbf{0}$. This equilibrium point is

(i) asymptotically stable if all eigenvalues of A have negative real parts. (In other words, the eigenvalues can be real and negative or they can be a complex conjugate pair with negative real parts.)

(ii) stable but not asymptotically stable if the eigenvalues of A are purely imaginary.

(iii) unstable if at least one eigenvalue of A has positive real part. (In other words, the eigenvalues can be real with at least one being positive or they can be a complex conjugate pair with positive real parts.)

Equilibrium points are sometimes characterized as being isolated or not isolated. A phase-plane equilibrium point is called an **isolated point** if it is the center of some small disk whose interior contains no other equilibrium points. The equilibrium point in Theorem 8.3 (as in all the examples in this section) is the *only* equilibrium point. It is therefore, an isolated equilibrium point.

We also note, with respect to Theorem 8.3, that the assumption A is invertible implies $\det(A) \neq 0$. Therefore, $\lambda = 0$ is not an eigenvalue of A. However, if A is not invertible, then $\mathbf{y}_e = \mathbf{0}$ is not the only equilibrium point (nor is it isolated). For example, suppose

$$A = \begin{bmatrix} \alpha & \beta \\ c\alpha & c\beta \end{bmatrix},$$

with α and β not both zero. Thus, $\det(A) = 0$ but A is not the zero matrix. One can show (see Exercise 33) that every point on the phase-plane line $\alpha x + \beta y = 0$ is an equilibrium point.

Theorem 8.3 applies equally well whether the (2×2) matrix A has distinct or repeated eigenvalues. Recall that if A has a repeated (real) eigenvalue $\lambda_1 = \lambda_2 = \lambda$, then solutions involving the function $te^{\lambda t}$, as well as $e^{\lambda t}$, are possible. If $\lambda < 0$, however, we know from calculus that $te^{\lambda t}$ is bounded for $t \geq 0$ and that $\lim_{t\to\infty} te^{\lambda t} = 0$. It follows that $\mathbf{y}_e = \mathbf{0}$ will always be an asymptotically stable equilibrium point when $\lambda < 0$.

In the higher dimensional case, stability characterization is somewhat more complicated. If A is a real invertible $(n \times n)$ matrix, then the unique equilibrium point $\mathbf{y}_e = \mathbf{0}$ of $\mathbf{y}' = A\mathbf{y}$ is asymptotically stable if all eigenvalues of A have negative real parts. The equilibrium point is unstable if at least one eigenvalue of A has positive real part. For $n \geq 4$, repeated complex conjugate pairs of purely imaginary eigenvalues, say $\lambda = \pm i\omega$, are possible. In this case, solutions of the form $t\cos\omega t$ and $t\sin\omega t$ are possible when the matrix A does not have n linearly independent eigenvectors; in that event, $\mathbf{y}_e = \mathbf{0}$ is an unstable equilibrium point.

EXERCISES

1. Assume that the phase-plane solution curves for a two-dimensional autonomous system consist of the family of concentric ellipses $x^2/4 + y^2 = C$, $C \geq 0$. Also assume the system has an isolated equilibrium point at the origin.

   (a) Apply the definition to show that the origin is a stable equilibrium point. In particular, given an $\varepsilon > 0$, determine a corresponding $\delta > 0$ so that all solutions starting within a circle of radius δ centered at the origin stay within the circle of radius ε centered at the origin for all $t \geq 0$. (The δ you determine should be expressed in terms of ε.)

   (b) Is the origin an asymptotically stable equilibrium point? Explain.

2. Assume that the phase-plane solution curves for a two-dimensional autonomous system consist of the family of hyperbolas $-x^2 + y^2 = C$, $C \geq 0$. The system has an isolated equilibrium point at the origin. Is this equilibrium point stable or unstable? Explain.

3. Consider the differential equation $x'' + \gamma x' + x = 0$, where γ is a real constant.

   (a) Rewrite the given scalar equation as a first order system, defining $y = x'$.

(b) Determine the values of γ for which the system is (i) asymptotically stable, (ii) stable but not asymptotically stable, (iii) unstable.

Exercises 4–15:

Each exercise lists a linear system $\mathbf{y}' = A\mathbf{y}$, where A is a real constant invertible (2×2) matrix. Use Theorem 8.3 to determine if the equilibrium point $\mathbf{y}_e = \mathbf{0}$ is

(a) asymptotically stable

(b) stable, but not asymptotically stable

(c) unstable

4. $\mathbf{y}' = \begin{bmatrix} -3 & -2 \\ 4 & 3 \end{bmatrix}\mathbf{y}$

5. $\mathbf{y}' = \begin{bmatrix} 5 & -14 \\ 3 & -8 \end{bmatrix}\mathbf{y}$

6. $\mathbf{y}' = \begin{bmatrix} 0 & -2 \\ 2 & 0 \end{bmatrix}\mathbf{y}$

7. $\mathbf{y}' = \begin{bmatrix} 1 & 4 \\ -1 & 1 \end{bmatrix}\mathbf{y}$

8. $x' = -7x - 3y$
$y' = 5x + y$

9. $x' = 9x + 5y$
$y' = -7x - 3y$

10. $x' = -3x - 5y$
$y' = 2x - y$

11. $x' = 9x - 4y$
$y' = 15x - 7y$

12. $x' = -13x - 8y$
$y' = 15x + 9y$

13. $x' = 3x - 2y$
$y' = 5x - 3y$

14. $x' = x - 5y$
$y' = x - 3y$

15. $x' = -3x + 3y$
$y' = x - 5y$

Exercises 16–23:

Each exercise lists the general solution of a linear system of the form

$$x' = a_{11}x + a_{12}y,$$
$$y' = a_{21}x + a_{22}y,$$

where $a_{11}a_{22} - a_{12}a_{21} \neq 0$. Determine whether the equilibrium point $\mathbf{y}_e = \mathbf{0}$ is

(a) asymptotically stable,

(b) stable but not asymptotically stable,

(c) unstable.

16. $x = c_1e^{-2t} + c_2e^{3t}$
$y = c_1e^{-2t} + 2c_2e^{3t}$

17. $x = c_1e^{2t} + c_2e^{3t}$
$y = c_1e^{2t} + 2c_2e^{3t}$

18. $x = c_1e^{-2t} + c_2e^{-4t}$
$y = c_1e^{-2t} + 2c_2e^{-4t}$

19. $x = c_1e^{t}\cos 2t + c_2e^{t}\sin 2t$
$y = -c_1e^{t}\sin 2t + c_2e^{t}\cos 2t$

20. $x = c_1\cos 2t + c_2\sin 2t$
$y = -c_1\sin 2t + c_2\cos 2t$

21. $x = c_1e^{-2t}\cos 2t + c_2e^{-2t}\sin 2t$
$y = -c_1e^{-2t}\sin 2t + c_2e^{-2t}\cos 2t$

22. $x = c_1e^{-2t} + c_2e^{3t}$
$y = c_1e^{-2t} - c_2e^{3t}$

23. $x = c_1e^{-2t} + c_2e^{-3t}$
$y = c_1e^{-2t} - c_2e^{-3t}$

24. Consider the nonhomogeneous linear system $\mathbf{y}' = A\mathbf{y} + \mathbf{g}_0$, where A is a real invertible (2×2) matrix and $\mathbf{g}_0$ is a real (2×1) constant vector.

(a) Determine the unique equilibrium point, $\mathbf{y}_e$, of this system.

(b) Show how Theorem 8.3 can be used to determine the stability properties of this equilibrium point. [Hint: Adopt the change of dependent variable $\mathbf{z}(t) = \mathbf{y}(t) - \mathbf{y}_e$.]

Exercises 25–28:

Locate the unique equilibrium point of the given nonhomogeneous system and determine the stability properties of this equilibrium point. Is it asymptotically stable, stable but not asymptotically stable, or unstable?

25. $\mathbf{y}' = \begin{bmatrix} -2 & 1 \\ 1 & -2 \end{bmatrix}\mathbf{y} + \begin{bmatrix} -4 \\ 2 \end{bmatrix}$

26. $x' = y + 2$
$y' = -x + 1$

27. $\mathbf{y}' = \begin{bmatrix} 3 & 2 \\ -4 & -3 \end{bmatrix}\mathbf{y} + \begin{bmatrix} -2 \\ 2 \end{bmatrix}$

28. $x' = -x + y + 1$
$y' = -10x + 5y + 2$

Exercises 29–32:

Higher Dimensional Systems In each exercise, locate all equilibrium points for the given autonomous system. Determine whether these equilibrium point(s) are asymptotically stable, stable but not asymptotically stable, or unstable.

29. $y_1' = 2y_1 + y_2 + y_3$
$y_2' = y_1 + y_2 + 2y_3$
$y_3' = y_1 + 2y_2 + y_3$

30. $\dfrac{d}{dt}\begin{bmatrix} y_1 \\ y_2 \\ y_3 \end{bmatrix} = \begin{bmatrix} 1 & -1 & 0 \\ 0 & -1 & 2 \\ 0 & 0 & -1 \end{bmatrix}\begin{bmatrix} y_1 \\ y_2 \\ y_3 \end{bmatrix} + \begin{bmatrix} 2 \\ 0 \\ 3 \end{bmatrix}$

31. $\dfrac{d}{dt}\begin{bmatrix} y_1 \\ y_2 \\ y_3 \\ y_4 \end{bmatrix} = \begin{bmatrix} -3 & -5 & 0 & 0 \\ 2 & -1 & 0 & 0 \\ 0 & 0 & 0 & 2 \\ 0 & 0 & -2 & 0 \end{bmatrix}\begin{bmatrix} y_1 \\ y_2 \\ y_3 \\ y_4 \end{bmatrix}$

32. $y_1' = y_2 - 1$
$y_2' = y_1 + 2$
$y_3' = -y_3 + 1$
$y_4' = -y_4$

33. Let A be a real (2×2) matrix. Assume that A has eigenvalues λ_1 and λ_2 and consider the linear homogeneous system $\mathbf{y}' = A\mathbf{y}$.

(a) Prove that if λ_1 and λ_2 are both nonzero, then $\mathbf{y}_e = \mathbf{0}$ is an isolated equilibrium point.

(b) Suppose that eigenvalue $\lambda_1 \neq 0$ but that $\lambda_2 = 0$ with corresponding eigenvector $\begin{bmatrix} \beta \\ -\alpha \end{bmatrix}$. Show that all points on the phase-plane line $\alpha x + \beta y = 0$ are equilibrium points. (In this case, $\mathbf{y}_e = \mathbf{0}$ is not an isolated equilibrium point.)

34. Consider the linear system

$$\mathbf{y}' = \begin{bmatrix} -1 & \alpha \\ \alpha & -1 \end{bmatrix}\mathbf{y},$$

where α is a real constant.

(a) What information can be obtained about the eigenvalues of the coefficient matrix simply by examining its structure?

(b) For what value(s) of the constant α is the equilibrium point $\mathbf{y}_e = \mathbf{0}$ an isolated equilibrium point? For what value(s) of the constant α is the equilibrium point $\mathbf{y}_e = \mathbf{0}$ not isolated?

(c) In the case where $\mathbf{y}_e = \mathbf{0}$ is not an isolated equilibrium point, what is the equation of the phase-plane line of equilibrium points?

(d) Is it possible in this example for $\mathbf{y}_e = \mathbf{0}$ to be an isolated equilibrium point that is stable but not asymptotically stable? Explain.

(e) For what values of the constant α, if any, is the equilibrium point $\mathbf{y}_e = \mathbf{0}$ an isolated asymptotically stable equilibrium point? For what values of the constant α, if any, is the equilibrium point $\mathbf{y}_e = \mathbf{0}$ an unstable equilibrium point?

35. Let $A = \begin{bmatrix} 1 & a_{12} \\ a_{21} & a_{22} \end{bmatrix}$ be a real (2×2) matrix. Assume that

$$A\begin{bmatrix} 1 \\ 2 \end{bmatrix} = \begin{bmatrix} 1 & a_{12} \\ a_{21} & a_{22} \end{bmatrix}\begin{bmatrix} 1 \\ 2 \end{bmatrix} = 2\begin{bmatrix} 1 \\ 2 \end{bmatrix}$$

and that the origin is *not* an isolated equilibrium point of the system $\mathbf{y}' = A\mathbf{y}$. Determine the constants a_{12}, a_{21}, and a_{22}.

8.5 Linearization and the Local Picture

We now consider nonlinear autonomous systems and a technique, known as linearization, for investigating the stability properties of such systems. The stability results cited in Theorem 8.3 for the linear system $\mathbf{y}' = A\mathbf{y}$ will be useful for the nonlinear equations treated in this section. Later, in Section 8.6, we will examine the phase-plane geometry of the linear two-dimensional system $\mathbf{y}' = A\mathbf{y}$ in more detail.

Nonlinear Systems

Although we are going to focus on the case $n = 2$, the ideas are applicable to general n-dimensional autonomous systems. Let

$$\mathbf{y}_e = \begin{bmatrix} x_e \\ y_e \end{bmatrix}$$

be an equilibrium solution of $\mathbf{y}' = \mathbf{f}(\mathbf{y})$. In component form, $\mathbf{y}' = \mathbf{f}(\mathbf{y})$ is given by

$$\begin{aligned} x' &= f(x, y), \\ y' &= g(x, y). \end{aligned} \tag{1}$$

For a nonlinear system such as (1), it is usually impossible to obtain explicit solutions. In the absence of explicit solutions we look for approximations or simplifications that provide qualitative insight into the stability properties of the equilibrium point $\mathbf{y}_e$.

Linearization

From the definition of stability, we know that the issue of equilibrium point stability is ultimately determined by the behavior of solutions very close to the equilibrium point. If nearby solutions diverge from the equilibrium point, it is an unstable equilibrium point. If all nearby solutions can be suitably confined, the equilibrium point is stable (and perhaps asymptotically stable).

To address the issue of equilibrium point stability, we begin with the observation that if the point (x, y) is near the equilibrium point (x_e, y_e), then the first few terms of their Taylor series expansions will yield good approximations to

$f(x,y)$ and $g(x,y)$:

$$f(x,y) = f(x_e,y_e) + \frac{\partial f(x_e,y_e)}{\partial x}(x - x_e) + \frac{\partial f(x_e,y_e)}{\partial y}(y - y_e) + \cdots,$$
$$g(x,y) = g(x_e,y_e) + \frac{\partial g(x_e,y_e)}{\partial x}(x - x_e) + \frac{\partial g(x_e,y_e)}{\partial y}(y - y_e) + \cdots. \tag{2}$$

With regard to (2), we make the following observations:

(i) Since (x_e, y_e) is an equilibrium point, $f(x_e,y_e) = g(x_e,y_e) = 0$. Thus, the first term on the right-hand side of each equation in (2) vanishes.

(ii) The error we make in truncating the series [retaining only the linear terms shown on the right-hand sides of (2)] can usually be bounded by a multiple of $\|\mathbf{y} - \mathbf{y}_e\|^2 = (x - x_e)^2 + (y - y_e)^2$.

If the Taylor expansion (2) is used in differential equation (1), we can write the system in matrix form as

$$\mathbf{y}'(t) = \begin{bmatrix} \dfrac{\partial f(x_e,y_e)}{\partial x} & \dfrac{\partial f(x_e,y_e)}{\partial y} \\ \dfrac{\partial g(x_e,y_e)}{\partial x} & \dfrac{\partial g(x_e,y_e)}{\partial y} \end{bmatrix} (\mathbf{y}(t) - \mathbf{y}_e) + \cdots. \tag{3}$$

Note that the (2×2) coefficient matrix in (3) is a *constant* matrix since the partial derivatives are evaluated at the equilibrium point. [In vector calculus, the matrix of first order partial derivatives in (3) is called the **Jacobian matrix** of $\mathbf{f}(\mathbf{y})$.]

Linearization is based upon the following two ideas:

1. Since $\mathbf{y}_e$ is a constant vector, the term $\mathbf{y}'(t)$ on the left-hand side of (3) can be replaced by $(\mathbf{y}(t) - \mathbf{y}_e)'$.
2. If we consider solutions close enough to the equilibrium point for the purpose of determining its stability characteristics, the higher order terms in (3) are typically small in norm relative to the linear term, $\mathbf{y}(t) - \mathbf{y}_e$. We will neglect these higher-order terms.

We introduce a new dependent variable, $\mathbf{z}(t) = \mathbf{y}(t) - \mathbf{y}_e$. Retaining only the linear term in (3) leads to the corresponding **linearized system**

$$\mathbf{z}'(t) = \begin{bmatrix} \dfrac{\partial f(x_e,y_e)}{\partial x} & \dfrac{\partial f(x_e,y_e)}{\partial y} \\ \dfrac{\partial g(x_e,y_e)}{\partial x} & \dfrac{\partial g(x_e,y_e)}{\partial y} \end{bmatrix} \mathbf{z}(t). \tag{4}$$

Note that equation (4) is a homogeneous constant coefficient linear system and that $\mathbf{z} = \mathbf{0}$ is an equilibrium point of the linear system. The stability properties of $\mathbf{z} = \mathbf{0}$ are easy to analyze since we can explicitly find the general solution of equation (4) using the techniques of Chapter 6; these properties are summarized in Theorem 8.3.

The underlying premise of linearization is a heuristic argument that the stability properties of $\mathbf{z} = \mathbf{0}$ for the linear system (4) should be the same as the stability properties of $\mathbf{y} = \mathbf{y}_e$ for the original nonlinear system (1). We illustrate linearization in the next example. Then, we address the question, "When does linearization work?"

EXAMPLE 1

Develop the linearized-system approximation for each of the equilibrium points of the nonlinear autonomous system

$$x' = \frac{1}{2}\left(1 - \frac{1}{2}x - \frac{1}{2}y\right)x,$$

$$y' = \frac{1}{4}\left(1 - \frac{1}{3}x - \frac{2}{3}y\right)y.$$

Also, determine the stability characteristics of the linearized system in each case.

Solution: This is the competing species model considered in Example 4 of Section 8.2. Recall that this system has four equilibrium solutions,

$$\mathbf{y}_e^{(1)} = \begin{bmatrix} 0 \\ 0 \end{bmatrix}, \qquad \mathbf{y}_e^{(2)} = \begin{bmatrix} 0 \\ \frac{3}{2} \end{bmatrix}, \qquad \mathbf{y}_e^{(3)} = \begin{bmatrix} 2 \\ 0 \end{bmatrix}, \qquad \mathbf{y}_e^{(4)} = \begin{bmatrix} 1 \\ 1 \end{bmatrix}.$$

For this system, the Jacobian matrix is given by

$$\begin{bmatrix} \dfrac{\partial f(x,y)}{\partial x} & \dfrac{\partial f(x,y)}{\partial y} \\ \dfrac{\partial g(x,y)}{\partial x} & \dfrac{\partial g(x,y)}{\partial y} \end{bmatrix} = \begin{bmatrix} \dfrac{1}{2} - \dfrac{1}{2}x - \dfrac{1}{4}y & -\dfrac{1}{4}x \\ -\dfrac{1}{12}y & \dfrac{1}{4} - \dfrac{1}{12}x - \dfrac{1}{3}y \end{bmatrix}. \tag{5}$$

The Jacobian matrix (5) must be evaluated at each of the equilibrium points in order to obtain the coefficient matrix of the appropriate linearized system. At the equilibrium point (0, 0), the coefficient matrix is

$$\left.\begin{bmatrix} \dfrac{1}{2} - \dfrac{1}{2}x - \dfrac{1}{4}y & -\dfrac{1}{4}x \\ -\dfrac{1}{12}y & \dfrac{1}{4} - \dfrac{1}{12}x - \dfrac{1}{3}y \end{bmatrix}\right|_{x=y=0} = \begin{bmatrix} \dfrac{1}{2} & 0 \\ 0 & \dfrac{1}{4} \end{bmatrix}.$$

The linearized system is therefore

$$\mathbf{z}' = \begin{bmatrix} \dfrac{1}{2} & 0 \\ 0 & \dfrac{1}{4} \end{bmatrix}\mathbf{z}.$$

[Observe, for this case, that $\mathbf{z}(t) = \mathbf{y}(t) - \mathbf{y}_e^{(1)} = \mathbf{y}(t)$.] Since the eigenvalues of the coefficient matrix ($\lambda = 1/2$ and $\lambda = 1/4$) are positive, we conclude from Theorem 8.3 that $\mathbf{z} = \mathbf{0}$ is an unstable equilibrium solution of the linearized system. Following the heuristic argument above, we anticipate that $\mathbf{y}_e^{(1)}$ the corresponding equilibrium point of the nonlinear system is also an unstable equilibrium point.

The remaining three equilibrium points can be analyzed in the same manner. The results for all four equilibrium points are summarized in Table 8.1. ▲

TABLE 8.1

Equilibrium Point	$\mathbf{z}(t)$	Linearized System Coefficient Matrix	Eigenvalues of the Coefficient Matrix	Stability Properties of the Linearized System
$(0, 0)$	$\begin{bmatrix} x(t) \\ y(t) \end{bmatrix}$	$\begin{bmatrix} \frac{1}{2} & 0 \\ 0 & \frac{1}{4} \end{bmatrix}$	$\frac{1}{2}, \frac{1}{4}$	Unstable
$\left(0, \frac{3}{2}\right)$	$\begin{bmatrix} x(t) \\ y(t) - \frac{3}{2} \end{bmatrix}$	$\begin{bmatrix} \frac{1}{8} & 0 \\ -\frac{1}{8} & -\frac{1}{4} \end{bmatrix}$	$\frac{1}{8}, -\frac{1}{4}$	Unstable
$(2, 0)$	$\begin{bmatrix} x(t) - 2 \\ y(t) \end{bmatrix}$	$\begin{bmatrix} -\frac{1}{2} & -\frac{1}{2} \\ 0 & \frac{1}{12} \end{bmatrix}$	$-\frac{1}{2}, \frac{1}{12}$	Unstable
$(1, 1)$	$\begin{bmatrix} x(t) - 1 \\ y(t) - 1 \end{bmatrix}$	$\begin{bmatrix} -\frac{1}{4} & -\frac{1}{4} \\ -\frac{1}{12} & -\frac{1}{6} \end{bmatrix}$	$\frac{-5 - \sqrt{13}}{24}, \frac{-5 + \sqrt{13}}{24}$	Stable

Observe that the stability properties obtained by studying the linearized systems in Table 8.1 are consistent with the phase-plane direction field portrait of Figure 8.4 in Section 8.2. The three equilibria designated as unstable in Table 8.1 are the ones that appear to have direction field arrows pointing away from them in Figure 8.4. The fourth equilibrium point, designated as stable in Table 8.1, appears to have all trajectories moving toward it in Figure 8.4.

While Example 1 suggests that linearization is a useful tool for predicting stability, we need more definitive results in order to justify placing our trust in linearization. These results are the topic of the next subsection.

When Does Linearization Work?

We restrict our attention to the two-dimensional autonomous system $\mathbf{y}' = \mathbf{f}(\mathbf{y})$. As before, let $\mathbf{y}_e$ be an equilibrium solution,

$$\mathbf{y}_e = \begin{bmatrix} x_e \\ y_e \end{bmatrix}.$$

We'll proceed as before by introducing a new dependent variable that shifts the equilibrium point to $\mathbf{0}$ and by rewriting $\mathbf{y}' = \mathbf{f}(\mathbf{y})$ in a form that explicitly exhibits the linearized part.

Let $\mathbf{z}(t) = \mathbf{y}(t) - \mathbf{y}_e$ and let A denote the Jacobian matrix evaluated at $\mathbf{y}_e$,

$$A = \begin{bmatrix} \dfrac{\partial f(x_e, y_e)}{\partial x} & \dfrac{\partial f(x_e, y_e)}{\partial y} \\ \dfrac{\partial g(x_e, y_e)}{\partial x} & \dfrac{\partial g(x_e, y_e)}{\partial y} \end{bmatrix}. \tag{6}$$

The system $\mathbf{y}' = \mathbf{f}(\mathbf{y})$ can be rewritten (without any approximation) as

$$\mathbf{z}'(t) = \mathbf{f}(\mathbf{z}(t) + \mathbf{y}_e) = A\mathbf{z}(t) + [\mathbf{f}(\mathbf{z}(t) + \mathbf{y}_e) - A\mathbf{z}(t)]. \tag{7}$$

We now define $\mathbf{g}(\mathbf{z}(t)) = \mathbf{f}(\mathbf{z}(t) + \mathbf{y}_e) - A\mathbf{z}(t)$. With this, equation (7) becomes

$$\mathbf{z}'(t) = A\mathbf{z}(t) + \mathbf{g}(\mathbf{z}(t)). \tag{8}$$

Written in this form, we see that $\mathbf{g}(\mathbf{z}(t))$ represents a nonlinear perturbation of the linearized system, $\mathbf{z}'(t) = A\mathbf{z}(t)$. The linearization approximation amounts to discarding the nonlinear term, $\mathbf{g}(\mathbf{z}(t))$.

If the behavior of the linearized system $\mathbf{z}' = A\mathbf{z}$ is going to be qualitatively similar to that of the nonlinear system (8) near the equilibrium point $\mathbf{z} = \mathbf{0}$, it seems clear that $\mathbf{g}(\mathbf{z})$ must be suitably "small" near $\mathbf{z} = \mathbf{0}$. In other words, the linear part of (8), $A\mathbf{z}$, must control the basic behavior of solutions near the equilibrium point. We will now describe such a class of nonlinear systems.

A two-dimensional autonomous system $\mathbf{y}' = \mathbf{f}(\mathbf{y})$ is called an **almost linear system at an equilibrium point $\mathbf{y}_e$** if

(i) $\mathbf{f}(\mathbf{y})$ is a continuous vector-valued function whose component functions possess continuous partial derivatives in an open region of the phase plane containing the equilibrium point, $\mathbf{y}_e$.

(ii) The matrix

$$A = \begin{bmatrix} \dfrac{\partial f(x_e, y_e)}{\partial x} & \dfrac{\partial f(x_e, y_e)}{\partial y} \\ \dfrac{\partial g(x_e, y_e)}{\partial x} & \dfrac{\partial g(x_e, y_e)}{\partial y} \end{bmatrix}$$

is invertible.

(iii) The perturbation function $\mathbf{g}(\mathbf{z}) = \mathbf{f}(\mathbf{z} + \mathbf{y}_e) - A\mathbf{z}$ is such that

$$\lim_{\|\mathbf{z}\| \to 0} \frac{\|\mathbf{g}(\mathbf{z})\|}{\|\mathbf{z}\|} = 0. \tag{9}$$

REMARKS:

1. The perturbation function $\mathbf{g}(\mathbf{z}) = \mathbf{f}(\mathbf{z} + \mathbf{y}_e) - A\mathbf{z}$ inherits the continuity and differentiability properties assumed for $\mathbf{f}$. Thus, $\mathbf{g}(\mathbf{z})$ is continuous with continuous partial derivatives in an open region of the $\mathbf{z}$-plane containing the origin, $\mathbf{z} = \mathbf{0}$.

2. Limit (9) establishes how "small" the nonlinear perturbation must be near the equilibrium point; the norm of the perturbation must tend to zero faster than the norm of $\|\mathbf{z}\|$ as $\mathbf{z}$ approaches the origin.
3. Since matrix A is invertible, $\mathbf{z} = \mathbf{0}$ is the only equilibrium point of the linearized problem $\mathbf{z}' = A\mathbf{z}$. It is also clear that $\mathbf{g}(\mathbf{0}) = \mathbf{0}$ [since $\mathbf{f}(\mathbf{y}_e) = \mathbf{0}$]. Therefore, $\mathbf{z} = \mathbf{0}$ is an equilibrium point of the nonlinear system $\mathbf{z}' = A\mathbf{z} + \mathbf{g}(\mathbf{z})$. In addition, it can be shown that the assumptions made in (i)–(iii) imply that $\mathbf{z} = \mathbf{0}$ is an isolated equilibrium point of $\mathbf{z}' = A\mathbf{z} + \mathbf{g}(\mathbf{z})$. Therefore, $\mathbf{y} = \mathbf{y}_e$ is an isolated equilibrium point of the original system $\mathbf{y}' = \mathbf{f}(\mathbf{y})$. ▌

Theorem 8.4

Let $\mathbf{y}' = \mathbf{f}(\mathbf{y})$ be a two-dimensional autonomous system that is almost linear at an equilibrium point, $\mathbf{y} = \mathbf{y}_e$. Let $\mathbf{z}' = A\mathbf{z}$ be the corresponding linearized system.

(i) If $\mathbf{z} = \mathbf{0}$ is an asymptotically stable equilibrium point of $\mathbf{z}' = A\mathbf{z}$, then $\mathbf{y} = \mathbf{y}_e$ is an asymptotically stable equilibrium point of $\mathbf{y}' = \mathbf{f}(\mathbf{y})$.

(ii) If $\mathbf{z} = \mathbf{0}$ is an unstable equilibrium point of $\mathbf{z}' = A\mathbf{z}$, then $\mathbf{y} = \mathbf{y}_e$ is an unstable equilibrium point of $\mathbf{y}' = \mathbf{f}(\mathbf{y})$.

(iii) If $\mathbf{z} = \mathbf{0}$ is a stable (but not asymptotically stable) equilibrium point of $\mathbf{z}' = A\mathbf{z}$, then no conclusions can be drawn about the stability properties of equilibrium point $\mathbf{y} = \mathbf{y}_e$.

The proof of Theorem 8.4 can be found in more advanced treatments of differential equations, such as Coddington and Levinson.[3] However, the assertions made in Theorem 8.4 should strike you as reasonable. When the linearized system is asymptotically stable, both eigenvalues of the coefficient matrix A have negative real parts. We know, therefore, that the norm of the solution of the linearized system, $\|\mathbf{z}(t)\|$, is exponentially decreasing to zero. If the nonlinear perturbation is sufficiently weak, we might expect this behavior to persist in the nonlinear system (8).

Similarly, if the linearized system is unstable, at least one of the two eigenvalues of A has a positive real part. The linearized system therefore has some solutions that grow exponentially in norm. In this case, we might expect instability to persist when the nonlinear perturbation $\mathbf{g}$ is sufficiently weak. Finally, if the linearized system is stable but not asymptotically stable, then the eigenvalues of A form a purely imaginary complex conjugate pair. In this case, the linearized system is sitting on the fence between stability and instability. It is possible for the nonlinear perturbation (however small) to tip the balance either way—causing the nonlinear system to be stable or causing it to be unstable. Example 3, found later in this section, illustrates this last point [and thereby proves condition (iii) of Theorem 8.4].

---

[3]Earl A. Coddington and Norman Levinson, *Theory of Ordinary Differential Equations* (Malabar, FL: R. E. Krieger, 1984).

EXAMPLE 2

Consider again the nonlinear system discussed in Example 1,

$$x' = \frac{1}{2}\left(1 - \frac{1}{2}x - \frac{1}{2}y\right)x,$$
$$y' = \frac{1}{4}\left(1 - \frac{1}{3}x - \frac{2}{3}y\right)y. \tag{10}$$

Use Theorem 8.4 to determine the stability properties of equilibrium point (1, 1).

Solution: We begin by making the change of dependent variable, $\mathbf{z}(t) = \mathbf{y}(t) - \mathbf{y}_e$:

$$z_1(t) = x(t) - 1 \quad \text{and} \quad z_2(t) = y(t) - 1.$$

With this change of variables, we can rewrite system (10) as

$$\mathbf{z}' = \begin{bmatrix} -\frac{1}{4} & -\frac{1}{4} \\ -\frac{1}{12} & -\frac{1}{6} \end{bmatrix} \mathbf{z} + \begin{bmatrix} -\frac{1}{4}(z_1^2 + z_1 z_2) \\ -\frac{1}{12}(z_1 z_2 + 2z_2^2) \end{bmatrix}. \tag{11}$$

We also know (see Table 8.1) that the linearized system

$$\mathbf{z}' = \begin{bmatrix} -\frac{1}{4} & -\frac{1}{4} \\ -\frac{1}{12} & -\frac{1}{6} \end{bmatrix} \mathbf{z}$$

has an asymptotically stable equilibrium point at $\mathbf{z} = \mathbf{0}$.

Theorem 8.4 asserts that $\mathbf{z} = \mathbf{0}$ will be an asymptotically stable equilibrium point of nonlinear system (11) (and therefore $\mathbf{y}_e$ is an asymptotically stable equilibrium point of the original system) if we can show the system is almost linear at the equilibrium point $\mathbf{y}_e$. Note that the first two conditions of the definition are clearly satisfied. So, to apply Theorem 8.4, all we need to do is establish the limit (9):

$$\frac{\|\mathbf{g}(\mathbf{z})\|}{\|\mathbf{z}\|} \to 0 \quad \text{as} \quad \|\mathbf{z}\| \to 0,$$

where (see equation (11))

$$\mathbf{g}(\mathbf{z}) = \begin{bmatrix} -\frac{1}{4}(z_1^2 + z_1 z_2) \\ -\frac{1}{12}(z_1 z_2 + 2z_2^2) \end{bmatrix}.$$

In order to calculate the quotient $\|\mathbf{g}(\mathbf{z})\|/\|\mathbf{z}\|$, it is convenient to introduce polar coordinates $z_1 = r\cos\theta$ and $z_2 = r\sin\theta$. Under this change of variables we see that $\|\mathbf{z}\| = r$ and

$$\mathbf{g}(\mathbf{z}) = \begin{bmatrix} -\frac{r^2}{4}(\cos^2\theta + \sin\theta\cos\theta) \\ -\frac{r^2}{12}(\sin\theta\cos\theta + 2\sin^2\theta) \end{bmatrix}.$$

Thus,

$$\frac{\|\mathbf{g}(\mathbf{z})\|}{\|\mathbf{z}\|} = \frac{\|\mathbf{g}(\mathbf{z})\|}{r}$$

$$= r\sqrt{\left(-\frac{1}{4}(\cos^2\theta + \sin\theta\cos\theta)\right)^2 + \left(-\frac{1}{12}(\sin\theta\cos\theta + 2\sin^2\theta)\right)^2}$$

$$< r\sqrt{\left(\frac{1}{4}(1+1)\right)^2 + \left(\frac{1}{12}(1+2)\right)^2} = \frac{\sqrt{5}}{4}r.$$

Since the right-hand side of this inequality vanishes as $r \to 0$, limit (9) is verified and we conclude that the nonlinear system is almost linear at equilibrium point $(1, 1)$. By Theorem 8.4, this equilibrium point is an asymptotically stable equilibrium point of the given nonlinear system. ▲

Polar Coordinates as Dependent Variables

It is often useful to employ polar coordinates when studying the stability properties of two-dimensional systems. We define new dependent variables $r(t)$ and $\theta(t)$ by means of the relations

$$z_1(t) = r(t)\cos(\theta(t)), \qquad z_2(t) = r(t)\sin(\theta(t)) \tag{12a}$$

$$r(t) = \sqrt{z_1^2(t) + z_2^2(t)}, \qquad \tan(\theta(t)) = \frac{z_2(t)}{z_1(t)}. \tag{12b}$$

We can transform system (8), $\mathbf{z}'(t) = A\mathbf{z}(t) + \mathbf{g}(\mathbf{z}(t))$, into a new system of differential equations in these polar variables.

The motivation for this transformation is the fact that stability properties of the equilibrium point $\mathbf{y} = \mathbf{y}_e$ (or, equivalently, $\mathbf{z} = \mathbf{0}$) depend on how the distance of the solution point $\mathbf{z}(t)$ from the origin varies with time. The radial variable, $r(t) = \|\mathbf{z}(t)\|$, is this distance.

Let the Jacobian matrix A be represented as

$$A = \begin{bmatrix} a_{11} & a_{12} \\ a_{21} & a_{22} \end{bmatrix}.$$

Then, substituting (12a) into $\mathbf{z}'(t) = A\mathbf{z}(t) + \mathbf{g}(\mathbf{z}(t))$, we obtain

$$r'\cos\theta - r\theta'\sin\theta = a_{11}r\cos\theta + a_{12}r\sin\theta + g_1(r\cos\theta, r\sin\theta),$$
$$r'\sin\theta + r\theta'\cos\theta = a_{21}r\cos\theta + a_{22}r\sin\theta + g_2(r\cos\theta, r\sin\theta).$$

We can write the transformed system in matrix form as

$$\begin{bmatrix} \cos\theta & -r\sin\theta \\ \sin\theta & r\cos\theta \end{bmatrix} \begin{bmatrix} r' \\ \theta' \end{bmatrix} = \begin{bmatrix} a_{11} & a_{12} \\ a_{21} & a_{22} \end{bmatrix} \begin{bmatrix} r\cos\theta \\ r\sin\theta \end{bmatrix} + \begin{bmatrix} g_1(r\cos\theta, r\sin\theta) \\ g_2(r\cos\theta, r\sin\theta) \end{bmatrix}. \tag{13}$$

The desired system of differential equations for the polar variables is obtained by multiplying both sides of (13) by

$$\begin{bmatrix} \cos\theta & -r\sin\theta \\ \sin\theta & r\cos\theta \end{bmatrix}^{-1} = \begin{bmatrix} \cos\theta & \sin\theta \\ -r^{-1}\sin\theta & r^{-1}\cos\theta \end{bmatrix}.$$

We illustrate the use of the polar coordinates transformation in the next example. As previously noted, this example establishes condition (iii) of Theorem 8.4.

EXAMPLE 3

Consider the nonlinear autonomous system

$$\begin{aligned} x' &= y + \alpha x(x^2 + y^2), \\ y' &= -x + \alpha y(x^2 + y^2), \end{aligned} \tag{14}$$

where α is a constant. Note that this system has $\mathbf{y}_e = \mathbf{0}$ as an equilibrium point for any choice of the parameter α.

Although $\mathbf{y}_e = \mathbf{0}$ is a stable equilibrium point of the linearized system, it is not asymptotically stable. We will show that if $\alpha = 1$, $\mathbf{y}_e = \mathbf{0}$ is an unstable equilibrium point of the nonlinear system, whereas if $\alpha = -1$, $\mathbf{y}_e = \mathbf{0}$ is a stable equilibrium point of the nonlinear system. This example, therefore, establishes assertion (iii) of Theorem 8.4.

To study the stability properties of $\mathbf{y}_e = \mathbf{0}$, we set $\mathbf{z}(t) = \mathbf{y}(t) - \mathbf{y}_e = \mathbf{y}(t)$ and note that the system can be written as $\mathbf{z}' = A\mathbf{z} + \mathbf{g}(\mathbf{z})$, where

$$A = \begin{bmatrix} 0 & 1 \\ -1 & 0 \end{bmatrix}, \qquad \mathbf{g}(\mathbf{z}) = \begin{bmatrix} \alpha x(x^2 + y^2) \\ \alpha y(x^2 + y^2) \end{bmatrix} = \alpha r^3 \begin{bmatrix} \cos\theta \\ \sin\theta \end{bmatrix}. \tag{15}$$

The eigenvalues of matrix A are $\pm i$, and we know from Example 3 in Section 8.4 that the associated linearized system is stable but not asymptotically stable at $\mathbf{z} = \mathbf{0}$. Note that the nonlinear system is almost linear at $\mathbf{0}$. In particular, it follows from (15) that the quotient,

$$\frac{\|\mathbf{g}(\mathbf{z})\|}{\|\mathbf{z}\|} = \frac{\|\mathbf{g}(\mathbf{z})\|}{r} = |\alpha| r^2,$$

vanishes as $r \to 0$.

Introducing polar coordinate variables $r(t)$ and $\theta(t)$ as in (12), it follows (see Exercise 27) that equation (14) transforms into the following simple decoupled system of equations for the new dependent variables,

$$\begin{aligned} r' &= \alpha r^3, \\ \theta' &= -1. \end{aligned} \tag{16}$$

Both of these equations are easy to solve; the differential equation for the radial variable $r(t)$ is a first order separable differential equation. The differential equation for the angular variable $\theta(t)$ can be solved by antidifferentiation. If we include initial conditions for the original system (14) of the form

$$\begin{bmatrix} x(0) \\ y(0) \end{bmatrix} = \begin{bmatrix} x_0 \\ y_0 \end{bmatrix},$$

we arrive at corresponding initial conditions for the transformed system (16):

$$\begin{aligned} r(0) &= r_0 = \sqrt{x_0^2 + y_0^2}, \\ \theta(0) &= \tan^{-1}\left(\frac{y_0}{x_0}\right). \end{aligned}$$

We are interested in the case where $r_0 \neq 0$ and, subject to the initial conditions, the solution of the transformed system (16) is

$$r(t) = \frac{r_0}{\sqrt{1 - 2r_0^2 \alpha t}}, \qquad \theta(t) = -t + \theta_0. \tag{17}$$

Suppose we now consider two choices for the parameter α, namely $\alpha = 1$ and $\alpha = -1$. For the case $\alpha = 1$, $r(t)$ becomes unbounded as t approaches $(1/2)r_0^2$. Therefore, since all nonzero solutions eventually move unboundedly far from the equilibrium point, $\mathbf{z} = \mathbf{0}$, the origin is an unstable equilibrium point of the nonlinear system. In other words, the linearized system is stable (but not asymptotically stable) and the nonlinear system is unstable at the origin.

If $\alpha = -1$, however, we see from (17) that

$$0 \le r(t) \le r_0 \quad \text{and} \quad \lim_{t\to\infty} r(t) = 0. \tag{18}$$

In this case, the origin is an asymptotically stable equilibrium point of the nonlinear system. The linearized system is stable (but not asymptotically stable) and the nonlinear system is asymptotically stable at $\mathbf{y} = \mathbf{0}$.

Therefore, we have established assertion (iii) of Theorem 8.4: If the linearized system is stable (but not asymptotically stable), nothing can be inferred about the stability properties of the nonlinear system.

In both cases, as time increases and the radial variable changes, the angular variable decreases at a constant rate. The solution point is moving clockwise around the origin with unit angular velocity. Solution point behaviors corresponding to the two cases, $\alpha = \pm 1$, are illustrated in Figure 8.13. ▲

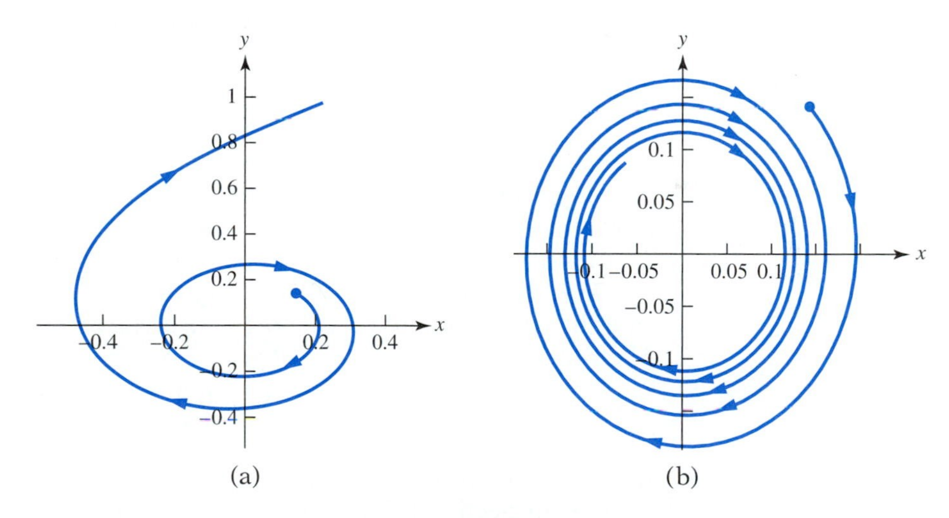

FIGURE 8.13

(a) The phase-plane plot of the solution $\mathbf{z}(t)$ in Example 3, with $\alpha = 1$, $r_0 = 0.2$, and $\theta_0 = \pi/4$ in (17), and where $0 \le t \le 12$. For this case, $\mathbf{z} = \mathbf{0}$ is an unstable equilibrium point for the nonlinear system (16); the solution point moves away from $\mathbf{z} = \mathbf{0}$ as t increases. (b) The phase-plane plot of the solution $\mathbf{z}(t)$ in Example 3, with $\alpha = -1$ and $0 \le t \le 30$. In this case $\mathbf{z} = \mathbf{0}$ is an asymptotically stable equilibrium point for the nonlinear system (16); the solution point moves toward $\mathbf{z} = \mathbf{0}$ as t increases.

The Pendulum Revisited

As a final example we consider the stability properties of the pendulum in the context of the ideas developed earlier in this chapter. Recall that, with $g = l$, the pendulum is described by the nonlinear system

$$\begin{aligned} x' &= y, \\ y' &= -\sin x. \end{aligned} \tag{19}$$

There are basically two equilibrium configurations,

$$\mathbf{y}_e = \begin{bmatrix} 0 \\ 0 \end{bmatrix} \quad \text{and} \quad \mathbf{y}_e = \begin{bmatrix} \pi \\ 0 \end{bmatrix}.$$

The first corresponds to the pendulum resting in the vertically downward position, and the second corresponds to the pendulum resting in an inverted position. It can be shown (Exercise 11) that system (19) is an almost linear system at both equilibrium points; thus we can apply Theorem 8.4.

At the equilibrium point $(0, 0)$, the coefficient matrix of the linearized system is

$$A = \begin{bmatrix} 0 & 1 \\ -1 & 0 \end{bmatrix}.$$

The eigenvalues of this matrix are $\pm i$, and thus the linearized system is stable but not asymptotically stable at $(0, 0)$. Therefore, Theorem 8.4 provides no information about stability of the pendulum equation at $(0, 0)$.

At $(\pi, 0)$, the coefficient matrix of the linearized system is

$$A = \begin{bmatrix} 0 & 1 \\ 1 & 0 \end{bmatrix}.$$

The eigenvalues of this matrix are ± 1 and so the linearized system is unstable (since one of the eigenvalues is positive). Therefore, Theorem 8.4 tells us that $(\pi, 0)$ is an unstable equilibrium point for the pendulum equation.

There is another way of deducing stability information. In Section 8.3 we saw that the pendulum leads to a conservative system, and we derived an explicit formula,

$$\frac{1}{2}y^2 - \cos x = E, \tag{20}$$

for a pendulum trajectory having energy per unit mass equal to E [see equation (8) in Section 8.3]. Figure 8.14(a) is similar to Figure 8.9 of Section 8.3, showing representative trajectories obtained by graphing equation (20) for various values of E, $E \geq -1$. As the value of the constant E decreases toward the equilibrium point value of $E = -1$, the corresponding phase-plane trajectories form a nested family of "ellipse-like" closed curves; see Figure 8.14(a). The uniqueness property of solutions prevents these closed curves from intersecting. We expect, therefore, that the origin is a stable equilibrium point of the nonlinear system.

A mathematical argument verifying that $(0, 0)$ is a stable equilibrium point can be developed along the following lines. Given any $\varepsilon > 0$, construct a circle of radius ε centered at the origin. For a value of energy E sufficiently close to -1, we can find a closed trajectory lying entirely within this circle. Let this value of energy be E_ε. Now we choose $\delta > 0$ sufficiently small so that a circle of radius δ lies within this closed trajectory. This choice of δ will work as far as

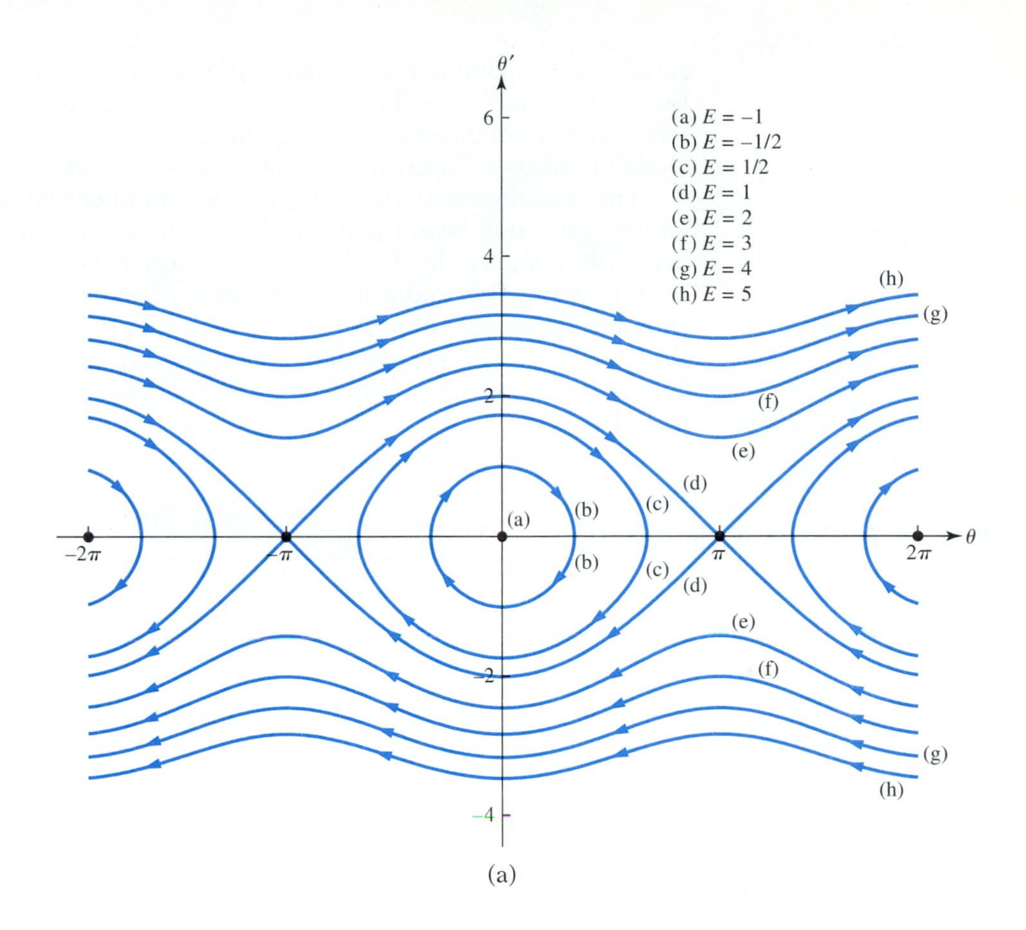

(a)

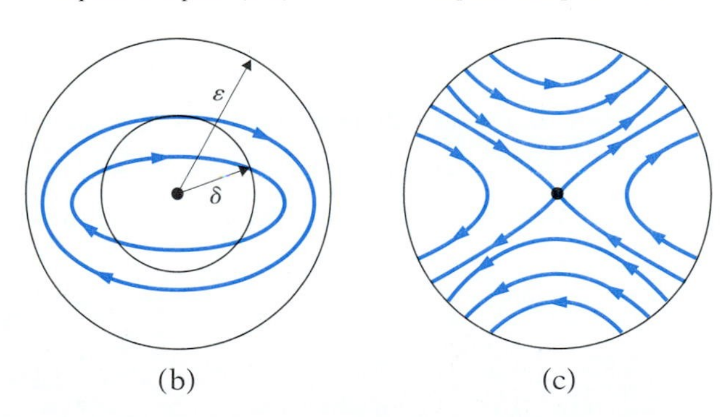

(b) (c)

FIGURE 8.14

(a) Some of the phase-plane trajectories for the pendulum equation (19). Note that the separatrices approach the unstable equilibrium point $(\pi, 0)$. (b) Any trajectory originating within the circle of radius δ must stay within the trajectory having energy E_ε and must therefore remain within the circle of radius ε. (c) Every sufficiently small circle centered at $(\pi, 0)$ has at least one trajectory that eventually exits the circle. In fact, only the equilibrium solution remains inside the circle for all $t > 0$.

satisfying the stability definition is concerned—all solutions originating within the circle of radius δ will remain within the closed trajectory of energy E_ε since solutions cannot intersect. Consequently the trajectories will remain within the circle of radius ε. Figure 8.14(b) illustrates these ideas.

These geometrical ideas also provide another way of seeing that $(\pi, 0)$ is an unstable equilibrium point of the nonlinear system. In particular, see Figure 8.14(c), any circle of radius $\delta > 0$ centered at $(\pi, 0)$ must contain portions of the separatrices and portions of some of the other trajectories that correspond to large-scale motions of the pendulum. Solution points originating on the latter trajectories eventually exit any sufficiently small circle of radius ε centered at $(\pi, 0)$. Therefore, the definition fails and the equilibrium point $(\pi, 0)$ is unstable.

EXERCISES

Exercises 1–9:

In each exercise, the given system is an almost linear system at each of its equilibrium points.

(a) Find the (real) equilibrium points of the given system.

(b) As in Example 2, find the corresponding linearized system, $\mathbf{z}' = A\mathbf{z}$ at each equilibrium point.

(c) What, if anything, can be inferred about the stability properties of the equilibrium point(s) by using Theorem 8.4?

1. $x' = x^2 + y^2 - 32$
$y' = y - x$

2. $x' = x^2 + 9y^2 - 9$
$y' = x$

3. $x' = 1 - x^2$
$y' = x^2 + y^2 - 2$

4. $x' = x - y - 1$
$y' = x^2 - y^2 + 1$

5. $x' = (x - 2)(y - 3)$
$y' = (x + 2y)(y - 1)$

6. $x' = (x - y)(y + 1)$
$y' = (x + 2)(y - 4)$

7. $x' = (x - 2y)(y + 4)$
$y' = 2x - y$

8. $x' = xy - 1$
$y' = (x + 4y)(x - 1)$

9. $x' = y^2 - x$
$y' = x^2 - y$

10. Complete the stability analysis of the competing species model

$$x' = \frac{1}{2}\left(1 - \frac{1}{2}x - \frac{1}{2}y\right)x,$$
$$y' = \frac{1}{4}\left(1 - \frac{1}{3}x - \frac{2}{3}y\right)y.$$

Specifically, repeat the analysis of Example 2 at each of the other three equilibrium points, $(0, 0)$, $(0, 3/2)$, and $(2, 0)$ to determine the stability properties of the nonlinear system at these points. (Assume the system is almost linear at each of these equilibrium points.)

11. Consider the system encountered in the study of pendulum motion,

$$x' = y,$$
$$y' = -\sin x,$$

at its equilibrium points $(0, 0)$ and $(\pi, 0)$.

(a) Let $z_1 = x, z_2 = y$. Show that the system becomes

$$\mathbf{z}' = \begin{bmatrix} 0 & 1 \\ -1 & 0 \end{bmatrix}\mathbf{z} + \begin{bmatrix} 0 \\ z_1 - \sin z_1 \end{bmatrix}.$$

(b) Let $z_1 = x - \pi, z_2 = y$. Show that the system becomes

$$\mathbf{z}' = \begin{bmatrix} 0 & 1 \\ 1 & 0 \end{bmatrix}\mathbf{z} - \begin{bmatrix} 0 \\ z_1 - \sin z_1 \end{bmatrix}.$$

(c) Show the system is almost linear at both equilibrium points. [Hint: One approach is to use Taylor's theorem and polar coordinates.]

Exercises 12–20:

Each exercise lists a nonlinear system $\mathbf{z}' = A\mathbf{z} + \mathbf{g}(\mathbf{z})$, where A is a constant (2×2) invertible matrix and $\mathbf{g}(\mathbf{z})$ is a (2×1) vector function. In each of the exercises, $\mathbf{z} = \mathbf{0}$ is an equilibrium point of the nonlinear system.

(a) Identify A and $\mathbf{g}(\mathbf{z})$.

(b) Calculate $\|\mathbf{g}(\mathbf{z})\|$.

(c) Is $\lim_{\|\mathbf{z}\| \to 0} \|\mathbf{g}(\mathbf{z})\| / \|\mathbf{z}\| = 0$? Is $\mathbf{z}' = A\mathbf{z} + \mathbf{g}(\mathbf{z})$ an almost linear system at $\mathbf{z} = \mathbf{0}$?

(d) If the system is almost linear, use Theorem 8.4 to choose one of the three statements:

(i) $\mathbf{z} = \mathbf{0}$ is an asymptotically stable equilibrium point.

(ii) $\mathbf{z} = \mathbf{0}$ is an unstable equilibrium point.

(iii) No conclusion can be drawn by using Theorem 8.4.

12. $z_1' = 9z_1 - 4z_2 + z_2^2$
$z_2' = 15z_1 - 7z_2$

13. $z_1' = 5z_1 - 14z_2 + z_1 z_2$
$z_2' = 3z_1 - 8z_2 + z_1^2 + z_2^2$

14. $z_1' = -3z_1 + z_2 + z_1^2 + z_2^2$
$z_2' = 2z_1 - 2z_2 + (z_1^2 + z_2^2)^{1/3}$

15. $z_1' = -z_1 + 3z_2 + z_2 \cos\sqrt{z_1^2 + z_2^2}$
$z_2' = -z_1 - 5z_2 + z_1 \cos\sqrt{z_1^2 + z_2^2}$

16. $z_1' = -2z_1 + 2z_2 + z_1 z_2 \cos z_2$
$z_2' = z_1 - 3z_2 + z_1 z_2 \sin z_2$

17. $z_1' = 2z_2 + z_2^2$
$z_2' = -2z_1 + z_1 z_2$

18. $z_1' = -3z_1 - 5z_2 + z_1 e^{-\sqrt{z_1^2 + z_2^2}}$
$z_2' = 2z_1 - z_2 + z_2 e^{-\sqrt{z_1^2 + z_2^2}}$

19. $z_1' = 9z_1 + 5z_2 + z_1 z_2$
$z_2' = -7z_1 - 3z_2 + z_1^2$

20. $z_1' = 2z_1 + 2z_2$
$z_2' = -5z_1 - 2z_2 + z_1^2$

21. Consider the autonomous system

$$x' = -x + xy + y,$$
$$y' = x - xy - 2y.$$

This is the reduced system for the chemical reaction discussed in Exercise 15 of Section 8.1 with $a(t) = x(t)$, $c(t) = y(t)$, $e_0 = 1$, and all rate constants set equal to 1.

(a) Show that this system has a single equilibrium point, $(x_e, y_e) = (0, 0)$.

(b) Determine the linearized system $\mathbf{z}' = A\mathbf{z}$ and analyze its stability properties.

(c) Show that the system is an almost linear system at equilibrium point $(0, 0)$.

(d) Use Theorem 8.4 to determine the equilibrium properties of the given nonlinear system at (0, 0).

22. Consider the nonlinear scalar differential equation $x'' = 1 - (1+x)^{3/2}$. An equation having this structure arises in modeling the bobbing motion of a floating parabolic trough (see the Chapter 4 Extended Problem).

(a) Let $y = x'$ and rewrite the given scalar equation as an equivalent first order system.

(b) Show that the system has a single equilibrium point at $(x_e, y_e) = (0, 0)$.

(c) Determine the linearized system $\mathbf{z}' = A\mathbf{z}$ and analyze its stability properties.

(d) Assume that the system is an almost linear system at equilibrium point (0, 0). Does Theorem 8.4 provide any information about the stability properties of the nonlinear system obtained in part (a)? Explain.

23. Consider again the differential equation of Exercise 22, $x'' = 1 - (1+x)^{3/2}$.

(a) In Section 8.3, equations having this structure were shown to have a conservation law. Derive this conservation law; it will have the form $y^2/2 + F(x) = C$, where $y = x'$.

(b) Let $C = 1/2$, $3/4$, and 1 in the conservation law derived in part (a). Plot the corresponding phase-plane solution trajectories, using computational software. Are these phase-plane trajectories consistent with a bobbing motion? Assuming these plots typify the behavior of solution trajectories near the origin, do they suggest that the origin is a stable or unstable equilibrium point for the nonlinear system? Explain.

24. Each of the autonomous nonlinear systems fails to satisfy the hypotheses of Theorem 8.4 at the equilibrium point (0, 0). Explain why.

(a) $x' = x - y + xy$
$y' = -x + y + 2x^2y^2$

(b) $x' = x - 2y - x^{2/3}$
$y' = x + y + 2y^{1/3}$

25. Polar Coordinates Consider the system $\mathbf{z}' = A\mathbf{z} + \mathbf{g}(\mathbf{z})$, where

$$A = \begin{bmatrix} a_{11} & a_{12} \\ a_{21} & a_{22} \end{bmatrix} \quad \text{and} \quad \mathbf{g}(\mathbf{z}) = \begin{bmatrix} g_1(\mathbf{z}) \\ g_2(\mathbf{z}) \end{bmatrix}.$$

Show that adopting polar coordinates $z_1(t) = r(t)\cos(\theta(t))$ and $z_2(t) = r(t)\sin(\theta(t))$ transforms the system $\mathbf{z}' = A\mathbf{z} + \mathbf{g}(\mathbf{z})$ into

$$r' = r[a_{11}\cos^2\theta + a_{22}\sin^2\theta + (a_{12} + a_{21})\sin\theta\cos\theta] + [g_1\cos\theta + g_2\sin\theta],$$
$$\theta' = [a_{21}\cos^2\theta - a_{12}\sin^2\theta + (a_{22} - a_{11})\sin\theta\cos\theta] + r^{-1}[-g_1\sin\theta + g_2\cos\theta].$$

26. Use the polar equations derived in Exercise 25 to show that if

$$a_{11} = a_{22}, \quad a_{21} = -a_{12}, \quad g_1(\mathbf{z}) = z_1 h\left(\sqrt{z_1^2 + z_2^2}\right), \quad g_2(\mathbf{z}) = z_2 h\left(\sqrt{z_1^2 + z_2^2}\right)$$

for some function h, then the polar equations uncouple into

$$r' = a_{11}r + rh(r),$$
$$\theta' = a_{21}.$$

Note that the radial equation is a separable differential equation and the angle equation can be solved by antidifferentiation.

27. Consider the system $x' = y + \alpha x[x^2 + y^2]$, $y' = -x + \alpha y[x^2 + y^2]$. Introduce polar coordinates and use the results of Exercises 25 and 26 to derive differential equations for $r(t)$ and $\theta(t)$. Solve these differential equations and then form $x(t)$ and $y(t)$.

Exercises 28–29:

Introduce polar coordinates and transform the given initial value problem into an equivalent initial value problem for the polar variables. Solve the polar initial value problem, and use the polar solution to obtain the solution of the original initial value problem. If the solution exists at time $t = 1$, evaluate it. If not, explain why.

28. $x' = x + x\sqrt{x^2 + y^2}, \quad x(0) = 1$
$y' = y + y\sqrt{x^2 + y^2}, \quad y(0) = \sqrt{3}$

29. $x' = y - x\ln[x^2 + y^2], \quad x(0) = e/\sqrt{2}$
$y' = -x - y\ln[x^2 + y^2], \quad y(0) = e/\sqrt{2}$

8.6 Two-Dimensional Linear Systems

In this section we examine the phase-plane behavior of solutions of the linear system $\mathbf{y}' = A\mathbf{y}$, where A is a (2×2) real invertible matrix. Since A is invertible, $\mathbf{y} = \mathbf{0}$ is the only equilibrium solution of $\mathbf{y}' = A\mathbf{y}$.

We have two principal reasons for studying this phase-plane behavior. First, $\mathbf{y}' = A\mathbf{y}$ is an important and intrinsically interesting system, one that we can thoroughly understand. Second, as we saw in the preceding section, such systems arise whenever we linearize about an equilibrium point, zooming in to study the behavior of a nonlinear autonomous system close to an equilibrium point.

By studying the eigenvalues and the phase-plane geometry of the associated eigenvectors at an equilibrium point, we can often sketch a good local picture—one that gives a qualitative description of system trajectory behavior near the equilibrium point. Such local pictures complement the large-scale overview provided by the direction field. Taken together, they provide a good overall view of system behavior.

To illustrate the ideas, we again consider the competing species problem that has served as a vehicle for discussion throughout this chapter.

EXAMPLE 1

In this example we use linearized system approximations to develop local pictures of system behavior near each of the equilibrium points of the nonlinear system

$$\begin{aligned} x' &= \frac{1}{2}\left(1 - \frac{1}{2}x - \frac{1}{2}y\right)x, \\ y' &= \frac{1}{4}\left(1 - \frac{1}{3}x - \frac{2}{3}y\right)y. \end{aligned} \qquad (1)$$

System (1) has four equilibrium points, $(0, 0)$, $(0, 3/2)$, $(2, 0)$, and $(1, 1)$. We now focus on the equilibrium point $(0, 3/2)$ in order to illustrate how the local picture at an equilibrium point complements the global picture provided by the direction field. Using

$$\mathbf{z}(t) = \begin{bmatrix} x(t) \\ y(t) - \dfrac{3}{2} \end{bmatrix},$$

(continued)

(continued)

the linearized system at (0, 3/2) is

$$\mathbf{z}' = \begin{bmatrix} \frac{1}{8} & 0 \\ -\frac{1}{8} & -\frac{1}{4} \end{bmatrix} \mathbf{z}.$$

The general solution of this linear system is

$$\mathbf{z}(t) = c_1 e^{t/8} \begin{bmatrix} 3 \\ -1 \end{bmatrix} + c_2 e^{-t/4} \begin{bmatrix} 0 \\ 1 \end{bmatrix}. \tag{2}$$

We can use the eigenpair information in (2) to sketch the qualitative behavior of solution trajectories of $\mathbf{z}' = A\mathbf{z}$. In turn, these sketches provide qualitative information about solutions of the original nonlinear system near the equilibrium point (0, 3/2).

To begin, consider the special case where $c_2 = 0$ and $c_1 \neq 0$. In this case,

$$z_1(t) = 3c_1 e^{t/8} \quad \text{and} \quad z_2(t) = -c_1 e^{t/8}. \tag{3}$$

We see from (3) that these solutions lie on the **z**-plane line

$$z_2 = -\frac{1}{3} z_1.$$

Since $e^{t/8}$ increases as t increases, we also see from (3) that both $z_1(t)$ and $z_2(t)$ increase in magnitude as t increases. Therefore, solution points originating on this line remain on this line and move away from the origin as t increases, as shown in the direction field plot in Figure 8.15(a). Similarly, consider the companion case where $c_1 = 0$ and $c_2 \neq 0$ in equation (2). In this case, solution points lie on the phase-plane line $z_1 = 0$ and approach the origin as t increases since $e^{-t/4}$ decreases as t increases (see Figure 8.15(a)).

Solution point behavior in these two special cases enables us to determine qualitatively the general phase-plane characteristics of (2). To explain, consider the general case where $c_1 \neq 0$, $c_2 \neq 0$. For sufficiently small values of t, both exponential functions are roughly comparable in size and both terms in the general solution influence solution point behavior. However, as t increases, the term

$$c_1 e^{t/8} \begin{bmatrix} 3 \\ -1 \end{bmatrix}$$

becomes increasingly dominant and all solution trajectories approach the phase-plane line $z_2 = -(1/3)z_1$ as an asymptote. Therefore, we obtain the phase-plane behavior shown in Figure 8.15(a).

A similar analysis can be used to study behavior near the other equilibrium points, (0, 0), (2, 0), and (1, 1). In all cases, the eigenvalues of the linearized system coefficient matrix are real and distinct. The eigenvectors determine lines through the **z**-plane origin on which solution points either travel toward the origin if the corresponding eigenvalue is negative or away from the origin if the eigenvalue is positive. Using this behavior as a guide, we can sketch in the qualitative behavior of solution points originating elsewhere in the plane. This qualitative behavior is shown in Figures 8.15(b)–(d). ▲

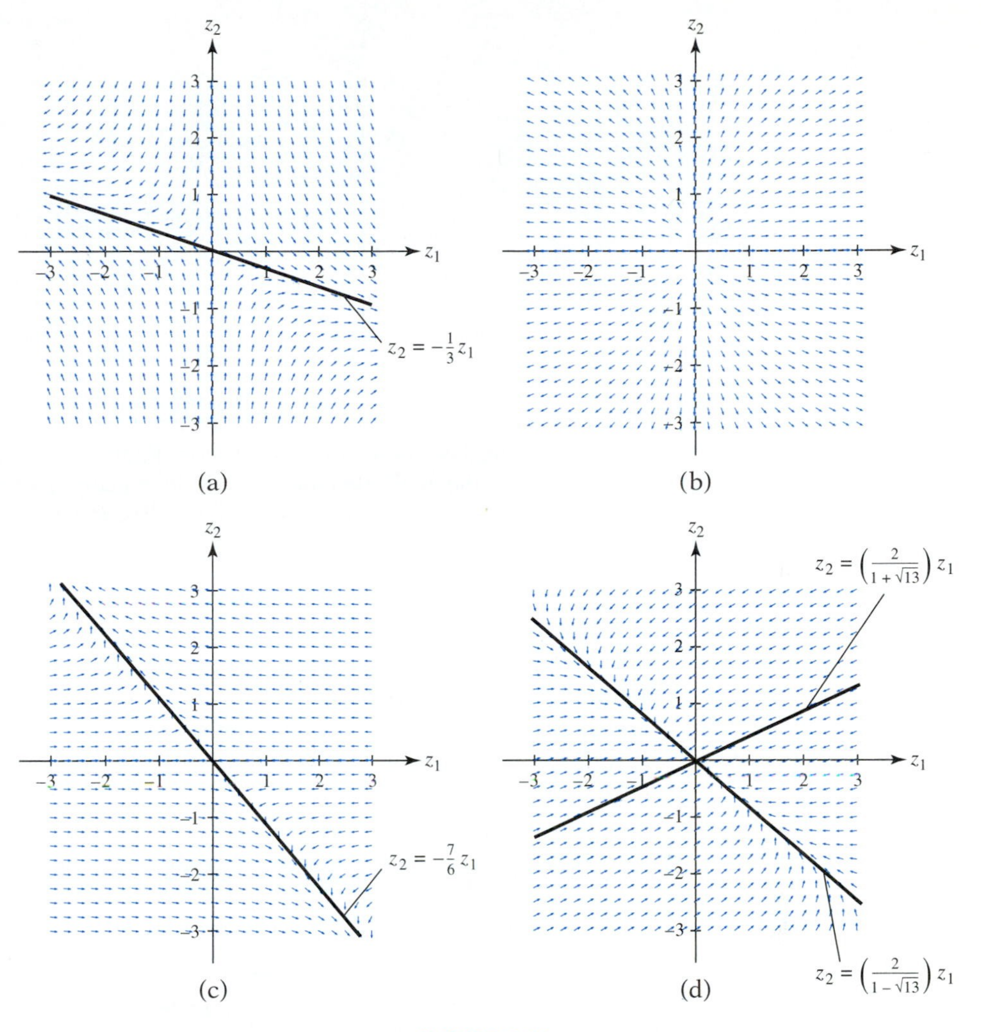

FIGURE 8.15

Direction fields for the various linearizations $\mathbf{z}' = A\mathbf{z}$ associated with the nonlinear system (2). Each $\mathbf{z}$-plane direction field corresponds to an equilibrium point of the nonlinear system. The equilibrium points are (a) $(0, 3/2)$, (b) $(0, 0)$, (c) $(2, 0)$, (d) $(1, 1)$.

The four $\mathbf{z}$-plane-phase portraits in Figure 8.15, when positioned at the corresponding $\mathbf{y}$-plane equilibrium points, provide local pictures that are complementary to and consistent with the large-scale overview developed in Section 8.2. This is illustrated in Figure 8.16 on the next page, where we have superimposed the local equilibrium pictures from Figure 8.15 upon Figure 8.6 from Section 8.2. Attention is restricted to the first quadrant, since the dependent variable $\mathbf{y}(t)$ has components that represent (nonnegative) populations.

Classifying Equilibrium Points

In Example 1, the coefficient matrix of the linearized system at each of the four equilibrium points had real, distinct eigenvalues. In two cases, the eigenvalues

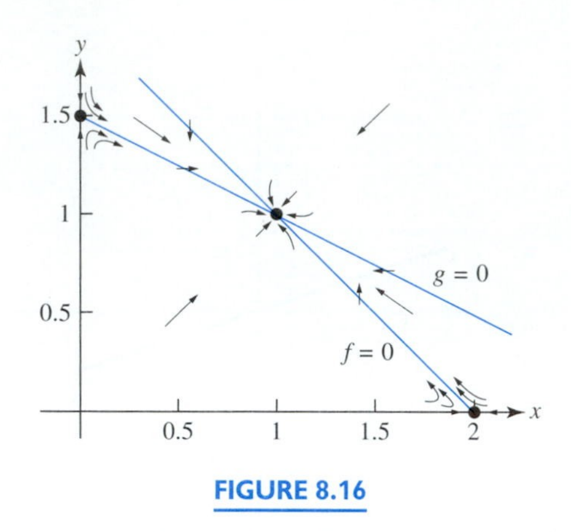

FIGURE 8.16

The local equilibrium pictures from Figure 8.15, superimposed upon the qualitative picture developed earlier in Section 8.2 (see Figure 8.6 of Section 8.2).

had the same sign; in the other two, the eigenvalues had opposite signs. If A is an invertible (2×2) matrix, other possibilities exist for the (nonzero) pair of eigenvalues. The eigenvalues might be real and repeated. They might be a complex conjugate pair with nonzero real part or they might be a purely imaginary complex conjugate pair. Table 8.2 enumerates the various possibilities and the names assigned to them.

TABLE 8.2

Classification of the Equilibrium Point at the Origin for $\mathbf{y}' = A\mathbf{y}$

Eigenvalues of A	Type of Equilibrium Point	Stability Characteristics of the Linear System
Real eigenvalues λ_1, λ_2, where		
$\lambda_1 \le \lambda_2 < 0$	Node	Asymptotically stable
$0 < \lambda_1 \le \lambda_2$	Node	Unstable
Real eigenvalues λ_1, λ_2 where		
$\lambda_1 < 0 < \lambda_2$	Saddle point	Unstable
Complex eigenvalues where		
$\lambda_{1,2} = \alpha \pm i\beta, \quad \alpha < 0$	Spiral point	Asymptotically stable
$\lambda_{1,2} = \alpha \pm i\beta, \quad \alpha > 0$	Spiral point	Unstable
Complex eigenvalues where		
$\lambda_{1,2} = \pm i\beta, \quad \beta \ne 0$	Center	Stable but not asymptotically stable

The "node" designation is often divided into two subcategories. If matrix A has two equal (real) eigenvalues and is a scalar multiple of the (2×2) identity matrix, the equilibrium point is called a **proper node**. In all other cases (when A has equal real eigenvalues but only one linearly independent eigenvector, or when A has unequal real eigenvalues of the same sign) the equilibrium point is called an **improper node**.

Figures 8.15 and 8.17–8.19 provide some examples of phase-plane behavior at nodes and saddle points. The following three examples illustrate typical behavior at a proper node, a spiral point, and a center.

EXAMPLE 2

Proper Node

Consider the linear system

$$\mathbf{y}' = \begin{bmatrix} \alpha & 0 \\ 0 & \alpha \end{bmatrix} \mathbf{y}, \qquad \alpha \neq 0.$$

The origin is a proper node since the eigenvalues are real and equal ($\lambda_1 = \lambda_2 = \alpha$), and the coefficient matrix is a nonzero multiple of the identity matrix. The phase-plane behavior of trajectories is easily recognized if we adopt polar coordinates. Let

$$x = r\cos\theta \quad \text{and} \quad y = r\sin\theta.$$

With this change of variables, the component equations transform into the differential equations for the polar variables

$$r' = \alpha r,$$
$$\theta' = 0.$$

As time increases, the solution points move on rays, since the polar angle θ remains constant. If $\alpha < 0$, solutions approach the origin and the origin is an asymptotically stable equilibrium point. If $\alpha > 0$, solution points move outward along the rays and the origin is an unstable equilibrium point. Note that the rays (the trajectories themselves) are independent of the value of α. The parameter α only governs how quickly solution points move inward or outward along the rays. Figure 8.17 depicts behavior for the case $\alpha > 0$. ▲

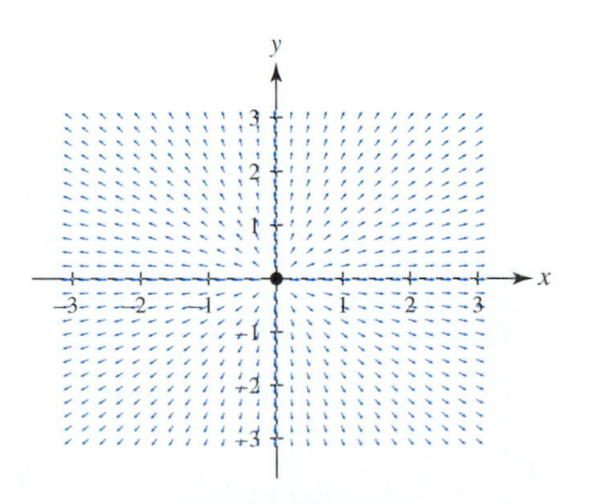

FIGURE 8.17

The origin is a proper node for the system in Example 2. Since $\alpha > 0$ in this example, the origin is an unstable equilibrium point.

EXAMPLE 3

Spiral Point

Consider the linear system

$$\mathbf{y}' = \begin{bmatrix} -1 & -1 \\ 1 & -1 \end{bmatrix} \mathbf{y}.$$

The eigenvalues of the coefficient matrix are the complex conjugate pair $\lambda_1 = -1 + i$, $\lambda_2 = -1 - i$. According to Table 8.2, the origin is an asymptotically stable spiral point. The behavior of solutions, as well as the reason for the terminology "spiral point," can be clearly seen when we change to polar coordinates. For this system, we obtain the following pair of differential equations for the polar variables:

$$r' = -r,$$
$$\theta' = 1.$$

Let the initial conditions be $r(0) = r_0$ and $\theta(0) = \theta_0$. Then the solutions are

$$r(t) = r_0 e^{-t}, \qquad \theta(t) = t + \theta_0.$$

Thus, as time increases, a solution point spirals inward toward the origin. Its distance from the origin decreases at an exponential rate while it moves counterclockwise about the origin. This behavior is shown in Figure 8.18. ▲

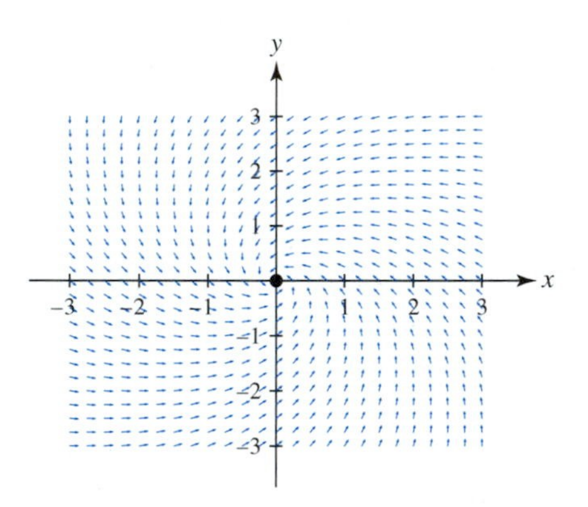

FIGURE 8.18

The origin is a spiral point for the system in Example 3. Since the eigenvalues of A have real part -1, the origin is asymptotically stable. A solution point follows a trajectory that spirals in toward the origin.

EXAMPLE 4

Center

Consider the linear system

$$\mathbf{y}' = \begin{bmatrix} -4 & 5 \\ -5 & 4 \end{bmatrix} \mathbf{y}.$$

The eigenvalues of the coefficient matrix are the purely imaginary complex conjugate pair $\lambda_1 = 3i$, $\lambda_2 = -3i$. According to Table 8.2, the origin is classified as a stable center. Phase-plane behavior is shown in Figure 8.19.

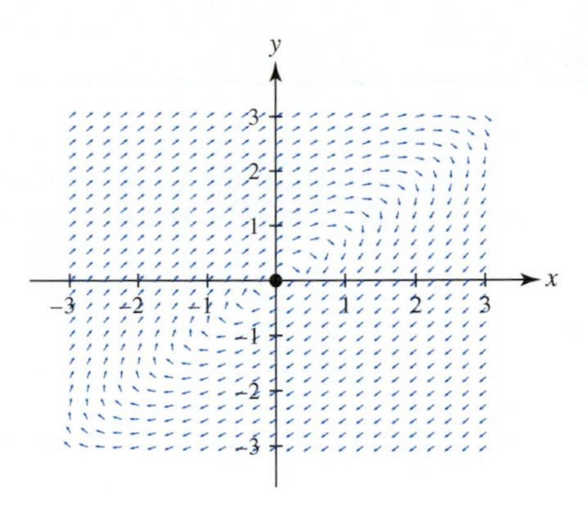

FIGURE 8.19

The origin is a center for the system in Example 4 since the eigenvalues of A are purely imaginary. The origin is a stable equilibrium point, but it is not asymptotically stable. The solution points follow elliptical trajectories about the origin.

One way to derive the equations for the elliptical trajectories in Figure 8.19 is to change to polar coordinates. For this linear system, the differential equations for the polar variables are

$$r' = -4r\cos 2\theta,$$
$$\theta' = -5 + 4\sin 2\theta.$$

Notice that θ is a decreasing function of t. Therefore, an inverse function exists and we can view r as a function of θ. Using the chain rule, we have

$$\frac{dr}{d\theta} = \frac{dr}{dt}\frac{dt}{d\theta} = -4r\cos 2\theta \frac{1}{-5 + 4\sin 2\theta},$$

or

$$\frac{dr}{d\theta} = \frac{4\cos 2\theta}{5 - 4\sin 2\theta} r.$$

This equation is a first order linear differential equation. Assuming an initial condition of $r = r_0$ when $\theta = 0$, we find the solution

$$r = \frac{r_0}{\sqrt{1 - 0.8\sin\theta}}.$$

Note that r is a periodic function of θ, with period π. Since θ is a decreasing function of t, the solution points (r, θ) move clockwise around the closed elliptical trajectories as shown in Figure 8.19. ▲

An alternative derivation of the trajectory equations is outlined in Exercise 31. This approach leads to equations in terms of the original x, y phase-plane variables.

EXERCISES

Exercises 1–5:

In each exercise, the eigenpairs of a (2×2) matrix A are given where both eigenvalues are real. Consider the phase-plane solution trajectories of the linear system $\mathbf{y}' = A\mathbf{y}$, where

$$\mathbf{y}(t) = \begin{bmatrix} x(t) \\ y(t) \end{bmatrix}.$$

(a) Use Table 8.2 to classify the type and stability characteristics of the equilibrium point at $\mathbf{y} = \mathbf{0}$.

(b) Sketch the two phase-plane lines defined by the eigenvectors. If an eigenvector is $\begin{bmatrix} u_1 \\ u_2 \end{bmatrix}$, the line of interest is $u_2 x - u_1 y = 0$. Solution trajectories originating on such a line stay on the line, and move toward the origin as time increases if the corresponding eigenvalue is negative and away from the origin if the eigenvalue is positive.

(c) Sketch appropriate direction field arrows on both lines. Use this information to sketch a representative trajectory in each of the four phase-plane regions having these lines as boundaries. Indicate the direction of motion of the solution point on each trajectory.

(d) Determine the coefficient matrix A. [Hint: Use the given information to construct a fundamental matrix $\Psi(t)$ and use the fact that $\Psi'(t) = A\Psi(t)$.]

1. $\lambda_1 = 2,\ \mathbf{x}_1 = \begin{bmatrix} 1 \\ 1 \end{bmatrix};\quad \lambda_2 = -1,\ \mathbf{x}_2 = \begin{bmatrix} 1 \\ -1 \end{bmatrix}$

2. $\lambda_1 = 1,\ \mathbf{x}_1 = \begin{bmatrix} 1 \\ 2 \end{bmatrix};\quad \lambda_2 = 2,\ \mathbf{x}_2 = \begin{bmatrix} 2 \\ -1 \end{bmatrix}$

3. $\lambda_1 = 2,\ \mathbf{x}_1 = \begin{bmatrix} 2 \\ 0 \end{bmatrix};\quad \lambda_2 = 1,\ \mathbf{x}_2 = \begin{bmatrix} 0 \\ 2 \end{bmatrix}$

4. $\lambda_1 = -2,\ \mathbf{x}_1 = \begin{bmatrix} 1 \\ 0 \end{bmatrix};\quad \lambda_2 = -1,\ \mathbf{x}_2 = \begin{bmatrix} 1 \\ 1 \end{bmatrix}$

5. $\lambda_1 = 1,\ \mathbf{x}_1 = \begin{bmatrix} 1 \\ 0 \end{bmatrix};\quad \lambda_2 = -1,\ \mathbf{x}_2 = \begin{bmatrix} 2 \\ 1 \end{bmatrix}$

Exercises 6–20:

In each exercise, consider the linear system $\mathbf{y}' = A\mathbf{y}$, where A is a constant invertible (2×2) matrix; therefore, $\mathbf{y} = \mathbf{0}$ is the unique (isolated) equilibrium point.

(a) Determine the eigenvalues of the coefficient matrix A.

(b) Use Table 8.2 to classify the type and stability characteristics of the equilibrium point at the phase-plane origin. If the equilibrium point is a node, designate it as either a proper node or an improper node.

6. $\mathbf{y}' = \begin{bmatrix} 1 & -6 \\ 1 & -4 \end{bmatrix}\mathbf{y}$

7. $\mathbf{y}' = \begin{bmatrix} 6 & -10 \\ 2 & -3 \end{bmatrix}\mathbf{y}$

8. $\mathbf{y}' = \begin{bmatrix} -6 & 14 \\ -2 & 5 \end{bmatrix}\mathbf{y}$

9. $\mathbf{y}' = \begin{bmatrix} 1 & 2 \\ -5 & -1 \end{bmatrix}\mathbf{y}$

10. $\mathbf{y}' = \begin{bmatrix} -1 & 1 \\ -1 & -1 \end{bmatrix}\mathbf{y}$

11. $\mathbf{y}' = \begin{bmatrix} 1 & -6 \\ 2 & -6 \end{bmatrix}\mathbf{y}$

12. $\mathbf{y}' = \begin{bmatrix} 2 & -3 \\ 3 & 2 \end{bmatrix}\mathbf{y}$ **13.** $\mathbf{y}' = \begin{bmatrix} -2 & -4 \\ 5 & 2 \end{bmatrix}\mathbf{y}$ **14.** $\mathbf{y}' = \begin{bmatrix} 7 & -24 \\ 2 & -7 \end{bmatrix}\mathbf{y}$

15. $\mathbf{y}' = \begin{bmatrix} -1 & 8 \\ -1 & 5 \end{bmatrix}\mathbf{y}$ **16.** $\mathbf{y}' = \begin{bmatrix} -2 & 1 \\ -1 & -2 \end{bmatrix}\mathbf{y}$ **17.** $\mathbf{y}' = \begin{bmatrix} 2 & 4 \\ -4 & -6 \end{bmatrix}\mathbf{y}$

18. $\mathbf{y}' = \begin{bmatrix} 3 & 0 \\ 0 & 3 \end{bmatrix}\mathbf{y}$ **19.** $\mathbf{y}' = \begin{bmatrix} 1 & 2 \\ -8 & 1 \end{bmatrix}\mathbf{y}$ **20.** $\mathbf{y}' = \begin{bmatrix} -1 & -2 \\ 2 & 3 \end{bmatrix}\mathbf{y}$

21. Consider the linear system $\mathbf{y}' = A\mathbf{y}$. Four direction fields are shown. Determine which of the four coefficient matrices listed corresponds to each of the direction fields shown.

(a) $A_1 = \begin{bmatrix} -2 & 1 \\ 1 & -2 \end{bmatrix}$ (b) $A_2 = \begin{bmatrix} 1 & 2 \\ -2 & -1 \end{bmatrix}$

(c) $A_3 = \begin{bmatrix} 2 & 1 \\ -1 & -2 \end{bmatrix}$ (d) $A_4 = \begin{bmatrix} 1 & 2 \\ -2 & 1 \end{bmatrix}$

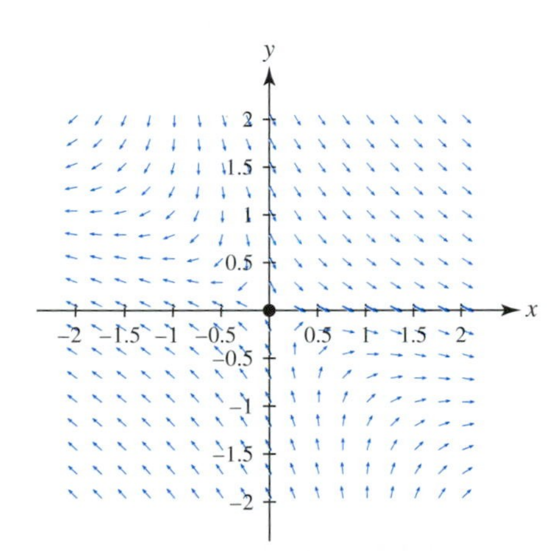

Direction Field 1

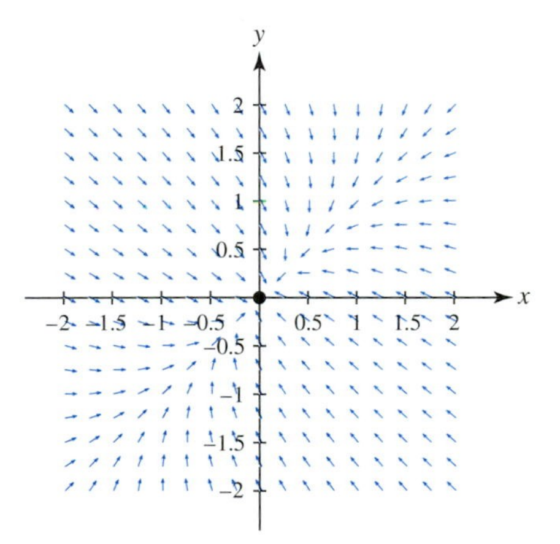

Direction Field 2

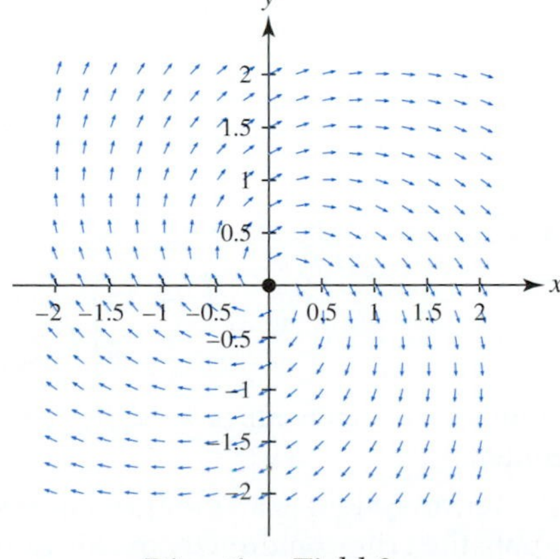

Direction Field 3

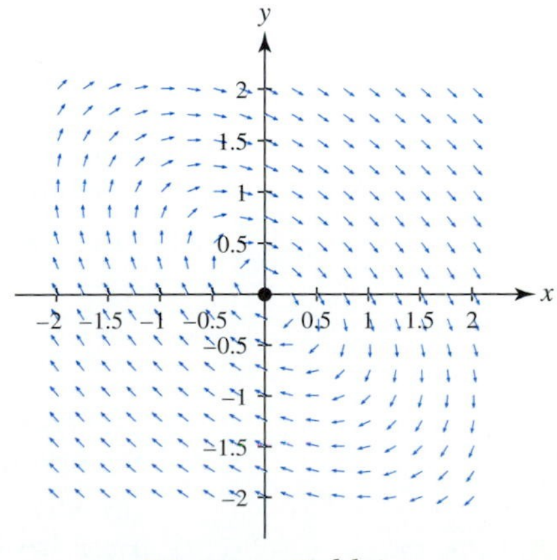

Direction Field 4

Figure for Exercise 21

Exercises 22–25:

Use the information given about the nature of the equilibrium point at the origin to determine the value or range of permissible values for the unspecified entry in the coefficient matrix.

22. The origin is a center for the linear system $\mathbf{y}' = \begin{bmatrix} 2 & 3 \\ -3 & \alpha \end{bmatrix} \mathbf{y}$; determine α.

23. Given $\mathbf{y}' = \begin{bmatrix} -4 & \alpha \\ -2 & 2 \end{bmatrix} \mathbf{y}$, for what values of α (if any) can the origin be an asymptotically stable spiral point?

24. The origin is an asymptotically stable proper node of $\mathbf{y}' = \begin{bmatrix} -2 & 0 \\ \alpha & -2 \end{bmatrix} \mathbf{y}$. What are the value(s) of α?

25. Given $\mathbf{y}' = \begin{bmatrix} 4 & -2 \\ \alpha & -4 \end{bmatrix} \mathbf{y}$, for what values of α (if any) can the origin be an (unstable) saddle point?

Exercises 26–29:

Locate the equilibrium point of the given nonhomogeneous linear system $\mathbf{y}' = A\mathbf{y} + \mathbf{g}_0$. [Hint: Introduce the change of dependent variable $\mathbf{z}(t) = \mathbf{y}(t) - \mathbf{y}_0$, where $\mathbf{y}_0$ is chosen so that the equation can be rewritten as $\mathbf{z}' = A\mathbf{z}$.] Use Table 8.2 to classify the type and stability characteristics of the equilibrium point.

26. $\mathbf{y}' = \begin{bmatrix} 1 & 4 \\ -1 & 1 \end{bmatrix} \mathbf{y} + \begin{bmatrix} 3 \\ 2 \end{bmatrix}$

27. $\mathbf{y}' = \begin{bmatrix} 6 & 5 \\ -7 & -6 \end{bmatrix} \mathbf{y} + \begin{bmatrix} 4 \\ -6 \end{bmatrix}$

28. $x' = 5x - 14y + 2$

$y' = 3x - 8y + 1$

29. $x' = -x + 2$

$y' = 2y - 4$

30. Let

$$A = \begin{bmatrix} a_{11} & a_{12} \\ a_{21} & a_{22} \end{bmatrix}$$

be a real invertible matrix, and consider the system $\mathbf{y}' = A\mathbf{y}$.

(a) What conditions must the matrix entries a_{ij} satisfy to make the equilibrium point $\mathbf{y}_e = \mathbf{0}$ a center?

(b) Assume that the equilibrium point at the origin is a center. Show that the system $\mathbf{y}' = A\mathbf{y}$ is a Hamiltonian system.

(c) Is the converse of the statement in part (b) true? In other words, if the system $\mathbf{y}' = A\mathbf{y}$ is a Hamiltonian system, does it necessarily follow that $\mathbf{y}_e = \mathbf{0}$ is a center? Explain.

31. Consider the linear system of Example 4,

$$\mathbf{y}' = \begin{bmatrix} -4 & 5 \\ -5 & 4 \end{bmatrix} \mathbf{y}.$$

The coefficient matrix has eigenvalues $\lambda_1 = 3i$, $\lambda_2 = -3i$; the equilibrium point at the origin is a center.

(a) Show that the linear system is a Hamiltonian system. Either use the results of Exercise 30 or apply the criterion directly to this example.

(b) Derive the conservation law for this system. The result, $(5/2)x^2 - 4xy + (5/2)y^2 = C > 0$, defines a family of ellipses. These ellipses are the trajectories on which the solution point moves as time changes.

(c) Plot the ellipses found in part (b) for $C = 1/4$, $1/2$, and 1. Indicate the direction in which the solution point moves on these ellipses.

Exercises 32–34:

A linear system is given in each exercise.

(a) Determine the eigenvalues of the coefficient matrix A.

(b) Use Table 8.2 to classify the type and stability characteristics of the equilibrium point at $\mathbf{y} = \mathbf{0}$.

(c) The given linear system is a Hamiltonian system. Derive the conservation law for this system.

32. $\mathbf{y}' = \begin{bmatrix} -2 & 1 \\ 5 & 2 \end{bmatrix}\mathbf{y}$

33. $x' = x + 3y$
$y' = -3x - y$

34. $\mathbf{y}' = \begin{bmatrix} 2 & 1 \\ 0 & -2 \end{bmatrix}\mathbf{y}$

8.7 Predator-Prey Population Models

By way of introduction, we pose a question. This question involves the familiar problem of a predatory population being introduced into a colony, either by accident or design. If the predator is undesirable, our goal may be to remove it from the colony. If the predator is desirable, our goal may be to establish a coexisting ecological balance between the predators and their prey. Which scenarios lead to predator eradication and which scenarios lead to predator-prey coexistence? What are the factors responsible for the desired outcome?

Mathematical Modeling

We develop a mathematical model of two-species predator-prey interaction to gain insight into answering the question posed above.[4] Let $P_1(t)$ and $P_2(t)$ represent the populations of predators and prey (respectively) in a colony at time t. Both populations change with time due to births, deaths, and harvesting. A "conservation of population" principle of the type discussed in Section 2.4 leads to differential equations having the following general structure,

$$\begin{aligned}\frac{dP_1}{dt} &= R_1P_1 - \mu_1P_1,\\ \frac{dP_2}{dt} &= R_2P_2 - \mu_2P_2.\end{aligned} \tag{1}$$

In equation (1), the term R_j represents the net birth rate per unit population of the jth species. We assume that R_1 and R_2 are functions of the populations, P_1 and P_2 (but not explicit functions of time t). The nonnegative constants μ_1 and μ_2 represent harvesting rates per unit population. Applying the ideas

[4]Important early work in developing and applying such models was done by Volterra. Vito Volterra (1860–1940) was an Italian mathematician and scientist noted for his work on functional calculus, partial differential equations, integral equations and mathematical biology. During his career, he held distinguished positions at the universities of Pisa, Turin, and Rome. In 1931, he was forced to leave the University of Rome after refusing to take an oath of allegiance to the Fascist government. The following year, he left Italy and spent the rest of his life abroad.

underlying the logistic population model developed in Section 3.5, we assume the rate functions R_1 and R_2 have the form

$$\begin{aligned} R_1 &= r_1(-1 - \alpha_1 P_1 + \beta_1 P_2), \\ R_2 &= r_2(1 - \beta_2 P_1 - \alpha_2 P_2), \end{aligned} \tag{2}$$

where the parameters r_j, α_j, and β_j are nonnegative constants, $j = 1, 2$.

What are we actually assuming in (2)? In the absence of prey for food (that is if $\beta_1 = 0$), the predator rate function would be $R_1 = -r_1(1 + \alpha_1 P_1) < 0$ and the predator population would continually decrease. The $\beta_1 P_2$ term embodies the beneficial aspects of the prey food supply upon the predator growth rate. The $-\alpha_1 P_1$ term allows for competition among the predators for the available food.

Consider now the prey rate function R_2. In the absence of predators and limitations upon the prey's food supply (that is, if $\beta_2 = 0$ and $\alpha_2 = 0$), the rate function would be $R_2 = r_2 > 0$. In that case, the prey population would grow exponentially whenever $r_2 > \mu_2$. The terms $-\beta_2 P_1$ and $-\alpha_2 P_2$ account for the negative effects of predation and limits on the prey's food supply, respectively. The predator-prey equations we use as our basic model are therefore

$$\begin{aligned} \frac{dP_1}{dt} &= r_1(-1 - \alpha_1 P_1 + \beta_1 P_2)P_1 - \mu_1 P_1, \\ \frac{dP_2}{dt} &= r_2(1 - \beta_2 P_1 - \alpha_2 P_2)P_2 - \mu_2 P_2. \end{aligned} \tag{3}$$

Managing a Predator Population

We now take up the question posed at the beginning of this section: How do we manage a predator population that has been introduced into a colony, whether by accident or design?

We assume the colony has resource limitations that exert a constraining influence on each of the predator and prey populations. We allow for the possibility of harvesting the predator population but not the prey population. Under these assumptions, system (3) becomes

$$\begin{aligned} \frac{dP_1}{dt} &= r_1\left(-1 - \frac{\mu}{r_1} - \alpha_1 P_1 + \beta_1 P_2\right) P_1, \\ \frac{dP_2}{dt} &= r_2\left(1 - \beta_2 P_1 - \alpha_2 P_2\right) P_2, \end{aligned} \tag{4}$$

where all the constants on the right-hand side of (4) are assumed positive with the possible exception of the harvesting rate μ, which we may allow to be zero.

From an ecological point of view, we want to know what combination of harvesting strategies and environmental factors will cause the predator population to

(a) die out, or

(b) achieve a coexisting balance with the prey population as time evolves.

Rephrasing in mathematical terms, we want to discover which relations among the constants in (4) will cause all solutions to:

(a) converge to an equilibrium value $(0, P_{2e})$, where $P_{2e} > 0$, or
(b) converge to an equilibrium value (P_{1e}, P_{2e}), where $P_{1e} > 0$, $P_{2e} > 0$.

Autonomous system (4) has at most three equilibrium points in the first quadrant of the phase plane,

$$\mathbf{P}_e^{(1)} = \begin{bmatrix} 0 \\ 0 \end{bmatrix}, \qquad \mathbf{P}_e^{(2)} = \begin{bmatrix} 0 \\ \dfrac{1}{\alpha_2} \end{bmatrix}, \qquad \mathbf{P}_e^{(3)} = \begin{bmatrix} \dfrac{-\alpha_2\left(1 + \dfrac{\mu}{r_1}\right) + \beta_1}{\alpha_1\alpha_2 + \beta_1\beta_2} \\[2ex] \dfrac{\beta_2\left(1 + \dfrac{\mu}{r_1}\right) + \alpha_1}{\alpha_1\alpha_2 + \beta_1\beta_2} \end{bmatrix}.$$

The equilibrium point $\mathbf{P}_e^{(1)}$ where $P_{1e} = P_{2e} = 0$ corresponds to neither species being present. The equilibrium point $\mathbf{P}_e^{(2)}$, where

$$P_{1e} = 0, \qquad P_{2e} = \frac{1}{\alpha_2}, \tag{5}$$

corresponds to a complete absence of predators. The third equilibrium point, $\mathbf{P}_e^{(3)}$, is found by solving the system of equations

$$\begin{aligned} -1 - \frac{\mu}{r_1} - \alpha_1 P_{1e} + \beta_1 P_{2e} &= 0, \\ 1 - \beta_2 P_{1e} - \alpha_2 P_{2e} &= 0. \end{aligned}$$

This equilibrium point is given by

$$P_{1e} = \frac{-\alpha_2\left(1 + \dfrac{\mu}{r_1}\right) + \beta_1}{\alpha_1\alpha_2 + \beta_1\beta_2}, \qquad P_{2e} = \frac{\beta_2\left(1 + \dfrac{\mu}{r_1}\right) + \alpha_1}{\alpha_1\alpha_2 + \beta_1\beta_2}. \tag{6}$$

By definition, the two species coexist when both P_{1e} and P_{2e} are positive. Thus, model (4) predicts that both species can coexist in equilibrium only if

$$\beta_1 > \alpha_2\left(1 + \frac{\mu}{r_1}\right). \tag{7}$$

If inequality (7) does not hold, then the two species cannot coexist in equilibrium and the only nontrivial equilibrium solution in the first quadrant is (5) wherein predators are absent.

Suppose our goal is to eradicate the predators. We see from (7) that we can eliminate the possibility of equilibrium predator-prey coexistence by sufficiently increasing $\alpha_2(1 + (\mu/r_1))$ and/or by decreasing β_1. Does this make sense? Increasing μ corresponds to increasing the harvesting rate of predators, while decreasing β_1 corresponds to somehow reducing the beneficial effects of the prey as food for the predators. It seems reasonable that either of these two strategies would be harmful to the predators.

What about increasing α_2, however? Recall that α_2 is the parameter modeling the constraining effects of the available colony resources upon the prey population. In the absence of predators, the equilibrium prey population $P_{2e} = 1/\alpha_2$ decreases as α_2 increases. Is it reasonable to conclude that we can adversely impact the predator population by indirectly constraining its food supply? This question will be examined further.

Table 8.3 summarizes the information that we can deduce from linearizing system (4). In either case (that is, when equilibrium coexistence is possible or when it is impossible), the origin is an unstable saddle point. If equilibrium coexistence is possible, then the equilibrium point $(0, 1/\alpha_2)$ is an unstable saddle point. If coexistence is impossible, however, this equilibrium point becomes an asymptotically stable node. In the case where equilibrium coexistence is possible, the equilibrium point with coordinate values (5) is either an asymptotically stable spiral point or a node.

The third equilibrium point in Table 8.3, corresponding to predators and prey coexisting, requires [as noted in equation (7)] that the numerator of P_{1e} is positive; that is,

$$\beta_1 > \alpha_2 \left(1 + \frac{\mu}{r_1}\right).$$

For this equilibrium point, the matrix A in Table 8.3 is given by

$$A = \begin{bmatrix} -\dfrac{\alpha_1 r_1 \left[\beta_1 - \alpha_2 \left(1 + \dfrac{\mu}{r_1}\right)\right]}{\alpha_1\alpha_2 + \beta_1\beta_2} & \dfrac{\beta_1 r_1 \left[\beta_1 - \alpha_2 \left(1 + \dfrac{\mu}{r_1}\right)\right]}{\alpha_1\alpha_2 + \beta_1\beta_2} \\ -\dfrac{\beta_2 r_2 \left[\alpha_1 + \beta_2 \left(1 + \dfrac{\mu}{r_1}\right)\right]}{\alpha_1\alpha_2 + \beta_1\beta_2} & -\dfrac{\alpha_2 r_2 \left[\alpha_1 + \beta_2 \left(1 + \dfrac{\mu}{r_1}\right)\right]}{\alpha_1\alpha_2 + \beta_1\beta_2} \end{bmatrix}. \tag{8}$$

The eigenvalues of A are

$$\lambda_{1,2} = \frac{(a_{11} + a_{22}) \pm \sqrt{(a_{11} - a_{22})^2 + 4a_{12}a_{21}}}{2}. \tag{9}$$

In (9), a_{ij} denotes the ij-entry of the matrix A.

The equilibrium-point analysis of system (4), summarized in Table 8.3, suggests that the desired predator-management objective can be achieved by appropriately adjusting the parameters. Figure 8.20 on page 530 reinforces this conclusion while further exploring the role of the parameter α_2. Recall that α_2 is a positive parameter whose size reflects the degree to which the colony's resources constrain prey growth. We want to focus on the role of this parameter. For simplicity, the parameters $r_1, r_2, \alpha_1, \beta_1, \beta_2$, and μ were all assigned the value 1, leading to the system

$$\begin{aligned} \frac{dP_1}{dt} &= (-2 - P_1 + P_2)P_1, \\ \frac{dP_2}{dt} &= (1 - P_1 - \alpha_2 P_2)P_2. \end{aligned} \tag{10}$$

The three phase-plane plots in Figure 8.20 correspond to different values of α_2. In Figure 8.20(a), $\alpha_2 = 0$. In this case, predator-prey coexistence is possible. All solution trajectories having both species initially present spiral in toward the asymptotically stable equilibrium point $(1, 3)$. In Figure 8.20(b), α_2 has been increased to $\alpha_2 = 9/20$. In this case the equilibrium point of the linearized system is a stable node. Here again, all solution trajectories of the nonlinear system having both species initially present approach the asymptotically stable equilibrium point at $(2/29, 60/29)$. Lastly, see Figure 8.20(c), when $\alpha_2 = 1$ coexistence

TABLE 8.3

Equilibrium Point	Linearized System $\mathbf{z}' = A\mathbf{z}$ $[\mathbf{z}(t) = \mathbf{P}(t) - \mathbf{P}_e]$	Eigenvalues	Stability Properties of System (4)
$P_{1e} = 0$ $P_{2e} = 0$	$A = \begin{bmatrix} -r_1 - \mu & 0 \\ 0 & r_2 \end{bmatrix}$	$\lambda_1 = -r_1 - \mu < 0$ $\lambda_2 = r_2 > 0$	Unstable
$P_{1e} = 0$ $P_{2e} = \dfrac{1}{\alpha_2}$	$A = \begin{bmatrix} r_1\left[-\left(1 + \dfrac{\mu}{r_1}\right) + \dfrac{\beta_1}{\alpha_2}\right] & 0 \\ -\dfrac{r_2\beta_2}{\alpha_2} & -r_2 \end{bmatrix}$	$\lambda_1 = r_1\left[-\left(1 + \dfrac{\mu}{r_1}\right) + \dfrac{\beta_1}{\alpha_2}\right]$ $\lambda_2 = -r_2 < 0$	Unstable if $\alpha_2\left(1 + \dfrac{\mu}{r_1}\right) < \beta_1$ Asymptotically stable if $\alpha_2\left(1 + \dfrac{\mu}{r_1}\right) > \beta_1$
$P_{1e} = \dfrac{\beta_1 - \alpha_2\left(1 + \dfrac{\mu}{r_1}\right)}{\alpha_1\alpha_2 + \beta_1\beta_2}$ $P_{2e} = \dfrac{\alpha_1 + \beta_2\left(1 + \dfrac{\mu}{r_1}\right)}{\alpha_1\alpha_2 + \beta_1\beta_2}$	A is given in (8) and we assume $\beta_1 > \alpha_2 \dfrac{1 + \mu}{r_1}$	λ_1, λ_2 are given in (9)	Asymptotically stable

is not possible. All solution trajectories having both species initially present approach the asymptotically stable equilibrium point $(0, 1)$ as $t \to \infty$. In this last case, the predator population tends toward extinction as time progresses. These direction field observations support our previous conclusions.

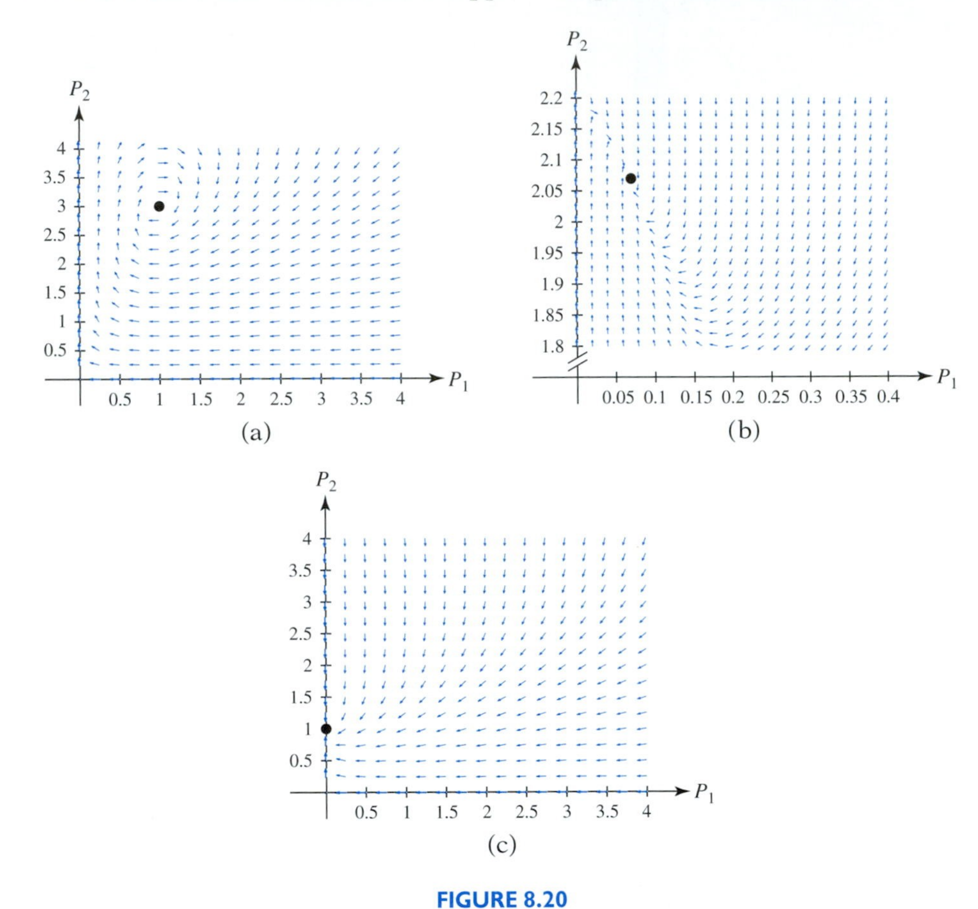

FIGURE 8.20

Portions of the direction field for the predator-prey equations (10), with different choices of α_2. (a) $\alpha_2 = 0$, (b) $\alpha_2 = 9/20$, (c) $\alpha_2 = 1$.

On one hand, the parametric study illustrated in Figure 8.20 indicates that our interpretation of the model's behavior seems correct. On the other hand, a model such as (4) is at best a gross simplification of reality. The trade-off in modeling is always one of reducing a problem to its essential features without "throwing away" reality. In particular, when model predictions seem counter-intuitive, we need to proceed with a healthy skepticism—both scrutinizing and refining the model to gain further confidence and insight.

EXERCISES

Exercises 1–4:

Assume the given autonomous system models the population dynamics of two species, x and y, within a colony.

(a) For each of the two species, answer the following questions.

(i) In the absence of the other species, does the remaining species' population continuously grow, decline toward extinction, or approach a nonzero equilibrium value as time evolves?

(ii) Is the presence of the other species in the colony beneficial, harmful, or a matter of indifference?

(b) Determine all equilibrium points lying in the first quadrant of the phase plane (including any lying on the coordinate axes).

(c) The given system is an almost linear system at the equilibrium point $(x, y) = (0, 0)$. Determine the stability properties of the system at $(0, 0)$.

1. $x' = x - x^2 - xy$
$y' = y - 3y^2 - \frac{1}{2}xy$

2. $x' = -x - x^2$
$y' = -y + xy$

3. $x' = x - x^2 - xy$
$y' = -y - y^2 + xy$

4. $x' = x - x^2 + xy$
$y' = y - y^2 + xy$

5. A scientist adopted the following mathematical model for a colony containing two species, x and y, that she is studying,

$$x' = r_1(1 + \alpha_1 x + \beta_1 y)x,$$
$$y' = r_2(1 + \beta_2 x + \alpha_2 y)y.$$

The following information is known:

(i) If only species x is present in the colony, any initial amount will vary with time as shown in figure (a). If only species y is present, any initial amount will vary as shown in figure (b).

(ii) If both species are initially present, $(x_e, y_e) = (2, 3)$ is an equilibrium point.

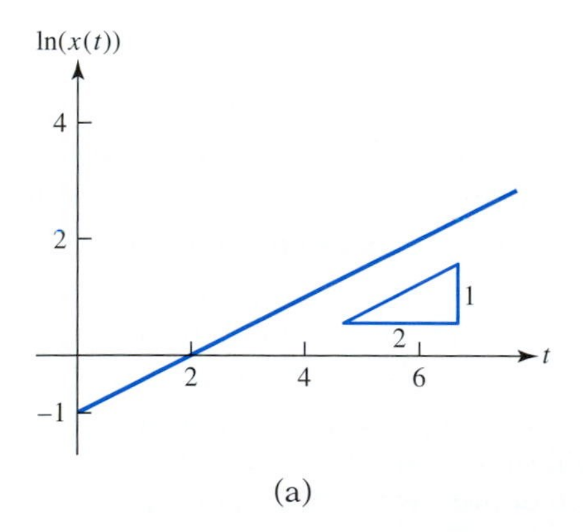

(a)

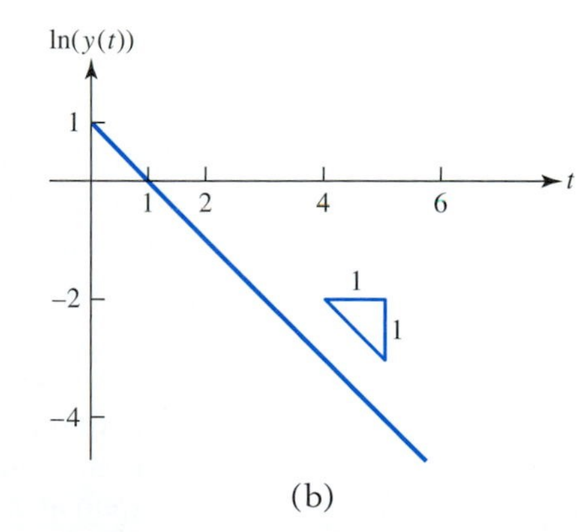

(b)

Figure for Exercise 5

(a) Determine the six constants $r_1, \alpha_1, \beta_1, r_2, \alpha_2, \beta_2$.

(b) How do the two populations relate to each other? Is population x beneficial, harmful, or indifferent to population y? Is population y beneficial, harmful, or indifferent to population x?

Exercises 6–7:

These exercises explore the question "When one of two species in a colony is desirable and the other is undesirable, is it better to use resources to nurture the growth of the desirable species or to harvest the undesirable one?"

Two Competing Species Let $x(t)$ and $y(t)$ represent the populations of two competing species, with $x(t)$ the desirable species. Assume that if resources are invested in promoting the growth of the desirable species, the population dynamics are given by

$$\begin{aligned} x' &= r(1 - \alpha x - \beta y)x + \mu x, \\ y' &= r(1 - \alpha y - \beta x)y. \end{aligned} \tag{11}$$

If resources are invested in harvesting the undesirable species, the dynamics are

$$\begin{aligned} x' &= r(1 - \alpha x - \beta y)x, \\ y' &= r(1 - \alpha y - \beta x)y - \mu y. \end{aligned} \tag{12}$$

In (12), r, α, β, and μ are positive constants. For simplicity, we assume the same parameter values for both species. For definiteness, assume that $\alpha > \beta$.

6. Consider system (11), which describes the strategy where resources are invested into nurturing the desirable species.

(a) Determine the four equilibrium points for the system.

(b) Show it is possible, by investing sufficient resources (that is, by making μ large enough), to prevent equilibrium coexistence of the two species.

(c) Assume that μ is large enough to preclude equilibrium coexistence of the two species. Compute the linearized system at each of the three physically relevant equilibrium points. Determine the stability characteristics of the linearized system at each of these equilibrium points.

(d) System (11) can be shown to be an almost linear system at each of the equilibrium points. Use this fact and the results of part (c) to infer the stability properties of system (11) at each of the three equilibrium points of interest.

(e) Sketch the direction field. Either use the information obtained from the nullclines and the general orientation of direction field arrows in different regions of the first quadrant, or adopt numerical parameter values consistent with the assumptions made and plot the direction field using computer software. Will a sufficiently aggressive nurturing of species x ultimately drive undesirable species y to extinction? If so, what is the limiting population of species x?

7. Consider system (12), which describes the strategy where resources are invested in harvesting the undesirable species. Again assume that $\alpha > \beta > 0$.

(a) Determine the four equilibrium points for the system.

(b) Show that it is possible, by investing sufficient resources (that is, by making μ large enough), to prevent equilibrium coexistence of the two species. In fact, if $\mu > r$, show there are only two physically relevant equilibrium points.

(c) Assume $\mu > r$. Compute the linearized system at each of the two physically relevant equilibrium points. Determine the stability characteristics of the linearized system at each of these equilibrium points.

(d) System (12) can be shown to be an almost linear system at each of the equilibrium points. Use this fact and the results of part (c) to infer the stability properties of system (12) at each of the three equilibrium points of interest.

(e) Sketch the direction field. Either use the information obtained from the nullclines and the general orientation of direction field arrows in different regions of the first quadrant or adopt numerical parameter values consistent with the assumptions made and plot the direction field using computer software. Will sufficiently aggressive harvesting of species y ultimately drive undesirable species y to extinction? If so, what is the limiting population of species x?

8. Compare the conclusions reached in Exercises 6 and 7. Assume we have sufficient resources to implement either strategy. What strategy will result in the larger number of desirable species x eventually being present: Promoting the desirable species or harvesting the undesirable one? Could this answer have been anticipated? In other words, suppose we assume that both strategies will lead to the eventual extinction of species y, with phase-plane trajectories approaching the equilibrium point where $x_e > 0$, $y_e = 0$. Will comparing the resulting one-species equilibrium values for x_e provide the answer?

9. Three species, designated as x, y, and z, inhabit a colony. Species x and y are two mutually competitive varieties of prey, while z is a predator that depends on x and y for sustenance. In the absence of the other two species, both x and y are known to evolve toward a nonzero equilibrium value as time increases. Species z, however, decreases exponentially toward extinction when both species of prey are absent. Assume that a mathematical model having the following structure is adopted to describe the population dynamics:

$$x' = \pm a_1 x \pm b_1 x^2 \pm c_1 xy \pm d_1 xz$$
$$y' = \pm a_2 y \pm b_2 y^2 \pm c_2 xy \pm d_2 yz$$
$$z' = \pm a_3 z \pm c_3 xz \pm d_3 yz$$

If we want the constants $a_1, b_1, c_1, a_2, \ldots, d_3$ to be nonnegative, use the information given to select the correct (plus or minus) sign in the model.

Exercises 10–11:

Infectious Disease Dynamics Consider a colony in which an infectious disease (such as the common cold) is present. The population consists of three "species" of individuals. Let s represent the susceptibles, healthy individuals capable of contracting the illness. Let i denote the infected individuals, and let r represent those who have recovered from the illness. For now, assume that those who have recovered from the illness are not permanently immunized but can become susceptible again. We also assume that the rate of infection is proportional to si, the product of the susceptible and infected populations. The differential equations describing the population dynamics are obtained by applying a conservation of population principle; the rate of change of a species equals the rate at which the species gains members minus the rate at which members are lost. We obtain the model

$$\begin{aligned} s' &= -\alpha si + \gamma r, \\ i' &= \alpha si - \beta i, \\ r' &= \beta i - \gamma r, \end{aligned} \tag{13}$$

where α, β, and γ are positive constants.

10. (a) Show that the system of equations (13) describes a population whose size remains constant in time. In particular, show that $s(t) + i(t) + r(t) = N$, a constant.

(b) How do the equations have to be modified if those who recover from the disease are, in fact, permanently immunized? Is $s(t) + i(t) + r(t)$ constant in this case? Support your answer by using the modified system of differential equations that you have formulated.

(c) Suppose those who recover from the disease are permanently immunized but that the disease is a serious one and some of the infected individuals perish. How does the system of equations you formulated in part (b) have to be further modified? Is $s(t) + i(t) + r(t)$ constant in this case? Show how your system of differential equations supports your assertion.

11. (a) Consider system (13). Use the fact that $s(t) + i(t) + r(t) = N$ to obtain a reduced system of two differential equations for the two dependent variables $s(t)$ and $i(t)$.

(b) For simplicity, set $\alpha = \beta = \gamma = 1$ and $N = 9$. Determine the equilibrium points of the reduced two-dimensional system.

(c) Determine the linearized system at each of the equilibrium points found in part (b). Use Table 8.2 to analyze the stability characteristics of each of these linearized systems.

(d) Assume that the nonlinear system is an almost linear system at each of the equilibrium points found in part (b). What are the stability characteristics of the nonlinear system at these points?

Extended Problem: Introduction of an Infectious Disease into a Predator-Prey Ecosystem

Problem Formulation

In this chapter we have discussed the following model of a predator-prey ecosystem,

$$\begin{aligned} x' &= -ax + bxy, \\ y' &= cy - dxy. \end{aligned} \tag{1}$$

In (1), $x(t)$ and $y(t)$ represent the populations of predator and prey, respectively, at time t. The terms a, b, c, and d are positive constants. The terms bxy and $-dxy$ account for the beneficial and detrimental impacts of predation upon the predator and prey populations.

We also discussed a model for the dynamics of an infectious disease within a population (see Exercise 10 of Section 8.7). In the present discussion, we will assume that infected individuals, when they have recovered, immediately become susceptible again. Therefore, we need not consider a "recovered" population as a separate entity. With this assumption the infectious disease model considered in Section 8.7 simplifies to

$$\begin{aligned} s' &= -\alpha si + \beta i, \\ i' &= \alpha si - \beta i. \end{aligned} \tag{2}$$

In (2), $s(t)$ and $i(t)$ are the populations of susceptible and infected individuals at time t, while α and β are positive constants. In the model (2), the total population, $s(t) + i(t)$, remains constant in time.

We will combine the ideas embodied in these two problems to model a situation where an infectious disease has been introduced into a predator-prey colony. The goal is to determine the behavior of the colony.

Assume the following additional facts.

(i) The disease is benign to the prey; that is, the prey are "carriers." The relative birth rate for infected prey remains the same as that for healthy, susceptible prey.

(ii) The disease is debilitating and ultimately fatal for predators. Once a predator is infected, it can basically be assumed to be deceased. Therefore, we need only consider one population of predators, those that are susceptible.

(iii) The disease is spread among the prey by contact. We assume the rate of infection to be proportional to the product of susceptible and infected populations.

(iv) The predators make no distinction between susceptible and infected individuals in their consumption of prey.

(v) The predators contract the disease only by consumption of prey. The rate of predator infection is proportional to the product of infected prey and susceptible predators.

We obtain a model by dividing the prey population into two subgroups, healthy, susceptible prey and infected prey. Let $s(t)$ and $i(t)$ represent the populations of susceptible and infected prey, respectively, at time t. Let $x(t)$ denote the population of healthy, susceptible predators. The autonomous system that will model the ecosystem is

$$\begin{aligned} x' &= -ax + bxs - \delta xi, \\ s' &= cs - dsx - \alpha si + \beta i, \\ i' &= ci - dxi + \alpha si - \beta i. \end{aligned} \tag{3}$$

where the constants $a, b, c, d, \alpha, \beta$, and δ are all positive.

PROBLEM:

1. Explain the modeling role played by each term in the three differential equations. (For example, the term $-ax$ accounts for the fact that, in the absence of prey, the predator population would decrease and exponentially approach extinction.)
2. One question of interest is whether the model has a nontrivial equilibrium solution in the first octant of x, s, i-space. Is it possible to have an equilibrium state where nonzero populations of predators, susceptible prey, and infected prey will remain constant in time? As usual, we assume that the variables x, s, and i have been scaled so that one unit of population corresponds to a large number of actual individuals. Assume the following values for the constants in equation (3),

$$a = 1, \quad b = 1, \quad c = 1, \quad d = 1, \quad \alpha = \frac{1}{2}, \quad \beta = 1, \quad \delta = 1.$$

   With this, equation (3) becomes

$$\begin{aligned} x' &= -x + xs - xi, \\ s' &= s - sx - \frac{1}{2}si + i, \\ i' &= i - xi + \frac{1}{2}si - i = -xi + \frac{1}{2}si. \end{aligned} \tag{4}$$

   (a) Observe that the point $(x_e, s_e, i_e) = (1, 2, 1)$ is an equilibrium solution of autonomous system (4). Show that this point is the only equilibrium point in the first octant of x, s, i-space where all three components are strictly positive.

   (b) To analyze the stability characteristics of the equilibrium solution, we linearize the system about the equilibrium point:

$$\begin{aligned} f_1(x, s, i) &= -x + xs - xi, \\ f_2(x, s, i) &= s - sx - \frac{1}{2}si + i, \\ f_3(x, s, i) &= -xi + \frac{1}{2}si. \end{aligned} \tag{5}$$

   The coefficient matrix of the linearized system is the (3×3) constant matrix

$$A = \begin{bmatrix} \dfrac{\partial f_1}{\partial x} & \dfrac{\partial f_1}{\partial s} & \dfrac{\partial f_1}{\partial i} \\[2ex] \dfrac{\partial f_2}{\partial x} & \dfrac{\partial f_2}{\partial s} & \dfrac{\partial f_2}{\partial i} \\[2ex] \dfrac{\partial f_3}{\partial x} & \dfrac{\partial f_3}{\partial s} & \dfrac{\partial f_3}{\partial i} \end{bmatrix}, \tag{6}$$

where all the partial derivatives in (6) are evaluated at $(x_e, s_e, i_e) = (1, 2, 1)$. Determine the entries in the coefficient matrix A. Without performing any further calculations, answer the following questions:

(i) Must the matrix A have at least one real eigenvalue?

(ii) Is it possible for A to have exactly two real eigenvalues?

(iii) Is the matrix A real and symmetric? Does it possess any special structure to suggest that it must have three real eigenvalues?

Now use computer software to compute the three eigenvalues of A.

It can be shown, see Coddington and Levinson,[5] that the (isolated) equilibrium point $(x_e, s_e, i_e) = (1, 2, 1)$ of nonlinear system (4) is

(a) asymptotically stable if all three eigenvalues have negative real parts.

(b) unstable if at least one eigenvalue has a positive real part.

Can either of these results be applied in this case to determine the stability properties of the equilibrium point? If so, describe the stability properties of this equilibrium point.

3. We now investigate the behavior of system (4) numerically, using Euler's method. (Euler's method, as discussed in Section 6.9 for linear systems, can be applied to nonlinear systems as well. In Chapter 9, we will discuss numerical methods that are more efficient than Euler's method.) Let $y_1(t) = x(t)$, $y_2(t) = s(t)$, $y_3(t) = i(t)$, and, using (5), define the vector functions $\mathbf{y}(t)$ and $\mathbf{f}(t)$ by

$$\mathbf{y}(t) = \begin{bmatrix} y_1(t) \\ y_2(t) \\ y_3(t) \end{bmatrix} \quad \text{and} \quad \mathbf{f}(\mathbf{y}) = \begin{bmatrix} f_1(y_1, y_2, y_3) \\ f_2(y_1, y_2, y_3) \\ f_3(y_1, y_2, y_3) \end{bmatrix}.$$

If h denotes the step size and $\mathbf{y}_n$ denotes the numerical approximation of the exact solution at $t_n = t_0 + nh$, Euler's method becomes

$$\mathbf{y}_{n+1} = \mathbf{y}_n + h\mathbf{f}(\mathbf{y}_n), \qquad n = 0, 1, 2, \ldots .$$

Assume the ecosystem is in the state $(x(0), s(0), i(0)) = (0.95, 1.90, 0.80)$ at time $t = 0$; Note that this state has initial values relatively close to the equilibrium values $(1, 2, 1)$. Use Euler's method with

$$\mathbf{y}_0 = \begin{bmatrix} 0.95 \\ 1.90 \\ 0.80 \end{bmatrix}$$

and step size $h = 0.02$ to obtain a numerical approximation of the initial value problem [system (4) and the given initial condition] on the interval $0 \le t \le 10$. Plot graphs of the three solution components. Describe qualitatively what is happening to the ecosystem over this time interval. What do you think will happen as time continues to increase?

[5]Earl A. Coddington and Norman Levinson, *Theory of Ordinary Differential Equations* (Malabar, FL: R. E. Krieger, 1984).

9

Numerical Methods

CHAPTER OVERVIEW

Introduction

We introduced a simple numerical method, Euler's method, in Section 3.8. In this chapter we discuss ways of improving Euler's method and we also begin a general study of numerical methods for solving ordinary differential equations.

Section 9.1 introduces Heun's method and the modified Euler's method as possible ways to improve the accuracy of Euler's method. Building on those ideas, Section 9.2 introduces Taylor series methods as well as the concepts of local truncation error and the order of a method.

Taylor series methods, while accurate, are not computationally friendly. This leads to a discussion in Section 9.3 of Runge-Kutta methods which are computationally friendly and (essentially) just as accurate as Taylor series methods.

The problem of interest is the first order scalar initial value problem

$$y' = f(t, y), \qquad y(t_0) = y_0, \qquad t_0 \le t \le t_0 + T. \tag{1}$$

We assume that problem (1) has a unique solution on the given t-interval. Our goal is to develop algorithms that generate accurate approximations to the solution, $y(t)$.

A numerical method frequently begins by imposing a partition of the form $t_0 < t_1 < t_2 < \cdots < t_{N-1} < t_N = t_0 + T$ on the t-interval $[t_0, t_0 + T]$. Often this partition is uniformly spaced—that is, the partition points are defined by

$$t_n = t_0 + nh, \qquad n = 0, 1, 2, \ldots, N \quad \text{where} \quad h = \frac{T}{N}.$$

The partition spacing, $h = T/N$, is called the **step length** or the **step size**. At each partition point, t_n, the numerical algorithm generates an approximation, y_n, to the exact solution value, $y(t_n)$. A **numerical solution** of the differential equation consists of the points $\{(t_0, y_0), (t_1, y_1), \dots, (t_N, y_N)\}$, where

$$y_n \approx y(t_n), \qquad n = 0, 1, \dots, N.$$

Note that the initial condition provides us with an exact starting point (t_0, y_0). A "good" numerical algorithm is one that generates subsequent points (t_n, y_n) that lie "close" to their exact solution counterparts, $(t_n, y(t_n))$ for $n = 1, 2, \dots, N$. The terms "good" and "close," while intuitively clear, will be made precise later.

Figure 9.1 displays the exact solution of the initial value problem $y' = y^2, y(0) = 1$ on the interval $0 \le t \le 0.95$ and a pair of numerical approximations corresponding to different step lengths h. [The exact solution, $y(t) = (1 - t)^{-1}$, does not exist for $t \ge 1$.]

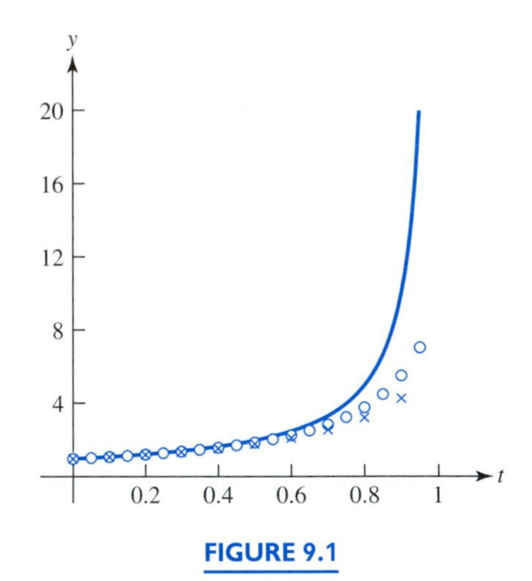

FIGURE 9.1

The initial value problem $y' = y^2, y(0) = 1$ has solution $y(t) = (1 - t)^{-1}, t < 1$. The solid curve is the graph of $y(t)$ for $0 \le t \le 0.95$. The points marked with a circle represent a numerical solution using step length $h = 0.05$ and the points marked with an $\times$ represent a numerical solution with $h = 0.1$. The numerical solutions were generated by Euler's method. As is usually the case, the smaller step length generates approximations that are more accurate.

Numerical Solutions for Systems of Differential Equations

Restricting our attention to the scalar first order initial value problem (1) may seem to be overly restrictive. This, however, is not the case. The algorithms we develop for the first order scalar problem extend directly to first order systems. And, as you have seen, first order systems basically encompass all the differential equations we have considered so far.

In this chapter we concentrate on first order scalar problems because they possess the virtues of relative simplicity and ease of visualization. In particular, we can graph and compare the exact and numerical solutions. To further simplify the development, we restrict our discussions to uniformly spaced partitions with a step size denoted by h.

The computational aspects of Euler's method were treated earlier. The algorithm was introduced and applied to first order scalar problems in Section 3.8 and extended to first order linear systems in Section 6.9. Euler's method will serve as our starting point in the next section, where we briefly review the method and begin to explore ways of improving it.

9.1 Euler's Method, Heun's Method, and the Modified Euler's Method

As we saw in Section 3.8, Euler's method develops a numerical solution of the initial value problem

$$y' = f(t, y), \qquad y(t_0) = y_0 \tag{1}$$

using the algorithm

$$y_{n+1} = y_n + hf(t_n, y_n), \qquad n = 0, 1, 2, \dots, N-1. \tag{2}$$

There are several different ways to derive Euler's method. In Section 3.8, we used a geometric approach based on direction fields. In this section we introduce two other ways of looking at Euler's method. While they are variations on a basic theme, they both provide useful insights when we are looking for ways to improve Euler's method.

Approximating the Integral Equation

Let $y(t)$ denote the exact solution of initial value problem (1). For now, we restrict our attention to the interval $t_n \le t \le t_{n+1}$. Assume we do not know the exact solution, $y(t)$, but that we have already calculated approximations y_k of $y(t_k)$, for $k = 0, 1, \dots, n$ (see Figure 9.2). Our goal is to find the next approximation, y_{n+1}, of $y(t_{n+1})$.

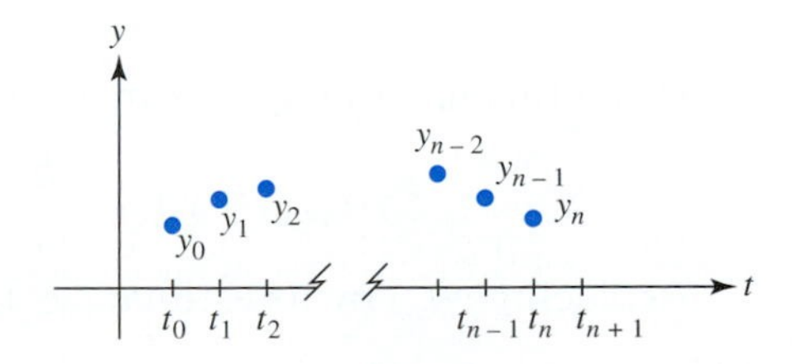

FIGURE 9.2

Let $y(t)$ denote the exact solution of initial value problem (1). While we do not know $y(t)$, we assume we have calculated approximations, y_k, to $y(t_k)$ for $k = 0, 1, \dots, n$. Our goal is to find the next approximation, y_{n+1}, to $y(t_{n+1})$.

Consider differential equation (1) and its exact solution, $y(t)$. Integrating both sides of equation (1) over the interval $[t_n, t_{n+1}]$, we obtain

$$\int_{t_n}^{t_{n+1}} y'(s)\,ds = \int_{t_n}^{t_{n+1}} f(s, y(s))\,ds.$$

By the fundamental theorem of calculus, the left-hand integral is

$$\int_{t_n}^{t_{n+1}} y'(s)\,ds = y(t_{n+1}) - y(t_n).$$

Therefore we obtain an equation for $y(t_{n+1})$,

$$y(t_{n+1}) = y(t_n) + \int_{t_n}^{t_{n+1}} f(s, y(s))\,ds. \tag{3}$$

We cannot use equation (3) computationally because we do not know $f(s, y(s))$, $t_n \le s \le t_{n+1}$. Suppose, however, that the step length h is small enough so that $f(s, y(s))$ is nearly constant over the interval $t_n \le s \le t_{n+1}$. In this case, we can approximate the integral reasonably well by the left Riemann sum,

$$\int_{t_n}^{t_{n+1}} f(s, y(s))\,ds \approx hf(t_n, y(t_n)). \tag{4}$$

Using approximation (4) in equation (3), we obtain

$$y(t_{n+1}) \approx y(t_n) + hf(t_n, y(t_n)).$$

Replacing $y(t_n)$ by the previously calculated estimate y_n, we are led to Euler's method,

$$y(t_{n+1}) \approx y_n + hf(t_n, y_n) \equiv y_{n+1}.$$

Therefore we can view Euler's method as a left Riemann sum approximation of integral equation (3).

Looked at in this light, we might seek to improve Euler's method by asking: "Are there better numerical integration schemes than approximation (4)?" The trapezoidal rule is one such numerical integration scheme. Using the trapezoidal rule, the integral in (3) is approximated by

$$\int_{t_n}^{t_{n+1}} f(s, y(s))\,ds \approx \frac{h}{2}\left[f(t_n, y(t_n)) + f(t_{n+1}, y(t_{n+1}))\right].$$

Using this integral approximation in (3), we obtain

$$y(t_{n+1}) \approx y(t_n) + \frac{h}{2}\left[f(t_n, y(t_n)) + f(t_{n+1}, y(t_{n+1}))\right].$$

Replacing $y(t_n)$ by its estimate y_n leads to

$$y(t_{n+1}) \approx y_n + \frac{h}{2}\left[f(t_n, y_n) + f(t_{n+1}, y(t_{n+1}))\right]. \tag{5}$$

At first glance, it appears we have made matters worse since the unknown $y(t_{n+1})$ appears on the right-hand side of (5), in the term $f(t_{n+1}, y(t_{n+1}))$. Approximation (5), if used as it stands, leads to an implicit algorithm with a nonlinear equation that has to be solved for $y(t_{n+1})$. Suppose, however, that we use Euler's method to approximate the unknown value $y(t_{n+1})$ on the right-hand side of (5). Doing so, we obtain the approximation

$$y(t_{n+1}) \approx y_n + \frac{h}{2}\left[f(t_n, y_n) + f(t_{n+1}, y_n + hf(t_n, y_n))\right],$$

which yields the explicit iteration

$$y_{n+1} = y_n + \frac{h}{2}\left[f(t_n, y_n) + f(t_{n+1}, y_n + hf(t_n, y_n))\right], \qquad n = 0, 1, 2, \ldots, N-1. \tag{6}$$

Algorithm (6) is often called **Heun's method** or the **improved Euler's method**.

Another simple numerical integration scheme is the modified Euler's method, in which the integrand is approximated over the interval $t_n \le t \le t_{n+1}$ by its midpoint value. If we use the midpoint rule to approximate the integral in (3) and again use Euler's method to approximate the unknown value $y(t)$ at the midpoint $t = t_n + h/2$, we obtain the algorithm

$$y_{n+1} = y_n + hf\left(t_n + \frac{h}{2}, y_n + \frac{h}{2}f(t_n, y_n)\right), \qquad n = 0, 1, 2, \ldots, N-1. \tag{7}$$

Algorithm (7) is known as the **modified Euler's method**. [There is no universal agreement on the names of algorithms (6) and (7).[1]]

Although algorithms (6) and (7) appear somewhat complicated, they are relatively easy to implement, since computers can readily evaluate compositions of functions. However, you may rightly ask whether algorithm (6) is, as one of its names implies, an improvement on Euler's method. If so, how do we quantitatively describe this improvement? The same question applies to (7) and we address it in Section 9.2. For now, we content ourselves with an example that compares Euler's method with algorithms (6) and (7) for a particular initial value problem.

EXAMPLE 1

Consider the initial value problem

$$y' = y^2, \qquad y(0) = 1, \qquad 0 \le t \le 0.95.$$

Using a step length of $h = 0.05$, compare the results of Euler's method (2), Heun's method (6) and the modified Euler's method (7). [The exact solution is $y(t) = 1/(1-t), t < 1$.]

Solution: For this example, $t_0 = 0$, $T = 0.95$, $N = T/h = 19$ and $f(t, y) = y^2$. Table 9.1 on the next page lists the results. Note that algorithms (6) and (7) do, in fact, represent an improvement over Euler's method. ▲

The relationship between numerical methods for solving differential equations and numerical integration schemes is a reciprocal one. Every numerical integration technique suggests an algorithm for the initial value problem—this is the approach we used in obtaining algorithms (6) and (7) from equation (3).

---

[1]We are using names found in Peter Henrici, *Discrete Variable Methods in Ordinary Differential Equations* (New York: Wiley, 1962).

TABLE 9.1

The results of Example 1. These results are typical and, as is usually the case, algorithms (6) and (7) give better approximations to $y(t_n)$ than does Euler's method. As is also typical, algorithms (6) and (7) have comparable accuracy. [Note: For this particular initial value problem, Heun's method (6) yields slightly better approximations than the modified Euler's method (7). For other examples, (7) may give slightly better approximations than (6).]

t_n	Euler's method	Heun's method	Modified Euler's method	Exact solution
0.0000	1.0000	1.0000	1.0000	1.0000
0.0500	1.0500	1.0526	1.0525	1.0526
0.1000	1.1051	1.1109	1.1109	1.1111
0.1500	1.1662	1.1762	1.1761	1.1765
0.2000	1.2342	1.2495	1.2493	1.2500
0.2500	1.3104	1.3326	1.3323	1.3333
0.3000	1.3962	1.4275	1.4271	1.4286
0.3500	1.4937	1.5370	1.5363	1.5385
0.4000	1.6052	1.6645	1.6636	1.6667
0.4500	1.7341	1.8151	1.8137	1.8182
0.5000	1.8844	1.9954	1.9934	2.0000
0.5500	2.0620	2.2153	2.2124	2.2222
0.6000	2.2745	2.4894	2.4850	2.5000
0.6500	2.5332	2.8402	2.8333	2.8571
0.7000	2.8541	3.3049	3.2935	3.3333
0.7500	3.2614	3.9488	3.9289	4.0000
0.8000	3.7932	4.8975	4.8597	5.0000
0.8500	4.5126	6.4264	6.3449	6.6667
0.9000	5.5308	9.2615	9.0471	10.0000
0.9500	7.0603	15.9962	15.2001	20.0000

Conversely, an algorithm for the initial value problem gives rise to a corresponding numerical integration scheme. To see why, consider the initial value problem $y' = f(t), y(t_0) = 0$. The solution is simply $y(t) = \int_{t_0}^{t} f(s)\,ds$. Therefore, any numerical method used to solve this initial value problem gives rise to an approximation of the integral. In particular, Euler's method, Heun's method, and the modified Euler's method, when applied to the initial value problem $y' = f(t), y(t_0) = 0$, reduce to a left Riemann sum, the trapezoidal rule, and the midpoint rule, respectively.

Approximating the Taylor Series Expansion

This subsection presents one last derivation of Euler's method. Since $y(t)$ is the solution of initial value problem (1) and is assumed to exist on the interval $t_0 \le t \le t_0 + T$, we know $y(t)$ is differentiable on that interval. Assume for now that the solution is not only differentiable, but that it can be expanded in a

Taylor series at $t = t_n$ where the Taylor series converges in an interval containing $[t_n, t_n + h]$. Therefore, we can express $y(t_{n+1})$ as

$$y(t_{n+1}) = y(t_n + h) = y(t_n) + y'(t_n)h + \frac{y''(t_n)}{2!}h^2 + \frac{y'''(t_n)}{3!}h^3 + \cdots. \tag{8}$$

Truncating the series (8) after two terms, we obtain the approximation

$$y(t_{n+1}) \approx y(t_n) + y'(t_n)h. \tag{9}$$

Since $y'(t_n) = f(t_n, y(t_n))$, we can rewrite approximation (9) as

$$y(t_{n+1}) \approx y(t_n) + f(t_n, y(t_n))h.$$

Replacing $y(t_n)$ in this approximation by its estimate y_n, we are once more led to Euler's method:

$$y(t_{n+1}) \approx y_n + f(t_n, y_n)h \equiv y_{n+1}.$$

Thus, we can obtain Euler's method by truncating the Taylor series (8) after two terms. Viewed in this light, we might seek to improve Euler's method by retaining more terms of the Taylor series—truncating after three terms, or four terms, and so forth. We investigate this possibility in Section 9.2.

Taylor series will be the focus of our discussion throughout the remainder of this chapter. Our goal is to develop accurate and "computationally friendly" algorithms; that is, algorithms that are easy to implement on a computer and that approximate enough terms of the Taylor series to provide sufficient accuracy.

EXERCISES

Most exercises in this chapter require a computer or programmable calculator.

Exercises 1–5:
In each exercise,

(a) Solve the initial value problem analytically, using an appropriate solution technique.

(b) For the given initial value problem, write the Heun's method algorithm,

$$y_{n+1} = y_n + \frac{h}{2}[f(t_n, y_n) + f(t_{n+1}, y_n + hf(t_n, y_n))].$$

(c) For the given initial value problem, write the modified Euler's method algorithm,

$$y_{n+1} = y_n + hf\left(t_n + \frac{h}{2}, y_n + \frac{h}{2}f(t_n, y_n)\right).$$

(d) Use a step size $h = 0.1$. Compute the first three approximations, y_1, y_2, y_3 using the method in part (b).

(e) Use a step size $h = 0.1$. Compute the first three approximations, y_1, y_2, y_3 using the method in part (c).

(f) For comparison, calculate and list the exact solution values $y(t_1), y(t_2), y(t_3)$.

1. $y' = 2t - 1,\quad y(1) = 0$ **2.** $y' = -y,\quad y(0) = 1$ **3.** $y' = -ty,\quad y(0) = 1$

4. $y' = -y + t,\quad y(0) = 0$ **5.** $y^2y' + t = 0,\quad y(0) = 1$

Exercises 6–9:

In each exercise,

(a) Find the exact solution of the given initial value problem.

(b) As in Example 1, use a step size of $h = 0.05$ for the given initial value problem. Compute 20 steps of Euler's method, Heun's method, and the modified Euler's method. Compare the numerical values obtained at $t = 1$ by calculating the error $|y(1) - y_{20}|$.

6. $y' = 1 + y^2, \; y(0) = -1$

7. $y' = -\dfrac{t}{y}, \; y(0) = 3$

8. $y' + 2y = 4, \; y(0) = 3$

9. $y' + 2ty = 0, \; y(0) = 2$

Exercises 10–14:

In each exercise, the given iteration is the result of applying Euler's method, Heun's method, or the modified Euler's method to an initial value problem of the form

$$y' = f(t, y), \qquad y(t_0) = y_0, \qquad t_0 \le t \le t_0 + T.$$

Identify the numerical method and determine t_0, T, and $f(t, y)$.

10. $y_{n+1} = y_n + h(y_n + t_n^2 y_n^3), \; y_0 = 1$
$t_n = 2 + nh, \; h = 0.02, \; n = 0, 1, 2, \ldots, 49$

11. $y_{n+1} = y_n + \dfrac{h}{2}[t_n y_n^2 + 2 + (t_n + h)(y_n + h(t_n y_n^2 + 1))^2], \; y_0 = 2$
$t_n = 1 + nh, \; h = 0.05, \; n = 0, 1, 2, \ldots, 99$

12. $y_{n+1} = y_n + h\left(t_n + \dfrac{h}{2}\right)\sin^2\left(y_n + \dfrac{h}{2} t_n \sin^2(y_n)\right), \; y_0 = 1$

$t_n = nh, \; h = 0.01, \; n = 0, 1, 2, \ldots, 199$

13. $y_{n+1} = y_n\left(1 + \dfrac{h}{t_n^2 + y_n^2}\right), \; y_0 = -1$

$t_n = 2 + nh, \; h = 0.01, \; n = 0, 1, 2, \ldots, 99$

14. $y_{n+1} = y_n + h\left[\sin\left(t_n + \dfrac{h}{2} + y_n + \dfrac{h}{2}\sin(t_n + y_n)\right)\right], \; y_0 = 1$

$t_n = -1 + nh, \; h = 0.05, \; n = 0, 1, 2, \ldots, 199$

15. Let h be a fixed positive step size and let λ be a nonzero constant. Suppose we apply Heun's method or the modified Euler's method to the initial value problem $y' = \lambda y, y(t_0) = y_0$, using this step size h. Show, in either case, that

$$y_k = \left(1 + h\lambda + \frac{(h\lambda)^2}{2!}\right) y_{k-1} \text{ and hence } y_k = \left(1 + h\lambda + \frac{(h\lambda)^2}{2!}\right)^k y_0, \qquad k = 1, 2, \ldots.$$

Exercises 16–17:

Assume a tank having a capacity of 200 gal initially contains 90 gal of fresh water. At time $t = 0$, a salt solution begins to flow into the tank at a rate of 6 gal/min and the well-stirred mixture flows out at a rate of 1 gal/min. Assume that the inflow concentration fluctuates in time, with the inflow concentration given by $c(t) = 2 - \cos(\pi t)$ lb/gal, where t is in minutes. Formulate the appropriate initial value problem for $Q(t)$, the amount of salt (lb) in the tank at time t. Our objective is to approximately determine the amount of salt in the tank when the tank contains 100 gal of liquid.

16. (a) Formulate the initial value problem.

(b) Obtain a numerical solution using Heun's method with a step size $h = 0.05$.

(c) What is your estimate of $Q(t)$ when the tank contains 100 gal?

17. (a) Formulate the initial value problem.

(b) Obtain a numerical solution using the modified Euler's method with a step size $h = 0.05$.

(c) What is your estimate of $Q(t)$ when the tank contains 100 gal?

Exercises 18–19:

Let $P(t)$ denote the population of a certain colony, measured in millions of members. Assume that $P(t)$ is the solution of the initial value problem,

$$P' = 0.1\left(1 - \frac{P}{3}\right)P + M(t), \qquad P(0) = P_0,$$

where time t is measured in years. Let $M(t) = e^{-t}$. Therefore, the colony experiences a migration influx that is initially strong but soon tapers off. Let $P_0 = 1/2$, that is, the colony had 500,000 members at time $t = 0$. Our objective is to estimate the colony size after two years.

18. Obtain a numerical solution of this problem, using Heun's method with a step size $h = 0.05$. What is your estimate of colony size at the end of two years?

19. Obtain a numerical solution of this problem, using the modified Euler's method with a step size $h = 0.05$. What is your estimate of colony size at the end of two years?

Practical Error Estimation In most applications of numerical methods, as in Exercises 16–19, an exact solution is unavailable to use as a benchmark. Therefore, it is natural to ask: "How accurate is our numerical solution?" For example, how accurate are the solutions obtained in Exercises 16fs–19 using the step size $h = 0.05$? The next exercise provides some insight.

20. Suppose we apply Heun's method or the modified Euler's method to the initial value problem $y' = f(t, y), y(t_0) = y_0$ and we use a fixed step size h. It can be shown, for most initial value problems and for h sufficiently small, that the error at a fixed point $t = t^*$ is proportional to h^2. That is, let n be a positive integer, let $h = (t^* - t_0)/n$, and let y_n denote the method's approximation to $y(t^*)$ using step size h. Then,

$$\lim_{h\to 0} \frac{y(t^*) - y_n}{h^2} = C, \qquad C \neq 0.$$

As a consequence of this limit, reducing a sufficiently small step size by 1/2 will reduce the error by approximately 1/4. In particular, let $\hat{y}_{2n}$ denote the method's approximation to $y(t^*)$ using step size $h/2$. Then, for most initial value problems, we expect that $y(t^*) - \hat{y}_{2n} \approx [y(t^*) - y_n]/4$. Rework Example 1, using Heun's method and step sizes of $h = 0.05$, $h = 0.025$, and $h = 0.0125$.

(a) Compare the three numerical solutions at $t = 0.05, 0.10, 0.15, \dots, 0.95$. Are the errors reduced by about 1/4 when the step size is reduced by 1/2? [Since the solution becomes unbounded as t approaches 1 from the left, the expected error reduction may not materialize near $t = 1$.]

(b) Suppose the exact solution is not available. How can the Heun's method solutions obtained using different step sizes be used to estimate the error? [Hint: Assuming that

$$y(t^*) - \hat{y}_{2n} \approx \frac{[y(t^*) - y_n]}{4},$$

derive an expression for $y(t^*) - \hat{y}_{2n}$ that only involves $\hat{y}_{2n}$ and y_n.]

(c) Test the error monitor derived in part (b) on the initial value problem in Example 1.

9.2 Taylor Series Methods

In Section 9.1 we saw we could obtain Euler's method by truncating the Taylor series for the solution $y(t)$ after the first two terms of the expansion; recall equation (9) in Section 9.1. That discussion also led us to anticipate that Euler's method can be improved by retaining more terms of the Taylor series.

In this section we describe how such an improvement of Euler's method is carried out. In addition, we use the Taylor series expansion as a basis for quantifying the accuracy of numerical algorithms. We begin this section with some preliminaries:

- First, we present Theorem 9.1. This theorem gives conditions that guarantee the solution of an initial value problem has a convergent Taylor series expansion.
- We then present Theorem 9.2, Taylor's theorem. This theorem from calculus enables us to measure the error that arises when we truncate a Taylor series.

Once these preliminary results are in place, we shall use Taylor series as a basis for systematically developing algorithms of increasing accuracy. These Taylor series algorithms, can (in principle) be made as accurate as we wish. They are not, however, computationally friendly. We combine accuracy with ease of implementation in Section 9.3, when we discuss Runge-Kutta methods.

Preliminaries

We begin with two definitions and then present a theorem guaranteeing that the solution of initial value problem (1),

$$y' = f(t, y), \qquad y(t_0) = y_0, \tag{1}$$

can be expanded in a Taylor series that converges in a neighborhood of the point t_0.

A function $y(t)$, defined on an open interval containing the point $\bar{t}$, is said to be **analytic** at $t = \bar{t}$ if it has a Taylor series expansion,

$$y(t) = \sum_{n=0}^{\infty} a_n (t - \bar{t})^n, \tag{2a}$$

that converges in an interval $\bar{t} - \delta < t < \bar{t} + \delta$ where $\delta > 0$. It is shown in calculus that if $y(t)$ is analytic at $t = \bar{t}$, then $y(t)$ has derivatives of all orders in the interval $(\bar{t} - \delta, \bar{t} + \delta)$. Moreover, the coefficients of the Taylor series are given by

$$a_n = \frac{y^{(n)}(\bar{t})}{n!}, \qquad n = 0, 1, 2, \ldots. \tag{2b}$$

In general, a function $y(t)$ is said to be **analytic in the interval** $a < t < b$ if it is analytic at every point $\bar{t}$ in this interval.

Consider the function $f(t, y)$ appearing on the right-hand side of differential equation (1). In the context of differential equation (1), $f(t, y)$ is understood to represent $f(t, y(t))$, where $y(t)$ is the unknown solution of interest. In the next definition, however, we view f as a function of two independent variables, t and y.

Let $f(t, y)$ be a function defined in an open region R of the ty-plane containing the point $(\bar{t}, \bar{y})$. The function $f(t, y)$ is said to be **analytic** at $(\bar{t}, \bar{y})$ if it has a two-variable Taylor series expansion,

$$f(t, y) = \sum_{m=0}^{\infty} \sum_{n=0}^{\infty} b_{mn}(t - \bar{t})^m (y - \bar{y})^n, \tag{3a}$$

that converges in a neighborhood N_ρ of $(\bar{t}, \bar{y})$,

$$N_\rho = \left\{(t, y) : \sqrt{(t - \bar{t})^2 + (y - \bar{y})^2} < \rho\right\}.$$

We say that $f(t, y)$ is **analytic in a region** R if it is analytic at every point $(\bar{t}, \bar{y})$ in R. The coefficients b_{mn} can be evaluated in terms of the function f and its partial derivatives, evaluated at $(\bar{t}, \bar{y})$; the two-variable Taylor series expansion has the form

$$\begin{aligned} f(t, y) &= f(\bar{t}, \bar{y}) + f_t(\bar{t}, \bar{y})(t - \bar{t}) + f_y(\bar{t}, \bar{y})(y - \bar{y}) \\ &+ \frac{1}{2}\left(f_{tt}(\bar{t}, \bar{y})(t - \bar{t})^2 + 2f_{ty}(\bar{t}, \bar{y})(t - \bar{t})(y - \bar{y}) + f_{yy}(\bar{t}, \bar{y})(y - \bar{y})^2\right) + \cdots. \end{aligned} \tag{3b}$$

The Existence of Analytic Solutions

It is natural to ask whether analyticity of $f(t, y)$ guarantees analyticity of the solution of initial value problem (1). An affirmative answer is contained in Theorem 9.1, which can be regarded as a refinement of Theorem 3.1. A proof of Theorem 9.1 can be found in Birkhoff and Rota.[2]

Theorem 9.1

Let R denote the rectangle defined by $a < t < b, \alpha < y < \beta$. Let $f(t, y)$ be a function defined and analytic in R and suppose that (t_0, y_0) is a point in R. Then there is a t-interval (c, d) containing t_0 in which there exists a unique analytic solution of the initial value problem

$$y' = f(t, y), \qquad y(t_0) = y_0.$$

EXAMPLE 1

Consider the initial value problem

$$y' = y^2 + t^2, \qquad y(t_0) = y_0.$$

Here, the function $f(t, y) = y^2 + t^2$ is a polynomial in the variables t and y and is therefore analytic in the entire ty-plane. Hence, the region R can be assumed to be the entire ty-plane. Theorem 9.1 guarantees the existence of a unique analytic solution $y(t)$ in an interval of the form $t_0 - \delta < t < t_0 + \delta$ for some $\delta > 0$. Note that the theorem does not tell us the value of δ, only that such a positive δ exists.

(continued)

[2] Garrett Birkhoff and Gian-Carlo Rota, *Ordinary Differential Equations*, 4th edition (New York: Wiley, 1989).

(continued)

Since the solution is analytic in $t_0 - \delta < t < t_0 + \delta$, we know $y(t)$ has the form

$$y(t) = \sum_{n=0}^{\infty} \frac{y^{(n)}(t_0)}{n!}(t - t_0)^n, \qquad t_0 - \delta < t < t_0 + \delta. \blacktriangle$$

We assume throughout this chapter that the hypotheses of Theorem 9.1 are satisfied. This theorem assures us that an analytic solution $y(t)$ exists on some interval of the form $t_0 - \delta < t < t_0 + \delta$. As noted in Example 1, however, Theorem 9.1 does not tell us the size of δ. Since we are interested in generating a numerical solution on an interval of the form $t_0 \le t \le t_0 + T$, we shall also assume that the interval of interest, $[t_0, t_0 + T]$, lies within the existence interval, $(t_0 - \delta, t_0 + \delta)$. Given this assumption, we can expand solution $y(t)$ in a Taylor series about any point $\bar{t}$ lying in the interval of interest. It is important to realize, however, that in practical computations involving nonlinear differential equations there is no a priori guarantee that the solution exists on a designated interval of interest, $[t_0, t_0 + T]$.

Using the Differential Equation to Compute the Taylor Series Coefficients

When Euler's method was developed in Section 3.8, we based the development on the fact that the differential equation determines the direction field. In particular, if we evaluate f at a point $(\bar{t}, \bar{y})$ in the ty-plane, then the value $f(\bar{t}, \bar{y})$ tells us the slope of the solution curve passing through $(\bar{t}, \bar{y})$.

We now show that the differential equation determines much more. In particular, suppose that a solution curve, $y(t)$, passes through the point $(\bar{t}, \bar{y})$. We will see that $f(t, y)$ and its partial derivatives, evaluated at $(\bar{t}, \bar{y})$, can be used to calculate all the derivatives of $y(t)$,

$$y^{(n)}(\bar{t}), \qquad n = 1, 2, 3, \ldots.$$

In turn, see equations (2a) and (2b), these derivative evaluations completely determine the Taylor series expansion of the solution $y(t)$.

In particular, we know the identity $y'(t) = f(t, y(t))$ holds for t in a neighborhood of $\bar{t}$. Therefore,

$$\begin{aligned} y'(\bar{t}) &= f(\bar{t}, y(\bar{t})) \\ &= f(\bar{t}, \bar{y}). \end{aligned}$$

We find higher derivatives by differentiating the identity $y'(t) = f(t, y(t))$. For example,

$$y''(t) = \frac{d}{dt} y'(t) = \frac{d}{dt} f(t, y(t)). \tag{4a}$$

As in Section 3.3, we can use the chain rule to calculate the derivative in equation (4a),

$$\frac{df}{dt} = \frac{\partial f}{\partial t} + \frac{\partial f}{\partial y}\frac{dy}{dt} = \frac{\partial f}{\partial t} + \frac{\partial f}{\partial y} f = f_t + f_y f. \tag{4b}$$

Once the partial derivatives in equation (4b) are computed, we substitute the function $y(t)$ for the second independent variable y, obtaining

$$y''(t) = f_t(t, y(t)) + f_y(t, y(t))f(t, y(t)).$$

Using the fact that $y(\bar{t}) = \bar{y}$, we find

$$y''(\bar{t}) = f_t(\bar{t}, \bar{y}) + f_y(\bar{t}, \bar{y})f(\bar{t}, \bar{y}). \tag{5}$$

Equation (5) determines the concavity of the solution curve at the point $(\bar{t}, \bar{y})$, just as $y'(\bar{t}) = f(\bar{t}, \bar{y})$ determines the slope of the solution curve at $(\bar{t}, \bar{y})$.

This differentiation process can be continued to compute higher derivatives of the solution at $(\bar{t}, \bar{y})$. To simplify the notation, we continue to use subscripts to denote partial derivatives but do not explicitly indicate the ultimate evaluation of the various functions at $(\bar{t}, \bar{y})$. Thus,

$$\begin{aligned} y'' &= f_t + f_y f, \\ y''' &= \frac{d}{dt}\left[f_t + f_y f\right] = \left[f_{tt} + f_{ty}f + (f_{yt} + f_{yy}f)f + f_y(f_t + f_y f)\right]. \end{aligned} \tag{6}$$

It is possible, in principle, to continue this differentiation process and compute as many derivatives of $y(t)$ at $t = \bar{t}$ as desired. It is clear from (6), however, that the computations may quickly become cumbersome.

The next example illustrates, however, that when the differential equation has a simple structure, it may be relatively easy to calculate higher derivatives.

EXAMPLE 2

Consider the initial value problem

$$y' = y^2, \qquad y(0) = 1.$$

Evaluate the derivatives $y'(0), y''(0), y'''(0)$, and $y^{(4)}(0)$.

Solution: In this case, $f(t, y) = y^2$ is a polynomial in y. Therefore, Theorem 9.1 applies and we know the solution $y(t)$ is an analytic function of t in the open interval $(-\delta, \delta)$, for some $\delta > 0$. Since $y'(t) = y^2(t)$, the chain rule yields

$$\begin{aligned} y''(t) &= [y^2(t)]' = 2y(t)y'(t) = 2y^3(t), \\ y'''(t) &= [2y^3(t)]' = 6y^2(t)y'(t) = 6y^4(t), \\ y^{(4)}(t) &= [6y^4(t)]' = 24y^3(t)y'(t) = 24y^5(t). \end{aligned}$$

Therefore, $y(0) = 1$, $y'(0) = 1$, $y''(0) = 2$, $y'''(0) = 6 = 3!$, and $y^{(4)}(0) = 24 = 4!$ Given these derivative values, the first few terms in the Taylor series expansion of $y(t)$ are

$$\begin{aligned} y(t) &= y(0) + y'(0)t + \frac{y''(0)}{2!}t^2 + \frac{y'''(0)}{3!}t^3 + \frac{y^{(4)}(0)}{4!}t^4 + \cdots \\ &= 1 + t + t^2 + t^3 + t^4 + \cdots. \end{aligned}$$

We recognize this expansion as a geometric series that converges to the exact solution,

$$y(t) = \frac{1}{1-t}, \qquad -1 < t < 1.$$

For this initial value problem, we find (after the fact) that $\delta = 1$. ▲

Taylor Series Methods

The preceding discussion shows how to calculate higher derivatives of the solution $y(t)$ of the initial value problem

$$y' = f(t, y), \qquad y(t_0) = y_0.$$

We can use these ideas to improve Euler's method. Let y_n be an approximation to $y(t_n)$, where t_n and $t_{n+1} = t_n + h$ are in the interval $t_0 \le t \le t_N$. As in equation (8) of Section 9.1, we have

$$y(t_{n+1}) = y(t_n) + y'(t_n)h + \frac{y''(t_n)}{2!}h^2 + \frac{y'''(t_n)}{3!}h^3 + \cdots.$$

Truncating this expansion after p terms, we obtain the approximation

$$y(t_{n+1}) \approx y(t_n) + y'(t_n)h + \frac{y''(t_n)}{2!}h^2 + \frac{y'''(t_n)}{3!}h^3 + \cdots + \frac{y^{(p)}(t_n)}{p!}h^p. \tag{7}$$

As we saw in equation (6), the Taylor series coefficients, $y'(t_n), y''(t_n), y'''(t_n), \ldots$, can be expressed in terms of f and its partial derivatives evaluated at $(t_n, y(t_n))$. For instance, with $p = 1$, (7) becomes

$$y(t_{n+1}) \approx y(t_n) + f(t_n, y(t_n))h.$$

Similarly, for $p = 2$, we obtain from (7)

$$y(t_{n+1}) \approx y(t_n) + f(t_n, y(t_n))h + \left[f_t(t_n, y(t_n)) + f_y(t_n, y(t_n))f(t_n, y(t_n))\right]\frac{h^2}{2!}.$$

We find similar approximations when $p \ge 3$. In order to use these approximations for computations, we replace $y(t_n)$ by its estimate, y_n. The algorithms we obtain in this manner are collectively referred to as **Taylor series methods**. We use the term **Taylor series method of order *p*** to identify the Taylor series method obtained from approximation (7). The Taylor series methods of orders 1, 2, and 3 are as follows:

Taylor Series Method of Order 1 (Euler's Method)

$$y_{n+1} = y_n + hf(t_n, y_n), \qquad n = 0, 1, \ldots, N-1. \tag{8a}$$

Taylor Series Method of Order 2

$$y_{n+1} = y_n + hf(t_n, y_n) + \frac{h^2}{2!}\left[f_t(t_n, y_n) + f_y(t_n, y_n)f(t_n, y_n)\right], \qquad n = 0, 1, \ldots, N-1. \tag{8b}$$

Taylor Series Method of Order 3

$$\begin{aligned} y_{n+1} = y_n &+ hf(t_n, y_n) + \frac{h^2}{2!}\left[f_t(t_n, y_n) + f_y(t_n, y_n)f(t_n, y_n)\right] \\ &+ \frac{h^3}{3!}\Big[f_{tt}(t_n, y_n) + 2f_{ty}(t_n, y_n)f(t_n, y_n) + f_{yy}(t_n, y_n)f^2(t_n, y_n) \\ &\qquad + f_y(t_n, y_n)f_t(t_n, y_n) + f_y^2(t_n, y_n)f(t_n, y_n)\Big], \qquad n = 0, 1, \ldots, N-1. \end{aligned} \tag{8c}$$

It is cumbersome to write out all the terms of the general pth order Taylor series method except for small values of p. In order to simplify the notation when discussing Taylor series methods, it is common to denote a pth order

Taylor series method as

$$y_{n+1} = y_n + hy'_n + \frac{h^2}{2!}y''_n + \frac{h^3}{3!}y'''_n + \cdots + \frac{h^p}{p!}y_n^{(p)}, \qquad n = 0, 1, \ldots, N-1. \quad \textbf{(9)}$$

We are using the name "pth order Taylor series method" to denote method (9). The term *order* has a precise meaning that is given later in this section. Once we state the formal definition of order, however, we will see that method (9) is properly called a pth order Taylor series method.

EXAMPLE 3

Consider the initial value problem

$$y' = y^2, \qquad y(0) = 1.$$

Using $h = 0.05$, execute 19 steps of the Taylor series method of order p, for $p = 1, 2, 3$, and 4. Do the results improve as p increases?

Solution: As we saw in Example 2, $y'' = 2y^3$, $y''' = 6y^4$, and $y^{(4)} = 24y^5$. The Taylor series methods of orders 1, 2, 3, and 4 are, respectively,

$$\begin{aligned}
y_{n+1} &= y_n + hy_n^2,\\
y_{n+1} &= y_n + hy_n^2 + h^2y_n^3,\\
y_{n+1} &= y_n + hy_n^2 + h^2y_n^3 + h^3y_n^4,\\
y_{n+1} &= y_n + hy_n^2 + h^2y_n^3 + h^3y_n^4 + h^4y_n^5.
\end{aligned}$$

Table 9.2 on the next page illustrates how the Taylor series method estimates improve as the order increases. ▲

Example 3 illustrates (for the special case of the differential equation $y' = y^2$) how the Taylor series method of order p becomes more accurate as p increases. We are now ready to make the concept of order precise and to discuss why we expect that higher order methods are usually more accurate than lower order methods.

Taylor's Theorem

In this subsection we consider the error made when we truncate a Taylor series. Theorem 9.2, known as Taylor's theorem, gives a convenient way of estimating the resulting truncation error. A proof of Taylor's theorem can be found in most calculus books.

Theorem 9.2

Let $y(t)$ be analytic at $t = \bar{t}$ where the Taylor series expansion (2) converges in the interval $\bar{t} - \delta < t < \bar{t} + \delta$. Let m be a positive integer and let t be in the interval $(\bar{t} - \delta, \bar{t} + \delta)$. Then

$$\begin{aligned}
y(t) &= y(\bar{t}) + y'(\bar{t})(t - \bar{t}) + \frac{y''(\bar{t})}{2!}(t - \bar{t})^2 + \cdots\\
&\quad + \frac{y^{(m)}(\bar{t})}{m!}(t - \bar{t})^m + \frac{y^{(m+1)}(\xi)}{(m+1)!}(t - \bar{t})^{m+1},
\end{aligned} \qquad \textbf{(10)}$$

where ξ is some point lying between $\bar{t}$ and t.

TABLE 9.2

In this table we designate the results of the pth order Taylor series method as "order p" for $p = 1, 2, 3, 4$ and the value of the exact solution at $t = t_n$ as $y(t_n)$. As anticipated, the results improve when we retain more terms in the Taylor series expansion; that is, as the order p increases.

	Taylor Series Methods				
t_n	Order 1	Order 2	Order 3	Order 4	$y(t_n)$
0.0000	1.0000	1.0000	1.0000	1.0000	1.0000
0.0500	1.0500	1.0525	1.0526	1.0526	1.0526
0.1000	1.1051	1.1108	1.1111	1.1111	1.1111
0.1500	1.1662	1.1759	1.1764	1.1765	1.1765
0.2000	1.2342	1.2491	1.2500	1.2500	1.2500
0.2500	1.3104	1.3320	1.3333	1.3333	1.3333
0.3000	1.3962	1.4266	1.4285	1.4286	1.4286
0.3500	1.4937	1.5357	1.5383	1.5385	1.5385
0.4000	1.6052	1.6626	1.6664	1.6666	1.6667
0.4500	1.7341	1.8123	1.8178	1.8182	1.8182
0.5000	1.8844	1.9914	1.9994	2.0000	2.0000
0.5500	2.0620	2.2095	2.2212	2.2221	2.2222
0.6000	2.2745	2.4805	2.4984	2.4999	2.5000
0.6500	2.5332	2.8264	2.8543	2.8569	2.8571
0.7000	2.8541	3.2822	3.3281	3.3328	3.3333
0.7500	3.2614	3.9093	3.9894	3.9987	4.0000
0.8000	3.7932	4.8227	4.9756	4.9963	5.0000
0.8500	4.5126	6.2661	6.5980	6.6536	6.6667
0.9000	5.5308	8.8443	9.7297	9.9301	10.0000
0.9500	7.0603	14.4850	17.8859	19.1273	20.0000

In Theorem 9.2, the polynomial

$$P_m(t) = y(\bar{t}) + y'(\bar{t})(t-\bar{t}) + \frac{y''(\bar{t})}{2!}(t-\bar{t})^2 + \cdots + \frac{y^{(m)}(\bar{t})}{m!}(t-\bar{t})^m$$

is referred to as the **Taylor polynomial of degree m**. The term

$$R_m(t) = \frac{y^{(m+1)}(\xi)}{(m+1)!}(t-\bar{t})^{m+1}$$

is the **remainder** and it measures the error made in approximating $y(t)$ by the Taylor polynomial, $P_m(t)$. When we consider the errors of a numerical method, the role of $\bar{t}$ is typically played by t_n and the generic point t lies in the interval $t_n \le t \le t_{n+1}$.

One-Step Methods and the Local Truncation Error

The methods we have considered thus far (Euler's method, Heun's method, the modified Euler's method, and Taylor series methods) are classified as "one-step methods." In general, a **one-step method** has the form

$$y_{n+1} = y_n + h\phi(t_n, y_n; h), \qquad n = 0, 1, 2, \ldots, N-1. \tag{11}$$

These methods are called one step because they use only the most recently computed point, (t_n, y_n), to compute the next point, (t_{n+1}, y_{n+1}). [By contrast, a multistep method uses multiple back values (t_n, y_n), (t_{n-1}, y_{n-1}), (t_{n-2}, y_{n-2}), ..., (t_{n-k}, y_{n-k}) to compute (t_{n+1}, y_{n+1}).[3] We restrict our consideration to one-step methods.]

In equation (11), the term $\phi(t_n, y_n; h)$ is called an **increment function**. Different increment functions define different one-step methods. For instance, Euler's method, $y_{n+1} = y_n + hf(t_n, y_n)$, is a one-step method with increment function

$$\phi(t_n, y_n; h) = f(t_n, y_n).$$

Heun's method is a one-step method with increment function

$$\phi(t_n, y_n; h) = \frac{1}{2}\left[f(t_n, y_n) + f(t_n + h, y_n + hf(t_n, y_n))\right].$$

EXAMPLE 4

Write the second order Taylor series method in the form of a one-step method and identify the increment function $\phi(t_n, y_n; h)$.

Solution: From equation (8b), the second order Taylor series method has the form

$$y_{n+1} = y_n + h\left(f(t_n, y_n) + \frac{h}{2!}\left[f_t(t_n, y_n) + f_y(t_n, y_n)f(t_n, y_n)\right]\right).$$

Thus,

$$\phi(t_n, y_n; h) = f(t_n, y_n) + \frac{h}{2!}\left[f_t(t_n, y_n) + f_y(t_n, y_n)f(t_n, y_n)\right]. \; \blacktriangle$$

A quantity known as the local truncation error is one of the keys to understanding and assessing the accuracy of one-step methods. Let $y(t)$ denote the unique solution of the initial value problem $y' = f(t, y)$, $y(t_0) = y_0$, and assume $y(t)$ exists on the interval of interest, $[t_0, t_0 + T]$. Let t_n and $t_{n+1} = t_n + h$ lie in the interval $[t_0, t_0 + T]$. For a given one-step method (11) we define the quantity T_{n+1} by

$$y(t_{n+1}) = y(t_n) + h\phi(t_n, y(t_n); h) + T_{n+1}. \tag{12}$$

The quantities T_{n+1}, $n = 0, 1, \ldots, N-1$ are called local truncation errors. A **local truncation error**[4] measures how much a single step of the numerical method misses the true solution value, $y(t_{n+1})$, given that the numerical method starts on the solution curve at the point $(t_n, y(t_n))$.

---

[3] John D. Lambert, *Numerical Methods for Ordinary Differential Systems* (Chichester, England: Wiley, 1991).

[4] There is no universal agreement about the definition of local truncation errors. Some texts express the quantity T_{n+1} in equation (12) as $h\tau_{n+1} = T_{n+1}$ and refer to τ_{n+1} as a local truncation error. However, no matter how local truncation errors are defined, there is universal agreement on the definition of the "order" of a one-step method, as given in the next subsection in equation (15).

EXAMPLE 5

Derive an expression for the local truncation errors of Euler's method.

Solution: Since Euler's method is given by $y_{n+1} = y_n + hf(t_n, y_n)$, the local truncation errors are defined by

$$y(t_{n+1}) = y(t_n) + hf(t_n, y(t_n)) + T_{n+1}, \tag{13}$$

where $y(t)$ is the unique solution of the initial value problem $y' = f(t, y), y(t_0) = y_0$. However, $f(t_n, y(t_n)) = y'(t_{n+1})$ and so equation (13) can be expressed as

$$y(t_{n+1}) = y(t_n) + y'(t_n)h + T_{n+1}. \tag{14a}$$

By Taylor's theorem, we can also write

$$y(t_{n+1}) = y(t_n) + y'(t_n)h + \frac{y''(\xi)}{2!}h^2, \tag{14b}$$

where $t_n < \xi < t_{n+1}$. Comparing (14a) and (14b), we see that

$$T_{n+1} = \frac{y''(\xi)}{2!}h^2, \tag{14c}$$

where ξ is some point in the t-interval $t_n < t < t_{n+1}$.

For later use, we note from (14c) that

$$\max_{0 \le n \le N-1} \left| T_{n+1} \right| \le Kh^2, \tag{14d}$$

where $K = \max_{t_0 \le t \le t_0+T} |y''(t)|/2!$. ▲

The Order of a Numerical Method

We now define the order of a numerical method and show that the terminology "Taylor series method of order p" is appropriate. We say that a one-step method has **order p** if there are positive constants K and h_0 such that

> For any point t_n in the interval $[t_0, t_0 + T - h_0]$ and any step size h satisfying $0 < h \le h_0$, we have
>
> $$|T_{n+1}| \le Kh^{p+1}. \tag{15}$$

Note that, in inequality (15), the constant K does not depend on the index n. From inequality (14d) of Example 5, we see that Euler's method has order $p = 1$. Similar arguments show that the Taylor series methods (8b) and (8c) have orders 2 and 3, respectively. In general, the Taylor series method

$$y_{n+1} = y_n + hy'_n + \frac{h^2}{2!}y''_n + \frac{h^3}{3!}y'''_n + \cdots + \frac{h^p}{p!}y^{(p)}_n,$$

has order p; this is consistent with our prior use of the term.

The order of a numerical method is a measure of how well the method replicates the Taylor expansion of the solution. A numerical method of order p has local truncation errors that satisfy $|T_{n+1}| \le Kh^{p+1}$. Noting Taylor's theorem, therefore, it follows that a pth order one-step method correctly replicates the Taylor series up to and including the term of order h^p.

The Global Error

The size of the local truncation error for a numerical method tells us how far we deviate from $y(t_{n+1})$, the true solution evaluated at $t = t_{n+1}$, if we were to take a single step of the method starting on the solution curve at the point $(t_n, y(t_n))$. However, except for the first step of the method [when we start at the initial point $(t_0, y(t_0)) = (t_0, y_0)$] we do not expect to take steps that begin on the solution curve. In this sense, the local truncation error is not a quantity that we can calculate without knowing the true solution of the initial value problem. We are using the concept of the local truncation error to define the order of a numerical method and to establish the convergence of numerical methods.

In practical computations, we are primarily interested in the **global errors**,

$$y(t_n) - y_n, \quad \text{for} \quad n = 0, 1, \ldots, N, \tag{16}$$

where $y(t)$ is the true solution of the initial value problem and y_n is the numerical method's estimate to $y(t_n)$.

When discussing local truncation errors and global errors, it is convenient to introduce the **"Big *O*" order symbol** (also known as the **Landau symbol**). This symbol is frequently used to characterize inequalities such as (15). We use the notation

$$q(h) = O(h^r), \qquad h \to 0 \quad \text{or simply} \quad q(h) = O(h^r)$$

to mean there exists some positive constant K such that $|q(h)| \le Kh^r$ for all positive, sufficiently small h. Thus, inequality (15) can be written as

$$T_{n+1} = O(h^{p+1}).$$

Note that the order of a numerical method, p, is one integer less than the order of the local truncation error. For example, from equation (14d), the local truncation error of Euler's method is $O(h^2)$ and therefore we say that Euler's method is a first order method.

In an appendix to Section 9.3, we state a theorem that shows how (for the types of problems and numerical methods we are considering) the order of the numerical method and the size of the global errors are related. In particular, there is a positive constant M such that the global errors for a pth order method satisfy the inequality

$$\max_{0 \le n \le N} |y(t_n) - y_n| \le Mh^p. \tag{17}$$

Inequality (17) tells us how the global errors are reduced when h is reduced. If we are using a pth order method and if we reduce the step size h by 1/2, then we anticipate that the global errors will be reduced by $(1/2)^p$.

EXAMPLE 6

We again consider the example

$$y' = y^2, \qquad y(0) = 1, \qquad 0 \le t \le 0.95.$$

Use Euler's method to generate numerical solutions first using step size $h_1 = 0.05$ and then using step size $h_2 = 0.025$. From (17) with $p = 1$, we expect the global errors to be reduced by approximately $1/2$ when h is reduced by $1/2$. Compare the global errors at $t = 0.05, 0.10, 0.15, \ldots, 0.95$. Does it appear

(continued)

(continued)

that the errors resulting from the smaller step size are about half the size of the errors of the larger step?

Solution: The results are listed in Table 9.3. The column headed E_1 gives the global errors $y(t_k) - y_k$, made using $h_1 = 0.05$. Similarly, the column headed E_2 lists the global errors, at the same values of t, made using $h_2 = 0.025$. As predicted by (17), the ratios of E_2 to E_1 (given in the column headed E_2/E_1) are close to 0.5 for smaller values of t. The ratios tend to deviate from 0.5 as the values t_k approach $t = 1$, where the exact solution has a vertical asymptote. ▲

TABLE 9.3

The results of Example 6. Note, as predicted by (17), that $E_2 \approx E_1/2$.

t_k	E_1 ($h = 0.05$)	E_2 ($h = 0.025$)	E_2/E_1
0.0500	0.0026	0.0014	0.5191
0.1000	0.0060	0.0031	0.5206
0.1500	0.0103	0.0054	0.5223
0.2000	0.0158	0.0083	0.5242
0.2500	0.0230	0.0121	0.5264
0.3000	0.0324	0.0171	0.5290
0.3500	0.0448	0.0238	0.5320
0.4000	0.0614	0.0329	0.5355
0.4500	0.0841	0.0454	0.5396
0.5000	0.1156	0.0630	0.5446
0.5500	0.1603	0.0883	0.5507
0.6000	0.2255	0.1259	0.5583
0.6500	0.3239	0.1840	0.5680
0.7000	0.4793	0.2782	0.5805
0.7500	0.7386	0.4412	0.5973
0.8000	1.2068	0.7491	0.6207
0.8500	2.1541	1.4111	0.6551
0.9000	4.4692	3.1700	0.7093
0.9500	12.9397	10.4052	0.8041

The Need for Computationally Friendly Algorithms

Taylor series expansions provide a clear blueprint for how to improve the accuracy of a numerical algorithm and Taylor series methods implement this blueprint. The exercises develop such algorithms for a variety of problems. In specific cases, as in Examples 1 and 2, the computations might not be overly difficult. In other cases, as the order of the algorithm increases, the computations rapidly become unwieldy and the possibility of mistakes in programming the numerical method grows as well. Moreover, Taylor series methods are problem specific; the various partial derivatives of $f(t, y)$ must be recomputed every time we are given a new differential equation.

For these reasons, a Taylor series method is not very attractive as an all-purpose method for solving initial value problems. The challenge is to develop algorithms that replicate the desired number of terms in the Taylor series expansion (thereby achieving the desired accuracy) but do not require calculation of partial derivatives. In particular, we want algorithms that only require evaluations of the function f.

Heun's method and the modified Euler's method, developed in Section 9.1, provide insight as to how these goals might be achieved using compositions of functions. Computers can evaluate functions with relative ease and compositions of functions, while they might look formidable to us, are also evaluated with relative ease on the machine. Nested compositions of functions, such as those used in Heun's method and the modified Euler's method, form the basis of Runge-Kutta methods that are discussed in Section 9.3. As you will see, Runge-Kutta methods achieve the accuracy of Taylor series methods, but in a computationally friendly way.

EXERCISES

Exercises 1–10:

Assume, for the given differential equation, that $y(0) = 1$.

(a) Use the differential equation itself to determine the values $y'(0), y''(0), y'''(0), y^{(4)}(0)$ and form the Taylor polynomial

$$P_4(t) = y(0) + y'(0)t + \frac{y''(0)}{2!}t^2 + \frac{y'''(0)}{3!}t^3 + \frac{y^{(4)}(0)}{4!}t^4.$$

(b) Verify that the given function is the solution of the initial value problem consisting of the differential equation and initial condition $y(0) = 1$.

(c) Evaluate both the exact solution $y(t)$ and $P_4(t)$ at $t = 0.1$. What is the error $E(0.1) = y(0.1) - P_4(0.1)$? [Note that $E(0.1)$ is the local truncation error incurred in using a Taylor series method of order 4 to step from $t_0 = 0$ to $t_1 = 0.1$ using step size $h = 0.1$.]

1. $y' = -y + 2;\ \ y(t) = 2 - e^{-t}$

2. $y' = 2ty;\ \ y(t) = e^{t^2}$

3. $y' = ty^2;\ \ y(t) = \left(1 - \frac{t^2}{2}\right)^{-1}$

4. $y' = t^2 + y;\ \ y(t) = 3e^t - (t^2 + 2t + 2)$

5. $y' = y^{1/2};\ \ y(t) = \left(1 + \frac{t}{2}\right)^2$

6. $y' = ty^{-1};\ \ y(t) = \sqrt{1 + t^2}$

7. $y' = y + \sin t;\ \ y(t) = \dfrac{3e^t - \cos t - \sin t}{2}$

8. $y' = y^{3/4};\ \ y(t) = \left(1 + \frac{t}{4}\right)^4$

9. $y' = 1 + y^2;\ \ y(t) = \tan\left(t + \frac{\pi}{4}\right)$

10. $y' = -4t^3y;\ \ y(t) = e^{-t^4}$

Results analogous to Theorem 9.1 guaranteeing the existence of analytic solutions can be established for higher order scalar problems and first order systems. The development of higher order numerical methods for such problems will be addressed in Section 9.3. Exercises 11–14 illustrate how a series expansion of the solution of a higher order scalar problem can be obtained from the differential equation itself. For example, consider the initial value problem $y'' = f(t, y, y'), y(t_0) = y_0, y'(t_0) = y_0'$. From the equation, we have $y''(t_0) = f(t_0, y_0, y_0')$. Differentiating the identity $y''(t) = f(t, y(t), y'(t))$ allows us to obtain $y'''(t_0)$ and then $y^{(4)}(t_0)$ and so forth.

Exercises 11–14:

In each exercise

(a) Obtain the fifth degree Taylor polynomial approximation of the solution,

$$P_5(t) = y(t_0) + y'(t_0)(t - t_0) + \frac{y''(t_0)}{2!}(t - t_0)^2 + \cdots + \frac{y^{(5)}(t_0)}{5!}(t - t_0)^5.$$

(b) If the exact solution is given, calculate the error at $t = t_0 + 0.1$,

11. $y'' - 3y' + 2y = 0,\ \ y(0) = 1,\ \ y'(0) = 0;\ \ t_0 = 0.$
[The exact solution is $y(t) = 2e^t - e^{2t}$.]

12. $y'' - y' = 0,\ \ y(1) = 1,\ \ y'(1) = 2;\ \ t_0 = 1.$
[The exact solution is $y(t) = -1 + 2e^{(t-1)}$.]

13. $y''' - y' = 0,\ \ y(0) = 1,\ \ y'(0) = 2,\ \ y''(0) = 0;\ \ t_0 = 0.$
[The exact solution is $y(t) = 1 + e^t - e^{-t}$.]

14. $y'' + y + y^3 = 0,\ \ y(0) = 1,\ \ y'(0) = 0;\ \ t_0 = 0.$

Exercises 15–18:

In each exercise, determine the largest positive integer r such that $q(h) = O(h^r)$. [Hint: Determine the first nonvanishing term in the Maclaurin expansion of q.]

15. $q(h) = \sin 2h$

16. $q(h) = 2h + h^3$

17. $q(h) = 1 - \cos h$

18. $q(h) = e^h - (1 + h)$

19. Give an example of functions f and g such that $f(h) = O(h), g(h) = O(h)$ but $(f + g)(h) = O(h^2)$.

Exercises 20–23:

For the given initial value problem,

(a) Execute 20 steps of the Taylor series method of order p for $p = 1, 2, 3$. Use step size $h = 0.05$.

(b) In each exercise, the exact solution is given. List the errors of the Taylor series method calculations at $t = 1$.

20. $y' = \dfrac{t}{y + 1},\ \ y(0) = 1.$ The exact solution is $y(t) = -1 + \sqrt{t^2 + 4}$.

21. $y' = 2ty^2,\ \ y(0) = -1.$ The exact solution is $y(t) = \dfrac{-1}{1 + t^2}$.

22. $y' = \dfrac{1}{2y},\ \ y(0) = 1.$ The exact solution is $y(t) = \sqrt{1 + t}$.

23. $y' = \dfrac{1 + y^2}{1 + t},\ \ y(0) = 0.$ The exact solution is $y(t) = \tan(\ln(1 + t))$.

Exercises 24–27:

Assume that a pth order Taylor series method is used to solve an initial value problem. When the step size h is reduced by 1/2, we expect the global error to be reduced by $1/2^p$.

Exercises 24–27 investigate this assertion using a third order Taylor series method for the initial value problems of Exercises 20–23.

Use the third order Taylor series method to numerically solve the given initial value problem for $0 \le t \le 1$. Let E_1 denote the global error at $t = 1$ with step size $h = 0.05$ and E_2 the error at $t = 1$ when $h = 0.025$. Calculate the error ratio E_2/E_1. Is the ratio close to $1/8$?

24. $y' = \dfrac{t}{y+1}, \quad y(0) = 1$

25. $y' = 2ty^2, \quad y(0) = -1$

26. $y' = \dfrac{1}{2y}, \quad y(0) = 1$

27. $y' = \dfrac{1+y^2}{1+t}, \quad y(0) = 0$

9.3 Runge-Kutta Methods

In this section we discuss Runge-Kutta[5] methods as a way of numerically solving the initial value problem

$$y' = f(t, y), \qquad y(t_0) = y_0. \tag{1}$$

Runge-Kutta methods are based on Taylor series methods, but they use nested compositions of function evaluations instead of the partial derivatives of $f(t, y)$ required by a Taylor series method. In theory, one can achieve any desired level of accuracy using the Runge-Kutta approach.

Heun's method and the modified Euler's method are two familiar algorithms that use the Runge-Kutta philosophy of evaluating compositions of functions. For instance, Heun's method has the form

$$y_{n+1} = y_n + \frac{h}{2}\left[f(t_n, y_n) + f(t_n + h, y_n + hf(t_n, y_n))\right], \qquad n = 0, 1, 2, \ldots, N-1.$$

Heun's method is easy to implement—in order to take a step, we need only evaluate the function $f(t, y)$ at the current estimate (t_n, y_n) and at the point $(t_n + h, y_n + hf(t_n, y_n))$. Moreover, as is shown in Example 1, Heun's method is a second order method. In contrast, a second order Taylor series method requires the calculation of two partial derivatives, $f_t(t_n, y_n)$ and $f_y(t_n, y_n)$, in order to make a step with comparable second order accuracy.

EXAMPLE 1

Calculate the order of Heun's method,

$$y_{n+1} = y_n + \frac{h}{2}\left[f(t_n, y_n) + f(t_n + h, y_n + hf(t_n, y_n))\right]. \tag{2}$$

Solution: Let $y(t)$ denote the unique solution of the initial value problem (1). To determine the order of the one-step method (2), we need to find an expression for the local truncation errors, T_{n+1} [recall equation (12) in Section 9.2].

(continued)

[5] Carle David Tolmé Runge (1856–1927) was a German scientist whose initial interest in pure mathematics was eventually supplanted by an interest in spectroscopy and applied mathematics. During his career, he held positions at universities in Hanover and Gottingen.

Martin Wilhelm Kutta (1867–1944) held positions at Munich, Jena, Aachen, and Stuttgart. In addition to the Runge-Kutta method (1901), he is remembered for his work in the study of airfoils.

(continued)

Assume that we apply Heun's method starting on the exact solution curve at t_n; that is, with $y_n = y(t_n)$. To determine the local truncation error, we must first unravel the composition $f(t_n + h, y_n + hf)$ [where functions without arguments will be assumed to be evaluated at (t_n, y_n)]. Expanding $f(t_n + h, y_n + hf)$ in a Taylor series about (t_n, y_n), we obtain

$$f(t_n + h, y_n + hf) = f + (f_t + f_y f)h + \frac{1}{2}(f_{tt} + 2f_{ty}f + f_{yy}f^2)h^2 + O(h^3). \quad (3)$$

Using this expansion in (2)

$$\begin{aligned} y_{n+1} &= y_n + \frac{h}{2}\left[f + f + (f_t + f_y f)h + \frac{1}{2}(f_{tt} + 2f_{ty}f + f_{yy}f^2)h^2 + O(h^3)\right] \\ &= y_n + fh + (f_t + f_y f)\frac{h^2}{2} + (f_{tt} + 2f_{ty}f + f_{yy}f^2)\frac{h^3}{4} + O(h^4). \end{aligned} \quad (4)$$

We now compare this expansion with the Taylor series of the exact solution, $y(t_{n+1})$. Using the fact that $y(t_n) = y_n$ and using the expressions for $y'(t_n), y''(t_n), y'''(t_n)$ derived in Section 9.2, we have

$$y(t_{n+1}) = y_n + fh + (f_t + f_y f)\frac{h^2}{2} + (f_{tt} + 2f_{ty}f + f_{yy}f^2 + f_y f_t + f_y^2 f)\frac{h^3}{6} + O(h^4). \quad (5)$$

Comparing expansions (4) and (5), we see that they agree up to and including the $O(h^2)$ terms but that the $O(h^3)$ term in the Heun method expansion does not correctly replicate the $O(h^3)$ term in the Taylor series of the exact solution. Therefore, the local truncation error of Heun's method is $T_{n+1} = O(h^3)$ and hence Heun's method is second order. ▲

Second Order Runge-Kutta Methods

To generalize the approach suggested by Heun's method, we choose a set of points $(\theta_i, \gamma_i), i = 1, 2, \dots, k$, that lie in the ty-plane, in the vertical strip bounded by the lines $t = t_n$ and $t = t_{n+1}$. As Figure 9.3 suggests, these points sample

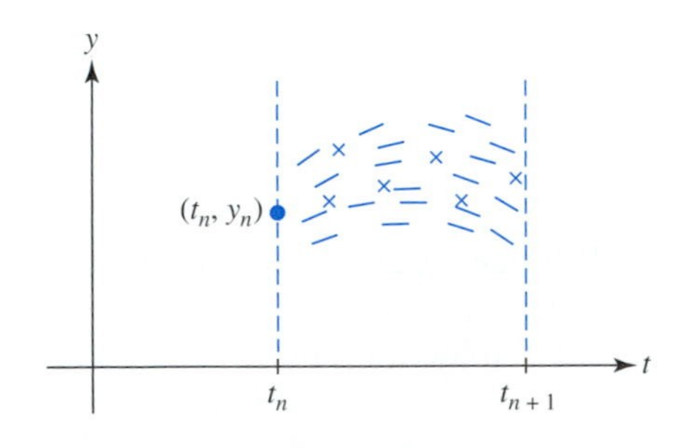

FIGURE 9.3

A portion of the direction field for $y' = f(t, y)$ near our latest estimate (t_n, y_n). We use a weighted sum of direction field evaluations at the points marked "×," to evolve the numerical solution from t_n to t_{n+1}.

the direction field in the vicinity of the point (t_n, y_n). To formalize this idea of sampling the direction field, consider a one-step method

$$y_{n+1} = y_n + h\phi(t_n, y_n; h), \tag{6}$$

where the increment function is defined by

$$\phi(t_n, y_n; h) \equiv A_1 f(\theta_1, \gamma_1) + A_2 f(\theta_2, \gamma_2) + \cdots + A_k f(\theta_k, \gamma_k). \tag{7}$$

The constants $A_1, A_2, \ldots, A_k$ are the **weights** of the method (6). Thus, the increment function is a weighted sum of direction field slopes. For a fixed integer k, the local truncation error is reduced by selecting weights, $\{A_i\}_{i=1}^k$, and direction field sampling points, $\{(\theta_i, \gamma_i)\}_{i=1}^k$, so that the method (6) replicates as many terms in the Taylor series expansion of the local solution as possible.

When $k = 2$, method (6) has the form

$$y_{n+1} = y_n + h[A_1 f(\theta_1, \gamma_1) + A_2 f(\theta_2, \gamma_2)]. \tag{8}$$

We need to choose the sampling points (θ_1, γ_1) and (θ_2, γ_2), and weights A_1 and A_2. Since we are viewing the term $A_1 f(\theta_1, \gamma_1) + A_2 f(\theta_2, \gamma_2)$ as an average slope, we want the sampling points to be near (t_n, y_n) and to be representative of the direction field between $t = t_n$ and $t = t_{n+1}$. A reasonable choice for one of the sampling points is $(\theta_1, \gamma_1) = (t_n, y_n)$. For a second point, our previous study suggests that we should sample somewhere along the "Euler line"—the line of slope $f(t_n, y_n)$ that passes through the point (t_n, y_n). Thus, as a second sampling point, we choose

$$(\theta_2, \gamma_2) = (t_n + \alpha h, y_n + \alpha h f(t_n, y_n))$$

where α is a constant, $0 < \alpha \leq 1$. See Figure 9.4.

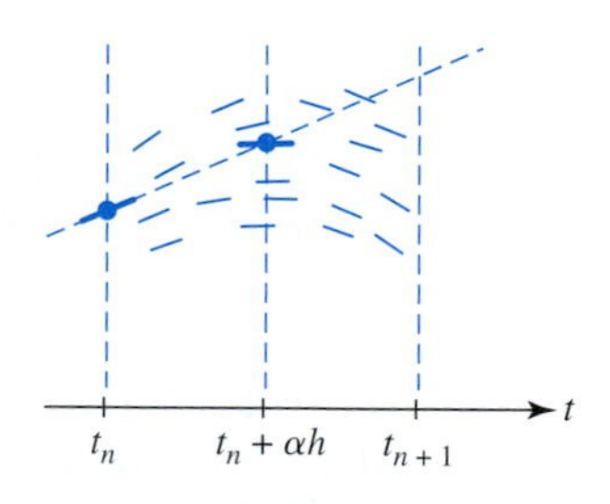

FIGURE 9.4

Given the two-sample method (8), we generally choose one sample at (t_n, y_n) and the second somewhere along the "Euler line," at $(t_n + \alpha h, y_n + \alpha h f(t_n, y_n))$ where α is a constant, $0 < \alpha \leq 1$.

With the choices shown in Figure 9.4, method (8) has the form

$$y_{n+1} = y_n + h\left[A_1 f(t_n, y_n) + A_2 f(t_n + \alpha h, y_n + \alpha h f(t_n, y_n))\right]. \tag{9}$$

We now need to select weights A_1 and A_2. Since the right-hand side of equation (9) is a function of h, it makes sense to expand the right-hand side in powers of h, with the objective of choosing the weights so that y_{n+1} matches a Taylor series method through as many powers of h as possible.

Expanding the right-hand side of (9) using a two-variable Taylor series,

$$\begin{aligned} y_{n+1} &= y_n + h[A_1 f(t_n, y_n) + A_2 f(t_n + \alpha h, y_n + \alpha h f(t_n, y_n))] \\ &= y_n + h[A_1 f(t_n, y_n) + A_2 \{f(t_n, y_n) + f_t(t_n, y_n)\alpha h \\ &\qquad\qquad + f_y(t_n, y_n)\alpha h f(t_n, y_n) + O(h^2)\}] \\ &= y_n + h(A_1 + A_2) f(t_n, y_n) + h^2 \alpha A_2 \left[f_t(t_n, y_n) + f_y(t_n, y_n) f(t_n, y_n) \right] + O(h^3). \end{aligned}$$

We now attempt to match this expansion with the Taylor series method of order two (denoted by y^T_{n+1}),

$$y^T_{n+1} = y_n + hf(t_n, y_n) + \frac{h^2}{2} \left[f_t(t_n, y_n) + f_y(t_n, y_n) f(t_n, y_n) \right].$$

Our objective is to select the parameters at our disposal, A_1, A_2, and α, so as to maximize the agreement between the expansions of y_{n+1} and y^T_{n+1}. Comparing the two expansions, we see we can obtain agreement through terms of order h^2 if A_1, A_2, and α satisfy the equations

$$\begin{aligned} A_1 + A_2 &= 1, \\ \alpha A_2 &= \frac{1}{2}. \end{aligned} \tag{10}$$

Once we satisfy these constraints, the method (9) matches the second order Taylor series method up through terms of order h^2 and therefore, like the second order Taylor series method, has an $O(h^3)$ local truncation error. [This is the best we can do with method (9). It is impossible to select A_1, A_2, and α to match the terms of the third order Taylor series method.]

In (10) we have a system of two (nonlinear) equations in three unknowns. This system has infinitely many solutions,

$$A_2 = \frac{1}{2\alpha} \quad \text{and} \quad A_1 = 1 - \frac{1}{2\alpha}, \tag{11}$$

with $0 < \alpha \le 1$. Since α represents the fraction of the step we move along the Euler line to the second sampling point,

$$(\theta_2, \gamma_2) = (t_n + \alpha h, y_n + \alpha h f(t_n, y_n)),$$

there are two "natural" choices for α, namely $\alpha = 1/2$ and $\alpha = 1$. If $\alpha = 1$ in equation (11), then $A_1 = 1/2$ and $A_2 = 1/2$. With this choice, method (9) reduces to Heun's method,

$$y_{n+1} = y_n + \frac{h}{2} \left[f(t_n, y_n) + f(t_n + h, y_n + hf(t_n, y_n)) \right].$$

If $\alpha = 1/2$ in equation (11), then $A_1 = 0$ and $A_2 = 1$. With this choice, method (9) reduces to the modified Euler's method,

$$y_{n+1} = y_n + hf\left(t_n + \frac{h}{2}, y_n + \frac{hf(t_n, y_n)}{2} \right).$$

R-stage Runge-Kutta Methods

In general, a Runge-Kutta method has the form

$$y_{n+1} = y_n + h\phi(t_n, y_n; h), \tag{12a}$$

where the increment function, $\phi(t_n, y_n; h)$, is given by

$$\phi(t_n, y_n; h) = \sum_{j=1}^{R} A_j K_j(t_n, y_n). \tag{12b}$$

In (12b), the terms A_j are constants (the weights) and the terms $K_j(t_n, y_n)$ are direction field samples, usually called **stages**. The stages are defined sequentially as follows:

$$K_1(t_n, y_n) = f(t_n, y_n)$$

$$K_j(t_n, y_n) = f(t_n + \alpha_j h, y_n + h\sum_{i=1}^{j-1} \beta_{ji} K_i(t_n, y_n)), \qquad j = 2, 3, \ldots, R, \tag{12c}$$

where $0 < \alpha_j \le 1$ and where $\beta_{j,1} + \beta_{j,2} + \cdots + \beta_{j,j-1} = \alpha_j$.

Method (12) is called an ***R*-stage Runge-Kutta method**. A Runge-Kutta method can be viewed as a "staged" sampling process. That is, for each j, we choose a value α_j that determines the t-coordinate of the jth sampling point. Then, see (12c), the y-coordinate of the jth sampling point is determined using the prior stages. In this sense, the sampling process is recursive. In (12c), the constraint $0 < \alpha_j \le 1$ means that all sampling points lie between $t = t_n$ and $t = t_{n+1}$. While this description of an R-stage Runge-Kutta method may seem complicated, the format of equation (12) makes programming a Runge-Kutta method very simple, see Figures 9.7 and 9.8 on page 568.

An example of a three-stage Runge-Kutta method is

$$\begin{aligned}
y_{n+1} &= y_n + \frac{h}{6}(K_1 + 4K_2 + K_3) \\
K_1 &= f(t_n, y_n), \\
K_2 &= f\left(t_n + \frac{1}{2}h, y_n + \frac{1}{2}hK_1\right), \\
K_3 &= f(t_n + h, y_n - hK_1 + 2hK_2).
\end{aligned} \tag{13}$$

It is not difficult to show that method (13) has order three; it matches the third order Taylor series method up through terms of order h^3 but not the order h^4 term.

As we saw in equations (9) and (11), there are infinitely many two-stage, second order Runge-Kutta methods. Similarly, there is an infinite two-parameter family of three-stage, third order Runge-Kutta methods, see Exercises 31–34. Likewise, when the parameters in (12) are chosen properly, there are four-stage, fourth order Runge-Kutta methods. One of the most popular fourth order Runge-Kutta methods is

$$\begin{aligned}
y_{n+1} &= y_n + \frac{h}{6}(K_1 + 2K_2 + 2K_3 + K_4), \\
K_1 &= f(t_n, y_n), \\
K_2 &= f\left(t_n + \frac{h}{2}, y_n + \frac{h}{2}K_1\right), \\
K_3 &= f\left(t_n + \frac{h}{2}, y_n + \frac{h}{2}K_2\right), \\
K_4 &= f(t_n + h, y_n + hK_3).
\end{aligned} \tag{14}$$

Viewing algorithm (14) geometrically, we can envision it as being formed in the following way. First we calculate K_1, the slope of the tangent line at starting point (t_n, y_n). We proceed a half-step along this tangent line to locate the direction field point at which slope K_2 is evaluated. We use this new slope K_2 to define another line through (t_n, y_n). Proceeding, in turn, a half step along this new line locates the point that determines slope K_3. Finally we proceed a full step from (t_n, y_n) along the line having slope K_3 to determine the point at which slope K_4 is evaluated. The appropriately weighted average of these four slopes defines the algorithm.

Runge-Kutta Methods for Systems

The discussion in this chapter has focused on the scalar initial value problem

$$y' = f(t, y), \qquad y(t_0) = y_0.$$

As mentioned in the Introduction to this chapter, however, the ideas developed and the ensuing methods extend naturally to first order systems. Consider the initial value problem

$$\mathbf{y}' = \mathbf{f}(t, \mathbf{y}), \qquad \mathbf{y}(t_0) = \mathbf{y}_0, \tag{15}$$

where

$$\mathbf{y}(t) = \begin{bmatrix} y_1(t) \\ y_2(t) \\ \vdots \\ y_m(t) \end{bmatrix}, \qquad \mathbf{y}_0 = \begin{bmatrix} y_0^{(1)} \\ y_0^{(2)} \\ \vdots \\ y_0^{(m)} \end{bmatrix}$$

and

$$\mathbf{f}(t, \mathbf{y}) = \begin{bmatrix} f_1(t, \mathbf{y}) \\ f_2(t, \mathbf{y}) \\ \vdots \\ f_m(t, \mathbf{y}) \end{bmatrix} = \begin{bmatrix} f_1(t, y_1, y_2, \ldots, y_m) \\ f_2(t, y_1, y_2, \ldots, y_m) \\ \vdots \\ f_m(t, y_1, y_2, \ldots, y_m) \end{bmatrix}.$$

The concept of an analytic function developed in Section 9.2 can be extended to the vector-valued functions $\mathbf{y}(t)$ and $\mathbf{f}(t, \mathbf{y})$. Theorem 9.1 can be extended to give analogous conditions sufficient for the existence of an analytic solution of (15) on an interval of the form $t_0 - \delta < t < t_0 + \delta$ for some $\delta > 0$.

We saw in Section 6.9 how Euler's method can be extended to systems. The higher order Runge-Kutta methods likewise extend naturally to initial value problems such as (15). For example, the system counterpart of algorithm (14) is

$$\begin{aligned}
\mathbf{y}_{n+1} &= \mathbf{y}_n + \frac{h}{6}\left[\mathbf{K}_1 + 2\mathbf{K}_2 + 2\mathbf{K}_3 + \mathbf{K}_4\right], \\
\mathbf{K}_1 &= \mathbf{f}(t_n, \mathbf{y}_n), \\
\mathbf{K}_2 &= \mathbf{f}\left(t_n + \frac{h}{2}, \mathbf{y}_n + \frac{h}{2}\mathbf{K}_1\right), \\
\mathbf{K}_3 &= \mathbf{f}\left(t_n + \frac{h}{2}, \mathbf{y}_n + \frac{h}{2}\mathbf{K}_2\right), \\
\mathbf{K}_4 &= \mathbf{f}(t_n + h, \mathbf{y}_n + h\mathbf{K}_3).
\end{aligned} \tag{16}$$

The Damped Pendulum

The next example illustrates how we can apply a Runge-Kutta method to a first order system of the form

$$\mathbf{y}' = \mathbf{f}(t, \mathbf{y}), \qquad \mathbf{y}(t_0) = \mathbf{y}_0.$$

EXAMPLE 2

Consider a pendulum whose motion is influenced not only by its weight but also by a resistive or damping force. The mathematical formulation of this problem leads to an initial value problem involving a scalar second order nonlinear differential equation. We rewrite this scalar second order problem as an equivalent problem for a first order (nonlinear) system and then use the fourth order Runge-Kutta method (16), with a step size of $h = 0.05$, to obtain a numerical solution.

Problem Formulation The pendulum is formed by a mass m attached to a rod of length l (see Figure 9.5). We neglect the mass of the rod. As the pendulum moves, it is acted upon by the force of gravity and also by a damping force, which acts to retard the pendulum motion. We assume this damping force is

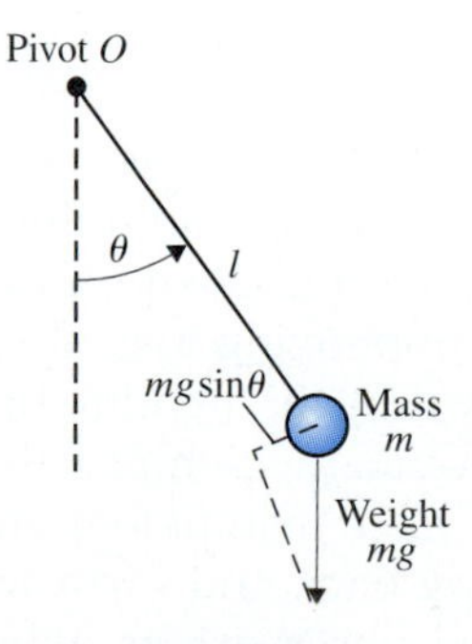

FIGURE 9.5

The damped pendulum described in Example 2.

(continued)

(continued)

proportional to the angular velocity of the pendulum and acts in the tangential direction as shown. The resulting motion is described by

$$ml^2\theta'' = -mgl\sin\theta - \kappa l\theta' \quad \text{or} \quad \theta'' + \frac{\kappa}{ml}\theta' + \frac{g}{l}\sin\theta = 0,$$

where κ is a positive damping constant. We complete the formulation by specifying both θ and θ' at the initial time of interest, say $t = 0$. These two constants give the initial position and initial angular velocity of the pendulum. For the purposes of this example, we adopt the numerical values

$$\frac{\kappa}{ml} = 0.2\ \mathrm{s}^{-1}, \qquad \frac{g}{l} = 1\ \mathrm{s}^{-2}, \qquad \theta(0) = 0\ \mathrm{rad}, \qquad \theta'(0) = 3\ \mathrm{rad/s}.$$

With this, the initial value problem of interest becomes

$$\theta'' + 0.2\theta' + \sin\theta = 0, \qquad \theta(0) = 0, \qquad \theta'(0) = 3.$$

The differential equation is recast as a first order system by defining

$$y_1(t) = \theta(t), \qquad y_2(t) = \theta'(t) \quad \text{and} \quad \mathbf{y}(t) = \begin{bmatrix} y_1(t) \\ y_2(t) \end{bmatrix}.$$

The initial value problem becomes

$$\mathbf{y}' = \begin{bmatrix} y_2 \\ -\sin y_1 - 0.2y_2 \end{bmatrix}, \qquad \mathbf{y}(0) = \begin{bmatrix} 0 \\ 3 \end{bmatrix}. \tag{17}$$

We will solve initial value problem (17) numerically using algorithm (16).

What Should We Expect? Before embarking on a numerical solution, it's usually worthwhile to bring any available physical insights to bear that will help determine what to expect. We know that, generally, solutions of nonlinear initial value problems do not exist on arbitrarily large time intervals. However, because of the nature of the pendulum motion it describes, we expect the exact solution of (17) to exist on an arbitrarily large time interval.

We saw in Chapter 8 that, in the absence of damping, a pendulum starting at $\theta(0) = 0$ with $\theta'(0) = 2$ has just enough energy to reach the inverted position (in the limit as $t \to \infty$). In our case, the initial angular velocity is greater, since $\theta'(0) = 3$. Damping, however, retards the motion and causes the pendulum to lose energy. If damping is not too large, we expect the pendulum to go past the inverted position at least once. If damping is large enough, however, the accompanying loss of energy will more than offset the increase in initial energy and the pendulum will not reach the inverted position. It's not clear at the outset which possibility will occur. In any event, the pendulum eventually will have insufficient energy to reach the inverted position and it will simply swing back and forth with decreasing amplitude as time increases. Based on these observations, what do you expect the graphs of $\theta(t)$ and $\theta'(t)$ to look like?

Interpreting the Results Figure 9.6 shows the results of the numerical simulation. Note that the graph of $y_1(t) = \theta(t)$ increases from zero to a maximum of

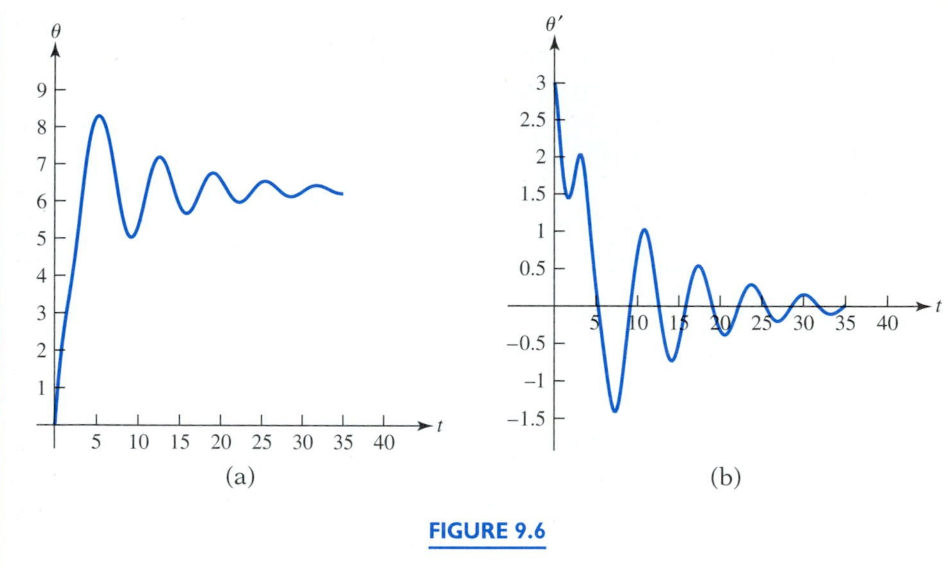

FIGURE 9.6

(a) The graph of $y_1(t) = \theta(t)$. (b) The graph of $y_2(t) = \theta'(t)$.

about 8.29 rad. Since $2\pi \approx 6.28$, the graph tells us that the pendulum makes one complete counterclockwise revolution, rotating an additional 2 rad $\approx 115°$ beyond the vertically downward position before falling back, beginning to swing back and forth with decreasing amplitude as time progresses. The graph has a horizontal asymptote of 2π since the pendulum approaches the vertically downward rest position as $t \to \infty$.

Is the graph of $y_2(t) = \theta'(t)$ consistent with this physical interpretation? What does the initial minimum and maximum (for $t > 0$) of this graph correspond to? Should they occur while the dependent variable (angular velocity) is positive? Should the zero crossings of this graph occur at the critical points of $y_1(t) = \theta(t)$? Should the maxima of $y_2(t) = \theta'(t)$ occur when the pendulum is in the vertically downward position? Should the graph of $y_2(t) = \theta'(t)$ have a horizontal asymptote of zero? Subjecting your numerical solution to simple commonsense checks such as these is an important final step. ▲

Coding a Runge-Kutta Method

We conclude this section with a short discussion about the practical aspects of writing a program to implement a Runge-Kutta method. We list in Figures 9.7 and 9.8 (on the next page) the program used to generate the numerical solution of Example 2. This particular code was written in MATLAB, but the principles are the same for any programming language.

Note first that no matter what numerical method we decide to use for the initial value problem

$$\mathbf{y}' = \mathbf{f}(t, \mathbf{y}), \qquad \mathbf{y}(t_0) = \mathbf{y}_0,$$

we need to write a subprogram (or module) that evaluates $\mathbf{f}(t, \mathbf{y})$. Such a module is listed in Figure 9.8, for the initial value problem of Example 2. Figure 9.7 lists

```
%
%  Set the initial conditions for the
%    initial value problem of Example 2
%
t=0;
y=[0,3]';
h=0.05;
output=[t,y(1),y(2)];
%
%  Execute the fourth order Runge-Kutta method
%    on the interval [0, 30]
%
for i=1:600
   ttemp=t;
   ytemp=y;
   k1=f(ttemp,ytemp);
   ttemp=t+h/2;
   ytemp=y+(h/2)*k1;
   k2=f(ttemp,ytemp);
   ttemp=t+h/2;
   ytemp=y+(h/2)*k2;
   k3=f(ttemp,ytemp);
   ttemp=t+h;
   ytemp=y+h*k3;
   k4=f(ttemp,ytemp);
   y=y+(h/6)*(k1+2*k2+2*k3+k4);
   t=t+h;
   output=[output;t,y(1),y(2)];
end
```

FIGURE 9.7

A Runge-Kutta code for the initial value problem in Example 2.

```
function yp=f(t,y)
yp=zeros(2,1);
yp(1)=y(2);
yp(2)=-sin(\hspace*{1.25pt}y(1))-0.2*y(2);
```

FIGURE 9.8

A function subprogram that evaluates $\mathbf{f}(t, \mathbf{y})$ for the differential equation of Example 2.

a MATLAB program that executes 600 steps of the fourth order Runge-Kutta method (16), for the initial value problem of Example 2.

Observe that the code listed in Figure 9.7 stays as close as possible to the notation and format of the fourth order Runge-Kutta method (16). In general, it is always a good idea to use variable names (such as k_1, and k_2) that match the names in the algorithm. Beyond the choice of variable names, the code in Figure 9.8 also mimics the steps of algorithm (16) as closely as possible. Adhering to such conventions make programs easier to read and debug.

EXERCISES

Exercises 1–10:

We reconsider the initial value problems studied in the exercises of Section 9.2. The solution of the differential equation satisfying initial condition $y(0) = 1$ is given.

(a) Carry out one step of the third order Runge-Kutta method (13) using a step size $h = 0.1$, obtaining a numerical approximation of the exact solution at $t = 0.1$.

(b) Carry out one step of the fourth order Runge-Kutta method (14) using a step size $h = 0.1$, obtaining a numerical approximation of the exact solution at $t = 0.1$.

(c) Examine the exact solution. Should either or both of the Runge-Kutta methods, in principle, yield an exact answer for the particular problem being considered? Explain.

(d) Compare the numerical values obtained in parts (a) and (b) with the exact solution evaluated at $t = 0.1$. Are the results consistent with your answer in part (c)? Is the error incurred using the four-stage algorithm less than the error of the three-stage calculation?

1. $y' = -y + 2;\ \ y(t) = 2 - e^{-t}$

2. $y' = 2ty;\ \ y(t) = e^{t^2}$

3. $y' = ty^2;\ \ y(t) = \dfrac{2}{2 - t^2}$

4. $y' = t^2 + y;\ \ y(t) = 3e^t - (t^2 + 2t + 2)$

5. $y' = \sqrt{y};\ \ y(t) = \left(1 + \dfrac{t}{2}\right)^2$

6. $y' = \dfrac{t}{y};\ \ y(t) = \sqrt{1 + t^2}$

7. $y' = y + \sin t;\ \ y(t) = \dfrac{3e^t - \cos t - \sin t}{2}$

8. $y' = y^{3/4};\ \ y(t) = \left(1 + \dfrac{t}{4}\right)^4$

9. $y' = 1 + y^2;\ \ y(t) = \tan\left(t + \dfrac{\pi}{4}\right)$

10. $y' = -4t^3y;\ \ y(t) = e^{-t^4}$

Exercises 11–16:

For the given initial value problem, an exact solution in terms of familiar functions is not available for comparison. If necessary, rewrite the problem as an initial value problem for a first order system. Implement one step of the fourth order Runge-Kutta method (14) using a step size $h = 0.1$ to obtain a numerical approximation of the exact solution at $t = 0.1$.

11. $y'' + ty' + y = 0,\ \ y(0) = 1,\ \ y'(0) = -1$

12. $\dfrac{d}{dt}\left(e^t \dfrac{dy}{dt}\right) + ty = 1,\ \ y(0) = 1,\ \ y'(0) = 2$

13. $\mathbf{y}' = \begin{bmatrix} 0 & t \\ e^t & 0 \end{bmatrix}\mathbf{y} + \begin{bmatrix} 1 \\ t \end{bmatrix},\ \ \mathbf{y}(0) = \begin{bmatrix} 2 \\ 1 \end{bmatrix}$

14. $\mathbf{y}' = \begin{bmatrix} -1 & t \\ 2 & 0 \end{bmatrix}\mathbf{y},\ \ \mathbf{y}(0) = \begin{bmatrix} -1 \\ 1 \end{bmatrix}$

15. $y''' - ty = 0,\ \ y(0) = 1,\ \ y'(0) = 0,\ \ y''(0) = -1$

16. $y'' + z + ty = 0$

$z' - y = t,\ \ y(0) = 1,\ \ y'(0) = 2,\ \ z(0) = 0$

Exercises 17–18:

One differential equation for which we can explicitly demonstrate the order of the Runge-Kutta algorithm is the linear homogeneous equation $y' = \lambda y$, where λ is a constant.

17. (a) Verify that the exact solution of $y' = \lambda y, y(t_0) = y_0$ is $y(t) = y_0 e^{\lambda(t-t_0)}$.

(b) Show, for the three-stage Runge-Kutta method (13), that

$$y(t_n) + h\phi(t_n, y(t_n); h) = y(t_n)\left[1 + \lambda h + \frac{(\lambda h)^2}{2!} + \frac{(\lambda h)^3}{3!}\right].$$

(c) Show that $y(t_{n+1}) = y(t_n)e^{\lambda h}$

(d) What is the order of the local truncation error?

18. Repeat the calculations of Exercise 17 using the four-stage Runge-Kutta method (14). In this case, show that the local truncation error is $O(h^5)$.

Exercises 19–22:

In these exercises, we ask you to use the fourth order Runge-Kutta method (14) to solve the problems in Exercises 20–23 of Section 9.2.

(a) For the given initial value problem, execute 20 steps of the method (14); use step size $h = 0.05$.

(b) The exact solution is given. Compare the numerical approximation y_{20} and the exact solution $y(t_{20}) = y(1)$.

19. $y' = \dfrac{t}{y+1}, \quad y(0) = 1.$ The exact solution is $y(t) = -1 + \sqrt{t^2 + 4}$.

20. $y' = 2ty^2, \quad y(0) = -1.$ The exact solution is $y(t) = \dfrac{-1}{1+t^2}$.

21. $y' = \dfrac{1}{2y}, \quad y(0) = 1.$ The exact solution is $y(t) = \sqrt{1+t}$.

22. $y' = \dfrac{1+y^2}{1+t}, \quad y(0) = 0.$ The exact solution is $y(t) = \tan\left(\ln(1+t)\right)$.

Exercises 23–25:

In each exercise,

(a) Verify that the given function is the solution of the initial value problem posed. If the initial value problem involves a higher order scalar differential equation, rewrite it as an equivalent initial value problem for a first order system.

(b) Execute the fourth order Runge-Kutta method (16) over the specified t-interval using step size $h = 0.1$ to obtain a numerical approximation of the exact solution. Tabulate the components of the numerical solution with their exact solution counterparts at the endpoint of the specified interval.

23. $y'' + 2y' + 2y = -2, \quad y(0) = 0, \quad y'(0) = 1; \quad y(t) = e^{-t}(\cos t + 2\sin t) - 1; \quad 0 \le t \le 2$

24. $\mathbf{y}' = \begin{bmatrix} -1 & \frac{1}{2} \\ \frac{1}{2} & -1 \end{bmatrix}\mathbf{y}, \quad \mathbf{y}(0) = \begin{bmatrix} 2 \\ 0 \end{bmatrix}; \quad \mathbf{y}(t) = \begin{bmatrix} e^{-t/2} + e^{-3t/2} \\ e^{-t/2} - e^{-3t/2} \end{bmatrix}; \quad 0 \le t \le 1,$

25. $t^2y'' - ty' + y = t^2, \quad y(1) = 2, \quad y'(1) = 2; \quad y(t) = t(t + 1 - \ln t); \quad 1 \le t \le 2$

Exercises 26–30:

These exercises ask you to use numerical methods to study the behavior of some scalar second order initial value problems. In each exercise, use the fourth order Runge-Kutta method (16) and step size $h = 0.05$ to solve the problem over the given interval.

26. $y'' + 4(1 + 3\tanh t)y = 0, \quad y(0) = 1, \quad y'(0) = 0; \quad 0 \le t \le 10.$

This problem might model the motion of a spring-mass system in which the mass is released from rest with a unit initial displacement at $t = 0$ and the spring stiffens as the motion progresses in time. Plot the numerical solutions for $y(t)$ and $y'(t)$. Since $\tanh t$ approaches 1 for large values of t, we might expect the solution to approximate a solution of $y'' + 16y = 0$ for time t sufficiently large. Do your graphs support this conjecture?

27. $y'' + y + y^3 = 0,\ \ y(0) = 0,\ \ y'(0) = 1;\quad 0 \le t \le 10.$

This nonlinear differential equation arose in modeling the motion of a bobbing sphere. We are interested in assessing the impact of the nonlinear y^3 term upon the motion. Plot the numerical solution for $y(t)$. If the nonlinear term were not present, the initial value problem would have solution $y(t) = \sin t$. On the same graph, plot the function $\sin t$. Does the nonlinearity increase or decrease the period of the motion? How do the amplitudes of the motion differ?

28. $\theta'' + \sin\theta = 0.2\sin t,\ \ \theta(0) = 0,\ \ \theta'(0) = 0;\quad 0 \le t \le 50.$

This nonlinear differential equation is used to model the forced motion of a pendulum initially at rest in the vertically downward position. For small angular displacements, the approximation $\sin\theta \approx \theta$ is often used in the differential equation. Note, however, that the solution of the resulting initial value problem $\theta'' + \theta = 0.2\sin t, \theta(0) = 0, \theta'(0) = 0$ is given by $\theta(t) = -0.1(\sin t - t\cos t)$, leading to pendulum oscillations that continue to grow in amplitude as time increases. Our goal is to determine how the nonlinear $\sin\theta$ term affects the motion. Plot the numerical solutions for $\theta(t)$ and $\theta'(t)$. Describe in simple terms what the pendulum is doing on the time interval considered.

29. $\theta'' + \sin\theta = 0,\ \ \theta(0) = 0,\ \ \theta'(0) = 2;\quad 0 \le t \le 20.$

This problem is used to model pendulum motion when the pendulum is initially in the vertically downward position with an initial angular velocity of 2 rad/s. For this conservative system, it was shown in Chapter 8 that $(\theta')^2 - 2\cos\theta = 2$. Therefore, the initial conditions have been chosen so that the pendulum will rotate upward in the positive (counterclockwise) direction, slowing down and approaching the vertically upward position as $t \to \infty$. The phase-plane solution point is moving on the separatrix; thus, loosely speaking, the exact solution is "moving on a knife's edge." If the initial velocity is slightly less, the pendulum will not reach the upright position but will reach a maximum value less than π and then proceed to swing back and forth. If the initial velocity is slightly greater, the pendulum passes through the vertically upright position and continues to rotate counterclockwise. What happens if we solve this problem numerically? Plot the numerical solutions for $\theta(t)$ and $\theta'(t)$. Interpret in simple terms what the numerical solution is saying about the pendulum motion on the time interval considered. Does the numerical solution conserve energy?

30. $mx'' + \dfrac{2k\delta}{\pi}\tan\left(\dfrac{\pi x}{2\delta}\right) = F(t),\ \ x(0) = 0,\ \ x'(0) = 0;\quad 0 \le t \le 15.$

This problem was used to model a nonlinear spring-mass system (see Exercise 14 in Section 8.1). The motion is assumed to occur on a frictionless horizontal surface. In this equation, m is the mass of the object attached to the spring, $x(t)$ is the horizontal displacement of the mass from the unstretched equilibrium position, and δ is the length that the spring can contract or elongate. The spring restoring force has vertical asymptotes at $x = \pm\delta$. Time t is in seconds.

Let $m = 100$ kg, $\delta = 0.15$, and $k = 100$ n/m. Assume that the spring-mass system is initially at rest with the spring at its unstretched length. At time $t = 0$ a force of large amplitude but short duration is applied:

$$F(t) = \begin{cases} F_0 \sin\pi t, & 0 \le t \le 1 \\ 0, & 1 < t < 15 \end{cases} \quad \text{newtons.}$$

Solve the problem numerically for the two cases $F_0 = 4$ n and $F_0 = 40$ n. Plot the corresponding displacements on the same graph. How do they differ?

Exercises 31–34:

Third Order Runge-Kutta methods As given in equation (12), the form of a three-stage Runge-Kutta method is

$$y_{n+1} = y_n + h[A_1K_1(t_n, y_n) + A_2K_2(t_n, y_n) + A_3K_3(t_n, y_n)] \tag{18a}$$

where

$$\begin{aligned} K_1(t_n, y_n) &= f(t_n, y_n) \\ K_2(t_n, y_n) &= f(t_n + \alpha_2 h, y_n + h\beta_{2,1}K_1(t_n, y_n)) \\ K_3(t_n, y_n) &= f(t_n + \alpha_3 h, y_n + h[\beta_{3,1}K_1(t_n, y_n) + \beta_{3,2}K_2(t_n, y_n)]) \end{aligned} \tag{18b}$$

and where, see equation (12c), $0 < \alpha_2 \le 1$, $0 < \alpha_3 \le 1$, $\beta_{2,1} = \alpha_2$, $\beta_{3,1} + \beta_{3,2} = \alpha_3$. It can be shown (see Lambert[6]) that this three-stage Runge-Kutta method has order 3 if the following four equations are satisfied

$$\begin{aligned} A_1 + A_2 + A_3 &= 1 \\ \alpha_2 A_2 + \alpha_3 A_3 &= \frac{1}{2} \\ \alpha_2^2 A_2 + \alpha_3^2 A_3 &= \frac{1}{3} \\ \alpha_2 \beta_{3,2} A_3 &= \frac{1}{6} \end{aligned} \tag{19}$$

One way to find a solution of this system of four nonlinear equations is first to select values for α_2 and α_3. [Note that α_2 and α_3 determine the t coordinate of the sampling points defining $K_2(t_n, y_n)$ and $K_3(t_n, y_n)$, respectively.] Once α_2 and α_3 are chosen, the first three equations in (19) can be solved for A_1, A_2, and A_3. The parameters α_2 and α_3 are nonzero; if they are distinct, then there are unique values A_1, A_2, and A_3 that satisfy the first three equations. Having A_1, A_2, and A_3, you can determine $\beta_{3,2}$ from the fourth equation and $\beta_{3,1}$ from the condition $\beta_{3,1} + \beta_{3,2} = \alpha_3$.

31. Consider a three-stage Runge-Kutta method. Show that if the first equation in system (19) holds, then the method has order at least 1.

32. Consider a three-stage Runge-Kutta method. Show that if the first and second equations in system (19) hold, then the method has order at least 2. [Note: In order to obtain order 3, the last two equations in (19) must hold as well.]

33. Determine the values of α_2 and α_3 that give rise to the three-stage third order Runge-Kutta method (13). Then solve equation (19) and verify that Runge-Kutta method (13) results.

34. (a) Verify that the choice of $\alpha_2 = 7/10$ and $\alpha_3 = 2/10$ leads to another solution of (19) having the same weights A_1, A_2, and A_3 as (13).

(b) Use the values from part (a) to form another three stage, third order Runge-Kutta method. Test this method on $y' = t/(y+1), y(0) = 1$, using step size $h = 0.05$. Compute the error at $t = 1$ (the exact solution is $y(t) = -1 + \sqrt{t^2 + 4}$).

[6] John D. Lambert, *Numerical Methods for Ordinary Differential Systems* (Chichester, England: Wiley, 1991).

Appendix 1: Convergence of One-Step Methods

In this appendix we state a theorem that guarantees convergence of the one-step method,

$$y_{n+1} = y_n + h\phi(t_n, y_n; h), \qquad n = 0, 1, 2, \ldots, N-1. \tag{1}$$

The convergence theorem, Theorem 9.3, applies to an initial value problem

$$y' = f(t, y), \qquad y(t_0) = y_0,$$

where we assume the function $f(t, y)$ satisfies a Lipschitz condition in y. We now define the concept of a Lipschitz condition.

Let $f(t, y)$ be a function defined on the rectangle R given by $a < t < b$, $\alpha < y < \beta$. The function f is said to satisfy a **Lipschitz condition in *y*** if there is a positive constant K such that

Whenever (t, y_1) and (t, y_2) are two points in R, then

$$|f(t, y_1) - f(t, y_2)| \le K|y_1 - y_2|. \tag{2}$$

The constant K in (2) is called a Lipschitz constant.

Note that Lipschitz constants are not unique; if a particular constant K can be used in inequality (2), then so can $K + \alpha$ for any positive constant α. A Lipschitz condition is not an overly restrictive assumption; if the partial derivative $f_y(t, y)$ exists on R, then (by the mean value theorem)

$$f(t, y_1) - f(t, y_2) = f_y(t, y^*)(y_1 - y_2),$$

where y^* is some value between y_1 and y_2. Thus, if we know that $|f_y(t, y)| \le K$ for all (t, y) in R, then the Lipschitz condition (2) holds where the assumed bound on $|f_y(t, y)|$ serves as a Lipschitz constant K.

Theorem 9.3

Consider the initial value problem

$$y' = f(t, y), \qquad y(t_0) = y_0, \tag{3}$$

where $f(t, y)$ is analytic and Lipschitz continuous in the vertical infinite strip defined by $a < t < b$, $-\infty < y < \infty$. Assume that $a < t_0 < t_0 + T < b$.

Let $y_{n+1} = y_n + h\phi(t_n, y_n; h)$ be a pth order one-step method and let $h = T/N$. Assume that for all step sizes less than some h_0 the increment function ϕ, when applied to the initial value problem (3) satisfies a Lipschitz condition in y with Lipschitz constant L. Then

$$\max_{0 \le n \le N} |y(t_n) - y_n| = O(h^p). \tag{4}$$

In words, conclusion (4) says that the global error can be bounded by some constant multiple of h^p as long as $0 < h \le h_0$. Also note that we are asking for a Lipschitz condition to hold on a vertical infinite strip. This rather restrictive condition simplifies the theorem since it ensures that initial value problem (3) has a unique solution on $[t_0, t_0 + T]$ and that $\phi(t_n, y_n; h)$ is defined for all points t_n in $t_0 \le t \le t_0 + T$.

Although Theorem 9.3 was stated for the scalar problem, a similar result can be established for a system of differential equations.

Appendix 2: Stability of One-Step Methods

Theorem 9.3 shows that we can, in principle, achieve arbitrarily good accuracy by using a one-step method with a sufficiently small step size h. We now consider the opposite situation and show that results sometimes become disastrously bad if we inadvertently use a step size that is just a little too large.

In particular, numerical methods for initial value problems are subject to difficulties of "stability." When a numerical method is applied to a given differential equation, it can happen that there is a sharp division between a step size h that is too large and that produces terrible results and a step size h that is small enough to produce acceptable results. Such a stability boundary is illustrated in Figures 9.9 and 9.10. In each case, we used Euler's method to solve the initial value problem

$$y'' + 251y' + 250y = 500 \cos 5t, \qquad y(0) = 10, \qquad y'(0) = 0. \tag{1}$$

This differential equation might model the forced vibrations of a spring-mass-dashpot system. (For the coefficients chosen, the spring constant and the damping coefficient per unit mass are relatively large.) As we know from Section 4.11, once the initial transients die out, the solution, $y(t)$, should tend toward a periodic steady-state solution. This expected behavior is exhibited by the results in Figure 9.9, but not by those in Figure 9.10. The only difference between the two computations is that the results in Figure 9.9 were obtained using a step size of $h = 1/130$ whereas the results in Figure 9.10 were obtained using a slightly larger step size of $h = 1/120$.

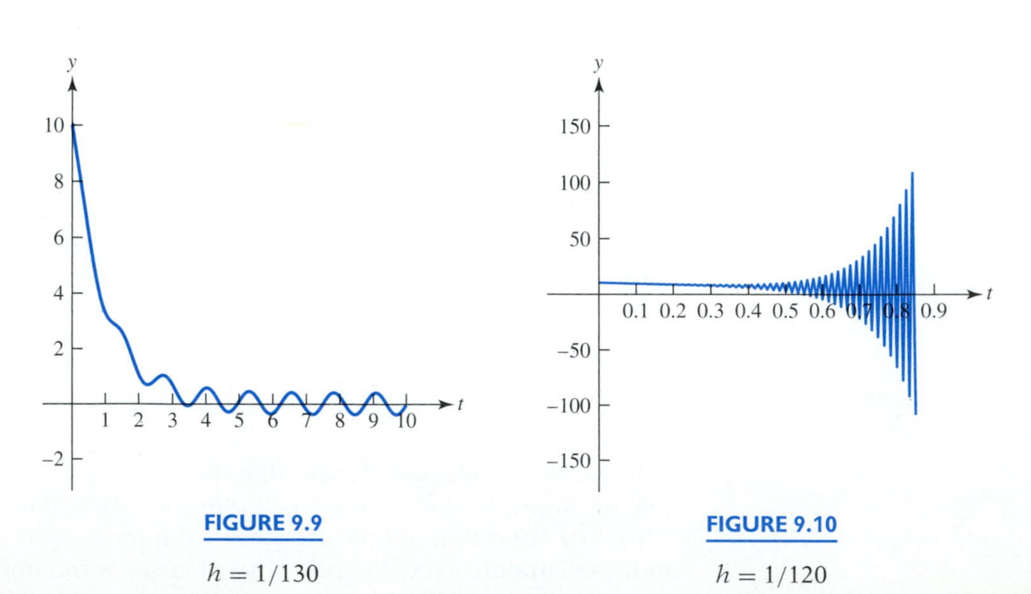

FIGURE 9.9

$h = 1/130$

FIGURE 9.10

$h = 1/120$

The sharp division between the accurate results of Figure 9.9 and the terrible results of Figure 9.10 can be explained by examining the behavior of Euler's method when it is applied to the homogeneous initial value problem $\mathbf{y}' = A\mathbf{y}, \mathbf{y}(0) = \mathbf{y}_0$. Assume that A is a (2×2) constant matrix, with distinct eigenvalues, λ_1 and λ_2, and corresponding eigenvectors, $\mathbf{u}_1$ and $\mathbf{u}_2$. Since the eigenvalues are distinct, the eigenvectors are linearly independent. Therefore, the initial condition can be represented as

$$\mathbf{y}_0 = \alpha_1 \mathbf{u}_1 + \alpha_2 \mathbf{u}_2 \tag{2}$$

for some constants α_1 and α_2. When applied to the initial value problem $\mathbf{y}' = A\mathbf{y}, \mathbf{y}(0) = \mathbf{y}_0$, Euler's method takes the form $\mathbf{y}_n = \mathbf{y}_{n-1} + hA\mathbf{y}_{n-1}$, or

$$\mathbf{y}_n = (I + hA)\mathbf{y}_{n-1}, \qquad n = 1, 2, \ldots. \tag{3}$$

It follows from (3) that $\mathbf{y}_n = (I + hA)^n \mathbf{y}_0, n = 1, 2, \ldots$. Using (2), it follows that

$$\begin{aligned} \mathbf{y}_n = (I + hA)^n \mathbf{y}_0 &= (I + hA)^n (\alpha_1 \mathbf{u}_1 + \alpha_2 \mathbf{u}_2) \\ &= \alpha_1 (I + hA)^n \mathbf{u}_1 + \alpha_2 (I + hA)^n \mathbf{u}_2. \end{aligned}$$

By Exercises 37 and 38 in Section 6.5, $(I + hA)^n \mathbf{u}_j = (1 + h\lambda_j)^n \mathbf{u}_j, j = 1, 2$. Thus,

$$\mathbf{y}_n = (I + hA)^n \mathbf{y}_0 = \alpha_1 (1 + h\lambda_1)^n \mathbf{u}_1 + \alpha_2 (1 + h\lambda_2)^n \mathbf{u}_2. \tag{4}$$

If the eigenvalues λ_1 and λ_2 are both negative, then the exact solution $\mathbf{y}(t)$ tends to $\mathbf{0}$ as t increases. Thus, the output from Euler's method [the sequence $\mathbf{y}_n$ in equation (4)] should also tend to $\mathbf{0}$ as t increases.

Assume that λ_1 and λ_2 are both negative and that $\lambda_1 < \lambda_2 < 0$. If α_1 and α_2 are both nonzero, having $\mathbf{y}_n \to \mathbf{0}$ as $n \to \infty$ requires $|1 + h\lambda_1| < 1$ and $|1 + h\lambda_2| < 1$. These two inequalities reduce to

$$-2 < h\lambda_1 < 0 \quad \text{and} \quad -2 < h\lambda_2 < 0.$$

Therefore, to obtain $\mathbf{y}_n \to \mathbf{0}$ as $n \to \infty$, we need to use a step size h that satisfies the inequality $h < -2/\lambda_1$.

When we write the homogeneous second order equation $y'' + 251y' + 250y = 0$ as a first order system $\mathbf{y}' = A\mathbf{y}, \mathbf{y}(0) = \mathbf{y}_0$, we find

$$A = \begin{bmatrix} 0 & 1 \\ -250 & -251 \end{bmatrix} \quad \text{and} \quad \lambda_1 = -250, \qquad \lambda_2 = -1.$$

Therefore, for this problem, the critical step size is $\overline{h} = -2/(-250) = 1/125$. If we apply Euler's method to the homogeneous problem $y'' + 251y' + 250y = 0$, we expect to see results qualitatively similar to those shown in Figure 9.10 when we use a step size h where $h > \overline{h} = 1/125$.

We now return to initial value problem (1), which we represent as $\mathbf{y}' = A\mathbf{y} + \mathbf{g}(t)$:

$$\mathbf{y}' = \begin{bmatrix} 0 & 1 \\ -250 & -251 \end{bmatrix} \mathbf{y} + (500 \cos 5t) \begin{bmatrix} 0 \\ 1 \end{bmatrix}, \qquad \mathbf{y}(0) = \begin{bmatrix} 10 \\ 0 \end{bmatrix}.$$

Applying Euler's method to this problem, we obtain

$$\begin{aligned} \mathbf{y}_{n+1} &= \mathbf{y}_n + h[A\mathbf{y}_n + \mathbf{g}(t_n)] \\ &= [I + hA]^{n+1} \mathbf{y}_0 + h \sum_{j=0}^{n} [I + hA]^j \mathbf{g}(t_{n-j}). \end{aligned}$$

Let us represent the vectors $\mathbf{y}_0$ and $\mathbf{g}(t_{n-j})$ in terms of the eigenvectors of A:

$$\mathbf{y}_0 = \alpha_1\mathbf{u}_1 + \alpha_2\mathbf{u}_2 \quad \text{and} \quad \mathbf{g}(t_{n-j}) = (500\cos 5t_{n-j})[\beta_1\mathbf{u}_1 + \beta_2\mathbf{u}_2].$$

With this, we see that Euler's method produces

$$\begin{aligned}\mathbf{y}_{n+1} &= \alpha_1(1+h\lambda_1)^{n+1}\mathbf{u}_1 + \alpha_2(1+h\lambda_2)^{n+1}\mathbf{u}_2 \\ &\quad + 500h\beta_1\sum_{j=0}^{n}\cos(t_{n-j})(1+h\lambda_1)^j\mathbf{u}_1 + 500h\beta_2\sum_{j=0}^{n}\cos(t_{n-j})(1+h\lambda_2)^j\mathbf{u}_2.\end{aligned}$$

Thus, as we saw with the homogeneous problem $\mathbf{y}' = A\mathbf{y}$, $\mathbf{y}(0) = \mathbf{y}_0$, if we do not choose a step size h such that $|1+h\lambda_1| < 1$ and $|1+h\lambda_2| < 1$, Euler's method will produce results qualitatively similar to those in Figure 9.10.

The ideas regarding stability that we have discussed in relation to Euler's method apply to one-step methods in general.

Extended Problem: Projectile Motion

At some initial time, a projectile (such as a meteorite) is traveling above Earth; assume that its position and velocity at that instant are known. We consider a highly idealized model in which the only force acting upon the projectile is the gravitational force exerted by Earth. Given this assumption, the projectile's trajectory lies in the plane containing the projectile's initial position vector and initial velocity vector. For simplicity, we assume this plane is the xy-plane. In our model the projectile eventually strikes the surface of Earth. Our goal is to determine where and when the impact occurs.

The dynamics of the projectile can be described by the equations

$$x''(t) = \frac{-Gm_e x(t)}{[x^2(t)+y^2(t)]^{3/2}}, \qquad y''(t) = \frac{-Gm_e y(t)}{[x^2(t)+y^2(t)]^{3/2}}, \tag{1}$$

where G is the universal gravitational constant and m_e is the mass of Earth. The center of Earth is at the origin and we let R_e denote the radius of Earth. The values of these constants are taken to be

$$G = 6.673(10^{-11})\,\frac{\text{m}^3}{\text{kg}\cdot\text{s}^2}, \qquad m_e = 5.976(10^{24})\text{ kg}, \qquad R_e = 6371\text{ km} = 6.371\times 10^6\text{ m}.$$

According to equation (1), the projectile dynamics are governed by a pair of coupled nonlinear second order differential equations. We will solve the problem numerically.

1. ***Setting Up the Problem*** The problem geometry and the nature of the force acting on the projectile suggest the use of polar coordinates and we will adopt this approach. Let

$$x(t) = r(t)\cos(\theta(t)), \qquad y(t) = r(t)\sin(\theta(t)). \tag{2}$$

Show that the equations in (1) transform into the following pair of equations for the polar variables

$$r'' - (\theta')^2 r = -\frac{Gm_e}{r^2}, \qquad \theta'' + 2\frac{r'}{r}\theta' = 0. \tag{3}$$

2. ***Initial Conditions*** Assume that the projectile is launched at time $t = 0$ at a point above Earth's surface as shown in the figure. Thus, $r(0) = R_0 > R_e, \theta(0) = 0$.

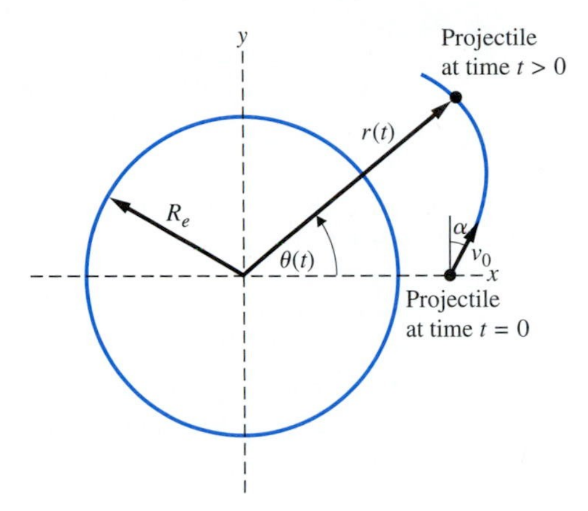

The initial conditions for the projectile whose trajectory is described by equation (3)

The initial speed v_0 and the angle α defining the orientation of the initial velocity vector are also shown on the figure. Show that

$$r'(0) = v_0 \sin\alpha \quad \text{and} \quad \theta'(0) = \frac{v_0}{R_0}\cos\alpha. \tag{4}$$

3. ***Scaled Variables*** When performing the numerical calculations, we want to deal with variables whose magnitudes are comparable to unity. To achieve this we will adopt Earth's radius, R_e, as the unit of length and the hour as the unit of time. For bookkeeping purposes, let

$$T = 3600\,\frac{\text{s}}{\text{hr}}.$$

Define the scaled variables

$$\rho(t) = \frac{r(t)}{R_e} \quad \text{and} \quad \tau = \frac{t}{T}.$$

Thus, points on Earth's surface correspond to $\rho = 1$ while 3600 sec corresponds to one unit of time on the τ-scale. Note, from the chain rule, that $d/dt = (d\tau/dt)(d/d\tau) = (1/T)(d/d\tau)$. Show that the initial value problem posed by (3) and (4) transforms into the following problem

$$\begin{aligned} &\frac{d^2\rho}{d\tau^2} - \left(\frac{d\theta}{d\tau}\right)^2 \rho = -19.985\frac{1}{\rho^2}, \qquad \rho(0) = \frac{R_0}{R_e}, \quad \frac{d\rho(0)}{d\tau} = \frac{v_0 T}{R_0}\sin\alpha \\ &\frac{d^2\theta}{d\tau^2} + 2\left(\frac{\frac{d\rho}{d\tau}}{\rho}\right)\frac{d\theta}{d\tau} = 0, \qquad \theta(0) = 0, \quad \frac{d\theta(0)}{d\tau} = \frac{v_0 T}{R_0}\cos\alpha \end{aligned} \tag{5}$$

The constant $Gm_eT^2/R_e^3 = 19.985$ has units of hr^{-2}.

Problem Specification and Solution Assume that the projectile is initially 9000 km above the surface of Earth with a speed $v_0 = 2000$ m/s and angle $\alpha = 10^\circ$.

4. Translate the assumed information into initial conditions for problem (5).

5. Recast initial value problem (5) as an initial value problem for a first order system, where

$$y_1(\tau) = \rho, \quad y_2(\tau) = \frac{d\rho}{d\tau}, \quad y_3(\tau) = \theta, \quad y_4(\tau) = \frac{d\theta}{d\tau}, \quad \text{and} \quad \mathbf{y}(\tau) = \begin{bmatrix} y_1(\tau) \\ y_2(\tau) \\ y_3(\tau) \\ y_4(\tau) \end{bmatrix}.$$

6. Solve this problem using a fourth order Runge-Kutta method and a step size $h = 0.005$. The projectile will strike Earth when $\rho = 1$. Run the program on a τ-interval sufficiently large to achieve this condition. [Hint: Gradually build up the size of the τ-interval. If too large an interval is used at the outset, the numerical solution will "blow up."]
7. Determine the polar coordinates of the impact point and the time of impact.
8. Suppose that the point $\rho = 1, \theta = 0$ corresponds to the point where the equator intersects the prime meridian, while the point $\rho = 1, \theta = \pi/2$ corresponds to the north pole. Use a globe to determine the approximate location of impact.

10

Series Solutions of Linear Differential Equations

CHAPTER OVERVIEW

Introduction

In this chapter attention is focused mainly on problems involving second order linear differential equations with variable coefficients,

$$\begin{aligned} &y'' + p(t)y' + q(t)y = 0, \qquad a < t < b \\ &y(t_0) = y_0, \qquad y'(t_0) = y'_0, \end{aligned} \tag{1}$$

where $a < t_0 < b$. In most of the cases we consider, the coefficient functions $p(t)$ and $q(t)$ are rational functions (that is, ratios of polynomials). For the moment, however, we make no assumptions about $p(t)$ and $q(t)$.

In Chapter 9, we discussed techniques for generating numerical approximations to the solution of problems such as (1). In this chapter, we look for solutions of (1) that have the form of a **power series**,

$$y(t) = \sum_{n=0}^{\infty} a_n(t-t_0)^n = a_0 + a_1(t-t_0) + a_2(t-t_0)^2 + a_3(t-t_0)^3 + \cdots. \tag{2}$$

To find such a power series solution of (1), we must answer several questions:

1. What properties must the coefficient functions $p(t)$ and $q(t)$ possess to guarantee that a power series solution (2) does, in fact, exist for initial value problem (1)?
2. If a power series solution of initial value problem (1) exists, how do we compute the coefficients $\{a_n\}_{n=0}^{\infty}$?
3. For what values of t does the resulting power series converge?

To answer these questions, we begin in Section 10.1 with a review of power series and the properties of functions defined by a power series. In Section 10.2 we introduce the concepts of ordinary points and singular points. We also state Theorem 10.1, which guarantees existence and uniqueness of a power series solution in some neighborhood of an ordinary point.

In Section 10.3 we consider the Euler equation and solution techniques for this equation. Building on the solution techniques for Euler equations, we define the notion of a regular singular point and, in Sections 10.4 and 10.5, discuss the method of Frobenius as a method for solving equations in the neighborhood of a regular singular point.

Applications

Second order linear differential equations with variable coefficients arise in many applications such as the one discussed in the next example.

EXAMPLE 1

We revisit a physical problem that gives rise to a linear variable-coefficient differential equation (see Exercise 21 in Section 4.4). The horizontally mounted rotating tube of length l shown in Figure 10.1 serves as a crude model of a centrifuge. A particle of mass m is initially placed in the frictionless tube at a distance r_0 from the pivot. At time $t = 0$, the tube begins to rotate with a constant (positive) angular acceleration α. As the tube rotates, the particle migrates

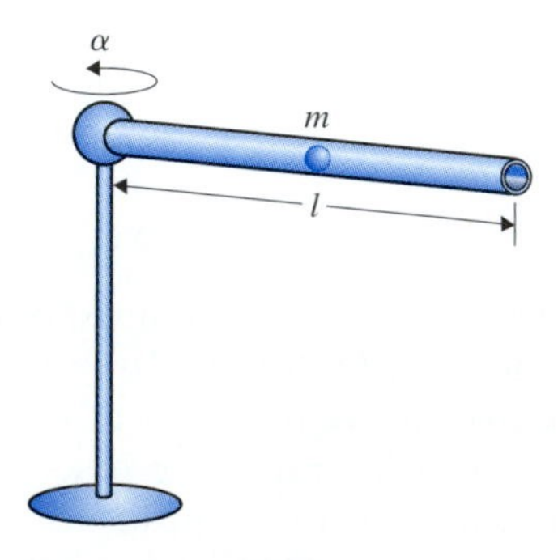

FIGURE 10.1

A particle of mass m can slide freely in a tube of length l. The particle is initially at rest and at a distance r_0 from the pivot. The tube begins to rotate with a constant angular acceleration α. When will the particle exit the tube?

radially outward and eventually exits the tube. How does the radial distance of the particle from the pivot, $r(t)$, vary with time?

To obtain a mathematical formulation of the problem, we apply Newton's second law of motion in the radial direction. Since no radial forces act upon the particle, its radial acceleration must be zero. However, if $r(t)$ and $\theta(t)$ represent the polar coordinates of the particle in the plane of the rotating tube, it was shown in Section 4.4 that the radial acceleration is

$$a_r(t) = r''(t) - r(t)(\theta'(t))^2. \tag{3}$$

In our case, $\theta'' = \alpha$ and (since the tube begins to rotate from rest) $\theta'(0) = 0$. Therefore, $\theta'(t) = \alpha t$ and the initial value problem of interest becomes

$$r'' - (\alpha t)^2 r = 0, \qquad r(0) = r_0, \qquad r'(0) = 0. \tag{4}$$

In equation (4), $r'(0) = 0$ since the particle is initially at rest.

Our goal is to develop a solution of initial value problem (4) in the form

$$r(t) = \sum_{n=0}^{\infty} a_n t^n$$

and we will do so in Section 10.2. ▲

10.1 Review of Power Series

Power series are at the center of the discussions in this chapter. We briefly review some facts that will be used later.

For any fixed value of t, the power series (1),

$$\sum_{n=0}^{\infty} a_n (t - t_0)^n, \tag{1}$$

is an infinite series of constants. If, for a fixed value of t, the sequence of partial sums,

$$S_N = \sum_{n=0}^{N} a_n (t - t_0)^n, \tag{2}$$

approaches a limit as $N \to \infty$, we say that the series **converges**. If the series does not converge, we say it **diverges**. Note that the power series (1) always converges for $t = t_0$ since $S_N = a_0$ for all N. In general, we want to know those values of t for which the power series converges and those values for which it diverges.

Convergence Possibilities

There are three distinct possibilities for the power series (1):

(i) The series $\sum_{n=0}^{\infty} a_n (t - t_0)^n$ might converge only at $t = t_0$ and diverge for all $t \neq t_0$.

(ii) The series might converge for all t in an interval of the form $|t - t_0| < R$, for some $0 < R < \infty$ and diverge for all t satisfying $|t - t_0| > R$. The number R is called the **radius of convergence**.

(iii) The series might converge for all t, $-\infty < t < \infty$.

It can be shown that every power series of the form (1) falls into exactly one of these three categories. [It is customary to say that the radius of convergence is $R = 0$ in case (i) and $R = \infty$ in case (iii).]

The power series (1) is said to be **absolutely convergent** at a value t if the infinite series

$$\sum_{n=0}^{\infty} |a_n||t - t_0|^n$$

converges. As the terminology suggests, absolute convergence implies convergence. The converse is not true, however, and convergence need not imply absolute convergence. In case (ii), it can be shown that the power series is absolutely convergent in the interval $|t - t_0| < R$ and, in case (iii), the power series is absolutely convergent for $-\infty < t < \infty$.

The **ratio test** is frequently used to test for absolute convergence of an infinite series. By way of reviewing the ratio test, consider the infinite series

$$\sum_{n=0}^{\infty} c_n, \tag{3}$$

and suppose that

$$\lim_{n\to\infty} \left| \frac{c_{n+1}}{c_n} \right| = L, \tag{4}$$

where L is either a nonnegative finite constant or $+\infty$. If $L < 1$, then the infinite series (3) is absolutely convergent. If $L > 1$ the infinite series (3) is divergent. If $L = 1$, the ratio rest is inconclusive.

The ratio test can be used to determine the radius of convergence of a power series, as we see in Example 1.

EXAMPLE 1

Use the ratio test to determine the radius of convergence of the power series

$$\sum_{n=1}^{\infty} nt^n = t + 2t^2 + 3t^3 + \cdots .$$

Solution: The power series clearly converges at $t = 0$. Applying the ratio test at an arbitrary value t, $t \neq 0$, we find

$$\lim_{n\to\infty} \left| \frac{(n+1)t^{n+1}}{nt^n} \right| = \lim_{n\to\infty} \left(1 + \frac{1}{n}\right) |t| = |t|.$$

Therefore, by the ratio test, the power series converges if $|t| < 1$ and diverges if $|t| > 1$. The radius of convergence is $R = 1$. ▲

Suppose a power series has a finite and positive radius of convergence R. The preceding discussion says nothing about whether or not the power series converges at the points $t = t_0 \pm R$. In fact, no general statements can be made about convergence or divergence at these points which separate the open interval of absolute convergence from the semi-infinite intervals of divergence; each individual power series must be scrutinized. For instance, the power series

$$\sum_{n=0}^{\infty} nt^n$$

considered in Example 1 diverges at $t = \pm 1$. In general, a power series might converge absolutely, converge **conditionally** (that is, converge but not converge absolutely), or diverge at the point $t = t_0 + R$. The same statement can be made with regard to the point $t = t_0 - R$.

Operations with Power Series

Every power series defines a function $f(t)$. The domain of $f(t)$ is the set of t-values for which the series converges. For example, consider the function $f(t)$ defined by

$$\begin{aligned} f(t) &= \sum_{n=0}^{\infty} a_n (t - t_0)^n \\ &= a_0 + a_1(t - t_0) + a_2(t - t_0)^2 + a_3(t - t_0)^3 + \cdots . \end{aligned}$$

Assume that the power series defining $f(t)$ has radius of convergence R, where $R > 0$. The following results are established in calculus and say, roughly, that power series can be treated like polynomials with respect to the operations of addition, subtraction, multiplication, and division.

1. ***Power series can be added and subtracted.*** If $f(t)$ and $g(t)$ are given by

   $$f(t) = \sum_{n=0}^{\infty} a_n (t - t_0)^n \quad \text{and} \quad g(t) = \sum_{n=0}^{\infty} b_n (t - t_0)^n,$$

   with both series converging in $|t - t_0| < R$, then the sum and difference functions are given by

   $$(f \pm g)(t) = \sum_{n=0}^{\infty} (a_n \pm b_n)(t - t_0)^n,$$

   where the sum and difference both converge absolutely in $|t - t_0| < R$.

2. ***Power series can be multiplied.*** If $f(t)$ and $g(t)$ are given by

   $$f(t) = \sum_{n=0}^{\infty} a_n (t - t_0)^n \quad \text{and} \quad g(t) = \sum_{n=0}^{\infty} b_n (t - t_0)^n,$$

   with both series converging in $|t - t_0| < R$, then the product function, $(fg)(t)$, has power series representation

   $$(fg)(t) = \sum_{n=0}^{\infty} c_n (t - t_0)^n,$$

which likewise converges in $|t - t_0| < R$. Moreover, the coefficients c_n can be obtained by formally multiplying the power series for $f(t)$ and $g(t)$ as if they were polynomials and grouping terms. In other words,

$$\begin{aligned} &c_0 + c_1(t-t_0) + c_2(t-t_0)^2 + \cdots \\ &\quad = [a_0 + a_1(t-t_0) + a_2(t-t_0)^2 + \cdots][b_0 + b_1(t-t_0) + b_2(t-t_0)^2 + \cdots] \\ &\quad = a_0b_0 + (a_0b_1 + a_1b_0)(t-t_0) + (a_0b_2 + a_1b_1 + a_2b_0)(t-t_0)^2 + \cdots. \end{aligned}$$

Therefore,

$$c_0 = a_0b_0, \qquad c_1 = a_0b_1 + a_1b_0, \qquad c_2 = a_0b_2 + a_1b_1 + a_2b_0,$$

and, in general,

$$c_n = \sum_{i=0}^{n} a_i b_{n-i}, \qquad n = 0, 1, 2, \ldots.$$

The product power series $(fg)(t)$ is called the Cauchy[1] product.

3. ***In some cases power series can be divided.*** If $f(t)$ and $g(t)$ are given by

$$f(t) = \sum_{n=0}^{\infty} a_n(t-t_0)^n \quad \text{and} \quad g(t) = \sum_{n=0}^{\infty} b_n(t-t_0)^n,$$

with both converging in $|t - t_0| < R$ and if $g(t_0) = b_0 \neq 0$, then the quotient function $(f/g)(t)$ has a power series representation

$$(f/g)(t) = \sum_{n=0}^{\infty} d_n(t-t_0)^n,$$

which converges in some neighborhood of t_0. Again, we can determine the coefficients d_n by formally manipulating the power series as if they were polynomials. We have

$$\frac{a_0 + a_1(t-t_0) + a_2(t-t_0)^2 + \cdots}{b_0 + b_1(t-t_0) + b_2(t-t_0)^2 + \cdots} = d_0 + d_1(t-t_0) + d_2(t-t_0)^2 + \cdots$$

or, after multiplying by the denominator series,

$$\begin{aligned} &a_0 + a_1(t-t_0) + a_2(t-t_0)^2 + \cdots \\ &\quad = [b_0 + b_1(t-t_0) + b_2(t-t_0)^2 + \cdots][d_0 + d_1(t-t_0) + d_2(t-t_0)^2 + \cdots]. \end{aligned}$$

The coefficients d_n can be recursively determined by forming the Cauchy product of the two series on the right and solving the resulting hierarchy

[1] Augustin Louis Cauchy (1789–1857) was a scientific giant whose life was enmeshed in the political turmoil of early nineteenth-century France. He contributed to many areas of mathematics and science and is considered to be the founder of the theory of functions of a complex variable. Numerous terms in mathematics bear his name, such as the Cauchy integral theorem, the Cauchy-Riemann equations, and Cauchy sequences. His collected works, when published, filled 27 volumes.

of linear equations. We obtain

$$a_0 = b_0 d_0 \quad \text{and hence} \quad d_0 = \frac{a_0}{b_0},$$

$$a_1 = b_0 d_1 + b_1 d_0 \quad \text{and hence} \quad d_1 = \frac{a_1 - b_1 d_0}{b_0},$$

$$a_2 = b_0 d_2 + b_1 d_1 + b_2 d_0 \quad \text{and hence} \quad d_2 = \frac{a_2 - b_1 d_1 - b_2 d_0}{b_0},$$

$$\vdots$$

Notice how the statement made in the case of division differs from that made in the previous cases. In particular, even though the series for $f(t)$ and $g(t)$ converge in $|t - t_0| < R$, it does *not* necessarily follow that the quotient series also converges in $|t - t_0| < R$. All we can say in general is that the quotient series converges in some neighborhood of t_0.

4. *A function defined by a power series can be differentiated termwise.* Let $f(t)$ be given by the power series

$$f(t) = \sum_{n=0}^{\infty} a_n (t - t_0)^n, \tag{5}$$

which converges in $|t - t_0| < R$. The function $f(t)$ has derivatives of all orders on the interval $t_0 - R < t < t_0 + R$. We can obtain these derivatives by termwise differentiation of the original power series. That is,

$$f'(t) = \sum_{n=1}^{\infty} n a_n (t - t_0)^{n-1} = a_1 + 2a_2(t - t_0) + 3a_3(t - t_0)^2 + \cdots,$$

$$f''(t) = \sum_{n=2}^{\infty} n(n-1) a_n (t - t_0)^{n-2} = 2a_2 + 6a_3(t - t_0) + \cdots,$$

and so forth. Each of these derived series also converges absolutely in the interval $t_0 - R < t < t_0 + R$. The derived series can be used to express the coefficient a_n in terms of the nth derivative of $f(t)$ evaluated at $t = t_0$. In particular, by evaluating the derived series at $t = t_0$, we see that

$$f(t_0) = a_0, \quad f'(t_0) = a_1, \quad f''(t_0) = (2 \cdot 1)a_2, \quad f'''(t_0) = (3 \cdot 2 \cdot 1)a_3, \ldots.$$

In general, for $f(t)$ given by (5),

$$f^{(n)}(t_0) = n! a_n, \qquad n = 0, 1, 2, \ldots.$$

5. *Taylor series and Maclaurin series.* If $f(t)$ is a function defined by the power series (5), then $f^{(n)}(t_0) = n! a_n$, $n = 0, 1, 2, \ldots$. Conversely, if we are given a function $f(t)$ that is defined and infinitely differentiable on an interval $t_0 - R < t < t_0 + R$, then we can associate $f(t)$ with its formal **Taylor series**

$$\sum_{n=0}^{\infty} \frac{f^{(n)}(t_0)}{n!} (t - t_0)^n. \tag{6}$$

Recall from calculus that the Taylor series for $f(t)$ need not necessarily converge to $f(t)$. However, for most of the functions considered in this chapter, the Taylor series converges to $f(t)$, so that

$$f(t) = \sum_{n=0}^{\infty} \frac{f^{(n)}(t_0)}{n!}(t-t_0)^n, \qquad t_0 - R < t < t_0 + R.$$

If $t_0 = 0$, the Taylor series is usually referred to as a Maclaurin series. For later reference, we list the Maclaurin series for several functions. We also give the interval of absolute convergence for the series.

$$e^t = \sum_{n=0}^{\infty} \frac{t^n}{n!} = 1 + t + \frac{t^2}{2!} + \frac{t^3}{3!} + \frac{t^4}{4!} + \cdots, \qquad -\infty < t < \infty. \tag{7a}$$

$$\sin t = \sum_{n=0}^{\infty} (-1)^n \frac{t^{2n+1}}{(2n+1)!} = t - \frac{t^3}{3!} + \frac{t^5}{5!} - \frac{t^7}{7!} + \cdots, \qquad -\infty < t < \infty. \tag{7b}$$

$$\cos t = \sum_{n=0}^{\infty} (-1)^n \frac{t^{2n}}{(2n)!} = 1 - \frac{t^2}{2!} + \frac{t^4}{4!} - \frac{t^6}{6!} + \cdots, \qquad -\infty < t < \infty. \tag{7c}$$

$$\frac{1}{1-t} = \sum_{n=0}^{\infty} t^n = 1 + t + t^2 + t^3 + t^4 + \cdots, \qquad -1 < t < 1. \tag{7d}$$

$$\ln(1+t) = \sum_{n=1}^{\infty} (-1)^{n-1} \frac{t^n}{n} = t - \frac{t^2}{2} + \frac{t^3}{3} - \frac{t^4}{4} + \cdots, \qquad -1 < t < 1. \tag{7e}$$

$$\tan^{-1} t = \sum_{n=0}^{\infty} (-1)^n \frac{t^{2n+1}}{2n+1} = t - \frac{t^3}{3} + \frac{t^5}{5} - \frac{t^7}{7} + \cdots, \qquad -1 < t < 1. \tag{7f}$$

Note that these basic series can be used to find the Taylor series of certain simple compositions. For example, by (7d)

$$\frac{1}{3-t} = \frac{1}{1-(t-2)} = \sum_{n=0}^{\infty} (t-2)^n, \qquad |t-2| < 1$$

and

$$\frac{1}{1+4t^2} = \sum_{n=0}^{\infty} (-1)^n 4^n t^{2n}, \qquad |t| < \frac{1}{2}.$$

6. ***A function defined by a power series can be integrated termwise.*** Let $f(t)$ be given by the power series

$$f(t) = \sum_{n=0}^{\infty} a_n (t-t_0)^n,$$

which converges in $|t-t_0| < R$. The function $f(t)$ has antiderivatives defined on the interval $t_0 - R < t < t_0 + R$. We can obtain these antiderivatives by termwise integration of the original power series. For example,

$$\int_{t_0}^{t} f(s)\,ds = \int_{t_0}^{t} \sum_{n=0}^{\infty} a_n (s-t_0)^n\,ds = \sum_{n=0}^{\infty} a_n \int_{t_0}^{t} (s-t_0)^n\,ds = \sum_{n=0}^{\infty} a_n \frac{(t-t_0)^{n+1}}{n+1}.$$

The integrated series also converges absolutely on the interval $t_0 - R < t < t_0 + R$.

Uniqueness of the Power Series Representation of a Function

A power series representation of a function is unique. In particular, let

$$\sum_{n=0}^{\infty} a_n (t-t_0)^n \quad \text{and} \quad \sum_{n=0}^{\infty} b_n (t-t_0)^n$$

be two power series having the same radius of convergence R. If

$$\sum_{n=0}^{\infty} a_n (t-t_0)^n = \sum_{n=0}^{\infty} b_n (t-t_0)^n \quad \text{for all } t \text{ such that} \quad |t-t_0| < R,$$

then it follows that the coefficients must be equal; that is,

$$a_n = b_n, \qquad n = 0, 1, 2, \ldots .$$

As an important special case, if

$$\sum_{n=0}^{\infty} a_n (t-t_0)^n = 0 \quad \text{for all } t \text{ such that} \quad |t-t_0| < R,$$

then $a_n = 0, n = 0, 1, 2, \ldots$.

Power Series Solutions of Linear Differential Equations

The next example introduces the ideas associated with finding a power series solution of a linear differential equation. In the sections that follow, we will elaborate on the theoretical foundations of the method and point out some of the potential difficulties.

EXAMPLE 2

Consider the equation

$$y'' + \omega^2 y = 0,$$

where ω is a positive constant. Assuming this equation has a solution of the form

$$y(t) = \sum_{n=0}^{\infty} a_n t^n,$$

determine the coefficients, $a_0, a_1, a_2, \ldots$. Can you also determine the *general solution*?

(continued)

(continued)

Solution: We look for a solution of the form $y(t) = \sum_{n=0}^{\infty} a_n t^n$, assuming that the power series has a positive radius of convergence. The actual radius of convergence will be determined once we find the coefficients, $a_0, a_1, a_2, \ldots$.

Differentiating termwise, we obtain

$$y' = \sum_{n=1}^{\infty} a_n n t^{n-1} \quad \text{and} \quad y'' = \sum_{n=2}^{\infty} a_n n(n-1)t^{n-2}.$$

In these series, we have adjusted the lower limit in the summation to correspond to the first nonzero term in the series.

Substituting these expressions into the differential equation $y'' + \omega^2 y = 0$, we find

$$\sum_{n=2}^{\infty} a_n n(n-1)t^{n-2} + \omega^2 \sum_{n=0}^{\infty} a_n t^n = 0. \tag{8}$$

We want to combine the two summations in (8) and eventually use the consequences of uniqueness. In order to combine the two series, we adjust the summation index of the first series so that powers of t are in agreement.

In particular, to match the powers of t, we can make a change of index, $k = n - 2$, in the first series. Doing so, we see from (8) that the lower limit of $n = 2$ transforms to a new lower limit of $k = 0$ (the upper limits remain at ∞). Thus, the first series in (8) can be rewritten as

$$\sum_{n=2}^{\infty} a_n n(n-1)t^{n-2} = \sum_{k=0}^{\infty} a_{k+2}(k+2)(k+1)t^k = \sum_{n=0}^{\infty} a_{n+2}(n+2)(n+1)t^n. \tag{9}$$

In the last step we have used the fact that the summation index is a dummy index and can be called n instead of k. Using (9) in (8) leads to

$$\sum_{n=0}^{\infty} a_{n+2}(n+2)(n+1)t^n + \omega^2 \sum_{n=0}^{\infty} a_n t^n = 0,$$

or

$$\sum_{n=0}^{\infty} [a_{n+2}(n+2)(n+1) + \omega^2 a_n] t^n = 0. \tag{10}$$

Since equality (10) is assumed to hold in some interval containing the origin, each coefficient of the series must vanish. We obtain the following infinite set of equalities

$$(n+1)(n+2)a_{n+2} + \omega^2 a_n = 0, \qquad n = 0, 1, 2, \ldots,$$

or

$$a_{n+2} = -\frac{\omega^2}{(n+1)(n+2)} a_n, \qquad n = 0, 1, 2, 3, \ldots \tag{11}$$

The set of equations (11) is referred to as a **recurrence relation**. Solving for the unknown coefficients recursively allows us to find all the coefficients, $\{a_n\}_{n=0}^{\infty}$, in terms of the coefficients a_0 and a_1. In particular, from (11) we find

$$n = 0: \quad a_2 = -\frac{\omega^2}{2 \cdot 1} a_0 \qquad\qquad n = 1: \quad a_3 = -\frac{\omega^2}{3 \cdot 2} a_1$$

$$n = 2: \quad a_4 = -\frac{\omega^2}{4 \cdot 3} a_2 = \frac{\omega^4}{4!} a_0 \qquad\qquad n = 3: \quad a_5 = -\frac{\omega^2}{5 \cdot 4} a_3 = \frac{\omega^4}{5!} a_1$$

$$n = 4: \quad a_6 = -\frac{\omega^2}{6 \cdot 5} a_4 = -\frac{\omega^6}{6!} a_0 \qquad\qquad n = 5: \quad a_7 = -\frac{\omega^2}{7 \cdot 6} a_5 = -\frac{\omega^6}{7!} a_1$$

$$\vdots \qquad\qquad\qquad\qquad \vdots$$

The emerging pattern is clear. For an even index, $n = 2k$, we have

$$a_{2k} = (-1)^k \frac{\omega^{2k}}{(2k)!} a_0.$$

For an odd index, $n = 2k + 1$, we have

$$a_{2k+1} = (-1)^k \frac{\omega^{2k}}{(2k+1)!} a_1 = (-1)^k \frac{\omega^{2k+1}}{(2k+1)!} \left(\frac{a_1}{\omega}\right).$$

Therefore, we can write the solution of the differential equation $y'' + \omega^2 y = 0$ as

$$y(t) = a_0 \left[1 - \frac{(\omega t)^2}{2!} + \frac{(\omega t)^4}{4!} - \frac{(\omega t)^6}{6!} + \cdots\right] + \left(\frac{a_1}{\omega}\right) \left[\omega t - \frac{(\omega t)^3}{3!} + \frac{(\omega t)^5}{5!} - \frac{(\omega t)^7}{7!} + \cdots\right]. \tag{12}$$

The first series is the Maclaurin series expansion of $\cos \omega t$ (see equation (7c)). The second series is that of $\sin \omega t$ (see equation (7b)). Therefore, identifying arbitrary constants c_1 with a_0 and c_2 with a_1/ω, we obtain the general solution familiar from Chapter 4,

$$y(t) = c_1 \cos \omega t + c_2 \sin \omega t.$$

Applying the ratio test, you can verify that the radius of convergence is $R = \infty$ for both of these power series. ▲

In the sections that follow, the same basic manipulations will be used to obtain series solutions of differential equations having variable coefficients.

Shifting the Index of Summation

We frequently find it convenient, as in Example 2, to shift the index of summation so that the general term in a series is a constant multiple of t^n. For example, consider the function $f(t) = t^3[e^t - 1]$. Using equation (7a), we see the series for $f(t) = t^3[e^t - 1]$ has the form

$$f(t) = t^3[e^t - 1] = t^3 \sum_{n=1}^{\infty} \frac{t^n}{n!} = \sum_{n=1}^{\infty} \frac{t^{n+3}}{n!}.$$

In order to rewrite the Maclaurin series so that the general term involves t^n, we make the change of index, $k = n + 3$. With this shift of index, $n = k - 3$. Therefore, the terms in the summation are all of the form $t^k/(k-3)!$. We also must transform the limits of the summation. At the lower limit, $n = 1$ implies that $k = 4$. At the upper limit, $n = \infty$ also implies $k = \infty$. Thus, we can rewrite the series for $f(t) = t^3[e^t - 1]$ as

$$f(t) = \sum_{n=1}^{\infty} \frac{t^{n+3}}{n!} = \sum_{k=4}^{\infty} \frac{t^k}{(k-3)!} = \sum_{n=4}^{\infty} \frac{t^n}{(n-3)!}. \tag{13}$$

The summation index is a dummy index. In the last step of (13), therefore, we can replace k by n.

EXERCISES

Exercises 1–12:

As in Example 1, use the ratio test to find the radius of convergence, R, for the given power series.

1. $\displaystyle\sum_{n=0}^{\infty} \frac{t^n}{2^n}$
2. $\displaystyle\sum_{n=1}^{\infty} \frac{t^n}{n^2}$
3. $\displaystyle\sum_{n=0}^{\infty} (t-2)^n$
4. $\displaystyle\sum_{n=0}^{\infty} (3t-1)^n$
5. $\displaystyle\sum_{n=0}^{\infty} \frac{(t-1)^n}{n!}$
6. $\displaystyle\sum_{n=0}^{\infty} n!(t-1)^n$
7. $\displaystyle\sum_{n=1}^{\infty} \frac{(-1)^n t^n}{n}$
8. $\displaystyle\sum_{n=0}^{\infty} \frac{(-1)^n (t-3)^n}{4^n}$
9. $\displaystyle\sum_{n=1}^{\infty} (\ln n)(t+2)^n$
10. $\displaystyle\sum_{n=0}^{\infty} n^3(t-1)^n$
11. $\displaystyle\sum_{n=0}^{\infty} \frac{\sqrt{n}}{2^n}(t-4)^n$
12. $\displaystyle\sum_{n=1}^{\infty} \frac{(t-2)^n}{\arctan n}$

Exercises 13–16:

In each exercise, functions $f(t)$ and $g(t)$ are given. The functions $f(t)$ and $g(t)$ are defined by a power series that converges in $-R < t - t_0 < R$ where R is a positive constant. In each exercise, determine the largest value R such that $f(t)$ and $g(t)$ both converge in $-R < t - t_0 < R$. In addition,

(a) Write out the first six terms of the power series for $f(t)$ and $g(t)$.

(b) Write out the first six terms of the power series for $f(t) + g(t)$.

(c) Write out the first six terms of the power series for $f(t) - g(t)$.

(d) Write out the first six terms of the power series for $f'(t)$.

(e) Write out the first six terms of the power series for $f''(t)$.

13. $f(t) = \displaystyle\sum_{n=0}^{\infty} t^n$, $\quad g(t) = \displaystyle\sum_{n=0}^{\infty} n^2 t^n$
14. $f(t) = \displaystyle\sum_{n=0}^{\infty} nt^n$, $\quad g(t) = \displaystyle\sum_{n=0}^{\infty} (-1)^n nt^n$
15. $f(t) = \displaystyle\sum_{n=0}^{\infty} (-1)^n 2^n (t-1)^n$, $\quad g(t) = \displaystyle\sum_{n=0}^{\infty} (t-1)^n$
16. $f(t) = \displaystyle\sum_{n=0}^{\infty} 2^n (t+1)^n$, $\quad g(t) = \displaystyle\sum_{n=0}^{\infty} n(t+1)^n$

Exercises 17–23:

By shifting the index of summation as in equation (9) or (13), rewrite the given power series so that the general term involves t^n.

17. $\sum_{n=0}^{\infty} 2^n t^{n+2}$

18. $\sum_{n=0}^{\infty} (n+1)(n+2)t^{n+3}$

19. $\sum_{n=0}^{\infty} a_n t^{n+2}$

20. $\sum_{n=1}^{\infty} n a_n t^{n-1}$

21. $\sum_{n=2}^{\infty} n(n-1)a_n t^{n-2}$

22. $\sum_{n=0}^{\infty} (-1)^n a_n t^{n+3}$

23. $\sum_{n=0}^{\infty} (-1)^{n+1}(n+1)a_n t^{n+2}$

Exercises 24–27:

Using the information given in equation (7), write a Maclaurin series for the given function $f(t)$. Determine the radius of convergence of the series.

24. $f(t) = t^2[t - \sin t]$

25. $f(t) = 1 - \cos(3t)$

26. $f(t) = \dfrac{1}{1+2t}$

27. $f(t) = \dfrac{1}{1-t^2}$

28. Use series (7a) to determine the first four nonvanishing terms of the Maclaurin series for

(a) $\sinh t = \dfrac{e^t - e^{-t}}{2}$

(b) $\cosh t = \dfrac{e^t + e^{-t}}{2}$.

29. Consider the differential equation $y'' - \omega^2 y = 0$, where ω is a positive constant. As in Example 2, assume this differential equation has a solution of the form $y(t) = \sum_{n=0}^{\infty} a_n t^n$.

(a) Determine a recurrence relation for the coefficients $a_0, a_1, a_2, \ldots$.

(b) As in equation (12), express the general solution in the form

$$y(t) = a_0 y_1(t) + \left(\frac{a_1}{\omega}\right) y_2(t).$$

What are the functions $y_1(t)$ and $y_2(t)$? [Hint: Recall the series in Exercise 28.]

Exercises 30–35:

In each exercise,

(a) Use the given information to determine a power series representation of the function $y(t)$.

(b) Determine the radius of convergence of the series found in part (a).

(c) Where possible, use (7) to identify the function $y(t)$.

30. $y'(t) = \sum_{n=1}^{\infty} n t^{n-1} = 1 + 2t + 3t^2 + \cdots, \quad y(0) = 1$

31. $y'(t) = \sum_{n=0}^{\infty} \dfrac{(t-1)^n}{n!} = 1 + (t-1) + \dfrac{(t-1)^2}{2!} + \cdots, \quad y(1) = 1$

32. $y''(t) = \sum_{n=0}^{\infty} (-1)^n \dfrac{t^n}{n!} = 1 - t + \dfrac{t^2}{2!} - \dfrac{t^3}{3!} + \cdots, \quad y(0) = 1, \ y'(0) = -1$

33. $y'(t) = \sum_{n=2}^{\infty} (-1)^n \dfrac{(t-1)^n}{n!} = \dfrac{(t-1)^2}{2!} - \dfrac{(t-1)^3}{3!} + \dfrac{(t-1)^4}{4!} - \dfrac{(t-1)^5}{5!} + \cdots, \quad y(1) = 0$

34. $y(t) = \int_0^t f(s)\,ds$, where $f(s) = \sum_{n=0}^{\infty} (-1)^n s^{2n} = 1 - s^2 + s^4 - s^6 + \cdots$

35. $\int_0^t y(s)\,ds = \sum_{n=1}^{\infty} \dfrac{t^n}{n} = t + \dfrac{t^2}{2} + \dfrac{t^3}{3} + \cdots$

Exercises 36–41:

In each exercise, an initial value problem is given. Assume that the initial value problem has a solution of the form $y(t) = \sum_{n=0}^{\infty} a_n t^n$, where the series has a positive radius of convergence. Determine the first six coefficients, $a_0, a_1, a_2, a_3, a_4, a_5$. Note that $y(0) = a_0$ and that $y'(0) = a_1$. Thus, the initial conditions determine the arbitrary constants. In Exercises 40 and 41, the exact solution is given in terms of exponential functions. Check your answer by comparing it with the Maclaurin series of the exact solution.

36. $y'' - ty' - y = 0, \;\; y(0) = 1, \;\; y'(0) = -1$

37. $y'' + ty' - 2y = 0, \;\; y(0) = 0, \;\; y'(0) = 1$

38. $y'' + ty = 0, \;\; y(0) = 1, \;\; y'(0) = 2$

39. $y'' + (1+t)y' + y = 0, \;\; y(0) = -1, \;\; y'(0) = 1$

40. $y'' - 5y' + 6y = 0, \;\; y(0) = 1, \;\; y'(0) = 2, \;\; y(t) = e^{2t}$

41. $y'' - 2y' + y = 0, \;\; y(0) = 0, \;\; y'(0) = 2, \;\; y(t) = 2te^{t}$

10.2 Series Solutions Near an Ordinary Point

Consider the linear differential equation

$$y'' + p(t)y' + q(t)y = 0 \tag{1}$$

in an open interval containing the point t_0. Frequently, t_0 is the point where the initial conditions are imposed. For our present discussion, however, t_0 is an arbitrary but fixed point. We are interested in answering the question: When is it possible to represent the general solution of (1) in terms of power series that converge in some neighborhood of the point t_0?

Ordinary Points and Singular Points

Recall (see Section 9.2) that a function $f(t)$ is called **analytic at t_0** if $f(t)$ has a Taylor series expansion,

$$f(t) = \sum_{n=0}^{\infty} \frac{f^{(n)}(t_0)}{n!}(t - t_0)^n,$$

with radius of convergence R, where $R > 0$. For later use, we also recall:

> If $f(t)$ and $g(t)$ are analytic at t_0, then the functions $f(t) \pm g(t)$ and $f(t)g(t)$ are also analytic at t_0. Furthermore, the quotient $f(t)/g(t)$ is analytic at t_0 if $g(t_0) \neq 0$. Polynomial functions are analytic at all points. Rational functions are analytic at all points where the denominator polynomial is nonzero. (When discussing rational functions, we assume the denominator and numerator have no factors in common.) If the denominator is nonzero at t_0, the radius of convergence is equal to the distance from t_0 to the nearest zero (either real or complex) of the denominator. (See Figure 10.2.)

As we will show, the general solution of $y'' + p(t)y' + q(t)y = 0$ can be expressed in terms of power series that converge in a neighborhood of t_0 whenever both $p(t)$ and $q(t)$ are analytic at t_0. The point t_0 is called an **ordinary point**

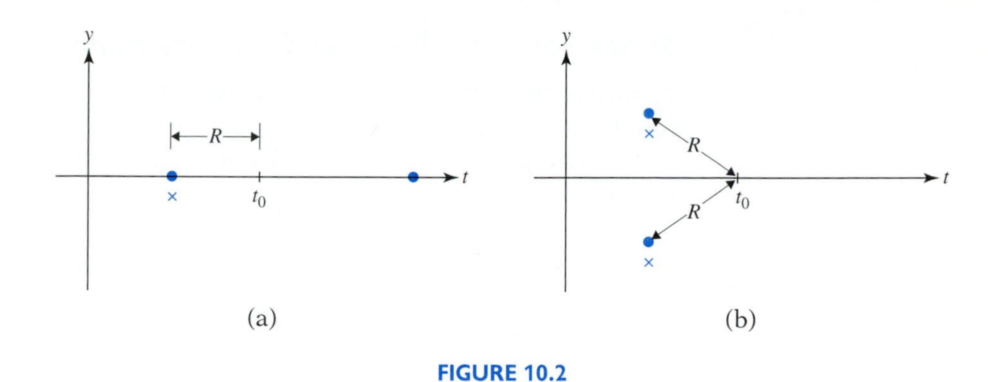

FIGURE 10.2

If $f(t) = P(t)/Q(t)$ is a rational function and if $Q(t_0) \neq 0$, then $f(t)$ is analytic at t_0. The radius of convergence of the power series is equal to the distance from t_0 to the nearest zero of $Q(t)$. (a) The nearest zero, marked $\times$, is real. (b) The nearest zeros, marked $\times$, are complex.

when both $p(t)$ and $q(t)$ are analytic at t_0. If $p(t)$ and/or $q(t)$ is not analytic at t_0, t_0 is called a **singular point**.

As noted in Section 10.1, if a function $f(t)$ is defined by a power series, $f(t) = \sum_{n=0}^{\infty} a_n(t - t_0)^n$, and if this power series has radius of convergence R where $R > 0$, then $f(t)$ has derivatives of all orders at $t = t_0$. Therefore, if some derivative of $f(t)$ fails to exist at a point t_0, then $f(t)$ cannot be analytic at t_0. To help identify ordinary points, we can use some facts noted earlier—sums, differences, and products of functions analytic at t_0 are again analytic at t_0. Quotients of functions analytic at t_0 are also analytic at t_0 if the denominator function is nonzero at t_0.

EXAMPLE 1

Consider the differential equation

$$(1 - t^2)y'' + (\tan t)y' + t^{5/3}y = 0,$$

in the open interval $-2 < t < 2$. Classify each point in this interval as an ordinary point or a singular point.

Solution: We first rewrite the equation in the form (1),

$$y'' + \frac{\tan t}{1 - t^2}y' + \frac{t^{5/3}}{1 - t^2}y = 0.$$

Therefore, the coefficient functions $p(t)$ and $q(t)$ are

$$p(t) = \frac{\tan t}{1 - t^2} \quad \text{and} \quad q(t) = \frac{t^{5/3}}{1 - t^2}.$$

The function $p(t)$ fails to be analytic at the points $t = \pm 1$ (where the denominator vanishes) and at $t = \pm\pi/2$ (where the graph of $y = \tan t$ has vertical asymptotes). The function $q(t)$ is not analytic at $t = \pm 1$ or at $t = 0$. [The numerator function, $y = t^{5/3}$, is continuous and has a continuous first derivative at $t = 0$. Its second derivative, however, does not exist at $t = 0$.] Thus, in the interval $-2 < t < 2$, the five points $t = 0, \pm 1, \pm\pi/2$ are singular points and all other points are ordinary points. ▲

Series Solutions Near an Ordinary Point

Theorem 10.1 shows that, in a neighborhood of an ordinary point, we can represent the general solution of equation (1) in terms of convergent power series. The proof of Theorem 10.1 is given in more advanced texts, such as Birkhoff and Rota.[2]

Theorem 10.1

Let $p(t)$ and $q(t)$ be analytic at t_0 and let R denote the smaller of the two radii of convergence of their respective Taylor series representations. Then the initial value problem

$$y'' + p(t)y' + q(t)y = 0, \qquad y(t_0) = y_0, \qquad y'(t_0) = y_0' \tag{2}$$

has a unique solution that is analytic in the interval $|t - t_0| < R$.

According to Theorem 10.1, if t_0 is an ordinary point, then initial value problem (2) has a power series solution of the form

$$y(t) = \sum_{n=0}^{\infty} a_n(t - t_0)^n = a_0 + a_1(t - t_0) + a_2(t - t_0)^2 + \cdots.$$

Note that the first two coefficients, a_0 and a_1, are determined by the initial conditions in (2), since $y(t_0) = a_0$ and $y_0'(t_0) = a_1$. Theorem 10.1 assures us that the power series we obtain by solving the recurrence relation for the remaining coefficients $a_2, a_3, \ldots$ converges in the interval $|t - t_0| < R$, where R is the smaller of the radii of convergence of the coefficient function $p(t)$ and $q(t)$. Note that Theorem 10.1 does not rule out the possibility that the power series for $y(t)$ may converge on a larger interval. This happens in some cases. The following corollary is a consequence of Theorem 10.1.

Corollary

Let $p(t)$ and $q(t)$ be analytic at t_0, and let R denote the smaller of the two radii of convergence of their respective Taylor series representations. The general solution of the differential equation

$$y'' + p(t)y' + q(t)y = 0 \tag{3}$$

can be expressed as

$$y(t) = \sum_{n=0}^{\infty} a_n(t - t_0)^n = a_0 y_1(t) + a_1 y_2(t),$$

where the constants a_0 and a_1 are arbitrary. The functions $y_1(t)$ and $y_2(t)$ form a fundamental set of solutions, analytic in the interval $|t - t_0| < R$.

---

[2]Garrett Birkhoff and Gian-Carlo Rota, *Ordinary Differential Equations*, 4th edition (New York: Wiley, 1989).

The solutions $y_1(t)$ and $y_2(t)$ forming the fundamental set can be obtained by adopting the particular initial conditions $y_1(t_0) = 1, y_1'(t_0) = 0$ and $y_2(t_0) = 0$, $y_2'(t_0) = 1$.

EXAMPLE 2

Consider the initial value problem

$$y'' + \frac{t+1}{t^3+t}y' + \frac{1}{t^2-4t+5}y = 0, \qquad y(2) = y_0, \qquad y'(2) = y_0'.$$

If $y(t) = \sum_{n=0}^{\infty} a_n(t-2)^n$ is the solution, determine a lower bound for the radius of convergence, R, of this series.

Solution: Since $t^3 + t = t(t^2+1)$, the coefficient function $p(t)$ has denominator zeros at $t = 0, t = -i$, and $t = i$. Likewise, the coefficient function $q(t)$ has denominator zeros at $t = 2 \pm i$. As illustrated in Figure 10.2, the radius of convergence, R_p, of the Taylor series expansion for $p(t)$ is equal to the distance from $t_0 = 2$ to the nearest denominator zero; that is, R_p is the smaller of $|2 \pm 0| = 2$ and $|2 - i| = \sqrt{5}$. Thus, $R_p = 2$. Similarly, the radius of convergence of the Taylor series for $q(t)$ is $R_q = |2 - (2 \pm i)| = 1$. Thus, by Theorem 10.1, the radius of convergence of the Taylor series for $y(t)$ is guaranteed to be no smaller than $R = 1$. ▲

When the coefficient functions of $y'' + p(t)y' + q(t)y = 0$ possess certain symmetries, some useful observations can be made; see Theorem 10.2.

Theorem 10.2

Consider the differential equation $y'' + p(t)y' + q(t)y = 0$.

(a) Let $p(t)$ be a continuous odd function defined on the domain $(-b, -a) \cup (a, b)$, where $a \geq 0$. Let $q(t)$ be a continuous even function defined on the same domain. If $f(t)$ is a solution of the differential equation on the interval $a < t < b$, then $f(-t)$ is a solution on the interval $(-b, -a)$.

(b) Let the coefficient functions $p(t)$ and $q(t)$ be analytic at $t = 0$ with a common radius of convergence $R > 0$. Let $p(t)$ be an odd function and $q(t)$ an even function. Then the differential equation has even and odd solutions that are analytic at $t = 0$ with radius of convergence R.

Recall the definitions of even and odd functions. We are assuming that $p(-t) = -p(t)$ and $q(-t) = q(t)$ for all t in $(-b, -a) \cup (a, b)$. The proof of Theorem 10.2 is outlined in Exercises 31–32.

In Example 2 of Section 10.1 we obtained a power series solution of $y'' + \omega^2 y = 0$ and observed that the ratio test could be used to show that each of the two series forming the general solution has an infinite radius of convergence. This fact is also an easy consequence of Theorem 10.1 since the differential equation $y'' + \omega^2 y = 0$ has coefficient functions $p(t) = 0$ and $q(t) = \omega^2$ that are analytic on $-\infty < t < \infty$. Moreover, since $p(t)$ is an odd function and $q(t)$ is an even function, it follows from Theorem 10.2 that even and odd solutions of this differential equation exist; they are $\cos \omega t$ and $\sin \omega t$, respectively.

EXAMPLE 3

Recall the centrifuge problem formulated in the Introduction,

$$r'' - \alpha^2 t^2 r = 0, \qquad r(0) = r_0, \qquad r'(0) = 0. \tag{4}$$

Find a Maclaurin series solution, $r(t) = \sum_{n=0}^{\infty} a_n t^n$, for the radial distance of the mass from the pivot.

Solution: For equation (4), $p(t) = 0$ and $q(t) = -\alpha^2 t^2$. Therefore, Theorem 10.1 assures us a Maclaurin series solution of the form $r(t) = \sum_{n=0}^{\infty} a_n t^n$ exists and that this series has an infinite radius of convergence.

Differentiating termwise, we obtain

$$r'(t) = \sum_{n=1}^{\infty} n a_n t^{n-1} \quad \text{and} \quad r''(t) = \sum_{n=2}^{\infty} n(n-1) a_n t^{n-2}.$$

Substituting these expressions into the differential equation $r'' - \alpha^2 t^2 r = 0$, we find

$$\sum_{n=2}^{\infty} n(n-1) a_n t^{n-2} - \alpha^2 t^2 \sum_{n=0}^{\infty} a_n t^n = 0,$$

or

$$\sum_{n=2}^{\infty} n(n-1) a_n t^{n-2} - \alpha^2 \sum_{n=0}^{\infty} a_n t^{n+2} = 0. \tag{5}$$

To obtain the desired recurrence relation, we shift summation indices so that each sum has coefficients multiplying t^n. We have

$$\sum_{n=2}^{\infty} n(n-1) a_n t^{n-2} = \sum_{m=0}^{\infty} (m+2)(m+1) a_{m+2} t^m = \sum_{n=0}^{\infty} (n+2)(n+1) a_{n+2} t^n$$

and

$$\sum_{n=0}^{\infty} a_n t^{n+2} = \sum_{m=2}^{\infty} a_{m-2} t^m = \sum_{n=2}^{\infty} a_{n-2} t^n.$$

Therefore, (5) becomes

$$\sum_{n=0}^{\infty} (n+2)(n+1) a_{n+2} t^n - \alpha^2 \sum_{n=2}^{\infty} a_{n-2} t^n = 0. \tag{6}$$

We cannot simply combine the two summations in (6) since the first sum begins at $n = 0$, while the second sum begins at $n = 2$. We can, however, separate the first two terms in the first sum and combine the rest with the second sum to obtain

$$(2)(1)a_2 + (3)(2)a_3 t + \sum_{n=2}^{\infty} [(n+2)(n+1) a_{n+2} - \alpha^2 a_{n-2}] t^n = 0.$$

Equating the coefficient of each power of t to zero leads to $a_2 = 0, a_3 = 0$, and

$$(n+2)(n+1)a_{n+2} - \alpha^2 a_{n-2} = 0, \qquad n = 2, 3, 4, \ldots.$$

Equivalently,

$$a_{n+2} = \frac{\alpha^2}{(n+2)(n+1)} a_{n-2}, \qquad n = 2, 3, 4, \ldots. \tag{7}$$

For recurrence relation (7), each nonzero even-indexed coefficient can be expressed as a multiple of a_0 while each nonzero odd-indexed coefficient can be expressed as a multiple of a_1. In particular

$$\begin{aligned} a_4 &= \frac{\alpha^2}{4\cdot 3} a_0 & a_5 &= \frac{\alpha^2}{5\cdot 4} a_1 \\ a_6 &= \frac{\alpha^2}{6\cdot 5} a_2 = 0 & a_7 &= \frac{\alpha^2}{7\cdot 6} a_3 = 0 \\ a_8 &= \frac{\alpha^2}{8\cdot 7} a_4 = \frac{\alpha^4}{8\cdot 7\cdot 4\cdot 3} a_0 & a_9 &= \frac{\alpha^2}{9\cdot 8} a_5 = \frac{\alpha^4}{9\cdot 8\cdot 5\cdot 4} a_1 \\ &\vdots & &\vdots \end{aligned} \tag{8}$$

Imposing the initial conditions, $a_0 = r(0) = r_0$ and $a_1 = r'(0) = 0$. Therefore, noting the pattern displayed in (8), we can write the solution of the initial value problem as

$$r(t) = \left[1 + \frac{\alpha^2 t^4}{4\cdot 3} + \frac{\alpha^4 t^8}{8\cdot 7\cdot 4\cdot 3} + \frac{\alpha^6 t^{12}}{12\cdot 11\cdot 8\cdot 7\cdot 4\cdot 3} + \cdots\right] r_0. \ \blacktriangle \tag{9}$$

When solving problems, we should make use of any checks that might easily be applied to the solutions we find. For this problem, two straightforward checks can be inferred from the differential equation itself:

(i) It follows from Theorem 10.2 that the differential equation has solutions that have even symmetry and solutions that have odd symmetry. [Note that the series solution $r(t)$ in (9) is an even function.]

(ii) Solutions of the differential equation are actually functions of $\alpha^{1/2}t$. To see this, consider a change of independent variable of the form $t = \beta s$, where β is a constant that we will choose to eliminate the parameter α from the differential equation. If we define $y(s)$ by $y(s) = r(\beta s)$, the differential equation transforms into

$$\frac{1}{\beta^2}\frac{d^2y}{ds^2} - \alpha^2\beta^2 s^2 y = 0.$$

If we choose β so that $1/\beta^2 = \alpha^2\beta^2$ or $\beta = \alpha^{-1/2}$, the differential equation becomes

$$\frac{d^2y}{ds^2} - s^2 y = 0.$$

We see, therefore, that solutions are functions of $s = \alpha^{1/2}t$. Note that the series solution $r(t)$ in (9) is a function of $\alpha^{1/2}t$.

In itself, the fact that solution (9) satisfies these two checks does not prove it is correct. It does, however, increase our confidence in its validity.

Figure 10.3 shows the solution plotted as a function of $s = \alpha^{1/2}t$ for an initial value of $r_0 = 0.01$. The figure shows both the numerical solution of the initial value problem and a polynomial approximation, $P_{12}(s)$, obtained by truncating series (9)

$$P_{12}(s) = \left[1 + \frac{s^4}{4 \cdot 3} + \frac{s^8}{8 \cdot 7 \cdot 4 \cdot 3} + \frac{s^{12}}{12 \cdot 11 \cdot 8 \cdot 7 \cdot 4 \cdot 3}\right](0.01).$$

In the interval $0 \le s \le 4$, the polynomial approximation agrees reasonably well with the numerical solution. To give the plot a physical context, think of a 1 m tube in which the particle is initially at rest 1 cm from the pivot. If the tube is set into motion with a constant angular acceleration of 1 rad/s^2, then $s = t$ and we see from Figure 10.3 that the particle exits the tube after roughly 3.5 s.

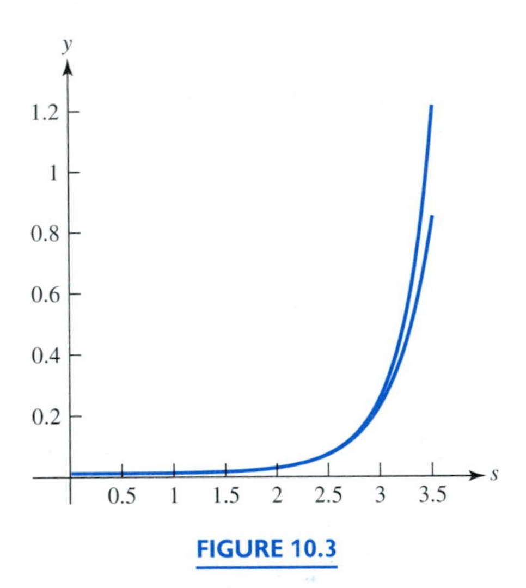

FIGURE 10.3

The graphs show (i) a numerical solution of the initial value problem in Example 3, and (ii) the first four terms of the Maclaurin series solution (9). In this figure, the independent variable is $s = \alpha^{1/2}t$. The initial position of the particle is $r_0 = 0.01$ m.

Polynomial Solutions

Some of the second order linear differential equations that arise in mathematical and scientific applications (such as the Legendre equation, the Hermite equation, and the Chebyshev equation) have polynomial solutions. The next example treats the Chebyshev equation. Other equations are considered in the exercises.

EXAMPLE 4

The Chebyshev[3] differential equation is

$$(1-t^2)y'' - ty' + \mu^2 y = 0, \tag{10}$$

where μ is a constant. Find a Maclaurin series solution of (10). Show that if μ is an integer, the Chebyshev differential equation (10) has a polynomial solution.

Solution: Rewriting the equation as

$$y'' - \frac{t}{1-t^2}y' + \frac{\mu^2}{1-t^2}y = 0,$$

we see that $t = \pm 1$ are singular points. All other points are ordinary points. In addition, we deduce from the structure of the differential equation itself that it possesses solutions with even symmetry and solutions with odd symmetry, see Theorem 10.2.

By Theorem 10.1, the general solution of the Chebyshev equation can be represented in terms of Maclaurin series that converge in $(-1, 1)$. Let $y(t) = \sum_{n=0}^{\infty} a_n t^n$. Substitution into differential equation (10) leads to

$$(1-t^2)\sum_{n=2}^{\infty} a_n n(n-1)t^{n-2} - t\sum_{n=1}^{\infty} a_n nt^{n-1} + \mu^2 \sum_{n=0}^{\infty} a_n t^n = 0. \tag{11}$$

Equation (11) can be rewritten as

$$\sum_{n=2}^{\infty} a_n n(n-1)t^{n-2} - \sum_{n=0}^{\infty} [a_n n(n-1) + na_n - \mu^2 a_n]t^n = 0,$$

or, after adjusting the index in the first summation and collecting terms,

$$\sum_{n=0}^{\infty} [a_{n+2}(n+2)(n+1) - a_n(n^2 - \mu^2)]t^n = 0.$$

The recurrence relation is therefore

$$a_{n+2} = \frac{n^2 - \mu^2}{(n+1)(n+2)}a_n, \qquad n = 0, 1, 2, \ldots \tag{12}$$

The recurrence relation determines all the even-indexed coefficients to be multiples of a_0 and all the odd-indexed coefficients to be multiples of a_1. The general solution is

$$\begin{aligned} y(t) = a_0 &\left(1 + \frac{-\mu^2}{2}t^2 + \frac{-(4-\mu^2)\mu^2}{24}t^4 + \cdots\right) \\ &+ a_1\left(t + \frac{1-\mu^2}{6}t^3 + \frac{(1-\mu^2)(9-\mu^2)}{120}t^5 + \cdots\right) \end{aligned} \tag{13}$$

(continued)

[3]Pafnuty Lvovich Chebyshev (1821–1894) was appointed to the University of St. Petersburg in 1847. He contributed to many areas of mathematics and science, including number theory, mechanics, probability theory, special functions, and the calculation of geometric volumes.

(continued)

The ratio test in conjunction with recurrence relation (12) can be used to show that the power series in (13) have radius of convergence $R = 1$ (except in the case when they terminate after a finite number of terms).

If μ is an even integer, we see from (12) that all the even coefficients having index greater than μ will vanish. For example, if $\mu = 4$, then $a_6 = 0, a_8 = 0, \ldots$. Thus, when $\mu = 4$, recurrence relation (12) leads to

$$a_2 = -8a_0, \qquad a_4 = -a_2 = 8a_0, \qquad a_6 = a_8 = a_{10} = \cdots = 0.$$

From this, we see that the 4th degree polynomial $P(t) = a_0[1 - 8t^2 + 8t^4]$ is a solution of the Chebyshev equation. Similarly, if μ is an odd positive integer, then we obtain an odd polynomial solution of degree μ. These polynomial solutions, generated as μ ranges over the nonnegative integers, are known as **Chebyshev polynomials of the first kind**. The Nth degree Chebyshev polynomial of the first kind is usually denoted $T_N(t)$. The first few Chebyshev polynomials are

$$T_0(t) = 1, \qquad T_1(t) = t, \qquad T_2(t) = 2t^2 - 1, \qquad T_3(t) = 4t^3 - 3t,$$
$$T_4(t) = 8t^4 - 8t^2 + 1.$$

The Chebyshev polynomials are normalized (that is, the arbitrary constant is selected) so that $T_N(1) = 1$. ▲

REMARKS:

1. Even though the differential equation has singular points at $t = \pm 1$, the Chebyshev polynomial solutions are well behaved at these points. The polynomial solutions are analytic with infinite radius of convergence. It is important to remember that solutions need not necessarily behave badly at singular points.
2. Chebyshev polynomials find important application in the design of antenna arrays and electrical filters. Consider, for example, the low-pass filtering problem illustrated in Figure 10.4. We want to build an electrical network having the power transfer function shown in Figure 10.4(a).

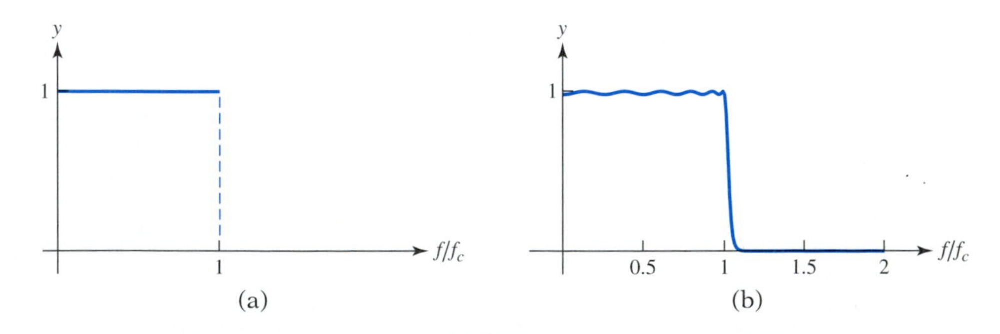

FIGURE 10.4

(a) The graph of an ideal power transfer function. (b) The graph of an approximation of the form (14) to the function graphed in (a).

Energy at all frequencies less than the cutoff frequency f_c should pass through the network unscathed while the passage of energy at all frequencies above f_c should be completely blocked. The problem, however, is that the network elements we have available to build the network only allow us to realize power transfer functions that are rational functions of frequency. The design problem is to find a rational function that closely approximates the ideal behavior in Figure 10.4(a). ∎

Chebyshev polynomials are particularly suited for such problems because they possess an "equal ripple" property (see Exercise 27). The polynomial $T_N(t)$ oscillates between ± 1 for t in the range $-1 \le t \le 1$ and grows monotonically in magnitude when $|t| \ge 1$. Therefore, a power transfer function of the form

$$\frac{1}{1 + \varepsilon T_N^2\left(\frac{f}{f_c}\right)} \tag{14}$$

with positive constant ε sufficiently small and integer N sufficiently large, can serve as a good rational approximation to the ideal power transfer function. Figure 10.4(b) illustrates the particular choice $\varepsilon = 0.02$, $N = 12$.

EXERCISES

Exercises 1–6:

Identify all the singular points of $y'' + p(t)y' + q(t)y = 0$ in the interval $-10 < t < 10$.

1. $y'' + (\sec t)y' + \dfrac{t}{t^2 - 4}y = 0$
2. $y'' + t^{2/3}y' + (\sin t)y = 0$
3. $(1 - t^2)y'' + ty' + (\csc t)y = 0$
4. $(\sin 2t)y'' + e^t y' + \dfrac{t}{25 - t^2}y = 0$
5. $(1 + \ln|t|)y'' + y' + (1 + t^2)y = 0$
6. $y'' + \dfrac{t}{1 + |t|}y' + (\tan t)y = 0$

Exercises 7–12:

In each exercise, $t = t_0$ is an ordinary point of $y'' + p(t)y' + q(t)y = 0$. Apply Theorem 10.1 to determine a value $R > 0$ such that the given initial value problem is guaranteed to have a unique solution that is analytic in the interval $t_0 - R < t < t_0 + R$.

7. $y'' + \dfrac{1}{1 + 2t}y' + \dfrac{t}{1 - t^2}y = 0, \quad t_0 = 0$
8. $(1 - 9t^2)y'' + 4y' + ty = 0, \quad t_0 = 1$
9. $y'' + \dfrac{1}{4 - 3t}y' + \dfrac{3t}{5 + 30t}y = 0, \quad t_0 = -1$
10. $y'' + \dfrac{1}{1 + 4t^2}y' + \dfrac{t}{4 + t}y = 0, \quad t_0 = 0$
11. $y'' + \dfrac{1}{1 + 3(t - 2)}y' + (\sin t)y = 0, \quad t_0 = 2$
12. $y'' + \dfrac{t + 3}{1 + t^2}y' + t^2 y = 0, \quad t_0 = 1$

Exercises 13–21:

In each exercise, $t = 0$ is an ordinary point of $y'' + p(t)y' + q(t)y = 0$.

(a) Find the recurrence relation that defines the coefficients of the power series solution $y(t) = \sum_{n=0}^{\infty} a_n t^n$.

(b) As in equation (13), find the first three nonzero terms in each of two linearly independent solutions.

(c) State the interval $-R < t < R$ on which Theorem 10.1 guarantees convergence.

(d) Does Theorem 10.2 indicate that the differential equation has solutions that are even and odd?

13. $y'' + ty' + y = 0$ **14.** $y'' + 2ty' + 3y = 0$ **15.** $(1 + t^2)y'' + ty' + 2y = 0$

16. $y'' - 5y' + 6y = 0$ **17.** $y'' - 4y' + 4y = 0$ **18.** $(1 + t)y'' + y = 0$

19. $(3 + t)y'' + 3ty' + y = 0$ **20.** $(2 + t^2)y'' + 4y = 0$ **21.** $y'' + t^2y = 0$

Exercises 22–25:

In each exercise, $t = 1$ is an ordinary point of $y'' + p(t)y' + q(t)y = 0$.

(a) Find the recurrence relation that defines the coefficients of the power series solution $y(t) = \sum_{n=0}^{\infty} a_n (t - 1)^n$.

(b) Find the first three nonzero terms in each of two linearly independent solutions.

(c) State the interval $-R < t - 1 < R$ on which Theorem 10.1 guarantees convergence.

22. $y'' + (t - 1)y' + y = 0$ **23.** $y'' + y = 0$

24. $(t - 2)y'' + y' + y = 0$ **25.** $y'' + y' + (t - 2)y = 0$

26. Recall Chebyshev's equation from Example 4, $(1 - t^2)y'' - ty' + \mu^2 y = 0$. As you saw in Example 4, this equation has a polynomial solution, $T_N(t)$, when $\mu = N$ is a nonnegative integer. Using recurrence relation (12), find $T_5(t)$ and $T_6(t)$.

27. The Equal Ripple Property of Chebyshev Polynomials Consider the Chebyshev differential equation $(1 - t^2)y'' - ty' + N^2 y = 0$, where N is a nonnegative integer.

(a) Show by substitution that the function $y(t) = \cos(N \arccos(t))$ is a solution for $-1 < t < 1$.

(b) Show, for $N = 0, 1, 2$ that the function $\cos(N \arccos(t))$ is a polynomial in t and that $T_N(t) = \cos(N \arccos(t))$. This result holds, in fact, for all nonnegative integers N. It can be shown that

$$T_N(t) = \begin{cases} \cos(N \arccos(t)), & -1 \le t \le 1 \\ \cosh(N \operatorname{arccosh}(t)), & 1 \le |t|. \end{cases}$$

(c) Use a computer graphics package to plot $T_N(t)$ for $N = 2, 5$, and 8 and for $-1.2 \le t \le 1.2$.

(d) What serves as a bound for $|T_N(t)|$ when $-1 \le t \le 1$? What is the behavior of $T_N(t)$ when $1 \le |t|$?

28. Legendre's Equation Legendre's equation is $(1 - t^2)y'' - 2ty' + \mu(\mu + 1)y = 0$. By Theorem 10.1, this equation has a power series solution of the form $y(t) = \sum_{n=0}^{\infty} a_n t^n$ which is guaranteed to be absolutely convergent in the interval $-1 < t < 1$. As in Example 4,

(a) Find the recurrence relation for the coefficients of the power series.

(b) Argue, when $\mu = N$ is a nonnegative integer, that Legendre's equation has a polynomial solution, $P_N(t)$.

(c) Show, by direct substitution, that the Legendre polynomials $P_0(t) = 1$ and $P_1(t) = t$ satisfy Legendre's equation for $\mu = 0$ and $\mu = 1$, respectively.

(d) Use the recurrence relation and the requirement that $P_n(1) = 1$ to determine the next four Legendre polynomials, $P_2(t), P_3(t), P_4(t), P_5(t)$.

29. Hermite's Equation Hermite's equation is $y'' - 2ty' + 2\mu y = 0$. By Theorem 10.1, this equation has a power series solution of the form $y(t) = \sum_{n=0}^{\infty} a_n t^n$ which is guaranteed to be absolutely convergent in the interval $-\infty < t < \infty$. As in Example 4,

(a) Find the recurrence relation for the coefficients of the power series.

(b) Argue that, when $\mu = N$ is a nonnegative integer, Hermite's equation has a polynomial solution, $H_N(t)$.

(c) Show, by direct substitution, that the Hermite polynomials $H_0(t) = 1$ and $H_1(t) = 2t$ satisfy Hermite's equation for $\mu = 0$ and $\mu = 1$, respectively.

(d) Use the recurrence relation and the requirement that $H_n(t) = 2^n t^n + \cdots$ to determine the next four Hermite polynomials, $H_2(t), H_3(t), H_4(t), H_5(t)$.

30. Consider the differential equation $y'' + p(t)y' + q(t)y = 0$. In some cases we may be able to find a power series solution of the form $y(t) = \sum_{n=0}^{\infty} a_n (t - t_0)^n$ even when t_0 is not an ordinary point. In other cases, there is no power series solution.

(a) The point $t = 0$ is a singular point of $ty'' + y' - y = 0$. Nevertheless, find a nontrivial power series solution, $y(t) = \sum_{n=0}^{\infty} a_n t^n$, of this equation.

(b) The point $t = 0$ is a singular point of $t^2 y'' + y = 0$. Show the only solution of this equation having the form $y(t) = \sum_{n=0}^{\infty} a_n t^n$ is the trivial solution.

Exercises 31 and 32 outline the proof of parts (a) and (b) of Theorem 10.2, respectively. In each exercise, consider the differential equation $y'' + p(t)y' + q(t)y = 0$, where p and q are continuous on the domain $(-b, -a) \cup (a, b)$, $a \geq 0$.

31. Let $f(t)$ be a solution on the interval (a, b).

(a) Let t lie in the interval $(-b, -a)$ and set $\tau = -t$, so that $a < \tau < b$. Show that if

$$\frac{d^2 f(\tau)}{d\tau^2} + p(\tau)\frac{df(\tau)}{d\tau} + q(\tau)f(\tau) = 0, \qquad a < \tau < b,$$

then

$$\frac{d^2 f(-t)}{dt^2} - p(-t)\frac{df(-t)}{dt} + q(-t)f(-t) = 0, \qquad -b < t < -a.$$

(b) Use the fact that p and q are odd and even functions (respectively) to show that $f(-t)$ is a solution of the given differential equation on the interval $-b < t < -a$.

32. Now let p and q be analytic at $t = 0$ with a common radius of convergence $R > 0$, where p is an odd function and q is an even function.

(a) Let $f_1(t)$ and $f_2(t)$ be solutions of the given differential equation, satisfying initial conditions $f_1(0) = 1, f_1'(0) = 0, f_2(0) = 0, f_2'(0) = 1$. What does Theorem 10.1 say about the solutions $f_1(t)$ and $f_2(t)$?

(b) Use the results of Exercise 31 to show that $f_1(-t)$ and $f_2(-t)$ are also solutions on the interval $-R < t < R$.

(c) Form the functions $f_e(t) = [f_1(t) + f_1(-t)]/2$ and $f_o(t) = [f_2(t) - f_2(-t)]/2$. Show that $f_e(t)$ and $f_o(t)$ are even and odd analytic solutions, respectively, on the interval $-R < t < R$.

(d) Show that $f_e(t)$ and $f_o(t)$ are nontrivial solutions by showing that $f_e(t) = f_1(t)$ and $f_o(t) = f_2(t)$. [Hint: Use the fact that solutions of initial value problems are unique.]

(e) Show that the solutions $f_e(t)$ and $f_o(t)$ form a fundamental set of solutions.

Exercises 33–38:

Suppose a linear differential equation $y'' + p(t)y' + q(t)y = 0$ satisfies the hypotheses of Theorem 10.2(b), on the interval $-\infty < t < \infty$. Then, by Exercise 32, we can assume the general solution of $y'' + p(t)y' + q(t)y = 0$ has the form

$$y(t) = c_1 y_e(t) + c_2 y_o(t), \tag{15}$$

where $y_e(t)$ is an even solution of $y'' + p(t)y' + q(t)y = 0$ and $y_o(t)$ is an odd solution. In each of the following exercises, determine whether Theorem 10.2(b) can be used to guarantee that the given differential equation has a general solution of the form (15). If your answer is No, explain why the equation fails to satisfy the hypotheses of Theorem 10.2(b).

33. $y'' + (\sin t)y' + t^2 y = 0$ **34.** $y'' + (\cos t)y' + ty = 0$ **35.** $y'' + t^2 y = 0$

36. $y'' + y' + t^2 y = 0$ **37.** $y'' + ty = 0$ **38.** $y'' + e^t y' + y = 0$

39. Consider the differential equation $y'' + ay' + by = 0$, where a and b are constants. For what values of a and b will the differential equation have nontrivial solutions that are odd and even?

40. Consider the initial value problem $(1 + t^2)y'' + y = 0, y(0) = 1, y'(0) = 0$.

(a) Show, by Theorem 10.1, that this initial value problem is guaranteed to have a unique solution of the form $y(t) = \sum_{n=0}^{\infty} a_n t^n$, where the series converges absolutely in $-1 < t < 1$.

(b) The recurrence relation has the form $r(n)a_{n+2} = s(n)a_n$ where $r(n)$ and $s(n)$ are quadratic polynomials. Therefore,

$$\left|\frac{a_{n+2}}{a_n}\right| = \left|\frac{s(n)}{r(n)}\right|.$$

Show that this ratio tends to 1 as n tends to ∞ and conclude, therefore, that the series solution of the initial value problem diverges for $1 < |t|$.

(c) Note, however, that the coefficient functions $p(t) = 0$ and $q(t) = (1 + t^2)^{-1}$ are continuous for $-\infty < t < \infty$ and hence, by Theorem 4.1, a unique solution exists for all values t. Do the conclusions of Theorem 10.1 and Theorem 4.1 contradict one another? Explain.

10.3 The Euler Equation

The second order linear homogeneous equation

$$t^2 y'' + \alpha t y' + \beta y = 0, \qquad t \neq 0, \tag{1}$$

is known as the **Euler equation** (it is also known as the **Cauchy-Euler** or the **equidimensional** equation). In equation (1), α and β are real constants. Note the special structure of the differential equation; the power of each monomial coefficient matches the order of the derivative that it multiplies; for example, t^2 multiplies y''. This special structure enables us to obtain an explicit representation for the general solution. In this section we present two related approaches to deriving the general solution.

If we rewrite the Euler equation in the form $y'' + p(t)y' + q(t)y = 0$, then

$$p(t) = \frac{\alpha}{t} \quad \text{and} \quad q(t) = \frac{\beta}{t^2}.$$

The coefficients $p(t)$ and $q(t)$ are analytic at every point except $t = 0$. Ignoring the trivial case where α and β are both zero, we see that $t = 0$ is a singular point and all other values of t are ordinary points. Note as well that $p(t)$ and $q(t)$ are not continuous at $t = 0$. Therefore, the basic existence-uniqueness result, Theorem 4.1, alerts us to possible problems—solutions may or may not exist at $t = 0$.

The Euler equation arises in a variety of applications. An example is the problem of determining the time-independent (or steady-state) temperature distribution within a circular geometry (such as the interior of a circular pipe) from a knowledge of the temperature on the boundary; see the Extended Problem at the end of this chapter. The Euler equation is also of interest because it serves as a prototype or model equation. In Section 10.4 we will define a special type of singular point, called a regular singular point; our understanding of the Euler equation will serve as the basis for studying regular singular points.

The General Solution of the Euler Equation

If $f(t)$ is a solution of the Euler equation (1), then so is $f(-t)$ (see Theorem 10.2 in Section 10.2). Therefore, we assume our interval of interest is $t > 0$. Once the general solution is obtained for the interval $t > 0$, we can obtain the general solution in $t < 0$ by replacing t with $-t$.

We present two approaches to deriving the general solution of (1). In retrospect, you will see that the two approaches are closely related. Nevertheless, both points of view are worthy of consideration.

Solutions of the Form $y(t) = t^\lambda$

The special structure of the Euler equation makes it possible to find solutions of the form $y(t) = t^\lambda$, where λ is a constant to be determined and where $t > 0$ is assumed. Substitution of this trial form into equation (1) leads to

$$t^2\lambda(\lambda - 1)t^{\lambda-2} + \alpha t\lambda t^{\lambda-1} + \beta t^\lambda = [\lambda(\lambda - 1) + \alpha\lambda + \beta]t^\lambda = 0, \qquad t > 0. \qquad (2)$$

Since t^λ is not identically zero on $0 < t < \infty$ for any real or complex value λ, we must have

$$\lambda^2 + (\alpha - 1)\lambda + \beta = 0, \qquad (3)$$

or

$$\lambda_{1,2} = \frac{-(\alpha - 1) \pm \sqrt{(\alpha - 1)^2 - 4\beta}}{2}. \qquad (4)$$

In other words, $y(t) = t^\lambda$ is a solution of the Euler equation whenever λ is a root of equation (3). We refer to equation (3) as the **characteristic equation** or the **indicial equation**. There are three possibilities for the roots of the characteristic equation:

1. There are **two real distinct roots** if the discriminant, $(\alpha - 1)^2 - 4\beta$, is positive. In this case, the Wronskian of the solution set is

$$W(t) = \begin{vmatrix} t^{\lambda_1} & t^{\lambda_2} \\ \lambda_1 t^{\lambda_1 - 1} & \lambda_2 t^{\lambda_2 - 1} \end{vmatrix} = (\lambda_2 - \lambda_1)t^{\lambda_1 + \lambda_2 - 1}.$$

Since $W(t)$ is nonzero for $t > 0$, the functions $y_1(t) = t^{\lambda_1}$ and $y_2(t) = t^{\lambda_2}$ form a fundamental set. The general solution of equation (1) is $y(t) = c_1 t^{\lambda_1} + c_2 t^{\lambda_2}$, $t > 0$, where c_1 and c_2 are arbitrary constants. As previously mentioned, we can obtain the general solution for $t < 0$ by replacing t by $-t$; the general solution for $t < 0$ is $y(t) = c_1(-t)^{\lambda_1} + c_2(-t)^{\lambda_2}$, where c_1 and c_2 again denote arbitrary constants. Since $-t > 0$ when $t < 0$, the general solution can be expressed in the form

$$y(t) = c_1|t|^{\lambda_1} + c_2|t|^{\lambda_2}, \qquad t \neq 0. \tag{5}$$

2. There is **one real repeated root** if $(\alpha - 1)^2 - 4\beta = 0$. In this case, one solution is $y_1(t) = t^{\lambda}$, $t > 0$, where $\lambda = -(\alpha - 1)/2$. We can use reduction of order (see Section 4.5) to find a second linearly independent solution, $y_2(t) = t^{\lambda}\ln t$, $t > 0$ (see Exercise 1 of the present section). The general solution in the repeated root case is therefore

$$y(t) = c_1|t|^{\lambda} + c_2|t|^{\lambda}\ln|t|, \qquad t \neq 0. \tag{6}$$

3. There are **complex conjugate roots** if $(\alpha - 1)^2 - 4\beta < 0$. For brevity, let

$$\lambda_{1,2} = \frac{-(\alpha - 1) \pm i\sqrt{4\beta - (\alpha - 1)^2}}{2} = \gamma \pm i\delta.$$

In this case,

$$t^{\gamma + i\delta} = e^{(\gamma + i\delta)\ln t} = e^{\gamma \ln t} e^{i\delta \ln t} = t^{\gamma} e^{i\delta \ln t}, \qquad t > 0.$$

Hence, by Euler's formula (see Section 4.6),

$$t^{\gamma + i\delta} = t^{\gamma}[\cos(\delta \ln t) + i \sin(\delta \ln t)], \qquad t > 0. \tag{7}$$

From equation (7) we obtain two real-valued solutions, $y_1(t) = t^{\gamma}\cos(\delta \ln t)$ and $y_2(t) = t^{\gamma}\sin(\delta \ln t)$, which can be shown to form a fundamental set on $0 < t < \infty$ (see Exercise 2). Therefore, the general solution is

$$y(t) = c_1|t|^{\gamma}\cos(\delta \ln|t|) + c_2|t|^{\gamma}\sin(\delta \ln|t|), \qquad t \neq 0. \tag{8}$$

EXAMPLE 1

Find the general solution of each of the Euler equations

(a) $t^2y'' - 2ty' + 2y = 0$ (b) $t^2y'' + 5ty' + 4y = 0$ (c) $t^2y'' + 3ty' + 5y = 0$

Solution:

(a) For this equation, $\alpha = -2$ and $\beta = 2$. Looking for solutions of the form t^{λ}, $t > 0$ leads to the condition $\lambda^2 - 3\lambda + 2 = (\lambda - 1)(\lambda - 2) = 0$. The general solution is therefore

$$y(t) = c_1 t + c_2 t^2, \qquad -\infty < t < \infty. \tag{9}$$

Since the general solution involves integer powers of t, we can dispense with the absolute value signs. Note further that solution (9) is valid for all values of t, including the singular point $t = 0$. For this equation, every solution is well behaved at $t = 0$.

(b) For this differential equation, $\alpha = 5$ and $\beta = 4$. The characteristic equation $\lambda^2 + 4\lambda + 4 = 0$ has real repeated roots, $\lambda_1 = \lambda_2 = -2$. The general solution is

$$y(t) = c_1 t^{-2} + c_2 t^{-2}\ln|t|, \qquad t \neq 0.$$

(c) For this equation, $\alpha = 3$ and $\beta = 5$. The characteristic equation $\lambda^2 + 2\lambda + 5 = 0$ has complex conjugate roots $\lambda_{1,2} = -1 \pm 2i$. The general solution is

$$y(t) = c_1 t^{-1} \cos(2\ln|t|) + c_2 t^{-1} \sin(2\ln|t|), \qquad t \neq 0.$$

Neither equation (b) nor (c) has solutions that exist at $t = 0$. ▲

Change of Independent Variable

After a change of independent variable, the Euler equation can be transformed into a new homogeneous constant coefficient equation. The techniques of Chapter 4 can be used to solve it.

Again consider the Euler equation $t^2y'' + \alpha ty' + \beta y = 0$ for $t > 0$. We introduce a new independent variable $z = \ln t$ or equivalently, $t = e^z$. The t-interval $0 < t < \infty$ transforms into the z-interval $-\infty < z < \infty$.

Let $y(t) = y(e^z) \equiv Y(z)$. Then, using the chain rule,

$$\frac{dy(t)}{dt} = \frac{dy(e^z)}{dz}\frac{dz}{dt} = \frac{dY(z)}{dz}\left(\frac{dt}{dz}\right)^{-1} = e^{-z}\frac{dY(z)}{dz}.$$

Therefore, $t\,(dy(t)/dt)$ transforms into

$$e^z e^{-z}\frac{dY(z)}{dz} = \frac{dY(z)}{dz} \tag{10}$$

while $t^2\,(d^2y(t)/dt^2)$ transforms into

$$\frac{d^2Y(z)}{dz^2} - \frac{dY(z)}{dz}. \tag{11}$$

Under this change of independent variable, the Euler equation (1) transforms into the constant coefficient equation

$$Y'' + (\alpha - 1)Y' + \beta Y = 0. \tag{12}$$

Therefore, we can solve the Euler equation (1) for $t > 0$ by

1. making the change of independent variable $t = e^z$,
2. solving the constant coefficient equation (12) using solution procedures developed in Chapter 4, and then
3. using the inverse map to obtain the desired solution, $y(t) = Y(\ln t)$.

Thus, looking for solutions of $t^2y'' + \alpha ty' + \beta y = 0$ having the form $y(t) = t^\lambda$ is equivalent to looking for solutions of $Y'' + (\alpha - 1)Y' + \beta Y = 0$ having the form $Y(z) = e^{\lambda z}$. The characteristic equation (3) results in either case.

Each of the two solution approaches has its utility. The first approach, looking for solutions of the form $|t|^\lambda$, serves as a guide in the next section when we study the behavior of solutions of differential equations in a neighborhood of certain types of singular points. However, the change of independent variable approach is useful because it permits us to use the techniques developed in Chapter 4.

As the next example shows, the change of variables approach can be applied to nonhomogeneous Euler equations.

EXAMPLE 2

Obtain the general solution of

$$4t^2y'' + 8ty' + y = t + \ln t, \qquad t > 0,$$

by making the change of independent variable, $t = e^z$. As before, let $y(e^z) \equiv Y(z)$ and, using (10) and (11), rewrite the left-hand side of the given differential equation in terms of $Y(z)$ and derivatives of $Y(z)$.

Solution: Under the change of variables $t = e^z$, the nonhomogeneous term transforms into $e^z + z$. Thus, using (10) and (11), the given differential equation becomes $4[Y'' - Y'] + 8Y' + Y = e^z + z$, or

$$4Y'' + 4Y' + Y = e^z + z, \tag{13}$$

where $Y' = dY/dz$. As seen in Chapter 4, the solution of (13) is the sum of a complementary solution, $Y_C(z)$, and a particular solution, $Y_P(z)$. The complementary solution is

$$Y_C(z) = c_1e^{-z/2} + c_2ze^{-z/2}. \tag{14}$$

A particular solution $Y_P(z)$ can be obtained using the method of undetermined coefficients (see Section 4.9). We find

$$Y_P(z) = \frac{1}{9}e^z + z - 4.$$

The general solution is therefore

$$Y(z) = c_1e^{-z/2} + c_2ze^{-z/2} + \frac{1}{9}e^z + z - 4.$$

Using $z = \ln t$ to convert to the original independent variable t, we have

$$y(t) = c_1t^{-1/2} + c_2t^{-1/2}\ln t + \frac{1}{9}t + \ln t - 4, \qquad t > 0. \blacktriangle$$

Generalizations of the Euler Equation

There are two natural ways to generalize the Euler equation. One such generalization is given in equation (15),

$$(t - t_0)^2y'' + (t - t_0)\alpha y' + \beta y = 0. \tag{15}$$

In equation (15) t_0 is a point in the interval $-\infty < t < \infty$. To solve equation (15) for $t > t_0$, we assume a solution of the form $y(t) = (t - t_0)^\lambda$. [We could also adopt the change of independent variable $t - t_0 = e^z$.]

Another natural generalization is a higher order version of the Euler equation, such as

$$t^3y''' + \alpha t^2y'' + \beta ty' + \gamma y = 0. \tag{16}$$

Examples of each of these generalizations are given in the exercises.

EXERCISES

1. Consider the Euler equation $t^2y'' - (2\alpha - 1)ty' + \alpha^2 y = 0$.

   (a) Show that the characteristic equation has a repeated root $\lambda_1 = \lambda_2 = \alpha$. One solution is therefore $y(t) = t^\alpha, t > 0$.

   (b) Use the method of reduction of order (Section 4.5) to obtain a second linearly independent solution for the interval $0 < t < \infty$.

2. Let $y_1(t) = t^\gamma \cos(\delta \ln t)$ and $y_2(t) = t^\gamma \sin(\delta \ln t), t > 0$, where δ and γ are real constants with $\delta \neq 0$. Solutions of this form arise when the characteristic equation has complex roots. Compute the Wronskian of this pair of solutions and show that they form a fundamental set of solutions on $0 < t < \infty$.

Exercises 3–18:

Identify the singular point and denote it by t_0. Find the general solution that is valid for all t in the interval $-\infty < t < \infty$, except for the singular point.

3. $t^2y'' - 4ty' + 6y = 0$
4. $t^2y'' - 6y = 0$
5. $t^2y'' - 3ty' + 4y = 0$
6. $t^2y'' - ty' + 5y = 0$
7. $t^2y'' - 3ty' + 29y = 0$
8. $t^2y'' - 5ty' + 9y = 0$
9. $t^2y'' + ty' + 9y = 0$
10. $t^2y'' + 3ty' + y = 0$
11. $t^2y'' + 3y' + 17y = 0$
12. $y'' + \dfrac{11}{t}y' + \dfrac{25}{t^2}y = 0$
13. $y'' + \dfrac{5}{t}y' + \dfrac{40}{t^2}y = 0$
14. $t^2y'' - 2ty' = 0$
15. $(t-1)^2y'' - (t-1)y' - 3y = 0$
16. $(t-1)^2y'' + 3(t-1)y' + 17y = 0$
17. $(t+2)^2y'' + 6(t+2)y' + 6y = 0$
18. $(t-2)^2y'' + (t-2)y' + 4y = 0$

Exercises 19–21:

A Euler equation $(t - t_0)^2y'' + \alpha(t - t_0)y' + \beta y = 0$ is known to have the given general solution. What are the constants t_0, α, and β?

19. $y(t) = c_1(t+2) + c_2\dfrac{1}{(t+2)^2}, \quad t \neq -2$
20. $y(t) = c_1 + c_2 \ln|t-1|, \quad t \neq 1$
21. $y(t) = c_1 t^2 \cos(\ln|t|) + c_2 t^2 \sin(\ln|t|), \quad t \neq 0$

Exercises 22–23:

A nonhomogeneous Euler equation, $t^2y'' + \alpha ty' + \beta y = g(t)$ is known to have the given general solution. Determine the constants α and β and the function $g(t)$.

22. $y(t) = c_1t^2 + c_2t^{-1} + 2t + 1, \quad t > 0$
23. $y(t) = c_1t^2 + c_2t^3 + \ln t, \quad t > 0$

Exercises 24–29:

Find the general solution of the given equation for $0 < t < \infty$. [Hint: You can, as in Example 2, use the change of variable $t = e^z$.]

24. $t^2y'' - 2y = 2$
25. $t^2y'' - ty' + y = t^{-1}$
26. $t^2y'' + ty' + 9y = 10t$
27. $t^2y'' - 6y = 10t^{-2} - 6$
28. $t^2y'' - 4ty' + 6y = 3\ln t$
29. $t^2y'' + 8ty' + 10y = 36(t + t^{-1})$

Exercises 30–33:

Solve the given initial value problem. What is the interval of existence of the solution?

30. $t^2y'' - ty' - 3y = 8t + 6, \quad y(1) = 1, \quad y'(1) = 3$
31. $t^2y'' - 5ty' + 5y = 10, \quad y(1) = 4, \quad y'(1) = 6$

32. $t^2y'' + 3ty' + y = 8t + 9,\ \ y(-1) = 1,\ \ y'(-1) = 0$

33. $t^2y'' + 3ty' + y = 2t^{-1},\ \ y(1) = -2,\ \ y'(1) = 1$

34. Consider the third order equation $t^3y''' + \alpha t^2y'' + \beta ty' + \gamma y = 0$, $t > 0$. Make the change of independent variable $t = e^z$ and let $Y(z) = y(e^z)$. Derive the third order equation satisfied by Y.

Exercises 35–38:

Obtain the general solution of the given differential equation for $0 < t < \infty$.

35. $t^3y''' + 3t^2y'' - 3ty' = 0$

36. $t^3y''' + ty' - y = 0$

37. $t^3y''' + 3t^2y'' + ty' = 8t^2 + 12$

38. $t^3y''' + 6t^2y'' + 7ty' + y = 2 + \ln t$

10.4 Solutions Near a Regular Singular Point and the Method of Frobenius

We introduced the term *singular point* in Section 10.2 to denote a point t where at least one of the coefficient functions of $y'' + p(t)y' + q(t)y = 0$ fails to be analytic. Near a singular point, the possible types of solution behavior are diverse and complicated.

In this section and in Section 10.5, we restrict our attention to a particular type of singular point, one known as a *regular singular point*. As we shall see, definitive statements can be made about the behavior of solutions near regular singular points. Many of the important equations of mathematical physics, such as the Euler equation, Bessel's equation, and Legendre's equation, possess regular singular points.

Regular Singular Points

To introduce the definition of a regular singular point, we begin with the Euler equation,

$$t^2y'' + \alpha ty' + \beta y = 0,$$

where α and β are constants, not both zero. Since the coefficient functions are

$$p(t) = \frac{\alpha}{t} \quad \text{and} \quad q(t) = \frac{\beta}{t^2},$$

$t = 0$ is a singular point and all other points are ordinary points. The Euler equation serves as a model or prototype for a differential equation having a regular singular point.

For the Euler equation, the functions $tp(t)$ and $t^2q(t)$ are analytic at the singular point $t = 0$ because $tp(t) = \alpha$ and $t^2q(t) = \beta$ are constant functions. A constant function is a very simple kind of analytic function; the definition of a regular singular point, which we now introduce, allows $tp(t)$ and $t^2q(t)$ to be general analytic functions. In this definition, we also generalize the location of the singular point, allowing it to occur at any point t_0, not just at $t = 0$.

Let t_0 be a singular point of the differential equation

$$y'' + p(t)y' + q(t)y = 0.$$

The point t_0 is a **regular singular point** if both of the functions

$$(t - t_0)p(t) \quad \text{and} \quad (t - t_0)^2 q(t)$$

are analytic at t_0. A singular point that is not a regular singular point is called an **irregular singular point**.

If $(t - t_0)p(t)$ and/or $(t - t_0)^2 q(t)$ are indeterminate at $t = t_0$ but the limits $\lim_{t \to t_0}(t - t_0)p(t)$ and $\lim_{t \to t_0}(t - t_0)^2 q(t)$ exist, then we are tacitly assuming that the functions being considered are defined to equal these limits at $t = t_0$. In many important cases, the functions $p(t)$ and $q(t)$ are rational functions (that is, ratios of polynomials). As previously noted, a rational function is analytic at every point where the denominator polynomial is nonzero. In this case, to show that a singular point t_0 is a regular singular point, it suffices to show that the limits $\lim_{t \to t_0}(t - t_0)p(t)$ and $\lim_{t \to t_0}(t - t_0)^2 q(t)$ both exist. Demonstrating that these two limits exist also suffices when, at a singular point t_0, both $p(t)$ and $q(t)$ are quotients of functions analytic at t_0.

EXAMPLE 1

For each differential equation, identify the singular points and classify them as regular or irregular singular points. In these equations, ν and μ are constants.

(a) (Bessel's equation[4]) $\quad t^2 y'' + t y' + (t^2 - \nu^2) y = 0$

(b) (Legendre's equation[5]) $\quad (1 - t^2) y'' - 2t y' + [\nu(\nu + 1) - \mu^2 (1 - t^2)^{-1}] y = 0$

(c) $\quad (t - 1) y'' + \left(\tan \dfrac{\pi}{2} t\right) y' + t^{5/3} y = 0, \qquad -2 < t < 2$

Solution:

(a) For Bessel's equation,

$$p(t) = \frac{1}{t} \quad \text{and} \quad q(t) = \frac{t^2 - \nu^2}{t^2}.$$

Therefore, $t = 0$ is a singular point. All other points are ordinary points. Because $p(t)$ and $q(t)$ are rational functions, we need only determine whether $\lim_{t \to 0} t p(t)$ and $\lim_{t \to 0} t^2 q(t)$ exist in order to establish analyticity. Since

$$\lim_{t \to 0} t p(t) = 1 \quad \text{and} \quad \lim_{t \to 0} t^2 q(t) = -\nu^2,$$

it follows that $t = 0$ is a regular singular point.

(continued)

[4]Friedrich Wilhelm Bessel (1784–1846) was a German scientist who is noted for important contributions to the fields of astronomy, celestial mechanics, and mathematics. His mathematical analysis of what is now known as the Bessel function arose during his studies of planetary motion. Bessel's achievements become even more remarkable when one realizes that his formal education ended at age 14.

[5]Adrien-Marie Legendre (1752–1833) was a French scientist whose research interests included projectile dynamics, celestial mechanics, number theory, and analysis. What are today called Legendre polynomials appeared in a 1784 paper on celestial mechanics. Legendre authored influential textbooks on Euclidean geometry and number theory.

(continued)

(b) For Legendre's equation,

$$p(t) = \frac{-2t}{(1-t^2)} = \frac{-2t}{(1-t)(1+t)}$$

and

$$q(t) = \frac{\nu(\nu+1)}{(1-t^2)} - \frac{\mu^2}{(1-t^2)^2} = \frac{\nu(\nu+1)(1-t^2)-\mu^2}{(1-t)^2(1+t)^2}.$$

Therefore, the points $t = \pm 1$ are singular points. All other points are ordinary points. We first consider the singular point $t = 1$. Since $p(t)$ and $q(t)$ are rational functions, we check the two limits

$$\lim_{t\to 1}(t-1)p(t) = \lim_{t\to 1}\frac{2t}{1+t} = 1$$

and

$$\lim_{t\to 1}(t-1)^2 q(t) = \lim_{t\to 1}\frac{\nu(\nu+1)(1-t^2)-\mu^2}{(1+t)^2} = \frac{-\mu^2}{4}.$$

Since both limits exist, $t = 1$ is a regular singular point. The analysis of the other singular point, $t = -1$, is very similar and we find that $t = -1$ is also a regular singular point.

(c) We have

$$p(t) = \frac{\tan\frac{\pi}{2}t}{t-1} = \frac{\sin\frac{\pi}{2}t}{(t-1)\cos\frac{\pi}{2}t} \quad \text{and} \quad q(t) = \frac{t^{5/3}}{t-1}.$$

Note that $p(t)$, viewed as $p(t) = \sin(\pi t/2)/(t-1)\cos(\pi t/2)$, is a quotient of functions analytic in $-2 < t < 2$. Therefore, $p(t)$ is analytic at all points in the interval $-2 < t < 2$ except $t = 1$ (where the denominator is zero). The function $q(t)$ fails to be analytic at $t = 0$ (where $t^{5/3}$ is not analytic) and $t = 1$ (where the denominator is zero). Consider first the singular point $t = 0$; $t^2 q(t) = (t-1)^{-1}t^{11/3}$ is not analytic at $t = 0$ and therefore $t = 0$ is an irregular singular point.

Next, consider the singular point $t = 1$. In this case, $p(t)$ and $q(t)$ are both quotients of functions analytic at $t = 1$. Therefore, $t = 1$ is a regular singular point if both of the limits exist:

$$\lim_{t\to 1}(t-1)p(t) \quad \text{and} \quad \lim_{t\to 1}(t-1)^2 q(t).$$

The first limit,

$$\lim_{t\to 1}(t-1)p(t) = \lim_{t\to 1}\tan\frac{\pi}{2}t,$$

does not exist. Therefore, $t = 1$ is also an irregular singular point. ▲

The Method of Frobenius

Just as the Euler equation served as a model for defining regular singular points, the general solution of the Euler equation will serve to introduce the method of Frobenius.[6] This method prescribes the type of solution to look for near a regular singular point.

For simplicity, consider solution behavior near $t = 0$. Suppose we know that the differential equation

$$y'' + p(t)y' + q(t)y = 0 \tag{1a}$$

has a regular singular point at $t = 0$. Then $tp(t)$ and $t^2q(t)$ are analytic at $t = 0$. Let the Maclaurin series for $tp(t)$ and $t^2q(t)$ be

$$tp(t) = \alpha_0 + \alpha_1 t + \alpha_2 t^2 + \cdots = \sum_{n=0}^{\infty} \alpha_n t^n$$

and

$$t^2q(t) = \beta_0 + \beta_1 t + \beta_2 t^2 + \cdots = \sum_{n=0}^{\infty} \beta_n t^n,$$

where both series converge in some interval $-R < t < R$. We next multiply differential equation (1a) by t^2, to obtain

$$t^2y'' + t[tp(t)]y' + [t^2q(t)]y = 0. \tag{1b}$$

Inserting the Maclaurin series for $tp(t)$ and $t^2q(t)$, we arrive at

$$t^2y'' + t[\alpha_0 + \alpha_1 t + \alpha_2 t^2 + \cdots]y' + [\beta_0 + \beta_1 t + \beta_2 t^2 + \cdots]y = 0. \tag{2}$$

The method of Frobenius can be motivated by a heuristic argument. Very close to the singular point $t = 0$, we expect that

$$\alpha_0 + \alpha_1 t + \alpha_2 t^2 + \cdots \approx \alpha_0 \quad \text{and} \quad \beta_0 + \beta_1 t + \beta_2 t^2 + \cdots \approx \beta_0. \tag{3}$$

If we use approximations (3) in equation (1b), we recover the Euler equation. Thus, we reason: If the two differential equations are nearly the same near $t = 0$, shouldn't the solutions themselves have similar behavior near $t = 0$?

Recall that solutions of the Euler equation were obtained by looking for solutions of the form $y(t) = |t|^{\lambda}, t \neq 0$. For definiteness, we consider $t > 0$. The method of Frobenius consists in looking for solutions in which the factor t^{λ} is multiplied by an infinite series. In other words, near the regular singular point $t = 0$, we look for solutions of $y'' + p(t)y' + q(t)y = 0$ that have the form

$$y(t) = t^{\lambda}[a_0 + a_1 t + a_2 t^2 + \cdots] = \sum_{n=0}^{\infty} a_n t^{\lambda+n}. \tag{4}$$

In representation (4), λ is a constant (possibly complex-valued) that is to be determined, along with the constants $a_0, a_1, \ldots$. Also, since λ has not been specified, we can assume without any loss of generality that $a_0 \neq 0$. [That is, there must be a "first nonzero term" in series (4) and we simply take that term to be $a_0 t^{\lambda}$.]

[6]Ferdinand Georg Frobenius (1849–1917), a German mathematician, served on the faculty at the University of Zurich and the University of Berlin. He is remembered for his contributions to group theory and differential equations.

REMARK: When $y'' + p(t)y' + q(t)y = 0$ has a regular singular point at $t = 0$ we usually, for convenience, consider solutions in the region $t > 0$. When the coefficient functions possess the symmetries discussed in Theorem 10.2, solutions in $t < 0$ can be obtained by simply replacing t with $-t$ in the solutions found for $t > 0$. In the general case, when such symmetries are not present, we can make the change of independent variable $\tau = -t$ and consider the differential equation

$$\frac{d^2\bar{y}(\tau)}{d\tau^2} - p(-\tau)\frac{d\bar{y}(\tau)}{d\tau} + q(-\tau)\bar{y}(\tau) = 0, \qquad \tau > 0.$$

Once the solution $\bar{y}(\tau)$ is obtained for $\tau > 0$, we obtain the desired solution $y(t)$ for $t < 0$ by setting $y(t) = \bar{y}(-t)$.

Implementing the Method of Frobenius

Substituting (4) into differential equation (1b) creates the following three terms:

$$t^2y'' = t^2\sum_{n=0}^{\infty} a_n(\lambda+n)(\lambda+n-1)t^{\lambda+n-2} = \sum_{n=0}^{\infty} a_n(\lambda+n)(\lambda+n-1)t^{\lambda+n} \tag{5a}$$

$$t[tp(t)]y' = t\left[\sum_{m=0}^{\infty}\alpha_m t^m\right]\left[\sum_{n=0}^{\infty} a_n(\lambda+n)t^{\lambda+n-1}\right]$$

$$= \left[\sum_{m=0}^{\infty}\alpha_m t^m\right]\left[\sum_{n=0}^{\infty} a_n(\lambda+n)t^{\lambda+n}\right] \tag{5b}$$

$$[t^2q(t)]y = \left[\sum_{m=0}^{\infty}\beta_m t^m\right]\left[\sum_{n=0}^{\infty} a_n t^{\lambda+n}\right] \tag{5c}$$

The three terms in (5) must be added and the sum equated to zero. The series products are computed using the Cauchy product defined in Section 10.2 and, as before, coefficients of like powers of t are grouped together. We obtain

$$\begin{aligned} t^2y'' + t[tp(t)]y' + [t^2q(t)]y = {} & [\lambda(\lambda-1) + \alpha_0\lambda + \beta_0]a_0t^{\lambda} + [\lambda(\lambda+1)a_1 \\ & + \alpha_0a_1(\lambda+1) + \alpha_1a_0\lambda + \beta_0a_1 + \beta_1a_0]t^{\lambda+1} \\ & + [(\lambda+2)(\lambda+1)a_2 + \alpha_0a_2(\lambda+2) + \alpha_1a_1(\lambda+1) \\ & + \alpha_2a_0\lambda + \beta_0a_2 + \beta_1a_1 + \beta_2a_0]t^{\lambda+2} + \cdots, \quad t > 0. \end{aligned} \tag{6}$$

Equating the right-hand side of (6) to zero and invoking the uniqueness property of power series representations, it follows that the coefficient of each power of t must vanish. Setting the first coefficient equal to zero leads to

$$[\lambda^2 + (\alpha_0 - 1)\lambda + \beta_0]a_0 = 0. \tag{7}$$

Since $a_0 \neq 0$, it follows that representation (4) is a solution of the differential equation only if the exponent λ is a root of the equation $F(\lambda) = 0$, where

$$F(\lambda) \equiv \lambda^2 + (\alpha_0 - 1)\lambda + \beta_0. \tag{8}$$

The equation $F(\lambda) = 0$ is called the **characteristic equation** or the **indicial equation**. (The latter term is used more frequently in this context and we will do so as well.) Note that this equation (8) is consistent with the heuristic argument used to motivate representation (4). The equation $F(\lambda) = 0$ is precisely the indicial equation for the Euler equation that results from using the two approximations, $tp(t) \approx \alpha_0$ and $t^2q(t) \approx \beta_0$.

Once we choose a root λ of the indicial equation, we set the coefficients of the higher powers of t equal to zero in (6). This gives us a recurrence relation for finding the coefficients $a_1, a_2, a_3, \ldots$ in terms of a_0. For instance, setting the coefficient of $t^{\lambda+1}$ equal to zero, we have

$$[(\lambda + 1)^2 + (\alpha_0 - 1)(\lambda + 1) + \beta_0]a_1 + [\alpha_1\lambda + \beta_1]a_0 = 0,$$

or

$$a_1 = -\frac{[\alpha_1\lambda + \beta_1]a_0}{F(\lambda + 1)}. \tag{9a}$$

Similarly, knowing a_1, we determine a_2 from

$$[(\lambda + 2)^2 + (\alpha_0 - 1)(\lambda + 2) + \beta_0]a_2 + [\alpha_1(\lambda + 1) + \beta_1]a_1 + [\alpha_2\lambda + \beta_2]a_0 = 0$$

or

$$a_2 = -\frac{[\alpha_1(\lambda + 1) + \beta_1]a_1 + [\alpha_2\lambda + \beta_2]a_0}{F(\lambda + 2)}. \tag{9b}$$

Note the difference between the procedure for constructing the general solution near an ordinary point and the method of Frobenius for constructing the general solution near a regular singular point. For example, if $t = 0$ is an ordinary point, then we look for a solution of the form

$$y(t) = \sum_{n=0}^{\infty} a_n t^n.$$

The recurrence relation obtained determines the coefficients a_n in terms of two of these coefficients, typically a_0 and a_1. Since a_0 and a_1 are arbitrary, both members of a fundamental set of solutions are obtained concurrently.

If $t = 0$ is a regular singular point, however, the method of Frobenius leads to the indicial equation, $F(\lambda) = 0$. Let the roots of this equation be denoted by λ_1 and λ_2. The recurrence relation found using (6) is used twice, first with $\lambda = \lambda_1$ and then with $\lambda = \lambda_2$. In this way we seek two linearly independent solutions,

$$y_1(t) = \sum_{n=0}^{\infty} a_n^{(1)} t^{\lambda_1 + n} \quad \text{and} \quad y_2(t) = \sum_{n=0}^{\infty} a_n^{(2)} t^{\lambda_2 + n},$$

where the coefficients $a_1^{(1)}, a_2^{(1)}, a_3^{(1)}, \ldots$ and $a_1^{(2)}, a_2^{(2)}, a_3^{(2)}, \ldots$ are generated (respectively) in terms of arbitrary constants $a_0^{(1)}$ and $a_0^{(2)}$.

There are cases where the method of Frobenius, as outlined in the foregoing text, does not produce the general solution. An obvious case occurs when the two roots of the indicial equation are equal. In this repeated-root case, the foregoing procedure yields only one member of the fundamental set. A second, less obvious, case occurs when the indicial equation possesses two (real) roots that differ by an integer. These special cases will be discussed (and the entire method summarized) in the next section. We conclude with an example in which the procedure previously outlined can be used to obtain the general solution.

EXAMPLE 2

Use the method of Frobenius to obtain the general solution of

$$2t^2y'' - ty' + (1+t^2)y = 0, \quad \text{for} \quad t > 0.$$

Solution: The point $t = 0$ is a singular point. Since $p(t)$ and $q(t)$ are rational functions and since $\lim_{t\to 0} tp(t) = -1/2$ and $\lim_{t\to 0} t^2q(t) = 1/2$, it follows that $t = 0$ is a regular singular point.

Using the method of Frobenius, we look for solutions of the form

$$y(t) = \sum_{n=0}^{\infty} a_n t^{\lambda+n}, \qquad t > 0.$$

Substituting this series into the differential equation leads to

$$\sum_{n=0}^{\infty} a_n(\lambda+n)(\lambda+n-1)t^{\lambda+n} - \frac{1}{2}\sum_{n=0}^{\infty} a_n(\lambda+n)t^{\lambda+n} + \frac{1}{2}\sum_{n=0}^{\infty} a_n t^{\lambda+n} + \frac{1}{2}\sum_{n=0}^{\infty} a_n t^{\lambda+n+2} = 0.$$

Rewriting the last series as $\sum_{n=0}^{\infty} a_n t^{\lambda+n+2} = \sum_{n=2}^{\infty} a_{n-2} t^{\lambda+n}$ and combining terms where possible, we obtain

$$\sum_{n=0}^{\infty}\left[(\lambda+n)(\lambda+n-1) - \frac{1}{2}(\lambda+n) + \frac{1}{2}\right] a_n t^{\lambda+n} + \sum_{n=2}^{\infty} \frac{1}{2} a_{n-2} t^{\lambda+n} = 0, \tag{10}$$

or

$$\left[\lambda^2 - \frac{3}{2}\lambda + \frac{1}{2}\right] a_0 t^{\lambda} + \left[(\lambda+1)^2 - \frac{3}{2}(\lambda+1) + \frac{1}{2}\right] a_1 t^{\lambda+1} + \sum_{n=2}^{\infty}\left[\left\{(\lambda+n)^2 - \frac{3}{2}(\lambda+n) + \frac{1}{2}\right\} a_n + \frac{1}{2}a_{n-2}\right] t^{\lambda+n} = 0.$$

The indicial equation, $F(\lambda) = \lambda^2 - (3/2)\lambda + (1/2) = 0$, has roots $\lambda_1 = 1/2$ and $\lambda_2 = 1$. We now examine the recurrence relations associated with each of these roots.

Case I: Let $\lambda = 1/2$ in (10). We set the coefficient of $t^{\lambda+n} = t^{(1/2)+n}$ equal to zero for each value of n and use $a_n^{(1)}$ to denote the coefficients associated with this value of λ. The coefficient $a_0^{(1)}$ is arbitrary. The coefficient multiplying $t^{3/2}$ must be zero and thus

$$\left[\left(\frac{1}{2}+1\right)^2 - \frac{3}{2}\left(\frac{1}{2}+1\right) + \frac{1}{2}\right] a_1^{(1)} = F\left(\frac{3}{2}\right) a_1^{(1)} = 0.$$

Since $F(3/2) \neq 0$, it follows that $a_1^{(1)} = 0$. As for the remaining coefficients, we have

$$\left[\left(\frac{1}{2}+n\right)^2 - \frac{3}{2}\left(\frac{1}{2}+n\right) + \frac{1}{2}\right] a_n^{(1)} + \frac{1}{2}a_{n-2}^{(1)} = 0, \qquad n = 2, 3, 4, \ldots,$$

and so

$$a_n^{(1)} = -\frac{a_{n-2}^{(1)}}{2F\left(\dfrac{1}{2}+n\right)}, \qquad n = 2, 3, 4, \ldots. \tag{11}$$

Since $F((1/2)+n) \neq 0$ for all $n \geq 2$, recurrence relation (11) is well defined. The coefficients $a_n^{(1)}$, for n even, can ultimately be expressed in terms of $a_0^{(1)}$. The coefficients $a_n^{(1)}$, for n odd, are all zero because $a_1^{(1)} = 0$. Thus, we find a solution $y_1(t)$,

$$y_1(t) = a_0^{(1)}\left[t^{1/2} - \frac{1}{6}t^{5/2} + \frac{1}{168}t^{9/2} - \cdots\right], \qquad t > 0.$$

Case 2: Let $\lambda = 1$ in (10). We repeat the sequence of computations just completed with this new value of λ and with the coefficients now denoted by $a_n^{(2)}$. As in Case 1, it follows that $a_1^{(2)} = 0$ since $F(2) \neq 0$. In general,

$$\left[(1+n)^2 - \frac{3}{2}(1+n) + \frac{1}{2}\right] a_n^{(2)} + \frac{1}{2}a_{n-2}^{(2)} = 0, \qquad n = 2, 3, 4, \ldots,$$

and thus

$$a_n^{(2)} = -\frac{a_{n-2}^{(2)}}{2F(1+n)}, \qquad n = 2, 3, 4, \ldots. \tag{12}$$

The solution obtained is

$$y_2(t) = a_0^{(2)}\left[t - \frac{1}{10}t^3 + \frac{1}{360}t^5 - \cdots\right]. \tag{13}$$

It is clear that the two solutions obtained are linearly independent on $t > 0$ since one solution is not a constant multiple of the other. Therefore, the two solutions form a fundamental set, and the general solution is

$$\begin{aligned} y(t) = a_0^{(1)}&\left[t^{1/2} - \frac{1}{6}t^{5/2} + \frac{1}{168}t^{9/2} - \cdots\right] \\ &+ a_0^{(2)}\left[t - \frac{1}{10}t^3 + \frac{1}{360}t^5 - \cdots\right], \qquad t > 0. \end{aligned} \tag{14}$$

Note that the differential equation possesses the symmetries discussed in Theorem 10.2; that is, $p(t) = -2/t$ is an odd function and $q(t) = (1+t^2)/t^2$ is an even function. Therefore, to find a solution for $t < 0$, we need only replace t by $-t$ in (14). As a final observation, recurrence relations (11) and (12), together with the ratio test, can be used to show that the two series in general solution (14) converge absolutely in $0 < t < \infty$. ▲

EXERCISES

Exercises 1–10:

In each exercise, find the singular points (if any) and classify them as regular or irregular.

1. $ty'' + (\cos t)y' + y = 0$

2. $t^2y'' + (\sin t)y' + y = 0$

3. $(t^2 - 1)y'' + (t - 1)y' + y = 0$

4. $(t^2 - 1)^2y'' + (t + 1)y' + y = 0$

5. $t^2y'' + (1 - \cos t)y' + y = 0$

6. $|t|y'' + y' + y = 0$

7. $(1 - e^t)y'' + y' + y = 0$

8. $(4 - t^2)y'' + (t + 2)y' + (4 - t^2)^{-1}y = 0$

9. $(1 - t^2)^{1/3}y'' + y' + ty = 0$

10. $y'' + y' + t^{1/3}y = 0$

Exercises 11–13:

In each exercise, determine the polynomial $P(t)$ of smallest degree that causes the given differential equation to have the stated properties.

11. $y'' + \dfrac{\sin 2t}{P(t)}y' + y = 0$ There is a regular singular point at $t = 0$, irregular singular points at $t = \pm 1$. All other points are ordinary points.

12. $y'' + \dfrac{1}{t}y' + \dfrac{1}{P(t)}y = 0$ There is a regular singular point at $t = 0$. All other points are ordinary points.

13. $y'' + \dfrac{1}{tP(t)}y' + \dfrac{1}{t^3}y = 0$ There are irregular singular points at $t = 0$ and $t = \pm 1$. All other points are ordinary points.

Exercises 14–15:

In each exercise, the exponent n in the given differential equation is a nonnegative integer. Determine the possible values of n (if any) for which
(a) $t = 0$ is a regular singular point.

(b) $t = 0$ is an irregular singular point.

14. $y'' + \dfrac{1}{t^n}y' + \dfrac{1}{1 + t^2}y = 0$

15. $y'' + \dfrac{1}{\sin t}y' + \dfrac{1}{t^n}y = 0$

Exercises 16–23:

In each exercise,

(a) Verify that $t = 0$ is a regular singular point.

(b) Find the indicial equation.

(c) Find the recurrence relation.

(d) Find the first three nonzero terms of the series solution, for $t > 0$, corresponding to the larger root of the indicial equation. If there are fewer than three nonzero terms, give the corresponding exact solution.

16. $2t^2y'' - ty' + (t + 1)y = 0$

17. $4t^2y'' + 4ty' + (t - 1)y = 0$

18. $16t^2y'' + t^2y' + 3y = 0$

19. $t^2y'' + ty' + (t - 9)y = 0$

20. $ty'' + (t + 2)y' - y = 0$

21. $t^2y'' + 3ty' + (2t + 1)y = 0$

22. $t^2y'' + t(t - 1)y' - 3y = 0$

23. $ty'' + (t - 2)y' + y = 0$

Exercises 24–27:

In each exercise,

(a) Verify that $t = 0$ is a regular singular point.

(b) Find the indicial equation.

(c) Find the first three terms of the series solution, for $t > 0$, corresponding to the larger root of the indicial equation.

24. $t^2y'' - (2\sin t)y' + (2+t)y = 0$

25. $ty'' - 4y' + e^t y = 0$

26. $(\sin t)y'' - y' + y = 0$

27. $(1 - e^t)y'' + (1/2)y' + y = 0$

10.5 The Method of Frobenius Continued: Special Cases and a Summary

There are two important special cases where the method of Frobenius (as described in the previous section) may not yield the general solution of $y'' + p(t)y' + q(t)y = 0$ near a regular singular point. These special cases arise when the indicial equation has two real roots that are equal or that differ by an integer. We will use Bessel's equation as a vehicle to illustrate these cases. We then conclude this section by summarizing, for all cases, the structure of the general solution near a regular singular point.

The General Solution of Bessel's Equation

Consider Bessel's equation,

$$t^2y'' + ty' + (t^2 - \nu^2)y = 0, \qquad t > 0,$$

where ν is a real nonnegative constant. As we saw in Section 10.4, $t = 0$ is a regular singular point. In addition, note that the coefficient functions

$$p(t) = \frac{1}{t} \quad \text{and} \quad q(t) = \frac{t^2 - \nu^2}{t^2}$$

are odd and even functions, respectively. Therefore, by Theorem 10.2, we can obtain solutions for $t < 0$ by finding solutions for $t > 0$ and replacing t with $-t$.

We apply the method of Frobenius to Bessel's equation. Substituting $y(t) = \sum_{n=0}^{\infty} a_n t^{\lambda+n}$ leads to

$$t^2 \sum_{n=0}^{\infty} a_n(\lambda+n)(\lambda+n-1)t^{\lambda+n-2} + t\sum_{n=0}^{\infty} a_n(\lambda+n)t^{\lambda+n-1} + t^2\sum_{n=0}^{\infty} a_n t^{\lambda+n} - \nu^2 \sum_{n=0}^{\infty} a_n t^{\lambda+n} = 0,$$

or

$$\sum_{n=0}^{\infty} [(\lambda+n)(\lambda+n-1) + (\lambda+n) - \nu^2]a_n t^{\lambda+n} + \sum_{n=2}^{\infty} a_{n-2} t^{\lambda+n} = 0.$$

Rewriting this equation, we obtain

$$[\lambda^2 - \nu^2]a_0 t^{\lambda} + [(\lambda+1)^2 - \nu^2]a_1 t^{\lambda+1} + \sum_{n=2}^{\infty} [\{(\lambda+n)^2 - \nu^2\}a_n + a_{n-2}]t^{\lambda+n} = 0, \qquad t > 0. \tag{1}$$

Without loss of generality, we assume $a_0 \neq 0$. The indicial equation is

$$F(\lambda) = \lambda^2 - \nu^2 = 0, \tag{2}$$

and has roots $\lambda_1 = \nu$ and $\lambda_2 = -\nu$. From (1) we also obtain the equations

$$F(\lambda + 1)a_1 = 0 \tag{3a}$$

and

$$F(\lambda + n)a_n + a_{n-2} = 0. \tag{3b}$$

Equation (3b) leads to the recurrence relation

$$a_n = \frac{-a_{n-2}}{F(\lambda + n)}, \qquad n = 2, 3, 4, \ldots. \tag{3c}$$

The special cases that arise can be illustrated by selecting particular values for the constant ν.

(i) ***Equal roots*** $(\lambda_1 = \lambda_2)$. Consider Bessel's equation with $\nu = 0$. In this case, $\lambda_1 = \lambda_2 = 0$. Since the root is repeated, the method of Frobenius gives us only one member of the fundamental set of solutions; the second linearly independent solution has a structure different from $\sum_{n=0}^{\infty} a_n t^{\lambda+n}$.

Since $\nu = 0$, the indicial polynomial reduces to $F(\lambda) = \lambda^2$. Therefore, $F(1)$ is nonzero and we see from (3a) that $a_1 = 0$. From (3c), we have

$$a_n = \frac{-a_{n-2}}{n^2}, \qquad n = 2, 3, 4, \ldots. \tag{4}$$

Since $a_1 = 0$, it follows from (4) that all odd-indexed coefficients are zero. The even-indexed coefficients can be expressed as multiples of a_0:

$$a_2 = \frac{-1}{2^2}a_0, \qquad a_4 = \frac{-1}{4^2}a_2 = \frac{(-1)^2}{2^4(2!)^2}a_0, \qquad a_6 = \frac{-1}{6^2}a_4 = \frac{(-1)^3}{2^6(3!)^2}a_0,$$

and, in general,

$$a_{2n} = \frac{(-1)^n}{2^{2n}(n!)^2}a_0, \qquad n = 0, 1, 2, \ldots.$$

The solution we thus obtain is

$$y_1(t) = a_0\left[1 - \frac{t^2}{4} + \frac{t^4}{64} - \frac{t^6}{2304} + \cdots\right] = a_0\left[1 + \sum_{n=1}^{\infty} \frac{(-1)^n t^{2n}}{2^{2n}(n!)^2}\right] \equiv a_0 J_0(t), \tag{5a}$$

where

$$J_0(t) = 1 + \sum_{n=1}^{\infty} \frac{(-1)^n t^{2n}}{2^{2n}(n!)^2}. \tag{5b}$$

The function $J_0(t)$ is called the **Bessel function of the first kind of order zero**. [We began our analysis by looking for solutions in $t > 0$. However, by the ratio test, we see that series (5a) defines a solution valid for $-\infty < t < \infty$. Also note that $J_0(t)$ is an even function of t, a fact consistent with Theorem 10.2.]

How do we obtain a second linearly independent solution? Having one solution, we could, in principle, use the method of reduction of order (see Section 4.5) to construct a second solution. We shall, however, simply state a form of the second solution that is commonly used in applications—the **Bessel function of the second kind of order zero** (also called Weber's function). The Bessel function of the second kind of order zero is given by

$$Y_0(t) = \frac{2}{\pi}\left[\gamma + \ln\frac{t}{2}\right]J_0(t) + \frac{2}{\pi}\sum_{n=1}^{\infty}(-1)^{n+1}k_n\frac{\left(\frac{t^2}{4}\right)^n}{(n!)^2}, \qquad t > 0. \quad \textbf{(6)}$$

In (6), $k_1 = 1$ and, in general,

$$k_n = 1 + \frac{1}{2} + \frac{1}{3} + \frac{1}{4} + \cdots + \frac{1}{n}, \qquad n = 2, 3, \ldots.$$

The constant γ in equation (6) is known as the **Euler-Mascheroni constant**[7] and is defined by the limit

$$\gamma = \lim_{n\to\infty}\left[\sum_{j=1}^{n}\frac{1}{j} - \ln n\right] \approx 0.5772\ldots.$$

Note the structure of $Y_0(t)$; it can ultimately be expressed as a constant multiple of $J_0(t)\ln t = y_1(t)\ln t$ added to a series of the form $\sum_{n=0}^{\infty} b_n t^n = \sum_{n=0}^{\infty} b_n t^{n+\lambda_2}$, $t > 0$. Since $J_0(0) = 1$, we see that $Y_0(t)$ has a logarithmic singularity at $t = 0$. If we recall the heuristic argument used to motivate the method of Frobenius, the presence of the function $\ln t$ in (6) and the corresponding logarithmic singularity at $t = 0$ is not surprising since the general solution of the Euler equation having repeated roots $\lambda_1 = \lambda_2 = 0$ is $y(t) = c_1 + c_2 \ln t$, $t > 0$. Figure 10.5 on the next page shows graphs of $J_0(t)$ and $Y_0(t)$.

(ii) ***Roots differing by unity ($\lambda_1 - \lambda_2 = 1$).*** As an illustration of this case, let $\nu = 1/2$ in Bessel's equation. The indicial equation,

$$F(\lambda) = \lambda^2 - \frac{1}{4}$$

has roots $\lambda_1 = 1/2$ and $\lambda_2 = -1/2$ and therefore, $\lambda_1 - \lambda_2 = 1$. Consider first the larger root, $\lambda_1 = 1/2$, which corresponds to a solution of the form

$$y_1(t) = \sum_{n=0}^{\infty} a_n^{(1)} t^{(1/2)+n}, \qquad t > 0.$$

Without loss of generality, we assume $a_0^{(1)} \neq 0$. Since $F(\lambda_1 + 1) = F(3/2) \neq 0$, it follows from (3a) that $a_1^{(1)} = 0$. From (3c), the recurrence

[7] Lorenzo Mascheroni (1750–1800) was an ordained priest, poet, and teacher of mathematics and physics. In 1786 he became professor of algebra and geometry at the University of Pavia and later became its rector. In 1790 Mascheroni correctly calculated the first 19 decimal places of Euler's constant. This accomplishment has caused his name to be linked with the constant.

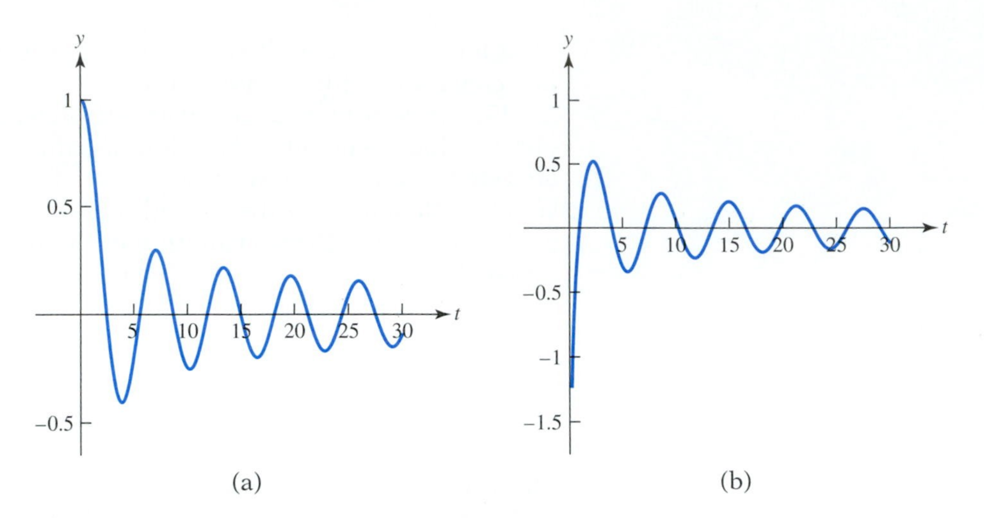

FIGURE 10.5

(a) The graph of $J_0(t)$, the Bessel function of the first kind of order zero. (b) The graph of $Y_0(t)$, the Bessel function of the second kind of order zero. Note that $J_0(t)$ is defined for all t, whereas $Y_0(t)$ has a logarithmic singularity at the regular singular point $t = 0$.

relation is

$$a_n^{(1)} = \frac{-a_{n-2}^{(1)}}{\left[\left(n + \frac{1}{2}\right)^2 - \frac{1}{4}\right]} = \frac{-a_{n-2}^{(1)}}{[n(n+1)]}, \qquad n = 2, 3, 4, \ldots. \tag{7}$$

Equation (7) allows us to determine all the even-indexed coefficients as multiples of $a_0^{(1)}$ and implies that all the odd-indexed coefficients are zero. Solving recurrence relation (7) leads to the solution:

$$\begin{aligned} y_1(t) &= a_0^{(1)}\left[t^{1/2} - \frac{t^{5/2}}{3!} + \frac{t^{9/2}}{5!} - \frac{t^{13/2}}{7!} + \cdots\right] \\ &= a_0^{(1)} t^{-1/2}\left[t - \frac{t^3}{3!} + \frac{t^5}{5!} - \frac{t^7}{7!} + \cdots\right] \\ &= a_0^{(1)} t^{-1/2} \sin t, \qquad t > 0. \end{aligned} \tag{8}$$

Consider now the smaller root, $\lambda_2 = -1/2$, where we look for a solution of the form

$$y_2(t) = \sum_{n=0}^{\infty} a_n^{(2)} t^{-(1/2)+n}, \qquad t > 0.$$

As before, $F(\lambda) = \lambda^2 - 1/4$. This time, however,

$$F(\lambda_2 + n) = \left(n - \frac{1}{2}\right)^2 - \frac{1}{4} = n(n-1),$$

and we see that $F(\lambda_2 + 1) = F(\lambda_1) = F(1/2) = 0$. Therefore, see equation (3a), $a_1^{(2)}$ need not be zero. From (3c), the recurrence relation

$$a_n^{(2)} = \frac{-a_{n-2}^{(2)}}{\left[\left(n - \frac{1}{2}\right)^2 - \frac{1}{4}\right]} = \frac{-a_{n-2}^{(2)}}{[n(n-1)]}, \qquad n = 2, 3, 4, \ldots, \tag{9}$$

allows us to determine all even-indexed coefficients as multiples of $a_0^{(2)}$ and all odd-indexed coefficients as multiples of $a_1^{(2)}$. We find

$$a_{2k}^{(2)} = \frac{(-1)^k}{(2k)!} a_0^{(2)} \quad \text{and} \quad a_{2k+1}^{(2)} = \frac{(-1)^k}{(2k+1)!} a_1^{(2)}, \qquad k = 1, 2, 3, \ldots. \tag{10}$$

The resulting solution $y_2(t)$ has the form

$$\begin{aligned} y_2(t) &= a_0^{(2)} t^{-1/2} \left[1 - \frac{t^2}{2!} + \frac{t^4}{4!} - \frac{t^6}{6!} + \cdots\right] \\ &\quad + a_1^{(2)} t^{-1/2} \left[t - \frac{t^3}{3!} + \frac{t^5}{5!} - \frac{t^7}{7!} + \cdots\right] \\ &= a_0^{(2)} t^{-1/2} \cos t + a_1^{(2)} t^{-1/2} \sin t, \qquad t > 0. \end{aligned} \tag{11}$$

We therefore obtain a second linearly independent solution, $t^{-1/2}\cos t$, added to a multiple of the solution, $t^{-1/2}\sin t$, previously obtained. In this example, the method of Frobenius has produced both members of a fundamental set of solutions.

The **Bessel functions of order one-half** are defined to be

$$J_{1/2}(t) = \sqrt{\frac{2}{\pi t}} \sin t \quad \text{and} \quad J_{-1/2}(t) = \sqrt{\frac{2}{\pi t}} \cos t, \qquad t > 0.$$

The general solution of Bessel's equation in this case is usually expressed as

$$y(t) = c_1 J_{1/2}(t) + c_2 J_{-1/2}(t). \tag{12}$$

Note that $J_{1/2}(t)$ is well behaved near $t = 0$ since

$$\lim_{t \to 0^+} \frac{\sin t}{\sqrt{t}} = 0.$$

By contrast, $\lim_{t \to 0^+} J_{-1/2}(t) = +\infty$. Figure 10.6 on the next page shows the behavior of these two functions.

In the special case where $\nu = 1/2$, it is also easy to obtain the general solution (12) by a change of independent variable. If we set $y(t) = t^{-1/2}u(t), t > 0$, Bessel's equation, $t^2y'' + ty' + (t^2 - (1/4))y = 0$, transforms into

$$t^{3/2}(u'' + u) = 0$$

or simply, $u'' + u = 0, t > 0$. The general solution is

$$u(t) = \tilde{c}_1 \cos t + \tilde{c}_2 \sin t,$$

and thus (12) follows readily.

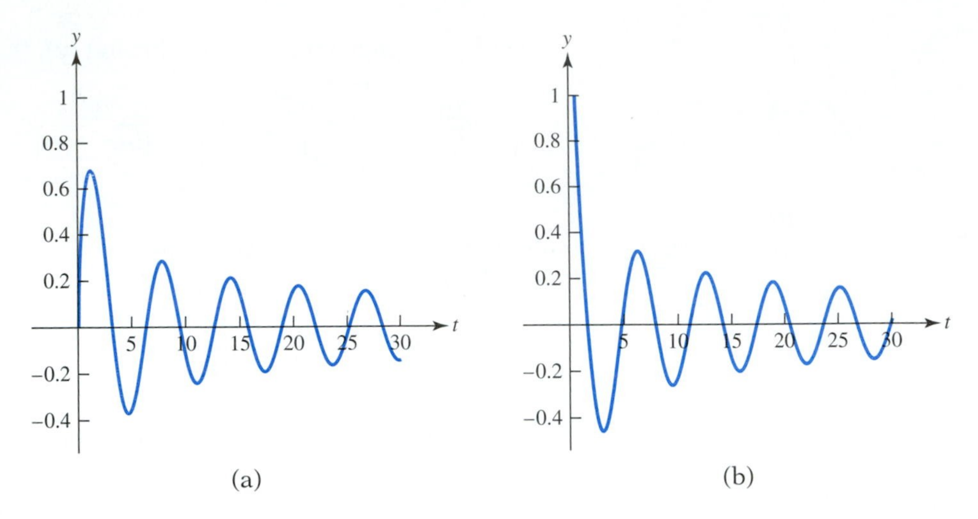

FIGURE 10.6

(a) The graph of the Bessel function $J_{1/2}(t)$.
(b) The graph of the Bessel function $J_{-1/2}(t)$.

(iii) ***Roots differing by an integer greater than 1 ($\lambda_1 - \lambda_2 = N \geq 2$).*** For this example, we let $\nu = M \geq 1$ in Bessel's equation. The two roots of the indicial equation, $\lambda^2 - M^2 = 0$, are $\lambda_1 = M$ and $\lambda_2 = -M$; therefore, the roots differ by an integer greater than 1 since $\lambda_1 - \lambda_2 = 2M$ and $2M \geq 2$. In this case, the method of Frobenius provides us with a solution corresponding to the larger root, $\lambda_1 = M$. It will fail, however, when we try to apply the method to the smaller root, $\lambda_2 = -M$.

Consider first the larger root, $\lambda_1 = M$. We look for a solution of the form

$$y_1(t) = \sum_{n=0}^{\infty} a_n^{(1)} t^{M+n}, \qquad t > 0.$$

Without loss of generality, we assume $a_0^{(1)} \neq 0$. Since $F(\lambda_1 + 1) = F(M+1) \neq 0$, we know from equation (3a) that $a_1^{(1)} = 0$. Using equation (3c), we obtain the recurrence relation

$$a_n^{(1)} = \frac{-a_{n-2}^{(1)}}{[(n+M)^2 - M^2]} = \frac{-a_{n-2}^{(1)}}{[n(n+2M)]}, \qquad n = 2, 3, 4, \ldots. \tag{13}$$

Equation (13) allows us to determine all the even-indexed coefficients as multiples of $a_0^{(1)}$ and tells us the odd-indexed coefficients are zero. Therefore, we obtain the solution

$$y_1(t) = a_0^{(1)} t^M \left\{ 1 - \frac{t^2}{2^2(M+1)} + \frac{t^4}{2^4 2!(M+2)(M+1)} - \frac{t^6}{2^6 3!(M+3)(M+2)(M+1)} + \cdots + \frac{(-t^2)^k M!}{2^{2k} k!(M+k)!} + \cdots \right\}. \tag{14}$$

Choosing $a_0^{(1)} = (1/2^M)M!$ in (14) leads to the **Bessel function of the first kind of order M**:

$$J_M(t) = \left(\frac{t}{2}\right)^M \sum_{k=0}^{\infty} \frac{\left(\dfrac{-t^2}{4}\right)^k}{k!(M+k)!}. \tag{15}$$

Note that $J_M(t)$ is analytic at $t = 0$ and the series (15) has an infinite radius of convergence. The function $J_M(t)$ vanishes at $t = 0$ when $M \ge 1$; $J_M(t)$ is an even function when M is an even integer and an odd function when M is an odd integer.

Suppose we now consider the smaller root, $\lambda_2 = -M$, and look for a solution of the form

$$y_2(t) = \sum_{n=0}^{\infty} a_n^{(2)} t^{-M+n}.$$

Assume, without loss of generality, that $a_0^{(2)} \ne 0$. Since $M \ge 1$, $F(\lambda_2 + 1) = F(-M+1) \ne 0$ and $a_1^{(2)} = 0$. The difficulty arises when we try to use equation (3c) to evaluate $a_2^{(2)}, a_4^{(2)}, a_6^{(2)}, \ldots$ in terms of $a_0^{(2)}$. Using $\lambda = -M$ and setting $n = 2k$, the recurrence relation (3c) becomes

$$a_{2k}^{(2)} = \frac{-a_{2k-2}^{(2)}}{4k(k-M)}, \qquad k = 1, 2, 3, \ldots. \tag{16}$$

The trouble occurs when we try to evaluate (16) for $k = M$. The breakdown of recurrence relation (16) at $k = M$ tells us that the assumed form of the solution is incorrect and that the second linearly independent solution does not have the structure assumed by the method of Frobenius.

The second linearly independent solution of Bessel's equation customarily used is called the **Bessel function of the second kind of order M** and is denoted by $Y_M(t)$. It is defined as

$$\begin{aligned} Y_M(t) = \frac{2}{\pi}\left[\gamma + \ln\frac{t}{2}\right] J_M(t) &- \frac{\left(\dfrac{t}{2}\right)^{-M}}{\pi} \sum_{k=0}^{M-1} \frac{(M-k-1)!}{k!}\left(\frac{t^2}{4}\right)^k \\ &- \frac{\left(\dfrac{t}{2}\right)^{M}}{\pi} \sum_{n=0}^{\infty} (-1)^n \left(k_{M+n} + k_n\right) \frac{\left(-\dfrac{t^2}{4}\right)^n}{n!(M+n)!}, \qquad t > 0, \end{aligned} \tag{17}$$

where the constants k_n are defined as $k_0 = 0$, $k_n = 1 + 2^{-1} + 3^{-1} + \cdots + n^{-1}$, $n \ge 1$, and γ is the Euler-Mascheroni constant.

The details of expression (17) are admittedly complicated. Note, however, the general structure; $Y_M(t)$ can be expressed as a multiple of $J_M(t)\ln t$ added to a series of the form $\sum_{n=0}^{\infty} b_n t^{n-M} = \sum_{n=0}^{\infty} b_n t^{n+\lambda_2}$, $t > 0$. Also note the behavior of $Y_M(t)$. The first term in (17) is well behaved near $t = 0$ since $J_M(t)$ can be bounded by a constant multiple of $|t^M|$ near $t = 0$ and hence

$$\lim_{t \to 0^+} J_M(t) \ln \frac{t}{2} = 0.$$

Similarly, the last term in (17) behaves like a multiple of t^M near $t = 0$. The middle term in (17), however, behaves like a multiple of t^{-M} near $t = 0$; consequently $Y_M(t)$ has an algebraic singularity at the origin. Figure 10.7 displays graphs of $J_1(t)$ and $Y_1(t)$.

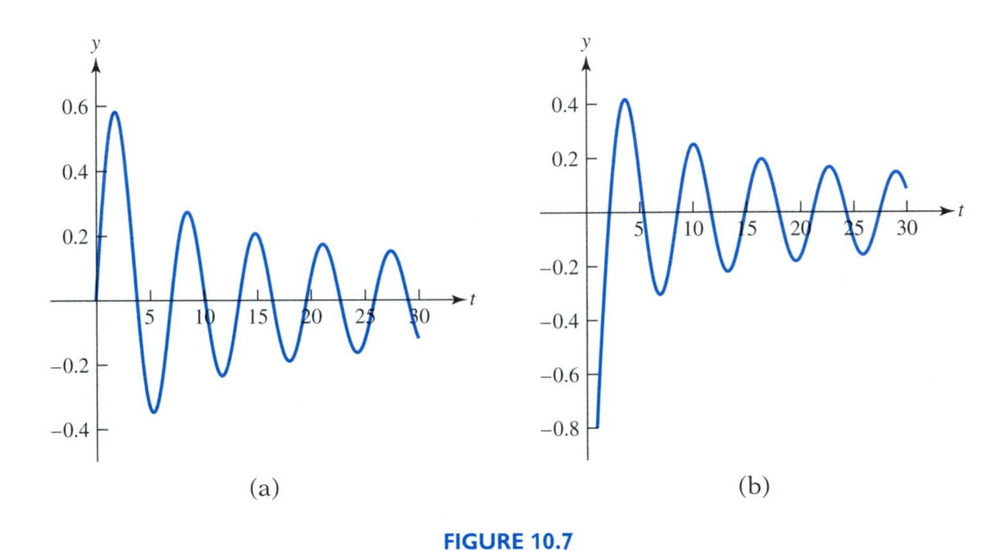

FIGURE 10.7

(a) The graph of $J_1(t)$. (b) The graph of $Y_1(t)$.

REMARK: For reasonably large values of t (say $t > 5$) all of the Bessel functions plotted in Figures 10.5–10.7 seem to behave like damped sinusoids; that is, sine or cosine waves having amplitudes that decrease with increasing t. To see why this behavior might be expected, note that the change of dependent variable $y(t) = t^{-1/2}u(t)$ transforms Bessel's equation, $t^2y'' + ty' + (t^2 - \nu^2)y = 0$, into

$$t^{3/2}u'' + \left[t^{3/2} - t^{-1/2}\left(\nu^2 - \frac{1}{4}\right)\right] u = 0,$$

or, equivalently,

$$u'' + \left[1 + \frac{\left(\nu^2 - \frac{1}{4}\right)}{t^2}\right] u = 0, \qquad t > 0.$$

For a fixed value of ν and for large t we have $(\nu^2 - (1/4))/t^2 \ll 1$. Under these circumstances, it seems reasonable to make the approximation

$$u'' + \left[1 + \frac{\left(\nu^2 - \frac{1}{4}\right)}{t^2}\right] u \approx u'' + u.$$

Thus, we anticipate for large t that $u(t) \approx c_1 \sin t + c_2 \cos t = R\cos(t-\delta)$ and hence

$$y(t) \approx \frac{R\cos(t-\delta)}{t^{1/2}}.$$

This approximation is examined in Exercise 26. ▮

The General Solution Near a Regular Singular Point

Our study of Bessel's equation for different values of ν provided some examples of the behavior of solutions near a regular singular point. Although these examples are representative, they are neither exhaustive nor completely general since Bessel's equation possesses a specific structure that is not necessarily present in the general case. The examples, however, illustrate the following summary describing the general behavior of solutions of $y'' + p(t)y' + q(t)y = 0$. We give the summary without proof.

Summary

Consider the differential equation $y'' + p(t)y' + q(t)y = 0$, where we assume $t = 0$ is a regular singular point. Let $tp(t)$ and $t^2q(t)$ be real-valued analytic functions with Maclaurin series

$$tp(t) = \sum_{n=0}^{\infty} \alpha_n t^n \quad \text{and} \quad t^2q(t) = \sum_{n=0}^{\infty} \beta_n t^n$$

that converge in $-R < t < R$. The corresponding indicial equation is

$$\lambda^2 + (\alpha_0 - 1)\lambda + \beta_0 = (\lambda - \lambda_1)(\lambda - \lambda_2) = 0.$$

The roots λ_1 and λ_2 are either real or they form a complex conjugate pair. In the event that λ_1 and λ_2 are real, we assume $\lambda_1 \ge \lambda_2$.

Then, in either of the intervals $(-R, 0)$ or $(0, R)$:

(a) There exists a solution having the form

$$y_1(t) = |t|^{\lambda_1} \sum_{n=0}^{\infty} a_n t^n, \qquad a_0 \neq 0 \tag{18}$$

where the series converges at least in $|t| < R$.

(b) The form of the second linearly independent solution, $y_2(t)$, depends on the difference $\lambda_1 - \lambda_2$.

(i) If $\lambda_1 - \lambda_2$ is not an integer, then

$$y_2(t) = |t|^{\lambda_2} \sum_{n=0}^{\infty} b_n t^n, \qquad b_0 \neq 0, \tag{19}$$

where the series converges at least in $|t| < R$.

(ii) If $\lambda_1 = \lambda_2$, then

$$y_2(t) = y_1(t)\ln|t| + |t|^{\lambda_2} \sum_{n=0}^{\infty} c_n t^n, \tag{20}$$

where the series converges at least in $|t| < R$.

(iii) If $\lambda_1 - \lambda_2$ equals a positive integer, then

$$y_2(t) = Cy_1(t)\ln|t| + |t|^{\lambda_2}\sum_{n=0}^{\infty} d_n t^n, \qquad d_0 \neq 0, \tag{21}$$

where C is a constant, possibly zero (if $C = 0$, there is no logarithmic term). Moreover, the series converges at least in $|t| < R$.

REMARKS:

1. We have assumed, for convenience, that $t = 0$ is a regular singular point. In general, if $t = t_0$ is a regular singular point, then the results in the Summary are valid when $t - t_0$ replaces t in the formulas.
2. When the roots λ_1 and λ_2 form a complex conjugate pair, for instance, $\lambda_1 = \gamma + i\delta$, $\lambda_2 = \gamma - i\delta$, the difference $\lambda_1 - \lambda_2 = 2\delta i$ is purely imaginary and thus not equal to zero or a positive integer. In that case, the two complex-valued solutions

$$y_1(t) = |t|^{\gamma+i\delta}\sum_{n=0}^{\infty} a_n t^n \quad \text{and} \quad y_2(t) = |t|^{\gamma-i\delta}\sum_{n=0}^{\infty} b_n t^n$$

obtained using the recurrence relations can be used to create an equivalent real-valued fundamental set. Using Euler's formula,

$$|t|^{\gamma+i\delta} = |t|^{\gamma}[\cos(\delta\ln|t|) + i\sin(\delta\ln|t|)].$$

3. In our study of Bessel's equation we considered two cases in which the roots differed by a positive integer. When $\lambda_1 - \lambda_2 = 1$, we found that the general solution did not involve a logarithmic term while when $\lambda_1 - \lambda_2 = 2M > 1$ we found that such a term was present, namely the term $J_M(t)\ln t$ in expression (17) for $Y_M(t)$. In a particular application where λ_1 and λ_2 differ by a positive integer, it seems advisable to look first for a second solution of the form given by (21) with $C = 0$. If the recurrence relation does not break down and a second linearly independent solution can be obtained in this way, then all is well. If not, one can always resort to the method of Reduction of Order as an (admittedly nontrivial) alternative. ❚

As we mentioned in the Introduction to this chapter, linear differential equations with variable coefficients play an important role in applications. Because of their importance, the functions that emerge as solutions of these equations, usually collectively referred to as special functions, have been studied exhaustively. There are books devoted to a particular special function[8] as well as general handbooks for special functions.[9] Modern software packages have many of these functions available as "built-in" functions—as accessible as the

[8] See, for example, George N. Watson, *A Treatise on the Theory of Bessel Functions*, 2nd ed. (Cambridge: Cambridge University Press, 1966).

[9] Milton Abramowitz and Irene A. Stegun, editors, *Handbook of Mathematical Functions with Formulas, Graphs, and Mathematical Tables* (New York: Dover Publications, 1970) and Wilhelm Magnus, Fritz Oberhettinger, and Raj Pal Soni, *Formulas and Theorems for the Special Functions of Mathematical Physics* (Berlin and New York: Springer-Verlag, 1966).

familiar exponential, logarithmic, and trigonometric functions. In that sense these so-called special functions are happily becoming less and less special.

Linear differential equations with variable coefficients, such as Bessel's equation and Legendre's equation, arise when the technique called separation of variables is applied to the partial differential equations that model certain physical problems. The extended problem at the end of this chapter studies steady-state heat conduction between concentric cylinders and provides a brief introduction into this circle of ideas.

EXERCISES

Exercises 1–12:

In each exercise,

(a) Verify that the given differential equation has a regular singular point at $t = 0$.

(b) Determine the indicial equation and its two roots. (These roots are often called the **exponents at the singularity**.)

(c) Determine the recurrence relation for the series coefficients.

(d) Consider the interval $t > 0$. If the two exponents obtained in (c) are unequal and do not differ by an integer, determine the first two nonzero terms in the series for each of the two linearly independent solutions. If the exponents are equal or differ by an integer, obtain the first two nonzero terms in the series for the solution having the larger exponent.

(e) When the given differential equation is put in the form $y'' + p(t)y' + q(t)y = 0$, note that $tp(t)$ and $t^2q(t)$ are polynomials. Do the series, whose initial terms were found in part (d), converge for all t, $0 < t < \infty$? Explain.

1. $2ty'' - (1 + t)y' + 2y = 0$

2. $2ty'' + 5y' + 3ty = 0$

3. $3t^2y'' - ty' + (1 + t)y = 0$

4. $6t^2y'' + ty' + (1 - t)y = 0$

5. $t^2y'' - 5ty' + (9 + t^2)y = 0$

6. $4t^2y'' + 8ty' + (1 + 2t)y = 0$

7. $t^2y'' - 2ty' + (2 + t)y = 0$

8. $ty'' + 4y' - 2ty = 0$

9. $t^2y'' + ty' - (1 + t^2)y = 0$

10. $t^2y'' + 5ty' + (4 - t^2)y = 0$

11. $t^2y'' + ty' - (16 + t)y = 0$

12. $8t^2y'' + 6ty' - (1 - t)y = 0$

Exercises 13–16:

In each exercise

(a) Determine all singular points of the given differential equation and classify them as regular or irregular singular points.

(b) At each regular singular point, determine the indicial equation and the exponents at the singularity.

13. $(t^3 + t)y'' - (1 + t)y' + y = 0$

14. $t^2y'' + (\sin 3t)y' + (\cos t)y = 0$

15. $(t^2 - 4)^2y'' - y' + y = 0$

16. $t^2(1 - t)^{1/3}y'' + ty' - y = 0$

17. The Legendre differential equation $(1 - t^2)y'' - 2ty' + \alpha(\alpha + 1)y = 0$ has regular singular points at $t = \pm 1$; all other points are ordinary points.

(a) Determine the indicial equation and the exponent at the singularity $t = 1$.

(b) Assume that $\alpha \neq 0, 1$. Find the first three nonzero terms of the series solution in powers of $t - 1$ for $t - 1 > 0$. [Hint: Rewrite the coefficient functions in powers of $t - 1$. For example, $1 - t^2 = -(t - 1)(t + 1) = -(t - 1)((t - 1) + 2)$.]

(c) What is an exact solution of the differential equation when $\alpha = 1$?

18. The Chebyshev differential equation $(1 - t^2)y'' - ty' + \alpha^2 y = 0$ has regular singular points at $t = \pm 1$; all other points are ordinary points.

(a) Determine the indicial equation and the exponent at the singularity $t = 1$.

(b) Assume α is nonzero and not an integer multiple of $1/2$. Find two linearly independent solutions for $t - 1 > 0$. [Use the hint in Exercise 17.]

(c) On what interval of the form $0 < t - 1 < R$ do the solutions found in part (b) converge?

(d) What is an exact solution of the differential equation when $\alpha = 1/2$?

19. The Laguerre[10] differential equation $ty'' + (1 - t)y' + \alpha y = 0$ has a regular singular point at $t = 0$.

(a) Determine the indicial equation and show that the roots are $\lambda_1 = \lambda_2 = 0$.

(b) Find the recurrence relation. Show that if $\alpha = N$, where N is a nonnegative integer, then the series solution reduces to a polynomial. Obtain the polynomial solution when $N = 5$. The polynomial solutions of this differential equation, when properly normalized, are called **Laguerre polynomials**.

(c) Is the polynomial obtained in part (b) for $\alpha = N = 5$ an even function, an odd function, or neither? Would you expect even and odd solutions of the differential equation based upon its structure and the conclusions of Theorem 10.2? Explain.

Exercises 20–23:

In each exercise, use the stated information to determine the unspecified coefficients in the given differential equation.

20. $t^2 y'' + t(\alpha + 2t)y' + (\beta + t^2)y = 0$

$t = 0$ is a regular singular point. The roots of the indicial equation at $t = 0$ are $\lambda_1 = 1$ and $\lambda_2 = 2$.

21. $t^2 y'' + \alpha t y' + (\beta + t - t^3)y = 0$

$t = 0$ is a regular singular point. The roots of the indicial equation at $t = 0$ are $\lambda_1 = 1 + 2i$ and $\lambda_2 = 1 - 2i$.

22. $t^2 y'' + \alpha t y' + (2 + \beta t)y = 0$

$t = 0$ is a regular singular point. One root of the indicial equation at $t = 0$ is $\lambda = 2$. The recurrence relation for the series solution corresponding to this root is

$$(n^2 + n)a_n - 4a_{n-1} = 0, \quad n = 1, 2, \ldots.$$

23. $ty'' + (1 + \alpha t)y' + \beta t y = 0$

The recurrence relation for a series solution is

$$n^2 a_n - (n - 1)a_{n-1} + 3a_{n-2} = 0, \quad n = 2, 3, \ldots.$$

24. Modified Bessel Equation The differential equation $t^2 y'' + ty' - (t^2 + \nu^2)y = 0$ is known as the modified Bessel equation. Its solutions, usually denoted by $I_\nu(t)$ and $K_\nu(t)$, are called **modified Bessel functions**. This equation arises when solving certain partial differential equations involving cylindrical coordinates.

(a) Do you anticipate that the modified Bessel equation will possess solutions that are even and odd functions of t? Explain.

(b) The point $t = 0$ is a regular singular point of the modified Bessel equation; all other points are ordinary points. Determine the indicial equation for the singularity at $t = 0$ and find the exponents at the singularity.

[10]Edmond Laguerre (1834–1886) attended the Ecole Polytechnique in Paris and returned there after ten years of service as a French artillery officer. He worked in the areas of analysis and geometry and is best remembered for his study of the polynomials that bear his name.

(c) Obtain the recurrence relation for the modified Bessel equation. How do the exponents and recurrence relation for this equation compare with their counterparts for Bessel's equation?

25. Consider the modified Bessel equation for $t > 0$ and for the special case $\nu = 1/2$, $t^2y'' + ty' - (t^s f\, 2 + (1/4))y = 0$.

(a) Define a new dependent variable $u(t)$ by the relation $y(t) = t^{-1/2}u(t)$. Show that $u(t)$ satisfies the differential equation $u'' - u = 0$.

(b) Show that the differential equation has a fundamental set of solutions

$$\frac{\sinh t}{\sqrt{t}}, \qquad \frac{\cosh t}{\sqrt{t}}, \qquad t \neq 0.$$

26. This exercise asks you to use computational software to show that Bessel functions behave like $R\cos(t - \delta)/\sqrt{t}$ for appropriate choices of constants R and δ and for t large enough. We restrict attention to $J_0(t)$.

(a) Locate the abscissa of the first maximum of $J_0(t)$ in $t > 0$; call this point t_m. Since $J_0'(t) = -J_1(t)$, this point can be found by applying a root-finding routine to $J_1(t)$.

(b) Evaluate the constants R and δ by setting $t_m - \delta = 2\pi$ and $R = \sqrt{t_m}\, J_0(t_m)$.

(c) Plot the two functions $J_0(t)$ and $R\cos(t - \delta)/\sqrt{t}$ on the same graph for $t_m \leq t \leq 50$. How do the two graphs compare?

EXTENDED PROBLEM: STEADY-STATE HEAT FLOW BETWEEN CONCENTRIC CYLINDERS

This exercise gives you a brief glimpse into one application involving variable coefficient linear differential equations.

Consider the two concentric cylinders shown in Figure 10.8. The inner cylinder has radius $a > 0$ while the outer cylinder has radius $b > a$. Assume that these cylinders represent the inner and outer surfaces of a pipe and that the pipe itself is designed to function as part of a simple cooling system. Heat, or thermal energy, is drawn from the region exterior to the pipe by the presence of coolant flowing within the pipe. We assume that the coolant is "well stirred" and that the inner surface of the pipe is maintained at

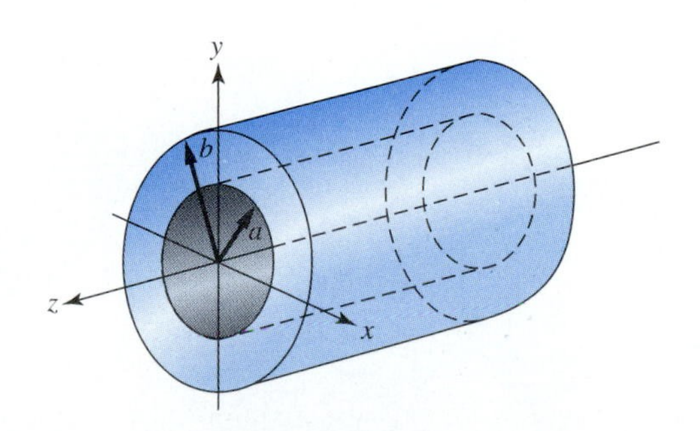

FIGURE 10.8

The concentric cylinders represent the inner and outer surfaces of a pipe. Heat is drawn from the exterior region by the presence of a coolant flowing through the pipe.

the coolant temperature. Suppose we know the temperature of the region outside the pipe as well as the temperature of the coolant. For a fixed geometry (that is, for fixed values of the radii a and b), we would like to know the rate at which heat is drawn from the exterior region. In addition, for a fixed coolant temperature, it would be interesting to know how the heat transfer varies as a function of the two pipe radii.

Mathematical Formulation

In general, a temperature reading depends on where and when the reading is taken. Thus, if T represents temperature (measured perhaps, in degrees Centigrade), then $T = T(x, y, z, t)$.

We shall assume, however, that our cooling system is operating in a steady-state mode; that is, system operation has "settled down" to the point where the temperature everywhere remains constant in time. In this case, T is a function of position only, $T = T(x, y, z)$. We further assume that temperature does not change in the axial or z direction; thus, in fact, $T = T(x, y)$. This is an assumption that you can rightly object to since, as the coolant flows along, it draws in thermal energy from the exterior region. Therefore, coolant temperature should gradually rise as a function of the axial variable z. However, if the coolant flow rate is sufficiently fast and the axial length of the system is sufficiently short, our approximation is a reasonable idealization.

Because of the cylindrical geometry, we introduce polar coordinates. Let $x = r\cos\theta$ and $y = r\sin\theta$, as in Figure 10.9. We view temperature as a function of the polar variables, $T = T(r, \theta)$. The domain of interest is the annular region between the cylinders, described in polar coordinates by $a \le r \le b$, $0 \le \theta < 2\pi$. We assume that the temperature at the outer radius of the pipe is known; that is,

$$T(b, \theta) = T_b(\theta), \qquad 0 \le \theta < 2\pi,$$

where T_b is a known function of angle θ. Likewise, we assume the temperature at the inner pipe radius is a known constant; that is,

$$T(a, \theta) = T_a, \qquad 0 \le \theta < 2\pi.$$

Since the coolant is to draw heat from the exterior region, we expect to have $T_b(\theta) > T_a$, $0 \le \theta < 2\pi$.

The problem is to determine the rate at which the coolant draws heat through the pipe from the exterior region. We solve this problem by determining the temperature

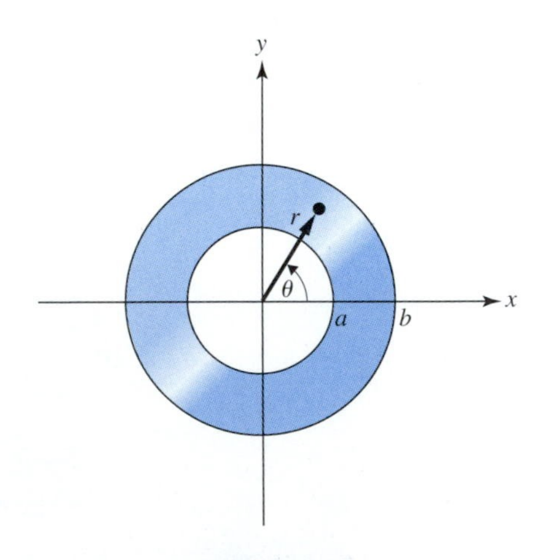

FIGURE 10.9

A cross section of the pipe in Figure 10.8. We introduce polar coordinates into the problem because of the cylindrical geometry.

$T(r, \theta)$ within the annular cross section of the pipe and then using this information to compute the required heat flow.

Within the annular region of the pipe, the steady-state temperature must be a solution of a partial differential equation known as Laplace's equation. In our case, the temperature $T(r, \theta)$ is a function of the two polar variables r and θ. In polar coordinates, Laplace's equation is

$$\frac{\partial^2 T}{\partial r^2} + \frac{1}{r}\frac{\partial T}{\partial r} + \frac{1}{r^2}\frac{\partial^2 T}{\partial \theta^2} = 0. \tag{1}$$

To determine the steady-state temperature within the annular pipe region we must solve partial differential equation (1), subject to the boundary conditions

$$T(a, \theta) = T_a, \qquad T(b, \theta) = T_b(\theta), \qquad 0 \le \theta < 2\pi. \tag{2}$$

We consider two different possibilities for the temperature, $T_b(\theta)$, at the outer radius. In each of these two cases, we can reduce the problem to that of solving an ordinary differential equation where the radial variable r is the independent variable. Imposing the boundary constraints on the solution at the inner and outer radii enables us to evaluate the two arbitrary constants in the general solution. It should be emphasized that the type of problem we ultimately solve in this exercise is not an initial value problem since constraints are imposed at two different values of the independent variable. The problems are two-point boundary value problems but we have the mathematical tools to obtain the solution.

Once we find the temperature distribution within the annular region,

$$T(r, \theta), \qquad a \le r \le b, \qquad 0 \le \theta < 2\pi,$$

we will compute the heat transfer rate into the coolant at the inner radius a. Heat flows "downhill," from hotter to cooler regions. Moreover, the rate of heat flow is proportional to the temperature gradient—the steeper the gradient, the greater the rate of heat transfer. At a point on the inner pipe radius, the rate of heat transfer per unit surface area is given by

$$\kappa \frac{\partial T(a, \theta)}{\partial r}, \tag{3}$$

where κ is a positive constant (known as the thermal conductivity) that depends on the nature of the pipe material. Better thermal conductors have larger values of κ. Equation (3) should make intuitive sense to you. If $\partial T/\partial r > 0$ when evaluated at a point on the inner pipe boundary, then heat "flows downhill" from that point into the coolant. From (3), it follows that the rate of heat transfer per unit axial length of the pipe can be found by integrating (3) around the inner pipe radius, obtaining

$$\int_0^{2\pi} \kappa \frac{\partial T(a, \theta)}{\partial r} a \, d\theta = \kappa a \int_0^{2\pi} \frac{\partial T(a, \theta)}{\partial r} \, d\theta. \tag{4}$$

PROBLEM 1:

Assume that the temperature at the outer pipe radius is constant; that is

$$T_b(\theta) = T_b, \qquad 0 \le \theta < 2\pi.$$

In this case, since neither boundary condition varies with angle, we expect the temperature in the annular pipe region likewise to be independent of θ. Assume a solution of the form $T = T(r), a \le r \le b$.

(a) Substitute $T = T(r)$ into Laplace's equation (1), obtaining

$$\frac{d^2 T(r)}{dr^2} + \frac{1}{r}\frac{dT(r)}{dr} = 0, \qquad a < r < b. \tag{5}$$

Note that (5) can be viewed as an Euler equation, $r^2T'' + rT' = 0$. (It can also be solved as a first order linear equation for T' followed by antidifferentiation.) Obtain the general solution of (5) and impose the two boundary constraints

$$T(a) = T_a, \qquad T(b) = T_b. \tag{6}$$

The constraints in (6) will determine the two arbitrary constants in the general solution.

(b) Compute dT/dr and evaluate the integral (4).

(c) Is the answer obtained in (b) consistent with common sense? Specifically:

(i) What happens to the heat transfer rate if the temperature differential $T_b - T_a$ is increased?

(ii) What happens if the heat transfer rate is calculated across an arbitrary radius $r = \rho, a < \rho < b$ instead of at the inner radius a? Referring to the figure, calculate

$$\int_0^{2\pi} \kappa \frac{\partial T(\rho, \theta)}{\partial r} \rho \, d\theta.$$

Is the answer you obtain different from that obtained in (b)? Should it be different?

(iii) How does the answer obtained in (b) depend upon the pipe radii a and b? What happens to the heat transfer rate if the pipe radii are changed in such a way that the quotient b/a remains constant?

(d) Let $T_a = 40°\text{F}$, $T_b = 120°\text{F}$, $a = 1"$, $b = 1.5"$. For these parameter values, plot $T(r)$ versus r for $a \le r \le b$. Do the maximum and minimum temperatures occur where you expect them to occur?

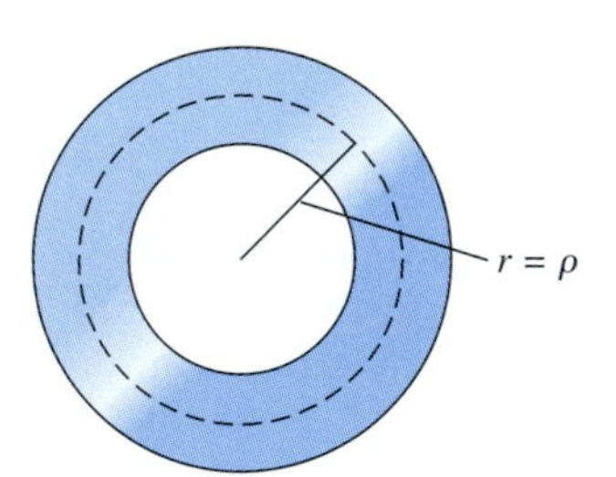

Heat transfer across an arbitrary cylinder of radius $r = \rho$

Figure for Problem 1

PROBLEM 2:

Assume now that the temperature distribution at the outer pipe radius is not constant; rather, assume

$$T(b, \theta) = [1 + \alpha \sin\theta]T_b, \qquad 0 \le \theta < 2\pi,$$

where T_b is the same positive constant as in Problem 1 and where $0 \le \alpha < 1$. Thus, some portions of the outer pipe surface are hotter and other portions cooler than in Problem 1; the outer radius temperature varies between a maximum of $[1 + \alpha]T_b$ and a minimum of $[1 - \alpha]T_b$. Do you expect the heat transfer rate to differ from that obtained in Problem 1 or not?

(a) Assume a solution of Laplace's equation of the form

$$T(r,\theta) = T_0(r) + T_1(r)\sin\theta \qquad \textbf{(7)}$$

within the annular pipe region. The unknown functions $T_0(r)$ and $T_1(r)$ must be determined. Substitute (7) into (1), obtaining

$$\left[\frac{d^2T_0(r)}{dr^2} + \frac{1}{r}\frac{dT_0(r)}{dr}\right] + \left[\frac{d^2T_1(r)}{dr^2} + \frac{1}{r}\frac{dT_1(r)}{dr} - \frac{1}{r^2}T_1(r)\right]\sin\theta = 0, \qquad \textbf{(8)}$$

$$a < r < b, \qquad 0 \le \theta < 2\pi.$$

Assume for the moment that the radial variable has an arbitrary but fixed value and view (8) as a function of polar angle θ. Since the set of functions $\{1, \sin\theta\}$ is linearly independent in $0 \le \theta < 2\pi$, equation (8) implies that

$$\frac{d^2T_0(r)}{dr^2} + \frac{1}{r}\frac{dT_0(r)}{dr} = 0$$

and

$$\frac{d^2T_1(r)}{dr^2} + \frac{1}{r}\frac{dT_1(r)}{dr} - \frac{1}{r^2}T_1(r) = 0.$$

for that particular value of r. However, since r is assumed to be arbitrary, these equations must hold for $a < r < b$. Find the general solution for each of these two Euler equations.

(b) Apply the boundary constraints (2). In particular, we have

$$T(a,\theta) = T_0(a) + T_1(a)\sin\theta = T_a$$
$$T(b,\theta) = T_0(b) + T_1(b)\sin\theta = T_b + T_b\alpha\sin\theta, \qquad 0 \le \theta < 2\pi.$$

Use the same linear independence argument as used in (a) to obtain boundary conditions for the functions T_0 and T_1. Impose these boundary conditions on the general solutions of the Euler equations obtained in (a) and determine $T(r,\theta)$.

(c) Again determine the heat transfer rate at the inner pipe radius. Is the heat transfer rate the same as that obtained in Problem 1? Was it actually necessary to explicitly calculate $T_1(r)$?

(d) Let $T_a = 40°$F, $T_b = 120°$F, $a = 1$", $b = 1.5$" as in Problem 1 and let $\alpha = 0.3$. We consider how the temperature varies as a function of angle midway between the inner and outer pipe radii. Let $r = 1.25$" and plot $T(1.25, \theta)$, $0 \le \theta \le 2\pi$. What are the maximum and minimum temperatures for this value of radius? Do the maximum and minimum temperatures occur where you expect them to occur?

REMARK: The problem consisting of Laplace's equation (1) and boundary constraints (2) can be shown to have a unique solution. Therefore, by assuming a solution having a certain form and then showing that this assumed form actually satisfies both (1) and (2), we have found the unique solution of the problem.

APPENDIX
Review of Matrices and Vectors

A **matrix** is a rectangular array of numbers. It is typically described in terms of the number of its rows and columns. For example, the matrix A represented below has two rows and three columns and is referred to as a (2×3) matrix:

$$A = \begin{bmatrix} a_{11} & a_{12} & a_{13} \\ a_{21} & a_{22} & a_{23} \end{bmatrix}.$$

As the notation indicates, a_{ij} refers to the number located in the ith row and jth column within the array A. A matrix having a single row, a $(1 \times n)$ matrix for example, is often referred to as a row matrix or **row vector**. Similarly, a matrix having a single column [an $(m \times 1)$ matrix] is called a column matrix or **column vector**. A matrix having equal numbers of rows and columns [an $(n \times n)$ matrix] is a **square matrix**.

Matrices and Systems of Equations

Matrices and vectors arise naturally when we study systems of linear equations, such as

$$\begin{aligned} a_{11}x_1 + a_{12}x_2 + a_{13}x_3 &= b_1 \\ a_{21}x_1 + a_{22}x_2 + a_{23}x_3 &= b_2 \\ a_{31}x_1 + a_{32}x_2 + a_{33}x_3 &= b_3. \end{aligned} \tag{1}$$

We can compactly represent the information contained within system (1) in terms of two matrices, the (3×3) square matrix of coefficients A and the (3×1) column vector $\mathbf{b}$ of right-hand sides:

$$A = \begin{bmatrix} a_{11} & a_{12} & a_{13} \\ a_{21} & a_{22} & a_{23} \\ a_{31} & a_{32} & a_{33} \end{bmatrix}, \qquad \mathbf{b} = \begin{bmatrix} b_1 \\ b_2 \\ b_3 \end{bmatrix}. \tag{2}$$

Matrices are more than simply notational shorthand, however. They become mathematical entities in their own right once the concept of matrix equality is defined and matrix operations, such as matrix addition, scalar multiplication, and matrix multiplication, are introduced. We can represent systems such as (1) as a matrix equation $A\mathbf{x} = \mathbf{b}$, where $\mathbf{x}$ is the (3×1) column vector of unknowns

$$\mathbf{x} = \begin{bmatrix} x_1 \\ x_2 \\ x_3 \end{bmatrix}.$$

Equality of Matrices; Matrix Operations

Matrix Equality

The question of matrix equality can be considered only when comparing two matrices of the same size; that is, matrices having the same number of rows and columns. Two $(m \times n)$ matrices A and B are equal if every entry of the A array equals the corresponding entry of the B array. In particular, if A and B have entries a_{ij} and b_{ij}, respectively, then $A = B$ if $a_{ij} = b_{ij}$ for $1 \le i \le m$, $1 \le j \le n$.

Scalar Multiplication

Numbers (real or complex) are referred to as scalars. If α is a scalar and A is an $(m \times n)$ matrix having entries a_{ij}, then the matrix αA is the $(m \times n)$ matrix formed by multiplying each entry of A by the number α. For example,

$$(-2)\begin{bmatrix} 1 & 3 \\ 0 & -1 \end{bmatrix} = \begin{bmatrix} -2 & -6 \\ 0 & 2 \end{bmatrix}.$$

Matrix Addition

Matrix addition is defined only for matrices of the same size. If A and B are $(m \times n)$ matrices, then $A + B$ is the $(m \times n)$ matrix formed by adding each entry of the matrix A to the corresponding entry of the matrix B. For example,

$$\begin{bmatrix} 1 & -2 & 0 \\ 2 & 3 & -1 \end{bmatrix} + \begin{bmatrix} 2 & 1 & 1 \\ -2 & 2 & -2 \end{bmatrix} = \begin{bmatrix} 3 & -1 & 1 \\ 0 & 5 & -3 \end{bmatrix}.$$

Matrix subtraction is defined in terms of scalar multiplication and matrix addition. If A and B are two $(m \times n)$ matrices, then $A - B = A + (-1)B$. If the entries of A and B are a_{ij} and b_{ij}, respectively, then the entries of $A - B$ are $a_{ij} - b_{ij}$, $1 \le i \le m$, $1 \le j \le n$.

Matrix Multiplication

The definition of matrix multiplication arises because we want to represent the left-hand side of system (1) as the matrix product $A\mathbf{x}$. Therefore, we define

$$A\mathbf{x} = \begin{bmatrix} a_{11} & a_{12} & a_{13} \\ a_{21} & a_{22} & a_{23} \\ a_{31} & a_{32} & a_{33} \end{bmatrix} \begin{bmatrix} x_1 \\ x_2 \\ x_3 \end{bmatrix} = \begin{bmatrix} a_{11}x_1 + a_{12}x_2 + a_{13}x_3 \\ a_{21}x_1 + a_{22}x_2 + a_{23}x_3 \\ a_{31}x_1 + a_{32}x_2 + a_{33}x_3 \end{bmatrix} = \begin{bmatrix} \sum_{j=1}^{3} a_{1j}x_j \\ \sum_{j=1}^{3} a_{2j}x_j \\ \sum_{j=1}^{3} a_{3j}x_j \end{bmatrix}. \tag{3}$$

In (3), the product of the (3×3) matrix A and the (3×1) vector $\mathbf{x}$ is a (3×1) vector, $A\mathbf{x}$. In general, if A is an $(m \times n)$ matrix having entries a_{ij} and $\mathbf{x}$ is an $(n \times 1)$ vector having entries x_j, then $A\mathbf{x}$ is defined to be the $(m \times 1)$ vector whose ith entry is equal to $a_{i1}x_1 + a_{i2}x_2 + \cdots + a_{in}x_n$. (In other words, the ith entry of $A\mathbf{x}$ is formed by taking the dot product of the ith row of A with the vector $\mathbf{x}$.) For example,

$$\begin{bmatrix} 1 & -2 & 0 \\ 2 & 3 & -1 \end{bmatrix} \begin{bmatrix} -1 \\ 2 \\ -3 \end{bmatrix} = \begin{bmatrix} (1)(-1) + (-2)(2) + (0)(-3) \\ (2)(-1) + (3)(2) + (-1)(-3) \end{bmatrix} = \begin{bmatrix} -5 \\ 7 \end{bmatrix}.$$

As another example, consider the system of equations

$$\begin{aligned} 3x_1 + 5x_2 &= 3 \\ 2x_1 + 4x_2 &= 1. \end{aligned} \tag{4a}$$

Using the definition of multiplication of a matrix times a vector, this system of equations can be written as $A\mathbf{x} = \mathbf{b}$, where

$$A = \begin{bmatrix} 3 & 5 \\ 2 & 4 \end{bmatrix}, \qquad \mathbf{x} = \begin{bmatrix} x_1 \\ x_2 \end{bmatrix}, \qquad \mathbf{b} = \begin{bmatrix} 3 \\ 1 \end{bmatrix}. \tag{4b}$$

The definition of matrix-vector multiplication can be extended to matrix-matrix products of the form $C = AB$, where A is an $(m \times n)$ matrix and B is an $(n \times p)$ matrix. Note that, in this product, the number of columns in the first matrix A equals the number of rows of the second matrix B. The resulting product matrix AB is an $(m \times p)$ matrix.

To define the product AB, we first view the $(n \times p)$ matrix B as being composed of p column vectors stacked side by side. We adopt the notation

$$B = \begin{bmatrix} b_{11} & b_{12} & \cdots & b_{1p} \\ b_{21} & b_{22} & & b_{2p} \\ \vdots & & & \vdots \\ b_{n1} & b_{n2} & \cdots & b_{np} \end{bmatrix} = [\mathbf{B}_1, \mathbf{B}_2, \ldots, \mathbf{B}_p],$$

where

$$\mathbf{B}_1 = \begin{bmatrix} b_{11} \\ b_{21} \\ \vdots \\ b_{n1} \end{bmatrix}, \quad \mathbf{B}_2 = \begin{bmatrix} b_{12} \\ b_{22} \\ \vdots \\ b_{n2} \end{bmatrix}, \quad \ldots, \quad \mathbf{B}_p = \begin{bmatrix} b_{1p} \\ b_{2p} \\ \vdots \\ b_{np} \end{bmatrix}.$$

The representation $B = [\mathbf{B}_1, \mathbf{B}_2, \ldots, \mathbf{B}_p]$ is referred to as the **column form of a matrix**. In terms of this notation, we define the matrix product AB to be

$$AB = A[\mathbf{B}_1, \mathbf{B}_2, \ldots, \mathbf{B}_p] = [A\mathbf{B}_1, A\mathbf{B}_2, \ldots, A\mathbf{B}_p].$$

According to this definition, the jth column of the product matrix is $A\mathbf{B}_j$, where $A\mathbf{B}_j$ is the product of the $(m \times n)$ matrix A and the $(n \times 1)$ column vector consisting of the jth column of B. For example,

$$\begin{bmatrix} 1 & 2 \\ -1 & 0 \end{bmatrix}\begin{bmatrix} 3 & -3 \\ 4 & -2 \end{bmatrix} = \begin{bmatrix} ((1)(3)+(2)(4)) & ((1)(-3)+(2)(-2)) \\ ((-1)(3)+(0)(4)) & ((-1)(-3)+(0)(-2)) \end{bmatrix} = \begin{bmatrix} 11 & -7 \\ -3 & 3 \end{bmatrix}.$$

In this example, $m = n = p = 2$. The product matrix is a (2×2) matrix.

When multiplying matrices, it is important to realize that generally $AB \neq BA$. For instance, consider the case where A is an $(m \times n)$ matrix and B is an $(n \times p)$ matrix. In this event, the matrix product AB is defined and the resulting product is an $(m \times p)$ matrix. The matrix product BA, however, is not defined unless $p = m$. If $p = m$, both matrix products, AB and BA, are defined. The product AB is a square $(m \times m)$ matrix, though, while the product BA is a square $(n \times n)$ matrix. In this case, we cannot even consider the question of matrix equality unless both matrices have the same size; that is, we must have $m = n$.

If A and B are both square $(n \times n)$ matrices, the matrix products AB and BA are both defined and are both $(n \times n)$ matrices; in this case we can entertain the question of matrix equality. However, even when A and B are $(n \times n)$, it is still generally true that $AB \neq BA$. For example, let

$$A = \begin{bmatrix} 1 & -1 \\ 0 & 2 \end{bmatrix} \quad \text{and} \quad B = \begin{bmatrix} 0 & 3 \\ 1 & -1 \end{bmatrix}.$$

Even though A and B are each (2×2), we still see that $AB \neq BA$, since

$$AB = \begin{bmatrix} -1 & 4 \\ 2 & -2 \end{bmatrix} \quad \text{and} \quad BA = \begin{bmatrix} 0 & 6 \\ 1 & -3 \end{bmatrix}.$$

As illustrated above, matrix multiplication is not **commutative**. It is, however, **associative**. In particular, suppose that matrices A, B, and C are $(m \times n)$, $(n \times p)$, and $(p \times q)$, respectively. Then, we can form the matrix products $(AB)C$ and $A(BC)$. In the first case, we form the product AB and then multiply this product matrix on the right by C. In the second case, we form the matrix product BC and then multiply the result on the left by A. It can be shown that the same $(m \times q)$ matrix results in each case; that is, $(AB)C = A(BC)$.

The Matrix Transpose Operation

Let A be an $(m \times n)$ matrix. The **transpose** of A, denoted by A^T, is the $(n \times m)$ matrix obtained by interchanging the rows and columns of A. Thus, the element in the ith row and jth column of A^T is a_{ji}, the element in the jth row and ith column of A. For example,

$$\text{if} \quad A = \begin{bmatrix} 1 & -1 & 2 \\ 3 & 0 & 4 \end{bmatrix}, \quad \text{then} \quad A^T = \begin{bmatrix} 1 & 3 \\ -1 & 0 \\ 2 & 4 \end{bmatrix}.$$

Some properties of the transpose operation are the following:

(a) If the matrix product AB is defined, then $(AB)^T = B^T A^T$.

(b) If the matrix sum $A + B$ is defined, then $(A + B)^T = A^T + B^T$.

(c) $(A^T)^T = A$.

Diagonal Matrices, Triangular Matrices, and the Identity Matrix

An $(n \times n)$ matrix $D = (d_{ij})$ is called a **diagonal matrix** if $d_{ij} = 0$ when $i \neq j$. For example,

$$D = \begin{bmatrix} 1 & 0 & 0 & 0 \\ 0 & -2 & 0 & 0 \\ 0 & 0 & 0 & 0 \\ 0 & 0 & 0 & 3 \end{bmatrix}$$

is a (4×4) diagonal matrix. In general, if $A = (a_{ij})$ is any $(n \times n)$ matrix, then the entries a_{ii} are called the **main diagonal entries** of A.

An $(n \times n)$ matrix $U = (u_{ij})$ is called an **upper triangular matrix** if $u_{ij} = 0$ when $i > j$. In other words, U is upper triangular if all the entries below the main diagonal are zero. For example,

$$U = \begin{bmatrix} -1 & 2 & 5 \\ 0 & 1 & 4 \\ 0 & 0 & 3 \end{bmatrix}$$

is a (3×3) upper triangular matrix. Similarly, an $(n \times n)$ matrix L is called a **lower triangular matrix** if all its entries above the main diagonal are zero. A matrix that is both upper triangular and lower triangular is a diagonal matrix. Also note that the transpose of an upper triangular matrix is a lower triangular matrix and conversely.

The $(n \times n)$ **identity matrix** I is a diagonal matrix with all diagonal elements equal to 1. For example,

$$I = \begin{bmatrix} 1 & 0 & 0 \\ 0 & 1 & 0 \\ 0 & 0 & 1 \end{bmatrix}$$

is the (3×3) identity matrix. One can show that if A is any $(n \times n)$ matrix and I is the $(n \times n)$ identity matrix, then $AI = IA = A$. Consequently, the identity matrix plays the same role in matrix multiplication that the number 1 plays when multiplying numbers.

Matrix Inverses

Given a scalar equation $ax = b$, where $a \neq 0$, we can solve for the unknown x by multiplying both sides of the equation by a^{-1}, where a^{-1} is a number such that $(a^{-1})(a) = (a)(a^{-1}) = 1$. For example, if $5x = 2$, then

$$(1/5)(5x) = ((1/5) \cdot 5)x = x = (1/5)(2) = 2/5.$$

Consider now $A\mathbf{x} = \mathbf{b}$, the matrix formulation of a system of linear equations, such as (1). In such a formulation, A is the $(n \times n)$ matrix of coefficients, $\mathbf{x}$ is the $(n \times 1)$ vector of unknowns and $\mathbf{b}$ is the $(n \times 1)$ vector of right-hand sides. If we can find an $(n \times n)$ matrix, call it A^{-1}, such that $A^{-1}A = I$, then we can solve the system of equations by multiplying both sides of the equation on the left by A^{-1}. That is, $A^{-1}(A\mathbf{x}) = (A^{-1}A)\mathbf{x} = I\mathbf{x} = \mathbf{x} = A^{-1}\mathbf{b}$. If such a matrix A^{-1} exists, we refer to it as the **inverse** of A and we say that A is **invertible**. It can be shown that if A^{-1} exists, then it is unique. Moreover, when A^{-1} exists, A and A^{-1} commute: $A^{-1}A = AA^{-1} = I$.

An example of a matrix and its inverse is

$$A = \begin{bmatrix} 3 & 5 \\ 2 & 4 \end{bmatrix}, \qquad A^{-1} = \frac{1}{2}\begin{bmatrix} 4 & -5 \\ -2 & 3 \end{bmatrix}. \tag{5}$$

The matrix A in (5) is the coefficient matrix for the system of two equations, (4a). As we saw earlier, this system can be represented as $A\mathbf{x} = \mathbf{b}$; see equation (4b). From our discussion of inverses, it follows that the solution of $A\mathbf{x} = \mathbf{b}$ is given by $\mathbf{x} = A^{-1}\mathbf{b}$:

$$\mathbf{x} = \begin{bmatrix} x_1 \\ x_2 \end{bmatrix} = A^{-1}\mathbf{b} = \frac{1}{2}\begin{bmatrix} 4 & -5 \\ -2 & 3 \end{bmatrix}\begin{bmatrix} 3 \\ 1 \end{bmatrix} = \begin{bmatrix} 7/2 \\ -3/2 \end{bmatrix}.$$

[Inserting the values $x_1 = 7/2$ and $x_2 = -3/2$ into equation (4a) confirms that the process of forming $\mathbf{x} = A^{-1}\mathbf{b}$ has indeed led to the solution of the system.]

Shortly, we will answer two fundamental questions relative to matrix inverses:

(a) How can we determine if a square matrix A is invertible?

(b) If matrix A is known to be invertible, how do we compute A^{-1}?

Determinants

If A is a square matrix, the **determinant** of A, denoted by $\det[A]$, is a number that we associate with A. The determinant provides a succinct answer to one of the questions posed earlier. In particular:

$$\text{A square matrix } A \text{ is invertible if and only if } \det[A] \neq 0. \tag{6}$$

A procedure for computing determinants can be described recursively—that is, the determinant of a square matrix of order n is evaluated in terms of determinants of square matrices of order $n - 1$. We restrict our discussion to determinants of (2×2) and (3×3) matrices since they are the ones of principal interest in this text.

If A is a (2×2) matrix, we define $\det[A]$ by

$$\det[A] = \det\left(\begin{bmatrix} a_{11} & a_{12} \\ a_{21} & a_{22} \end{bmatrix}\right) = a_{11}a_{22} - a_{12}a_{21}.$$

If A is a (3×3) matrix, we define $\det[A]$ by means of the formulas

$$\det[A] = \sum_{j=1}^{3} (-1)^{i+j} a_{ij} M_{ij} \quad \text{or} \quad \det[A] = \sum_{i=1}^{3} (-1)^{i+j} a_{ij} M_{ij}. \tag{7}$$

In (7), M_{ij} is the determinant of the (2×2) matrix obtained by deleting the ith row and jth column of A. In the first formula, the sum is taken over the column index j; the row index i can be chosen to be any fixed integer between 1 and 3.

In the second formula, the sum is taken over the row index i; the column index j can be chosen to be any fixed integer between 1 and 3. These determinant calculations are sometimes referred to as an "expansion along the ith row" or an "expansion along the jth column," respectively. The remarkable fact is that the same number, det[A], is obtained for all possible choices of row and column expansions. From (7) we see that if a particular row or column of A has several zero elements, the effort of computing a determinant can be reduced by performing the expansion along that row or column.

As an example to illustrate a determinant computation, let A denote the matrix

$$A = \begin{bmatrix} 3 & 2 & 1 \\ 2 & -1 & 3 \\ 4 & 0 & 1 \end{bmatrix}.$$

Expanding along the first row, $\det[A] = a_{11}M_{11} - a_{12}M_{12} + a_{13}M_{13} = 3M_{11} - 2M_{12} + M_{13}$. Therefore,

$$\begin{aligned}\det[A] &= 3\det\left(\begin{bmatrix} -1 & 3 \\ 0 & 1 \end{bmatrix}\right) - 2\det\left(\begin{bmatrix} 2 & 3 \\ 4 & 1 \end{bmatrix}\right) + \det\left(\begin{bmatrix} 2 & -1 \\ 4 & 0 \end{bmatrix}\right) \\ &= (3)(-1) - (2)(-10) + (1)(4) = 21.\end{aligned}$$

For comparison, we also expand along the second column, obtaining

$$\begin{aligned}\det[A] &= -a_{12}M_{12} + a_{22}M_{22} - a_{32}M_{32} = -2M_{12} + (-1)M_{22} - (0)M_{32} \\ &= -2\det\left(\begin{bmatrix} 2 & 3 \\ 4 & 1 \end{bmatrix}\right) - \det\left(\begin{bmatrix} 3 & 1 \\ 4 & 1 \end{bmatrix}\right) = (-2)(-10) - (1)(-1) = 21.\end{aligned}$$

[By (6), the matrix A is invertible since $\det[A] \neq 0$.]

Properties of Determinants

We list without proof some of the properties of determinants. In this list, A and B are $(n \times n)$ matrices, while $\mathbf{b}$ and $\mathbf{x}$ are $(n \times 1)$ vectors.

(a) The equation $A\mathbf{x} = \mathbf{b}$ has a unique solution if and only if $\det[A] \neq 0$.

(b) The matrix A is invertible if and only if $\det[A] \neq 0$.

(c) $\det[AB] = \det[A]\det[B]$.

(d) $\det[A^T] = \det[A]$.

(e) If A is invertible, then $\det[A^{-1}] = 1/\det[A]$.

(f) If B is obtained from A by multiplying all elements of any row (or column) of A by a number α, then $\det[B] = \alpha \det[A]$.

(g) If B is obtained from A by interchanging any two rows (or columns) of A, then $\det[B] = -\det[A]$.

(h) If B is obtained from A by adding a scalar multiple of one row (or column) of A to another row (or column) of A, then $\det[B] = \det[A]$.

(i) If any row (or column) of A is a scalar multiple of another row (or column) of A, then $\det[A] = 0$. In particular, if all entries in any row (or column) of A are zero, then $\det[A] = 0$.

(j) If A is an upper (or lower) triangular matrix, then $\det[A] = a_{11}a_{22} \cdots a_{nn}$. That is, the determinant of a triangular matrix is the product of its diagonal entries.

Solving a System of Linear Equations—Gaussian Elimination and Cramer's Rule

One well-known method for solving a system $A\mathbf{x} = \mathbf{b}$ is the process known as Gaussian elimination. Gaussian elimination is carried out using the following three **elementary row operations**:

(a) Interchanging any two rows of the matrix

(b) Multiplying all elements of any row by a nonzero constant

(c) Multiplying the elements of a row by a nonzero constant and adding the result to another row.

The process of Gaussian elimination is often organized by using an augmented matrix. In particular, given an $(n \times n)$ system $A\mathbf{x} = \mathbf{b}$, we form the $(n \times (n+1))$ **augmented matrix**, $[A, \mathbf{b}]$, and use elementary row operations to transform (if possible) the augmented matrix to a matrix of the form $[I, \hat{\mathbf{b}}]$. If we can achieve the form $[I, \hat{\mathbf{b}}]$ using elementary row operations, then the unique solution of $A\mathbf{x} = \mathbf{b}$ is given by $\mathbf{x} = \hat{\mathbf{b}}$.

When we have several systems to solve, $A\mathbf{x} = \mathbf{b}_j, j = 1, 2, \ldots, p$, we can solve them all at the same time if we apply elementary row operations to the $(n \times (n+p))$ augmented matrix $[A, \mathbf{b}_1, \mathbf{b}_2, \ldots, \mathbf{b}_p]$. If it is possible to transform this augmented matrix to the form $[I, \hat{\mathbf{b}}_1, \hat{\mathbf{b}}_2, \ldots, \hat{\mathbf{b}}_p]$, then the solution of $A\mathbf{x} = \mathbf{b}_j$ is $\mathbf{x} = \hat{\mathbf{b}}_j, j = 1, 2, \ldots, p$.

If the $(n \times n)$ system $A\mathbf{x} = \mathbf{b}$ is not too large, we might also consider using **Cramer's rule** instead of Gaussian elimination to solve the system. To explain Cramer's rule for solving $A\mathbf{x} = \mathbf{b}$, let C_k denote the matrix obtained by replacing the kth column of A by the right-hand side, $\mathbf{b}$. Furthermore, let us assume that $\det[A] \neq 0$ so that the system $A\mathbf{x} = \mathbf{b}$ has a unique solution. Then, the kth component of the solution is given by

$$x_k = \frac{\det[C_k]}{\det[A]}, \qquad k = 1, 2, \ldots, n.$$

Computing Matrix Inverses

If A is a (2×2) invertible matrix, then there is a simple formula for A^{-1}. In particular,

$$\text{if} \quad A = \begin{bmatrix} a_{11} & a_{12} \\ a_{21} & a_{22} \end{bmatrix} \quad \text{where} \quad \det[A] \neq 0, \quad \text{then} \quad A^{-1} = \frac{1}{\det[A]} \begin{bmatrix} a_{22} & -a_{12} \\ -a_{21} & a_{11} \end{bmatrix}.$$

The validity of this formula can be checked directly by showing that $A^{-1}A = AA^{-1} = I$. For square invertible matrices of size (3×3) or larger, there is no simple expression for A^{-1}.

We now describe an approach to calculating A^{-1}. If A is a (3×3) invertible matrix, the task of computing A^{-1} is tantamount to finding an invertible matrix X such that $AX = I$. If the unknown inverse matrix X is expressed in column form as $X = [\mathbf{x}_1, \mathbf{x}_2, \mathbf{x}_3]$ and the (3×3) identity matrix I is expressed in column form as $I = [\mathbf{e}_1, \mathbf{e}_2, \mathbf{e}_3]$, then the equation $AX = I$ has the form

$$AX = A[\mathbf{x}_1, \mathbf{x}_2, \mathbf{x}_3] = [A\mathbf{x}_1, A\mathbf{x}_2, A\mathbf{x}_3] = [\mathbf{e}_1, \mathbf{e}_2, \mathbf{e}_3].$$

Thus, solving $AX = I$ is equivalent to solving the three systems $A\mathbf{x}_j = \mathbf{e}_j$, $j = 1, 2, 3$.

As we saw earlier, we can use Gaussian elimination to solve these three systems at the same time if we begin with the (3×6) augmented matrix $[A, \mathbf{e}_1, \mathbf{e}_2, \mathbf{e}_3] = [A, I]$. If we can use elementary row operations to transform $[A, I]$ to a matrix of the form $[I, \hat{B}]$, then $\hat{B}$ is the inverse of A.

For example, consider the matrix

$$A = \begin{bmatrix} 2 & 4 & 4 \\ -2 & -3 & -3 \\ 2 & 5 & 6 \end{bmatrix}.$$

We begin with the augmented matrix

$$[A, I] = \begin{bmatrix} 2 & 4 & 4 & 1 & 0 & 0 \\ -2 & -3 & -3 & 0 & 1 & 0 \\ 2 & 5 & 6 & 0 & 0 & 1 \end{bmatrix}.$$

We next apply the following sequence of elementary row operations: Add row 1 to row 2, add -1 times row 1 to row 3, add -4 times row 2 to row 1, add -1 times row 2 to row 3, add -1 times row 3 to row 2, multiply row 1 by $1/2$. The result of these row operations is

$$[I, \hat{B}] = \begin{bmatrix} 1 & 0 & 0 & -3/2 & -2 & 0 \\ 0 & 1 & 0 & 3 & 2 & -1 \\ 0 & 0 & 1 & -2 & -1 & 1 \end{bmatrix}.$$

Thus, we find that A^{-1} exists and is given by

$$A^{-1} = \begin{bmatrix} -3/2 & -2 & 0 \\ 3 & 2 & -1 \\ -2 & -1 & 1 \end{bmatrix}.$$

Linear Independence of Column Vectors

Let A be an $(n \times n)$ matrix that we represent in column form as $A = [\mathbf{A}_1, \mathbf{A}_2, \ldots, \mathbf{A}_n]$. Let $\mathbf{x}$ be an $(n \times 1)$ column vector having components $x_j, j = 1, \ldots, n$. From (3) we can show that

$$A\mathbf{x} = x_1\mathbf{A}_1 + x_2\mathbf{A}_2 + \cdots + x_n\mathbf{A}_n.$$

In other words, the matrix-vector product $A\mathbf{x}$ can be represented as a linear combination of the columns of the matrix A, with the coefficients of the linear combination being the components of the column vector $\mathbf{x}$.

Suppose we are given a set of $(n \times 1)$ column vectors $\{\mathbf{v}_1, \mathbf{v}_2, \ldots, \mathbf{v}_n\}$. To test whether this set is linearly independent, we form the equation

$$c_1\mathbf{v}_1 + c_2\mathbf{v}_2 + \cdots + c_n\mathbf{v}_n = \mathbf{0},$$

and ask, "Is the only solution of this equation the trivial one, $c_1 = 0$, $c_2 = 0, \ldots, c_n = 0$?" If the answer is "yes," the set of vectors is **linearly independent**; if the answer is "no," the set is **linearly dependent**.

We can view the equation $c_1\mathbf{v}_1 + c_2\mathbf{v}_2 + \cdots + c_n\mathbf{v}_n = \mathbf{0}$ as the column form version of the matrix equation $V\mathbf{c} = \mathbf{0}$, where $V = [\mathbf{v}_1, \mathbf{v}_2, \ldots, \mathbf{v}_n]$. The question of linear independence, therefore, depends on whether the matrix V has an inverse. If V is invertible, then the only solution of $V\mathbf{c} = \mathbf{0}$ is $\mathbf{c} = \mathbf{0}$ and the set of vectors is linearly independent. If V^{-1} does not exist, it can be shown that the equation $V\mathbf{c} = \mathbf{0}$ has a nonzero solution and the set of vectors is linearly dependent. Hence, in terms of determinants, we can say

$$\{\mathbf{v}_1, \mathbf{v}_2, \ldots, \mathbf{v}_n\} \quad \text{is linearly independent if and only if} \quad \det[\mathbf{v}_1, \mathbf{v}_2, \ldots, \mathbf{v}_n] \neq 0.$$

Answers to Odd-Numbered Exercises

CHAPTER 1

Section 1.1, page 6

1. Order is 2.
3. Order is 1.
5. (b) $C = 2e^{-1}$
7. (b) $C_1 = 3,\ C_2 = 1$
9. $c = 0$ and $c = 1$
11. $r = 1$ and $r = 2$
13. $y = e^{2t} + e^{-2t} = 2\cosh 2t$
15. $y = 3e^{-2t}$
17. $m = -2, y_0 = 1$
19. $t_{\text{impact}} = \sqrt{2y_0/g},\quad v_{\text{impact}} = -\sqrt{2gy_0}$
21. $\varepsilon = \pi/4$

Section 1.2, page 12

1. (a) Autonomous
   (b) $y = 1$
   (c)

3. (a) Autonomous
   (b) $y = 0,\ \ y = \pm\pi,\ \ y = \pm 2\pi, \ldots$
   (c)

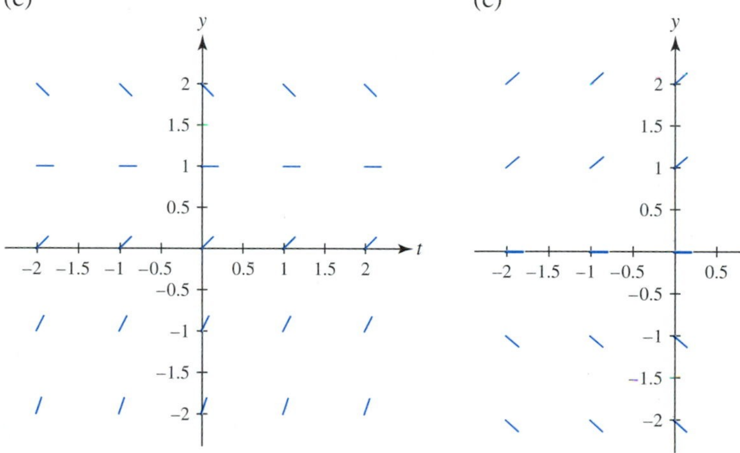

5. (a) Autonomous
   (b) There are none.
   (c)

7. (a) The requested isoclines are the lines $y = 2, y = 1$, and $y = 0$.
(b)

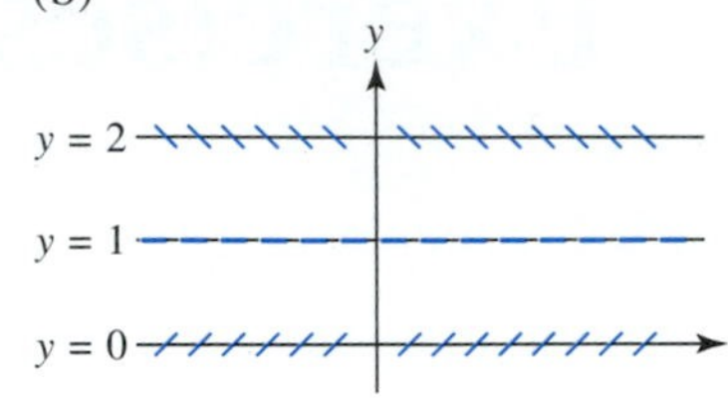

9. (a) The requested isoclines are the hyperbolas $y^2 - t^2 = -1, y^2 - t^2 = 0$, and $y^2 - t^2 = 1$.
(b)

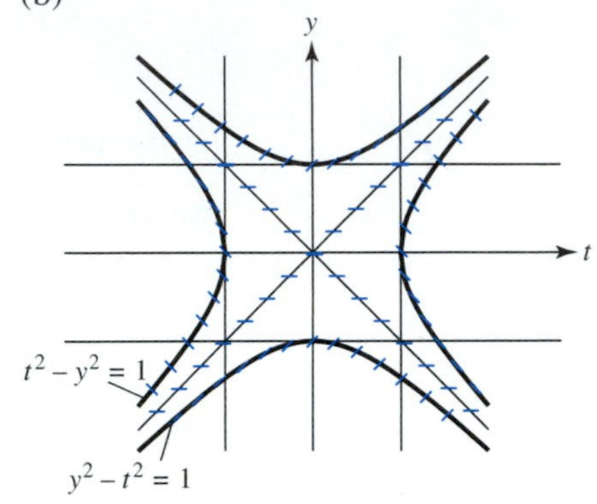

11. One possibility is $y' = -(y - 1)^2$.

13. One possibility is $y' = \sin(2\pi y)$.

15. (f)

17. (b)

19. (e)

CHAPTER 2

Section 2.1, page 18

1. Linear and nonhomogeneous
3. Nonlinear
5. Nonlinear
7. Nonlinear
9. Linear and nonhomogeneous
11. (a) $-\infty < t < \infty$ (b) $-\infty < t < \infty$ (c) $-\infty < t < \infty$
13. (a) $3 < t < \infty$ (b) $-2 < t < 2$ (c) $-2 < t < 2$ (d) $-\infty < t < -2$ (e) $-2 < t < 2$
15. $p(t) = -2t$ and $y_0 = 3$
17. $y(t) \equiv 0, a < t < b$

Section 2.2, page 23

1. (a) $y = Ce^{-3t}$ (b) $y = -3e^{-3t}$
3. (a) $y = Ce^{t^2}$ (b) $y = 3e^{-1}e^{t^2} = 3e^{(t^2-1)}$
5. (a) $y = Ct^{-4}$ (b) $y = t^{-4}$
7. (a) $y = Ce^{\sin 2t}$ (b) $y = -2e^{\sin 2t}$
9. (a) $y = Ce^{t^3+3t}$ (b) $y = 4e^{t^3+3t-4}$
11. (a) #2 (b) #3 (c) #1
13. $\alpha = 2$ and $y_0 = 1/4$
15. $y = 3 + e^{-t^2}$
17. $y_0 = 2$ and $c = 1$
19. (a) #2, $y(0) = e^2$ (b) #1, $y(0) = 1$ (c) #4, $y(0) = e$ (d) #3, $y(0) = 1$
21. (a) $y_0 = 1$ and $n = 3$ (b) $y(-1) = e^{-1/4}$

Section 2.3, page 33

1. $y = 0.5 + Ce^{-2t}$
3. $y = te^{-2t} + Ce^{-2t}$
5. $y = (1/4)t^2 + Ct^{-2}$
7. $y = t - 1 + Ce^{-t}$
9. $y = -2 + 3e^{3t}$
11. $y = (e^t - e^{-3t/2})/5$
13. $y = -3 - e^{-0.5\sin t}$
15. $y = (3t + 4 - 3t^{-3})/12, -\infty < t < 0$
17. $p(t) = 2,\ g(t) = 2t + 3$
19. $p(t) = t^{-1},\ g(t) = t^{-1}$
21. $g(t) = 2e^t + \sin t + \cos t,\ y_0 = -1$
23. $\lim_{t\to\infty} y(t) = -1$
25. $y = \begin{cases} 1 + 2e^{-1+\cos t}, & 0 \le t \le \pi \\ -1 + 2e^{1+\cos t} + 2e^{-1+\cos t}, & \pi \le t \le 2\pi \end{cases}$

27. $y = \begin{cases} 3e^{-t^2+t}, & 0 \le t \le 1 \\ 3, & 1 \le t \le 3 \\ t, & 3 \le t \le 4 \end{cases}$

29. $y = e^{t^2}\left[2 + \frac{\sqrt{\pi}}{2}\operatorname{erf}(t)\right]$

Section 2.4, page 42

1. $P(30) = 5000e^{1.5} = 22408.45$
3. (a) $t = (\ln 2)/(\ln 1.06) \approx 11.9$ years (b) $t = (\ln 2)/(2 \ln 1.03) \approx 11.72$ years
   (c) $t = (\ln 2)/(0.06) \approx 11.55$ years
5. (a) $P'_B = (0.04 + 0.004t)P_B,\ P_B(0) = A_0$ (b) $P_B = A_0 \exp(0.04t + 0.002t^2)$
   (c) $t = 10$ years
7. $r = 0.07428\ldots$
9. $t = (5 \ln 2)/(\ln 1.1) \approx 36.36$ days
11. $P = 104{,}000$
13. (a) For Strategy I, $M_{\mathrm{I}} = kP_0$. For Strategy II, $M_{\mathrm{II}} = (e^k - 1)P_0$.
   (b) $\text{Profit}_{\mathrm{I}} = 500{,}000(0.3172)(0.75) = \$118{,}950$
   $\text{Profit}_{\mathrm{II}} = 500{,}000(e^{0.3172} - 1)(0.6) \approx \$111{,}983$
15. (a) $Q_0 = 100(10/3)^{1/3} \approx 149.4$ mg (b) $\tau = (\ln 2)/k \approx 1.73$ months
   (c) $t = (-\ln 0.01)/k \approx 11.48$ months
17. Approximately 42.78 mg/yr
19. $t = (\tau \ln 100/99)/(\ln 2) \approx 58$ million years
21. Contact angle is $\theta_2 - \theta_1 = 2\pi + 2\pi + \pi = 5\pi$. $T_2 = e^{5\pi/10}(100)(9.8) \approx 4714$ N
23. Let $\sin\alpha = 3/5$. The contact angles are: for T_2, $\alpha + \pi/2$; for T_3, $3\alpha + \pi/2$; for T_4, $4\alpha + \pi/2$. Therefore, $T_2 \approx 155.7$ lb, $T_3 \approx 201.4$ lb, $T_4 \approx 229.1$ lb.

Section 2.5, page 52

1. (a) $Q(10) = 20(1 - e^{-0.3}) \approx 5.18$ lb
   (b) $\lim_{t\to\infty} Q(t) = 20$ and the limiting concentration is 0.2 lb/gal.
3. (a) $r = 10 \ln(9/4) \approx 8.11$ gal/min
   (b) It is not possible to achieve this objective.
5. (a) $t = 600$ min
   (b) $c(300) = Q(300)/V(300) = 197.5/400 \approx 0.494$ lb/gal
   (c) $0.5 - (40/700)(1/49) \approx 0.4988$ lb/gal
7. (a) $Q'_A = -1000(Q_A/500{,}000),\ Q_A(0) = 1000$
   $Q'_B = 1000(Q_A/500{,}000) - 1000(Q_B/200{,}000),\ Q_B(0) = 0$
   (b) $Q_A(t) = 1000e^{-t/500}$ lb, $Q_B(t) = (2000/3)(e^{-t/500} - e^{-t/200})$ lb
   (c) The maximum value is attained at $t = (1000/3) \ln 2.5 \approx 305.4$ hr.
   (d) It requires 4056 hours or approximately 169 days.
9. $f(t) = 3 + \sin t$ and therefore, $\tau = \int_0^t (3 + \sin s)\,ds = 1 + 3t - \cos t$.
   $dQ/d\tau = 0.5 - (1/200)Q,\ Q(0) = 10.$
   $Q(\tau) = 100 - 90e^{-\tau/200} = 100 - 90e^{-(1+3t-\cos t)/200}$.
11. (a) In this case, $Q(t)$ changes both by decay and by flow out of the lake. Therefore, $Q(t + \Delta t) - Q(t) \approx -kQ(t)\Delta t - r(Q(t)/V)\Delta t$. Thus, the initial value problem is $Q' = -(k + (r/V))Q,\ Q(0) = 5$. For this problem, $k = (\ln 2)/18,\ r = 60{,}000$, $V = 1{,}200{,}000$ where t is in hours. (b) $t \approx 104.06$ hours
13. The serving time is approximately 35 minutes after removal from the oven.
15. The times are the same.

CHAPTER 3

Section 3.1, page 64

1. (a) $f(t,y) = (1 - 2t\cos y)/3$ (b) $f_y(t,y) = (2t\sin y)/3$ (c) The entire ty-plane
3. (a) $f(t,y) = -2t/(1+y^2)$ (b) $f_y(t,y) = 4ty/(1+y^2)^2$ (c) The entire ty-plane
5. (a) $f(t,y) = -ty^{1/3} + \tan t$ (b) $f_y(t,y) = -(1/3)ty^{-2/3}$
   (c) $-\pi/2 < t < \pi/2, \quad 0 < y < \infty$
7. (a) $f(t,y) = (2+\tan t)/\cos y$ (b) $f_y(t,y) = (2+\tan t)\sec y \tan y$
   (c) $-\pi/2 < t < \pi/2, \quad -\pi/2 < y < \pi/2$
9. One such example is $y' = [t(t-4)(y+1)(y-2)]^{-1}, y(1) = 0$
11. $\bar{y}(t) = 2/\sqrt{1-(t-1)}$. Therefore, $\bar{y}(0) = \sqrt{2}$.
13. (a) $z_1(t) = y(t+2)$. Therefore, $z_1(-5) = y(-3) = 2$.
    (b) $z_2(t) = y(t-2)$. Therefore, $z_2(3) = y(1) = 0$.

Section 3.2, page 71

1. (a) $y^2 = 4 - 2\cos t; \quad y = -\sqrt{4 - 2\cos t}$ (b) $-\infty < t < \infty$
3. (a) $y^2 + 2y + 2(t-1) = 0; \quad y = -1 + \sqrt{2-t}$ (b) $-\infty < t \le 2$
5. (a) $y^{-2} + t^2 = 1/4; \quad y = 2/\sqrt{1-4t^2}$ (b) $-1/2 < t < 1/2$
7. (a) $\tan^{-1} y = t - \pi/2; \quad y = \tan(t - \pi/2)$ (b) $0 < t < \pi$
9. (a) $|(y+1)/(y-1)| = 3e^{t^2}; \quad y = (3e^{t^2} - 1)/(3e^{t^2} + 1)$ (b) $-\infty < t < \infty$
11. (a) $e^y = e^t + e - 1; \quad y = \ln(e^t + e - 1)$ (b) $-\infty < t < \infty$
13. (a) $\tan y = e^{-t}; \quad y = \tan^{-1}(e^{-t})$ (b) $-\infty < t < \infty$
15. (a) $ye^y = 2e^2 + (1 - (t-2)^2)/2$; there is no explicit solution.
    (b) Approximately $-3.5 < t < 7.5$
17. $\alpha = 2/3, n = 3, y_0 = 1$
19. $(1+y)e^y y' + 2t - \cos t = 0, y(0) = 0$
21. (a) $S + K\ln S = -\alpha t + S_0 + K\ln S_0$ (b) $K \approx 1.769, \alpha \approx 0.759$ (c) $t \approx 10.41$
23. $y = -2 + \tan((t^2/2) - \pi/4), -\sqrt{3\pi/2} < t < \sqrt{3\pi/2}$
25. The half life is $\tau = 3/(2kQ_0^2)$ and it does depend on Q_0.
27. (a) It is nonlinear and separable.
    (b) The two curves are the same and are
    (c) The two curves are different and are

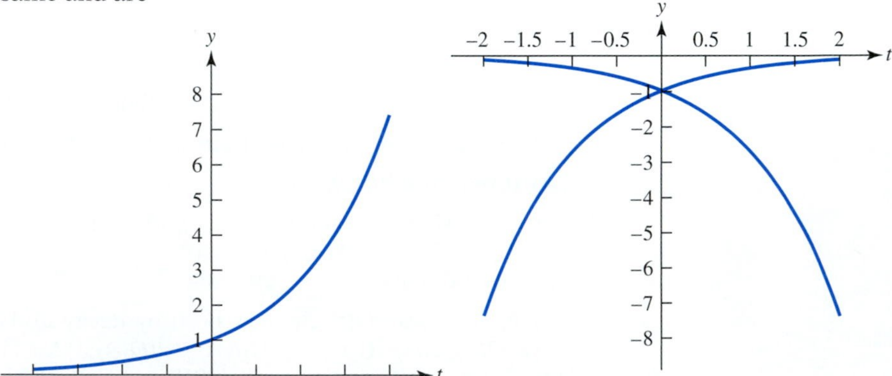

29. The appropriate graph is shifted 3 units to the right.

31. $f(y) = 3/(\sin y + y\cos y)$. An implicit solution is $y\sin y = t^3 - 1$.

Section 3.3, page 80

1. The equation is separable and exact. $y^2 = 6 + 4t - 2t^3$; $\quad y = -\sqrt{6 + 4t - 2t^3}$
3. The equation is separable and exact. $\tan^{-1} y = t^3 + t + \pi/4$; $\quad y = \tan(t^3 + t + \pi/4)$
5. The equation is exact. $e^{t+y} + t^3 + y^2 = 1$
7. The equation is exact. $2y\sin(t + y) + 2ty + t^2 + y^2 = 0$
9. $M(t, y) = 2ty + (y^3/3)\cos t + m(t)$; $m(t)$ is arbitrary.
11. $M(t, y) = 3t^2y + e^t, N(t, y) = t^3 + 2y, y_0 = \pm 2$

Section 3.4, page 83

1. (a) The equation is separable and therefore exact as well.

   (b) Using $v = y^{-1}$ we obtain $v' + 2v = 1, v(0) = 1$. Solving for v and transforming back, $y = 2/(1 + e^{-2t})$.

   (c) $-\infty < t < \infty$
3. (a) The equation is neither separable nor exact.

   (b) Using $v = y^{-1}$ we obtain $v' - v = -e^t, v(-1) = -1$. Solving for v and transforming back, $y = -1/((t + 1)e^t + e^{t+1})$.

   (c) $-(1 + e) < t < \infty$
5. (a) The equation is neither separable nor exact.
   (b) Using $v = y^3$ we obtain $tv' + 3v = 3t^3, v(1) = 1$. Solving for v and transforming back, $y = (0.5(t^3 + t^{-3}))^{1/3}$.
   (c) $0 < t < \infty$
7. Using $u = e^y$ we obtain $u' - (2/t)u = 1$ which has solution $u = -t + Ct^2$. Therefore, after imposing the initial condition, $y = \ln(2t^2 - t)$.
9. $y_0 = 3$ and $q(t) = e^t$

Section 3.5, page 87

1. Using $v = P^{-1}$ we obtain $v' + rv = r/P_e, v(0) = 1/P_0$. Solving for v and transforming back, $P = P_0P_e/(P_0 - (P_0 - P_e)e^{-rt})$.
3. (a) $P^2 - P_eP - (M/r)P_e = 0$

   (b) Let $P_e = 1$ and define $x = M/r$. For a fixed value of x, the equilibrium solutions are given by $P = (1 \pm \sqrt{1 + 4x})/2$. The graph is

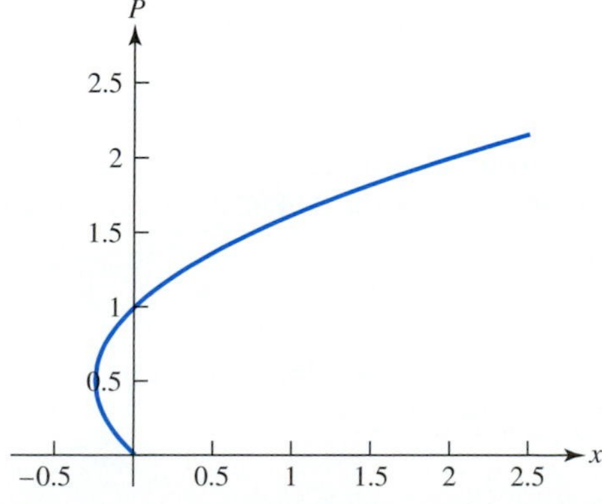

   There are two nonnegative equilibrium solutions when $-1/4 < x \leq 0$ and one when $x > 0$.

5. $P = (3 + e^{-t/2})/(4 + 4e^{-t/2})$
7. $P = P_0/(P_0 - (P_0 - 1)e^{-s(t)})$ where $s(t) = 0.5(t + (1 - \cos 2\pi t)/\pi)$
11. (a) Note that $(A - B)' = 0$. Hence, $A(t) - B(t)$ is constant at 3 moles.
    (b) $A' = kA(3 - A), A(0) = 5$
    (c) $A(4) = 3.195$ moles and $B(4) = 0.195$ moles

Section 3.6, page 94

1. $v = -(mg/k)(1 - e^{-kt/m})$ and therefore, $t = (m/k) \ln 2$
3. Terminal velocity is $-\sqrt{mg/\kappa}$ and therefore we need $\kappa = 0.930$ lb-sec^2/ft^2.
5. (a) $\kappa \approx 0.247$ lb-sec^2/ft^2 (b) 562.4 ft
7. $y(t_m) = \int_0^{t_m} v(t)\,dt = -(mg/k)t_m + (m/k)[v_0 + (mg/k)][1 - e^{-(k/m)t_m}]$
9. (a) 187.3 ft/sec or approximately 127.7 mph
    (b) 128.18 ft/sec or approximately 87.4 mph
11. Assuming the drag chute acts instantaneously, it takes about 5.56 seconds for the dragster to slow to 50 mph. In the delayed case, it takes about 6.25 seconds.

Section 3.7, page 100

1. The transformed equation is $dv/dx = -(k/m)x^2$. Thus, $v(x) = -(k/m)(x^3/3) + v_0$, leading to a stopping position of $x_f = [3(m/k)v_0]^{1/3}$.
3. The transformed equation is $mv(dv/dx) = -ke^{-x}$. If $v_0^2 \geq 2k/m$, then $x_f = \infty$. Otherwise, $x_f = -\ln[1 - mv_0^2/(2k)]$.
5. $v = v_0 e^{-(\kappa/m)d}$
7. From Newton's second law of motion, $-mv(dv/dx) = xF/\sqrt{x^2 + h^2}$. Therefore, $v = -\left[(2F/m)\left(\sqrt{D^2 + h^2} - \sqrt{(D^2/9) + h^2}\right)\right]^{1/2}$
9. (a) $mv(dv/dx) + \kappa_0 xv^2 = 0$
    (b) $\kappa_0 = (2m/d^2) \ln 100$
11. 2044 m/s
13. 7.39 rad/sec

Section 3.8, page 108

1. (a) $y = t^2 - t$
    (b) $y_{k+1} = y_k + h(2t_k - 1)$
    (c) $y_1 = 0.1, y_2 = 0.22, y_3 = 0.36$
    (d) $y(t_1) = 0.11, y(t_2) = 0.24, y(t_3) = 0.39$
3. (a) $y = e^{-t^2/2}$
    (b) $y_{k+1} = y_k - ht_k y_k$
    (c) $y_1 = 1, y_2 = 0.99, y_3 = 0.9702$
    (d) $y(t_1) = 0.9950\ldots, y(t_2) = 0.9802\ldots, y(t_3) = 0.9559\ldots$
5. (a) $y = 1/(1 - t)$
    (b) $y_{k+1} = y_k + hy_k^2$
    (c) $y_1 = 1.1, y_2 = 1.221, y_3 = 1.370\ldots$
    (d) $y(t_1) = 1.111\ldots, y(t_2) = 1.25, y(t_3) = 1.428\ldots$
7. $t_0 = 1, y_0 = 1,\ n = 2, \alpha = -2$

9. 

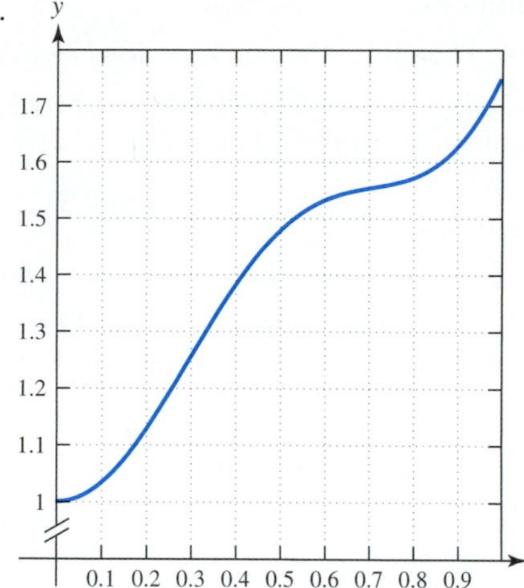

11. Using $h = 0.01$,

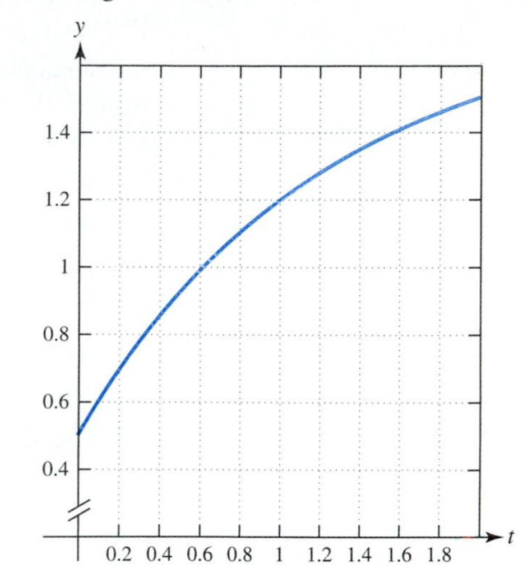

CHAPTER 4

Introduction, page 120

1. (a) $Y \approx 1.603$ ft

   (b) $\omega^2 = (\rho_l g)/(\rho L) = (1996.8)/(100)$. Therefore, $\omega = 4.468\ldots$ and $y(t) = -0.25\cos\omega t \approx -0.25\cos 4.47t$

   (c) Maximum depth $= Y + |y_0| \approx 1.853$ ft

3. (a) Drum #1 bobs more rapidly. (b) Drum #1 bobs more rapidly.

5. (a) $my'' = mg - \rho_l Vg$, or $\rho V y'' = \rho Vg - \rho_l Vg$. Therefore, $y'' = (1 - \rho_l/\rho)g$.

   (b) $y = (1 - \rho_l/\rho)(gt^2/2) + y_0' t + y_0$

Section 4.1, page 123

1. $-\infty < t < \infty$

3. $-\infty < t < -1$

5. An example is $y'' + y = 0, y(0) = 1, y'(0) = 3$.

7. An example is $y'' + (1+t)^{-1}y' + (t-5)^{-1}y = 0, y(0) = 1, y'(0) = 3$.

9. No, because Theorem 4.1 guarantees that a unique solution exists on $-4 < t < 4$.

11. (a) Note that $y''(0) = 1$. Therefore, the solution is decreasing and concave up at $t = 0$. Graph B (b) Graph D (c) Graph A (d) Graph C

Section 4.2, page 133

1. (a) Both are solutions.

   (b) $W(t) = -8$. Therefore, the two functions form a fundamental set.

   (c) The general solution is $y = c_1 e^{2t} + c_2(2e^{-2t})$. The solution of the initial value problem is $y = e^{-2t}$.

3. (a) $y_1(t)$ is not a solution, but $y_2(t)$ is.

5. (a) Both are solutions.

   (b) $W(t) = e^{4t}$. Therefore, the two functions form a fundamental set.

   (c) The general solution is $y = c_1 e^{2t} + c_2 te^{2t}$. The solution of the initial value problem is $y = 2e^{2t} - 4te^{2t}$.

7. (a) Both are solutions.

   (b) $W(t) = \sqrt{3}/4$. Therefore, the two functions form a fundamental set.

   (c) The general solution is $y = c_1 \sin[(t/2) + (\pi/3)] + c_2 \sin[(t/2) - (\pi/3)]$. The solution of the initial value problem is $y = 2\sin[(t/2) + (\pi/3)] + 2\sin[(t/2) - (\pi/3)]$ or $y = 2\sin(t/2)$.

9. (a) Both are solutions.

   (b) $W(t) = -t^{-1} \ln 3$. Therefore, the two functions form a fundamental set.

   (c) The general solution is $y = c_1 \ln t + c_2 \ln 3t$. The solution of the initial value problem is $y = 18 \ln t - 9 \ln 3t$ or equivalently, $y = 9 \ln(t/3)$.

11. (a) Both are solutions.

    (b) $W(t) = 4t$. Therefore, the two functions form a fundamental set.

    (c) The general solution is $y = c_1 t^3 + c_2(-t^{-1})$. The solution of the initial value problem is $y = -0.5t^3 + 0.5t^{-1}$.

13. (a) Both are solutions.

    (b) $W(t) = -3$. Therefore, the two functions form a fundamental set.

    (c) The general solution is $y = c_1(t+1) + c_2(-t+2)$. The solution of the initial value problem is $y = (t+1) + 2(-t+2)$ or, $y = 5 - t$.

15. (a) $y_1(t)$ is not a solution; $y_2(t)$ is not a solution.

17. (b) $c_1 = 2 + \ln 3$, $c_2 = 1$

19. $\alpha = 0$, $\beta = -9$

21. $W(2) = 4e^{-3/2}$

23. $W(4) = -3$

Section 4.3, page 142

1. (a) $W(1) = 0$. Therefore, the functions do not form a fundamental set.

   (b) No; because $-0.5y_1(t)$ and $y_2(t)$ satisfy the same initial conditions, Theorem 4.1 says they are the same functions.

3. (a) $W(0) = 1$. Therefore, the functions do form a fundamental set.

   (b) Yes, by Theorem 4.7.

5. (b) $y_1(1) = e^2$, $y_1'(1) = 2e^2$, $y_2(1) = e^{-2}$, $y_2'(1) = -2e^{-2}$

   (c) $W(1) = -4$. Therefore, the functions form a fundamental set.

7. (b) $y_1(1) = 0$, $y_1'(1) = 3$, $y_2(1) = 2$, $y_2'(1) = 0$

   (c) $W(1) = -6$. Therefore, the functions form a fundamental set.

9. (b) $y_1(2) = e^{-6}$, $y_1'(2) = -3e^{-6}$, $y_2(2) = 1$, $y_2'(2) = -3$

   (c) $W(2) = 0$. Therefore, the functions do not form a fundamental set.

11. $[\bar{y}_1, \bar{y}_2] = [y_1, y_2]A$, where $A = \begin{bmatrix} 2 & 1 \\ -1 & 1 \end{bmatrix}$. Since the determinant of A is nonzero, Theorem 4.8 guarantees that $\{\bar{y}_1, \bar{y}_2\}$ is a fundamental set.

13. $[\bar{y}_1, \bar{y}_2] = [y_1, y_2]A$, where $A = \begin{bmatrix} 0 & 2 \\ 1 & -1 \end{bmatrix}$. Since the determinant of A is nonzero, Theorem 4.8 guarantees that $\{\bar{y}_1, \bar{y}_2\}$ is a fundamental set.

15. Linearly dependent since $2f_1 - f_2 = 0$.

17. Linearly independent.

19. Linearly dependent since $-f_1 + 2f_2 + f_3 = 0$.

21. (a) Linearly dependent. (b) Linearly dependent. (c) Linearly independent.

27. $A = \begin{bmatrix} 2 & 1 \\ -1 & 1 \end{bmatrix}$

Section 4.4, page 149

1. (a) The general solution is $y = c_1 e^{-2t} + c_2 e^{t}$.
   (b) The solution of the initial value problem is $y = 2e^{-2t} + e^{t}$.
   (c) $\lim_{t\to-\infty} y(t) = \infty$ and $\lim_{t\to\infty} y(t) = \infty$
3. (a) The general solution is $y = c_1 e^{t} + c_2 e^{3t}$.
   (b) The solution of the initial value problem is $y = -2e^{t} + e^{3t}$.
   (c) $\lim_{t\to-\infty} y(t) = 0$ and $\lim_{t\to\infty} y(t) = \infty$
5. (a) The general solution is $y = c_1 e^{-t} + c_2 e^{t}$.
   (b) The solution of the initial value problem is $y = e^{-t}$.
   (c) $\lim_{t\to-\infty} y(t) = \infty$ and $\lim_{t\to\infty} y(t) = 0$
7. (a) The general solution is $y = c_1 e^{-3t} + c_2 e^{-2t}$.
   (b) The solution of the initial value problem is $y = -e^{-3t} + 2e^{-2t}$.
   (c) $\lim_{t\to-\infty} y(t) = -\infty$ and $\lim_{t\to\infty} y(t) = 0$
9. (a) The general solution is $y = c_1 e^{-2t} + c_2 e^{2t}$.
   (b) The solution of the initial value problem is $y = 0e^{-2t} + 0e^{2t}$, or $y(t) \equiv 0$.
   (c) $\lim_{t\to-\infty} y(t) = 0$ and $\lim_{t\to\infty} y(t) = 0$
11. (a) The general solution is $y = c_1 + c_2 e^{1.5t}$.
   (b) The solution of the initial value problem is $y = 3 + 0e^{1.5t}$ or $y(t) \equiv 3$.
   (c) $\lim_{t\to-\infty} y(t) = 3$ and $\lim_{t\to\infty} y(t) = 3$
13. (a) The general solution is $y = c_1 e^{(-2-\sqrt{2})t} + c_2 e^{(-2+\sqrt{2})t}$.
   (b) The solution of the initial value problem is $y = -\sqrt{2}\, e^{(-2-\sqrt{2})t} + \sqrt{2}\, e^{(-2+\sqrt{2})t}$.
   (c) $\lim_{t\to-\infty} y(t) = -\infty$ and $\lim_{t\to\infty} y(t) = 0$
15. (a) The general solution is $y = c_1 e^{-t/\sqrt{2}} + c_2 e^{t/\sqrt{2}}$.
   (b) The solution of the initial value problem is $y = -2e^{-t/\sqrt{2}}$.
   (c) $\lim_{t\to-\infty} y(t) = -\infty$ and $\lim_{t\to\infty} y(t) = 0$
17. (a) $y_2(t) = e^{3t}$ (b) $\alpha = -2,\ \ \beta = -3$ (c) $y = e^{-t} + 2e^{3t}$
19. $y = c_1 e^{2t} + c_2 e^{3t} + c_3$
21. (a) $r = 0.5(r_0 - \Omega^{-1} r'_0)e^{-\Omega t} + 0.5(r_0 + \Omega^{-1} r'_0)e^{\Omega t}$ (b) Approximately 1.45 seconds
23. (a) $k/m = 3\Omega/2$ (b) $r(t) = (2r'_0/5\Omega)[e^{0.5\Omega t} - e^{-2\Omega t}]$

Section 4.5, page 157

1. (a) The general solution is $y = c_1 e^{-t} + c_2 t e^{-t}$.
   (b) The solution of the initial value problem is $y = te^{-(t-1)}$.
   (c) $\lim_{t\to-\infty} y(t) = -\infty$ and $\lim_{t\to\infty} y(t) = 0$
3. (a) The general solution is $y = c_1 e^{-3t} + c_2 t e^{-3t}$.
   (b) The solution of the initial value problem is $y = (2 + 4t)e^{-3t}$.
   (c) $\lim_{t\to-\infty} y(t) = -\infty$ and $\lim_{t\to\infty} y(t) = 0$

5. (a) The general solution is $y = c_1 e^{t/2} + c_2 t e^{t/2}$.
   (b) The solution of the initial value problem is $y = (-6 + 2t)e^{(t-1)/2}$.
   (c) $\lim_{t\to-\infty} y(t) = 0$ and $\lim_{t\to\infty} y(t) = \infty$
7. (a) The general solution is $y = c_1 e^{t/4} + c_2 t e^{t/4}$.
   (b) The solution of the initial value problem is $y = (-4 + 4t)e^{t/4}$.
   (c) $\lim_{t\to-\infty} y(t) = 0$ and $\lim_{t\to\infty} y(t) = \infty$
9. (a) The general solution is $y = c_1 e^{5t/2} + c_2 t e^{5t/2}$.
   (b) The solution of the initial value problem is $y = (2 + t)e^{2.5(t+2)}$.
   (c) $\lim_{t\to-\infty} y(t) = 0$ and $\lim_{t\to\infty} y(t) = \infty$
11. $\alpha = -1/2,\ y_0 = 0,\ y_0' = 4,\ y(t) = 4te^{-t/2}$
13. $y(t) = (2 - t)e^{-t/2}$ and therefore, $y(0) = 2, y'(0) = -2$
15. (a) $y_2(t) = t \ln|t|,\ \ t \neq 0$
    (b) $W(t) = t$. The Wronskian is nonzero on $-\infty < t < 0$ and $0 < t < \infty$.
    (c) $p(t) = -1/t,\ q(t) = 1/t^2$; continuous on $-\infty < t < 0$ and $0 < t < \infty$
17. (a) $y_2(t) = (t + 1)^3$
    (b) $W(t) = (t + 1)^4$. The Wronskian is nonzero on $-\infty < t < -1$ and $-1 < t < \infty$.
    (c) $p(t) = -4/(t + 1),\ q(t) = 6/(t + 1)^2$; continuous on $-\infty < t < -1$ and $-1 < t < \infty$
19. (a) $y_2(t) = 1/(t - 2)^2$
    (b) $W(t) = -4/(t - 2)$. The Wronskian is nonzero on $-\infty < t < 2$ and $2 < t < \infty$.
    (c) $p(t) = 1/(t - 2),\ q(t) = -4/(t - 1)^2$; continuous on $-\infty < t < 2$ and $2 < t < \infty$

Section 4.6, page 166

1. (a) $1 + \sqrt{3}i$ (b) $-2 + 2i$ (c) $-1 - 2i$ (d) $-\left(\sqrt{6} + \sqrt{2}i\right)/8$ (e) $-2 + 2\sqrt{3}i$
3. (a) $\lambda = \pm 2i$ (b) $y = c_1 \cos 2t + c_2 \sin 2t$ (c) $y = -0.5 \cos 2t - 2 \sin 2t$
5. (a) $\lambda = \pm i/3$ (b) $y = c_1 \cos(t/3) + c_2 \sin(t/3)$ (c) $y = 2\sqrt{3}\cos(t/3) + 2\sin(t/3)$
7. (a) $\lambda = \left(-1 \pm i\sqrt{3}\right)/2$
   (b) $y = c_1 e^{-t/2} \cos\left(\sqrt{3}t/2\right) + c_2 e^{-t/2} \sin\left(\sqrt{3}t/2\right)$
   (c) $y = -2e^{-t/2} \cos\left(\sqrt{3}t/2\right) - 2\sqrt{3}e^{-t/2} \sin\left(\sqrt{3}t/2\right)$
9. (a) $\lambda = (-1 \pm i)/3$
   (b) $y = c_1 e^{-t/3} \cos(t/3) + c_2 e^{-t/3} \sin(t/3)$
   (c) $y = -e^{-(t-3\pi)/3} \sin(t/3)$
11. (a) $\lambda = \sqrt{2} \pm i$
    (b) $y = c_1 e^{\sqrt{2}t} \cos t + c_2 e^{\sqrt{2}t} \sin t$
    (c) $y = -0.5e^{\sqrt{2}t} \cos t + 1.5\sqrt{2}e^{\sqrt{2}t} \sin t$
13. $a = 0,\ b = 1,\ y_0 = (\sqrt{2} - 2)/2,\ y_0' = (\sqrt{2} + 2)/2$
15. $a = 4,\ b = 5,\ y_0 = 1,\ y_0' = -3$
17. $a = 0,\ b = \pi^2,\ y_0 = -1,\ y_0' = -\sqrt{3}\pi$

19. $y = \sqrt{2}\cos(\pi t - 7\pi/4)$

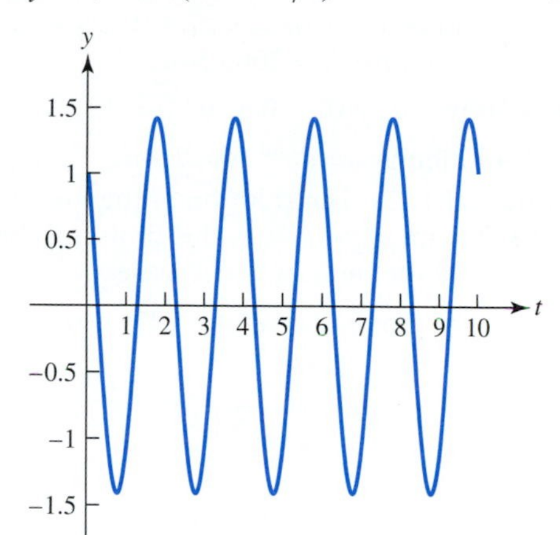

21. $y = 2e^{-t}\cos(t - 2\pi/3)$

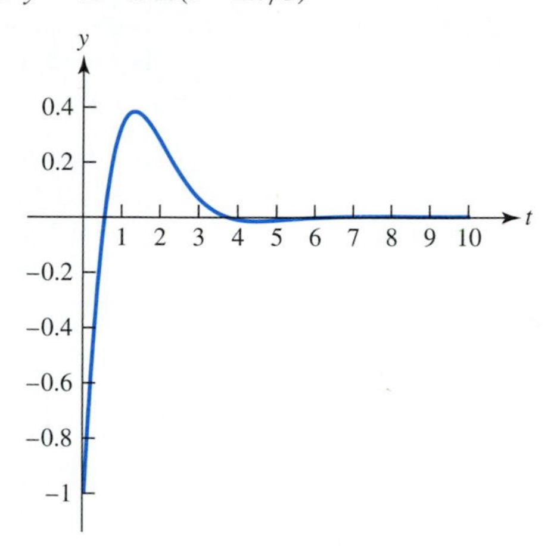

23. $a = 0,\ b = \pi^2/4,\ y_0 = 2,\ y_0' = 0$

25. $a = 0,\ b = 4,\ y_0 = 0.5\cos(5\pi/6),\ y_0' = \sin(5\pi/6)$

29. $y = c_1 e^{it} + c_2 e^{-5it}$

Section 4.7, page 179

1.

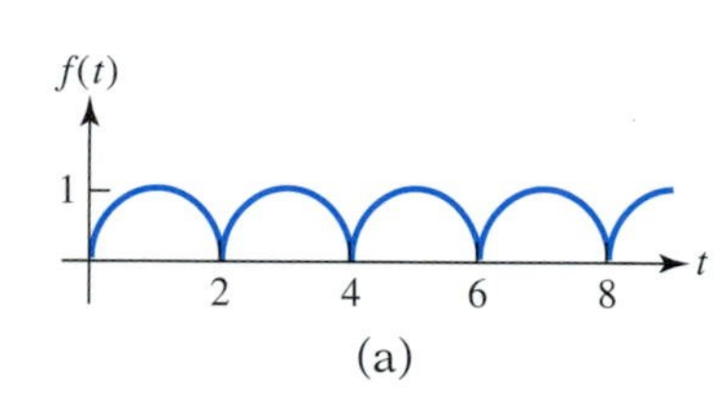

(a)

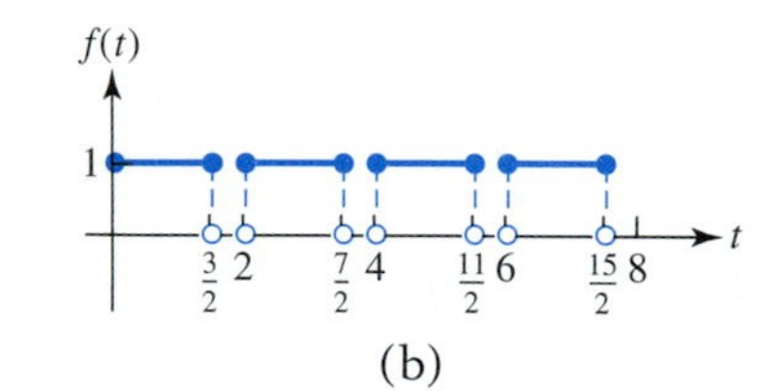

(b)

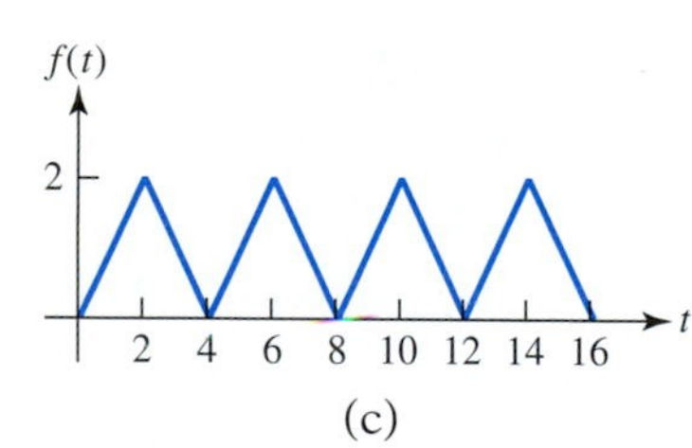

(c)

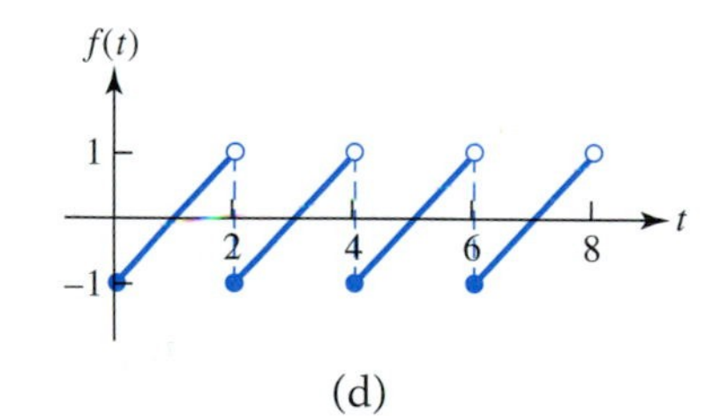

(d)

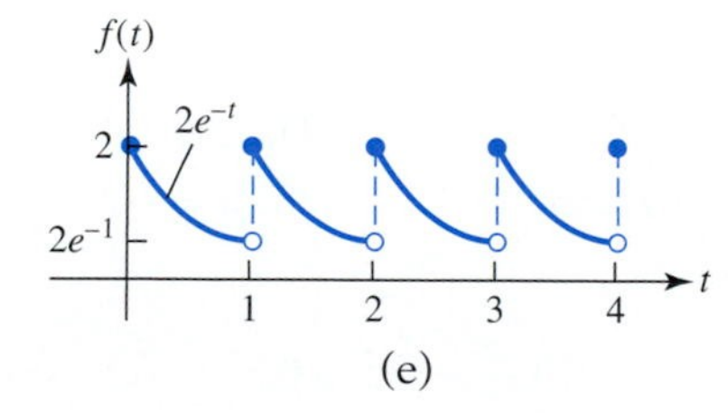

(e)

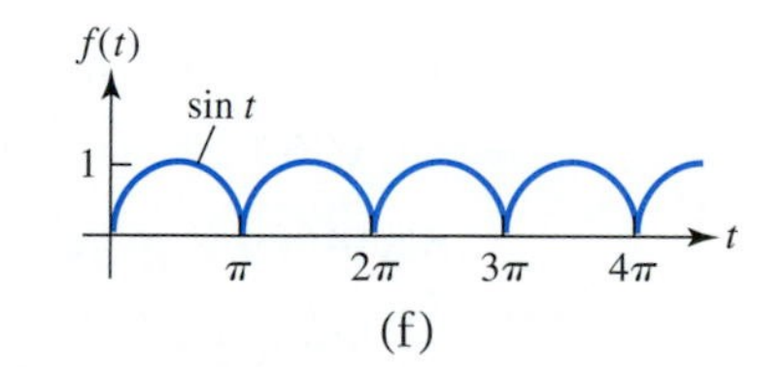

(f)

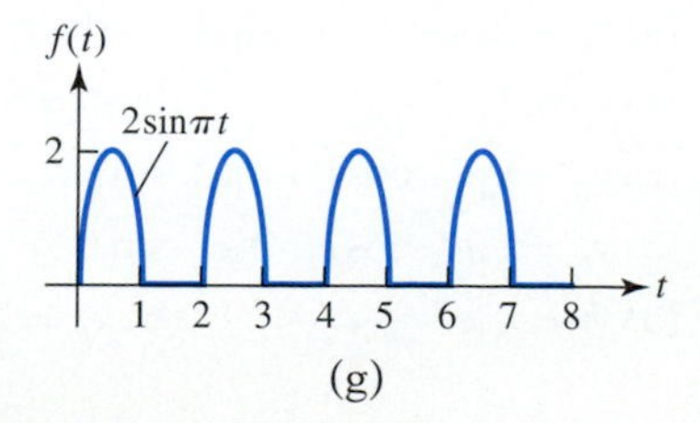

(g)

3. The initial value problem is $my'' + ky = 0,\ y(0) = 0,\ y'(0) = 2$. The general solution is $y = c_1 \cos \omega t + c_2 \sin \omega t$, where $\omega = \sqrt{k/m}$. The solution of the initial value problem is $y = (2/\omega) \sin \omega t$. Therefore, $k = 2000$ N/m.

5. (a) $10y'' + 7y' + 100y = 0,\ y(0) = 0.5,\ y'(0) = 1$

   (b) The general solution is $y = c_1 e^{\alpha t} \cos \beta t + c_2 e^{\alpha t} \sin \beta t$ where $\alpha = -0.35$ and $\beta = \sqrt{0.49 - 40}/2 \approx 3.1428$. The solution of the initial value problem requires $c_1 = 0.5, c_2 \approx 0.3739$. $\lim_{t\to\infty} y(t) = 0$; this limit is to be expected since damping dissipates energy causing the motion to decrease.

   (c) Zooming in on the graph, τ is about 5.25 seconds.

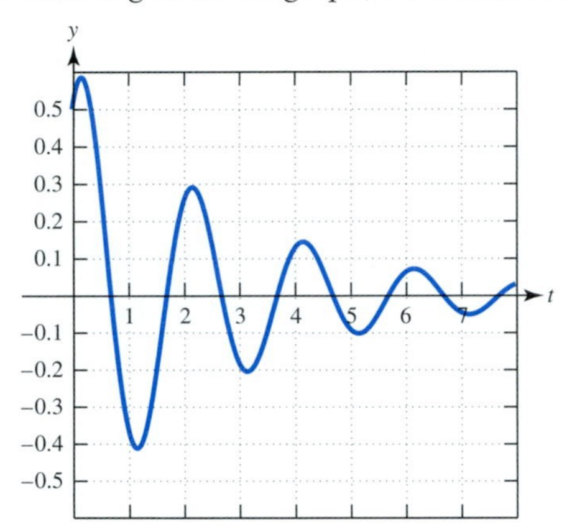

7. (a) $y(t) = (\lambda_2 e^{\lambda_1 t} - \lambda_1 e^{\lambda_2 t})y_0/(\lambda_2 - \lambda_1)$ (c) $\lim_{\gamma\to\infty} y(t) = y_0$. As damping increases, motion is suppressed; the system "locks up" and tends to remain at the initial displacement.

9. $k = 32.67\ldots$ N/m, $\gamma = 0.5$ kg/s

Section 4.8, page 186

1. (b) $y_C = c_1 e^{-t} + c_2 e^{3t}$ (c) $y = 1.5e^{-t} + 0.5e^{3t} + 3t - 1$
3. (b) $y_C = c_1 e^{-t} + c_2 e^{2t}$ (c) $y = e^{-t} - 3e^{2t} + 2e^{4t}$
5. (b) $y_C = c_1 e^{-t} + c_2$ (c) $y = 2e^{-(t-1)} + t^2 - 2t$
7. (b) $y_C = c_1 \cos t + c_2 \sin t$ (c) $y = -\cos t - 2 \sin t + 2t + \cos 2t$
9. (b) $y_C = c_1 e^t \cos t + c_2 e^t \sin t$ (c) $y = -5e^t \cos t - 5e^t \sin t + 5(t+1)^2$
11. (b) $y_C = c_1 e^t + c_2 te^t$ (c) $y = -2e^t + 4te^t + t^2 e^t/2$
15. $y_P = 2u_2 + (2/3)u_3$
17. $g(t) = 5e^{2t} + t^2 - 2t - 2$
19. $g(t) = 3e^t + 6t$
21. $g(t) = (2|t| - 1)\cos t - \sin^2 t$
23. $\alpha = 1,\ \beta = 0,\ g(t) = 2 + 2t$
25. $\alpha = -2,\ \beta = 2,\ g(t) = e^t - 2\cos t + \sin t$

Section 4.9, page 195

1. (a) $y_C = c_1 e^{-2t} + c_2 e^{2t}$ (b) $y_P = -t^2 - 0.5$ (c) $y = c_1 e^{-2t} + c_2 e^{2t} - t^2 - 0.5$
3. (a) $y_C = c_1 \cos t + c_2 \sin t$ (b) $y_P = 4e^t$ (c) $y = c_1 \cos t + c_2 \sin t + 4e^t$
5. (a) $y_C = c_1 e^{2t} + c_2 te^{2t}$ (b) $y_P = (1/2)t^2 e^{2t}$
7. (a) $y_C = c_1 e^{-t} \cos t + c_2 e^{-t} \sin t$ (b) $y_P = (t^3 - 3t^2 + 3t)/2$
9. (a) $y_C = c_1 e^{-t} \cos t + c_2 e^{-t} \sin t$ (b) $y_P = e^{-t} + (\cos t + 2 \sin t)/5$
11. (a) $y_C = c_1 e^{t/2} + c_2 e^{2t}$ (b) $y_P = 2te^{t/2}$

13. (a) $y_C = c_1 e^{t/3} + c_2 t e^{t/3}$ (b) $y_P = (1/6)t^3 e^{t/3}$
15. (a) $y_C = c_1 e^{-2t}\cos t + c_2 e^{-2t}\sin t$ (b) $y_P = 2e^{-2t} + (\cos t + \sin t)/8$
17. (a) $y_C = c_1 \cos 3t + c_2 \sin 3t$
    (b) $y_P = t(A_2t^2 + A_1t + A_0)\cos 3t + t(B_2t^2 + B_1t + B_0)\sin 3t + C\cos t + D\sin t$
19. (a) $y_C = c_1 e^t \cos t + c_2 e^t \sin t$
    (b) $y_P = Ae^{-t}\cos 2t + Be^{-t}\sin 2t + C_1t + C_0 + e^{-t}(D_1t + D_0)\cos t + e^{-t}(E_1t + E_0)\sin t$
21. (a) $y_C = c_1 \cos 2t + c_2 \sin 2t$
    (b) $y_P = At\cos 2t + Bt\sin 2t + C + D\cos 4t + E\sin 4t$
23. $\alpha = -1,\ \beta = -2,\ y = c_1e^{-t} + c_2e^{2t} - 2t + 1$
25. $\alpha = 4,\ \beta = 4,\ y = c_1e^{-2t} + c_2te^{-2t} - (4\cos t - 3\sin t)/5$
27. $\alpha = 2,\ \beta = 5,\ y = c_1e^{-t}\cos 2t + c_2e^{-t}\sin 2t + 2e^{-t}$
29. $\alpha = 0,\ \beta = -4$
31. (a) Figure C (b) Figure E (c) Figure A (d) Figure B (e) Figure D
35. (a) $y_P = -(i/2)e^{it}$ 37. (a) $y_P = -(i/4)te^{2it}$ 39. (a) $y_P = 0.2(1-2i)e^{(1+i)t}$

Section 4.10, page 203

1. (a) $y_C = c_1 \cos 2t + c_2 \sin 2t$
   (b) $y_P = -(t/2)\cos 2t + (1/8)\sin 4t\cos 2t + (1/4)\sin^3 2t$
3. (a) $y_C = c_1e^t + c_2t^2e^t$ (b) $y_P = t + 1$
5. (a) $y_C = c_1e^{-t} + c_2e^t$ (b) $y_P = -(1/4)e^t + (t/2)e^t$
7. (a) $y_C = c_1e^t + c_2te^t$ (b) $y_P = (t^2/2)e^t$
9. (a) $y_C = c_1 \sin t + c_2 t\sin t$ (b)$y_P = (t^3/6)\sin t$
11. (a) $y_C = c_1t + c_2e^t$ (b) $y_P = [(t^2/2) - t]e^t$
13. (a) $y_C = c_1(t-1)^2 + c_2(t-1)^3$ (b) $y_P = (3t-2)/6$
17. $\alpha = 0,\ \beta = -1,\ y_0 = 1,\ y_0' = -1$

Section 4.11, page 214

3. (b) $y'' + 100y = 2e^{-t},\ y(0) = 0,\ y'(0) = 0,\ y = (2/101)[-\cos 10t + 0.1\sin 10t + e^{-t}]$
   (c) $|y|_{\max} \approx 0.035$ m

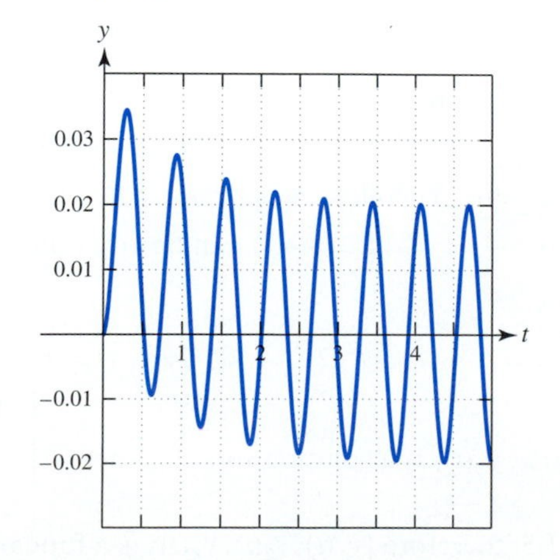

5. (b) $y'' + 100y = 2,\ \ y(0) = 0,\ \ y'(0) = 0, \quad 0 \le t \le \pi$
$y = (1/50)[1 - \cos 10t], \quad 0 \le t \le \pi$
$y'' + 100y = 0,\ \ y(\pi) = 0,\ \ y'(\pi) = 0, \quad \pi \le t$
$y(t) \equiv 0, \quad \pi \le t$

(c) $|y|_{\max} = 0.04$ m

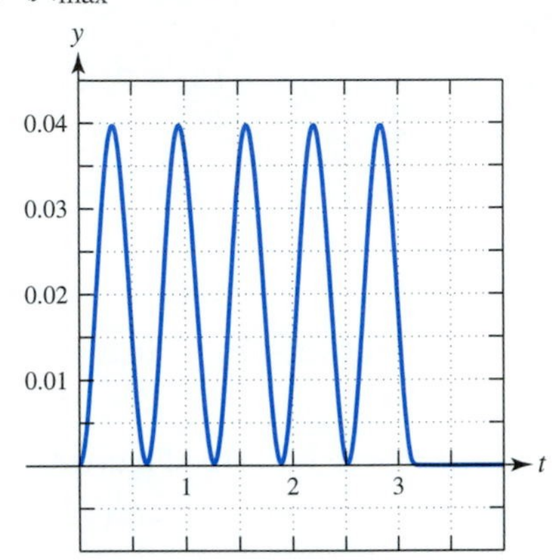

7. (a) $y = (1/60)e^{-2t}[9\cos 6t - 13\sin 6t] + (1/20)[-3\cos 8t + 4\sin 8t]$
(b) $\lim_{t\to\infty} y(t)$ does not exist. For large values of t, $y \approx (1/20)[-3\cos 8t + 4\sin 8t]$

9. (a) $y = (1/74)e^{-2t}[30\cos 6t + 5\sin 6t] + (1/74)[-30\cos 6t + 5\sin 6t]$
(b) $\lim_{t\to\infty} y(t)$ does not exist. For large values of t, $y \approx (1/74)[-30\cos 6t + 5\sin 6t]$

13. (b) Approximately 52.42 ft/sec (c) Approximately 85.86 ft

15. $I = (12/7)[\cos(t/2) - \cos 3t]$

17. $V = 2e^{-t}(1 - \cos t)$

CHAPTER 5

Section 5.1, page 223

7. $-1 < t < \pi/2$

9. $-\infty < t < \infty$

11. $\lambda = 0,\ \ \lambda = 1,\ \ \lambda = -1$

13. $\lambda = 1,\ \ \lambda = -1$

Section 5.2, page 229

1. (a) The general solution is $y = c_2t^2 + c_1t + c_0$.

3. $W(t) = e^{-t}$

5. $W(t) = 2t^{-3}$

7. $y = 2 - e^t + 2e^{-t}$

9. $y = -(1 + \pi) + 4t + \cos t + 3\sin t$

11. $y = 2 - t^2 + 4\ln t$

17. $W(t) = e^{3(t-1)}$

19. $W(t) = 3/t$

23. $v = c_1\cos t + c_2\sin t$. Therefore, $y = A\cos t + B\sin t + Ct + D$

25. (a) $\alpha = 4,\ \ \beta = -4$ (b) $\beta = -4$, α can be any value.

Section 5.3, page 235

1. $y_1(t) = 1,\ y_2(t) = t,\ y_3(t) = t^2/2$

3. $y_1(t) = 1,\ y_2(t) = t,\ y_3(t) = e^{-t} + t - 1$

5. (a) $\{\bar{y}_1(t), \bar{y}_2(t), \bar{y}_3(t)\}$ is a solution set. (b) $A = \begin{bmatrix} 0 & 1 & 2 \\ 1/2 & -1/2 & 1/2 \\ 1/2 & 1/2 & -1/2 \end{bmatrix}$
(c) $\det[A] = 1.5$; therefore $\{\bar{y}_1(t), \bar{y}_2(t), \bar{y}_3(t)\}$ is a fundamental set.

7. (a) $\bar{y}_3(t)$ is not a solution.

9. Linearly independent

11. Linearly dependent since $2y_1 - y_2 - y_3 = 0$.

13. Linearly independent

15. Linearly independent

Section 5.4, page 242

1. (a) $\lambda^3 - 4\lambda = 0$ (b) $\lambda = 0,\ \lambda = 2,\ \lambda = -2$ (c) $y = c_1 + c_2 e^{2t} + c_3 e^{-2t}$
3. (a) $\lambda^3 + \lambda^2 + 4\lambda + 4 = 0$ (b) $\lambda = -1,\ \lambda = 2i,\ \lambda = -2i$
(c) $y = c_1 e^{-t} + c_2 \cos 2t + c_3 \sin 2t$
5. (a) $16\lambda^4 + 8\lambda^2 + 1 = 0$ (b) $\lambda = i/2,\ \lambda = -i/2$
(c) $y = c_1 \cos(t/2) + c_2 t \cos(t/2) + c_3 \sin(t/2) + c_4 t \sin(t/2)$
7. (a) $\lambda^3 - 2\lambda^2 - \lambda + 2 = 0$ (b) $\lambda = 1,\ \lambda = -1,\ \lambda = 2$ (c) $y = c_1 e^t + c_2 e^{-t} + c_3 e^{2t}$
9. (a) $\lambda^3 + 4\lambda = 0$ (b) $\lambda = 0,\ \lambda = 2i,\ \lambda = -2i$ (c) $y = c_1 + c_2 \cos 2t + c_3 \sin 2t$
(d) $y = 2 - \cos 2t + 3 \sin 2t$

11. $a_0 = 0,\ a_1 = 0,\ a_2 = 9,\ a_3 = 0$

13. $a_0 = 1,\ a_1 = 0,\ a_2 = -2,\ a_3 = 0$

15. $a_0 = 1,\ a_1 = -4,\ a_2 = 6,\ a_3 = -4$

17. (a) $n = 5$ (b) $y_1(t), y_2(t), y_3(t)$ along with $y_4(t) = e^t \sin 2t, y_5(t) = e^{-t} \sin 2t$

19. (a) $n = 7$ (b) A fundamental set is $\{\sin t, \cos t, t\sin t, t\cos t, e^t, te^t, t^2e^t\}$.

21. $n = 1,\ a = 1$

23. $n = 4,\ a = 0$

25. $n = 3,\ a = -27$

Section 5.5, page 249

1. (a) $y_C = c_1 + c_2 e^t + c_3 e^{-t}$ (b) $y_P = e^{2t}/6$ (c) $y = c_1 + c_2 e^t + c_3 e^{-t} + e^{2t}/6$
3. (a) $y_C = c_1 + c_2 e^t + c_3 e^{-t}$ (b) $y_P = -2t^2$ (c) $y = c_1 + c_2 e^t + c_3 e^{-t} - 2t^2$
5. (a) $y_C = c_1 + c_2 t + c_3 e^{-t}$ (b) $y_P = 6te^{-t}$ (c) $y = c_1 + c_2 t + c_3 e^{-t} + 6te^{-t}$
7. (a) $y_C = c_1 + c_2 e^t + c_3 te^t$ (b) $y_P = 2t^2e^t + 2t + 0.5t^2$
(c) $y = c_1 + c_2 e^t + c_3 te^t + 2t^2e^t + 2t + 0.5t^2$
9. (a) $y_C = c_1 e^t + c_2 e^{-t/2} \cos(\sqrt{3}\,t/2) + c_3 e^{-t/2} \sin(\sqrt{3}\,t/2)$ (b) $y_P = (t/3)e^t$
(c) $y = c_1 e^t + c_2 e^{-t/2} \cos(\sqrt{3}\,t/2) + c_3 e^{-t/2} \sin(\sqrt{3}\,t/2) + (t/3)e^t$
11. (a) $y_C = c_1 + c_2 e^{2t} + c_3 te^{2t}$
(b) $y_P = A_0 t + A_1 t^2 + A_2 t^3 + A_3 t^4 + (B_0 t^2 + B_1 t^3 + B_2 t^4)e^{2t}$
13. (a) $y_C = c_1 e^{2t} + c_2 e^{-2t} + c_3 \cos 2t + c_4 \sin 2t$
(b) $y_P = (A_0 t + A_1 t^2)e^{2t} + (B_0 t + B_1 t^2)e^{-2t} + (C_0 t + C_1 t^2)\cos 2t + (D_0 t + D_1 t^2)\sin 2t$
15. (a) $y_C = c_1 e^{-t} + c_2 e^t + c_3 \cos t + c_4 \sin t$
(b) $y_P = (A_0 t + A_1 t^2)e^{-t} + (B_0 t + B_1 t^2)\cos t + (C_0 t + C_1 t^2)\sin t$
17. $a = -1,\ b = 4,\ c = -4,\ g(t) = -4t^2 + 8t - 2$
19. $a = -1,\ b = 0,\ c = 0,\ g(t) = 12t^4$
21. (b) $y = c_1 t + c_2 t^2 + c_3 t^4 - (16/21)\sqrt{t}$
23. $y = c_1 t + c_2 t^2 + c_3 t^4 - (1/2)t^3 - (1/8) - t^2 \ln t$

CHAPTER 6

Section 6.1, page 262

1. $$\begin{bmatrix} -3t^2 + 2t - 2 & 2t^2 + 3t \\ 4 & -3t^2 - 2t + 2 \end{bmatrix}$$

3. $$\begin{bmatrix} -1 \\ 1 \end{bmatrix}$$

5. $\det[A(t)B(t)] = \det[A(t)]\det[B(t)] = -t(t+1)(t+2)$

7. $A^{-1}(t) = \dfrac{1}{(t-4)(t+1)} \begin{bmatrix} t-3 & -2 \\ -2 & t \end{bmatrix}, \quad t \neq 4, t \neq -1$

9. $A(t)$ cannot be inverted for any value of t.

11. $\begin{bmatrix} 0 & 0 \\ -2 & 1 \end{bmatrix}$

13. $A(t)$ is defined for $-\infty < t < 0, 0 < t \leq 1$. $A'(t) = \begin{bmatrix} 0 & 1/t \\ (-1/2)(1-t)^{-1/2} & 3e^{3t} \end{bmatrix}$,

$A''(t) = \begin{bmatrix} 0 & -1/t^2 \\ (-1/4)(1-t)^{-3/2} & 9e^{3t} \end{bmatrix}, \quad -\infty < t < 0, 0 < t < 1.$

15. $P(t) = \begin{bmatrix} t^{-1} & t^2+1 \\ 4 & t^{-1} \end{bmatrix}, \quad \mathbf{g}(t) = \begin{bmatrix} t \\ 8t \ln t \end{bmatrix}$

17. $A(t) = \begin{bmatrix} 2 + \ln|t| & 2t^2+3 \\ 5t-4 & t^3-3 \end{bmatrix}$

19. $A(t) = \begin{bmatrix} t^2 & \sin t & 2t \\ 5t & \ln|t+1| & t^3 \end{bmatrix}$

Section 6.2, page 269

1. $\pi/2 < t < 3\pi/2$

3. $0 < t < \pi/2$

7. (a) $A = \begin{bmatrix} 4 & 1 \\ 1 & 4 \end{bmatrix}$ (b) $\mathbf{y} = c_1 \begin{bmatrix} e^{5t} \\ e^{5t} \end{bmatrix} + c_2 \begin{bmatrix} e^{3t} \\ -e^{3t} \end{bmatrix}$

9. $c_1 = 1, c_2 = 2$

11. $P(t) = \begin{bmatrix} 0 & 1 \\ -4 & -t^2 \end{bmatrix}, \quad \mathbf{G}(t) = \begin{bmatrix} 0 \\ \sin t \end{bmatrix}$

13. $P(t) = \begin{bmatrix} 0 & 1 & 0 \\ 0 & 0 & 1 \\ -e^{-t}\tan t & -t^{-1}e^{-t} & -5e^{-t} \end{bmatrix}, \quad \mathbf{G}(t) = \begin{bmatrix} 0 \\ 0 \\ e^{-t} \end{bmatrix}$

15. $y'' - 2y' + 3y = 2\cos 2t, \ \ y(-1) = 1, \ \ y'(-1) = 4$

17. $y^{(4)} - (y'')(y'' + \sin y) + y' = 0, \ \ y(1) = 0, \ \ y'(1) = 0, \ \ y''(1) = -1, \ \ y'''(1) = 2$

19. $P(t) = \begin{bmatrix} 0 & 1 & 0 & 0 \\ 4 & t^{-1} & -t & \sin t \\ 0 & 0 & 0 & 1 \\ 1 & 0 & 0 & -5 \end{bmatrix}, \quad \mathbf{G}(t) = \begin{bmatrix} 0 \\ e^{2t} \\ 0 \\ 0 \end{bmatrix}$

21. $P(t) = \begin{bmatrix} 0 & 1 & 0 & 0 \\ 4 & -3 & -5 & -2 \\ 0 & 0 & 0 & 1 \\ 5 & 6 & -2 & 1 \end{bmatrix}, \quad \mathbf{G}(t) = \begin{bmatrix} 0 \\ t^2 \\ 0 \\ -t \end{bmatrix}$

Section 6.3, page 277

1. (a) $\mathbf{y}' = \begin{bmatrix} 9 & -4 \\ 15 & -7 \end{bmatrix} \mathbf{y}$

3. (a) $\mathbf{y}' = \begin{bmatrix} 1 & 4 \\ -1 & 1 \end{bmatrix} \mathbf{y}$

5. (a) $\mathbf{y}' = \begin{bmatrix} 0 & 1 & 1 \\ -6 & -3 & 1 \\ -8 & -2 & 4 \end{bmatrix} \mathbf{y}$

7. (b) $W(t) = 4e^{2t}$ (c) $\mathbf{y}(t) = \begin{bmatrix} 2e^{3t} & 2e^{-t} \\ 3e^{3t} & 5e^{-t} \end{bmatrix} \begin{bmatrix} c_1 \\ c_2 \end{bmatrix}$

(d) $\mathbf{c} = \begin{bmatrix} 3/4 \\ -1/4 \end{bmatrix}, \quad \mathbf{y}(t) = (3/4) \begin{bmatrix} 2e^{3t} \\ 3e^{3t} \end{bmatrix} - (1/4) \begin{bmatrix} 2e^{-t} \\ 5e^{-t} \end{bmatrix} = \begin{bmatrix} (3e^{3t} - e^{-t})/2 \\ (9e^{3t} - 5e^{-t})/4 \end{bmatrix}$

9. (b) $W(t) \equiv 0$; therefore, these solutions do not form a fundamental set.

11. (b) $W(t) \equiv 1$ (c) $\mathbf{y}(t) = \begin{bmatrix} e^t & e^{-t} \\ -2e^t & -e^{-t} \end{bmatrix} \begin{bmatrix} c_1 \\ c_2 \end{bmatrix}$

(d) $\mathbf{c} = \begin{bmatrix} 2e^{-1} \\ -e \end{bmatrix}$, $\mathbf{y}(t) = 2e^{-1} \begin{bmatrix} e^t \\ -2e^t \end{bmatrix} - e \begin{bmatrix} e^{-t} \\ -e^{-t} \end{bmatrix} = \begin{bmatrix} 2e^{t-1} - e^{-(t+1)} \\ -4e^{t-1} + e^{-(t+1)} \end{bmatrix}$

13. (b) $W(t) = -t^2$ (c) $\mathbf{y}(t) = \begin{bmatrix} t^2 - 2t & t-1 \\ 2t & 1 \end{bmatrix} \begin{bmatrix} c_1 \\ c_2 \end{bmatrix}$

(d) $\mathbf{c} = \begin{bmatrix} 1 \\ -2 \end{bmatrix}$, $\mathbf{y}(t) = \begin{bmatrix} t^2 - 2t \\ 2t \end{bmatrix} - 2 \begin{bmatrix} t-1 \\ 1 \end{bmatrix} = \begin{bmatrix} t^2 - 4t + 2 \\ 2t - 2 \end{bmatrix}$

15. (b) $W(t) = -11e^t$ (c) $\mathbf{y}(t) = \begin{bmatrix} 5e^t & e^t & e^{-t} \\ -11e^t & 0 & -e^{-t} \\ 0 & 11e^t & 5e^{-t} \end{bmatrix} \begin{bmatrix} c_1 \\ c_2 \\ c_3 \end{bmatrix}$

(d) $\mathbf{c} = \begin{bmatrix} 1 \\ -1 \\ -1 \end{bmatrix}$, $\mathbf{y}(t) = \begin{bmatrix} 5e^t \\ -11e^t \\ 0 \end{bmatrix} - \begin{bmatrix} e^t \\ 0 \\ 11e^t \end{bmatrix} - \begin{bmatrix} e^{-t} \\ -e^{-t} \\ 5e^{-t} \end{bmatrix} = \begin{bmatrix} 4e^t - e^{-t} \\ -11e^t + e^{-t} \\ -11e^t - 5e^{-t} \end{bmatrix}$

17. (a) $W(t) = 2e^{6t}$ (b) 6

19. (a) $W(t) = -6e^{4t}$ (b) 4

21. $\alpha = -3$

Section 6.4, page 284

1. (a) The determinant is $2t - t^2$. (b) No (c) Yes
3. (a) The determinant is $2te^t - (t-1)\sin^2 t$. (b) No (c) Yes
5. Linearly independent
7. Linearly independent
9. Linearly independent
11. (a) The determinant is te^t. (b) No
   (c) $P(t) = \begin{bmatrix} 1 & -t+2 \\ 0 & 1/t \end{bmatrix}$, continuous on $-\infty < t < 0$ and $0 < t < \infty$
13. (b) $C = \begin{bmatrix} 1/2 & 1/2 \\ -1/2 & 1/2 \end{bmatrix}$ (c) $\hat{\Psi}(t)$ is also a fundamental matrix.
15. (b) $C = \begin{bmatrix} 0 & 0 \\ 2 & 0 \end{bmatrix}$ (c) $\hat{\Psi}(t)$ is not a fundamental matrix.
17. $C = \begin{bmatrix} \frac{1}{2} & \frac{1}{2} \\ \frac{1}{2} & -\frac{1}{2} \end{bmatrix}$

Section 6.5, page 294

1. (a) $(\lambda_1, \mathbf{x}_1) = (2, \mathbf{x}_1)$ is an eigenpair. $(\lambda_2, \mathbf{x}_2) = (3, \mathbf{x}_2)$ is an eigenpair.
   (b) $\mathbf{y}_1(t) = e^{2t} \begin{bmatrix} 1 \\ -1 \end{bmatrix}$, $\mathbf{y}_2(t) = e^{3t} \begin{bmatrix} -2 \\ 1 \end{bmatrix}$
   (c) $W(t) = -e^{5t}$ and therefore they form a fundamental set.

3. (a) $\mathbf{x}_1$ is not an eigenvector. $(\lambda_2, \mathbf{x}_2) = (1, \mathbf{x}_2)$ is an eigenpair.

   (b) $\mathbf{y}_2(t) = e^t \begin{bmatrix} 1 \\ -2 \end{bmatrix}$

   (c) A single solution does not constitute a fundamental set for this system.

5. (a) $(\lambda_1, \mathbf{x}_1) = (-1, \mathbf{x}_1)$ is an eigenpair. $(\lambda_2, \mathbf{x}_2) = (1, \mathbf{x}_2)$ is an eigenpair.

   (b) $\mathbf{y}_1(t) = e^{-t} \begin{bmatrix} 1 \\ -1 \end{bmatrix}, \quad \mathbf{y}_2(t) = e^t \begin{bmatrix} 2 \\ 2 \end{bmatrix}$

   (c) $W(t) \equiv 4$ and therefore they form a fundamental set.

7. $\mathbf{x} = \begin{bmatrix} 1 \\ 2 \end{bmatrix}$ 9. $\mathbf{x} = \begin{bmatrix} 1 \\ 4 \end{bmatrix}$ 11. $\mathbf{x} = \begin{bmatrix} 1 \\ -2 \\ 1 \end{bmatrix}$ 13. $\mathbf{x} = \begin{bmatrix} -1 \\ -1 \\ 1 \end{bmatrix}$

15. $\lambda^2 - 10\lambda + 16 = 0$; the eigenvalues are $\lambda_1 = 8$ and $\lambda_2 = 2$.

17. $\lambda^2 - 8\lambda + 7 = 0$; the eigenvalues are $\lambda_1 = 1$ and $\lambda_2 = 7$.

19. $-\lambda^3 + 5\lambda^2 - 6\lambda = 0$; the eigenvalues are $\lambda_1 = 0$, $\lambda_2 = 2$, and $\lambda_3 = 3$.

21. $\mathbf{y}_1(t) = e^{2t} \begin{bmatrix} 1 \\ 2 \end{bmatrix}, \quad \mathbf{y}_2(t) = e^{-2t} \begin{bmatrix} 3 \\ 2 \end{bmatrix}; \quad \mathbf{y}(t) = \mathbf{y}_1(t) + \mathbf{y}_2(t)$

23. $\mathbf{y}_1(t) = e^{5t} \begin{bmatrix} 1 \\ 2 \end{bmatrix}, \quad \mathbf{y}_2(t) = e^{3t} \begin{bmatrix} 1 \\ 1 \end{bmatrix}; \quad \mathbf{y}(t) = 4\mathbf{y}_1(t) + 3\mathbf{y}_2(t)$

25. $\mathbf{y}_1(t) = e^{t} \begin{bmatrix} -1 \\ -1 \\ 1 \end{bmatrix}, \quad \mathbf{y}_2(t) = e^{2t} \begin{bmatrix} 1 \\ 0 \\ 0 \end{bmatrix}, \quad \mathbf{y}_3(t) = e^{3t} \begin{bmatrix} 1 \\ 1 \\ 0 \end{bmatrix}; \quad \mathbf{y}(t) = -\mathbf{y}_1(t) + \mathbf{y}_2(t) + 2\mathbf{y}_3(t)$

27. $\mathbf{y}_1(t) = e^{2t} \begin{bmatrix} 1 \\ 0 \\ 0 \end{bmatrix}, \quad \mathbf{y}_2(t) = e^{4t} \begin{bmatrix} 3 \\ 2 \\ 2 \end{bmatrix}, \quad \mathbf{y}_3(t) = e^{-2t} \begin{bmatrix} 3 \\ 4 \\ -8 \end{bmatrix}$

29. $\mathbf{y}_1(t) = e^{-t} \begin{bmatrix} -1 \\ 4 \\ 5 \end{bmatrix}, \quad \mathbf{y}_2(t) = e^{t} \begin{bmatrix} -1 \\ 2 \\ 3 \end{bmatrix}, \quad \mathbf{y}_3(t) = e^{2t} \begin{bmatrix} -2 \\ 2 \\ 1 \end{bmatrix}$

31. $x = 8$, $\lambda = -6$

39. (a) $\lambda_1 = -1, \quad \mathbf{x}_1 = \begin{bmatrix} 1 \\ 1 \end{bmatrix}, \quad \lambda_2 = -3, \quad \mathbf{x}_2 = \begin{bmatrix} 1 \\ -1 \end{bmatrix}, \quad \mathbf{y}(\tau) = c_1 e^{-\tau} \begin{bmatrix} 1 \\ 1 \end{bmatrix} + c_2 e^{-3\tau} \begin{bmatrix} 1 \\ -1 \end{bmatrix}$

   (b) Approximately 250.5 sec

Section 6.6, page 307

1. $\lambda_1 = 2 + i, \quad \mathbf{x}_1 = \begin{bmatrix} 1 \\ i \end{bmatrix}, \quad \lambda_2 = 2 - i, \quad \mathbf{x}_2 = \begin{bmatrix} 1 \\ -i \end{bmatrix}$

3. $\lambda_1 = 3 + 2i, \quad \mathbf{x}_1 = \begin{bmatrix} 3 + 2i \\ 1 \end{bmatrix}, \quad \lambda_2 = 3 - 2i, \quad \mathbf{x}_2 = \begin{bmatrix} 3 - 2i \\ 1 \end{bmatrix}$

5. $\lambda_1 = 1, \quad \mathbf{x}_1 = \begin{bmatrix} -3 \\ 0 \\ 1 \end{bmatrix}, \quad \lambda_2 = 1 + i, \quad \mathbf{x}_2 = \begin{bmatrix} -5 + i \\ -1 + i \\ 2 \end{bmatrix}, \quad \lambda_2 = 1 - i, \quad \mathbf{x}_2 = \begin{bmatrix} -5 - i \\ -1 - i \\ 2 \end{bmatrix}$

7. $\mathbf{y}_1(t) = e^{4t}\begin{bmatrix} 4\cos 2t \\ -\cos 2t - \sin 2t \end{bmatrix}, \quad \mathbf{y}_2(t) = e^{4t}\begin{bmatrix} 4\sin 2t \\ -\sin 2t + \cos 2t \end{bmatrix}$

9. $\mathbf{y}_1(t) = \begin{bmatrix} -\cos 2t + \sin 2t \\ \cos 2t \end{bmatrix}, \quad \mathbf{y}_2(t) = \begin{bmatrix} -\cos 2t - \sin 2t \\ \sin 2t \end{bmatrix}$

11. $\mathbf{y}_1(t) = e^{2t}\begin{bmatrix} 1 \\ 0 \\ -1 \end{bmatrix}, \quad \mathbf{y}_2(t) = e^{2t}\begin{bmatrix} -5\cos 3t - 3\sin 3t \\ 3\cos 3t - 3\sin 3t \\ 2\cos 3t \end{bmatrix}, \quad \mathbf{y}_3(t) = e^{2t}\begin{bmatrix} -5\sin 3t + 3\cos 3t \\ 3\sin 3t + 3\cos 3t \\ 2\sin 3t \end{bmatrix}$

13. $\mathbf{y}(t) = 4e^{2t}\begin{bmatrix} \cos t \\ -\sin t \end{bmatrix} + 7e^{2t}\begin{bmatrix} \sin t \\ \cos t \end{bmatrix}$

15. $\mathbf{y}(t) = 3e^{3t}\begin{bmatrix} 3\cos 2t - 2\sin 2t \\ \cos 2t \end{bmatrix} - 4e^{3t}\begin{bmatrix} 3\sin 2t + 2\cos 2t \\ \sin 2t \end{bmatrix}$

17. $\mathbf{y}(t) = -9e^{t}\begin{bmatrix} -3 \\ 0 \\ 1 \end{bmatrix} + (11/2)e^{t}\begin{bmatrix} -5\cos t - \sin t \\ -\cos t - \sin t \\ 2\cos t \end{bmatrix} + (13/2)e^{t}\begin{bmatrix} \cos t - 5\sin t \\ \cos t - \sin t \\ 2\sin t \end{bmatrix}$

23. $-\infty < \mu < 6$

25. There is no such value μ.

27. (a) $\mathbf{v}' = \begin{bmatrix} 0 & -1 & 1 \\ 1 & 0 & -1 \\ -1 & 1 & 0 \end{bmatrix}\mathbf{v}$ (b) $\mathbf{v}(t) \equiv c\begin{bmatrix} 1 \\ 1 \\ 1 \end{bmatrix}$

(c) $\mathbf{y}(t) = c_1\begin{bmatrix} 1 \\ 1 \\ 1 \end{bmatrix} + c_2\begin{bmatrix} 2\cos\sqrt{3}t \\ -\cos\sqrt{3}t + \sqrt{3}\sin\sqrt{3}t \\ -\cos\sqrt{3}t - \sqrt{3}\sin\sqrt{3}t \end{bmatrix} + c_3\begin{bmatrix} 2\sin\sqrt{3}t \\ -\sin\sqrt{3}t - \sqrt{3}\cos\sqrt{3}t \\ -\sin\sqrt{3}t + \sqrt{3}\cos\sqrt{3}t \end{bmatrix}$

(d) $\mathbf{y}(t) = v_0\begin{bmatrix} 1 \\ 1 \\ 1 \end{bmatrix}$

29. (a) $\mathbf{z}' = \begin{bmatrix} 0 & 1 \\ -k/m & 0 \end{bmatrix}\mathbf{z}, \quad \mathbf{z}(0) = \begin{bmatrix} y_0 \\ y_0' \end{bmatrix}$

(b) $\lambda_1 = i\omega, \quad \mathbf{x}_1 = \begin{bmatrix} 1 \\ i\omega \end{bmatrix}, \quad \lambda_2 = -i\omega, \quad \mathbf{x}_2 = \begin{bmatrix} 1 \\ -i\omega \end{bmatrix}$ where $\omega = \sqrt{k/m}$

(c) $\mathbf{z}(t) = \begin{bmatrix} \cos\omega t & \sin\omega t \\ -\omega\sin\omega t & \omega\cos\omega t \end{bmatrix}\begin{bmatrix} c_1 \\ c_2 \end{bmatrix}$

(d) $\mathbf{z}(t) = \begin{bmatrix} (y_0'/\omega)\sin\omega t + y_0\cos\omega t \\ y_0'\cos\omega t - \omega y_0\sin\omega t \end{bmatrix}$

Section 6.7, page 319

1. (a) $\lambda_1 = \lambda_2 = 2$, $\mathbf{x} = c\begin{bmatrix} 1 \\ 0 \end{bmatrix}$, geometric multiplicity is 1.

(b) $\mathbf{y}_1(t) = e^{2t}\begin{bmatrix} 1 \\ 0 \end{bmatrix}, \quad \mathbf{y}_2(t) = e^{2t}\left\{t\begin{bmatrix} 1 \\ 0 \end{bmatrix} + \begin{bmatrix} 0 \\ 1 \end{bmatrix}\right\}$ (c) $\mathbf{y}(t) = e^{2t}\begin{bmatrix} 1 - t \\ -1 \end{bmatrix}$

3. (a) $\lambda_1 = \lambda_2 = 6,\ \mathbf{x} = c\begin{bmatrix} 0 \\ 1 \end{bmatrix}$, geometric multiplicity is 1.

   (b) $\mathbf{y}_1(t) = e^{6t}\begin{bmatrix} 0 \\ 1 \end{bmatrix},\ \mathbf{y}_2(t) = e^{6t}\left\{t\begin{bmatrix} 0 \\ 1 \end{bmatrix} + \begin{bmatrix} 1/2 \\ 0 \end{bmatrix}\right\}$

   (c) $\mathbf{y}(t) = e^{6t}\begin{bmatrix} -2 \\ -4t \end{bmatrix}$

5. (a) $\lambda_1 = \lambda_2 = 3,\ \mathbf{x} = c\begin{bmatrix} 1 \\ 2 \end{bmatrix}$, geometric multiplicity is 1.

   (b) $\mathbf{y}_1(t) = e^{3t}\begin{bmatrix} 1 \\ 2 \end{bmatrix},\ \mathbf{y}_2(t) = e^{3t}\left\{t\begin{bmatrix} 1 \\ 2 \end{bmatrix} + \begin{bmatrix} 1/2 \\ 0 \end{bmatrix}\right\}$

   (c) $\mathbf{y}(t) = e^{3t}\begin{bmatrix} 1+t \\ 1+2t \end{bmatrix}$

7. (a) $\lambda_1 = \lambda_2 = 2,\ \mathbf{x} = c\begin{bmatrix} 1 \\ -1 \end{bmatrix}$, geometric multiplicity is 1.

   (b) $\mathbf{y}_1(t) = e^{2t}\begin{bmatrix} 1 \\ -1 \end{bmatrix},\ \mathbf{y}_2(t) = e^{2t}\left\{t\begin{bmatrix} 1 \\ -1 \end{bmatrix} + \begin{bmatrix} -1 \\ 0 \end{bmatrix}\right\}$

   (c) $\mathbf{y}(t) = e^{2t}\begin{bmatrix} 4-3t \\ -1+3t \end{bmatrix}$

9. (a) $\lambda_1 = \lambda_2 = \lambda_3 = 2,\ \mathbf{x} = c\begin{bmatrix} 1 \\ 0 \\ 0 \end{bmatrix}$, geometric multiplicity is 1.

   (b) $\lambda_1 = \lambda_2 = \lambda_3 = 2,\ \mathbf{x}_1 = \begin{bmatrix} 1 \\ 0 \\ 0 \end{bmatrix},\ \mathbf{x}_2 = \begin{bmatrix} 0 \\ 0 \\ 1 \end{bmatrix}$, geometric multiplicity is 2.

11. (a) $y_3 = c_3e^{2t}, y_2 = c_2e^{2t}, y_1 = c_1e^{2t} + c_2te^{2t}$

    (b) $\mathbf{y}(t) = \begin{bmatrix} e^{2t} & te^{2t} & 0 \\ 0 & e^{2t} & 0 \\ 0 & 0 & e^{2t} \end{bmatrix}\begin{bmatrix} c_1 \\ c_2 \\ c_3 \end{bmatrix}$; $W(t) = e^{6t}$ and therefore $\Psi(t)$ is a fundamental matrix.

13. $\lambda_1 = \lambda_2 = \lambda_3 = 5,\ \mathbf{x}_1 = c\begin{bmatrix} 0 \\ 0 \\ 1 \end{bmatrix}$ Thus, geometric multiplicity is 1 and the matrix does not have a full set of eigenvectors.

15. $\lambda_1 = \lambda_2 = \lambda_3 = 5,\ \mathbf{x}_1 = \begin{bmatrix} 1 \\ 0 \\ 0 \end{bmatrix},\ \mathbf{x}_2 = \begin{bmatrix} 0 \\ 1 \\ 0 \end{bmatrix}$ Thus, geometric multiplicity is 2 and the matrix does not have a full set of eigenvectors.

17. The eigenvalues are $\lambda = 2$ (algebraic multiplicity 2, geometric multiplicity 1) and $\lambda = 3$ (algebraic multiplicity 2, geometric multiplicity 1). A does not have a full set of eigenvectors.

19. The eigenvalues are $\lambda = 2$ (algebraic multiplicity 3, geometric multiplicity 3) and $\lambda = 3$ (algebraic multiplicity 1, geometric multiplicity 1). A does have a full set of eigenvectors.

21. $x = 9,\ y = 4$ 23. $x = \pm 1,\ y = 2$ 25. $x = 0,\ y = -2$

31. $\mathbf{y}_1(t) = e^{4t}\begin{bmatrix} 0 \\ 0 \\ 1 \end{bmatrix},\quad \mathbf{y}_2(t) = e^{4t}\left\{\begin{bmatrix} 0 \\ 1/3 \\ 0 \end{bmatrix} + t\begin{bmatrix} 0 \\ 0 \\ 1 \end{bmatrix}\right\},$

$$\mathbf{y}_3(t) = e^{4t}\left\{\begin{bmatrix} 1/6 \\ -1/18 \\ 0 \end{bmatrix} + t\begin{bmatrix} 0 \\ 1/3 \\ 0 \end{bmatrix} + (t^2/2)\begin{bmatrix} 0 \\ 0 \\ 1 \end{bmatrix}\right\}$$

33. $\mathbf{y}_1(t) = e^{2t}\begin{bmatrix} 1 \\ -1 \\ 0 \end{bmatrix},\quad \mathbf{y}_2(t) = e^{2t}\left\{\begin{bmatrix} -3/2 \\ 0 \\ -1/2 \end{bmatrix} + t\begin{bmatrix} 1 \\ -1 \\ 0 \end{bmatrix}\right\},$

$$\mathbf{y}_3(t) = e^{2t}\left\{\begin{bmatrix} -1/2 \\ 0 \\ -1/4 \end{bmatrix} + t\begin{bmatrix} -3/2 \\ 0 \\ -1/2 \end{bmatrix} + (t^2/2)\begin{bmatrix} 1 \\ -1 \\ 0 \end{bmatrix}\right\}$$

Section 6.8, page 331

1. (a) $\mathbf{y}_C(t) = c_1 e^{-3t}\begin{bmatrix} 1 \\ -1 \end{bmatrix} + c_2 e^{-t}\begin{bmatrix} 1 \\ 1 \end{bmatrix}$ (b) $\mathbf{y}_P(t) = \begin{bmatrix} 1 \\ 1 \end{bmatrix}$

   (d) $\mathbf{y}(t) = \begin{bmatrix} e^{-3t} + e^{-t} + 1 \\ -e^{-3t} + e^{-t} + 1 \end{bmatrix}$

3. (a) $\mathbf{y}_C(t) = c_1 e^{t}\begin{bmatrix} 1 \\ 1 \end{bmatrix} + c_2 e^{-t}\begin{bmatrix} -1 \\ 1 \end{bmatrix}$ (b) $\mathbf{y}_P(t) = \begin{bmatrix} 0 \\ -t \end{bmatrix}$

   (d) $\mathbf{y}(t) = \begin{bmatrix} 0.5e^{t} + 1.5e^{-t} \\ 0.5e^{t} - 1.5e^{-t} - t \end{bmatrix}$

5. (a) $\mathbf{y}_C(t) = c_1 e^{t}\begin{bmatrix} 1 \\ -2 \end{bmatrix} + c_2 e^{-t}\begin{bmatrix} 1 \\ -1 \end{bmatrix}$ (b) $\mathbf{y}_P(t) = \begin{bmatrix} 1.5\sin t - 0.5\cos t \\ -2\sin t \end{bmatrix}$

   (d) $\mathbf{y}(t) = \begin{bmatrix} -0.5e^{t} + e^{-t} + 1.5\sin t - 0.5\cos t \\ e^{t} - e^{-t} - 2\sin t \end{bmatrix}$

7. $\mathbf{y}_0 = \begin{bmatrix} 1 \\ e^{\pi/2} - 1 \end{bmatrix},\quad \mathbf{g}(t) = \begin{bmatrix} -2e^{t} \\ e^{t} + 2 \end{bmatrix}$

9. $P(t) = \begin{bmatrix} 1 & e^{t} \\ 0 & -1 \end{bmatrix}$ 11. $\mathbf{y}(t) \equiv \begin{bmatrix} 10 \\ -3 \end{bmatrix}$

13. $\mathbf{y}(t) \equiv \begin{bmatrix} 0 \\ 2 \end{bmatrix} + c\begin{bmatrix} 1 \\ 1 \end{bmatrix}$ 15. $\mathbf{y}(t) \equiv \begin{bmatrix} 2 \\ 0 \\ 0 \end{bmatrix} + c\begin{bmatrix} 0 \\ -1 \\ 1 \end{bmatrix}$

17. $\Psi(t) = \begin{bmatrix} \sin 2t & -\cos 2t - \sin 2t \\ \cos 2t & -\cos 2t + \sin 2t \end{bmatrix}$ 19. $\Psi(t) = \begin{bmatrix} -e^{-t} + 2e^{t} & 2e^{-t} - 2e^{t} \\ -e^{-t} + e^{t} & 2e^{-t} - e^{t} \end{bmatrix}$

21. $\mathbf{y}_C(t) = c_1 \begin{bmatrix} 1 \\ -1 \end{bmatrix} + c_2 e^{2t} \begin{bmatrix} 1 \\ 1 \end{bmatrix}$, $\mathbf{y}(t) = (1/4) \begin{bmatrix} -1 + e^{2t} + 2te^{2t} \\ 1 - e^{2t} + 2te^{2t} \end{bmatrix}$

23. $\mathbf{y}_C(t) = c_1 \begin{bmatrix} \sin t \\ \cos t \end{bmatrix} + c_2 \begin{bmatrix} -\cos t \\ \sin t \end{bmatrix}$, $\mathbf{y}(t) = \begin{bmatrix} 3\sin t - \cos t + 1 \\ 3\cos t + \sin t - 2 \end{bmatrix}$

25. (a) Yes, $\mathbf{Q}(t) = \begin{bmatrix} (2/3)cV \\ (1/3)cV \end{bmatrix}$

(b) $\mathbf{Q}_C(t) = \begin{bmatrix} e^{-(r/V)t} & e^{-3(r/V)t} \\ e^{-(r/V)t} & -e^{-3(r/V)t} \end{bmatrix} \begin{bmatrix} c_1 \\ c_2 \end{bmatrix}$

(c) $\mathbf{Q}_P(t) = \begin{bmatrix} (2/3)cV \\ (1/3)cV \end{bmatrix}$

(d) $\mathbf{Q}(t) = \dfrac{Vc}{6} \begin{bmatrix} -3e^{-(r/V)t} - e^{-3(r/V)t} + 4 \\ -3e^{-(r/V)t} + e^{-3(r/V)t} + 2 \end{bmatrix}$

(e) $\mathbf{Q}_\infty(t) = \begin{bmatrix} (2/3)c \\ (1/3)c \end{bmatrix}$

27. $\mathbf{v}_0 = (1/4) \begin{bmatrix} \sqrt{3} + 1 \\ \sqrt{3} - 1 \end{bmatrix}$

Section 6.9, page 344

1. (a) $\mathbf{y}_{k+1} = \mathbf{y}_k + h \left\{ \begin{bmatrix} 1 & 2 \\ 2 & 3 \end{bmatrix} \mathbf{y}_k + \begin{bmatrix} 1 \\ 1 \end{bmatrix} \right\}$, $\mathbf{y}_0 = \begin{bmatrix} -1 \\ 1 \end{bmatrix}$

(b) $t_k = kh$; $k = 0, 1, \ldots, 100$ in the case $h = 0.01$ and $T = 1$.

3. (a) $\mathbf{y}_{k+1} = \mathbf{y}_k + h \left\{ \begin{bmatrix} -t_k^2 & t_k \\ 2 - t_k & 0 \end{bmatrix} \mathbf{y}_k + \begin{bmatrix} 1 \\ t_k \end{bmatrix} \right\}$, $\mathbf{y}_0 = \begin{bmatrix} 2 \\ 0 \end{bmatrix}$

(b) $t_k = 1 + kh$; $k = 0, 1, \ldots, 300$ in the case $h = 0.01$ and $T = 3$.

5. (a) $\mathbf{y}_{k+1} = \mathbf{y}_k + h \left\{ \begin{bmatrix} -t_k^{-1} & \sin t_k \\ 1 - t_k & 1 \end{bmatrix} \mathbf{y}_k + \begin{bmatrix} 0 \\ t_k^2 \end{bmatrix} \right\}$, $\mathbf{y}_0 = \begin{bmatrix} 0 \\ 0 \end{bmatrix}$

(b) $t_k = 1 + kh$; $k = 0, 1, \ldots, 500$ in the case $h = 0.01$ and $T = 5$.

7. $\mathbf{y}_0 = \begin{bmatrix} 2 \\ 1 \end{bmatrix}$, $\mathbf{y}_1 = \begin{bmatrix} 2.04 \\ 1.09 \end{bmatrix}$, $\mathbf{y}_2 = \begin{bmatrix} 2.0814\ldots \\ 1.1833\ldots \end{bmatrix}$

9. $\mathbf{y}_0 = \begin{bmatrix} 0 \\ 0 \\ 1 \end{bmatrix}$, $\mathbf{y}_1 = \begin{bmatrix} 0.01 \\ 0.03 \\ 0.99 \end{bmatrix}$, $\mathbf{y}_2 = \begin{bmatrix} 0.02 \\ 0.0608 \\ 0.9808 \end{bmatrix}$

11. (a) $\mathbf{z}' = \begin{bmatrix} 0 & 1 \\ -1 & 0 \end{bmatrix} \mathbf{z} + \begin{bmatrix} 0 \\ t^{3/2} \end{bmatrix}$, $\mathbf{z}(0) = \begin{bmatrix} 1 \\ 0 \end{bmatrix}$

(b) $\mathbf{z}_{k+1} = \mathbf{z}_k + h \left\{ \begin{bmatrix} 0 & 1 \\ -1 & 0 \end{bmatrix} \mathbf{z}_k + \begin{bmatrix} 0 \\ t_k^{3/2} \end{bmatrix} \right\}$, $\mathbf{z}_0 = \begin{bmatrix} 1 \\ 0 \end{bmatrix}$

(c) $\mathbf{z}_0 = \begin{bmatrix} 1 \\ 0 \end{bmatrix}$, $\mathbf{z}_1 = \begin{bmatrix} 1 \\ -0.01 \end{bmatrix}$, $\mathbf{z}_2 = \begin{bmatrix} 0.9999\ldots \\ -0.0199\ldots \end{bmatrix}$

13. (a) $\mathbf{z}' = \begin{bmatrix} 0 & 1 & 0 \\ 0 & 0 & 1 \\ -t & -2 & 0 \end{bmatrix} \mathbf{z} + \begin{bmatrix} 0 \\ 0 \\ t+1 \end{bmatrix}, \quad \mathbf{z}(0) = \begin{bmatrix} 1 \\ -1 \\ 0 \end{bmatrix}$

(b) $\mathbf{z}_{k+1} = \mathbf{z}_k + h \left\{ \begin{bmatrix} 0 & 1 & 0 \\ 0 & 0 & 1 \\ -t_k & -2 & 0 \end{bmatrix} \mathbf{z}_k + \begin{bmatrix} 0 \\ 0 \\ t_k+1 \end{bmatrix} \right\}, \quad \mathbf{z}_0 = \begin{bmatrix} 1 \\ -1 \\ 0 \end{bmatrix}$

(c) $\mathbf{z}_0 = \begin{bmatrix} 1 \\ -1 \\ 0 \end{bmatrix}, \quad \mathbf{z}_1 = \begin{bmatrix} 0.99 \\ -1 \\ 0.03 \end{bmatrix}, \quad \mathbf{z}_2 = \begin{bmatrix} 0.98 \\ -0.9997 \\ 0.0600\ldots \end{bmatrix}$

17. $\mathbf{y}(1) - \bar{\mathbf{y}}_{2n} = \begin{bmatrix} 0.00768\ldots \\ 0.00584\ldots \end{bmatrix}, \quad \bar{\mathbf{y}}_{2n} - \mathbf{y}_n = \begin{bmatrix} 0.00762\ldots \\ 0.00577\ldots \end{bmatrix}$

19. $\mathbf{y}(1) - \bar{\mathbf{y}}_{2n} = \begin{bmatrix} -0.1718\ldots \\ -0.0624\ldots \end{bmatrix}, \quad \bar{\mathbf{y}}_{2n} - \mathbf{y}_n = \begin{bmatrix} -0.1693\ldots \\ -0.0583\ldots \end{bmatrix}$

21. (a) $\mathbf{Q}' = \begin{bmatrix} -15/(200-10t) & 5/(500-20t) \\ 15/(200-10t) & -35/(500-20t) \end{bmatrix} \mathbf{Q}, \quad \mathbf{Q}(0) = \begin{bmatrix} 40 \\ 40 \end{bmatrix}$

23. (a) $\mathbf{r}' = \begin{bmatrix} 0 & 1 \\ \alpha^2 t^2 & -\gamma \end{bmatrix} \mathbf{r}, \quad \mathbf{r}(0) = \begin{bmatrix} 2 \\ 0 \end{bmatrix}$

Section 6.10, page 356

1. $T = \begin{bmatrix} 1 & 2 \\ 1 & 1 \end{bmatrix}$ 3. $T = \begin{bmatrix} -1 & 1 \\ 1 & 1 \end{bmatrix}$ 5. $T = \begin{bmatrix} -1 & 1 \\ 1 & 1 \end{bmatrix}$ 7. $T = \begin{bmatrix} 0 & 1 \\ 1 & 1 \end{bmatrix}$

9. $\lambda_1 = 1$, algebraic multiplicity 2, geometric multiplicity 2. $\lambda_2 = 2$, algebraic multiplicity 1, geometric multiplicity 1.

$$T = \begin{bmatrix} -5 & 1 & -2 \\ 0 & 3 & -2 \\ 4 & 0 & 1 \end{bmatrix}$$

11. $\lambda_1 = 1$, algebraic multiplicity 2, geometric multiplicity 1. $\lambda_2 = 2$, algebraic multiplicity 1, geometric multiplicity 1. A is not diagonalizable.

13. $\lambda_1 = 1$, algebraic multiplicity 3, geometric multiplicity 1. A is not diagonalizable.

15. $\mathbf{z}' = \begin{bmatrix} 2 & 0 \\ 0 & 3 \end{bmatrix} \mathbf{z} + \begin{bmatrix} e^t \\ 2 \end{bmatrix}, \quad \mathbf{z}(t) = c_1 e^{2t} \begin{bmatrix} 1 \\ 0 \end{bmatrix} + c_2 e^{3t} \begin{bmatrix} 0 \\ 1 \end{bmatrix} + \begin{bmatrix} -e^t \\ -2/3 \end{bmatrix}; \; T = \begin{bmatrix} 3 & 2 \\ 2 & 1 \end{bmatrix}$

17. $\mathbf{z}' = \begin{bmatrix} 0 & 0 \\ 0 & 3 \end{bmatrix} \mathbf{z} + \begin{bmatrix} t-1 \\ 1 \end{bmatrix}, \quad \mathbf{z}(t) = c_1 \begin{bmatrix} 1 \\ 0 \end{bmatrix} + c_2 \begin{bmatrix} 0 \\ e^{3t} \end{bmatrix} + \begin{bmatrix} -t + 0.5t^2 \\ -1/3 \end{bmatrix}; \; T = \begin{bmatrix} 1 & 1 \\ -1 & 2 \end{bmatrix}$

19. $\mathbf{x} = c_1 \cos 2t \begin{bmatrix} -1 \\ 1 \end{bmatrix} + c_2 \sin 2t \begin{bmatrix} -1 \\ 1 \end{bmatrix} + c_3 \cos t \begin{bmatrix} -5 \\ 8 \end{bmatrix} + c_4 \sin t \begin{bmatrix} -5 \\ 8 \end{bmatrix}$

21. $\mathbf{x} = c_1 e^{-t} \begin{bmatrix} 1 \\ -3 \end{bmatrix} + c_2 e^{t} \begin{bmatrix} 1 \\ -3 \end{bmatrix} + c_3 \cos t \begin{bmatrix} 1 \\ -1 \end{bmatrix} + c_4 \sin t \begin{bmatrix} 1 \\ -1 \end{bmatrix}$

27. (a) $\lambda_1 = 100, \mathbf{x}_1 = \begin{bmatrix} 1 \\ 2 \end{bmatrix}, \quad \lambda_2 = 600, \mathbf{x}_2 = \begin{bmatrix} 2 \\ -1 \end{bmatrix}$

(c) $\mathbf{y}(t) = \dfrac{1}{100} \begin{bmatrix} 8\cos(10t) + 2\cos(10\sqrt{6}\,t) \\ 16\cos(10t) - \cos(10\sqrt{6}\,t) \end{bmatrix}$

Section 6.11, page 371

1. (a) $\Phi(t) = \begin{bmatrix} -4+5e^t & 4-4e^t \\ -5+5e^t & 5-4e^t \end{bmatrix}$ (b) $\mathbf{y}(2) = \begin{bmatrix} -4+5e^3 \\ -5+5e^3 \end{bmatrix}$

3. (a) $\Phi(t) = \dfrac{1}{6}\begin{bmatrix} e^t+5e^{7t} & -5e^t+5e^{7t} \\ -e^t+e^{7t} & 5e^t+e^{7t} \end{bmatrix}$ (b) $\dfrac{1}{6}\begin{bmatrix} -4e^{-1}+10e^{-7} \\ 4e^{-1}+2e^{-7} \end{bmatrix}$

5. (a) $\Phi(t,s) = \dfrac{1}{s^2}\begin{bmatrix} 2st-t^2 & -s^2t+st^2 \\ 2s-2t & -s^2+2st \end{bmatrix}$ (b) $\mathbf{y}(3) = \begin{bmatrix} -9 \\ -9 \end{bmatrix}$

9. (a) $A^n = I$ when n is even. (b) $A^n = A$ when n is odd. (c) $A^{-n} = I$ when n is even and $A^{-n} = A$ when n is odd.

11. One such matrix is $B = \begin{bmatrix} 9-8i & -6+6i \\ 12-12i & -8+9i \end{bmatrix}$

13. $A^3 = \begin{bmatrix} 24 & -16 \\ 32 & -24 \end{bmatrix}$

15. $\cos(\pi A) = \begin{bmatrix} 1/2 & 0 \\ 0 & 1/2 \end{bmatrix}$, $\sin(\pi A) = \begin{bmatrix} \sqrt{3}/2 & 0 \\ 0 & \sqrt{3}/2 \end{bmatrix}$

17. $\mathbf{y} = c_1 e^t \begin{bmatrix} 1 \\ -2 \end{bmatrix} + c_2 e^{-t}\begin{bmatrix} 1 \\ -1 \end{bmatrix} + \begin{bmatrix} 1 \\ 1 \end{bmatrix}$

19. $\mathbf{y}(t) = c_1 \begin{bmatrix} 1 \\ -2 \end{bmatrix} + c_2 e^{2t}\begin{bmatrix} 1 \\ -2 \end{bmatrix} + k_1 \begin{bmatrix} 1 \\ -1 \end{bmatrix} + k_2 e^{-2t}\begin{bmatrix} 1 \\ -1 \end{bmatrix}$

21. (a) $\lambda_1 = 0$, $\mathbf{x}_1 = \begin{bmatrix} 1 \\ 1 \\ 1 \end{bmatrix}$, $\lambda_2 = \dfrac{k}{m}$, $\mathbf{x}_2 = \begin{bmatrix} 1 \\ 0 \\ -1 \end{bmatrix}$, $\lambda_3 = \dfrac{3k}{m}$, $\mathbf{x}_2 = \begin{bmatrix} 1 \\ -2 \\ 1 \end{bmatrix}$

(b) $\mathbf{x} = T\mathbf{y}$ where $\mathbf{y} = \begin{bmatrix} d_1 t + c_1 \\ c_2\cos(\sqrt{k/m}\,t) + d_2\sin(\sqrt{k/m}\,t) \\ c_3\cos(\sqrt{3k/m}\,t) + d_3\sin(\sqrt{3k/m}\,t) \end{bmatrix}$

and $T = \begin{bmatrix} 1 & 1 & 1 \\ 1 & 0 & -2 \\ 1 & -1 & 1 \end{bmatrix}$.

CHAPTER 7

Section 7.1, page 388

1. $\mathcal{L}\{1\} = 1/s, \quad s > 0$

3. $\mathcal{L}\{te^{-t}\} = 1/(s+1)^2, \quad s > -1$

5. The Laplace transform does not exist.

7. $\mathcal{L}\{|t-1|\} = 2e^{-s}s^{-2} + s^{-1} - s^{-2}, \quad s > 0$

9. $\mathcal{L}\{f(t)\} = e^{-s}s^{-1}, \quad s > 0$

11. $\mathcal{L}\{f(t)\} = e^{-s}s^{-1} - e^{-2s}s^{-1}, \quad s \neq 0; \quad \mathcal{L}\{f(t)\} = 1, \quad s = 0$

15. (b) $\mathcal{L}\{t^2\} = 2/s^3$, $\mathcal{L}\{t^3\} = 2\cdot 3/s^4$, $\mathcal{L}\{t^4\} = 24/s^5$, $\mathcal{L}\{t^5\} = 120/s^6$, $s > 0$

(c) $\mathcal{L}\{t^m\} = m!/s^{m+1}, \quad s > 0$

17. $\mathcal{L}\{\sin\omega t\} = \omega/(s^2+\omega^2), \quad s > 0$

19. $\mathcal{L}\{\sin[\omega(t-2)]\} = (\omega\cos 2\omega - s\sin 2\omega)/(s^2+\omega^2), \quad s > 0$
21. $\mathcal{L}\{e^{-2t}\cos 4t\} = (s+2)/[(s+2)^2+16], \quad s > -2$
23. $R(s) = 5\mathcal{L}\{e^{-7t}\} + \mathcal{L}\{t\} + 2\mathcal{L}\{e^{2t}\} = 5/(s+7) + (1/s^2) + 2/(s-2), \quad s > 2$
25. f is continuous and exponentially bounded. $M = 1, a = 1$
27. f is continuous and exponentially bounded. $M = 1, a = 2$(since $\cosh 2t \le e^{2t}$)
29. f is piecewise continuous and exponentially bounded. $M = 1, a = 2$
31. f is neither continuous nor exponentially bounded.
33. The integral diverges.
35. The integral converges to 1/2.
37. $\mathcal{L}^{-1}\{F(s)\} = -2t + e^{-t}, \quad t \ge 0$
39. $\mathcal{L}^{-1}\{F(s)\} = e^{t} - e^{-t}, \quad t \ge 0$

Section 7.2, page 403

1. $F(s) = (s^2+2s+6)/s^3, \quad s > 0$
3. $F(s) = s^{-1} + 3/(9+s^2), \quad s > 0$
5. $F(s) = e^{-s}(2/s^3), \quad s > 0$
7. $F(s) = 2/(s+2)^2, \quad s > -2$
9. $F(s) = e^{-2s}(2s^{-2} + 4s^{-1}), \quad s > 0$
11. $F(s) = e^{3-s}/(s-3), \quad s > -3$
13. $\mathcal{L}^{-1}\{F(s)\} = 3 + 4t^3, \quad t \ge 0$
15. $\mathcal{L}^{-1}\{F(s)\} = 2e^{2t}\cos 3t, \quad t \ge 0$
17. $\mathcal{L}^{-1}\{F(s)\} = h(t-2)\sin[3(t-2)], \quad t \ge 0$
19. $\mathcal{L}^{-1}\{F(s)\} = (2/3)e^{t}[6\cos 3t - \sin 3t], \quad t \ge 0$
21. $\mathcal{L}^{-1}\{F(s)\} = 2(t-3)^4 h(t-3) + 4(t-5)^4 h(t-5), \quad t \ge 0$

23.

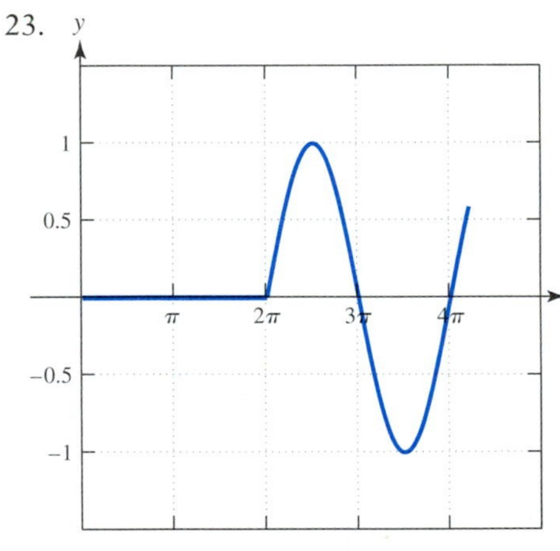

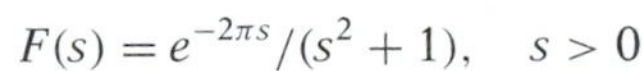
$F(s) = e^{-2\pi s}/(s^2+1), \quad s > 0$

25.

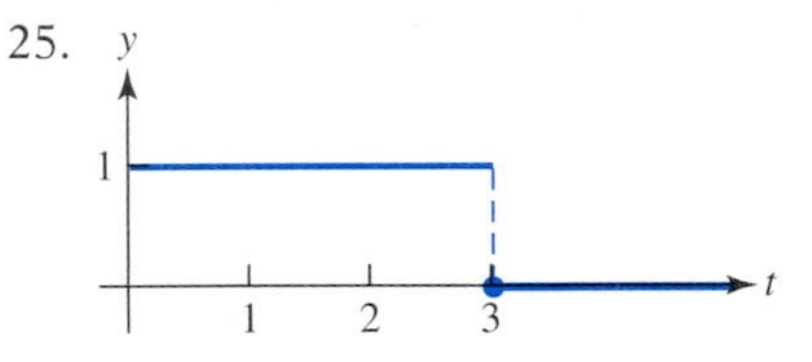

$F(s) = (1 - e^{-3s})/s$

27.

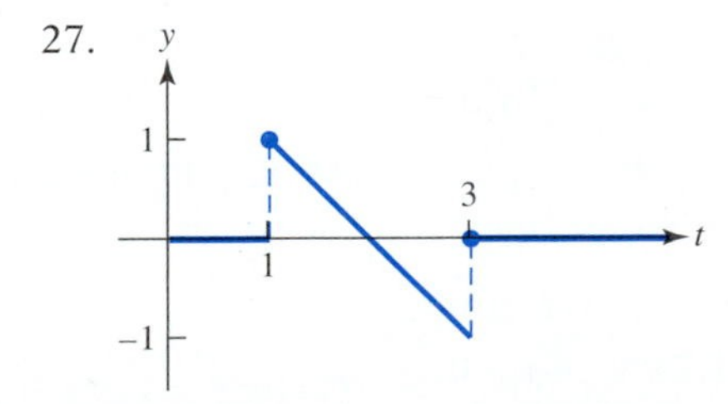

$F(s) = (e^{-s}(s-1) + e^{-3s}(s+1))/s^2$

29.

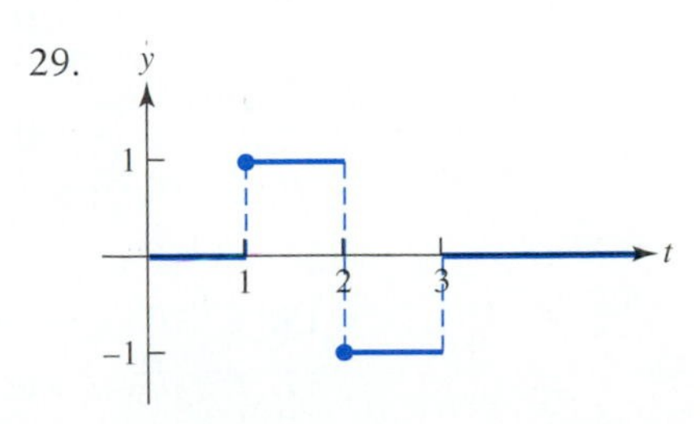

$F(s) = (e^{-s} - 2e^{-2s} + e^{-3s})/s$

31.

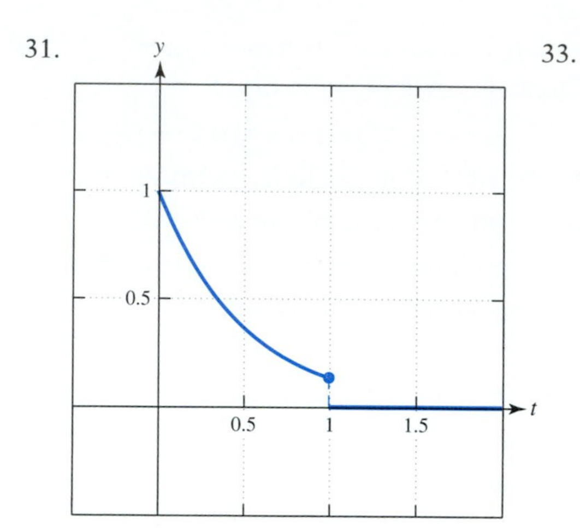

33.

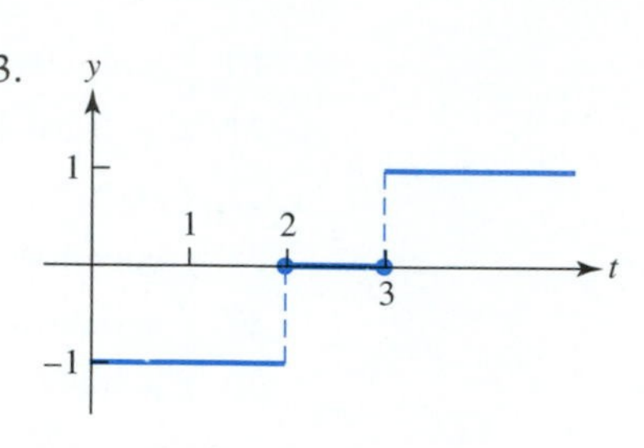

$F(s) = (e^{-2s} + e^{-3s} - 1)/s, \quad s > 0$

$F(s) = (1 - e^{-(s+2)})/(s + 2)$

35. $f(t) = h(t-1) + h(t-2) - 2h(t-3), \quad F(s) = (e^{-s} + e^{-2s} - 2e^{-3s})/s$

37. $f(t) = (1-t)[h(t) - h(t-1)] + (2-t)[h(t-1) - h(t-2)]$, and after expanding,
$f(t) = 1 - t + h(t-1) + (t-2)h(t-2), \quad t \geq 0$
$F(s) = (1/s) - (1/s^2) + (e^{-s}/s) + (e^{-2s}/s^2)$

39. $F(s) = (2/s) - [2/(s+2)], \quad \mathcal{L}^{-1}\{F(s)\} = 2 - 2e^{-2t}, \quad t \geq 0$

41. $F(s) = [10e^{-s}/(s-3)] - [10e^{-s}/(s-2)]$,
$\mathcal{L}^{-1}\{F(s)\} = 10h(t-1)[e^{3(t-1)} - e^{2(t-1)}], \quad t \geq 0$

43. $Y(s) = [1/(s-1)] + e^{12-4s}/[(s-3)(s-1)]$
$y(t) = \mathcal{L}^{-1}\{Y(s)\} = e^t + h(t-4)[e^{3t} - e^{t+8}]/2, \quad t \geq 0$

45. $Y(s) = s/[(s-1)(s^2 - 2s - 8)]$
$y(t) = \mathcal{L}^{-1}\{Y(s)\} = (-e^{-2t} - e^t + 2e^{4t})/9, \quad t \geq 0$

47. $G(s) = F(s)/s^2, \quad s > \max\{a, 0\}$ where $|f(t)| \leq Me^{at}$

49. (a) No, they differ at $t = 3$. (b) $F(s) = G(s) = (1 - e^{-3s})/s$

Section 7.3, page 412

1. $F(s) = \dfrac{A}{s-1} + \dfrac{B_2}{(s-2)^2} + \dfrac{B_1}{(s-2)}$

3. $F(s) = \dfrac{A_2}{s^2} + \dfrac{A_1}{s} + \dfrac{Bs + C}{(s+1)^2 + 9}$

5. $F(s) = \dfrac{A_2}{(s-3)^2} + \dfrac{A_1}{(s-3)} + \dfrac{B_2}{(s+3)^2} + \dfrac{B_1}{(s+3)}$

7. $F(s) = \dfrac{Bs + C}{(s+4)^2 + 1} + \dfrac{Ds + E}{(s+3)^2 + 4}$

9. $\mathcal{L}^{-1}\{F(s)\} = 2e^{3t}, \quad t \geq 0$

11. $\mathcal{L}^{-1}\{F(s)\} = 4\cos 3t + (5/3)\sin 3t, \quad t \geq 0$

13. $\mathcal{L}^{-1}\{F(s)\} = e^{-3t} + 2e^{-t}, \quad t \geq 0$

15. $\mathcal{L}^{-1}\{F(s)\} = 2 + \cos 2t + (1/2)\sin 2t, \quad t \geq 0$

17. $\mathcal{L}^{-1}\{F(s)\} = te^t + (1/2)t^2e^t, \quad t \geq 0$

19. $y(t) = 4e^{3t} - 3\cos 2t + 2\sin 2t$

21. $y(t) = e^{3t} + te^{3t}$

23. $y(t) = 2t + 2\cos 2t + 2\sin 2t$

25. $y(t) = \cos 2t - (1/4)t\cos 2t + (1/8)\sin 2t$

27. $y(t) = te^{-t} + (1/2)t^2e^{-t}$

29. $y(t) = t + \cos t - \sin t + h(t-2)[t - 4 + 2\cos(t-2) + \sin(t-2)]$

31. $\alpha = 0, \beta = -4, y_0 = 0, y_0' = 3$

Section 7.4, page 422

1. $F(s) = \dfrac{3(1-e^{-2s})^2}{s(1-e^{-4s})} = \dfrac{3(1-e^{-2s})}{s(1+e^{-2s})}$

3. $F(s) = \dfrac{-2e^{-4s} + 5e^{-2s} - 3}{s(1-e^{-4s})}$

5. $F(s) = (1-e^{-s})/[s^2(1+e^{-s})]$

7. $F(s) = [1 + (2e^{-s}/s) + (2e^{-s}/s^2) - (2/s^2)]/[s(1-e^{-2s})]$

9.

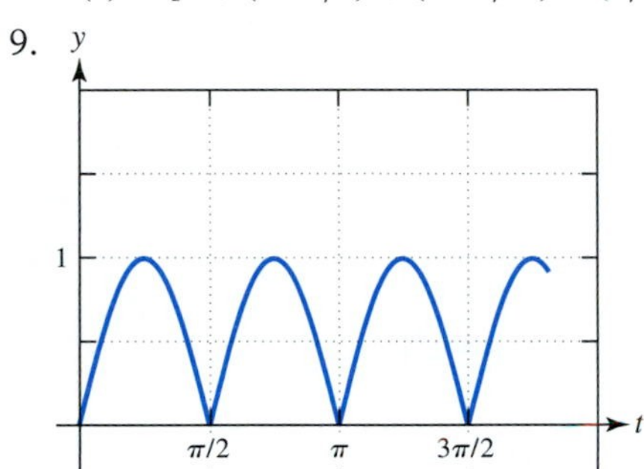

$$F(s) = \frac{2(1+e^{-\pi s/2})}{(s^2+4)(1-e^{-\pi s/2})}$$

11.

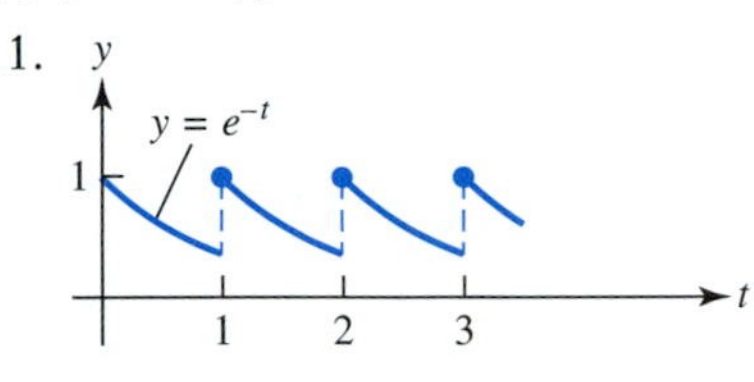

$$F(s) = \frac{1-e^{-(s+1)}}{(s+1)(1-e^{-s})}$$

13.

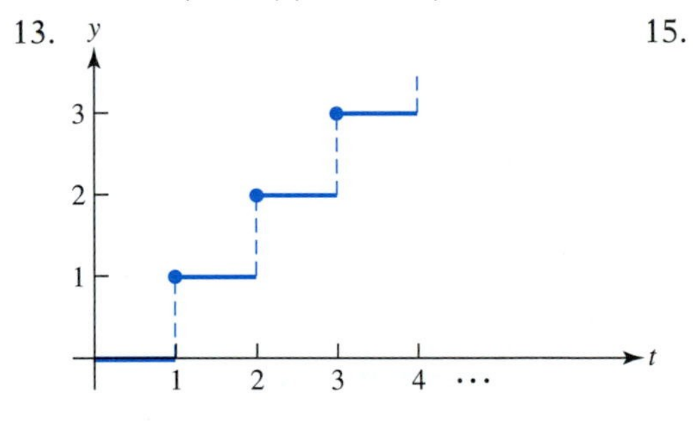

15.

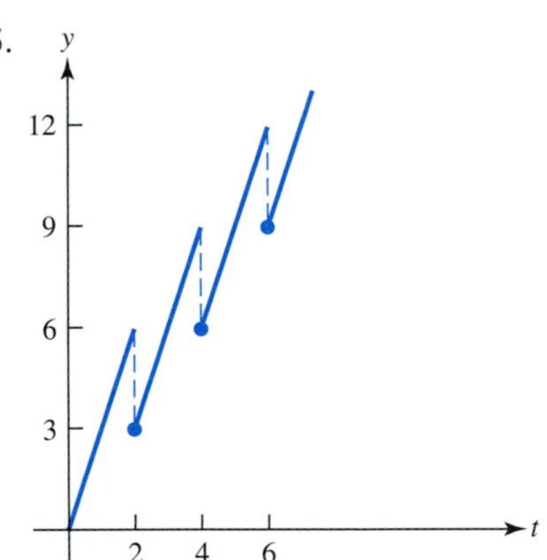

$$\mathcal{L}^{-1}\{F(s)\} = 3t - 3[h(t-2) + h(t-4) + h(t-6) + \cdots]$$

17. (a) $q' + 0.1q = 5c_i(t), \quad q(0) = 0, \quad$ where $c_i(t) = \begin{cases} 1, & 0 \le t < 1/2, \\ 0, & 1/2 \le t < 1, \end{cases} \quad c_i(t+1) = c_i(t)$

(b) $Q(s) = \dfrac{5(1-e^{-s/2})}{s(s+0.1)(1-e^{-s})}$

(c) $q(t) = \begin{cases} 50[1 - e^{-t/10} + e^{-(t-1/2)/10} - e^{-(t-1)/10}], & 1 \le t < 3/2 \\ 50[-e^{-t/10} + e^{-(t-1/2)/10} - e^{-(t-1)/10} + e^{-(t-3/2)/10}], & 3/2 \le t < 2 \end{cases}$

21. $\Phi(s) = 1/(s+1)^2$

23. (a) $\Phi(s) = 1/(s^2+4)$ (b) $Y(s) = 2/[s^3(s^2+4)]$

25. (a) $\Phi(s) = 1/(s+2)^2$ (b) $Y(s) = [1 - (s+1)e^{-s}]/[s^2(1-e^{-s})(s+2)^2]$

27. (a) $\Phi(s) = 1/[s(s^2+4)]$ (b) $Y(s) = 1/(s^2+4)^2$

29. $b = 0, c = 4, y_0 = 1, y_0' = 0$

Section 7.5, page 431

1. $\mathbf{Y}(s) = \begin{bmatrix} s/(s^2+1) \\ 1/s^2 \\ 1/(s-1)^2 \end{bmatrix}$

3. $\mathbf{Y}(s) = \begin{bmatrix} (2-se^{-2s})/s^2 \\ 2e^{-2s}/s \end{bmatrix}$

5. $\mathbf{Y}(s) = \begin{bmatrix} e^{-s}/(1+s^2) \\ 1/[(s-1)e] - 2/s^2 \end{bmatrix}$

7. $\mathcal{L}^{-1}\{\mathbf{Y}(s)\} = h(t-1)\begin{bmatrix} 1-\sin(t-1) \\ 2\sin(t-1) \end{bmatrix}$

9. $\mathbf{y}(t) = \begin{bmatrix} 4+e^t \\ 5+e^t \end{bmatrix}$

11. $\mathbf{y}(t) = \begin{bmatrix} t+e^t-e^{2t} \\ 1.5t-0.25+e^t-0.75e^{2t} \end{bmatrix}$

13. $\mathbf{y}(t) = e^t\begin{bmatrix} 2\cos 2t \\ -\sin 2t \end{bmatrix}$

15. $\mathbf{y}(t) = \begin{bmatrix} 2e^{3(t-1)} + 3e^{-2(t-1)} \\ 2e^{3(t-1)} + 8e^{-2(t-1)} \end{bmatrix}$

17. $\mathbf{y}(t) = (1/120)\begin{bmatrix} t^5-5t^4+20t^3 \\ t^5-5t^4+60t^2 \end{bmatrix}$

19. $\mathbf{y}(t) = \begin{bmatrix} 5e^{-t}-3e^t \\ -7e^{-t}+3e^t \\ -e^{-2t} \end{bmatrix}$

21. (a) $\lambda_1 = 6,\ \lambda_2 = 3$ (b) $A = \begin{bmatrix} 7 & -1 \\ 4 & 2 \end{bmatrix}$

23. $A = \begin{bmatrix} 6 & -3 \\ 8 & -5 \end{bmatrix}$

25. (b) $\mathbf{i}(t) = 0.25e^{-t}\begin{bmatrix} t^2 \\ t^2-2t \end{bmatrix}$

Section 7.6, page 442

3. $f*g = t^4/12$

5. $f*g = t - \sin t$

7. $f*g = (t^2/2) - 0.5h(t-1)(t-1)^2$

9. $\begin{bmatrix} t^3/6 \\ 1-\cos t \end{bmatrix}$

11.

$f*g = (t-2)h(t-2) - 2(t-3)h(t-3) + (t-4)h(t-4)$

13. $t^5/120$

15. $-t + (e^t - e^{-t})/2$

17. $n = 5, C = 1/24, \alpha = -1$

19. $y(t) = 2 - 2e^{-t} - 2t^2e^{-t}$

21. $y(t) = 6t$ or $y(t) = -6t$

23. $y(t) = 2 - 2e^{-t} - te^{-t}$

25. $y(t) = 1 + (t^4/24)$

Section 7.7, page 449

1. (a) $1+e^{-2}$ (b) 0 (c) $\begin{bmatrix} 1 \\ 0 \end{bmatrix}$ (d) $\begin{bmatrix} e^{-4}-2 \\ e^2+1 \\ 0 \end{bmatrix}$

3. One possible choice is $t_0 = 1/6$.

5.

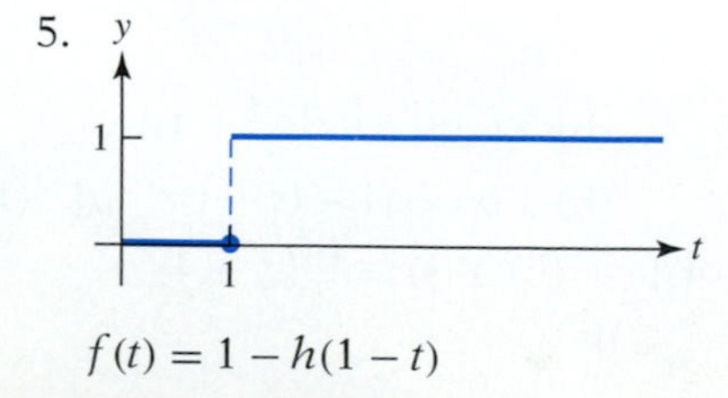

$f(t) = 1 - h(1-t)$

7.

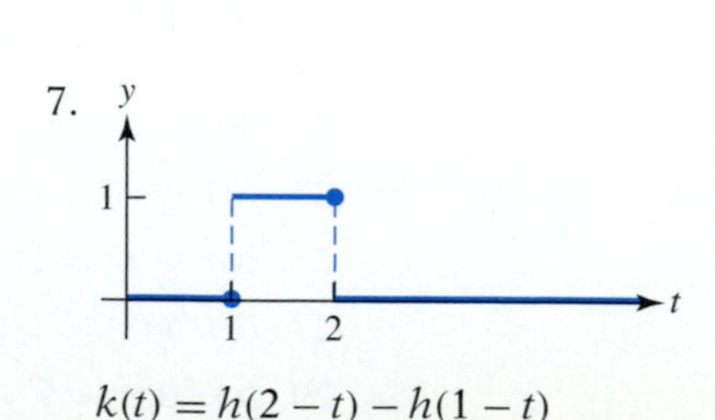

$k(t) = h(2-t) - h(1-t)$

9. (a) $y(t) = -1 + e^t$ (b) $\phi(t) = e^t$

11. (a) $y(t) = -t - 1 + e^t$ (b) $\phi(t) = e^t$

13.

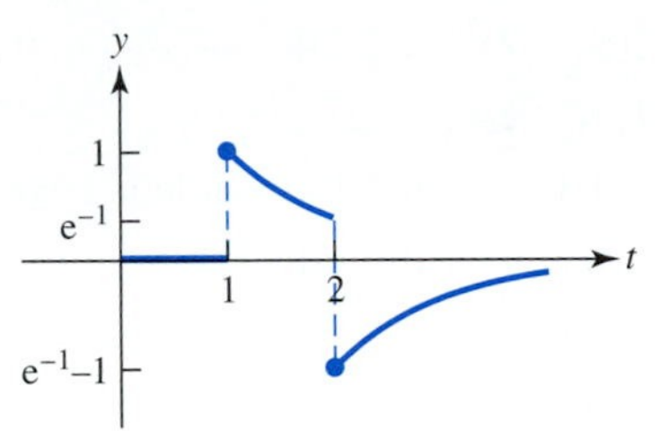

$y(t) = h(t-1)e^{-(t-1)} - h(t-2)e^{-(t-2)}$

15.

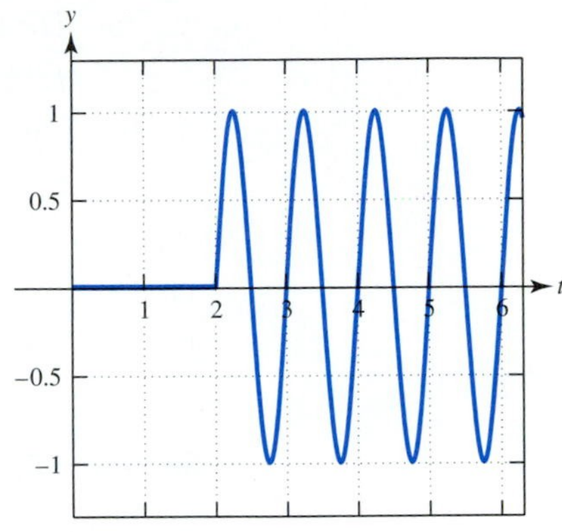

$y(t) = h(t-2)\sin[2\pi(t-2)]$

17.

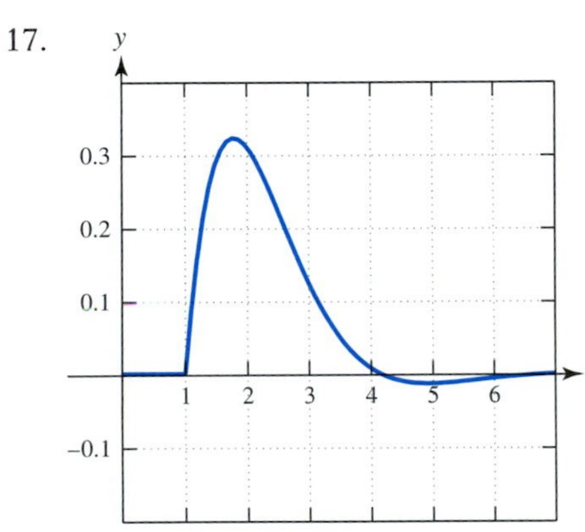

$y(t) = h(t-1)e^{-(t-1)}\sin(t-1)$

19. $\mathbf{y}(t) = \dfrac{h(t-1)}{2}\begin{bmatrix} 1 + e^{2(t-1)} \\ -1 + e^{2(t-1)} \end{bmatrix}$

CHAPTER 8

Section 8.1, page 462

1. (a) $\mathbf{y}' = \begin{bmatrix} y_2 \\ -ty_1 + \sin y_2 \end{bmatrix}$, $\mathbf{y}(0) = \begin{bmatrix} 0 \\ 1 \end{bmatrix}$

   (b) $\partial f_1/\partial y_1 = 0$, $\partial f_1/\partial y_2 = 1$,
   $\partial f_2/\partial y_1 = -t$, $\partial f_2/\partial y_2 = \cos y_2$

   (c) The hypotheses of Theorem 8.1 are satisfied everywhere.

3. (a) $\mathbf{y}' = \begin{bmatrix} y_2 \\ (e^{-t}/t) - 1/[(1 + y_1 + 2y_2)t] \end{bmatrix}$, $\mathbf{y}(2) = \begin{bmatrix} 2 \\ 1 \end{bmatrix}$

   (b) $\partial f_1/\partial y_1 = 0$, $\partial f_1/\partial y_2 = 1$,
   $\partial f_2/\partial y_1 = (1/t)(1 + y_1 + 2y_2)^{-2}$, $\partial f_2/\partial y_2 = (2/t)(1 + y_1 + 2y_2)^{-2}$

   (c) The hypotheses of Theorem 8.1 are satisfied except when $t = 0$ or $1 + y_1 + 2y_2 = 0$.

5. (a) $\mathbf{y}' = \begin{bmatrix} y_2 \\ y_3 \\ -2t^{1/3}/[(y_1 - 2)(y_3 + 2)] \end{bmatrix}$, $\mathbf{y}(2) = \begin{bmatrix} 0 \\ 2 \\ 2 \end{bmatrix}$

(b) $\partial f_1/\partial y_1 = 0, \quad \partial f_1/\partial y_2 = 1, \quad \partial f_1/\partial y_3 = 0$
$\partial f_2/\partial y_1 = 0, \quad \partial f_2/\partial y_2 = 0, \quad \partial f_2/\partial y_3 = 1$
$\partial f_3/\partial y_1 = 2t^{1/3}/[(y_1-2)^2(y_3+2)], \quad \partial f_3/\partial y_2 = 0,$
$\partial f_3/\partial y_3 = 2t^{1/3}/[(y_1-2)(y_3+2)^2]$
(c) The hypotheses of Theorem 8.1 are satisfied except when $y_1 = 2$ or $y_3 = -2$.

7. $y'' = e^{y'} + y' \tan y, \quad y(0) = 0, \quad y'(0) = 1$
9. $y''' = (y'y'' + t^2)^{1/2}, \quad y(1) = 1, \quad y'(1) = 1/2, \quad y''(1) = 3$
11. No, because the differential equation is nonlinear.
15. (b) $a' = -k_1 e_0 a + k_1 ac + k_1' c$
$c' = k_1 e_0 a - k_1 ac - (k_1' + k_2)c$

Section 8.2, page 477

1. (0, 0), (1, 1)
3. (1, −1), (3, −1)
5. (0, 0)
7. (2, 2), (2, −2), (−2, 2), (−2, −2)
9. (0, 1, 2), (−1, 1, 2)
11. $x' = y$
$y' = -x - x^3$
The only equilibrium point is (0, 0).
13. $x' = y$
$y' = 1 - x^2 - 2y/(1 + x^4)$
The only equilibrium points are (1, 0) and (−1, 0).
15. $x' = y$
$y' = z$
$z' = y^2 + (x^2 - 4)/(y^2 + 2)$
The only equilibrium points are (−2, 0, 0) and (2, 0, 0).
17. $\alpha = -1, \quad \beta = -1, \quad \gamma = 1/2, \quad \delta = 1/2$
19. $\alpha = 1, \quad \beta = 1, \quad \gamma = -2$
21. $\alpha = 8$
23. (b) The velocity vector is a constant multiple of the position vector. Therefore, the velocity vector is oriented along the line.
25. Direction Field B
27. Direction Field A
29. (a) $x' = y$
$y' = -x - x^3$
(c)

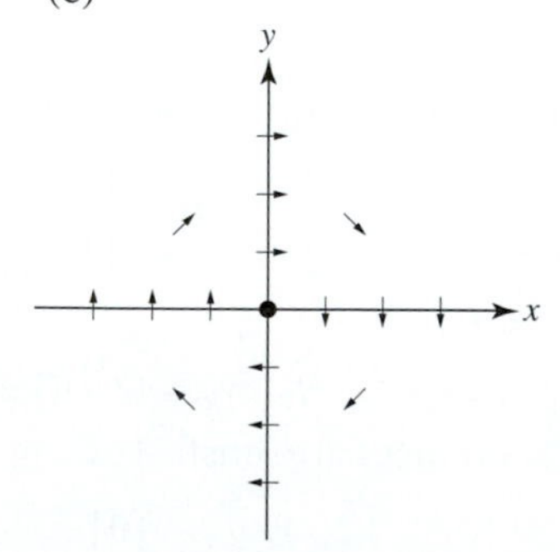

31. (a) $x' = y$
$y' = 1 - 2\sin^2 x$
(c)

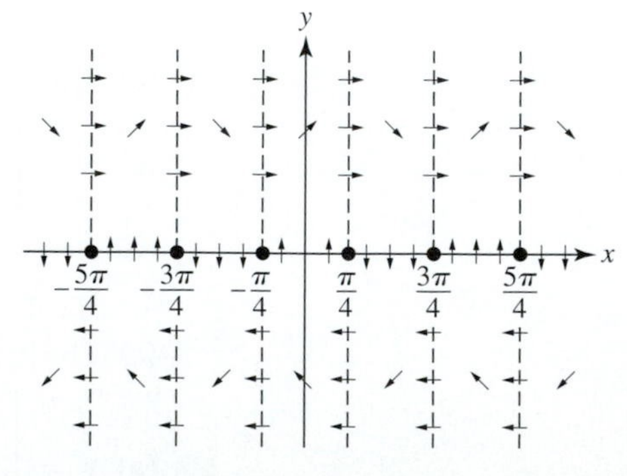

33. (a) equilibrium point (1, 1).

(b)

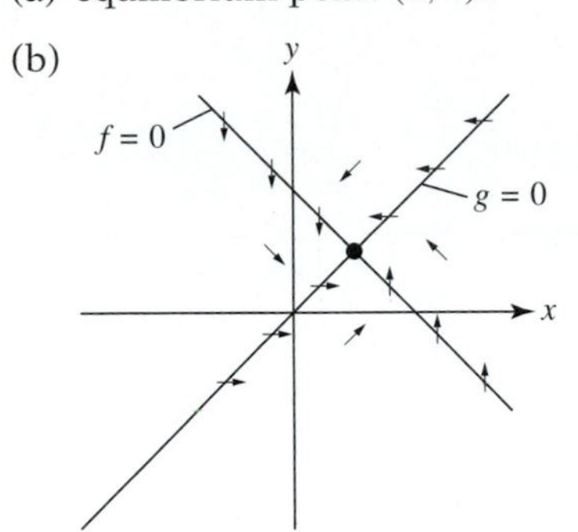

35. (a) equilibrium points (0, 0) and (2, −2)

(b)

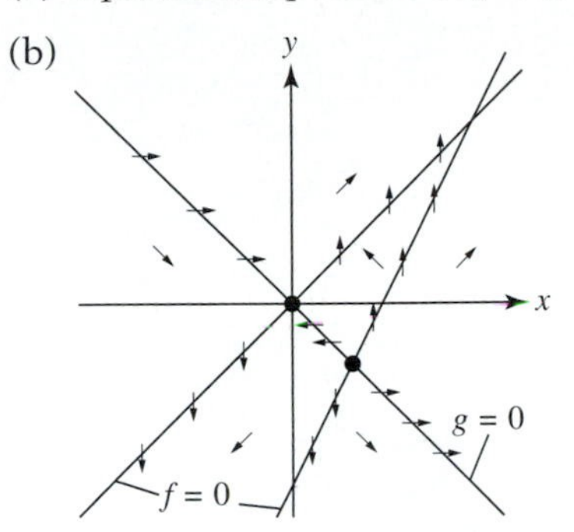

Section 8.3, page 487

1. (a) $0.5(x')^2 + 2x^2 = C$

(b) $x' = y$

$y' = -4x$

(c) $\mathbf{v}(1, 1) = \mathbf{i} - 4\mathbf{j}$

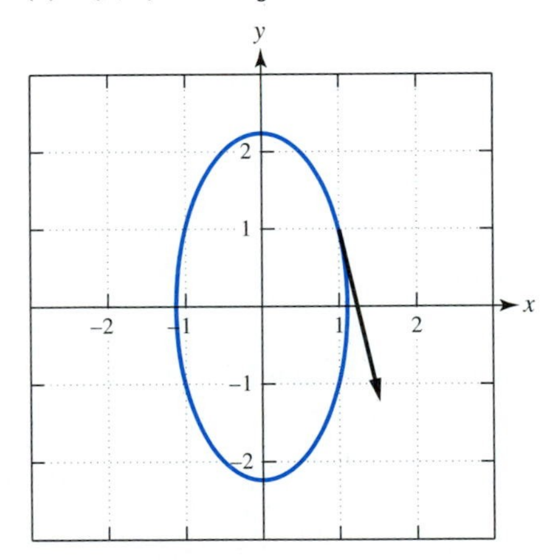

3. (a) $0.5(x')^2 + 0.25x^4 = C$

(b) $x' = y$

$y' = -x^3$

(c) $\mathbf{v}(1, 1) = \mathbf{i} - \mathbf{j}$

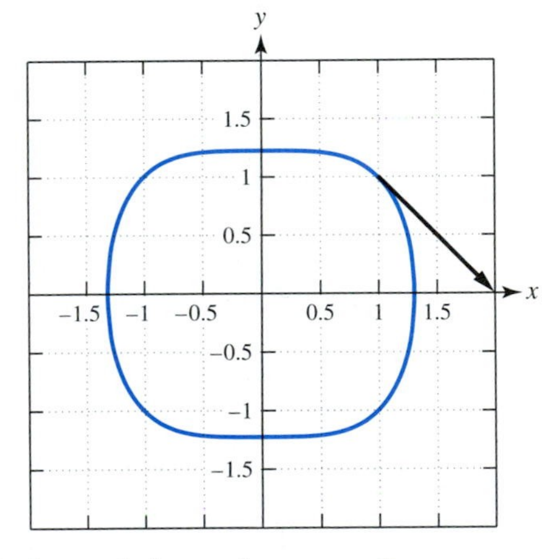

5. (a) $3(x')^2 + 2x^3 = C$

(b) $x' = y$

$y' = -x^2$

(c) $\mathbf{v}(1, 1) = \mathbf{i} - \mathbf{j}$

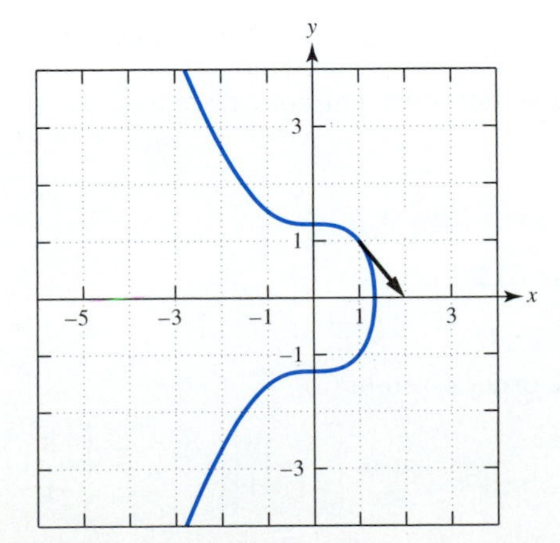

7. $2x'' - x^2 \sin x + 2x \cos x = 0$

9. (a)

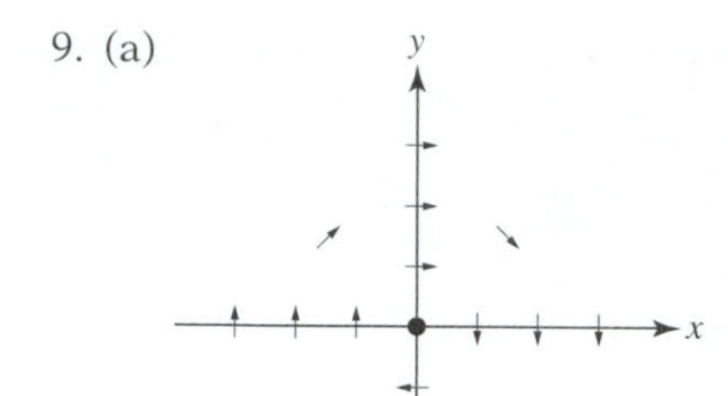

(c) $H(x, y) = 0.5(x^2 + y^2) + (0.25)x^4$

11. $0.5(y'')^2 + F(y') = C$, where $F(u)$ is an antiderivative of $f(u)$.

13. (b) $H(x, y) = 2xy$

15. (b) $H(x, y) = xy - x^2y + y - 2x^2$

17. (b) $H(x, y) = y^2 \cos x$

19. (b) $H(x, y) = -x^3 - y^2$

21. $H(x, y) = x^3y - \cos(2x + 3y)$

23. Not a Hamiltonian system

25. $H(x, y) = 0.5(y^2 - x^2) + x^3/3$

27. $H(x, y) = F(y) - G(x)$

29. $H(x, y) = 3F(y) - xy^2 - G(x) - x$

31. Yes, $K(x, y)$

Section 8.4, page 497

1. (a) $\delta = \varepsilon/2$ (b) Not asymptotically stable

3. (a) $x' = y$

$y' = -x - \gamma y$

(b) Asymptotically stable if $\gamma > 0$, stable if $\gamma = 0$, unstable if $\gamma < 0$

5. Asymptotically stable

7. Unstable

9. Unstable

11. Unstable

13. Stable but not asymptotically stable

15. Asymptotically stable

17. Unstable

19. Unstable

21. Asymptotically stable

23. Asymptotically stable

25. $\mathbf{y}_e = \begin{bmatrix} -2 \\ 0 \end{bmatrix}$, asymptotically stable

27. $\mathbf{y}_e = \begin{bmatrix} 2 \\ -2 \end{bmatrix}$, unstable

29. $\mathbf{y}_e = \begin{bmatrix} 0 \\ 0 \\ 0 \end{bmatrix}$, unstable

31. $\mathbf{y}_e = \begin{bmatrix} 0 \\ 0 \\ 0 \\ 0 \end{bmatrix}$, stable but not asymptotically stable

35. $a_{12} = 1/2, \quad a_{22} = 1, \quad a_{21} = 2$

Section 8.5, page 512

1. (a) $(4, 4)$, $(-4, -4)$

(b) Using the order given in (a),

$$\mathbf{z}' = \begin{bmatrix} 8 & 8 \\ -1 & 1 \end{bmatrix} \mathbf{z}, \quad \mathbf{z}' = \begin{bmatrix} -8 & -8 \\ -1 & 1 \end{bmatrix} \mathbf{z}.$$

(c) $(4, 4)$ and $(-4, -4)$ are each unstable equilibrium points.

3. (a) $(1,1), (-1,1), (-1,-1), (1,-1)$

   (b) Using the order given in (a),

$$\mathbf{z}' = \begin{bmatrix} -2 & 0 \\ 2 & 2 \end{bmatrix}\mathbf{z}, \quad \mathbf{z}' = \begin{bmatrix} 2 & 0 \\ -2 & 2 \end{bmatrix}\mathbf{z}, \quad \mathbf{z}' = \begin{bmatrix} 2 & 0 \\ -2 & -2 \end{bmatrix}\mathbf{z}, \quad \mathbf{z}' = \begin{bmatrix} -2 & 0 \\ 2 & -2 \end{bmatrix}\mathbf{z}.$$

   (c) $(1,1)$ is unstable, $(-1,1)$ is unstable, $(-1,-1)$ is unstable, $(1,-1)$ is asymptotically stable.

5. (a) $(2,1), (2,-1), (-6,3)$

   (b) Using the order given in (a),

$$\mathbf{z}' = \begin{bmatrix} -2 & 0 \\ 0 & 4 \end{bmatrix}\mathbf{z}, \quad \mathbf{z}' = \begin{bmatrix} -4 & 0 \\ -2 & -4 \end{bmatrix}\mathbf{z}, \quad \mathbf{z}' = \begin{bmatrix} 0 & -8 \\ 2 & 4 \end{bmatrix}\mathbf{z}.$$

   (c) $(2,1)$ is an unstable equilibrium point, $(2,-1)$ is an asymptotically stable equilibrium point, $(-6,3)$ is an unstable equilibrium point.

7. (a) $(0,0), (-2,-4)$

   (b) Using the order given in (a),

$$\mathbf{z}' = \begin{bmatrix} 4 & -8 \\ 2 & -1 \end{bmatrix}\mathbf{z}, \quad \mathbf{z}' = \begin{bmatrix} 0 & 6 \\ 2 & -1 \end{bmatrix}\mathbf{z}.$$

   (c) $(0,0)$ and $(-2,-4)$ are each unstable equilibrium points.

9. (a) $(0,0), (1,1)$

   (b) Using the order given in (a),

$$\mathbf{z}' = \begin{bmatrix} -1 & 0 \\ 0 & -1 \end{bmatrix}\mathbf{z}, \quad \mathbf{z}' = \begin{bmatrix} -1 & 2 \\ 2 & -1 \end{bmatrix}\mathbf{z}.$$

   (c) $(0,0)$ is an asymptotically stable equilibrium point, $(1,1)$ is an unstable equilibrium point.

13. (a) $A = \begin{bmatrix} 5 & -14 \\ 3 & -8 \end{bmatrix}, \quad \mathbf{g}(\mathbf{z}) = \begin{bmatrix} z_1 z_2 \\ z_1^2 + z_2^2 \end{bmatrix}$

   (c) The limit is 0. $\mathbf{z}' = A\mathbf{z} + \mathbf{g}(\mathbf{z})$ is an almost linear system.

   (d) $\mathbf{z} = \mathbf{0}$ is an asymptotically stable equilibrium point.

15. (a) $A = \begin{bmatrix} -1 & 3 \\ -1 & -5 \end{bmatrix}, \quad \mathbf{g}(\mathbf{z}) = \begin{bmatrix} z_2 \cos\sqrt{z_1^2 + z_2^2} \\ z_1 \cos\sqrt{z_1^2 + z_2^2} \end{bmatrix}$

   (c) The limit is 1. $\mathbf{z}' = A\mathbf{z} + \mathbf{g}(\mathbf{z})$ is not an almost linear system.

17. (a) $A = \begin{bmatrix} 0 & 2 \\ -2 & 0 \end{bmatrix}, \quad \mathbf{g}(\mathbf{z}) = \begin{bmatrix} z_2^2 \\ z_1 z_2 \end{bmatrix}$

   (c) The limit is 0. $\mathbf{z}' = A\mathbf{z} + \mathbf{g}(\mathbf{z})$ is an almost linear system.

   (d) No conclusion can be drawn using Theorem 8.4.

19. (a) $A = \begin{bmatrix} 9 & 5 \\ -7 & -3 \end{bmatrix}, \quad \mathbf{g}(\mathbf{z}) = \begin{bmatrix} z_1 z_2 \\ z_1^2 \end{bmatrix}$

   (c) The limit is 0. $\mathbf{z}' = A\mathbf{z} + \mathbf{g}(\mathbf{z})$ is an almost linear system.

   (d) $\mathbf{z} = \mathbf{0}$ is an unstable equilibrium point.

21. (b) $\mathbf{z}' = \begin{bmatrix} -1 & 1 \\ 1 & -2 \end{bmatrix}\mathbf{z}$, asymptotically stable at $\mathbf{z} = \begin{bmatrix} 0 \\ 0 \end{bmatrix}$.

   (d) The given nonlinear system is asymptotically stable at $(0,0)$.

23. (a) $0.5y^2 - x + (2/5)(1+x)^{5/2} = C$

(b) The graphs are consistent with bobbing motion. The origin appears to be a stable equilibrium point but not an asymptotically stable equilibrium point.

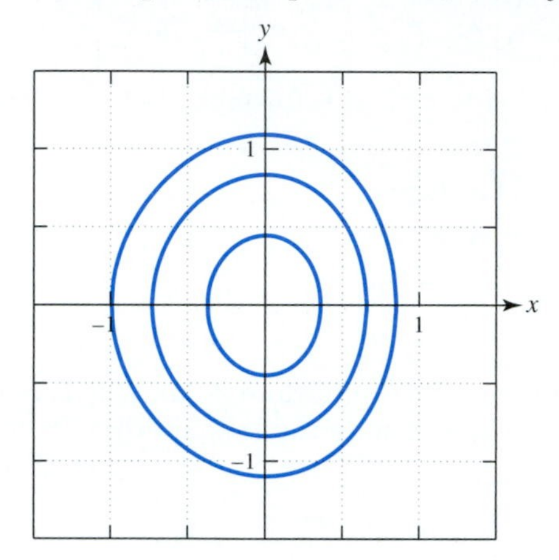

27. $x = [\cos(c_1 - t)]/\sqrt{2(c_2 - \alpha t)}$, $y = [\sin(c_1 - t)]/\sqrt{2(c_2 - \alpha t)}$; c_1 and c_2 are arbitrary constants.

29. $x(1) = \exp(e^{-2})\cos(\pi/4 - 1)$, $y(1) = \exp(e^{-2})\sin(\pi/4 - 1)$

Section 8.6, page 522

1. (a) Saddle point, unstable

(b)

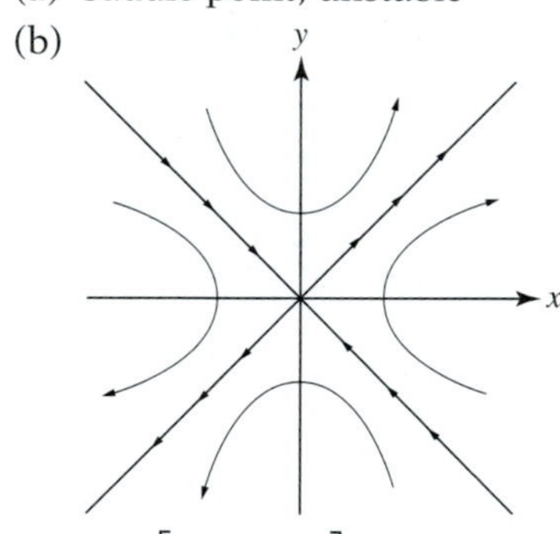

(c) $A = \begin{bmatrix} 1/2 & 3/2 \\ 3/2 & 1/2 \end{bmatrix}$

3. (a) Unstable improper node

(b)

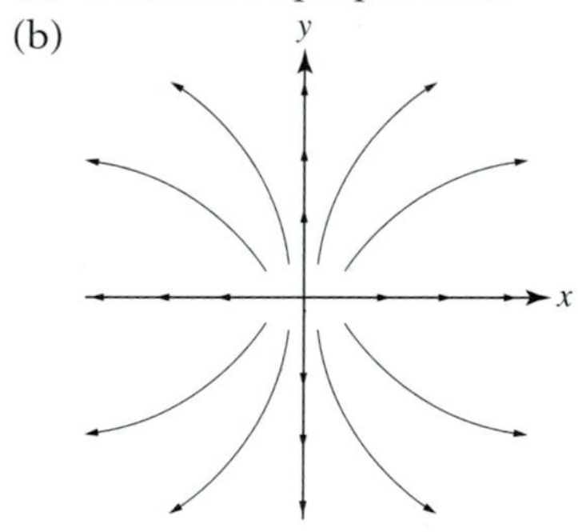

(c) $A = \begin{bmatrix} 2 & 0 \\ 0 & 1 \end{bmatrix}$

5. (a) Saddle point, unstable

(b)

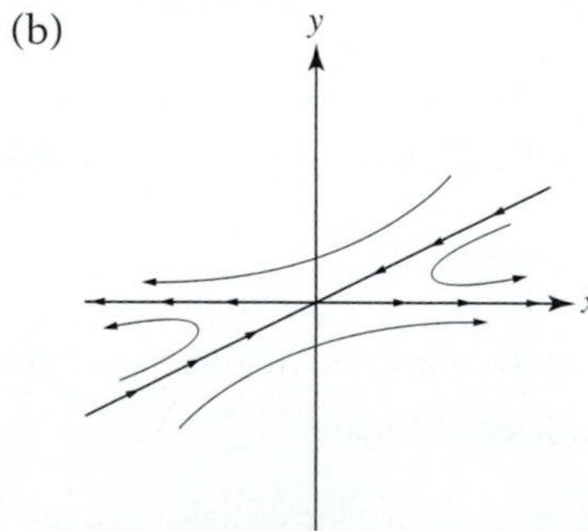

(c) $A = \begin{bmatrix} 1 & -4 \\ 0 & -1 \end{bmatrix}$

7. (a) $\lambda = 1,\ \lambda = 2$ (b) Unstable improper node

9. (a) $\lambda = 3i,\ \lambda = -3i$ (b) Center, stable but not asymptotically stable

11. (a) $\lambda = -2,\ \lambda = -3$ (b) Improper node, asymptotically stable

13. (a) $\lambda = 4i,\ \lambda = -4i$ (b) Center, stable but not asymptotically stable

15. (a) $\lambda = 1,\ \lambda = 3$ (b) Improper node, unstable

17. (a) $\lambda = -2,\ \lambda = -2$ (b) Improper node, asymptotically stable

19. (a) $\lambda = 1 + 4i,\ \lambda = 1 - 4i$ (b) Spiral point, unstable

21. (a) Direction Field 2 (b) Direction Field 4 (c) Direction Field 1
(d) Direction Field 3

23. $\alpha > 9/2$

25. $\alpha < 8$

27. $\mathbf{y}_e = \begin{bmatrix} 6 \\ -8 \end{bmatrix}$ is an unstable saddle point.

29. $\mathbf{y}_e = \begin{bmatrix} 2 \\ 2 \end{bmatrix}$ is an unstable saddle point.

31. (c) The solution point moves clockwise.

33. (a) $\lambda = \pm 2\sqrt{2}\,i$ (b) Center, stable but not asymptotically stable
(c) $H(x, y) = (3x^2 + 2xy + 3y^2)/2$

Section 8.7, page 530

1. (a) x approaches an equilibrium population of 1 in the absence of y. If y is present, it is harmful to x. y approaches an equilibrium population of 1/3 in the absence of x. If x is present, it is harmful to y.

   (b) $(0, 0)$, $(1, 0)$, $(0, 1/3)$, $(4/5, 1/5)$

   (c) The origin is an unstable proper node of the linearized system $\mathbf{z}' = \begin{bmatrix} 1 & 0 \\ 0 & 1 \end{bmatrix} \mathbf{z}$.

   Therefore, the given system is unstable at the origin.

3. (a) x approaches an equilibrium population of 1 in the absence of y. If y is present, it is harmful to x. y approaches an equilibrium population of 0 in the absence of x. If x is present, it is beneficial to y.

   (b) $(0, 0)$, $(1, 0)$

   (c) The origin is an unstable saddle point of the linearized system $\mathbf{z}' = \begin{bmatrix} 1 & 0 \\ 0 & -1 \end{bmatrix} \mathbf{z}$.

   Therefore, the given system is unstable at the origin.

5. (a) $\alpha_1 = 0, \alpha_2 = 0, r_1 = 1/2, r_2 = -1, \beta_1 = -1/3, \beta_2 = -1/2$

   (b) x is beneficial to y but y is harmful to x.

7. (a) $(0, 0)$, $(1/\alpha, 0)$, $(0, [r - \mu]/[\alpha r])$,
   $([\alpha r - \beta r + \beta\mu]/[\alpha^2 r - \beta^2 r], [\alpha r - \beta r - \alpha\mu]/[\alpha^2 r - \beta^2 r])$

   (c) $(0, 0)$ is an unstable saddle point, $(1/\alpha, 0)$ is an asymptotically stable improper node.

9. $x' = a_1 x - b_1 x^2 - c_1 xy - d_1 xz$

   $y' = a_2 y - b_2 y^2 - c_2 xy - d_2 yz$

   $z' = -a_3 z + c_3 xz + d_3 yz$

11. (a) $s' = -\alpha si - \gamma s - \gamma i + \gamma N$

   $i' = \alpha si - \beta i$

   (b) $(s, i) = (9, 0)$ and $(s, i) = (1, 4)$

(c) For $(9, 0)$, $\mathbf{z}' = \begin{bmatrix} -1 & -10 \\ 0 & 8 \end{bmatrix} \mathbf{z}$, unstable saddle point

For $(1, 4)$, $\mathbf{z}' = \begin{bmatrix} -5 & -2 \\ 4 & 0 \end{bmatrix} \mathbf{z}$, asymptotically stable spiral point

CHAPTER 9

Section 9.1, page 543

1. (a) $y = t^2 - t$
   (b) $y_{n+1} = y_n + (h/2)[(2t_n - 1) + (2t_{n+1} - 1)]$
   (c) $y_{n+1} = y_n + h[2(t_n + h/2) - 1]$
   (d) $y_1 = 0.11,\ \ y_2 = 0.24,\ \ y_3 = 0.39$
   (e) $y_1 = 0.11,\ \ y_2 = 0.24,\ \ y_3 = 0.39$
   (f) $y(t_1) = 0.11,\ \ y(t_2) = 0.24,\ \ y(t_3) = 0.39$
3. (a) $y = e^{-t^2/2}$
   (b) $y_{n+1} = y_n + (h/2)[-t_n y_n - t_{n+1}(y_n - h t_n y_n)]$
   (c) $y_{n+1} = y_n - h(t_n + 0.5h)(y_n - 0.5h t_n y_n)$
   (d) $y_1 = 0.9950,\ \ y_2 = 0.9801\ldots,\ \ y_3 = 0.9559\ldots$
   (e) $y_1 = 0.9950,\ \ y_2 = 0.9801\ldots,\ \ y_3 = 0.9558\ldots$
   (f) $y(t_1) = 0.9950\ldots,\ \ y(t_2) = 0.9801\ldots,\ \ y(t_3) = 0.9559\ldots$
5. (a) $y = (1 - 1.5t^2)^{1/3}$
   (b) $y_{n+1} = y_n + (h/2)[-t_n y_n^{-2} - t_{n+1}(y_n - h t_n y_n^{-2})^{-2}]$
   (c) $y_{n+1} = y_n - h(t_n + 0.5h)(y_n - 0.5h t_n y_n^{-2})^{-2}$
   (d) $y_1 = 0.9950,\ \ y_2 = 0.9796\ldots,\ \ y_3 = 0.9529\ldots$
   (e) $y_1 = 0.9950,\ \ y_2 = 0.9796\ldots,\ \ y_3 = 0.9530\ldots$
   (f) $y(t_1) = 0.9949\ldots,\ \ y(t_2) = 0.9795\ldots,\ \ y(t_3) = 0.9528\ldots$
7. (a) $y = \sqrt{9 - t^2}$
   (b) $\left|y(1) - y_{20}^{\text{Euler}}\right| = 0.009146\ldots,$
   $\left|y(1) - y_{20}^{\text{Heun}}\right| = 6.902 \times 10^{-7},\ \ \left|y(1) - y_{20}^{\text{modEul}}\right| = 1.375 \times 10^{-5}$
9. (a) $y = 2e^{-t^2}$
   (b) $\left|y(1) - y_{20}^{\text{Euler}}\right| = 0.01300\ldots,$
   $\left|y(1) - y_{20}^{\text{Heun}}\right| = 6.022 \times 10^{-4},\ \ \left|y(1) - y_{20}^{\text{modEul}}\right| = 3.329 \times 10^{-4}$
11. Heun's method, with $t_0 = 1, T = 5, f(t, y) = ty^2 + 1$
13. Euler's method, with $t_0 = 2, T = 1, f(t, y) = y/(t^2 + y^2)$
17. (a) The initial value problem is $Q' = 12 - 6\cos(\pi t) - Q/(90 + 5t), Q(0) = 0.$
    (c) About 23.75 lb
19. About 1.5003 million individuals

Section 9.2, page 557

1. (a) $y'(0) = 1, y''(0) = -1, y'''(0) = 1, y^{(4)}(0) = -1$
   $P_4(t) = 1 + t - (1/2)t^2 + (1/6)t^3 - (1/24)t^4$

   (c) $E(0.1) = 8.196 \times 10^{-8}$

3. (a) $y'(0) = 0, y''(0) = 1, y'''(0) = 0, y^{(4)}(0) = 6$
   $P_4(t) = 1 + (1/2)t^2 + (1/4)t^4$

   (c) $E(0.1) = 1.256 \times 10^{-7}$

5. (a) $y'(0) = 1, y''(0) = 1/2, y'''(0) = 0, y^{(4)}(0) = 0$
   $P_4(t) = 1 + t + (1/4)t^2$

   (c) $E(0.1) = 0$

7. (a) $y'(0) = 1, y''(0) = 2, y'''(0) = 2, y^{(4)}(0) = 1$
   $P_4(t) = 1 + t + t^2 + (1/3)t^3 + (1/24)t^4$

   (c) $E(0.1) = 8.615 \times 10^{-8}$

9. (a) $y'(0) = 2, y''(0) = 4, y'''(0) = 16, y^{(4)}(0) = 80$
   $P_4(t) = 1 + 2t + 2t^2 + (8/3)t^3 + (10/3)t^4$

   (c) $E(0.1) = 4.888 \times 10^{-5}$

11. (a) $P_5(t) = 1 - t^2 - t^3 - (7/12)t^4 - (1/4)t^5$

    (b) $E(0.1) = -8.867 \times 10^{-8}$

13. (a) $P_5(t) = 1 + 2t + (1/3)t^3 + (1/60)t^5$

    (b) $E(0.1) = 3.968 \times 10^{-11}$

15. $r = 1$

17. $r = 2$

21. At $t = 1$, the errors are: (order 1) $1.8054\ldots \times 10^{-3}$, (order 2) $-4.0475\ldots \times 10^{-4}$, (order 3) $6.8372\ldots \times 10^{-6}$.

23. At $t = 1$, the errors are: (order 1) $-7.2978\ldots \times 10^{-3}$, (order 2) $8.2708\ldots \times 10^{-4}$, (order 3) $-3.0262\ldots \times 10^{-5}$.

25. At $t = 1$, the error ratio is $E_2/E_1 = 8.4648\ldots \times 10^{-7}/6.8372\ldots \times 10^{-6} = 0.1238\ldots$.

27. At $t = 1$, the error ratio is
    $E_2/E_1 = -3.6501\ldots \times 10^{-6}/ -3.0262\ldots \times 10^{-5} = 0.1206\ldots$.

Section 9.3, page 569

1. (a) $y_1 = 1.09516666\ldots$ (b) $y_1 = 1.09516250\ldots$ (c) No
   (d) $y(t_1) = 1.09516258\ldots$

3. (a) $y_1 = 1.00503350\ldots$ (b) $y_1 = 1.00502513\ldots$ (c) No
   (d) $y(t_1) = 1.00502512\ldots$

5. (a) $y_1 = 1.10249901\ldots$

   (b) $y_1 = 1.10249998\ldots$

   (c) Yes, since methods of order p or higher yield exact results if the solution is a polynomial of degree p or less.

   (d) $y(t_1) = 1.10250000\ldots$ Note that the numerical results do not agree with our answer in part (c). This discrepancy is due to the finite precision of computer arithmetic.

7. (a) $y_1 = 1.11032909\ldots$ (b) $y_1 = 1.11033743\ldots$ (c) No
   (d) $y(t_1) = 1.11033758\ldots$

9. (a) $y_1 = 1.22304273\ldots$ (b) $y_1 = 1.22304891\ldots$ (c) No
(d) $y(t_1) = 1.22304888\ldots$

11. $\mathbf{y}_1 = \begin{bmatrix} 0.89534540\ldots \\ -1.08953454\ldots \end{bmatrix}$

13. $\mathbf{y}_1 = \begin{bmatrix} 2.10571842\ldots \\ 1.22088665\ldots \end{bmatrix}$

15. $\mathbf{y}_1 = \begin{bmatrix} 0.99500416\ldots \\ -0.09983354\ldots \\ -0.99501250\ldots \end{bmatrix}$

19. $y_{20} = 1.23606797\ldots,\ y(1) = 1.23606797\ldots$

21. $y_{20} = 1.41421356\ldots,\ y(1) = 1.41421356\ldots$

23. $y_{20} = -0.81020240\ldots,\ y(1) = -0.81019930\ldots$
$y'_{20} = -0.42549625\ldots,\ y'(1) = -0.42549942\ldots$

25. $y_{10} = 4.613705448\ldots,\ y(1) = 4.613705638\ldots$
$y'_{10} = 3.30685272\ldots,\ y'(1) = 3.30685281\ldots$

27.

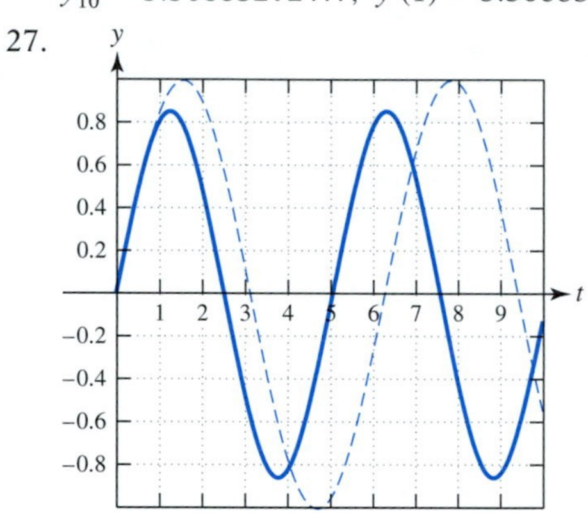

Solid is numerical solution, dotted is sin(*t*).

29.

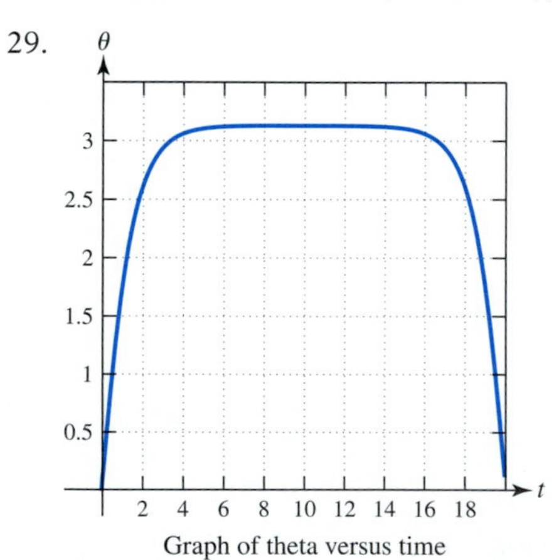

Graph of theta versus time

and

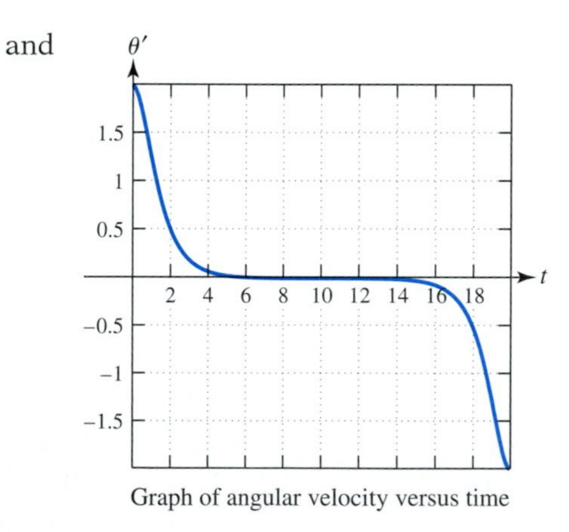

Graph of angular velocity versus time

CHAPTER 10

Section 10.1, page 590

1. $R = 2$
3. $R = 1$
5. $R = \infty$
7. $R = 1$
9. $R = 1$
11. $R = 2$

13. $R = 1$

(a) $f(t) = 1 + t + t^2 + t^3 + t^4 + t^5 + \cdots$
$g(t) = t + 4t^2 + 9t^3 + 16t^4 + 25t^5 + \cdots$

(b) $f(t) + g(t) = 1 + 2t + 5t^2 + 10t^3 + 17t^4 + 26t^5 + \cdots$

(c) $f(t) - g(t) = 1 - 3t^2 - 8t^3 - 15t^4 - 24t^5 + \cdots$

(d) $f'(t) = 1 + 2t + 3t^2 + 4t^3 + 5t^4 + 6t^5 + \cdots$

(e) $f''(t) = 2 + 6t + 12t^2 + 20t^3 + 30t^4 + 42t^5 + \cdots$

15. $R = 1/2$

(a) $f(t) = 1 - 2(t-1) + 4(t-1)^2 - 8(t-1)^3 + 16(t-1)^4 - 32(t-1)^5 + \cdots$
$g(t) = 1 + (t-1) + (t-1)^2 + (t-1)^3 + (t-1)^4 + (t-1)^5 + \cdots$

(b) $f(t) + g(t) = 2 - (t-1) + 5(t-1)^2 - 7(t-1)^3 + 17(t-1)^4 - 31(t-1)^5 + \cdots$

(c) $f(t) - g(t) = -3(t-1) + 3(t-1)^2 - 9(t-1)^3 + 15(t-1)^4 - 33(t-1)^5 + \cdots$

(d) $f'(t) = -2 + 8(t-1) - 24(t-1)^2 + 64(t-1)^3 - 160(t-1)^4 + \cdots$

(e) $f''(t) = 8 - 48(t-1) + 192(t-1)^2 - 640(t-1)^3 + \cdots$

17. $\sum_{n=2}^{\infty} 2^{n-2} t^n$

19. $\sum_{n=2}^{\infty} a_{n-2} t^n$

21. $\sum_{n=0}^{\infty} (n+2)(n+1)a_{n+2} t^n$

23. $-\sum_{n=2}^{\infty} (-1)^n (n-1) a_{n-2} t^n$

25. $f(t) = \sum_{n=1}^{\infty} (-1)^{n+1} (3t)^{2n}/(2n)!, R = \infty$

27. $f(t) = \sum_{n=0}^{\infty} t^{2n}, R = 1$

29. (a) $a_{n+2} = \dfrac{\omega^2}{(n+2)(n+1)} a_n$ (b) $y_1(t) = \cosh \omega t, y_2(t) = \sinh \omega t$

31. (a) $y(t) = \sum_{n=0}^{\infty} (t-1)^n/n!$ (b) $R = \infty$ (c) $y(t) = e^{t-1}$

33. (a) $y(t) = -\sum_{n=3}^{\infty} (-1)^n (t-1)^n/n!$ (b) $R = \infty$

(c) $y(t) = -e^{-(t-1)} + 1 - (t-1) + (t-1)^2/2$

35. (a) $y(t) = \sum_{n=0}^{\infty} t^n$ (b) $R = 1$ (c) $y(t) = 1/(1-t)$

37. $a_0 = 0,\ a_1 = 1,\ a_2 = 0,\ a_3 = 1/6,\ a_4 = 0,\ a_5 = -1/120$

39. $a_0 = -1,\ a_1 = 1,\ a_2 = 0,\ a_3 = -1/3,\ a_4 = 1/12,\ a_5 = 1/20$

41. $a_0 = 0,\ a_1 = 2,\ a_2 = 2,\ a_3 = 1,\ a_4 = 1/3,\ a_5 = 1/12$

Section 10.2, page 601

1. $\pm\pi/2,\ \pm 2,\ \pm 3\pi/2,\ \pm 5\pi/2$

3. $0,\ \pm 1,\ \pm\pi,\ \pm 2\pi,\ \pm 3\pi$

5. $0, \pm e^{-1}$

7. $R = 1/2$

9. $R = 5/6$

11. $R = 1/3$

13. (a) $a_{n+2} = -a_n/(n+2)$

(b) $y_1(t) = a_0[1 - (1/2)t^2 + (1/8)t^4 + \cdots]$
$y_2(t) = a_1[t - (1/3)t^3 + (1/15)t^5 + \cdots]$

(c) $R = \infty$

(d) Yes

15. (a) $a_{n+2} = -(n^2+2)a_n/[(n+2)(n+1)]$

(b) $y_1(t) = a_0[1 - t^2 + (1/2)t^4 + \cdots]$
$y_2(t) = a_1[t - (1/2)t^3 + (11/40)t^5 + \cdots]$

(c) $R = 1$

(d) Yes

17. (a) $(n+2)(n+1)a_{n+2} = 4(n+1)a_{n+1} - 4a_n$

(b) $y_1(t) = a_0[1 - 2t^2 - (8/3)t^3 + \cdots]$
$y_2(t) = a_1[t + 2t^2 + 2t^3 + \cdots]$

(c) $R = \infty$

(d) No

19. (a) $3(n+2)(n+1)a_{n+2} = -n(n+1)a_{n+1} - (3n+1)a_n$

(b) $y_1(t) = a_0[1 - (1/6)t^2 + (1/54)t^3 + \cdots]$
$y_2(t) = a_1[t - (2/9)t^3 + (1/27)t^4 + \cdots]$

(c) $R = 3$

(d) No

21. (a) $(n+2)(n+1)a_{n+2} = -a_{n-2}$

(b) $y_1(t) = a_0[1 - (1/12)t^4 + (1/672)t^8 + \cdots]$
$y_2(t) = a_1[t - (1/20)t^5 + (1/1440)t^9 + \cdots]$

(c) $R = \infty$

(d) Yes

23. (a) $(n+2)(n+1)a_{n+2} = -a_n$

(b) $y_1(t) = a_0[1 - (1/2)(t-1)^2 + (1/24)(t-1)^4 + \cdots]$
$y_2(t) = a_1[(t-1) - (1/6)(t-1)^3 + (1/120)(t-1)^5 + \cdots]$

(c) $R = \infty$

25. (a) $(n+2)(n+1)a_{n+2} = -(n+1)a_{n+1} + a_n - a_{n-1}$

(b) $y_1(t) = a_0[1 + (1/2)(t-1)^2 - (1/3)(t-1)^3 + \cdots]$
$y_2(t) = a_1[(t-1) - (1/2)(t-1)^2 + (1/3)(t-1)^3 + \cdots]$

(c) $R = \infty$

29. (a) $(n+2)(n+1)a_{n+2} = -2(\mu - n)a_n$

(d) $H_2(t) = 4t^2 - 2,\ \ H_3(t) = 8t^3 - 12t,\ \ H_4(t) = 16t^4 - 48t^2 + 12$
$H_5(t) = 32t^5 - 160t^3 + 120t$

33. Yes

35. Yes

37. No

39. $a = 0$, b is arbitrary.

Section 10.3, page 609

3. $y = c_1t^2 + c_2t^3$

5. $y = c_1t^2 + c_2t^2 \ln|t|$

7. $y = c_1t^2\cos(5\ln|t|) + c_2t^2\sin(5\ln|t|)$

9. $y = c_1\cos(3\ln|t|) + c_2\sin(3\ln|t|)$

11. $y = c_1t^{-1}\cos(4\ln|t|) + c_2t^{-1}\sin(4\ln|t|)$

13. $y = c_1t^{-2}\cos(6\ln|t|) + c_2t^{-2}\sin(6\ln|t|)$

15. $y = c_1(t-1)^{-1} + c_2(t-1)^3$

17. $y = c_1(t+2)^{-2} + c_2(t+2)^{-3}$

19. $t_0 = -2, \alpha = 2, \beta = -2$

21. $t_0 = 0, \alpha = -3, \beta = 5$

23. $\alpha = -4, \beta = 6, g(t) = -5 + 6\ln t$

25. The transformed equation is $Y'' - 2Y' + Y = e^{-z}$.
    $y = c_1 t + c_2 t \ln t + (1/4)t^{-1}$

27. The transformed equation is $Y'' - Y' - 6Y = 10e^{-2z} - 6$.
    $y = c_1 t^{-2} + c_2 t^3 + 1 - 2t^{-2} \ln t$

29. The transformed equation is $Y'' + 7Y' + 10Y = 36(e^z + e^{-z})$.
    $y = c_1 t^{-5} + c_2 t^{-2} + 2t + 9t^{-1}$

31. $y = t + t^5 + 2,\ -\infty < t < \infty$

33. $y = -2t^{-1} - t^{-1} \ln t + t^{-1}(\ln t)^2,\ 0 < t < \infty$

35. $y = c_1 + c_2 t^{-2} + c_3 t^2,\ t \neq 0$

37. $y = c_1 + c_2 \ln|t| + c_3(\ln|t|)^2 + t^2 + 2(\ln|t|)^3$

Section 10.4, page 618

1. $t = 0$ is a regular singular point.
3. $t = 1$ and $t = -1$ are regular singular points.
5. $t = 0$ is a regular singular point.
7. $t = 0$ is a regular singular point.
9. $t = 1$ and $t = -1$ are irregular singular points.
11. $P(t) = (t^2 - 1)^2 t^2$
13. $P(t) = (t - 1)^2 (t + 1)^2$
15. (a) $n = 0, 1, 2$ (b) $n > 2$
17. (b) $F(\lambda) = 4\lambda^2 - 1 = 0$
    (c) $F(\lambda + n)a_n = -a_{n-1}$
    (d) $y_1 = a_0[t^{1/2} - (1/8)t^{3/2} + (1/192)t^{5/2} + \cdots]$
19. (b) $F(\lambda) = \lambda^2 - 9 = 0$
    (c) $F(\lambda + n)a_n = -a_{n-1}$
    (d) $y_1 = a_0[t^3 - (1/7)t^4 + (1/112)t^5 + \cdots]$
21. (b) $F(\lambda) = \lambda^2 + 2\lambda + 1 = 0$
    (c) $F(\lambda + n)a_n = -2a_{n-1}$
    (d) $y_1 = a_0[t^{-1} - 2 + t + \cdots]$
23. (b) $F(\lambda) = \lambda^2 - 3\lambda = 0$
    (c) $F(\lambda + n)a_{n+1} = -(\lambda + n + 1)a_n$
    (d) $y_1 = a_0[t^3 - t^4 + (1/2)t^5 + \cdots]$
25. (b) $\lambda^2 - 5\lambda = 0$
    (c) $y_1 = a_0[t^5 - (1/6)t^6 - (5/84)t^7 + \cdots]$
27. (b) $\lambda(\lambda - 1.5) = 0$
    (c) $y_1 = a_0[t^{3/2} + (1/2)t^{5/2} - (17/96)t^{7/2} + \cdots]$

Section 10.5, page 629

1. (b) $\lambda(2\lambda - 3) = 0$
   (c) $[(\lambda + n)(2\lambda + 2n - 3)]a_n = (\lambda + n - 3)a_{n-1}$
   (d) $y_1 = a_0[1 + 2t - t^2]$
       $y_2 = a_0 t^{3/2}[1 - (1/10)t - (1/280)t^2 + \cdots]$
   (e) Yes

3. (b) $(\lambda - 1)(3\lambda - 1) = 0$
   (c) $[3(\lambda + n)^2 - 4(\lambda + n) + 1]a_n = -a_{n-1}$
   (d) $y_1 = a_0[t - (1/5)t^2 + (1/80)t^3 + \cdots]$
   $y_2 = a_0[t^{1/3} - t^{4/3} + (1/8)t^{7/3} + \cdots]$
   (e) Yes
5. (b) $(\lambda - 3)(\lambda - 3) = 0$
   (c) $[(\lambda + n)^2 - 6(\lambda + n) + 9]a_n = -a_{n-2}$
   (d) $y_1 = a_0 t^3[1 - (1/4)t^2 + (1/64)t^4 + \cdots]$
   (e) Yes
7. (b) $(\lambda - 2)(\lambda - 1) = 0$
   (c) $[(\lambda + n)^2 - 3(\lambda + n) + 2]a_n = -a_{n-1}$
   (d) $y_1 = a_0[t^2 - (1/2)t^3 + (1/12)t^4 + \cdots]$
   (e) Yes
9. (b) $(\lambda + 1)(\lambda - 1) = 0$
   (c) $[(\lambda + n)^2 - 1]a_n = a_{n-2}$
   (d) $y_1 = a_0[t + (1/8)t^3 + (1/192)t^4 + \cdots]$
   (e) Yes
11. (b) $(\lambda - 4)(\lambda + 4) = 0$
   (c) $[(\lambda + n)^2 - 16]a_n = a_{n-1}$
   (d) $y_1 = a_0[t^4 + (1/9)t^5 + (1/180)t^6 + \cdots]$
   (e) Yes
13. (a) $t = 0$ is a regular singular point.
   (b) $\lambda(\lambda - 2) = 0$
15. (a) $t = 2$ and $t = -2$ are irregular singular points.
17. (a) $\lambda^2 = 0$
   (b) $y_1 = a_0[1 + (1/2)(\alpha^2 + \alpha)(t - 1) + \cdots]$
   (c) $y_1 = a_0 t$
19. (a) $\lambda^2 = 0$ (b) $y_1 = a_0[1 - 5t + 5t^2 - (5/3)t^3 + (5/24)t^4 - (1/120)t^5]$
21. $\alpha = -1, \beta = 5$
23. $\alpha = -1, \beta = 3$

Index

Terms are defined on pages with bold numbers.